DATE DUE

DEMCO 38-296

DYNAMICS
OF
STRUCTURES

Second Edition

DYNAMICS OF STRUCTURES

Ray W. Clough

Professor of Civil Engineering
University of California, Berkeley

Joseph Penzien

International Civil Engineering
Consultants, Inc.

SECOND EDITION

McGraw-Hill, Inc.
New York St. Louis San Francisco Auckland Bogotá
Caracas Lisbon London Madrid Mexico City Milan
Montreal New Delhi San Juan Singapore
Sydney Tokyo Toronto

This book is printed on acid-free paper.

DYNAMICS OF STRUCTURES

4 5 6 7 8 9 0 DOC DOC 9 0 9 8 7 6

ISBN 0-07-011394-7

The editor was B. J. Clark;
the production supervisor was Kathryn Porzio.
R. R. Donnelley & Sons Company was printer and binder.
Type setting and drawing by:
Drawing and Editing Services Company, Taipei, Taiwan, R. O. C.

Library of Congress Cataloging-in-Publication Data

Clough, Ray W., (date).
    Dynamics of structures / Ray W. Clough, Joseph Penzien.
        p.   cm.
    Includes index.
    ISBN 0-07-011394-7
    1. Structural dynamics.   I. Penzien, Joseph   II. Title.
    TA654.C6   1993
    624.1'71—dc20   92-44039

CONTENTS

PART IV   RANDOM VIBRATIONS

PART V   EARTHQUAKE ENGINEERING

PREFACE

Since the first edition of this book was published in 1975, major advances have been made in the subject "Dynamics Of Structures." While it would be impossible to give a comprehensive treatment of all such changes in this second edition, those considered to be of most practical significance are included.

The general organization of text material remains unchanged from the first edition. It progresses logically from a treatment of single-degree-of-freedom systems to multi-degree-of-freedom discrete-parameter systems and then on to infinite-degree-of-freedom continuous systems. The concept of force equilibrium, which forms the basis of static analysis of structures, is retained so that the experienced engineer can easily make the transition to performing a dynamic analysis. It is essential therefore that the student of structural dynamics have a solid background in the theories of statics of structures, including matrix methods, and it is assumed that the readers of this text have such preparation.

The theoretical treatment in Parts I, II, and III is deterministic in nature because it makes use of dynamic loadings which are fully prescribed eventhough they may be highly irregular and transient with respect to time. The treatment of random vibrations in Part IV is however stochastic (random) in form since the loadings considered can be characterized only in a statistical manner. An understanding of basic probability theory is therefore an essential prerequisite to the study of this subject. Before proceeding with this study, it is recommended that the student take a full course on probability theory; however, if this has not been done, the brief treatment of probability concepts given in Chapter 20 can serve as minimum preparation.

The solution of a typical structural dynamics problem is considerably more complicated than its static counterpart due to the addition of inertia and damping to the elastic resistance forces and due to the time dependency of all force quantities. For most practical situations, the solution usually is possible only through the use of a high-speed digital computer, which has become the standard tool of the structural dynamicist. However, most of the problems in the text, which are intended to teach the fundamentals of dynamics, are quite simple in form allowing their solutions to be obtained using a hand calculator. Nevertheless, the student of dynamics of structures should have previously studied computer coding techniques and the associated analytical procedures. Such background will permit an early transition from solving dynamics problems by hand calculator to solving them on a PC computer using programs specially developed for this purpose. The program CAL-91, developed by Professor E. L. Wilson of the University of California, Berkeley, is such a program which has been used very effectively in teaching even the first course in Dynamics Of Structures. Instructors using this book are encouraged to implement such PC computer solutions into their courses so that more realistic problems can be considered.

A large number of example problems have been solved in the text to assist the reader in understanding the subject material. To fully master the analytical techniques, it is essential that the student solve many of the homework problems presented at the ends of chapters. They should be assigned sparingly however because dynamic-response analyses are notoriously time consuming. The authors have found that from one to four problems may constitute an adequate weekly assignment, depending on the subject matter and type of solution required. On this basis, the book includes many more problems than can be assigned during a one-year sequence of courses on structural dynamics.

The subject matter of this text can serve as the basis of a series of graduate-level courses. The first course could cover the material in Part I and a portion of that in Part II. The full extent of this coverage would depend, of course, upon whether the course is of quarter or semester duration. If of quarter duration, the material coverage in Parts I and II is sufficient to provide the basis of a sequence of two quarter courses and some material from Part III also could be included in the second course.

It is now generally expected that nearly all Masters-Degree students in structural engineering should have had at least the basic first-course in dynamics of structures and it is recommended that the advanced (fourth-year level) undergraduate student be provided on opportunity to take a similar course, eventhough its material coverage may be somewhat reduced.

The material in Part IV can serve as the subject matter of a basic course on random vibration which is needed in fully understanding practical applications of stochastic methods in various fields such as earthquake engineering, wind engineering, and ocean engineering. Many such applications are presented in Part V which treats the broad subject of earthquake engineering. A separate course is needed however to fully cover the material in Part V. Students taking either of these latter two courses should have a good background in deterministic dynamic analysis of structures and a reasonable maturity in mathematics.

This book has been written to serve not only as a textbook for college and university students, but to serve as a reference book for practicing engineers as well. The analytical formulations and techniques presented can serve effectively as the basis for continued development of new computer programs to be used by the engineer in designing and analyzing structures which function in dynamic environments.

In closing, the authors wish to express their sincere thanks and appreciation to the many individuals (students, faculty members, and practicing engineers) who have both directly and indirectly contributed to the content of this book. The number of such contributors is much too large however to attempt listing them by name.

One person most deserving of special recognition is Ms. Huey-Shu Ni who typed the entire text and, with assistance from her staff at Drawing and Editing Services, Ltd. in Taipei, Taiwan, prepared all of the figures. Her patient and congenial manner, which was always present over the many years of the book's preparation, is to be admired. The authors express to her their deep appreciation and thanks for a job superbly done.

Ray W. Clough
Joseph Penzien

LIST OF SYMBOLS

a	distance
a_0	Fourier coefficient, dimensionless frequency
a_n	Fourier coefficients, constants
A	area, constant
A_1, A_2	constants
b	distance, integer
b_0, b_n	Fourier coefficients constants
B	constant
c	damping coefficient
c^*	generalized damping coefficient
c_c	critical damping coefficient
c_{ij}	damping influence coefficients
C_n	normal mode generalized damping coefficients
CQC	complete quadratic combination
D	dynamics magnification factor
$\mathbf{D}$	dynamics matrix $= \mathbf{k}^{-1}\mathbf{m}$
DFT	discrete Fourier transform
DRV	derived Ritz vector
e	axial displacement
E	Young's modulus, energy release
$\mathbf{E}$	dynamic matrix $= \mathbf{D}^{-1}$
$E[\ ]$	expected value, ensemble average
E_D	damping energy loss/cycle
ED	epicentral distance
EI	flexural stiffness
f	natural cyclic frequency

$\tilde{f}_{ij}$	flexibility influence coefficients
$f_I,\ f_D,\ f_S$	inertial, damping, and spring forces, respectively
FD	focal depth
FFT	fast Fourier transform
g	acceleration of gravity
g_i	stress wave functions
G	shear modulus, complex constant
$G_1,\ G_2$	constants
$G_I,\ G_R$	real constants
$\overline{G}$	length of vector
$G(ia_0)$	boundary impedance function
GC	geological condition
h	height, plate thickness, time interval
$h_{ij}(t),\ h(t)$	unit impulse response functions
$\mathbf{H}_{ij}(i\overline{\omega}),\ \mathrm{H}(i\overline{\omega})$	complex frequency response functions
H_z	Hertz (frequency in $cycles/sec$)
i	integer
I	impulse, cross-section moment of inertia
$\mathbf{I}$	identity matrix
$\mathrm{I}_{ij}(i\overline{\omega})$	impedance function
IE	isolation effectiveness
Im	imaginary
j	integer, mass moment of inertia
$k,\ k_i$	spring constants
k^*	generalized spring constant
$\overline{k^*}$	combined generalized stiffness
$\hat{k}$	complex stiffness
$\tilde{k}_c,\ \tilde{k}_d$	effective stiffnesses
k_{ij}	stiffness influence coefficients
$\overline{k}_{ij}$	combined stiffness influence coefficients
$\mathbf{k}_G$	geometric stiffness
k_G^*	generalized geometric stiffness
$k_{G_{ij}}$	geometric stiffness influence coefficients
K_n	generalized stiffness of nth normal mode
$\hat{K}_n$	complex stiffness of nth normal mode
L	length
$\mathcal{L}$	earthquake-excitation factor
m	mass, integer

m_i	mass
m_{ij}	mass influence coefficients
m^*	generalized mass
$\overline{m}$	uniform mass/unit length
M	Richter magnitude, integer
$\boldsymbol{M}$	mass matrix for normal modes
M_n	generalized mass of nth normal mode
$M(t)$, $M(x,t)$	internal moment
MDOF	multi-degree of freedom
MF	magnification factor
MM	modified Mercalli scale
n	integer, constant
N	number of time increments, number of degrees of freedom, integer
N	axial load
N_{cr}	critical axial load
$N(x)$, $N(x,t)$	internal axial force
p , p_0	load
$\overline{p}$	uniform loading/unit length
p_{eff}	effective loading
$p(t)$	applied loading
$\boldsymbol{P}(t)$	load vector in time domain
$p^*(t)$	generalized loading
$p(x)$	probability density function
$p(x,y)$	joint probability density function
$p(x|y)$	conditional probability density function
$p(x_1, x_2, \cdots, x_m)$	multivariate probability density function
P	power
$\boldsymbol{P}(i\overline{\omega})$	load vector in frequency domain
$P(x)$	probability distribution function
P_n	complex amplitude coefficient
$P_n(t)$	generalized loading of nth normal mode in time domain
$\overline{\mathrm{P}}_n(i\overline{\omega})$	generalized loading of nth normal mode in frequency domain
PGA	peak ground acceleration
Pr	probability
$P(X)$, $P(X,Y)$	probability density functions
q_0 , q_i	constants, generalized coordinates
$q(x,t)$	applied axial loading

$Q_i(t)$	ith generalized forcing function
r	radius of gyration
Re	real
$R(t)$	response ratio
$R_x(\tau)$	autocorrelation function
$R_{xy}(\tau)$	cross-correlation function
s	constant
$S_a(\xi, \omega)$	spectral absolute acceleration response
$S_d(\xi, \omega)$	spectral relative displacement response
$S_{ii}(i\overline{\omega})$	power-spectral density functions
$S_{ij}(i\overline{\omega})$	cross-spectral density functions
$S_{pa}(\xi, \omega)$	pseudo-acceleration spectral response
$S_{pv}(\xi, \omega)$	pseudo-velocity spectral response
$S_v(\xi, \omega)$	spectral relative velocity response
$\mathbf{S}_1$	first-mode sweeping matrix
SC	soil conditions
SDOF	single degree of freedom
$SI(\xi)$	Housner's spectrum intensity
SM	source mechanism
SRSS	square root of the sum of squares
t, t_i	time
t_1	impulse duration
T	period of vibration, kinetic energy
$\mathbf{T}$	matrix of orthonormal eigenvectors
T_n	period of nth normal mode
T_p	period of motion
TR	transmissibility
u	displacement in x-direction
U	strain energy
v	displacement in y-direction
$\overline{v}$	dynamic displacement
v^t	total displacement
$\mathbf{v}(t)$	displacement in time domain
v_g, v_{g0}	ground displacement
$\ddot{v}_g(t)$	ground acceleration in time domain
$\ddot{\mathbf{V}}_g(i\overline{\omega})$	ground acceleration in frequency domain
v_{st}	static displacement
V	potential energy

$\mathbf{V}(i\overline{\omega})$	displacement in frequency domain
$V(x,t)$	internal shear force
V_a	apparent wave velocity
V_c , V_p , V_s	wave velocities
V_{ff}	free-field wave velocity
V_n	complex constant
w	displacement in z-direction
W	work, weight
W_{nc}	work by nonconservative forces
W_N	work by axial load N
x	space coordinate, random variable
$\overline{x}$	mean value of x
$\overline{x^2}$	mean square value of x
$x(t)$	random process
X	space coordinate, random variable
y	space coordinate
$y(t)$	random process
Y	random variable, space coordinate
$Y_n(t)$	generalized displacement of nth normal mode in time domain
$Y_n(i\overline{\omega})$	generalized displacement of nth normal mode in frequency domain
z	space coordinate
$z(t)$	generalized coordinate response in time domain
Z , Z_n , Z_0	generalized coordinates
$\mathbf{Z}(i\overline{\omega})$	generalized coordinate response in frequency domain
α	constant, non-dimensionless time parameter
β	frequency ratio
γ	integer, mass/unit area, unit weight
$\gamma_{ij}(i\overline{\omega})$	coherency functions
δ	log decrement, variation, residual
δe , δv , δZ	virtual displacements
δW_I	internal virtual work
δW_E	external virtual work
Δ	increment
Δ_{st}	static displacement
$\Delta\tilde{p}_d$	effective loading increment
Δt	time interval
$\Delta\omega$	frequency interval

ϵ	normal strain
ζ	time function, hysteretic damping coefficient
λ	wave length
λ_G	axial load factor
λ_i	Lagrange multiplier
λ_n	nth eigenvalue
θ	phase angle, slope, rotation
μ	ductility factor
μ_{ij}	covariances
ν	Poisson's ratio
ξ, ξ_n	damping ratios
ρ	vector amplitude, mass/unit volume
ρ_{xy}	correlation coefficient
σ	normal stress
σ_x	standard deviation
σ_x^2	variance
τ	time
ϕ	phase angle
ϕ_{ij}	modal displacement
$\boldsymbol{\phi}_n, \phi_n(x)$	nth mode shape
$\boldsymbol{\phi}$	mode shape matrix
ψ, ψ_n	generalized displacement functions
$\boldsymbol{\psi}_n$	generalized displacement vector
$\boldsymbol{\Psi}$	matrix of assumed made shapes
ω, ω_n	undamped natural circular frequencies
ω_D, ω_{Dn}	damped natural circular frequencies
$\overline{\omega}$	circular frequency of harmonic forcing function
$\chi(x)$	load distribution

CHAPTER

1

OVERVIEW OF STRUCTURAL DYNAMICS

1-1 FUNDAMENTAL OBJECTIVE OF STRUCTURAL DYNAMICS ANALYSIS

The primary purpose of this book is to present methods for analyzing the stresses and deflections developed in any given type of structure when it is subjected to an arbitrary dynamic loading. In one sense, this objective may be considered to be an extension of standard methods of structural analysis, which generally are concerned with static loading only, to permit consideration of dynamic loading as well. In this context, the static-loading condition may be looked upon merely as a special form of dynamic loading. However, in the analysis of a linear structure it is convenient to distinguish between the static and the dynamic components of the applied loading, to evaluate the response to each type of loading separately, and then to superpose the two response components to obtain their total effect. When treated thusly, the static and dynamic methods of analysis are fundamentally different in character.

For the purposes of this presentation, the term *dynamic* may be defined simply as time-varying; thus a dynamic load is any load of which its magnitude, direction, and/or position varies with time. Similarly, the structural response to a dynamic load, i.e., the resulting stresses and deflections, is also time-varying, or dynamic.

1

Two basically different approaches are available for evaluating structural response to dynamic loads: deterministic and nondeterministic. The choice of method to be used in any given case depends upon how the loading is defined. If the time variation of loading is fully known, even though it may be highly oscillatory or irregular in character, it will be referred to herein as a *prescribed dynamic loading*; and the analysis of the response of any specified structural system to a prescribed dynamic loading is defined as a deterministic analysis. On the other hand, if the time variation is not completely known but can be defined in a statistical sense, the loading is termed a *random dynamic loading*; and its corresponding analysis of response is defined as a nondeterministic analysis. The principal emphasis in this text is placed on development of methods of deterministic dynamic analysis; however, Part Four is devoted to presenting an introduction to nondeterministic methods of analysis and Part Five contains a chapter dealing with the application of nondeterministic methods of analysis in the field of earthquake engineering.

In general, structural response to any dynamic loading is expressed basically in terms of the displacements of the structure. Thus, a deterministic analysis leads directly to displacement time-histories corresponding to the prescribed loading history; other related response quantities, such as stresses, strains, internal forces, etc., are usually obtained as a secondary phase of the analysis. On the other hand, a nondeterministic analysis provides only statistical information about the displacements resulting from the statistically defined loading; corresponding information on the related response quantities are then generated using independent nondeterministic analysis procedures.

1-2 TYPES OF PRESCRIBED LOADINGS

Almost any type of structural system may be subjected to one form or another of dynamic loading during its lifetime. From an analytical standpoint, it is convenient to divide prescribed or deterministic loadings into two basic categories, periodic and nonperiodic. Some typical forms of prescribed loadings and examples of situations in which such loadings might be developed are shown in Fig. 1-1.

As indicated in this figure, a periodic loading exhibits the same time variation successively for a large number of cycles. The simplest periodic loading has the sinusoidal variation shown in Fig. 1-1a, which is termed *simple harmonic*; loadings of this type are characteristic of unbalanced-mass effects in rotating machinery. Other forms of periodic loading, e.g., those caused by hydrodynamic pressures generated by a propeller at the stern of a ship or by inertial effects in reciprocating machinery, frequently are more complex. However, by means of a Fourier analysis any periodic loading can be represented as the sum of a series of simple harmonic components; thus, in principle, the analysis of response to any periodic loading follows the same general procedure.

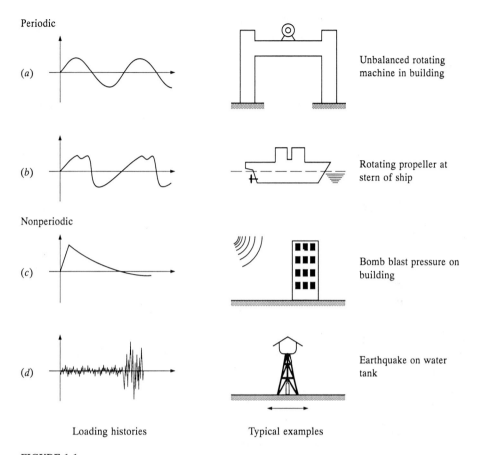

Periodic

(a)

Unbalanced rotating machine in building

(b)

Rotating propeller at stern of ship

Nonperiodic

(c)

Bomb blast pressure on building

(d)

Earthquake on water tank

Loading histories   Typical examples

FIGURE 1-1
Characteristics and sources of typical dynamic loadings: (*a*) simple harmonic; (*b*) complex; (*c*) impulsive; (*d*) long-duration.

Nonperiodic loadings may be either short-duration *impulsive* loadings or long-duration general forms of loads. A blast or explosion is a typical source of impulsive load; for such short-duration loads, special simplified forms of analysis may be employed. On the other hand, a general, long-duration loading such as might result from an earthquake can be treated only by completely general dynamic-analysis procedures.

1-3 ESSENTIAL CHARACTERISTICS OF A DYNAMIC PROBLEM

A structural-dynamic problem differs from its static-loading counterpart in two important respects. The first difference to be noted, by definition, is the time-varying nature of the dynamic problem. Because both loading and response vary with time, it is evident that a dynamic problem does not have a single solution, as a static problem does; instead the analyst must establish a succession of solutions corresponding to all times of interest in the response history. Thus a dynamic analysis is clearly more complex and time-consuming than a static analysis.

The second and more fundamental distinction between static and dynamic problems is illustrated in Fig. 1-2. If a simple beam is subjected to a static load p, as shown in Fig. 1-2a, its internal moments and shears and deflected shape depend only upon this load and they can be computed by established principles of force equilibrium. On the other hand, if the load $p(t)$ is applied dynamically, as shown in Fig. 1-2b, the resulting displacements of the beam depend not only upon this load but also upon inertial forces which oppose the accelerations producing them. Thus the corresponding internal moments and shears in the beam must equilibrate not only the externally applied force $p(t)$ but also the inertial forces resulting from the accelerations of the beam.

Inertial forces which resist accelerations of the structure in this way are the most important distinguishing characteristic of a structural-dynamics problem. In general, if the inertial forces represent a significant portion of the total load equilibrated by the internal elastic forces of the structure, then the dynamic character of the problem must be accounted for in its solution. On the other hand, if the motions are so slow that the inertial forces are negligibly small, the analysis of response for any desired instant of time may be made by static structural-analysis procedures even though the load and response may be time-varying.

1-4   METHODS OF DISCRETIZATION

Lumped-Mass Procedure

An analysis of the dynamic system in Fig. 1-2b is obviously made complicated by the fact that the inertial forces result from structural time-varying displacements which in turn are influenced by the magnitudes of inertial forces. This closed cycle of cause and effect can be attacked directly only by formulating the problem in terms

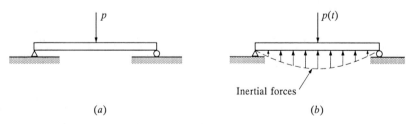

(a)   (b)

FIGURE 1-2
Basic difference between static and dynamic loads: (a) static loading; (b) dynamic loading.

of differential equations. Furthermore, because the mass of the beam is distributed continuously along its length, the displacements and accelerations must be defined for each point along the axis if the inertial forces are to be completely defined. In this case, the analysis must be formulated in terms of partial differential equations because position along the span as well as time must be taken as independent variables.

However, if one assumes the mass of the beam to be concentrated at discrete points as shown in Fig. 1-3, the analytical problem becomes greatly simplified because inertial forces develop only at these mass points. In this case, it is necessary to define the displacements and accelerations only at these discrete locations.

The number of displacement components which must be considered in order to represent the effects of all significant inertial forces of a structure may be termed the *number of dynamic degrees of freedom* of the structure. For example, if the three masses in the system of Fig. 1-3 are fully concentrated and are constrained so that the corresponding mass points translate only in a vertical direction, this would be called a three-degree-of-freedom (3 DOF) system. On the other hand, if these masses are not fully concentrated so that they possess finite rotational inertia, the rotational displacements of the three points will also have to be considered, in which case the system has 6 DOF. If axial distortions of the beam are significant, translation displacements parallel with the beam axis will also result giving the system 9 DOF. More generally, if the structure can deform in three-dimensional space, each mass will have 6 DOF; then the system will have 18 DOF. However, if the masses are fully concentrated so that no rotational inertia is present, the three-dimensional system will then have 9 DOF. On the basis of these considerations, it is clear that a system with continuously distributed mass, as in Fig. 1-2b, has an infinite number of degrees of freedom.

Generalized Displacements

The lumped-mass idealization described above provides a simple means of limiting the number of degrees of freedom that must be considered in conducting a dynamic analysis of an arbitrary structural system. The lumping procedure is most effective in treating systems in which a large proportion of the total mass actually is concentrated at a few discrete points. Then the mass of the structure which supports these concentrations can be included in the lumps, allowing the structure itself to be considered weightless.

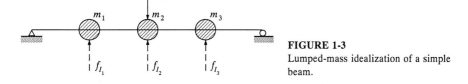

FIGURE 1-3
Lumped-mass idealization of a simple beam.

However, in cases where the mass of the system is quite uniformly distributed throughout, an alternative approach to limiting the number of degrees of freedom may be preferable. This procedure is based on the assumption that the deflected shape of the structure can be expressed as the sum of a series of specified displacement patterns; these patterns then become the displacement coordinates of the structure. A simple example of this approach is the trigonometric-series representation of the deflection of a simple beam. In this case, the deflection shape may be expressed as the sum of independent sine-wave contributions, as shown in Fig. 1-4, or in mathematical form,

$$v(x) = \sum_{n=1}^{\infty} b_n \sin \frac{n\pi x}{L} \tag{1-1}$$

In general, any arbitrary shape compatible with the prescribed support conditions of the simple beam can be represented by this infinite series of sine-wave components. The amplitudes of the sine-wave shapes may be considered to be the displacement coordinates of the system, and the infinite number of degrees of freedom of the actual beam are represented by the infinite number of terms included in the series. The advantage of this approach is that a good approximation to the actual beam shape can be achieved by a truncated series of sine-wave components; thus a 3 DOF approximation would contain only three terms in the series, etc.

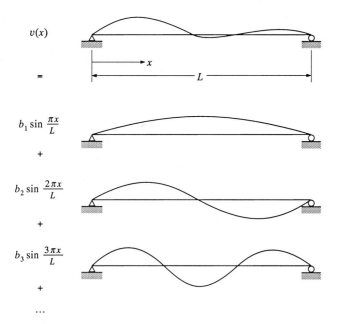

FIGURE 1-4
Sine-series representation of simple beam deflection.

This concept can be further generalized by recognizing that the sine-wave shapes used as the assumed displacement patterns were an arbitrary choice in this example. In general, any shapes $\psi_n(x)$ which are compatible with the prescribed geometric-support conditions and which maintain the necessary continuity of internal displacements may be assumed. Thus a generalized expression for the displacements of any one-dimensional structure might be written

$$v(x) = \sum_n Z_n \psi_n(x) \qquad (1\text{-}2)$$

For any assumed set of displacement functions $\psi(x)$, the resulting shape of the structure depends upon the amplitude terms Z_n, which will be referred to as *generalized coordinates*. The number of assumed shape patterns represents the number of degrees of freedom considered in this form of idealization. In general, better accuracy can be achieved in a dynamic analysis for a given number of degrees of freedom by using the shape-function method of idealization rather than the lumped-mass approach. However, it also should be recognized that greater computational effort is required for each degree of freedom when such generalized coordinates are employed.

The Finite-Element Concept

A third method of expressing the displacements of any given structure in terms of a finite number of discrete displacement coordinates, which combines certain features of both the lumped-mass and the generalized-coordinate procedures, has now become popular. This approach, which is the basis of the finite-element method of analysis of structural continua, provides a convenient and reliable idealization of the system and is particularly effective in digital-computer analyses.

The finite-element type of idealization is applicable to structures of all types: framed structures, which comprise assemblages of one-dimensional members (beams, columns, etc.); plane-stress, plate- and shell-type structures, which are made up of two-dimensional components; and general three-dimensional solids. For simplicity, only the one-dimensional type of structural components will be considered in the present discussion, but the extension of the concept to two- and three-dimensional structural elements is straightforward.

The first step in the finite-element idealization of any structure, e.g., the beam shown in Fig. 1-5, involves dividing it into an appropriate number of segments, or elements, as shown. Their sizes are arbitrary; i.e., they may be all of the same size or all different. The ends of the segments, at which they are interconnected, are called *nodal points*. The displacements of these nodal points then become the generalized coordinates of the structure.

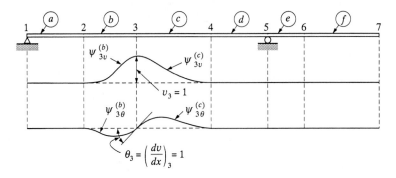

FIGURE 1-5
Typical finite-element beam coordinates.

The deflection shape of the complete structure can now be expressed in terms of these generalized coordinates by means of an appropriate set of assumed displacement functions using an expression similar to Eq. (1-2). In this case, however, the displacement functions are called *interpolation functions* because they define the shapes produced by specified nodal displacements. For example, Fig. 1-5 shows the interpolation functions associated with two degrees of freedom of nodal point 3, which produce transverse displacements in the plane of the figure. In principle, each interpolation function could be any curve which is internally continuous and which satisfies the geometric displacement condition imposed by the nodal displacement. For one-dimensional elements it is convenient to use the shapes which would be produced by these same nodal displacements in a uniform beam. It will be shown later in Chapter 10 that these interpolation functions are cubic hermitian polynomials.

Because the interpolation functions used in this procedure satisfy the requirements stated in the preceding section, it should be apparent that coordinates used in the finite-element method are just special forms of generalized coordinates. The advantages of this special procedure are as follows:

(1) The desired number of generalized coordinates can be introduced merely by dividing the structure into an appropriate number of segments.

(2) Since the interpolation functions chosen for each segment may be identical, computations are simplified.

(3) The equations which are developed by this approach are largely uncoupled because each nodal displacement affects only the neighboring elements; thus the solution process is greatly simplified.

In general, the finite-element approach provides the most efficient procedure for expressing the displacements of arbitrary structural configurations by means of a discrete set of coordinates.

1-5 FORMULATION OF THE EQUATIONS OF MOTION

As mentioned earlier, the primary objective of a deterministic structural-dynamic analysis is the evaluation of the displacement time-histories of a given structure subjected to a given time-varying loading. In most cases, an approximate analysis involving only a limited number of degrees of freedom will provide sufficient accuracy; thus, the problem can be reduced to the determination of the time-histories of these selected displacement components. The mathematical expressions defining the dynamic displacements are called the *equations of motion* of the structure, and the solution of these equations of motion provides the required displacement time-histories.

The formulation of the equations of motion of a dynamic system is possibly the most important, and sometimes the most difficult, phase of the entire analysis procedure. In this text, three different methods will be employed for the formulation of these equations, each having advantages in the study of special classes of problems. The fundamental concepts associated with each of these methods are described in the following paragraphs.

Direct Equilibration Using d'Alembert's Principle

The equations of motion of any dynamic system represent expressions of Newton's second law of motion, which states that the rate of change of momentum of any mass particle m is equal to the force acting on it. This relationship can be expressed mathematically by the differential equation

$$\mathbf{p}(t) = \frac{d}{dt}\left(m\frac{d\mathbf{v}}{dt}\right)$$
(1-3)

where $\mathbf{p}(t)$ is the applied force vector and $\mathbf{v}(t)$ is the position vector of particle mass m. For most problems in structural dynamics it may be assumed that mass does not vary with time, in which case Eq. (1-3) may be written

$$\mathbf{p}(t) = m\frac{d^2\mathbf{v}}{dt^2} \equiv m\,\ddot{\mathbf{v}}(t)$$
(1-3a)

where the dots represent differentiation with respect to time. Equation (1-3a), indicating that force is equal to the product of mass and acceleration, may also be written in the form

$$\mathbf{p}(t) - m\,\ddot{\mathbf{v}}(t) = \mathbf{0}$$
(1-3b)

in which case, the second term $m\ddot{\mathbf{v}}(t)$ is called the *inertial force* resisting the acceleration of the mass.

The concept that a mass develops an inertial force proportional to its acceleration and opposing it is known as *d'Alembert's principle*. It is a very convenient device in problems of structural dynamics because it permits the equations of motion to be expressed as equations of dynamic equilibrium. The force $\mathbf{p}(t)$ may be considered to include many types of force acting on the mass: elastic constraints which oppose displacements, viscous forces which resist velocities, and independently defined external loads. Thus if an inertial force which resists acceleration is introduced, the equation of motion is merely an expression of equilibration of all forces acting on the mass. In many simple problems, the most direct and convenient way of formulating the equations of motion is by means of such direct equilibrations.

Principle of Virtual Displacements

However, if the structural system is reasonably complex involving a number of interconnected mass points or bodies of finite size, the direct equilibration of all the forces acting in the system may be difficult. Frequently, the various forces involved may readily be expressed in terms of the displacement degrees of freedom, but their equilibrium relationships may be obscure. In this case, the principle of virtual displacements can be used to formulate the equations of motion as a substitute for the direct equilibrium relationships.

The principle of virtual displacements may be expressed as follows. If a system which is in equilibrium under the action of a set of externally applied forces is subjected to a virtual displacement, i.e., a displacement pattern compatible with the system's constraints, the total work done by the set of forces will be zero. With this principle, it is clear that the vanishing of the work done during a virtual displacement is equivalent to a statement of equilibrium. Thus, the response equations of a dynamic system can be established by first identifying all the forces acting on the masses of the system, including inertial forces defined in accordance with d'Alembert's principle. Then, the equations of motion are obtained by separately introducing a virtual displacement pattern corresponding to each degree of freedom and equating the work done to zero. A major advantage of this approach is that the virtual-work contributions are scalar quantities and can be added algebraically, whereas the forces acting on the structure are vectorial and can only be superposed vectorially.

Variational Approach

Another means of avoiding the problems of establishing the vectorial equations of equilibrium is to make use of scalar quantities in a variational form known as Hamilton's principle. Inertial and elastic forces are not explicitly involved in this principle; instead, variations of kinetic and potential energy terms are utilized. This formulation has the advantage of dealing only with purely scalar energy quantities, whereas the forces and displacements used to represent corresponding effects in the virtual-work procedure are all vectorial in character, even though the work terms themselves are scalars.

It is of interest to note that Hamilton's principle can also be applied to statics problems. In this case, it reduces to the well-known principle of minimum potential energy so widely used in static analyses.

It has been shown that the equation of motion of a dynamic system can be formulated by any one of three distinct procedures. The most straightforward approach is to establish directly the dynamic equilibrium of all forces acting in the system, taking account of inertial effects by means of d'Alembert's principle. In more complex systems, however, especially those involving mass and elasticity distributed over finite regions, a direct vectorial equilibration may be difficult, and work or energy formulations which involve only scalar quantities may be more convenient. The most direct of these procedures is based on the principle of virtual displacements, in which the forces acting on the system are evaluated explicitly but the equations of motion are derived by consideration of the work done during appropriate virtual displacements. On the other hand, the alternative energy formulation, which is based on Hamilton's principle, makes no direct use of the inertial or conservative forces acting in the system; the effects of these forces are represented instead by variations of the kinetic and potential energies of the system. It must be recognized that all three procedures are completely equivalent and lead to identical equations of motion. The method to be used in any given case is largely a matter of convenience and personal preference; the choice generally will depend on the nature of the dynamic system under consideration.

1-6 ORGANIZATION OF THE TEXT

This book, "Dynamics of Structures," has been written in five parts. Part One presents an extensive treatment of the single-degree-of-freedom (SDOF) system having only one independent displacement coordinate. This system is studied in great detail for two reasons: (1) the dynamic behavior of many practical structures can be expressed in terms of a single coordinate, so that this SDOF treatment applies directly in those cases, and (2) the response of complex linear structures can be expressed as the sum of the responses of a series of SDOF systems so that this same treatment once again applies to each system in the series. Thus, the SDOF analysis techniques provide the basis for treating the vast majority of structural-dynamic problems.

Part Two treats discrete-parameter multi-degree-of-freedom (MDOF) systems, i.e., systems for which their dynamic responses can be expressed in terms of a limited number of displacement coordinates. For the analysis of linearly elastic systems, procedures are presented for evaluating their properties in a free-vibration state, i.e., for evaluating normal mode shapes and corresponding frequencies. Then, two general methods for calculating the dynamic responses of these systems to arbitrarily specified loadings are given: (1) making use of mode-superposition in which total response is expressed as the sum of individual responses in the various normal modes of vibration, each of which can be determined by analysis procedures of the SDOF system, and (2) solving directly the MDOF equations of motion in their original coupled form. Finally, the variational formulation of the structural-dynamic problem is presented and step-by-step numerical integration techniques are formulated for solving directly both SDOF and MDOF equations of motion representing either linear or nonlinear systems.

Dynamic linearly elastic systems having continuously distributed properties are considered in Part Three. Such systems have an infinite number of degrees of freedom requiring that their equations of motion be written in the form of partial differential equations. However, it is shown that the mode-superposition procedure is still applicable to these systems and that practical solutions can be obtained by considering only a limited number of the lower modes of vibration.

Part Four covers the general topic of random vibrations of linear SDOF and MDOF systems. Since the loadings under consideration can be characterized only in a statistical sense, the corresponding responses are similarly characterized. To provide a basis for treating these systems, introductions to probability theory and stochastic processes are given.

Earthquake engineering, with special focus on structural response and performance, is the subject of Part Five. A very brief seismological background on the causes and characteristics of earthquakes is given, along with a discussion of the ground motions they produce. Methods are then given for evaluating the response of structures to these motions using both deterministic and nondeterministic procedures.

PART
I

SINGLE-
DEGREE-
OF-
FREEDOM
SYSTEMS

CHAPTER

2

ANALYSIS OF FREE VIBRATIONS

2-1 COMPONENTS OF THE BASIC DYNAMIC SYSTEM

The essential physical properties of any linearly elastic structural or mechanical system subjected to an external source of excitation or dynamic loading are its mass, elastic properties (flexibility or stiffness), and energy-loss mechanism or damping. In the simplest model of a SDOF system, each of these properties is assumed to be concentrated in a single physical element. A sketch of such a system is shown in Fig. 2-1a.

The entire mass m of this system is included in the rigid block which is constrained by rollers so that it can move only in simple translation; thus, the single displacement coordinate $v(t)$ completely defines its position. The elastic resistance to displacement is provided by the weightless spring of stiffness k, while the energy-loss mechanism is represented by the damper c. The external dynamic loading producing the response of this system is the time-varying force $p(t)$.

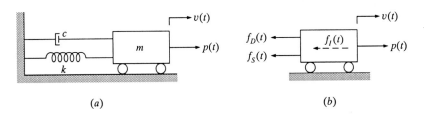

(a)   (b)

FIGURE 2-1
Idealized SDOF system: (a) basic components; (b) forces in equilibrium.

2-2   EQUATION OF MOTION OF THE BASIC DYNAMIC SYSTEM

The equation of motion for the simple system of Fig. 2-1a is most easily formulated by directly expressing the equilibrium of all forces acting on the mass using d'Alembert's principle. As shown in Fig. 2-1b, the forces acting in the direction of the displacement degree of freedom are the applied load $p(t)$ and the three resisting forces resulting from the motion, i.e., the inertial force $f_I(t)$, the damping force $f_D(t)$, and the spring force $f_S(t)$. The equation of motion is merely an expression of the equilibrium of these forces as given by

$$f_I(t) + f_D(t) + f_S(t) = p(t) \tag{2-1}$$

Each of the forces represented on the left hand side of this equation is a function of the displacement $v(t)$ or one of its time derivatives. The positive sense of these forces has been deliberately chosen to correspond with the negative-displacement sense so that they oppose a positive applied loading.

In accordance with d'Alembert's principle, the inertial force is the product of the mass and acceleration

$$f_I(t) = m\, \ddot{v}(t) \tag{2-2a}$$

Assuming a viscous damping mechanism, the damping force is the product of the damping constant c and the velocity

$$f_D(t) = c\, \dot{v}(t) \tag{2-2b}$$

Finally, the elastic force is the product of the spring stiffness and the displacement

$$f_S(t) = k\, v(t) \tag{2-2c}$$

When Eqs. (2-2) are introduced into Eq. (2-1), the equation of motion for this SDOF system is found to be

$$m\, \ddot{v}(t) + c\, \dot{v}(t) + k\, v(t) = p(t) \tag{2-3}$$

To introduce an alternative formulation procedure, it is instructive to develop this same equation of motion by a virtual-work approach. If the mass is given a virtual displacement δv compatible with the system's constraints, the total work done by the equilibrium system of forces in Fig. 2-1b must equal zero as shown by

$$-f_I(t)\, \delta v - f_D(t)\, \delta v - f_S(t)\, \delta v + p(t)\, \delta v = 0 \tag{2-4}$$

in which the negative signs result from the fact that the associated forces act opposite to the sense of the virtual displacement. Substituting Eqs. (2-2) into Eq. (2-4) and factoring out δv leads to

$$\left[-m\, \ddot{v}(t) - c\, \dot{v}(t) - k\, v(t) + p(t) \right] \delta v = 0 \tag{2-5}$$

Since δv is nonzero, the bracket quantity in this equation must equal zero, thus giving the same equation of motion as shown by Eq. (2-3). While a virtual-work formulation has no advantage for this simple system, it will be found very useful for more general types of SDOF systems treated subsequently.

2-3 INFLUENCE OF GRAVITATIONAL FORCES

Consider now the system shown in Fig. 2-2a, which is the system of Fig. 2-1a rotated through 90° so that the force of gravity acts in the direction of the displacement. In this case, the system of forces acting in the direction of the displacement degree of freedom is that set shown in Fig. 2-2b. Using Eqs. (2-2), the equilibrium of these forces is given by

$$m \, \ddot{v}(t) + c \, \dot{v}(t) + k \, v(t) = p(t) + W \qquad (2\text{-}6)$$

where W is the weight of the rigid block.

However, if the total displacement $v(t)$ is expressed as the sum of the static displacement Δ_{st} caused by the weight W plus the additional dynamic displacement $\bar{v}(t)$ as shown in Fig. 2-2c, i.e.,

$$v(t) = \Delta_{st} + \bar{v}(t) \qquad (2\text{-}7)$$

then the spring force is given by

$$f_S(t) = k \, v(t) = k \, \Delta_{st} + k \, \bar{v}(t) \qquad (2\text{-}8)$$

Introducing Eq. (2-8) into (2-6) yields

$$m \, \ddot{v}(t) + c \, \dot{v}(t) + k \, \Delta_{st} + k \, \bar{v}(t) = p(t) + W \qquad (2\text{-}9)$$

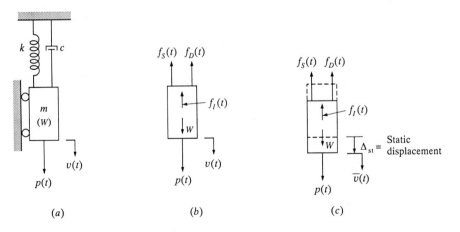

FIGURE 2-2
Influence of gravity on SDOF equilibrium.

and noting that $k \, \triangle_{\text{st}} = W$ leads to

$$m \, \ddot{\overline{v}}(t) + c \, \dot{\overline{v}}(t) + k \, \overline{v}(t) = p(t) \qquad (2\text{-}10)$$

Now by differentiating Eq. (2-7) and noting that $\triangle_{\text{st}}$ does not vary with time, it is evident that $\ddot{v}(t) = \ddot{\overline{v}}(t)$ and $\dot{v}(t) = \dot{\overline{v}}(t)$ so that Eq. (2-10) can be written

$$m \, \ddot{\overline{v}}(t) + c \, \dot{\overline{v}}(t) + k \, \overline{v}(t) = p(t) \qquad (2\text{-}11)$$

Comparison of Eqs. (2-11) and (2-3) demonstrates that the equation of motion expressed with reference to the static-equilibrium position of the dynamic system is not affected by gravity forces. For this reason, displacements in all future discussions will be referenced from the static-equilibrium position and will be denoted $v(t)$ (i.e., without the overbar); the displacements which are determined will represent dynamic response. Therefore, total deflections, stresses, etc. are obtained by adding the corresponding static quantities to the results of the dynamic analysis.

2-4   INFLUENCE OF SUPPORT EXCITATION

Dynamic stresses and deflections can be induced in a structure not only by a time-varying applied load, as indicated in Figs. 2-1 and 2-2, but also by motions of its support points. Important examples of such excitation are the motions of a building foundation caused by an earthquake or motions of the base support of a piece of equipment due to vibrations of the building in which it is housed. A simplified model of the earthquake-excitation problem is shown in Fig. 2-3, in which the horizontal ground motion caused by the earthquake is indicated by the displacement $v_g(t)$ of the structure's base relative to the fixed reference axis.

The horizontal girder in this frame is assumed to be rigid and to include all the moving mass of the structure. The vertical columns are assumed to be weightless and inextensible in the vertical (axial) direction, and the resistance to girder displacement provided by each column is represented by its spring constant $k/2$. The mass thus has a single degree of freedom, $v(t)$, which is associated with column flexure; the damper c provides a velocity-proportional resistance to the motion in this coordinate.

As shown in Fig. 2-3b, the equilibrium of forces for this system can be written as

$$f_I(t) + f_D(t) + f_S(t) = 0 \qquad (2\text{-}12)$$

in which the damping and elastic forces can be expressed as in Eqs. (2-2). However, the inertial force in this case is given by

$$f_I(t) = m \, \ddot{v}^t(t) \qquad (2\text{-}13)$$

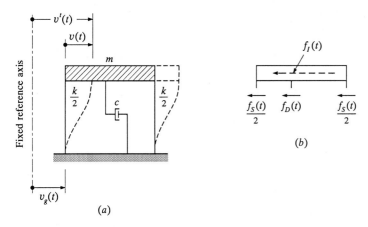

FIGURE 2-3
Influence of support excitation on SDOF equilibrium: (a) motion of system;
(b) equilibrium forces.

where $v^t(t)$ represents the total displacement of the mass from the fixed reference axis. Substituting for the inertial, damping, and elastic forces in Eq. (2-12) yields

$$m\,\ddot{v}^t(t) + c\,\dot{v}(t) + k\,v(t) = 0 \qquad (2\text{-}14)$$

Before this equation can be solved, all forces must be expressed in terms of a single variable, which can be accomplished by noting that the total motion of the mass can be expressed as the sum of the ground motion and that due to column distortion, i.e.,

$$v^t(t) = v(t) + v_g(t) \qquad (2\text{-}15)$$

Expressing the inertial force in terms of the two acceleration components obtained by double differentiation of Eq. (2-15) and substituting the result into Eq. (2-14) yields

$$m\,\ddot{v}(t) + m\,\ddot{v}_g(t) + c\,\dot{v}(t) + k\,v(t) = 0 \qquad (2\text{-}16)$$

or, since the ground acceleration represents the specified dynamic input to the structure, the same equation of motion can more conveniently be written

$$m\,\ddot{v}(t) + c\,\dot{v}(t) + k\,v(t) = -m\,\ddot{v}_g(t) \equiv p_{\text{eff}}(t) \qquad (2\text{-}17)$$

In this equation, $p_{\text{eff}}(t)$ denotes the effective support excitation loading; in other words, the structural deformations caused by ground acceleration $\ddot{v}_g(t)$ are exactly the same as those which would be produced by an external load $p(t)$ equal to $-m\,\ddot{v}_g(t)$. The negative sign in this effective load definition indicates that the effective force opposes the sense of the ground acceleration. In practice this has little significance inasmuch as the engineer is usually only interested in the maximum absolute value of $v(t)$; in this case, the minus sign can be removed from the effective loading term.

An alternative form of the equation of motion can be obtained by using Eq. (2-15) and expressing Eq. (2-14) in terms of $v^t(t)$ and its derivatives, rather than in terms of $v(t)$ and its derivatives, giving

$$m\,\ddot{v}^t(t) + c\,\dot{v}^t(t) + k\,v^t(t) = c\,\dot{v}_g(t) + k\,v_g(t) \tag{2-18}$$

In this formulation, the effective loading shown on the right hand side of the equation depends on the velocity and displacement of the earthquake motion, and the response obtained by solving the equation is total displacement of the mass from a fixed reference rather than displacement relative to the moving base. Solutions are seldom obtained in this manner, however, because the earthquake motion generally is measured in terms of accelerations and the seismic record would have to be integrated once and twice to evaluate the effective loading contributions due to the velocity and displacement of the ground.

2-5   ANALYSIS OF UNDAMPED FREE VIBRATIONS

It has been shown in the preceding sections that the equation of motion of a simple spring-mass system with damping can be expressed as

$$m\,\ddot{v}(t) + c\,\dot{v}(t) + k\,v(t) = p(t) \tag{2-19}$$

in which $v(t)$ represents the dynamic response (that is, the displacement from the static-equilibrium position) and $p(t)$ represents the effective load acting on the system, either applied directly or resulting from support motions.

The solution of Eq. (2-19) will be obtained by considering first the homogeneous form with the right side set equal to zero, i.e.,

$$m\,\ddot{v}(t) + c\,\dot{v}(t) + k\,v(t) = 0 \tag{2-20}$$

Motions taking place with no applied force are called free vibrations, and it is the free-vibration response of the system which now will be examined.

The free-vibration response that is obtained as the solution of Eq. (2-20) may be expressed in the following form:

$$v(t) = G\,\exp(st) \tag{2-21}$$

where G is an arbitrary complex constant and $\exp(st) \equiv e^{st}$ denotes the exponential function. In subsequent discussions it often will be convenient to use complex numbers in expressing dynamic loadings and responses; therefore it is useful now to briefly review the complex number concept.

Considering first the complex constant G, this may be represented as a vector plotted in the complex plane as shown in Fig. 2-4. This sketch demonstrates that the vector may be expressed in terms of its real and imaginary Cartesian components:

$$G = G_R + i\,G_I \qquad (2\text{-}22a)$$

or alternatively that it may be expressed in polar coordinates using its absolute value $\overline{G}$ (the length of the vector) and its angle θ, measured counterclockwise from the real axis:

$$G = \overline{G}\,\exp(i\,\theta) \qquad (2\text{-}22b)$$

In addition, from the trigonometric relations shown in the sketch, it is clear that Eq. (2-22a) also may be written

$$G = \overline{G}\,\cos\theta + i\,\overline{G}\,\sin\theta \qquad (2\text{-}22c)$$

Using this expression and noting that $\cos\theta = \sin\left(\theta + \frac{\pi}{2}\right)$ and $\sin\theta = -\cos\left(\theta + \frac{\pi}{2}\right)$ it is easy to show that multiplying a vector by i has the effect of rotating it counterclockwise in the complex plane through an angle of $\frac{\pi}{2}$ radians or 90 degrees. Similarly it may be seen that multiplying by $-i$ rotates the vector 90° clockwise. Now equating Eq. (2-22c) to Eq. (2-22b), and also noting that a negative imaginary component would be associated with a negative vector angle, leads to Euler's pair of equations that serve to transform from trigonometric to exponential functions:

$$\left.\begin{array}{l} \exp(i\theta) = \cos\theta + i\,\sin\theta \\[4pt] \exp(-i\theta) = \cos\theta - i\sin\theta \end{array}\right\} \qquad (2\text{-}23a)$$

Furthermore, Eqs. (2-23a) may be solved simultaneously to obtain the inverse form of Euler's equations:

$$\left.\begin{array}{l} \cos\theta = \frac{1}{2}\left[\exp(i\theta) + \exp(-i\theta)\right] \\[4pt] \sin\theta = -\frac{i}{2}\left[\exp(i\theta) - \exp(-i\theta)\right] \end{array}\right\} \qquad (2\text{-}23b)$$

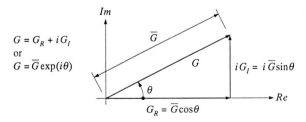

FIGURE 2-4
Complex constant representation in complex plane.

To derive a free-vibration response expression, Eq. (2-21) is substituted into Eq. (2-20), leading to

$$(m\,s^2 + c\,s + k)\,G\,\exp(st) = 0$$

and after dividing by $mG\exp(st)$ and introducing the notation

$$\omega^2 \equiv \frac{k}{m} \tag{2-24}$$

this expression becomes

$$s^2 + \frac{c}{m}\,s + \omega^2 = 0 \tag{2-25}$$

The two values of s that satisfy this quadratic expression depend on the value of c relative to the values of k and m; thus the type of motion given by Eq. (2-21) depends on the amount of damping in the system.

Considering now the undamped system for which $c = 0$, it is evident that the two values of s given by solving Eq.(2-25) are

$$s_{1,2} = \pm\,i\,\omega \tag{2-26}$$

Thus the total response includes two terms of the form of Eq. (2-21), as follows:

$$v(t) = G_1\,\exp(i\omega t) + G_2\,\exp(-i\omega t) \tag{2-27}$$

in which the two exponential terms result from the two values of s, and the complex constants G_1 and G_2 represent the (as yet) arbitrary amplitudes of the corresponding vibration terms.

We now establish the relation between these constants by expressing each of them in terms of its real and imaginary components:

$$G_1 = G_{1R} + i\,G_{1I} \quad ; \quad G_2 = G_{2R} + i\,G_{2I}$$

and by transforming the exponential terms to trigonometric form using Eqs. (2-23a), so that Eq. (2-27) becomes

$$v(t) = \big(G_{1R} + i\,G_{1I}\big)\,\big(\cos\omega t + i\,\sin\omega t\big) + \big(G_{2R} + i\,G_{2I}\big)\,\big(\cos\omega t - i\,\sin\omega t\big)$$

or after simplifying

$$\begin{aligned} v(t) = {}& (G_{1R} + G_{2R})\cos\omega t - (G_{1I} - G_{2I})\sin\omega t \\ & + i\left[(G_{1I} + G_{2I})\cos\omega t + (G_{1R} - G_{2R})\sin\omega t\right] \end{aligned} \tag{2-28}$$

However, this free-vibration response must be real, so the imaginary term (shown in square brackets) must be zero for all values of t, and this condition requires that

$$G_{1I} = -G_{2I} \equiv G_I \qquad\qquad G_{1R} = G_{2R} \equiv G_R$$

From this it is seen that G_1 and G_2 are a complex conjugate pair:

$$G_1 = G_R + i\,G_I \qquad\qquad G_2 = G_R - i\,G_I$$

and with these Eq. (2-27) becomes finally

$$v(t) = (G_R + i\,G_I)\,\exp(i\omega t) + (G_R - i\,G_I)\,\exp(-i\omega t) \qquad (2\text{-}29)$$

The response given by the first term of Eq. (2-29) is depicted in Fig. 2-5 as a vector representing the complex constant G_1 rotating in the counterclockwise direction with the angular velocity ω; also shown are its real and imaginary constants. It will be noted that the resultant response vector $(G_R + i\,G_I)\exp(i\omega t)$ leads vector $G_R\exp(i\omega t)$ by the phase angle θ; moreover it is evident that the response also can be expressed in terms of the absolute value, $\overline{G}$, and the combined angle $(\omega t + \theta)$. Examination of the second term of Eq. (2-29) shows that the response associated with it is entirely equivalent to that shown in Fig. 2-5 except that the resultant vector $\overline{G}\exp[-i(\omega t + \theta)]$ is rotating in the clockwise direction and the phase angle by which it leads the component $G_R\exp(-i\omega t)$ also is in the clockwise direction.

The two counter-rotating vectors $\overline{G}\exp[i(\omega t + \theta)]$ and $\overline{G}\exp[-i(\omega t + \theta)]$ that represent the total free-vibration response given by Eq. (2-29) are shown in Fig. 2-6;

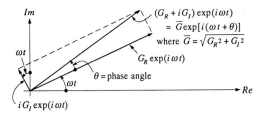

FIGURE 2-5
Portrayal of first term of Eq. (2-29).

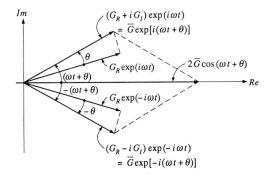

FIGURE 2-6
Total free-vibration response.

it is evident here that the imaginary components of the two vectors cancel each other leaving only the real vibratory motion

$$v(t) = 2\,\overline{G}\,\cos(\omega t + \theta) \tag{2-30}$$

An alternative for this real motion expression may be derived by applying the Euler transformation Eq. (2-23a) to Eq. (2-29), with the result

$$v(t) = A\,\cos\omega t + B\,\sin\omega t \tag{2-31}$$

in which $A = 2G_R$ and $B = -2G_I$. The values of these two constants may be determined from the initial conditions, that is, the displacement $v(0)$ and velocity $\dot{v}(0)$ at time $t = 0$ when the free vibration was set in motion. Substituting these into Eq. (2-31) and its first time derivative, respectively, it is easy to show that

$$v(0) = A = 2G_R \qquad \frac{\dot{v}(0)}{\omega} = B = -2G_I \tag{2-32}$$

Thus Eq. (2-31) becomes

$$v(t) = v(0)\,\cos\omega t + \frac{\dot{v}(0)}{\omega}\,\sin\omega t \tag{2-33}$$

This solution represents a simple harmonic motion (SHM) and is portrayed graphically in Fig. 2-7. The quantity ω, which we have identified previously as the angular velocity (measured in radians per unit of time) of the vectors rotating in the complex plane, also is known as the circular frequency. The cyclic frequency, usually referred to as the frequency of motion, is given by

$$f = \frac{\omega}{2\pi} \tag{2-34}$$

Its reciprocal

$$\frac{1}{f} = \frac{2\pi}{\omega} = T \tag{2-35}$$

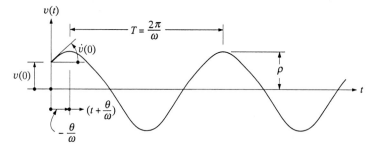

FIGURE 2-7
Undamped free-vibration response.

is the time required to complete one cycle and is called the period of the motion. Usually for structural and mechanical systems the period T is measured in seconds and the frequency is measured in cycles per second, commonly referred to as Hertz (Hz).

The motion represented by Eq. (2-33) and depicted by Fig. 2-7 also may be interpreted in terms of a pair of vectors, $v(0)$ and $\frac{\dot{v}(0)}{\omega}$ rotating counter-clockwise in the complex plane with angular velocity ω, as shown in Fig. 2-8. Using previously stated relations among the free-vibration constants and the initial conditions, it may be seen that Fig. 2-8 is equivalent to Fig. 2-5, but with double amplitude and with a negative phase angle to correspond with positive initial conditions. Accordingly, the amplitude $\rho = 2\overline{G}$, and as shown by Eq. (2-30) the free vibration may be expressed as

$$v(t) = \rho \, \cos(\omega t + \theta) \tag{2-36}$$

in which the amplitude is given by

$$\rho = \sqrt{[v(0)]^2 + \left[\frac{\dot{v}(0)}{\omega}\right]^2} \tag{2-37}$$

and the phase angle by

$$\theta = \tan^{-1}\left[\frac{-\dot{v}(0)}{\omega \, v(0)}\right] \tag{2-38}$$

2-6 DAMPED FREE VIBRATIONS

If damping is present in the system, the solution of Eq. (2-25) which defines the response is

$$s_{1,2} = -\frac{c}{2m} \pm \sqrt{\left(\frac{c}{2m}\right)^2 - \omega^2} \tag{2-39}$$

Three types of motion are represented by this expression, according to whether the quantity under the square-root sign is positive, negative, or zero. It is convenient to

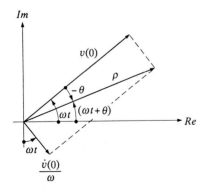

FIGURE 2-8
Rotating vector representation of undamped free vibration.

discuss first the case when the radical term vanishes, which is called the critically-damped condition.

Critically-Damped Systems

If the radical term in Eq. (2-39) is set equal to zero, it is evident that $c/2m = w$; thus, the critical value of the damping coefficient, c_c, is

$$c_c = 2mw \tag{2-40}$$

Then both values of s given by Eq. (2-39) are the same, i.e.,

$$s_1 = s_2 = -\frac{c_c}{2m} = -w \tag{2-41}$$

The solution of Eq. (2-20) in this special case must now be of the form

$$v(t) = (G_1 + G_2 t) \exp(-wt) \tag{2-42}$$

in which the second term must contain t since the two roots of Eq. (2-25) are identical. Because the exponential term $\exp(-wt)$ is a real function, the constants G_1 and G_2 must also be real.

Using the initial conditions $v(0)$ and $\dot{v}(0)$, these constants can be evaluated leading to

$$v(t) = \left[v(0) (1 - wt) + \dot{v}(0) t\right] \exp(-wt) \tag{2-43}$$

which is portrayed graphically in Fig. 2-9 for positive values of $v(0)$ and $\dot{v}(0)$. Note that this free response of a critically-damped system does not include oscillation about the zero-deflection position; instead it simply returns to zero asymptotically in accordance with the exponential term of Eq. (2-43). However, a single zero-displacement crossing would occur if the signs of the initial velocity and displacement were different from each other. A very useful definition of the critically-damped

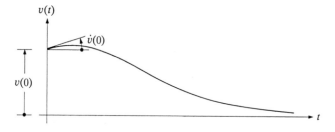

FIGURE 2-9
Free-vibration response with critical damping.

condition described above is that it represents the smallest amount of damping for which no oscillation occurs in the free-vibration response.

Undercritically-Damped Systems

If damping is less than critical, that is, if $c < c_c$ (i.e., $c < 2m\omega$), it is apparent that the quantity under the radical sign in Eq. (2-39) is negative. To evaluate the free-vibration response in this case, it is convenient to express damping in terms of a damping ratio ξ which is the ratio of the given damping to the critical value;

$$\xi \equiv \frac{c}{c_c} = \frac{c}{2m\omega} \tag{2-44}$$

Introducing Eq. (2-44) into Eq. (2-39) leads to

$$s_{1,2} = -\xi\omega \pm i\,\omega_D \tag{2-45}$$

where

$$\omega_D \equiv \omega\sqrt{1-\xi^2} \tag{2-46}$$

is the free-vibration frequency of the damped system. Making use of Eq. (2-21) and the two values of s given by Eq. (2-45), the free-vibration response becomes

$$v(t) = \left[G_1 \exp(i\omega_D t) + G_2 \exp(-i\omega_D t)\right]\exp(-\xi\omega t) \tag{2-47}$$

in which the constants G_1 and G_2 must be complex conjugate pairs for the response $v(t)$ to be real, i.e., $G_1 = G_R + iG_I$ and $G_2 = G_R - iG_I$ similar to the undamped case shown by Eq. (2-27).

The response given by Eq. (2-47) can be represented by vectors in the complex plane similar to those shown in Fig. 2-6 for the undamped case; the only difference is that the damped circular frequency ω_D must be substituted for the undamped circular frequency ω and the magnitudes of the vectors must be forced to decay exponentially with time in accordance with the term outside of the brackets, $\exp(-\xi\omega t)$.

Following the same procedure used in arriving at Eq. (2-31), Eq. (2-47) also can be expressed in the equivalent trigonometric form

$$v(t) = \left[A \cos\omega_D t + B \sin\omega_D t\right]\exp(-\xi\omega t) \tag{2-48}$$

where $A = 2G_R$ and $B = -2G_I$. Using the initial conditions $v(0)$ and $\dot{v}(0)$, constants A and B can be evaluated leading to

$$v(t) = \left[v(0)\cos\omega_D t + \left(\frac{\dot{v}(0) + v(0)\xi\omega}{\omega_D}\right)\sin\omega_D t\right]\exp(-\xi\omega t) \tag{2-49}$$

Alternatively, this response can be written in the form

$$v(t) = \rho \, \cos(\omega_D t + \theta) \, \exp(-\xi\omega t) \tag{2-50}$$

in which

$$\rho = \left\{ v(0)^2 + \left(\frac{\dot{v}(0) + v(0)\xi\omega}{\omega_D} \right)^2 \right\}^{1/2} \tag{2-51}$$

$$\theta = - \tan^{-1} \left(\frac{\dot{v}(0) + v(0)\xi\omega}{\omega_D \, v(0)} \right) \tag{2-52}$$

Note that for low damping values which are typical of most practical structures, $\xi < 20\%$, the frequency ratio ω_D/ω as given by Eq. (2-46) is nearly equal to unity. The relation between damping ratio and frequency ratio may be depicted graphically as a circle of unit radius as shown in Fig. 2-10.

A plot of the response of an undercritically-damped system subjected to an initial displacement $v(0)$ but starting with zero velocity is shown in Fig. 2-11. It is of

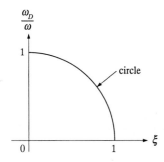

FIGURE 2-10
Relationship between frequency ratio and damping ratio.

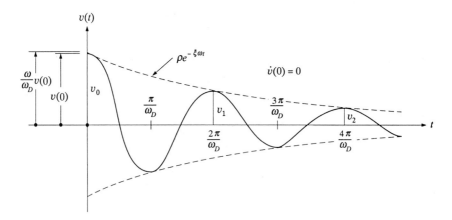

FIGURE 2-11
Free-vibration response of undercritically-damped system.

interest to note that the underdamped system oscillates about the neutral position, with a constant circular frequency ω_D. The rotating-vector representation of Eq. (2-47) is equivalent to Fig. 2-6 except that ω is replaced by ω_D and the lengths of the vectors diminish exponentially as the response damps out.

The true damping characteristics of typical structural systems are very complex and difficult to define. However, it is common practice to express the damping of such real systems in terms of equivalent viscous-damping ratios ξ which show similar decay rates under free-vibration conditions. Therefore, let us now relate more fully the viscous-damping ratio ξ to the free-vibration response shown in Fig. 2-11.

Consider any two successive positive peaks such as v_n and v_{n+1} which occur at times $n\left(\frac{2\pi}{\omega_D}\right)$ and $(n+1)\frac{2\pi}{\omega_D}$, respectively. Using Eq. (2-50), the ratio of these two successive values is given by

$$v_n/v_{n+1} = \exp(2\pi\xi\omega/\omega_D) \qquad (2\text{-}53)$$

Taking the natural logarithm (ln) of both sides of this equation and substituting $\omega_D = \omega\sqrt{1-\xi^2}$, one obtains the so-called logarithmic decrement of damping, δ, defined by

$$\delta \equiv \ln\frac{v_n}{v_{n+1}} = \frac{2\pi\xi}{\sqrt{1-\xi^2}} \qquad (2\text{-}54)$$

For low values of damping, Eq. (2-54) can be approximated by

$$\delta \doteq 2\pi\xi \qquad (2\text{-}55)$$

where the symbol $\doteq$ represents "approximately equal," thus,

$$\frac{v_n}{v_{n+1}} = \exp(\delta) \doteq \exp(2\pi\xi) = 1 + 2\pi\xi + \frac{(2\pi\xi)^2}{2!} + \cdots \qquad (2\text{-}56)$$

Sufficient accuracy is obtained by retaining only the first two terms in the Taylor's series expansion on the right hand side, in which case

$$\xi \doteq \frac{v_n - v_{n+1}}{2\pi\, v_{n+1}} \qquad (2\text{-}57)$$

To illustrate the accuracy of Eq. (2-57), the ratio of the exact value of ξ as given by Eq. (2-54) to the approximate value as given by Eq. (2-57) is plotted against the approximate value in Fig. 2-12. This graph permits one to correct the damping ratio obtained by the approximate method.

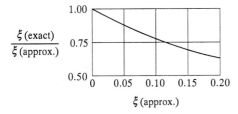

FIGURE 2-12
Damping-ratio correction factor to be applied to result obtained from Eq. (2-57).

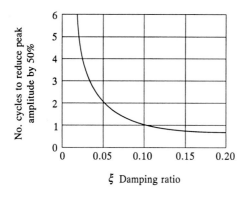

FIGURE 2-13
Damping ratio vs. number of cycles required to reduce peak amplitude by 50 percent.

For lightly damped systems, greater accuracy in evaluating the damping ratio can be obtained by considering response peaks which are several cycles apart, say m cycles; then

$$\ln \frac{v_n}{v_{n+m}} = \frac{2m\pi\xi}{\sqrt{1-\xi^2}} \tag{2-58}$$

which can be simplified for low damping to an approximate relation equivalent to Eq. (2-57):

$$\xi \doteq \frac{v_n - v_{n+m}}{2\,m\,\pi\,v_{n+m}} \tag{2-59}$$

When damped free vibrations are observed experimentally, a convenient method for estimating the damping ratio is to count the number of cycles required to give a 50 percent reduction in amplitude. The relationship to be used in this case is presented graphically in Fig. 2-13. As a quick rule of thumb, it is convenient to remember that for percentages of critical damping equal to 10, 5, and 2.5, the corresponding amplitudes are reduced by 50 percent in approximately one, two, and four cycles, respectively.

Example E2-1.   A one-story building is idealized as a rigid girder supported by weightless columns, as shown in Fig. E2-1. In order to evaluate the dynamic properties of this structure, a free-vibration test is made, in which the roof system (rigid girder) is displaced laterally by a hydraulic jack and then suddenly released. During the jacking operation, it is observed that a force of 20 $kips$ [9,072 kg] is required to displace the girder 0.20 in [0.508 cm]. After the instantaneous release of this initial displacement, the maximum displacement on the first return swing is only 0.16 in [0.406 cm] and the period of this displacement cycle is $T = 1.40$ sec.

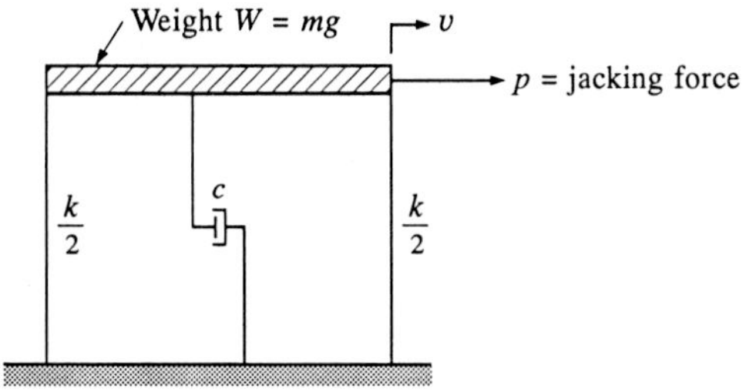

Weight $W = mg$

v

p = jacking force

$\dfrac{k}{2}$

c

$\dfrac{k}{2}$

FIGURE E2-1
Vibration test of a simple building.

From these data, the following dynamic behavioral properties are determined:

(1) Effective weight of the girder:

$$T = \frac{2\pi}{\omega} = 2\pi \sqrt{\frac{W}{gk}} = 1.40 \ sec$$

Hence

$$W = \left(\frac{1.40}{2\pi}\right)^2 gk = 0.0496 \frac{20}{0.2} 386 = 1,920 \ kips \quad [870.9 \times 10^3 \ kg]$$

where the acceleration of gravity is taken to be $g = 386 \ in/sec^2$

(2) Undamped frequency of vibration:

$$f = \frac{1}{T} = \frac{1}{1.40} = 0.714 \ Hz$$

$$\omega = 2\pi f = 4.48 \ rad/sec$$

(3) Damping properties:

Logarithmic decrement: $\quad \delta = \ln \dfrac{0.20}{0.16} = 0.223$

Damping ratio: $\qquad\quad \xi \doteq \dfrac{\delta}{2\pi} = 3.55\%$

Damping coefficient: $\quad c = \xi c_c = \xi \, 2m\omega = 0.0355 \dfrac{2(1,920)}{386} 4.48$

$$= 1.584 \ kips \cdot sec/in \quad [282.9 \ kg \cdot sec/cm]$$

Damped frequency: $\quad \omega_D = \omega \sqrt{1 - \xi^2} = \omega(0.999)^{1/2} \doteq \omega$

(4) Amplitude after six cycles:

$$v_6 = \left(\frac{v_1}{v_0}\right)^6 v_0 = \left(\frac{4}{5}\right)^6 (0.20) = 0.0524 \ in \quad [0.1331 \ cm]$$

Overcritically-Damped Systems

Although it is very unusual under normal conditions to have overcritically-damped structural systems, they do sometimes occur as mechanical systems; therefore, it is useful to carry out the response analysis of an overcritically-damped system to make this presentation complete. In this case having $\xi \equiv c/c_c > 1$, it is convenient to write Eq. (2-39) in the form

$$s_{1,2} = -\xi w \pm w\sqrt{\xi^2 - 1} = -\xi w \pm \hat{w} \qquad (2\text{-}60)$$

in which

$$\hat{w} \equiv w\sqrt{\xi^2 - 1} \qquad (2\text{-}61)$$

Substituting the two values of s given by Eq. (2-60) into Eq. (2-21) and simplifying leads eventually to

$$v(t) = [A \sinh \hat{w}t + B \cosh \hat{w}t] \exp(-\xi wt) \qquad (2\text{-}62)$$

in which the real constants A and B can be evaluated using the initial conditions $v(0)$ and $\dot{v}(0)$. It is easily shown from the form of Eq. (2-62) that the response of an overcritically-damped system is similar to the motion of a critically-damped system as shown in Fig. 2-9; however, the asymptotic return to the zero-displacement position is slower depending upon the amount of damping.

PROBLEMS

2-1. The weight W of the building of Fig. E2-1 is 200 *kips* and the building is set into free vibration by releasing it (at time $t = 0$) from a displacement of 1.20 *in*. If the maximum displacement on the return swing is 0.86 *in* at time $t = 0.64$ *sec*, determine:

    (a) the lateral spring stiffness k

    (b) the damping ratio ξ

    (c) the damping coefficient c

2-2. Assume that the mass and stiffness of the structure of Fig. 2-1a are as follows: $m = 2$ *kips* $\cdot$ *sec*^{2}/*in*, $k = 40$ *kips*/*in*. If the system is set into free vibration with the initial conditions $v(0) = 0.7$ *in* and $\dot{v}(0) = 5.6$ *in*/*sec*, determine the displacement and velocity at $t = 1.0$ *sec*, assuming:

    (a) $c = 0$ (undamped system)

    (b) $c = 2.8$ *kips* $\cdot$ *sec*/*in*

2-3. Assume that the mass and stiffness of the system of Fig. 2-1a are $m = 5$ *kips* $\cdot$ *sec*^{2}/*in* and $k = 20$ *kips*/*in*, and that it is undamped. If the initial displacement is $v(0) = 1.8$ *in*, and the displacement at $t = 1.2$ *sec* is also 1.8 *in*, determine:

    (a) the displacement at $t = 2.4$ *sec*

    (b) the amplitude of free vibration ρ

CHAPTER

3

RESPONSE TO HARMONIC LOADING

3-1 UNDAMPED SYSTEM

Complementary Solution

Assume the system of Fig. 2-1 is subjected to a harmonically varying load $p(t)$ of sine-wave form having an amplitude p_o and circular frequency $\overline{\omega}$ as shown by the equation of motion

$$m\,\ddot{v}(t) + c\,\dot{v}(t) + k\,v(t) = p_o\,\sin\overline{\omega}t \tag{3-1}$$

Before considering this viscously damped case, it is instructive to examine the behavior of an undamped system as controlled by

$$m\,\ddot{v}(t) + k\,v(t) = p_o\,\sin\overline{\omega}t \tag{3-2}$$

which has a complementary solution of the free-vibration form of Eq. (2-31)

$$v_c(t) = A\,\cos\omega t + B\,\sin\omega t \tag{3-3}$$

Particular Solution

The general solution must also include the particular solution which depends upon the form of dynamic loading. In this case of harmonic loading, it is reasonable to assume that the corresponding motion is harmonic and in phase with the loading; thus, the particular solution is

$$v_p(t) = C\,\sin\overline{\omega}t \tag{3-4}$$

33

in which the amplitude C is to be evaluated.

Substituting Eq. (3-4) into Eq. (3-2) gives

$$-m\bar{\omega}^2 C \sin\bar{\omega}t + kC \sin\bar{\omega}t = p_o \sin\bar{\omega}t \qquad (3\text{-}5)$$

Dividing through by $\sin\bar{\omega}t$ (which is nonzero in general) and by k and noting that $k/m = \omega^2$, one obtains after some rearrangement

$$C = \frac{p_o}{k}\left[\frac{1}{1-\beta^2}\right] \qquad (3\text{-}6)$$

in which β is defined as the ratio of the applied loading frequency to the natural free-vibration frequency, i.e.,

$$\beta \equiv \bar{\omega}\,/\,\omega \qquad (3\text{-}7)$$

General Solution

The general solution of Eq. (3-2) is now obtained by combining the complementary and particular solutions and making use of Eq. (3-6); thus, one obtains

$$v(t) = v_c(t) + v_p(t) = A\,\cos\omega t + B\,\sin\omega t + \frac{p_o}{k}\left[\frac{1}{1-\beta^2}\right]\sin\bar{\omega}t \qquad (3\text{-}8)$$

In this equation, the values of A and B depend on the conditions with which the response was initiated. For the system starting from rest, i.e., $v(0) = \dot{v}(0) = 0$, it is easily shown that

$$A = 0 \qquad\qquad B = -\frac{p_o\beta}{k}\left[\frac{1}{1-\beta^2}\right] \qquad (3\text{-}9)$$

in which case the response of Eq. (3-8) becomes

$$v(t) = \frac{p_o}{k}\left[\frac{1}{1-\beta^2}\right](\sin\bar{\omega}t - \beta\,\sin\omega t) \qquad (3\text{-}10)$$

where $p_o/k = v_{st}$ is the displacement which would be produced by the load p_o applied statically and $1/(1-\beta^2)$ is the magnification factor (MF) representing the amplification effect of the harmonically applied loading. In this equation, $\sin\bar{\omega}t$ represents the response component at the frequency of the applied loading; it is called the steady-state response and is directly related to the loading. Also $\beta\sin\omega t$ is the response component at the natural vibration frequency and is the free-vibration effect controlled by the initial conditions. Since in a practical case, damping will cause the last term to vanish eventually, it is termed the transient response. For this hypothetical undamped system, however, this term will not damp out but will continue indefinitely.

Response Ratio — A convenient measure of the influence of dynamic loading is provided by the ratio of the dynamic displacement response to the displacement produced by static application of load p_o, i.e.,

$$R(t) \equiv \frac{v(t)}{v_{st}} = \frac{v(t)}{p_o/k} \tag{3-11}$$

From Eq. (3-10) it is evident that the response ratio resulting from the sine-wave loading of an undamped system starting from rest is

$$R(t) = \left[\frac{1}{1 - \beta^2}\right](\sin \bar{\omega} t - \beta \sin \omega t) \tag{3-12}$$

It is informative to examine this response behavior in more detail by reference to Fig. 3-1. Figure 3-1a represents the steady-state component of response while Fig. 3-1b represents the so-called transient response. In this example, it is assumed that $\beta = 2/3$, that is, the applied loading frequency is two-thirds of the free-vibration frequency. The total response $R(t)$, i.e., the sum of both types of response, is shown in Fig. 3-1c. Two points are of interest: (1) the tendency for the two components

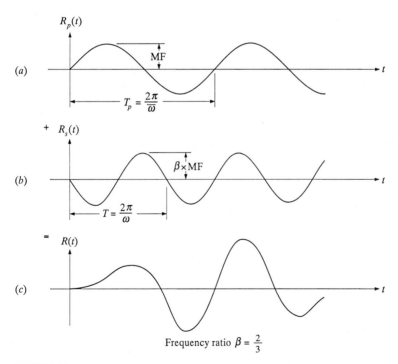

FIGURE 3-1
Response ratio produced by sine wave excitation starting from at-rest initial conditions:
(a) steady state; (b) transient; (c) total R(t).

to get in phase and then out of phase again, causing a "beating" effect in the total response; and (2) the zero slope of total response at time $t = 0$, showing that the initial velocity of the transient response is just sufficient to cancel the initial velocity of the steady-state response; thus, it satisfies the specified initial condition $\dot{v}(0) = 0$.

3-2 SYSTEM WITH VISCOUS DAMPING

Returning to the equation of motion including viscous damping, Eq. (3-1), dividing by m, and noting that $c/m = 2\,\xi\,\omega$ leads to

$$\ddot{v}(t) + 2\,\xi\,\omega\,\dot{v}(t) + \omega^2\,v(t) = \frac{p_o}{m}\,\sin\overline{\omega}t \qquad (3\text{-}13)$$

The complementary solution of this equation is the damped free-vibration response given by Eq. (2-48), i.e.,

$$v_c(t) = \left[A\,\cos\omega_D t + B\,\sin\omega_D t\right]\,\exp(-\xi\,\omega\,t) \qquad (3\text{-}14)$$

The particular solution to Eq. (3-13) is of the form

$$v_p(t) = G_1\,\cos\overline{\omega}t + G_2\,\sin\overline{\omega}t \qquad (3\text{-}15)$$

in which the cosine term is required as well as the sine term because, in general, the response of a damped system is not in phase with the loading.

Substituting Eq. (3-15) into Eq. (3-13) and separating the multiples of $\cos\overline{\omega}t$ from the multiples of $\sin\overline{\omega}t$ leads to

$$\left[-G_1\overline{\omega}^2 + G_2\overline{\omega}\,(2\xi\omega) + G_1\,\omega^2\right]\,\cos\overline{\omega}t$$

$$+ \left[-G_2\overline{\omega}^2 - G_1\overline{\omega}\,(2\xi\omega) + G_2\,\omega^2 - \frac{p_o}{m}\right]\,\sin\overline{\omega}t = 0 \qquad (3\text{-}16)$$

In order to satisfy this equation for all values of t, it is necessary that each of the two square bracket quantities equal zero; thus, one obtains

$$G_1\,(1 - \beta^2) + G_2\,(2\xi\beta) = 0$$

$$G_2\,(1 - \beta^2) - G_1\,(2\xi\beta) = \frac{p_o}{k} \qquad (3\text{-}17)$$

in which β is the frequency ratio given by Eq. (3-7). Solving these two equations simultaneously yields

$$G_1 = \frac{p_o}{k}\left[\frac{-2\xi\beta}{(1 - \beta^2)^2 + (2\xi\beta)^2}\right]$$

$$G_2 = \frac{p_o}{k}\left[\frac{1 - \beta^2}{(1 - \beta^2)^2 + (2\xi\beta)^2}\right] \qquad (3\text{-}18)$$

Introducing these expressions into Eq. (3-15) and combining the result with the complementary solution of Eq. (3-14), the total response is obtained in the form

$$v(t) = \left[A \cos \omega_D t + B \sin \omega_D t \right] \exp(-\xi \omega t)$$

$$+ \frac{p_o}{k} \left[\frac{1}{(1 - \beta^2)^2 + (2\xi\beta)^2} \right] \left[(1 - \beta^2) \sin \overline{\omega} t - 2\xi\beta \cos \overline{\omega} t \right] \qquad (3\text{-}19)$$

The first term on the right hand side of this equation represents the transient response, which damps out in accordance with $\exp(-\xi \omega t)$, while the second term represents the steady-state harmonic response, which will continue indefinitely. The constants A and B can be evaluated for any given initial conditions, $v(0)$ and $\dot{v}(0)$. However, since the transient response damps out quickly, it is usually of little interest; therefore, the evaluation of constants A and B will not be pursued here.

<u>Steady-State Harmonic Response</u> — Of great interest, however, is the steady-state harmonic response given by the second term of Eq. (3-19)

$$v_p(t) = \frac{p_o}{k} \left[\frac{1}{(1 - \beta^2)^2 + (2\xi\beta)^2} \right] \left[(1 - \beta^2) \sin \overline{\omega} t - 2\xi\beta \cos \overline{\omega} t \right] \qquad (3\text{-}20)$$

This steady-state displacement behavior can be interpreted easily by plotting two corresponding rotating vectors in the complex plane as shown in Fig. 3-2, where their components along the real axis are identical to the two terms in Eq. (3-20). The real component of the resultant vector, $-\rho i \exp[i(\overline{\omega}t - \theta)]$, gives the steady-state response in the form

$$v_p(t) = \rho \sin(\overline{\omega} t - \theta) \qquad (3\text{-}21)$$

having an amplitude

$$\rho = \frac{p_o}{k} \left[(1 - \beta^2)^2 + (2\xi\beta)^2 \right]^{-1/2} \qquad (3\text{-}22)$$

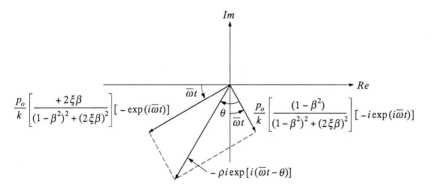

FIGURE 3-2
Steady-state displacement response.

and a phase angle, θ, by which the response lags behind the applied loading

$$\theta = \tan^{-1}\left[\frac{2\xi\beta}{1 - \beta^2}\right] \tag{3-23}$$

It should be understood that this phase angle is limited to the range $0 < \theta < 180°$.

The ratio of the resultant harmonic response amplitude to the static displacement which would be produced by the force p_o will be called the dynamic magnification factor D; thus

$$D \equiv \frac{\rho}{p_o/k} = \left[(1 - \beta^2)^2 + (2\xi\beta)^2\right]^{-1/2} \tag{3-24}$$

It is seen that both the dynamic magnification factor D and the phase angle θ vary with the frequency ratio β and the damping ratio ξ. Plots of D vs. β and θ vs. β are shown in Figs. 3-3 and 3-4, respectively, for discrete values of damping ratio, ξ.

At this point it is instructive to solve for the steady-state harmonic response once again using an exponential form of solution. Consider the general case of

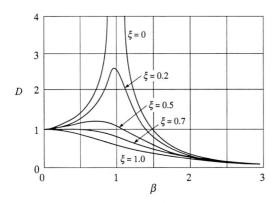

FIGURE 3-3
Variation of dynamic magnification factor with damping and frequency.

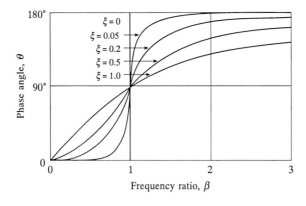

FIGURE 3-4
Variation of phase angle with damping and frequency.

harmonic loading expressed in exponential form:

$$\ddot{v}(t) + 2\,\xi\,\omega\,\dot{v}(t) + \omega^2\,v(t) = \frac{p_o}{m}\,\exp[i\,(\overline{\omega}t + \phi)] \tag{3-25}$$

where ϕ is an arbitrary phase angle in the harmonic loading function. In dealing with completely general harmonic loads, especially for the case of periodic loading where the excitation is expressed as a series of harmonic terms, it is essential to define the input phase angle for each harmonic; however, this usually is accomplished most conveniently by expressing the input in complex number form rather than by amplitude and phase angle. In this chapter only a single harmonic loading term will be considered; therefore, its phase angle is arbitrarily taken to be zero for simplicity, so it need not be included in the loading expression.

The particular solution of Eq. (3-25) and its first and second time derivatives are

$$v_p(t) = G\,\exp(i\overline{\omega}t)$$

$$\dot{v}_p(t) = i\overline{\omega}\,G\,\exp(i\overline{\omega}t) \tag{3-26}$$

$$\ddot{v}_p(t) = -\overline{\omega}^2\,G\,\exp(i\overline{\omega}t)$$

where G is a complex constant. To evaluate G, substitute Eqs. (3-26) into Eq. (3-25), cancel out the quantity $\exp(i\overline{\omega}t)$ common to each term, substitute $k\,\omega^2$ for m and β for $\overline{\omega}/\omega$, and solve for G yielding

$$G = \frac{p_o}{k}\left[\frac{1}{(1-\beta^2)+i\,(2\xi\beta)}\right] = \frac{p_o}{k}\left[\frac{(1-\beta^2)-i\,(2\xi\beta)}{(1-\beta^2)^2+(2\xi\beta)^2}\right] \tag{3-27}$$

Substituting this complex value of G into the first of Eqs. (3-26) and plotting the resulting two vectors in the complex plane, one obtains the representation shown in Fig. 3-5. Note that these two vectors and their resultant along with phase angle θ

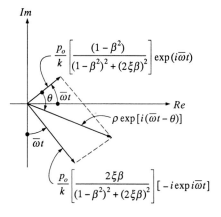

FIGURE 3-5
Steady-state response using viscous damping.

are identical to the corresponding quantities in Fig. 3-2, except that now the set of vectors has been rotated counterclockwise through 90 degrees. This difference in the figures corresponds to the phase angle difference between the harmonic excitations $-i\,(p_o/m)\,\exp(i\overline{\omega}t)$ and $(p_o/m)\,\exp(i\overline{\omega}t)$ producing the results of Figs. 3-2 and 3-5, respectively. Note that $(p_o/m)\,\sin\overline{\omega}t$ is the real part of $-i\,(p_o/m)\,\exp(i\overline{\omega}t)$.

It is of interest to consider the balance of forces acting on the mass under the above steady-state harmonic condition whereby the total response, as shown in Fig. 3-5, is

$$v_p(t) = \rho\,\exp[i\,(\overline{\omega}t - \theta)] \tag{3-28}$$

having an amplitude ρ as given by Eq. (3-22). Force equilibrium requires that the sum of the inertial, damping, and spring forces equal the applied loading

$$p(t) = p_o\,\exp(i\overline{\omega}t) \tag{3-29}$$

Using Eq. (3-28), these forces are

$$f_{I_p}(t) = m\,\ddot{v}_p(t) = -m\overline{\omega}^2\,\rho\,\exp[i\,(\overline{\omega}t - \theta)]$$

$$f_{D_p}(t) = c\,\dot{v}_p(t) = ic\overline{\omega}\,\rho\,\exp[i\,(\overline{\omega}t - \theta)] \tag{3-30}$$

$$f_{S_p}(t) = k\,v_p(t) = k\,\rho\,\exp[i\,(\overline{\omega}t - \theta)]$$

which along with the applied loading are shown as vectors in the complex plane of Fig. 3-6.   Also shown is the closed polygon of forces required for equilibrium in accordance with Eq. (2-1). Note that although the inertial, damping, and spring forces as given in Eqs. (3-30) are in phase with the acceleration, velocity, and displacement

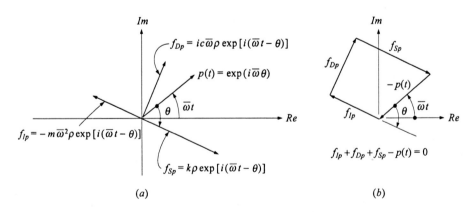

(a)

(b)

FIGURE 3-6
Steady-state harmonic forces using viscous damping: (a) complex plane representation; (b) closed force polygon representation.

motions, respectively, they actually oppose their corresponding motions in accordance with the sign convention given in Fig. 2-1b which was adopted in Eq. (2-1).

Example E3-1.   A portable harmonic-loading machine provides an effective means for evaluating the dynamic properties of structures in the field. By operating the machine at two different frequencies and measuring the resulting structural-response amplitude and phase relationship in each case, it is possible to determine the mass, damping, and stiffness of a SDOF structure. In a test of this type on a single-story building, the shaking machine was operated at frequencies of $\bar{\omega}_1 = 16 \ rad/sec$ and $\bar{\omega}_2 = 25 \ rad/sec$, with a force amplitude of $500 \ lb$ [$226.8 \ kg$] in each case. The response amplitudes and phase relationships measured in the two cases were

$$\rho_1 = 7.2 \times 10^{-3} \ in \quad [18.3 \times 10^{-3} \ cm] \qquad \cos\theta_1 = 0.966$$
$$\theta_1 = 15° \qquad\qquad\qquad\qquad\qquad\qquad\quad \sin\theta_1 = 0.259$$

$$\rho_2 = 14.5 \times 10^{-3} \ in \quad [36.8 \times 10^{-3} \ cm] \qquad \cos\theta_2 = 0.574$$
$$\theta_2 = 55° \qquad\qquad\qquad\qquad\qquad\qquad\quad \sin\theta_2 = 0.819$$

To evaluate the dynamic properties from these data, it is convenient to rewrite Eq. (3-22) as

$$\rho = \frac{P_o}{k}\frac{1}{1-\beta^2}\left\{\frac{1}{1+[2\xi\beta/(1-\beta^2)]^2}\right\}^{1/2} = \frac{P_o}{k}\frac{\cos\theta}{1-\beta^2} \qquad (a)$$

where the trigonometric function has been derived from Eq. (3-23). With further algebraic simplification this becomes

$$k(1-\beta^2) = k - \bar{\omega}^2\, m = \frac{P_o\,\cos\theta}{\rho}$$

Then introducing the two sets of test data leads to the matrix equation

$$\begin{bmatrix} 1 & -16^2 \\ 1 & -25^2 \end{bmatrix}\begin{bmatrix} k \\ m \end{bmatrix} = 500\ lb \begin{bmatrix} \frac{0.966}{7.2\times10^{-3}} \\ \frac{0.574}{14.5\times10^{-3}} \end{bmatrix}$$

which can be solved to give

$$k = 100 \times 10^3 \ lb/in \quad [17.8 \times 10^3 \ kg/cm]$$
$$m = 128.5 \ lb \cdot sec^2/in \quad [22.95 \ kg \cdot sec^2/cm]$$

Thus,

$$W = m\,g = 49.6 \times 10^3 \ lb \quad [22.5 \times 10^3 \ kg]$$

The natural frequency is given by

$$\omega = \sqrt{\frac{k}{m}} = 27.9 \; rad/sec$$

To determine the damping coefficient, two expressions for $\cos \theta$ can be derived from Eqs. (a) and (3-23). Equating these and solving for the damping ratio leads to

$$\xi = \frac{p_o \sin \theta}{2 \beta k \rho} = \frac{p_o \sin \theta}{c_c \overline{\omega} \rho}$$

Thus with the data of the first test

$$c = \xi c_c = \frac{500 \, (0.259)}{16 \, (7.2 \times 10^{-3})} = 1,125 \; lb \cdot sec/in \quad [200.9 \; kg \cdot sec/cm]$$

and the same result (within engineering accuracy) is given by the data of the second test. The damping ratio therefore is

$$\xi = \frac{c}{2 \, k/\omega} = \frac{1,125 \, (27.9)}{200 \times 10^3} = 15.7\%$$

3-3   RESONANT RESPONSE

From Eq. (3-12), it is apparent that the steady-state response amplitude of an undamped system tends toward infinity as the frequency ratio β approaches unity; this tendency can be seen in Fig. 3-3 for the case of $\xi = 0$. For low values of damping, it is seen in this same figure that the maximum steady-state response amplitude occurs at a frequency ratio slightly less than unity. Even so, the condition resulting when the frequency ratio equals unity, i.e., when the frequency of the applied loading equals the undamped natural vibration frequency, is called resonance. From Eq. (3-24) it is seen that the dynamic magnification factor under this condition ($\beta = 1$) is

$$D_{\beta=1} = \frac{1}{2 \xi} \tag{3-31}$$

To find the maximum or peak value of dynamic magnification factor, one must differentiate Eq. (3-24) with respect to β and solve the resulting expression for β obtaining

$$\beta_{\text{peak}} = \sqrt{1 - 2 \xi^2} \tag{3-32}$$

(which yields positive real values for damping ratios $\xi < 1/\sqrt{2}$), and then substitute this value of frequency ratio back into Eq. (3-24) giving

$$D_{\text{max}} = \frac{1}{2 \xi \sqrt{1 - \xi^2}} = \frac{1}{2 \xi} \frac{\omega}{\omega_D} \tag{3-33}$$

For typical values of structural damping, say $\xi < 0.10$, the difference between Eq. (3-33) and the simpler Eq. (3-31) is small, the difference being one-half of 1 percent for $\xi = 0.10$ and 2 percent for $\xi = 0.20$.

For a more complete understanding of the nature of the resonant response of a structure to harmonic loading, it is necessary to consider the general response Eq. (3-19), which includes the transient term as well as the steady-state term. At the resonant exciting frequency ($\beta = 1$), this equation becomes

$$v(t) = (A \, \cos \omega_D t + B \, \sin \omega_D t) \, \exp(-\xi \omega t) - \frac{P_o}{k} \frac{\cos \omega t}{2 \xi} \qquad (3\text{-}34)$$

Assuming that the system starts from rest $[v(0) = \dot{v}(0) = 0]$, the constants are

$$A = \frac{P_o}{k} \frac{1}{2 \xi} \qquad B = \frac{P_o}{k} \frac{\omega}{2 \omega_D} = \frac{P_o}{k} \frac{1}{2 \sqrt{1 - \xi^2}} \qquad (3\text{-}35)$$

Thus Eq. (3-34) becomes

$$v(t) = \frac{1}{2 \xi} \frac{P_o}{k} \left[\left(\frac{\xi}{\sqrt{1 - \xi^2}} \, \sin \omega_D t + \cos \omega_D t \right) \, \exp(-\xi \omega t) - \cos \omega t \right] \qquad (3\text{-}36)$$

For the amounts of damping to be expected in structural systems, the term $\sqrt{1 - \xi^2}$ is nearly equal to unity; in this case, this equation can be written in the approximate form

$$R(t) = \frac{v(t)}{P_o/k} \doteq \frac{1}{2 \xi} \left\{ \left[\exp(-\xi \omega t) - 1 \right] \, \cos \omega t + \xi \left[\exp(-\xi \omega t) \right] \, \sin \omega t \right\} \qquad (3\text{-}37)$$

For zero damping, this approximate equation is indeterminate; but when L'Hospital's rule is applied, the response ratio for the undamped system is found to be

$$R(t) = \frac{1}{2} \, (\sin \omega t - \omega t \, \cos \omega t) \qquad (3\text{-}38)$$

Plots of these equations are shown in Fig. 3-7. Note that because the terms containing $\sin \omega t$ contribute little to the response, the peak values in this figure build up linearly for the undamped case, changing by an amount π in each cycle; however, they build up in accordance with $(1/2\xi)[\exp(-\xi \omega t) - 1]$ for the damped case. This latter envelope function is plotted against frequency in Fig. 3-8 for discrete values of damping. It is seen that the buildup rate toward the steady-state level $1/2\xi$ increases with damping and that buildup to nearly steady-state level occurs in a relatively small number of cycles for values of damping in the practical range of interest; e.g., 14 cycles brings the response very close to the steady-state level for a case having 5 percent of critical damping.

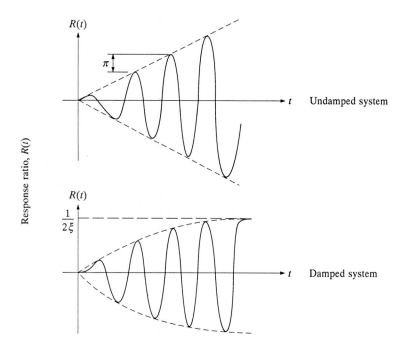

FIGURE 3-7
Response to resonant loading $\beta = 1$ for at-rest initial conditions.

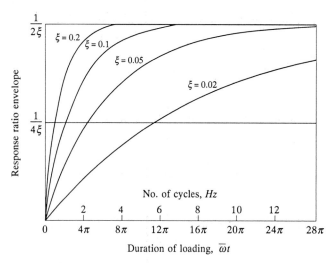

FIGURE 3-8
Rate of buildup of resonant response from rest.

3-4   ACCELEROMETERS AND DISPLACEMENT METERS

At this point it is convenient to discuss the fundamental principles on which the operation of an important class of dynamic measurement devices is based. These are seismic instruments, which consist essentially of a viscously damped oscillator as shown in Fig. 3-9.   The system is mounted in a housing which may be attached to the surface where the motion is to be measured. The response is measured in terms of the motion $v(t)$ of the mass relative to the housing.

The equation of motion for this system already has been shown in Eq. (2-17) to be

$$m\,\ddot{v}(t) + c\,\dot{v}(t) + k\,v(t) = -m\,\ddot{v}_g(t) \equiv p_{\mathrm{eff}}(t)$$

where $\ddot{v}_g(t)$ is the vertical acceleration of the housing support. Considering a harmonic support acceleration of the form $\ddot{v}_g(t) = \ddot{v}_{g0}\sin\overline{\omega}t$, so that $p_{\mathrm{eff}}(t) = -m\,\ddot{v}_{g0}\sin\overline{\omega}t$, the dynamic steady-state response amplitude of motion $v(t)$ is given by Eq. (3-22), i.e.,

$$\rho = \frac{m\,\ddot{v}_{g0}}{k}\,D \tag{3-39}$$

in which D as given by Eq. (3-24) is presented graphically in Fig. 3-3. Examination of this figure shows that for a damping ratio $\xi = 0.7$, the value of D is nearly constant over the frequency range $0 < \beta < 0.6$. Thus it is clear from Eq. (3-39) that the response indicated by this instrument is almost directly proportional to the support-acceleration amplitude for applied frequencies up to about six-tenths the natural frequency of the instrument ($\omega = 2\pi f = \sqrt{k/m}$). Hence, this type of instrument when properly damped will serve effectively as an accelerometer for relatively low frequencies; its range of applicability will be broadened by increasing its natural frequency relative to the exciting frequency, i.e., by increasing the stiffness of the spring and/or decreasing the mass. Calibration of an accelerometer is easily carried out by first placing the instrument with its axis of sensitivity vertically and then

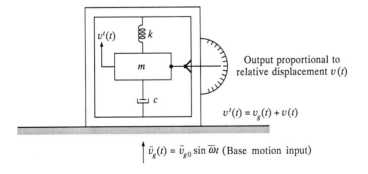

FIGURE 3-9
Schematic diagram of a typical seismometer.

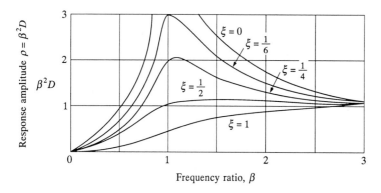

FIGURE 3-10
Response of seismometer to harmonic base displacement.

turning the instrument upside-down and recording the resulting change of response which corresponds to an acceleration twice that of gravity.

Consider now the response of the above described instrument subjected to a harmonic support displacement $v_g(t) = v_{g0} \sin \bar{\omega} t$. In this case, $\ddot{v}_g(t) = -\bar{\omega}^2 v_{g0} \sin \bar{\omega} t$ and the effective loading is $p_{\text{eff}} = m \bar{\omega} v_{g0} \sin \bar{\omega} t$. In accordance with Eq. (3-22), the relative-displacement response amplitude is

$$\rho = \frac{m \bar{\omega}^2 v_{g0}}{k} D = v_{g0} \beta^2 D \qquad (3\text{-}40)$$

A plot of the response function $\beta^2 D$ is presented in Fig. 3-10. In this case, it is evident that $\beta^2 D$ is essentially constant at frequency ratios $\beta > 1$ for a damping ratio $\xi = 0.5$. Thus, the response of a properly damped instrument is essentially proportional to the base-displacement amplitude for high-frequency support motions; i.e., it will serve as a displacement meter in measuring such motions. Its range of applicability for this purpose will be broadened by reducing the natural frequency, i.e., by reducing the spring stiffness and/or increasing the mass.

3-5 VIBRATION ISOLATION

Although the subject of vibration isolation is too broad to be discussed thoroughly here, the basic principles involved will be presented as they relate to two types of problems: (1) prevention of harmful vibrations in supporting structures due to oscillatory forces produced by operating equipment and (2) prevention of harmful vibrations in sensitive instruments due to vibrations of their supporting structures.

The first situation is illustrated in Fig. 3-11 where a rotating machine produces an oscillatory vertical force $p_o \sin \bar{\omega} t$ due to unbalance in its rotating parts. If the

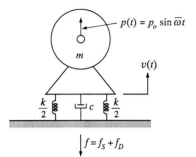

FIGURE 3-11
SDOF vibration-isolation system (applied loading).

machine is mounted on a SDOF spring-damper support system as shown, its steady-state relative-displacement response is given by

$$v_p(t) = \frac{p_o}{k} D \sin(\overline{\omega}t - \theta) \tag{3-41}$$

where D is defined by Eq. (3-24). This result assumes, of course, that the support motion induced by total reaction force $f(t)$ is negligible in comparison with the system motion relative to the support.

Using Eq. (3-41) and its first time derivative, the spring and damping reaction forces become

$$f_S(t) = k\, v(t) = p_o\, D\, \sin(\overline{\omega}t - \theta)$$

$$f_D(t) = c\, \dot{v}(t) = \frac{c\, p_o\, D\overline{\omega}}{k}\, \cos(\overline{\omega}t - \theta) = 2\,\xi\,\beta\, p_o\, D\, \cos(\overline{\omega}t - \theta) \tag{3-42}$$

Since these two forces are 90° out of phase with each other, it is evident that the amplitude of the total base reaction force is given by

$$f_{\max}(t) = [f_{S,\max}(t)^2 + f_{D,\max}(t)^2]^{1/2} = p_o\, D \left[1 + (2\xi\beta)^2\right]^{1/2} \tag{3-43}$$

Thus, the ratio of the maximum base force to the amplitude of the applied force, which is known as the transmissibility (TR) of the support system, becomes

$$\mathrm{TR} \equiv \frac{f_{\max}(t)}{p_o} = D\,\sqrt{1 + (2\xi\beta)^2} \tag{3-44}$$

The second type of situation in which vibration isolation is important is illustrated in Fig. 3-12, where the harmonic support motion $v_g(t)$ forces a steady-state relative-displacement response

$$v_p(t) = v_{g0}\,\beta^2\, D\, \sin(\overline{\omega}t - \theta) \tag{3-45}$$

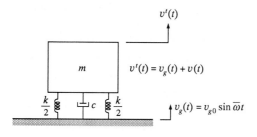

FIGURE 3-12
SDOF vibration-isolation system (support excitation).

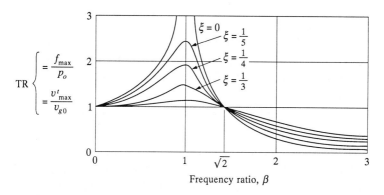

FIGURE 3-13
Vibration-transmissibility ratio (applied loading or support excitation).

in accordance with Eqs. (3-21) and (3-40). Adding this motion vectorially to the support motion $v_g(t) = v_{g0} \sin \bar{\omega} t$, the total steady-state response of mass m is given by

$$v^t(t) = v_{g0} \sqrt{1 + (2\xi\beta)^2} \, D \sin(\bar{\omega} t - \bar{\theta}) \tag{3-46}$$

in which the phase angle $\bar{\theta}$ is of no particular interest in the present discussion. Thus, if the transmissibility in this situation is defined as the ratio of the amplitude of total motion of the mass to the corresponding base-motion amplitude, it is seen that this expression for transmissibility is identical to that given by Eq. (3-44), i.e.,

$$\text{TR} \equiv \frac{v_{max}^t}{v_{g0}} = D \sqrt{1 + (2\xi\beta)^2} \tag{3-47}$$

Note that this transmissibility relation also applies to the acceleration ratio $(\ddot{v}_{max}^t / \ddot{v}_{gmax})$ because $\ddot{v}_{max}^t = \bar{\omega}^2 v_{max}^t$ and $\ddot{v}_{gmax} = \bar{\omega}^2 v_{g0}$.

Since the transmissibility relations given by Eqs. (3-44) and (3-47) are identical, the common relation expresses the transmissibility of vibration-isolation systems for both situations described above. This relation is plotted as a function of frequency ratio in Fig. 3-13 for discrete values of damping. Note that all curves pass through the same point at a frequency ratio of $\beta = \sqrt{2}$. Clearly because of this feature, increasing the damping when $\beta < \sqrt{2}$ increases the effectiveness of the vibration-isolation system, while increasing the damping when $\beta > \sqrt{2}$ decreases the effectiveness. Since the transmissibility values for $\beta > \sqrt{2}$ are generally much lower than those for $\beta < \sqrt{2}$, one should take advantage of operating in the higher frequency ratio range when it is practical to do so. This is not always possible, however, because in many cases the system must operate below $\beta = \sqrt{2}$ for some intervals of time, and in some cases even operate near the resonant condition $\beta = 1$. The following example illustrates such a condition:

Example E3-2.   Deflections sometimes develop in concrete bridge girders due to creep, and if the bridge consists of a long series of identical spans, these deformations will cause a harmonic excitation in a vehicle traveling over the bridge at constant speed. Of course, the springs and shock absorbers of the car are intended to provide a vibration-isolation system which will limit the vertical motions transmitted from the road to the occupants.

Figure E3-1 shows a highly idealized model of this type of system, in which the vehicle weight is $4,000$ lb $[1,814$ $kg]$ and its spring stiffness is defined by a test which showed that adding 100 lb $[45.36$ $kg]$ caused a deflection of 0.08 in $[0.203$ $cm]$. The bridge profile is represented by a sine curve having a wavelength (girder span) of 40 ft $[12.2$ $m]$ and a (single) amplitude of 1.2 in $[3.05$ $cm]$. From these data it is desired to predict the steady-state vertical motions in the car when it is traveling at a speed of 45 mph $[72.4$ $km/hr]$, assuming that the damping is 40 percent of critical.

The transmissibility for this case is given by Eq. (3-47); hence the amplitude of vertical motion is

$$v^t_{max} = v_{g0} \left[\frac{1 + (2\xi\beta)^2}{(1 - \beta^2)^2 + (2\xi\beta)^2} \right]^{1/2}$$

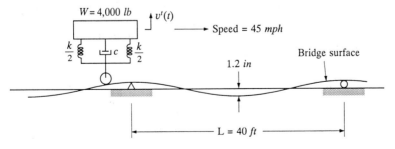

FIGURE E3-1
Idealized vehicle traveling over an uneven bridge deck.

When the car is traveling at $45\ mph = 66\ ft/sec$, the excitation period is

$$T_p = \frac{40\ ft}{66\ ft/sec} = 0.606\ sec$$

while the natural period of the vehicle is

$$T = \frac{2\pi}{\omega} = 2\pi\sqrt{\frac{W}{kg}} = 0.572\ sec$$

Hence $\beta = T/T_p = 0.572/0.606 = 0.944$, and with $\xi = 0.4$ the response amplitude is

$$v^t_{\max} = 1.2\ (1.642) = 1.97\ in \quad [5.00\ cm]$$

It also is of interest to note that if there were no damping in the vehicle ($\xi = 0$), the amplitude would be

$$v^t_{\max} = v_{g0}\left[\frac{1}{1 - \beta^2}\right] = \frac{1.2}{0.11} = 10.9\ in \quad [27.7\ cm]$$

This is beyond the spring range, of course, and thus has little meaning, but it does demonstrate the important function of shock absorbers in limiting the motions resulting from waviness of the road surface.

When designing a vibration-isolation system which will operate at frequencies above the critical value represented by $\beta = \sqrt{2}$, it is convenient to express the behavior of the SDOF system in terms of isolation effectiveness (IE) rather than transmissibility. This quantity is defined by

$$\text{IE} \equiv [1 - \text{TR}] \tag{3-48}$$

in which IE = 1 represents complete isolation approachable only as $\beta \to \infty$ and IE = 0 represents no isolation which takes place at $\beta = \sqrt{2}$. For values of β below this critical value, amplification of the motion of the mass takes place; thus, actual vibration isolation can take place only when the system functions at values of β greater than $\sqrt{2}$. In this case the isolation system should have as little damping as possible.

For small damping, the transmissibility given by Eq. (3-44) or Eq. (3-47), after substitution of Eq. (3-24), can be expressed by the approximate relation

$$\text{TR} \doteq 1/(\beta^2 - 1) \tag{3-49}$$

in which case the isolation effectiveness becomes

$$IE = (\beta^2 - 2) / (\beta^2 - 1) \tag{3-50}$$

Solving this relation for β^2, one obtains its inverse form

$$\beta^2 = (2 - IE) / (1 - IE) \tag{3-51}$$

Noting that $\beta^2 = \bar{\omega}^2/\omega^2 = \bar{\omega}^2 (m/k) = \bar{\omega}^2 (W/kg) = \bar{\omega}^2 (\triangle_{st}/g)$, where g is the acceleration of gravity and $\triangle_{st}$ is the static deflection produced by the dead weight W on its spring mounting, Eq. (3-51) can be expressed in the form

$$\bar{f} = \frac{\bar{\omega}}{2\pi} = \frac{1}{2\pi} \sqrt{\frac{g}{\triangle_{st}} \left[\frac{2 - IE}{1 - IE} \right]} \qquad 0 < IE < 1 \tag{3-52}$$

Frequency $\bar{f}$ measured in Hertz (*cycles/sec*), as derived from this expression, is plotted against the static deflection $\triangle_{st}$ in Fig. 3-14 for discrete values of isolation efficiency IE. Knowing the frequency of impressed excitation $\bar{f}$, one can determine directly from the curves in this figure the support-pad deflection $\triangle_{st}$ required to achieve any desired level of vibration isolation efficiency (IE), assuming, of course, that the isolation system has little damping. It is apparent that any isolation system must be very flexible to be effective.

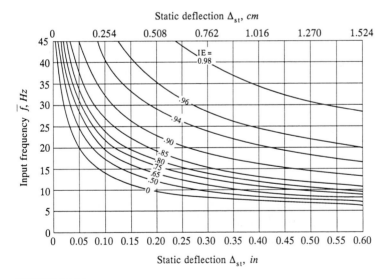

FIGURE 3-14
Vibration-isolation design chart.

Example   E3-3.   A reciprocating machine weighing $20,000 \; lb \; [9,072 \; kg]$ is known to develop a vertically oriented harmonic force of amplitude

500 *lb* [226.8 *kg*] at its operating speed of 40 *Hz*. In order to limit the vibrations excited in the building in which this machine is to be installed, it is to be supported by a spring at each corner of its rectangular base. The designer wants to know what support stiffness will be required of each spring to limit the total harmonic force transmitted from the machine to the building to 80 *lb* [36.3 *kg*].

The transmissibility in this case is TR $= 80/500 = 0.16$ which corresponds to an isolation efficiency of IE $= 1 - $ TR $= 0.84$. From Fig. 3-14 for $\overline{f} = 40$ *Hz* and IE $= 0.84$, one finds that Δ_{st} is about 0.045 *in* [0.114 *cm*]; thus, the required stiffness k of each spring is

$$k = \frac{W}{4 \, \Delta_{st}} = \frac{20}{(4)(0.045)} = 111 \; kips/in \quad [19,823 \; kg/cm]$$

3-6 EVALUATION OF VISCOUS-DAMPING RATIO

In the foregoing discussion of the dynamic response of SDOF systems, it has been assumed that the physical properties consisting of mass, stiffness, and viscous damping are known. While in most cases, the mass and stiffness can be evaluated rather easily using simple physical considerations or generalized expressions as discussed in Chapter 8, it is usually not feasible to determine the damping coefficient by similar means because the basic energy-loss mechanisms in most practical systems are seldom fully understood. In fact, it is probable that the actual energy-loss mechanisms are much more complicated than the simple viscous (velocity proportional) damping force that has been assumed in formulating the SDOF equation of motion. But it generally is possible to determine an appropriate equivalent viscous-damping property by experimental methods. A brief treatment of the methods commonly used for this purpose is presented in the following sections:

Free-Vibration Decay Method

This is the simplest and most frequently used method of finding the viscous-damping ratio ξ through experimental measurements. When the system has been set into free vibration by any means, the damping ratio can be determined from the ratio of two peak displacements measured over m consecutive cycles. As shown in Chapter 2, the damping ratio can be evaluated using

$$\xi = \frac{\delta_m}{2 \pi m \, (\omega/\omega_D)} \doteq \frac{\delta_m}{2 \pi m} \tag{3-53}$$

where $\delta_m \equiv \ln(v_n/v_{n+m})$ represents the logarithmic decrement over m cycles and ω and ω_D are the undamped and damped circular frequencies, respectively. For low values of damping, the approximate relation in Eq. (3-53) can be used which is only 2 percent in error when $\xi = 0.2$. A major advantage of this free-vibration method is that equipment and instrumentation requirements are minimal; the vibrations can be initiated by any convenient method and only the relative-displacement amplitudes need be measured. If the damping is truly of the linear viscous form as previously assumed, any set of m consecutive cycles will yield the same damping ratio through the use of Eq. (3-53). Unfortunately, however, the damping ratio so obtained often is found to be amplitude dependent, i.e., m consecutive cycles in the earlier portion of high-amplitude free-vibration response will yield a different damping ratio than m consecutive cycles in a later stage of much lower response. Generally it is found in such cases that the damping ratio decreases with decreasing amplitude of free-vibration response. Caution must be exercised in the use of these amplitude-dependent damping ratios for predicting dynamic response.

Resonant Amplification Method

This method of determining the viscous-damping ratio is based on measuring the steady-state amplitudes of relative-displacement response produced by separate harmonic loadings of amplitude p_o at discrete values of excitation frequency $\overline{\omega}$ over a wide range including the natural frequency. Plotting these measured amplitudes against frequency provides a frequency-response curve of the type shown in Fig. 3-15. Since the peak of the frequency-response curve for a typical low damped structure is quite narrow, it is usually necessary to shorten the intervals of the discrete frequencies

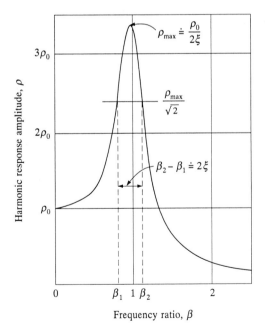

FIGURE 3-15
Frequency-response curve for moderately damped system.

in the neighborhood of the peak in order to get good resolution of its shape. As shown by Eqs. (3-32) and (3-33), the actual maximum dynamic magnification factor $D_{max} \equiv \rho_{max}/\rho_0$ occurs at the excitation frequency $\overline{\omega} = \omega \sqrt{1 - 2\xi^2}$ and is given by $D_{max} = 1/2\xi \sqrt{1 - \xi^2}$; however, for damping values in the practical range of interest, one can use the approximate relation $D_{max} \doteq D\,(\beta = 1) = 1/2\xi$. The damping ratio can then be determined from the experimental data using

$$\xi \doteq \rho_0/2 \, \rho_{max} \tag{3-54}$$

This method of determining the damping ratio requires only simple instrumentation to measure the dynamic response amplitudes at discrete values of frequency and fairly simple dynamic-loading equipment; however, obtaining the static displacement ρ_0 may present a problem because the typical harmonic loading system cannot produce a loading at zero frequency. As pointed out above, the damping ratio for practical systems is often amplitude dependent. In this case, the value of ξ obtained by Eq. (3-54) depends on the amplitude p_o of the applied harmonic loading. This dependency should be taken into consideration when specifying an appropriate value of ξ for dynamic analysis purposes.

Half-Power (Band-Width) Method

It is evident from Eq. (3-22), in which $(p_o/k) \equiv \rho_0$, that the frequency-response curve ρ vs. β shown in Fig. 3-15 has a shape which is controlled by the amount of damping in the system; therefore, it is possible to derive the damping ratio from many different properties of the curve. One of the most convenient of these is the half-power or band-width method whereby the damping ratio is determined from the frequencies at which the response amplitude ρ is reduced to the level $1/\sqrt{2}$ times its peak value ρ_{max}.

The controlling frequency relation is obtained by setting the response amplitude in Eq. (3-22) equal to $1/\sqrt{2}$ times its peak value given by Eq. (3-33), that is, by setting

$$\left[(1 - \beta^2)^2 + (2\xi\beta)^2\right]^{-1/2} = (1/\sqrt{2}) \left[1/2\xi \sqrt{1 - \xi^2}\right] \tag{3-55}$$

Squaring both sides of this equation and solving the resulting quadratic equation for β^2 gives

$$\beta_{1,2}^2 = 1 - 2\xi^2 \mp 2\xi \sqrt{1 - \xi^2} \tag{3-56}$$

which, for small values of damping in the practical range of interest, yields the frequency ratios

$$\beta_{1,2} \doteq 1 - \xi^2 \mp \xi \sqrt{1 - \xi^2} \tag{3-57}$$

Subtracting β_1 from β_2, one obtains

$$\beta_2 - \beta_1 = 2\xi \sqrt{1 - \xi^2} \doteq 2\xi \tag{3-58}$$

while adding β_1 and β_2 gives

$$\beta_2 + \beta_1 = 2\,(1 - \xi^2) \doteq 2 \qquad (3\text{-}59)$$

Combining Eqs. (3-58) and (3-59) yields

$$\xi = \frac{\beta_2 - \beta_1}{\beta_2 + \beta_1} = \frac{f_2 - f_1}{f_2 + f_1} \qquad (3\text{-}60)$$

where f_1 and f_2 are the frequencies at which the amplitudes of response equal $1/\sqrt{2}$ times the maximum amplitude. The use of either Eq. (3-58) or Eq. (3-60) in evaluating the damping ratio is illustrated in Fig. 3-15 where a horizontal line has been drawn across the curve at $1/\sqrt{2}$ times its peak value. It is evident that this method of obtaining the damping ratio avoids the need for obtaining the static displacement ρ_0; however, it does require that the frequency-response curve be obtained accurately at its peak and at the level $\rho_{max}/\sqrt{2}$.

To clarify why the above method is commonly referred to as the half-power method, consider the time-average power input provided by the applied loading, which must equal the corresponding average rate of energy dissipation caused by the damping force $F_D(t) = c\,\dot{v}(t)$. Under the steady-state harmonic condition at frequency $\bar{\omega}$ where the displacement response amplitude is ρ, the average rate of energy dissipation is

$$P_{avg} = \frac{c\bar{\omega}}{2\pi} \int_0^{2\pi/\bar{\omega}} \dot{v}(t)^2\, dt = c\bar{\omega}^2 \left[\frac{\bar{\omega}}{2\pi} \int_0^{2\pi/\bar{\omega}} v(t)^2\, dt \right] = \xi\, m\, \omega\, \bar{\omega}^2\, \rho^2 \qquad (3\text{-}61)$$

which shows that the corresponding average power input is proportional to $\beta^2\rho^2$; thus, since $\rho_1 = \rho_2 = \rho_{peak}/\sqrt{2}$, the average power inputs at frequency ratios β_1 and β_2 are

$$P_{\beta_1} = \left(\frac{\beta_1}{\beta_{peak}} \right)^2 \frac{P_{peak}}{2} \qquad\qquad P_{\beta_2} = \left(\frac{\beta_2}{\beta_{peak}} \right)^2 \frac{P_{peak}}{2} \qquad (3\text{-}62)$$

where β_{peak} is given by Eq. (3-32). While the average power input at β_1 is somewhat less than one-half the peak power input and the average power input at β_2 is somewhat greater, the mean value of these two averaged inputs is very close to one-half the peak average power input.

Example E3-4. Data from a frequency-response test of a SDOF system have been plotted in Fig. E3-2. The pertinent data for evaluating the damping ratio are shown. The sequence of steps in the analysis after the curve was plotted were as follows:

(1) Determine the peak response $= 5.67 \times 10^{-2}\ in$ $[14.4 \times 10^{-2}\ cm]$.

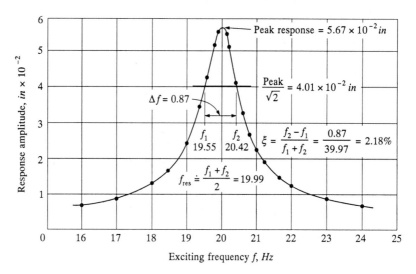

FIGURE E3-2
Frequency-response experiment to determine damping ratio.

(2) Construct a horizontal line at $1/\sqrt{2}$ times the peak level.

(3) Determine the two frequencies at which this horizontal line cuts the response curve; $f_1 = 19.55$, $f_2 = 20.42\ Hz$.

(4) The damping ratio is given by

$$\xi = \frac{f_2 - f_1}{f_2 + f_1} = 0.022$$

showing 2.2 percent of critical damping in the system.

Resonance Energy Loss Per Cycle Method

If instrumentation is available to measure the phase relationship between the input force and the resulting displacement response, the damping ratio can be evaluated from a steady-state harmonic test conducted only at resonance: $\beta = \frac{\overline{\omega}}{\omega} = 1$. This procedure involves establishing resonance by adjusting the input frequency until the displacement response is 90° out-of-phase with the applied loading. As shown in Fig. 3-6 for $\theta = 90°$, the applied loading is exactly balancing the damping force so that if the relationship between the applied loading and the resulting displacement is plotted for one loading cycle as shown in Fig. 3-16, the result can be interpreted as the damping force vs. displacement diagram. If the system truly possesses linear viscous damping, this diagram will be an ellipse as shown by the dashed line in this

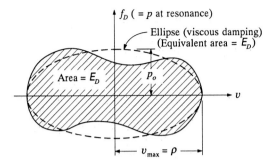

FIGURE 3-16
Actual and equivalent damping
energy per cycle.

figure. In this case, the damping ratio can be determined directly from the maximum damping force and the maximum velocity using the relation

$$p_o = f_{D_{max}} = c\,\dot{v}_{max} = 2\,\xi\,m\,\omega\,\dot{v}_{max} = 2\,\xi\,m\,\omega^2\,\rho \qquad (3\text{-}63)$$

or

$$\xi = p_o/2\,m\,\omega^2\,\rho \qquad (3\text{-}64)$$

If damping is not of the linear viscous form previously assumed but is of a nonlinear viscous form, the shape of the applied-force/displacement diagram obtained by the above procedure will not be elliptical; rather, it will be of a different shape as illustrated by the solid line in Fig. 3-16. In this case, the response $v(t)$ will be a distorted harmonic, even though the applied loading remains a pure harmonic. Nevertheless, the energy input per cycle, which equals the damping energy loss per cycle E_D, can be obtained as the area under the applied-force/displacement diagram. This permits one to evaluate an equivalent viscous-damping ratio for the corresponding displacement amplitude, which when used in the linear viscous form will dissipate the same amount of energy per cycle as in the real experimental case. This equivalent damping ratio is associated with an elliptical applied-force/displacement diagram having the same area E_D as the measured nonelliptical diagram. Making use of Eq. (3-61), this energy equivalence requires that

$$E_D = (2\pi/\omega)\,P_{avg} = (2\pi/\omega)\,(\xi_{eq}\,m\,\omega^3\,\rho^2) \qquad (3\text{-}65)$$

or

$$\xi_{eq} = E_D/(2\,\pi\,m\,\omega^2\,\rho^2) = E_D/(2\,\pi\,k\,\rho^2) \qquad (3\text{-}66)$$

The latter form of Eq. (3-66) is more convenient here because the stiffness of the structure can be measured by the same instrumentation used to obtain the energy loss per cycle, merely by operating the system very slowly at essentially static conditions. The static-force displacement diagram obtained in this way will be of the form shown

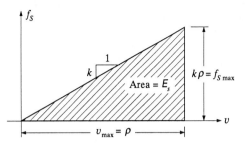

FIGURE 3-17
Elastic stiffness and strain energy.

in Fig. 3-17, if the structure is linearly elastic. The stiffness is obtained as the slope of the straight line curve.

3-7 COMPLEX-STIFFNESS DAMPING

Damping of the linear viscous form discussed above is commonly used because it leads to a convenient form of equation of motion. It has one serious deficiency, however; as seen from Eq. (3-61), the energy loss per cycle

$$E_D = (2\pi/\overline{\omega}) \, P_{\text{avg}} = 2\pi \, \xi \, m \, \omega \, \overline{\omega} \, \rho^2 \tag{3-67}$$

at a fixed amplitude ρ is dependent upon the excitation (or response) frequency $\overline{\omega}$. This dependency is at variance with a great deal of test evidence which indicates that the energy loss per cycle is essentially independent of frequency. It is desirable therefore to model the damping force so as to remove this frequency dependence. This can be accomplished by using the so-called "hysteretic" form of damping in place of viscous damping. Hysteretic damping may be defined as a damping force proportional to the displacement amplitude but in phase with the velocity, and for the case of harmonic motion it may be expressed as

$$f_D(t) = i \, \zeta \, k \, v(t) \tag{3-68}$$

where ζ is the hysteretic damping factor which defines the damping force as a function of the elastic stiffness force, and the imaginary constant i puts the force in phase with the velocity. It is convenient to combine the elastic and damping resistance into the complex stiffness $\hat{k}$ defined as

$$\hat{k} = k \, (1 + i \, \zeta) \tag{3-69}$$

leading to the following harmonic forced vibration equation of motion:

$$m \, \ddot{v}(t) + \hat{k} \, v(t) = p_o \, \exp(i\overline{\omega}t) \tag{3-70}$$

The particular (or steady-state) solution of Eq. (3-70) is

$$v_p(t) = G \ \exp(i\bar{\omega}t) \tag{3-71}$$

in which G is a complex constant, and the corresponding acceleration is given by

$$\ddot{v}_p(t) = -\bar{\omega}^2 \ G \ \exp(i\bar{\omega}t) \tag{3-72}$$

Substituting these expressions into Eq. (3-70) yields

$$\left[-m\bar{\omega}^2 + \hat{k} \right] G \ \exp(i\bar{\omega}t) = p_o \ \exp(i\bar{\omega}t)$$

from which the value of G is found to be

$$G = \cfrac{p_o}{k \left[-\cfrac{m}{k}\bar{\omega}^2 + (1 + i\,\zeta) \right]} = \frac{p_o}{k} \frac{1}{\left[(1 - \beta^2) + i\,\zeta \right]}$$

or in a more convenient complex form

$$G = \frac{p_o}{k} \left[\frac{(1 - \beta^2) - i\,\zeta}{(1 - \beta^2)^2 + \zeta^2} \right] \tag{3-73}$$

Substituting this into Eq. (3-71) finally gives the following expression for the steady-state response with hysteretic damping

$$v_p(t) = \frac{p_o}{k} \left[\frac{(1 - \beta^2) - i\,\zeta}{(1 - \beta^2)^2 + \zeta^2} \right] \exp(i\bar{\omega}t) \tag{3-74}$$

This response is depicted graphically by its two orthogonal vectors plotted in the complex plane of Fig. 3-18. The resultant of these two vectors gives the response in terms of a single-amplitude vector, namely

$$v_p(t) = \bar{\rho} \ \exp\left[i\bar{\omega}t - \bar{\theta} \right] \tag{3-75}$$

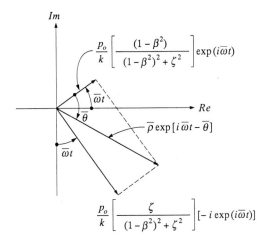

FIGURE 3-18
Steady-state displacement response using complex stiffness damping.

in which

$$\bar{p} = \frac{p_o}{k} \left[(1 - \beta^2)^2 + \zeta^2 \right]^{-1/2} \tag{3-76}$$

and the response phase angle is

$$\bar{\theta} = \tan^{-1} \left[\frac{\zeta}{(1 - \beta^2)} \right] \tag{3-77}$$

Comparing these three equations with Eqs. (3-28), (3-22), and (3-23), respectively, it is evident that the steady-state response provided by hysteretic damping is identical to that with viscous damping if the hysteretic damping factor has the value

$$\zeta = 2\xi\beta \tag{3-78}$$

In this case, the energy loss per cycle at a fixed amplitude $\bar{p}$ is dependent upon the excitation frequency $\bar{\omega}$ exactly as in the case of viscous damping. As will be shown subsequently, this frequency dependence can be removed by making the hysteretic damping factor ζ frequency independent. In doing so, it is convenient to use Eq. (3-78) and to adopt the factor given at resonance for which $\beta = 1$; thus the recommended hysteretic damping factor is $\zeta = 2\xi$, and the complex stiffness coefficient given by Eq. (3-69) becomes

$$\hat{k} = k \left[1 + i\,2\xi \right] \tag{3-79}$$

Then as shown by Eqs. (3-76) and (3-77), the response amplitude and phase angle, respectively, are

$$\bar{p} = \frac{p_0}{k} \left[(1 - \beta^2)^2 + (2\xi)^2 \right]^{-1/2} \tag{3-80}$$

$$\bar{\theta} = \tan^{-1} \left[\frac{2\xi}{(1 - \beta^2)} \right] \tag{3-81}$$

This response with hysteretic damping is identical to the viscous-damping response if the system is excited at resonance ($\beta = 1$). However, when $\beta \neq 1$, the two amplitudes differ in accordance with Eqs. (3-22) and (3-80) and the corresponding phase angles differ in accordance with Eqs. (3-23) and (3-81).

When the complex stiffness is defined in accordance with Eq. (3-69) and when $\zeta = 2\xi$, the damping force component under steady-state harmonic excitation is given by

$$f_D(t) = 2\,i\xi\,k\,\bar{p} \left[\exp(i\bar{\omega}t - \bar{\theta}) \right] \tag{3-82}$$

and the damping energy loss per cycle, E_D, can be obtained by integrating the instantaneous power loss

$$P(t) = f_D(t)\,\dot{v}_p(t) = 2\,\xi\,k\,\bar{\omega}\,\bar{p}^2 \left[-\exp(i\bar{\omega}t - \bar{\theta}) \right]^2 \tag{3-83}$$

over one cycle, with the final result

$$E_D = 2\pi \xi m \omega^2 \bar{p}^2 \tag{3-84}$$

It is evident that this energy loss per cycle at fixed amplitude $\bar{p}$ is independent of the excitation frequency, $\bar{\omega}$, thus it is consistent with the desired frequency-independent behavior; for this reason it is recommended that this form of hysteretic damping (complex stiffness damping) be used in most cases for general harmonic response analysis purposes.

PROBLEMS

3-1. Consider the basic structure of Fig. 2-1a with zero damping and subjected to harmonic excitation at the frequency ratio $\beta = 0.8$. Including both steady-state and transient effects, plot the response ratio $R(t)$. Evaluate the response at increments $\bar{\omega}\Delta t = 80°$ and continue the analysis for 10 increments.

3-2. Consider the basic system of Fig. 2-1a with the following properties: $m = 2\ kips \cdot sec^2/in$ and $k = 20\ kips/in$. If this system is subjected to resonant harmonic loading ($\bar{\omega} = \omega$) starting from "at rest" conditions, determine the value of the response ratio $R(t)$ after four cycles ($\bar{\omega}t = 8\pi$), assuming:
  (a) $c = 0$ [use Eq. (3-38)]
  (b) $c = 0.5\ kips \cdot sec/in$ [use Eq. (3-37)]
  (c) $c = 2.0\ kips \cdot sec/in$ [use Eq. (3-37)]

3-3. Consider the same vehicle and bridge structure of Example E3-2, except with the girder spans reduced to $L = 36\ ft$. Determine:
  (a) the vehicle speed required to induce resonance in the vehicle spring system.
  (b) the total amplitude of vertical motion v^t_{max} at resonance.
  (c) the total amplitude of vertical motion v^t_{max} at the speed of 45 mph.

3-4. A control console containing delicate instrumentation is to be located on the floor of a test laboratory where it has been determined that the floor slab is vibrating vertically with an amplitude of 0.03 in at 20 Hz. If the weight of the console is 800 lb, determine the stiffness of the vibration isolation system required to reduce the vertical-motion amplitude of the console to 0.005 in.

3-5. A sieving machine weighs 6,500 lb, and when operating at full capacity, it exerts a harmonic force on its supports of 700 lb amplitude at 12 Hz. After mounting the machine on spring-type vibration isolators, it was found that the harmonic force exerted on the supports had been reduced to a 50 lb amplitude. Determine the spring stiffness k of the isolation system.

3-6. The structure of Fig. P3-1a can be idealized by the equivalent system of Fig. P3-1b. In order to determine the values of c and k for this mathematical model, the concrete column was subjected to a harmonic load test as shown in Fig. P3-1c. When operating at a test frequency of $\bar{\omega} = 10$ $rads/sec$, the force-deflection (hysteresis) curve of Fig. P3-1d was obtained. From this data:

(a) determine the stiffness k.

(b) assuming a viscous damping mechanism, determine the apparent viscous damping ratio ξ and damping coefficient c.

(c) assuming a hysteretic damping mechanism, determine the apparent hysteretic damping factor ζ.

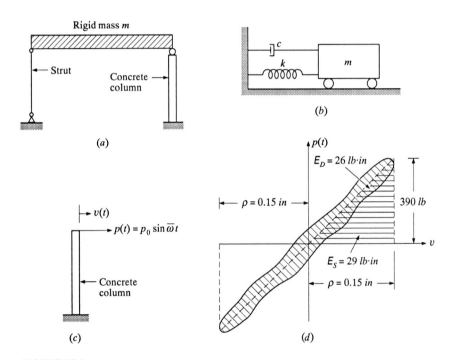

FIGURE P3-1

3-7. Suppose that the test of Prob. 3-6 were repeated, using a test frequency $\bar{\omega} = 20$ $rads/sec$, and that the force-deflection curve (Fig. P3-1d) was found to be unchanged. In this case:

(a) determine the apparent viscous damping values ξ and c.

(b) determine the apparent hysteretic damping factor ζ.

(c) Based on these two tests ($\bar{\omega} = 10$ and $\bar{\omega} = 20$ $rads/sec$), which type of damping mechanism appears more reasonable — viscous or hysteretic?

3-8. If the damping of the system of Prob. 3-6 actually were provided by a viscous damper as indicated in Fig. P3-1b, what would be the value of E_D obtained in a test performed at $\bar{\omega} = 20 \; rads/sec$?

RESPONSE
TO PERIODIC
LOADING

4-1 FOURIER SERIES EXPRESSIONS OF PERIODIC LOADING

Trigonometric Form

Because any periodic loading can be expressed as a series of harmonic loading terms, the response analysis procedures presented in Chapter 3 have a wide range of applicability. To treat the case of an arbitrary periodic loading of period T_p, as indicated in Fig. 4-1, it is convenient to express it in a Fourier series form with harmonic loading components at discrete values of frequency. The well-known trigonometric form of the Fourier series is given by

$$p(t) = a_0 + \sum_{n=1}^{\infty} a_n \, \cos \overline{\omega}_n t + \sum_{n=1}^{\infty} b_n \, \sin \overline{\omega}_n t \qquad (4\text{-}1)$$

in which

$$\overline{\omega}_n = n \, \overline{\omega}_1 = n \, \frac{2\,\pi}{T_p} \qquad (4\text{-}2)$$

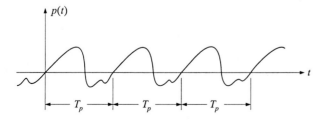

FIGURE 4-1
Arbitrary periodic loading.

and the harmonic amplitude coefficients can be evaluated using the expressions

$$a_0 = \frac{1}{T_p} \int_0^{T_p} p(t) \, dt$$

$$a_n = \frac{2}{T_p} \int_0^{T_p} p(t) \, \cos \overline{\omega}_n t \, dt \qquad n = 1, 2, 3, \cdots \qquad (4\text{-}3)$$

$$b_n = \frac{2}{T_p} \int_0^{T_p} p(t) \, \sin \overline{\omega}_n t \, dt \qquad n = 1, 2, 3, \cdots$$

When $p(t)$ is of arbitrary periodic form, the integrals in Eqs. (4-3) must be evaluated numerically. This can be done by dividing the period T_p into N equal intervals Δt ($T_p = N \, \Delta t$), evaluating the ordinates of the integrand in each integral at discrete values of $t = t_m = m \, \Delta t$ ($m = 0, 1, 2, \cdots, N$) denoted by $q_0, q_1, q_2, \cdots, q_N$, and then applying the trapezoidal rule of integration in accordance with

$$\int_0^{T_p} q(t) \, dt \doteq \Delta t \left[\frac{q_0}{2} + \left(\sum_{m=1}^{N-1} q_m \right) + \frac{q_N}{2} \right] \qquad (4\text{-}4)$$

In practical solutions, the beginning and end of the time period usually can be set so that the ordinates q_0 and q_N are equal to zero, in which case, Eq. (4-4) simplifies to

$$\int_0^{T_p} q(t) \, dt \doteq \Delta t \sum_{m=1}^{N-1} q_m \qquad (4\text{-}5)$$

The harmonic amplitude coefficients of Eq. (4-3) then may be expressed as

$$\left. \begin{array}{c} a_0 \\ a_n \\ b_n \end{array} \right\} = \frac{2\Delta t}{T_p} \sum_{m=1}^{N-1} q_m \qquad \text{where} \qquad q_m = \left\{ \begin{array}{c} \frac{1}{2} p(t_m) \\ p(t_m) \, \cos \overline{\omega}_n(m\Delta t) \\ p(t_m) \, \sin \overline{\omega}_n(m\Delta t) \end{array} \right\} \qquad (4\text{-}6)$$

Exponential Form

The exponential form of the Fourier series equivalent to Eq. (4-1) is obtained by substituting the inverse Euler relations, Eqs. (2-23b) (with $\overline{\omega}_n t$ replacing the angle θ):

$$\cos \overline{\omega}_n t = \frac{1}{2} \left[\exp(i\overline{\omega}_n t) + \exp(-i\overline{\omega}_n t) \right]$$

$$\sin \overline{\omega}_n t = -\frac{i}{2} \left[\exp(i\overline{\omega}_n t) - \exp(-i\overline{\omega}_n t) \right] \qquad (4\text{-}7)$$

into Eqs. (4-1) and (4-3) leading to

$$p(t) = \sum_{n=-\infty}^{\infty} P_n \, \exp(i\overline{\omega}_n t) \qquad (4\text{-}8)$$

in which the complex amplitude coefficients are given by

$$P_n = \frac{1}{T_p} \int_0^{T_p} p(t) \, \exp(-i\overline{\omega}_n t) \, dt \qquad n = 0, \pm 1, \pm 2, \cdots \qquad (4\text{-}9)$$

It should be noted that for each positive value of n in Eq. (4-8), say $n = +m$, there is a corresponding $n = -m$. From the form of Eq. (4-9) it is seen that P_m and P_{-m} are complex conjugate pairs which is a necessary condition for the imaginary parts in the corresponding terms of Eq. (4-8) to cancel each other.

Equation (4-8) may be evaluated numerically by the trapezoidal rule in a manner equivalent to that described above, i.e., by defining the function $q(t) \equiv p(t) \exp(-i\overline{\omega}_n t)$ at discrete values of $t = t_m = m\Delta t$, making the substitutions $\overline{\omega}_n = 2\pi n/T_p = 2\pi n/N\Delta t$ and $t_m = m\Delta t$, and assuming that $q_0 = q_N = 0$, leading finally to

$$P_n \doteq \frac{1}{N} \sum_{m=1}^{N-1} p(t_m) \, \exp\left(-i\,\frac{2\pi\,n\,m}{N}\right) \qquad n = 0, 1, 2, \cdots, (N-1) \qquad (4\text{-}10)$$

4-2 RESPONSE TO THE FOURIER SERIES LOADING

Having expressed the periodic loading as a series of harmonic terms, the response of a linear system to this loading may be obtained by simply adding up the responses to the individual harmonic loadings. In Chapter 3 [Eq. (3-10)], it was shown that the steady-state response produced in an undamped SDOF system by the nth sine-wave harmonic of Eq. (4-1) (after omitting the transient response term) is given by

$$v_n(t) = \frac{b_n}{k} \left[\frac{1}{1 - \beta_n^2} \right] \sin \overline{\omega}_n t \qquad (4\text{-}11)$$

where

$$\beta_n \equiv \overline{\omega}_n / \omega \qquad (4\text{-}12)$$

Similarly, the steady-state response produced by the nth cosine-wave harmonic in Eq. (4-1) is

$$v_n(t) = \frac{a_n}{k} \left[\frac{1}{1 - \beta_n^2} \right] \cos \overline{\omega}_n t \qquad (4\text{-}13)$$

Finally, the steady-state response to the constant load a_0 is the static deflection

$$v_0 = a_0 / k \qquad (4\text{-}14)$$

The total periodic response of the undamped structure then can be expressed as the sum of the individual responses to the loading terms in Eq. (4-1) as follows:

$$v(t) = \frac{1}{k} \left\{ a_0 + \sum_{n=1}^{\infty} \left[\frac{1}{1 - \beta_n^2} \right] (a_n \cos \overline{\omega}_n t + b_n \sin \overline{\omega}_n t) \right\} \qquad (4\text{-}15)$$

where the load-amplitude coefficients are given by Eqs. (4-3) or Eqs. (4-6).

To take account of viscous damping in evaluating the steady-state response of a SDOF system to periodic loading, it is necessary to substitute the damped-harmonic-response expressions of the form of Eq. (3-20) for the undamped expressions used above. In this case the total steady-state response is given by

$$v(t) = \frac{1}{k} \left(a_0 + \sum_{n=1}^{\infty} \left[\frac{1}{(1 - \beta_n^2)^2 + (2\xi\beta_n)^2} \right] \right.$$

$$\times \left\{ \left[2\,\xi\,a_n\,\beta_n + b_n\,(1 - \beta_n^2) \right]\,\sin\overline{\omega}_n t \right.$$

$$\left. \left. + \left[a_n\,(1 - \beta_n^2) - 2\,\xi\,b_n\,\beta_n \right]\,\cos\overline{\omega}_n t \right\} \right) \qquad (4\text{-}16)$$

Example   E4-1.   As an example of the response analysis of a periodically loaded structure, consider the system and loading shown in Fig. E4-1. The loading in this case consists of the positive portion of a simple sine function. The Fourier coefficients of Eq. (4-1) are found by using Eqs. (4-2) and (4-3) to obtain

$$a_0 = \frac{1}{T_p} \int_0^{T_p/2} p_0\,\sin\frac{2\pi t}{T_p}\,dt = \frac{p_0}{\pi}$$

$$a_n = \frac{2}{T_p} \int_0^{T_p/2} p_0\,\sin\frac{2\pi t}{T_p}\,\cos\frac{2\,\pi\,n\,t}{T_p}\,dt = \begin{cases} 0 & n \quad \text{odd} \\ \frac{p_0}{\pi}\left[\frac{2}{1-n^2}\right] & n \quad \text{even} \end{cases} \qquad (a)$$

$$b_n = \frac{2}{T_p} \int_0^{T_p/2} p_0\,\sin\frac{2\pi t}{T_p}\,\sin\frac{2\,\pi\,n\,t}{T_p}\,dt = \begin{cases} \frac{p_0}{2} & n = 1 \\ 0 & n > 1 \end{cases}$$

Substituting these coefficients into Eq. (4-15) leads to the following series expression for the periodic loading:

$$p(t) = \frac{p_0}{\pi} \left(1 + \frac{\pi}{2}\,\sin\overline{\omega}_1 t - \frac{2}{3}\,\cos 2\,\overline{\omega}_1 t - \frac{2}{15}\,\cos 4\,\overline{\omega}_1 t - \frac{2}{35}\,\cos 6\,\overline{\omega}_1 t + \cdots \right) \qquad (b)$$

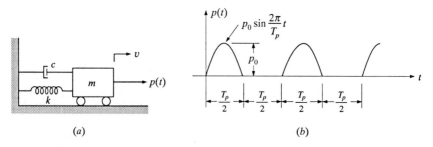

(a)   (b)

FIGURE E4-1
Example analysis of response to periodic loading: (a) SDOF system; (b) periodic loading.

in which $\bar{\omega}_1 = 2\pi / T_p$.

If it is now assumed that the structure of Fig. E4-1 is undamped, and if, for example, the period of loading is taken as four-thirds the period of vibration of the structure, i.e.,

$$\frac{T_p}{T} = \frac{\omega}{\bar{\omega}_1} = \frac{4}{3} \qquad\qquad \beta_n = \frac{n\bar{\omega}_1}{\omega} = \frac{3}{4} n \qquad\qquad (c)$$

the steady-state response given by Eq. (4-15) becomes

$$v(t) = \frac{p_0}{k\pi} \left(1 + \frac{8\pi}{7} \sin\bar{\omega}_1 t + \frac{8}{15} \cos 2\bar{\omega}_1 t + \frac{1}{60} \cos 4\bar{\omega}_1 t + \cdots \right) \qquad (d)$$

If the structure were damped, the analysis would proceed similarly, using Eq. (4-16) instead of Eq. (4-15).

If the periodic loading is expressed in terms of individual harmonics of the exponential form of Eq. (4-8), the nth harmonic steady-state response of the viscously damped SDOF system will be

$$v_n(t) = H_n P_n \exp(i\bar{\omega}_n t) \qquad\qquad (4\text{-}17)$$

where the complex loading coefficient P_n is given by Eq. (4-9) [or Eq. (4-10)] and where the complex frequency response coefficient H_n is given by Eq. (3-27) after dividing by the harmonic load amplitude, i.e.,

$$H_n = \frac{1}{k} \left[\frac{1}{(1 - \beta_n^2) + i\,(2\xi\beta_n)}\right] = \frac{1}{k} \left[\frac{(1 - \beta_n^2) - i\,(2\xi\beta_n)}{(1 - \beta_n^2)^2 + (2\xi\beta_n)^2}\right] \qquad (4\text{-}18)$$

Using the principle of superposition again, the total steady-state response of the SDOF system to the periodic loading of Eq. (4-8) is

$$v(t) = \sum_{n=-\infty}^{\infty} H_n P_n \exp(i\bar{\omega}_n t) \qquad\qquad (4\text{-}19)$$

Total response obtained by this equation will, of course, be the same as the total response obtained through Eq. (4-16).

4-3 PREVIEW OF FREQUENCY-DOMAIN ANALYSIS

It is useful at this time to point out that the above-described response analysis procedure for a SDOF system subjected to periodic loading contains all the essential elements of the "frequency-domain" method of analysis. That method is discussed extensively in Chapter 6, but its general concepts are evident in the preceding description. The first stage of the process, in which the Fourier coefficients of the periodic loading are evaluated, may be looked upon as the conversion of the applied loading expression from the time-domain to the frequency-domain form. In other words, the values $p_m = p(t_m)$ which express the applied load at a sequence of times, t_m, are replaced by the complex values $P_n = P(i\bar{\omega}_n)$ which express the harmonic load amplitudes at a specified sequence of frequencies, $\bar{\omega}_n$. These values constitute the frequency-domain expression of the loading.

In the second stage of the analysis, the SDOF response for any given frequency is characterized by the complex frequency response coefficient, H_n, which expresses the harmonic response amplitude due to a unit harmonic loading at the frequency $\bar{\omega}_n$. When this response coefficient is multiplied by the complex Fourier series coefficient P_n that expresses the harmonic input amplitude at that frequency, the result is the complex response amplitude, V_n, for that frequency. Thus the complete set of values V_n for all considered frequencies, $\bar{\omega}_n$, constitute the frequency-domain expression of system response.

In the final stage of the analysis, the frequency-domain response is converted back to the time domain by superposing the response components determined for all of the frequencies included in the Fourier series loading expression. In this superposition operation, it is necessary to evaluate all of the response harmonics at the same instants of time, t_m, recognizing the relative phase relationships associated with each frequency. When these response harmonics are added together, the final result is the time-domain expression of the response history, $v_m = v(t_m)$. As is explained in Chapter 6, the analysis task is made computationally feasible by a special computer technique known as the "Fast Fourier Transform" (FFT); but this brief description gives the essence of the frequency-domain procedure.

Example E4-2. Consider the periodic loading shown in Fig. E4-2. The corresponding Fourier coefficients P_n to be used in the SDOF response expression are

$$P_n = \frac{p_0}{T_p} \int_0^{T_p/2} \exp\left(-i\,\frac{2\pi n}{T_p}\,t\right) dt$$

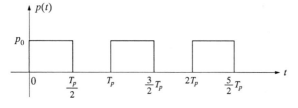

FIGURE E4-2
Rectangular-pulse-type periodic loading.

or

$$P_n = \frac{p_0}{T_p} \left(\frac{-T_p}{i\,2\pi\,n} \right) \left[\exp\left(-i\,\frac{2\pi n}{T_p}\,t \right) \right]_0^{T_p/2} = \begin{cases} p_0/2 & n = 0 \\ 0 & n \quad \text{odd} \\ -p_0\,i\,/\,\pi n & n \quad \text{even} \end{cases}$$

Making use of these coefficients and the values of H_n directly from Eq. (4-18), the total response of Eq. (4-19) is obtained.

PROBLEMS

4-1. Express the periodic loading shown in Fig. P4-1 as a Fourier series. Thus, determine the coefficients a_n and b_n by means of Eqs. (4-3) for the periodic loading given by

$$p(t) = p_0 \, \sin \frac{3\pi}{T_p} \, t \qquad\qquad (0 < t < 2\pi)$$

$$p(t) = 0 \qquad\qquad (2\pi < t < 3\pi)$$

Then write the loading in the series form of Eq. (4-1).

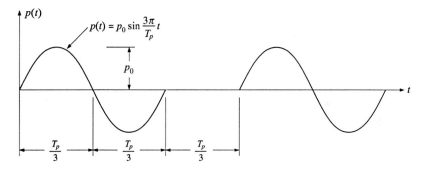

FIGURE P4-1

4-2. Repeat Prob. 4-1 for the periodic loading shown in Fig. P4-2.

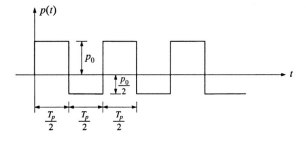

FIGURE P4-2

4-3. Solve the problem of Example E4-1, assuming that the structure is 10 percent critically damped.

4-4. Construct an Argand diagram similar to that of Fig. 3-6, showing to scale the applied load vector and the steady-state inertia, damping, and elastic resisting-force vectors. Assume the structure has 15 percent critical damping and is subjected to the harmonic loading term $p(t) = p_0 \exp[i\bar{\omega}t]$, where $\bar{\omega} = (6/5)\omega$ (i.e., $\beta = 6/5$). Construct the diagram for the time when $\bar{\omega}t = \pi/4$.

4-5. The periodic loading of Fig. P4-3 can be expressed by the sine series

$$p(t) = \sum_{n=1}^{\infty} b_n \, \sin\bar{\omega}_n t$$

where

$$b_n = -\frac{2p_0}{n\pi}(-1)^n$$

Plot the steady-state response of the structure of Fig. E4-1a to this loading for one full period, considering only the first four terms of the series and evaluating at time increments given by $\bar{\omega}_1\Delta t = 30°$.

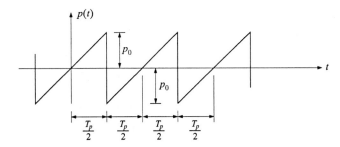

FIGURE P4-3

CHAPTER
5

RESPONSE TO IMPULSIVE LOADING

5-1 GENERAL NATURE OF IMPULSIVE LOADING

Another special class of dynamic loading of the SDOF system will now be considered, the impulsive load. Such a load consists of a single principal impulse of arbitrary form, as illustrated in Fig. 5-1, and generally is of relatively short duration. Impulsive or shock loads frequently are of great importance in the design of certain classes of structural systems, e.g., vehicles such as trucks or automobiles or traveling cranes. Damping has much less importance in controlling the maximum response of a structure to impulsive loads than for periodic or harmonic loads because the maximum response to a particular impulsive load will be reached in a very short time, before the damping forces can absorb much energy from the structure. For this reason only the undamped response to impulsive loads will be considered in this chapter.

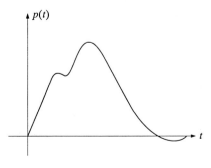

FIGURE 5-1
Arbitrary impulsive loading.

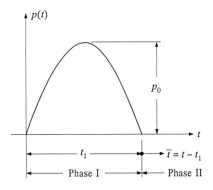

FIGURE 5-2
Half-sine-wave impulse.

5-2   SINE-WAVE IMPULSE

For impulsive loads which can be expressed by simple analytical functions, closed form solutions of the equations of motion can be obtained. For example, consider the single half-sine-wave impulse shown in Fig. 5-2. The response to such an impulse will be divided into two phases as shown, the first corresponding to the forced-vibration phase in the interval during which the load acts and the second corresponding to the free-vibration phase which follows.

Phase I — During this phase, the structure is subjected to the single half-sine-wave loading shown in Fig. 5-2. Assuming the system starts from rest, the undamped response-ratio time-history $R(t) \equiv v(t) / (p_0/k)$, including the transient as well as the steady-state term, is given by the simple harmonic load expression, Eq. (3-12). Introducing the nondimensional time parameter $\alpha \equiv t/t_1$ so that $\bar{\omega}t = \pi\alpha$ and $\omega t = \pi\alpha/\beta$, this equation can be written in the form

$$R(\alpha) = \left[\frac{1}{1-\beta^2}\right]\left[\sin\pi\alpha - \beta\sin\frac{\pi\alpha}{\beta}\right] \qquad 0 \leq \alpha \leq 1 \qquad (5\text{-}1)$$

where $\beta \equiv \bar{\omega}/\omega = T/2t_1$. This equation is, of course, valid only in Phase I corresponding to $0 \leq \alpha \leq 1$. Since it is indeterminate for $\beta = 1$, L'Hospital's rule must be applied to obtain a useable expression for this special case. Taking this action, one obtains [by analogy with Eq. (3-38)]

$$R(\alpha) = \frac{1}{2}\left[\sin\pi\alpha - \pi\alpha\cos\pi\alpha\right] \qquad \beta = 1 \qquad 0 \leq \alpha \leq 1 \qquad (5\text{-}2)$$

Phase II — The free-vibration motion which occurs during this phase, $t \geq t_1$, depends on the displacement $v(t_1)$ and velocity $\dot{v}(t_1)$ existing at the end of Phase I; in other words, in terms of the response ratio, it depends on the values of $R(1)$ and $\dot{R}(1)$ given by Eq. (5-1) and its first time derivative expression, respectively. Thus, using Eq. (2-33) in its response-ratio form this free-vibration response is shown to be

$$R(\alpha) = \left[\frac{-\beta}{1-\beta^2}\right] \left\{ \left(1+\cos\frac{\pi}{\beta}\right) \sin\left[\frac{\pi}{\beta}(\alpha-1)\right] + \left(\sin\frac{\pi}{\beta}\right) \cos\left[\frac{\pi}{\beta}(\alpha-1)\right]\right\}$$

$$\alpha \geq 1 \qquad (5\text{-}3)$$

in which $\left[\frac{\pi}{\beta}(\alpha-1)\right] = \omega(t-t_1)$. This expression, like Eq. (5-1), is indeterminate for $\beta = 1$, requiring once again the use of L'Hospital's rule leading to

$$R(\alpha) = \frac{\pi}{2} \cos\left[\pi(\alpha-1)\right] \qquad \beta = 1 \qquad \alpha \geq 1 \qquad (5\text{-}4)$$

Using Eqs. (5-1) and (5-2) for Phase I and Eqs. (5-3) and (5-4) for Phase II, response-ratio time-histories can be generated for discrete values of β as illustrated by the solid lines in Fig. 5-3. The values of β selected for this figure are 1/4, 1/3, 1/2, 1, and 3/2 which correspond to values of t_1/T equal to 2, 3/2, 1, 1/2, and 1/3, respectively. Also shown for comparison is the dashed line representing the quasi-static response ratio $[p(t)/k] / (p_0/k) = p(t)/p_0$ which has a peak value equal to unity. Notice that for $t_1/T = 1/2$ ($\beta = 1$), the maximum response at Point d occurs exactly at the end of Phase I. For any value of t_1/T less than 1/2 ($\beta > 1$), the maximum response will occur in Phase II; while for any value of t_1/T greater than 1/2 ($\beta < 1$), it will occur in Phase I. Clearly, the maximum value of response depends on the ratio of the load duration to the period of vibration of the structure, i.e., on the ratio $t_1/T = 1/2\,\beta$.

While it is very important to understand the complete time-history behavior as shown in Fig. 5-3, the engineer is usually only interested in the maximum value of

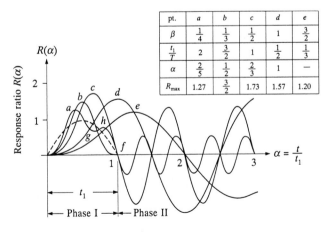

pt.	a	b	c	d	e
β	$\frac{1}{4}$	$\frac{1}{3}$	$\frac{1}{2}$	1	$\frac{3}{2}$
$\frac{t_1}{T}$	2	$\frac{3}{2}$	1	$\frac{1}{2}$	$\frac{1}{3}$
α	$\frac{2}{5}$	$\frac{1}{2}$	$\frac{2}{3}$	1	—
R_{max}	1.27	$\frac{3}{2}$	1.73	1.57	1.20

FIGURE 5-3
Response ratios due to half-sine pulse.

response as represented by Points a, b, c, d, and e. If a maximum value occurs in Phase I, the value of α at which it occurs can be determined by differentiating Eq. (5-1) with respect to α and equating to zero, thus obtaining

$$\frac{d\,R(\alpha)}{d\,\alpha} = \left[\frac{\pi}{1-\beta^2}\right]\left[\cos\pi\alpha - \cos\frac{\pi\alpha}{\beta}\right] = 0 \tag{5-5}$$

from which

$$\cos\pi\alpha = \cos\frac{\pi\alpha}{\beta} \tag{5-6}$$

This equation is satisfied when

$$\pi\alpha = \pm\frac{\pi\alpha}{\beta} + 2\pi n \qquad n = 0, \pm 1, \pm 2, \cdots \tag{5-7}$$

Solving for α gives

$$\alpha = \frac{2\beta n}{(\beta \pm 1)} \qquad n = 0, \pm 1, \pm 2, \cdots \tag{5-8}$$

which is valid, of course, only when the resulting values of α fall in Phase I, i.e., in the range $0 \leq \alpha \leq 1$. As previously shown, this condition is met only when $0 \leq \beta \leq 1$. To satisfy both of these conditions, it is necessary that the positive and negative values of n be used along with the plus and minus signs, respectively, in Eq. (5-8). Note that the zero value of n can be dropped from consideration as it yields $\alpha = 0$ which simply confirms that the zero-velocity initial condition has been satisfied.

To develop an understanding of the use of Eq. (5-8), let us consider the cases shown in Fig. 5-3. For the limit-value case $\beta = 1$, using the plus sign and $n = +1$, one obtains $\alpha = 1$ which when substituted into Eq. (5-2) yields $R(1) = \pi/2$. This corresponds to Point d in Fig. 5-3. When $\beta = 1/2$, Eq. (5-8) has only one valid solution, namely the solution using the plus sign and $n = +1$. The resulting α value is 2/3 which when substituted into Eq. (5-1) gives $R(2/3) = 1.73$ as shown by Point c. For $\beta = 1/3$, the plus-sign form of Eq. (5-8) gives $\alpha = 1/2$ and 1 when $n = +1$ and $+2$, respectively; when substituted into Eq. (5-1) these yield $R(1/2) = 3/2$ and $R(1) = 0$, as shown in Fig. 5-3 by Points b and f. Note that because $\dot{R}(1)$ is zero in this case, there is no free vibration in Phase II. For the case $\beta = 1/4$, two maxima (Points a and h) and one minimum (Point g) are clearly present in Phase I. Points a and h correspond to using the plus sign along with $n = +1$ and $+2$, respectively, giving $\alpha = 2/5$ and 4/5, while Point g corresponds to using the minus sign along with $n = -1$ giving $\alpha = 2/3$. It is now clear that using the plus sign in Eq. (5-8) along with positive values of n yields α-values for the maxima, while using the minus sign along with the negative values of n yields α-values for the minima. Substituting the above values of α into Eq. (5-1) gives $R(2/5) = 1.268$, $R(4/5) = 0.784$, and

$R(2/3) = 0.693$ corresponding to Points a, h, and g, respectively. If one examined additional cases by further reducing the value of β, the numbers of maxima and minima will continue to increase in Phase I with the largest of the maxima changing from the first (as in the case of $\beta = 1/4$) to the second, then to the third, etc. In the limit, as $\beta \to 0$, the response-ratio curve will approach the quasi-static response curve shown by the dashed line in Fig. 5-3 and $R_{\max}$ will approach unity.

Finally, consider the case $\beta = 3/2$ which has its maximum response in Phase II as indicated by Point e. It is not necessary in this case of free vibration to determine the value of α corresponding to maximum response because the desired maximum value is obtained directly by simply taking the vector sum of the two orthogonal components in Eq. (5-3) giving

$$R_{\max} = \left[\frac{-\beta}{1 - \beta^2} \right] \left[\left(1 + \cos \frac{\pi}{\beta} \right)^2 + \left(\sin \frac{\pi}{\beta} \right)^2 \right]^{1/2}$$

$$= \left[\frac{-\beta}{1 - \beta^2} \right] \left[2 \left(1 + \cos \frac{\pi}{\beta} \right) \right]^{1/2}$$

Finally using the trigonometric identity $\left[2 \left(1 + \cos \frac{\pi}{\beta} \right) \right]^{1/2} \equiv 2 \cos \frac{\pi}{2\beta}$ this may be written in the following simplified form:

$$R_{\max} = \left[\frac{-2\beta}{1 - \beta^2} \right] \cos \frac{\pi}{2\beta} \tag{5-9}$$

For the above case of $\beta = 3/2$, this expression gives $R_{\max} = 1.2$.

5-3 RECTANGULAR IMPULSE

A second example of the analysis of the response to an impulse load will now make use of the rectangular loading shown in Fig. 5-4. Again the response will be divided into the loading phase and the subsequent free-vibration phase.

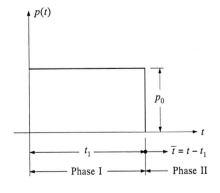

FIGURE 5-4
Rectangular impulse.

Phase I — The suddenly applied load which remains constant during this phase is called a step loading. The particular solution to the equation of motion for this case is simply the static deflection

$$v_p = p_0 / k \qquad\qquad R_p = 1 \qquad\qquad (5\text{-}10)$$

Using this result, the general response-ratio solution, in which the complementary free-vibration solution constants have been evaluated to satisfy the at-rest initial conditions, is easily found to be

$$R(\alpha) = \left[1 - \cos 2\pi \left(\frac{t_1}{T}\right)\alpha\right] \qquad 0 \le \alpha \le 1 \qquad (5\text{-}11)$$

where again $\alpha \equiv t/t_1$ so that $\omega t = 2\pi (t_1/T)\alpha$. The first maximum of this expression occurs when $(t_1/T)\alpha = 1/2$. If it is to occur exactly at the end of Phase I, i.e., $\alpha = 1$, then the ratio t_1/T must equal 1/2; in this case, from Eq. (5-11), $R(1/2) = 2$. As t_1/T continues to increase above 1/2, additional maxima will appear in Phase I each having the value $R_{max} = 2$. As t_1/T decreases from 1/2, no maximum can occur in Phase I in accordance with Eq. (5-11); instead the maximum response will occur in Phase II under the free-vibration condition.

Phase II — Using Eq. (2-33) in its response-ratio form and applying Eq. (5-11) to find $R(1)$ and $\dot{R}(1)$, the free vibration in this phase is given by

$$R(\alpha) = \left(1 - \cos 2\pi \frac{t_1}{T}\right) \cos\left[2\pi \frac{t_1}{T}(\alpha - 1)\right]$$

$$+ \left(\sin 2\pi \frac{t_1}{T}\right) \sin\left[2\pi \frac{t_1}{T}(\alpha - 1)\right] \qquad \alpha \ge 1 \qquad (5\text{-}12)$$

in which $\left[2\pi \frac{t_1}{T}(\alpha - 1)\right] = \omega (t - t_1)$. Taking the vector sum of the two orthogonal components in this expression gives

$$R_{max} = \left[\left(1 - \cos 2\pi \frac{t_1}{T}\right)^2 + \left(\sin 2\pi \frac{t_1}{T}\right)^2\right]^{1/2}$$

$$= \left[2\left(1 - \cos 2\pi \frac{t_1}{T}\right)\right]^{1/2} = 2 \sin \pi \frac{t_1}{T} \qquad (5\text{-}13)$$

showing that the maximum response to the rectangular impulse varies as a sine function for $0 \le t_1/T \le 1/2$.

5-4   TRIANGULAR IMPULSE

The last impulse loading to be analyzed in detail is the decreasing triangular impulse shown in Fig. 5-5.

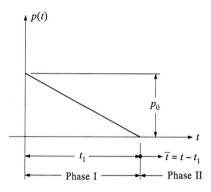

FIGURE 5-5
Triangular impulse.

Phase I — The loading during this phase is $p_0 \left(1 - \frac{t}{t_1}\right)$ for which it is easily demonstrated that the particular solution to the equation of motion, in its response-ratio form, is

$$R_p(t) = (1 - \alpha) \qquad\qquad 0 \leq \alpha \leq 1 \qquad\qquad (5\text{-}14)$$

in which $\alpha = \frac{t}{t_1}$. Combining this solution with the complementary free-vibration solution and evaluating its constants to satisfy the zero initial conditions, one finds

$$R(\alpha) = \left(\frac{1}{2\pi \frac{t_1}{T}}\right) \sin 2\pi \frac{t_1}{T} \alpha - \cos 2\pi \frac{t_1}{T} \alpha - \alpha + 1 \qquad 0 \leq \alpha \leq 1 \qquad (5\text{-}15)$$

Taking the first time derivative of this expression and setting it to zero, one can show that the first maximum will occur exactly at the end of Phase I (i.e., at $\alpha = 1$), when $t_1/T = 0.37101$. Substituting this value into Eq. (5-15) gives $R(0.37101) = 1$. For values of $t_1/T > 0.37101$, the maximum response will be in Phase I and can be obtained from Eq. (5-15) upon substitution of the proper α-value representing the zero-velocity condition.

Phase II — When $t_1/T < 0.37101$, the maximum response will be the free-vibration amplitude in Phase II. It is found in the same manner as in the previous cases by substituting $R(1)$ and $\dot{R}(1)$ obtained from Eq. (5-15) and its first time derivative expression, respectively, into the response-ratio form of the free-vibration response given by Eq. (2-33). The maximum response is then the vector sum of the two orthogonal components in the resulting free-vibration equation.

5-5 SHOCK OR RESPONSE SPECTRA

In the expressions derived above, the maximum response produced in an undamped SDOF structure by each type of impulsive loading depends only on the ratio of the impulse duration to the natural period of the structure, i.e., on the ratio t_1/T. Thus, it is useful to plot the maximum value of response ratio $R_{\max}$ as a function of

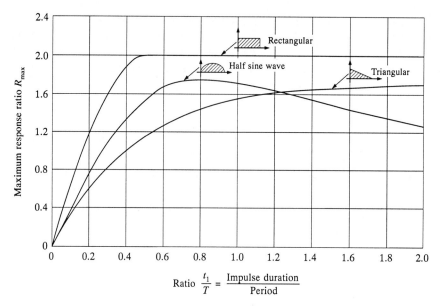

FIGURE 5-6
Displacement-response spectra (shock spectra) for three types of impulse.

t_1/T for various forms of impulsive loading. Such plots, shown in Fig. 5-6 for the three forms of loading treated above, are commonly known as displacement-response spectra, or merely as response spectra. Generally plots like these can be used to predict with adequate accuracy the maximum effect to be expected from a given type of impulsive loading acting on a simple structure.

These response spectra also serve to indicate the response of the structure to an acceleration pulse applied to its base. If the applied base acceleration is $\ddot{v}_g(t)$, it produces an effective impulsive loading $p_{\text{eff}} = -m\,\ddot{v}_g(t)$ [see Eq. (2-17)]. If the maximum base acceleration is denoted by $\ddot{v}_{g0}$, the maximum effective impulsive load is $p_{0,\text{max}} = -m\,\ddot{v}_{g0}$. The maximum response ratio can now be expressed as

$$R_{\text{max}} = \left| \frac{v_{\text{max}}}{m\,\ddot{v}_{g0}\,/\,k} \right| \tag{5-16}$$

in which only the absolute magnitude is generally of interest. Alternatively, this maximum response ratio can be written in the form

$$R_{\text{max}} = \left| \ddot{v}^t_{\text{max}}\,/\,\ddot{v}_{g0} \right| \tag{5-17}$$

where $\ddot{v}^t_{\text{max}}$ is the maximum total acceleration of the mass. This follows from the fact that in an undamped system, the product of the mass and the total acceleration must be equal in magnitude to the elastic spring force $k\,v_{max}$. Accordingly, it is evident that the response spectrum plots of Fig. 5-6 can be used to predict the maximum acceleration response of mass m to an impulsive acceleration as well as the maximum displacement response to impulsive loads. When used to predict response to base acceleration, the plots are generally referred to as shock spectra.

Example E5-1. As an example of the use of the above described response (or shock) spectra in evaluating the maximum response of a SDOF structure to an impulsive load, consider the system shown in Fig. E5-1, which represents a single-story building subjected to the triangular blast load. For the given weight and column stiffness of this structure, the natural period of vibration is

$$T = \frac{2\pi}{\omega} = 2\pi\sqrt{\frac{W}{kg}} = 2\pi\sqrt{\frac{600}{10,000\,(386)}} = 0.079\ sec$$

The ratio of impulse duration to natural period becomes

$$\frac{t_1}{T} = \frac{0.05}{0.079} = 0.63$$

and from Fig. 5-6, the maximum response ratio is $R_{max} = 1.33$. Thus, the maximum displacement will be

$$v_{max} = R_{max}\left(\frac{p_0}{k}\right) = 1.33\left(\frac{1,000}{10,000}\right) = 0.133\ in\quad [0.338\ cm]$$

and the maximum total elastic force developed is

$$f_{S,max} = k\,v_{max} = 10,000\,(1.33) = 1,330\ kips\quad [603,300\ kg]$$

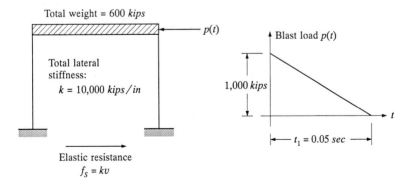

FIGURE E5-1
SDOF building subjected to blast load.

If the blast-pressure impulse had been only one-tenth as long $(t_1 = 0.005\ sec)$, the maximum response ratio for this impulse duration $(t_1/T = 0.063)$ would be only $R_{max} = 0.20$. Thus for an impulse of very short-duration, a large part of the applied load is resisted by the inertia of the structure, and the stresses produced are much smaller than those produced by loadings of longer duration.

It should be kept in mind that although the response (or shock) spectra described above have been developed for the undamped SDOF system, they can be used for damped systems as well since damping in the practical range of interest has little effect on the maximum response produced by short-duration impulsive loads.

5-6 APPROXIMATE ANALYSIS OF IMPULSIVE-LOAD RESPONSE

From a study of the response spectra presented in Fig. 5-6 and similar spectra for other forms of loadings, two general conclusions may be drawn concerning the response of structures to impulsive loadings:

(1) For long-duration loadings, for example, $t_1/T > 1$, the dynamic magnification factor depends principally on the rate of increase of the load to its maximum value. A step loading of sufficient duration produces a magnification factor of 2; a very gradual increase causes a magnification factor of 1.

(2) For short-duration loads, for example, $t_1/T < 1/4$, the maximum displacement amplitude v_{max} depends principally upon the magnitude of the applied impulse $I = \int_0^{t_1} p(t)\ dt$ and is not strongly influenced by the form of the loading impulse. The maximum response ratio R_{max} is, however, quite dependent upon the form of loading because it is proportional to the ratio of impulse area to peak-load amplitude, as may be noted by comparing the curves of Fig. 5-6 in the short-period range. Thus v_{max} is the more significant measure of response.

A convenient approximate procedure for evaluating the maximum response to a short-duration impulsive load, which represents a mathematical expression of this second conclusion, may be derived as follows. The impulse-momentum relationship for the mass m may be written

$$m\ \Delta \dot{v} = \int_0^{t_1} \left[p(t) - k\,v(t) \right]\ dt \qquad (5\text{-}18)$$

in which $\Delta \dot{v}$ represents the change of velocity produced by the loading. In this expression it may be observed that for small values of t_1 the displacement developed during the loading $v(t_1)$ is of the order of $(t_1)^2$ while the velocity change $\Delta \dot{v}$ is of the order of t_1. Thus since the impulse is also of the order of t_1, the elastic force term $k\,v(t)$ vanishes from the expression as t_1 approaches zero and is negligibly small for short-duration loadings.

On this basis, the approximate relationship may be used:

$$m\,\Delta\dot{v} \doteq \int_0^{t_1} p(t)\,dt \qquad (5\text{-}19)$$

or

$$\Delta\dot{v} = \frac{1}{m}\int_0^{t_1} p(t)\,dt \qquad (5\text{-}20)$$

The response after termination of loading is the free vibration

$$v(\bar{t}) = \frac{\dot{v}(t_1)}{\omega}\,\sin\omega\bar{t} + v(t_1)\,\cos\omega\bar{t}$$

in which $\bar{t} = t - t_1$. But since the displacement term $v(t_1)$ is negligibly small and the velocity $\dot{v}(t_1) = \Delta\dot{v}$, the following approximate relationship may be used:

$$v(\bar{t}) \doteq \frac{1}{m\omega}\left(\int_0^{t_1} p(t)\,dt\right)\sin\omega\bar{t} \qquad (5\text{-}21)$$

Example E5-2.   As an example of the use of this approximate formula, consider the response of the structure shown in Fig. E5-2 to the impulsive loading indicated. In this case, $\omega = \sqrt{kg/W} = 3.14\,rad/sec$ and $\int_0^{t_1} p(t)\,dt = 10\ kip\cdot sec$. The response then is approximately

$$v(\bar{t}) = \frac{10\,(386)}{2,000\,(3.14)}\,\sin\omega\bar{t}$$

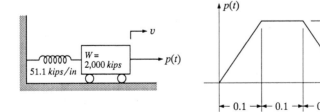

FIGURE E5-2
Approximate impulse-response analysis.

in which the acceleration of gravity is taken as $g = 386\ in/sec^2$ [$980.7\ cm/sec^2$]. The maximum response results when $\sin w\bar{t} = 1$, that is,

$$v_{max} \doteq 0.614\ in \quad [1.56\ cm]$$

The maximum elastic force developed in the spring, which is of major concern to the structural engineer, is

$$f_{S,max} = k\ v_{max} = 51.1\ (0.614) = 31.4\ kips \quad [14,240\ kg]$$

Since the period of vibration of this system is $T = 2\pi/w = 2\ sec$, the ratio of load duration to period is $t_1/T = 0.15$; thus, the approximate analysis in this case is quite accurate. In fact, the exact maximum response determined by direct integration of the equation of motion is $0.604\ in$ [$1.53\ cm$], and so the error in the approximate result is less than 2 percent.

PROBLEMS

5-1. Consider the basic dynamic system of Fig. 2-1a with the following properties: $W = 600\ lb$ ($m = W/g$) and $k = 1,000\ lb/in$. Assume that it is subjected to a half sine-wave impulse (Fig. 5-2) of amplitude $p_0 = 500\ lb$ and duration $t_1 = 0.15\ sec$. Determine:
   (a) The time at which the maximum response will occur.
   (b) The maximum spring force produced by this loading; check this result with that obtained by use of Fig. 5-6.

5-2. A triangular impulse that increases linearly from zero to the peak value is expressed as $p(t) = p_0(t/t_1)$ $(0 < t < t_1)$.
   (a) Derive an expression for the response of a SDOF structure to this loading, starting from "at rest" conditions.
   (b) Determine the maximum response ratio

$$R_{max} = \frac{v_{max}}{p_0/k}$$

resulting from this loading if $t_1 = 3\pi/w$.

5-3. A quarter cosine-wave impulse is expressed as

$$p(t) = p_0\ \cos \bar{w}t \qquad 0 < t < \frac{\pi}{2\bar{w}}$$

   (a) Derive an expression for the response to this impulse, starting from rest.
   (b) Determine the maximum response ratio

$$R_{max} = \frac{v_{max}}{p_0/k} \qquad if \ \ \bar{w} = w$$

5-4. The basic SDOF system of Fig. 2-1a, having the following properties, $k =$ 20 $kips/in$ and $m = 4\ kips \cdot sec^2/in$, is subjected to a triangular impulse of the form of Fig. 5-5 with $p_0 = 15\ kips$ and $t_1 = 0.15\ T$.

    (a) Using the shock spectra of Fig. 5-6, determine the maximum spring force $f_{S\max}$.

    (b) Using Eq. (5-21), compute approximately the maximum displacement and spring force; compare with the result of part a.

5-5. The water tank of Fig. P5-1a can be treated as a SDOF structure with the following properties: $m = 4\ kips \cdot sec^2/in$, $k = 40\ kips/in$. As a result of an explosion, the tank is subjected to the dynamic-load history shown in Fig. P5-1b. Compute approximately the maximum overturning moment M_0 at the base of the tower using Eq. (5-21) and evaluating the impulse integral by means of Simpson's rule:

$$\int p\ dt = \frac{\Delta t}{3}\ (p_0 + 4p_1 + 2p_2 + 4p_3 + p_4)$$

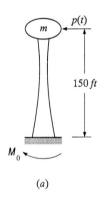

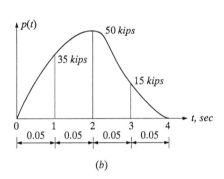

(a)   (b)

FIGURE P5-1

RESPONSE
TO GENERAL
DYNAMIC
LOADING:
SUPERPOSITION
METHODS

6-1 ANALYSIS THROUGH THE TIME DOMAIN

Formulation of Response Integral

<u>Undamped System</u> — The procedure described in Chapter 5 for approximating the response of an undamped SDOF structure to short-duration impulsive loads can be used as the basis for developing a formula for evaluating response to a general dynamic loading. Consider an arbitrary general loading $p(t)$ as illustrated in Fig. 6-1 and, for the moment, concentrate on the intensity of loading $p(\tau)$ acting at time $t = \tau$. This

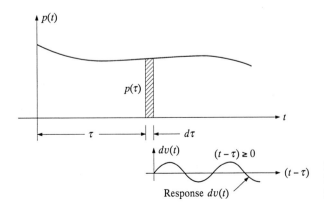

FIGURE 6-1
Derivation of the Duhamel integral (undamped).

87

loading acting during the interval of time $d\tau$ represents a very short-duration impulse $p(\tau)\,d\tau$ on the structure, so that Eq. (5-21) can be used to evaluate the resulting response. It should be noted carefully that although this equation is approximate for impulses of finite duration, it becomes exact as the duration of loading approaches zero. Thus, for the differential time interval $d\tau$, the response produced by the impulse $p(\tau)\,d\tau$ is exactly equal to

$$dv(t) = \frac{p(\tau)\,d\tau}{m\,\omega}\ \sin\omega\,(t-\tau) \qquad\qquad t \geq \tau \qquad\qquad (6\text{-}1)$$

In this expression, the term $dv(t)$ represents the time-history response to the differential impulse over the entire time $t \geq \tau$; it is not the change of v during a time interval dt.

The entire loading history can be considered to consist of a succession of such short impulses, each producing its own differential response of the form of Eq. (6-1). For this linearly elastic system, the total response can then be obtained by summing all the differential responses developed during the loading history, that is, by integrating Eq. (6-1) as follows:

$$v(t) = \frac{1}{m\,\omega}\ \int_0^t p(\tau)\ \sin\omega\,(t-\tau)\ d\tau \qquad\qquad t \geq 0 \qquad\qquad (6\text{-}2)$$

This relation, generally known as the Duhamel integral equation, can be used to evaluate the response of an undamped SDOF system to any form of dynamic loading $p(t)$; however, for arbitrary loadings the evaluation must be performed numerically using procedures described subsequently.

Equation (6-2) can also be expressed in the general convolution integral form:

$$v(t) = \int_0^t p(\tau)\ h(t-\tau)\ d\tau \qquad\qquad t \geq 0 \qquad\qquad (6\text{-}3)$$

in which the function
$$h(t-\tau) = \frac{1}{m\,\omega}\ \sin\omega\,(t-\tau) \qquad\qquad (6\text{-}4)$$

is known as the unit-impulse response function because it expresses the response of the SDOF system to a pure impulse of unit magnitude applied at time $t = \tau$. Generating response using the Duhamel or convolution integral is one means of obtaining response through the time domain. It is important to note that this approach may be applied only to linear systems because the response is obtained by superposition of individual impulse responses.

In Eqs. (6-1) and (6-2) it has been tacitly assumed that the loading was initiated at time $t = 0$ and that the structure was at rest at that time. For any other specified initial conditions $v(0) \neq 0$ and $\dot{v}(0) \neq 0$, the additional free-vibration response must be added to this solution; thus, in general

$$v(t) = \frac{\dot{v}(0)}{\omega} \sin \omega t + v(0) \cos \omega t + \frac{1}{m\omega} \int_0^t p(\tau) \sin \omega (t - \tau) \, d\tau \qquad (6\text{-}5)$$

Should the nonzero initial conditions be produced by known loading $p(t)$ for $t < 0$, the total response given by this equation could also be found through Eq. (6-2) by changing the lower limit of the integral from zero to minus infinity.

Under-Critically-Damped System — The derivation of the Duhamel integral equation which expresses the response of a viscously damped system to a general dynamic loading is entirely equivalent that for the undamped case, except that the free-vibration response initiated by the differential load impulse $p(\tau) \, d\tau$ experiences exponential decay. By expressing Eq. (2-49) in terms of $t - \tau$ rather than t, and substituting zero for $v(0)$ and $p(\tau) \, d\tau / m$ for $\dot{v}(0)$, one obtains the damped differential response

$$dv(t) = \left[\frac{p(\tau) \, d\tau}{m\omega_D} \sin \omega_D (t - \tau) \right] \exp\left[-\xi \omega (t - \tau) \right] \qquad t \geq \tau \qquad (6\text{-}6)$$

showing that the exponential decay begins as soon as the load $p(\tau)$ is applied. Summing these differential response terms over the loading interval $0 < \tau < t$ results in

$$v(t) = \frac{1}{m\omega_D} \int_0^t p(\tau) \sin \omega_D (t - \tau) \exp\left[-\xi \omega (t - \tau) \right] d\tau \qquad t \geq 0 \qquad (6\text{-}7)$$

which is the damped-response equivalent of Eq. (6-2).

When expressing Eq. (6-7) in terms of the convolution integral of Eq. (6-4), the damped unit-impulse response function

$$h(t - \tau) = \frac{1}{m\omega_D} \sin \omega_D (t - \tau) \exp\left[-\xi \omega (t - \tau) \right] \qquad (6\text{-}8)$$

must be used. If the initial conditions $v(0)$ and $\dot{v}(0)$ are not equal to zero, then the corresponding free-vibration response given by Eq. (2-49) must be added to Eq. (6-7).

Numerical Evaluation of Response Integral

Undamped System — If the applied-loading function $p(\tau)$ is of simple analytic form, then the integrals in Eqs. (6-2) and (6-7) can be evaluated directly. However, this usually is not possible in most practical cases as the loading is known only from experimental data. The response integrals must then be evaluated by numerical procedures.

To develop these procedures, use is made of the trigonometric identity

$$\sin(\omega t - \omega\tau) = \left[\sin\omega t \ \cos\omega\tau - \cos\omega t \ \sin\omega\tau\right] \tag{6-9}$$

so that Eq. (6-2), which assumes zero initial conditions, can be written as

$$v(t) = \sin\omega t \left[\frac{1}{m\omega} \int_0^t p(\tau) \ \cos\omega\tau \ d\tau\right] - \cos\omega t \left[\frac{1}{m\omega} \int_0^t p(\tau) \ \sin\omega\tau \ d\tau\right]$$

or

$$v(t) = \left[\overline{A}(t) \ \sin\omega t - \overline{B}(t) \ \cos\omega t\right] \tag{6-10}$$

where

$$\overline{A}(t) \equiv \frac{1}{m\omega} \int_0^t p(\tau) \ \cos\omega\tau \ d\tau \qquad \overline{B}(t) \equiv \frac{1}{m\omega} \int_0^t p(\tau) \ \sin\omega\tau \ d\tau \tag{6-11}$$

Numerical procedures, which can be used to evaluate $\overline{A}(t)$ and $\overline{B}(t)$, will now be described.

Consider first the numerical integration of $y(\tau) \equiv p(\tau) \ \cos\omega\tau$ as required to find $\overline{A}(t)$. For convenience of numerical calculation, the function $y(\tau)$ is evaluated at equal time increments $\Delta\tau$ as shown in Fig. 6-2, with the successive ordinates being identified by appropriate subscripts. The integral $\overline{A}_N \equiv \overline{A}(t = N \, \Delta t)$ can now be obtained approximately by summing these ordinates, after multiplying by weighting

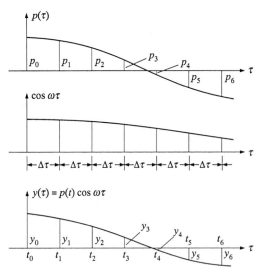

FIGURE 6-2
Formulation of numerical summation process for Duhamel integral.

factors that depend on the numerical integration scheme being used as follows:

Simple summation:

$$\overline{A}_N \doteq \tfrac{\Delta \tau}{m \, \omega} \left[y_0 + y_1 + y_2 + \cdots + y_{N-1} \right] \qquad\qquad N = 1, 2, 3, \cdots \quad (6\text{-}12a)$$

Trapezoidal rule:

$$\overline{A}_N \doteq \tfrac{\Delta \tau}{2 \, m \, \omega} \left[y_0 + 2 \, y_1 + 2 \, y_2 + \cdots + 2 \, y_{N-1} + y_N \right] \quad N = 1, 2, 3, \cdots \quad (6\text{-}12b)$$

Simpson's rule:

$$\overline{A}_N \doteq \tfrac{\Delta \tau}{3 \, m \, \omega} \left[y_0 + 4 \, y_1 + 2 \, y_2 + \cdots + 4 \, y_{N-1} + y_N \right] \quad N = 2, 4, 6, \cdots \quad (6\text{-}12c)$$

Using any one of these equations, $\overline{A}_N$ can be obtained directly for any specific value of N indicated. However, usually the entire time-history of response is required so that one must evaluate $\overline{A}_N$ for successive values of N until the desired time-history of response is obtained. For this purpose, it is more efficient to use these equations in their recursive forms:

Simple summation:

$$\overline{A}_N \doteq \overline{A}_{N-1} + \tfrac{\Delta \tau}{m \, \omega} \left[y_{N-1} \right] \qquad\qquad N = 1, 2, 3, \cdots \quad (6\text{-}13a)$$

Trapezoidal rule:

$$\overline{A}_N \doteq \overline{A}_{N-1} + \tfrac{\Delta \tau}{2 \, m \, \omega} \left[y_{N-1} + y_N \right] \qquad\qquad N = 1, 2, 3, \cdots \quad (6\text{-}13b)$$

Simpson's rule:

$$\overline{A}_N \doteq \overline{A}_{N-2} + \tfrac{\Delta \tau}{3 \, m \, \omega} \left[y_{N-2} + 4 \, y_{N-1} + y_N \right] \quad N = 2, 4, 6, \cdots \quad (6\text{-}13c)$$

such that $\overline{A}_0 = 0$.

Evaluation of $\overline{B}(t)$ in Eq. (6-10) can be carried out in the same manner, leading to expressions for $\overline{B}_N$ having exactly the same forms shown by Eqs. (6-13); however, in doing so, the definition of $y(\tau)$ must be changed to $y(\tau) \equiv p(\tau) \sin \omega \tau$ consistent with the second of Eqs. (6-11).

Having calculated the values of $\overline{A}_N$ and $\overline{B}_N$ for successive values of N, the corresponding values of response $v_N \equiv v \, (t = N \, \Delta \tau)$ are obtained using

$$v_N = \overline{A}_N \, \sin \omega t_N - \overline{B}_N \, \cos \omega t_N \qquad\qquad (6\text{-}14)$$

Example E6-1.   The dynamic response of a water tower subjected to a blast loading will now be presented to illustrate the above numerical procedure for obtaining undamped response through the time domain in accordance with Eq. (6-14). The idealizations of the structure and blast loading are shown in Fig. E6-1. For this system, the vibration frequency and period are

$$\omega = \sqrt{\frac{kg}{W}} = \sqrt{\frac{2,700 \, (32.2)}{96.6}} = 30 \; rad/sec \qquad T = \frac{2 \, \pi}{\omega} = 0.209 \; sec$$

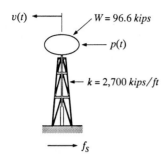

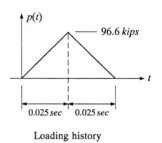

FIGURE E6-1
Water tower subjected to blast load.

The time increment used in the numerical integration is $\Delta\tau = 0.005$ *sec*, which corresponds to an angular increment in free vibrations of $\omega\,\Delta\tau = 0.15$ *rad* (probably a somewhat longer increment would give equally satisfactory results). In this analysis, Simpson's-rule summation as given by Eq. (6-13c) is used.

An evaluation of response over the first 10 time steps is presented in a convenient tabular format in Table E6-1. The operations performed in each column are generally apparent from the labels at the top; however, a few brief comments may be helpful as follows: (a) Columns (4) through (10) are used to evaluate $\overline{A}_N/F$ (where $F \equiv \Delta t/3m\,\omega$) in accordance with Eq. (6-13c) using $y_N \equiv p_N \cos \omega t_N$. (b) Columns (11) through (17) are used to evaluate $\overline{B}_N/F$ in accordance with its equivalence of Eq. (6-13c). (c) Columns (18) through (21) are used to evaluate v_N in accordance with Eq. (6-14). (d) The last column is used to evaluate the spring force $f_{S_N} = k\,v_N$. (e) The multiplication factor $M_2 = 1$ need not be shown in Table E6-1; however, it is entered for later comparison with $M_2 \neq 1$ as required in the damped-response solution.

Since the blast loading terminates at the end of the first 10 time steps, the values of $\overline{A}$ and $\overline{B}$ remain constant after time $t = 0.050$. If these constant values are designated $\overline{A}^*$ and $\overline{B}^*$, the free vibrations which follow the blast loading are given by

$$v(t) = \overline{A}^* \sin \omega t - \overline{B}^* \cos \omega t$$

in accordance with Eq. (6-10). The amplitude of this motion is

$$v_{\max} = \left[(\overline{A}^*)^2 + (\overline{B}^*)^2 \right]^{1/2}$$

In the above example, $\overline{A}^* = 1026\,F = 0.0190$ *ft* $[0.579$ *cm*$]$ and $\overline{B}^* = 956\,F = 0.0177$ *ft* $[0.539$ *cm*$]$ [see Columns (10) and (17) for $N = 10$] so that $v_{\max} = 0.0260$ *ft* $[0.792$ *cm*$]$ and $f_{S_{\max}} = 70.2$ *kips* $[31,840$ *kg*$]$.

TABLE E6-1
Numerical Duhamel integral analysis without damping

N	t_N (sec)	p_N (kips)	$\sin 30t_N$	$\cos 30t_N$	$y_N=(1)\times(3)$ (kips)	y_{N-1}	y_{N-2}	$M_1\times(5)$	$M_2\times[(6)+(9)]$	$\bar A_{N-2}/F$	$\bar A_N/F=(4)+(7)+(8)$	$y_N=(1)\times(2)$ (kips)	y_{N-1}	y_{N-2}	$M_1\times(12)$	$M_2\times[(13)+(16)]$	$\bar B_{N-2}/F$	$\bar B_N/F=(11)+(14)+(15)$	$(10)\times(2)$	$(17)\times(3)$	$(18)-(19)$	$v_N=F\times(20)$ (ft)	$f_{S_N}=k\times(21)$ (kips)
		(1)	(2)	(3)	(4)	(5)	(6)	(7)	(8)	(9)	(10)	(11)	(12)	(13)	(14)	(15)	(16)	(17)	(18)	(19)	(20)	(21)	(22)
0	0.000	0	0	1.000	0	—	—	—	—	—	0	0	—	—	—	—	—	0	0	0	0	0	0
1	0.005	19.32	0.149	0.989	19.1	0	—	—	—	—	—	2.88	0	—	—	—	—	—	—	—	—	—	—
2	0.010	38.64	0.296	0.955	36.9	19.1	0	76.4	0	0	113.3	11.4	2.88	0	11.5	0	0	22.9	33.5	21.9	11.6	0.0002	0.54
3	0.015	57.96	0.435	0.900	52.2	36.9	19.1	—	—	—	—	25.2	11.4	2.88	—	—	—	—	—	—	—	—	—
4	0.020	77.28	0.565	0.825	63.8	52.2	36.9	208.8	150.2	113.3	422.8	43.7	25.2	11.4	100.8	34.3	22.9	178.8	239	148	91	0.0017	4.60
5	0.025	96.60	0.682	0.732	70.7	63.8	52.2	—	—	—	—	65.9	43.7	25.2	—	—	—	—	—	—	—	—	—
6	0.030	77.28	0.783	0.622	48.1	70.7	63.8	282.8	486.6	422.8	817.5	60.5	65.9	43.7	263.6	222.5	178.8	546.6	640	340	300	0.0056	15.1
7	0.035	57.96	0.867	0.498	28.9	48.1	70.7	—	—	—	—	50.3	60.5	65.9	—	—	—	—	—	—	—	—	—
8	0.040	38.64	0.932	0.362	14.0	28.9	48.1	115.6	865.6	817.5	995.2	36.0	50.3	60.5	201.2	607.1	546.6	844.3	928	306	622	0.0115	31.0
9	0.045	19.32	0.976	0.219	4.23	14.0	28.9	—	—	—	—	18.9	36.0	50.3	—	—	—	—	—	—	—	—	—
10	0.050	0	0.997	0.0707	0	4.23	14.0	16.9	1009	995.2	1026	0	18.9	36.0	75.6	880.3	844.3	955.9	1023	67.6	955	0.0177	47.8

$$\omega = \sqrt{\frac{kg}{W}} = 30 \; rad/sec \qquad \Delta\tau = 0.0005 \; sec \qquad M_1 = 4 \qquad M_2 = 1 \qquad F \equiv \frac{\Delta\tau}{3m\omega} = 1.852 \times 10^{-5} \; ft/kip \qquad k = 2700 \; kips/ft$$

The above example solution could have been obtained by formal integration of the Duhamel integral since the loading is of very simple form; however, in practice, one would normally obtain the solution numerically as computer programs are readily available for this purpose.

Under-Critically-Damped System — For numerical evaluation of the response of a damped system, Eq. (6-7) can be written in a form similar to Eq. (6-10) as given by

$$v(t) = A(t) \sin \omega_D t - B(t) \cos \omega_D t \qquad (6\text{-}15)$$

in which

$$A(t) \equiv \frac{1}{m \omega_D} \int_0^t p(\tau) \frac{\exp(\xi \omega \tau)}{\exp(\xi \omega t)} \cos \omega_D \tau \, d\tau$$

$$(6\text{-}16)$$

$$B(t) \equiv \frac{1}{m \omega_D} \int_0^t p(\tau) \frac{\exp(\xi \omega \tau)}{\exp(\xi \omega t)} \sin \omega_D \tau \, d\tau$$

These integral expressions can be evaluated by an incremental summation procedure equivalent to that used previously for the undamped system, but one must now account for the exponential decay behavior caused by damping. To illustrate, the first of Eqs. (6-16) can be written in the approximate recursive forms

Simple summation:

$$A_N \doteq A_{N-1} \exp(-\xi \omega \, \Delta \tau) + \frac{\Delta \tau}{m \omega_D} y_{N-1} \exp(-\xi \omega \, \Delta \tau)$$

$$N = 1, 2, 3, \cdots \quad (6\text{-}17a)$$

Trapezoidal rule:

$$A_N \doteq A_{N-1} \exp(-\xi \omega \, \Delta \tau) + \frac{\Delta \tau}{2 m \omega_D} \left[y_{N-1} \exp(-\xi \omega \, \Delta \tau) + y_N \right]$$

$$N = 1, 2, 3, \cdots \quad (6\text{-}17b)$$

Simpson's rule:

$$A_N \doteq A_{N-2} \exp(-2 \xi \omega \, \Delta \tau)$$

$$+ \frac{\Delta \tau}{3 m \omega_D} \left[y_{N-2} \exp(-2 \xi \omega \, \Delta \tau) + 4 y_{N-1} \exp(-\xi \omega \, \Delta \tau) + y_N \right]$$

$$N = 2, 4, 6, \cdots \quad (6\text{-}17c)$$

which are equivalent to the undamped-response forms of Eqs. (6-13), but with the exponential decay terms added to account for damping. It should be recognized that in this damped case $y_1 \equiv p_1 \cos \omega_D t_1$, $y_2 \equiv p_2 \cos \omega_D t_2$, etc., differing from the undamped case where they were defined as $y_1 \equiv p_1 \cos \omega t$, $y_2 \equiv p_2 \cos \omega t$, etc. However, for small values of damping $\omega \doteq \omega_D$, so these latter terms generally can be used for the damped case as well.

The expressions for B_N are identical in form to those given for A_N in Eqs. (6-17); however, one must use $y_1 \equiv p_1 \sin \omega_D t$, $y_2 \equiv p_2 \sin \omega_D t$, etc.

Having calculated the values of A_N and B_N for successive values of N, the corresponding ordinates of response are obtained using

$$v_N = A_N \sin \omega_D t_N - B_N \cos \omega_D t_N \qquad (6\text{-}18)$$

The accuracy to be expected from any of the above numerical procedures depends, of course, on the duration of time interval $\Delta \tau$. In general, this duration must be selected short enough for both the load and the trigonometric functions used in the analysis to be well defined, and further, to provide the normal engineering accuracy, it should also satisfy the condition $\Delta \tau \leq T/10$. Clearly the accuracy and computational effort increase with the complexity of the numerical integration procedure. Usually, the increased accuracy obtained using Simpson's rule, rather than the simple summation or trapezoidal rule, justifies its use, even though it is more complex.

Example E6-2. To demonstrate how damping can be included in the numerical evaluation of the Duhamel integral, the response analysis of the system of Fig. E6-1 has been repeated using a damping ratio of 5 percent ($\xi = 0.05$). The integrals have been evaluated using the Simpson's-rule integration procedure given by Eq. (6-17c) and its counterpart expression for B_N. For this lightly damped system, the damped frequency has been assumed to be the same as the undamped frequency.

An evaluation of response over the first 18 time steps is presented in Table E6-2. The operations are quite apparent from the labels at the top; however, comments similar to those made for the undamped case may be helpful as follows: (a) Columns (4) through (10) are used to evaluate A_N/F, in accordance with Eq. (6-17c), using $y_N \equiv p_N \cos \omega t_N$. (b) Columns (11) through (17) are used to evaluate B_N/F in accordance with its equivalent form of Eq. (6-17c), using $y_N \equiv p_N \sin \omega t_N$. (c) Columns (18) through (21) are used to evaluate v_N in accordance with Eq. (6-18). (d) The last column is used to evaluate the spring force $f_{S_N} = k v_N$.

Notice that the numerical analysis procedure in Table E6-2 continues into the damped free-vibration Phase II ($t_N \geq 0.050$ sec) stopping at $t_N = 0.090$ sec. As shown in columns (21) and (22), the maximum response is reached very near to $t_N = 0.080$ sec with $v_{max} \doteq 0.024$ ft [0.732 cm] and $f_{S_{max}} \doteq 64.8$ kips [29,390 kg].

TABLE E6-2
Numerical Duhamel integral analysis including damping

N	t_N sec	p_N kips (1)	$\sin 30t_N$ (2)	$\cos 30t_N$ (3)	$y_N=(1)\times(3)$ kips (4)	y_{N-1} (5)	y_{N-2} (6)	$M_1\times(5)$ (7)	$M_2\times[(6)+(9)]$ (8)	$\dfrac{A_{N-2}}{F}$ (9)	$\dfrac{A_N}{F}=(4)+(7)+(8)$ (10)	$y_N=(1)\times(2)$ kips (11)	y_{N-1} (12)	y_{N-2} (13)	$M_1\times(12)$ (14)	$M_2\times[(13)+(16)]$ (15)	$\dfrac{B_{N-2}}{F}$ (16)	$\dfrac{B_N}{F}=(11)+(14)+(15)$ (17)	$(10)\times(2)$ (18)	$(17)\times(3)$ (19)	$(18)-(19)$ (20)	v_N $F\times(20)$ ft (21)	f_{S_N} $k\times(21)$ kips (22)
0	0.000	0	0	1.000	0	—	—	—	—	—	0	0	—	—	—	—	—	0	0	0	0	0	0
1	0.005	19.32	0.149	0.989	19.1	0	—	—	—	—	—	2.88	0	—	—	—	0	—	—	—	—	—	—
2	0.010	38.64	0.296	0.955	36.9	19.1	0	75.8	0	0	112.7	11.4	2.88	0	11.4	0	0	22.8	33.3	21.8	11.5	0.0002	0.58
3	0.015	57.96	0.435	0.900	52.2	36.9	19.1	—	—	—	—	25.2	11.4	2.88	—	—	—	—	—	—	—	—	—
4	0.020	77.28	0.565	0.825	63.8	52.2	36.9	207.2	147.4	112.7	418.4	43.7	25.2	11.4	100.0	33.7	22.8	177.4	236	146	90	0.0017	4.50
5	0.025	96.60	0.682	0.732	70.7	63.8	52.2	—	—	—	—	65.9	43.7	25.2	—	—	—	—	—	—	—	—	—
6	0.030	77.28	0.783	0.622	48.1	70.7	63.8	280.7	475.0	418.4	803.8	60.5	65.9	43.7	261.6	217.8	177.4	539.9	629	336	293	0.0054	14.65
7	0.035	57.96	0.867	0.498	28.9	48.1	70.7	—	—	—	—	50.3	60.5	65.9	—	—	—	—	—	—	—	—	—
8	0.040	38.64	0.932	0.362	14.0	28.9	48.1	114.7	839.1	803.8	967.8	36.0	50.3	60.5	199.7	591.4	539.9	827.1	902	299	603	0.0112	30.2
9	0.045	19.32	0.976	0.219	4.23	14.0	28.9	—	—	—	—	18.9	36.0	50.3	—	—	—	—	—	—	—	—	—
10	0.050	0	0.997	0.0707	0	4.23	14.0	16.8	967.1	967.8	983.9	0	18.9	36.0	75.0	850.1	827.1	925.1	981	65.4	915	0.0169	45.8
11	0.055	0	0.997	−0.079	0	0	4.23	—	—	—	—	0	0	18.9	—	—	—	—	—	—	—	—	—
12	0.060	0	0.974	−0.227	0	0	0	0	969.1	983.9	969.1	0	0	0	0	911.2	925.1	911.2	900	−206	1106	0.0205	55.4
13	0.065	0	0.929	−0.370	0	0	0	—	—	—	—	0	0	0	—	—	—	—	—	—	—	—	—
14	0.070	0	0.863	−0.505	0	0	0	0	954.6	969.1	0	0	0	0	0	897.5	911.2	897.5	824	−453	1277	0.0236	63.9
15	0.075	0	0.778	−0.628	0	0	0	—	—	—	—	0	0	0	—	—	—	—	—	—	—	—	—
16	0.080	0	0.675	−0.737	0	0	0	0	940.3	954.6	940.3	0	0	0	0	884.0	897.5	884.0	635	−651.5	1286	0.0238	64.3
17	0.085	0	0.558	−0.830	0	0	0	—	—	—	—	0	0	0	—	—	—	—	—	—	—	—	—
18	0.090	0	0.427	−0.904	0	0	0	0	926.2	940.3	926.2	0	0	0	0	870.7	884.0	870.7	395	−787	1182	0.0219	59.1

$\omega = \sqrt{\dfrac{kg}{W}} = 30 \, rad/sec$ $\Delta\tau = 0.005 \, sec$ $M_1 = 4\exp(-\xi\omega\Delta\tau) = 3.97$ $M_2 = \exp(-2\,\xi\omega\Delta\tau) = 0.985$ $F \equiv \dfrac{\Delta\tau}{3m\omega} = 1852 \times 10^{-5} \, ft/kip$ $k = 2700 \, kips/ft$

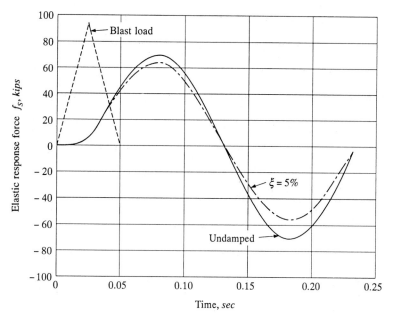

FIGURE E6-2
Response of water tower to blast load.

Plots of the time-histories of spring-force response for the undamped and damped cases treated in Examples E6-1 and E6-2, respectively, are shown in Fig. E6-2, along with the time-history of the blast loading. Notice that the maximum response for the damped case is only slightly less than that for the undamped case. This is clear evidence that damping in the practical range of interest has only a small influence on the maximum response produced by short-duration impulsive loads. One should realize, however, that damping has a very large influence on maximum response produced by arbitrary oscillatory loadings as will be shown later in Section 6-2. In such cases, it is essential that damping effects be included in the numerical solution of response.

6-2 ANALYSIS THROUGH THE FREQUENCY DOMAIN

As shown in Section 6-1, the time-domain analysis procedure can be used to determine the response of any linear SDOF system to any arbitrary loading, even if highly oscillatory. It is sometimes more convenient, however, to perform the analysis in the frequency domain as will be shown subsequently. Furthermore, when the equation of motion contains parameters which might be frequency dependent, such as stiffness k or damping c, the frequency-domain approach is much superior to the time-domain approach. The purpose of this section is to develop continuous and discrete integral formulations of the frequency-domain approach and then to establish numerical procedures for evaluating the response to arbitrary loadings.

Fourier Response Integral

The general frequency-domain approach is similar in concept to the periodic-load-analysis procedure as was mentioned in Chapter 4. Both procedures involve expressing the applied loading in terms of harmonic components, evaluating the response of the structure to each component, and then superposing the harmonic responses to obtain total structural response. However, to apply the periodic-loading technique to arbitrary loadings, it obviously is necessary to extend the Fourier series concept so that it will actually represent nonperiodic functions. In doing so, the Fourier series representations in exponential form, Eqs. (4-8) and (4-9), will be used.

Consider, for example, the arbitrary nonperiodic loading shown by the solid line in Fig. 6-3. If one attempted to represent this function by Eq. (4-8), after obtaining coefficients P_n by integrating Eq. (4-9) over an arbitrary time interval $0 < t < T_p$, the resulting representation would be a periodic function as shown in the figure by both the solid and dashed lines. It is apparent, however, that the spurious repetitive dashed-line loadings can be eliminated by letting $T_p \to \infty$. Toward this end, it is convenient to express Eqs. (4-8) and (4-9) in slightly modified forms by introducing the following:

$$\frac{1}{T_p} = \frac{\bar{\omega}_1}{2\pi} \equiv \frac{\Delta\bar{\omega}}{2\pi} \qquad n\bar{\omega}_1 = n\,\Delta\bar{\omega} \equiv \bar{\omega}_n \qquad P_n\,T_p \equiv P\,(i\bar{\omega}_n) \qquad (6\text{-}19)$$

Taking this action, the Fourier series expressions Eqs. (4-8) and (4-9) become

$$p(t) = \frac{\Delta\bar{\omega}}{2\pi} \sum_{n=-\infty}^{\infty} P(i\bar{\omega}_n)\,\exp(i\bar{\omega}_n t) \qquad (6\text{-}20)$$

$$P(i\bar{\omega}_n) = \int_{-T_p/2}^{T_p/2} p(t)\,\exp(-i\bar{\omega}_n t)\,dt \qquad (6\text{-}21)$$

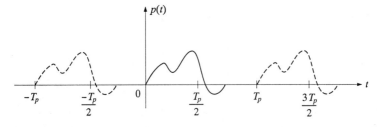

FIGURE 6-3
Arbitrary loading represented by Fourier series.

after taking advantage of the fact that the limits of the integral in Eq. (4-9) are arbitrary so long as they span exactly one loading period T_p.

If the loading period is extended to infinity ($T_p \to \infty$), the frequency increment becomes an infinitesimal ($\Delta\bar{\omega} \to d\bar{\omega}$) and the discrete frequencies $\bar{\omega}_n$ become a continuous function of $\bar{\omega}$. Thus, in the limit, the Fourier series expressions (6-20) and (6-21) take the integral forms

$$p(t) = \frac{1}{2\pi} \int_{-\infty}^{\infty} P(i\bar{\omega}) \exp(i\bar{\omega}t) \, d\bar{\omega} \tag{6-22}$$

$$P(i\bar{\omega}) = \int_{-\infty}^{\infty} p(t) \exp(-i\bar{\omega}t) \, dt \tag{6-23}$$

which are known as the inverse and direct Fourier transforms, respectively. Using the direct Fourier transform, the arbitrary loading $p(t)$ can be expressed as an infinite sum of harmonics having known complex amplitudes. The function $P(i\bar{\omega})/2\pi$ represents the complex amplitude intensity at frequency $\bar{\omega}$ per unit of $\bar{\omega}$. A necessary condition for the direct Fourier transform to exist is that the integral $\int_{-\infty}^{\infty} |p(t)| \, dt$ be finite. Clearly, this is satisfied as long as the loading $p(t)$ acts over a finite period of time.

By analogy with the Fourier series expression of Eq. (4-19), the total response $v(t)$ of the linearly viscously-damped SDOF system is

$$v(t) = \frac{1}{2\pi} \int_{-\infty}^{\infty} H(i\bar{\omega}) \, P(i\bar{\omega}) \exp(i\bar{\omega}t) \, d\bar{\omega} \tag{6-24}$$

The complex frequency response function $H(i\bar{\omega})$ takes the equivalent form of Eq. (4-18) as given by

$$H(i\bar{\omega}) = \frac{1}{k} \left[\frac{1}{(1 - \beta^2) + i\,(2\xi\beta)} \right] = \frac{1}{k} \left[\frac{(1 - \beta^2) - i\,(2\xi\beta)}{(1 - \beta^2)^2 + (2\xi\beta)^2} \right] \tag{6-25}$$

where $\beta \equiv \bar{\omega}/\omega$. Solution of Eq. (6-24) is known as obtaining response through the frequency domain.

Example E6-3. As an example of obtaining response through the frequency domain, consider the rectangular impulse loading of Fig. 5-4: $p(t) = p_0$ in the interval $0 < t < t_1$, with zero loading otherwise. The direct Fourier transform [Eq. (6-23)] of this load function is

$$P(i\bar{\omega}) = \frac{p_0}{-i\bar{\omega}} \left[\exp(-i\bar{\omega}t) - 1 \right] \tag{a}$$

Substituting this load expression together with the complex-frequency-response expression of Eq. (6-25) into Eq. (6-24) leads to response in the integral form

$$v(t) = \frac{i\bar{\omega}_D}{2\pi k} \left[\int_{-\infty}^{\infty} \frac{\exp[-i\omega\beta\,(t_1 - t)]}{\beta\,(\beta - \gamma_1)\,(\beta - \gamma_2)}\, d\beta - \int_{-\infty}^{\infty} \frac{\exp[i\omega\beta t]}{\beta\,(\beta - \gamma_1)\,(\beta - \gamma_2)}\, d\beta \right]$$

(b)

where

$$\gamma_1 = \xi i + \sqrt{1 - \xi^2} \qquad\qquad \gamma_2 = \xi i - \sqrt{1 - \xi^2} \qquad\qquad \text{(c)}$$

The two integrals of Eq. (b) can be evaluated by contour integration in the complex β plane, giving for the case $0 < \xi < 1$

$$v(t) = 0 \qquad\qquad\qquad\qquad\qquad\qquad\qquad\qquad\qquad\qquad t \le 0$$

$$v(t) = \frac{p_0}{k} \left[1 - \exp(-\xi\omega t) \left(\cos\omega_D t + \frac{\xi}{\sqrt{1 - \xi^2}} \sin\omega_D t \right) \right] \qquad 0 \le t \le t_1$$

$$v(t) = \frac{p_0}{k} \exp[-\xi\omega\,(t - t_1)]$$

$$\times \left\{ \left[\exp(-\xi\omega t_1) \left(\sin\omega_D t_1 - \frac{\xi}{\sqrt{1 - \xi^2}} \cos\omega_D t_1 \right) + \frac{\xi}{\sqrt{1 - \xi^2}} \right] \right.$$

$$\times \sin\omega_D\,(t - t_1) + \left[1 - \exp(-\xi\omega t_1) \left(\cos\omega_D t_1 + \frac{\xi}{\sqrt{1 - \xi^2}} \sin\omega_D t_1 \right) \right]$$

$$\left. \times \cos\omega_D\,(t - t_1) \right\} \qquad\qquad\qquad\qquad\qquad\qquad\qquad t \ge t_1 \qquad \text{(d)}$$

Formal application of the frequency-domain-analysis procedure, as illustrated in the above example, is limited to cases for which the Fourier integral transforms of the applied-loading functions are available, and even in these cases the evaluation of the integrals can be a tedious process. Thus to make the procedure practical, it is necessary to formulate it in terms of a numerical-analysis approach. This formulation requires (1) expressing both the direct and inverse Fourier integral transforms in convenient approximate forms and (2) developing effective and efficient numerical techniques for evaluating these forms.

Discrete Fourier Transforms (DFT)

To be practical, the Fourier integral transforms of Eqs. (6-22) and (6-23) must be expressed in their approximate Fourier series forms given by Eqs. (4-7) and (4-8) but, for effective evaluation, they should be converted to different, but equivalent, forms. First, consider Eq. (4-8) in its form

$$P_n = \frac{1}{T_p} \int_0^{T_p} p(t) \, \exp(-i\bar{\omega}_n t) \, dt \qquad n = 0, \pm 1, \pm 2, \cdots \qquad (6\text{-}26)$$

in which $\bar{\omega}_n = n \, \Delta\bar{\omega} = n \, 2\pi/T_p$. Dividing the period T_p into N equal intervals Δt ($T_p = N \, \Delta t$), evaluating the ordinates of the function $q(t) \equiv p(t) \exp(-i\bar{\omega}_n t)$ at discrete values of $t = t_m = m \, \Delta t$ ($m = 0, 1, 2, \cdots, N$) denoted by $q_0, q_1, q_2, \cdots, q_N$, and then applying the trapezoidal rule of integration as given in Eq. (6-12b), the above expression for P_n can written in the approximate form

$$P_n \doteq \frac{1}{N \, \Delta t} \left\{ \Delta t \left[\frac{q_0}{2} + \left(\sum_{m=1}^{N-1} q_m \right) + \frac{q_N}{2} \right] \right\} \qquad (6\text{-}27)$$

Considering times at the beginning and end of the loading period ($t = 0$ and $t = T_p$) so that $q_0 = q_N = 0$ as shown in Fig. 6-3, and making the substitutions $\bar{\omega}_n = 2\pi \, n/T_p = 2\pi \, n/N \, \Delta t$ and $t_m = m \, \Delta t$, Eq. (6-27) becomes

$$P_n \doteq \frac{1}{N} \sum_{m=1}^{N-1} p(t_m) \, \exp\left(-i\frac{2\pi \, n \, m}{N}\right) \qquad n = 0, 1, 2, \cdots, (N-1) \qquad (6\text{-}28)$$

Now consider Eq. (4-7) in its form

$$p(t) = \sum_{n=-\infty}^{\infty} P_n \, \exp(i\bar{\omega}_n t) \qquad (6\text{-}29)$$

It is convenient and acceptable, as far as engineering accuracy is concerned, to limit the discrete frequencies to a finite range $-M \, \Delta\bar{\omega} \le \bar{\omega}_n \le M \, \Delta\bar{\omega}$ corresponding to $-M \le n \le M$. Introducing these finite limits into Eq. (6-29), making the same substitutions mentioned above, and letting $t = t_m$, one obtains

$$p(t_m) \doteq \sum_{n=-M}^{M} P_n \, \exp\left(i\frac{2\pi \, n \, m}{N}\right) \qquad (6\text{-}30)$$

The summation in this equation can be separated into two parts; the first part summing from $-M$ to -1 and the second part summing from zero to M. The first part can be modified as

$$\sum_{n=-M}^{-1} P_n \, \exp\left(i\frac{2\pi \, n \, m}{N}\right) = \sum_{n=-M}^{-1} P_{n+N} \, \exp\left(i\frac{2\pi \, (n+N) \, m}{N}\right) \qquad (6\text{-}31)$$

since each term is periodic over intervals $n = N$. Substituting $Z \equiv n + N$ and $M \equiv (N - 1)/2$ into the right hand side of this equation and combining the resulting expression with the second part of Eq. (6-30) as defined above, the complete Eq. (6-30) becomes

$$p(t_m) \doteq \sum_{Z=M+1}^{2M} P_Z \exp\left(i\frac{2\pi Z m}{N}\right) + \sum_{n=0}^{M} P_n \exp\left(i\frac{2\pi n m}{N}\right) \qquad (6\text{-}32)$$

or

$$p(t_m) \doteq \sum_{n=0}^{2M} P_n \exp\left(i\frac{2\pi n m}{N}\right) \qquad (6\text{-}33)$$

Noting that $2M = N - 1$, Eq. (6-33) can be expressed as

$$p(t_m) \doteq \sum_{n=0}^{N-1} P_n \exp\left(i\frac{2\pi n m}{N}\right) \qquad m = 0, 1, 2, \cdots, N - 1 \qquad (6\text{-}34)$$

Equations (6-28) and (6-34) are the discrete Fourier transform (DFT) equations which are in convenient forms for solution by numerical procedures. It should be recognized that both of these transforms are periodic; Eq. (6-28) has a period $n = N$ and Eq. (6-34) a period $m = N$. Further, it should be recognized that the quantity $p(t_m)$ is simply the ordinate to the function $p(t)$ at time $t = t_m$; while P_n is the complex amplitude of the discrete harmonic at frequency $\overline{\omega}_n$.

Fast Fourier Transforms (FFTs)

The Fast Fourier Transform, which allows very efficient and accurate evaluations of the discrete Fourier transforms, is based on an algorithm developed by Cooley and Tukey.[1] Since the algorithm is used in exactly the same way in evaluating both the direct and inverse FFTs, let us now consider only the direct FFT, namely

$$A_n \equiv P_n N = \sum_{m=0}^{N-1} p_m \exp\left(-i\frac{2\pi n m}{N}\right) \qquad n = 1, 2, 3, \cdots, (N - 1) \qquad (6\text{-}35)$$

in which p_m denotes $p(t = t_m)$. A straightforward evaluation of this summation for all values of n requires N^2 complex multiplications. This number would be prohibitive for most practical solutions requiring large values of N, say $N > 1,000$, thus providing the incentive to develop the FFT algorithm.

---

[1] J. W. Cooley and J. W. Tukey, "An Algorithm for Machine Calculation of Complex Fourier Series," Math. Computation, Vol. 19, April 1965.

The FFT algorithm is based on letting $N = 2^\gamma$ where γ is an integer. In this case, each value of n and m in their common range from zero to $N - 1$ can be expressed in terms of binary coefficients as given by

$$n = 2^{\gamma-1} n_{\gamma-1} + 2^{\gamma-2} n_{\gamma-2} + \cdots + n_0$$
$$m = 2^{\gamma-1} m_{\gamma-1} + 2^{\gamma-2} m_{\gamma-2} + \cdots + m_0$$

(6-36)

in which each binary coefficient is either $+1$ or 0 depending upon the particular value of n or m being represented. Using these relations and letting $W_N \equiv \exp(-i2\pi/N)$, Eq. (6-35) can be expressed as

$$A(n_{\gamma-1}, n_{\gamma-2}, \cdots, n_0) = \sum_{m_0=0}^{1} \sum_{m_1=0}^{1} \cdots \sum_{m_{\gamma-1}=0}^{1} p_0(m_{\gamma-1}, m_{\gamma-2}, \cdots, m_0)\, W_N^{nm}$$

(6-37)

Note that each coefficient A_n for $n = 0, 1, 2, \cdots, N - 1$ is represented by $A(n_{\gamma_1}, n_{\gamma-2}, \cdots, n_0)$ and each load ordinate p_m for $m = 0, 1, 2, \cdots, N - 1$ is represented by $p_0(m_{\gamma-1}, m_{\gamma-2}, \cdots, m_0)$. The subscript zero has been added to p only to indicate the multiplier of the W_N^{nm} term in the first summation. The reason for introducing this addition to the notation will become apparent as the algorithm develops.

Consider now the term W_N^{nm} of Eq. (6-37) in the form

$$W_N^{nm} = W_N^{\left(2^{\gamma-1} n_{\gamma-1} + 2^{\gamma-2} n_{\gamma-2} + \cdots + n_0\right)\left(2^{\gamma-1} m_{\gamma-1} + 2^{\gamma-2} m_{\gamma-2} + \cdots + m_0\right)}$$

(6-38)

Making use of $W_N^{(a+b)} = W_N^a W_N^b$, this equation can be modified to

$$W_N^{nm} = W_N^{\left(2^{\gamma-1} n_{\gamma-1} + 2^{\gamma-2} n_{\gamma-2} + \cdots + n_0\right)\left(2^{\gamma-1} m_{\gamma-1}\right)}$$
$$\times W_N^{\left(2^{\gamma-1} n_{\gamma-1} + 2^{\gamma-2} n_{\gamma-2} + \cdots + n_0\right)\left(2^{\gamma-2} m_{\gamma-2}\right)}$$
$$\times \cdots \times W_N^{\left(2^{\gamma-1} n_{\gamma-1} + 2^{\gamma-2} n_{\gamma-2} + \cdots + n_0\right) m_0}$$

(6-39)

Let us now examine each individual W_N term on the right hand side of this equation separately. The first term can be written

$$W_N^{\left(2^{\gamma-1} n_{\gamma-1} + 2^{\gamma-2} n_{\gamma-2} + \cdots + n_0\right)\left(2^{\gamma-1} m_{\gamma-1}\right)}$$
$$= W_N^{2^\gamma \left(2^{\gamma-2} n_{\gamma-1} m_{\gamma-1}\right)} \times W_N^{2^\gamma \left(2^{\gamma-3} n_{\gamma-2} m_{\gamma-1}\right)} \times \cdots$$
$$\times W_N^{2^\gamma \left(2 n_1 m_{\gamma-1}\right)} \times W_N^{2^{\gamma-1} (n_0 m_{\gamma-1})}$$
$$= W_N^{2^{\gamma-1} (n_0 m_{\gamma-1})}$$

(6-40)

since each W_N term of the form

$$W_N^{2^\gamma \text{(integer)}} = \left[\exp\left(-i\frac{2\pi}{N}\right)\right]^{N \text{(integer)}} = 1 \qquad (6\text{-}41)$$

Writing the second term similarly and making use of Eq. (6-41), one finds that

$$W_N^{\left(2^{\gamma-1} n_{\gamma-1} + 2^{\gamma-2} n_{\gamma-2} + \cdots + n_0\right)\left(2^{\gamma-2} m_{\gamma-2}\right)} = W_N^{2^{\gamma-2}(2n_1+n_0)m_{\gamma-2}} \qquad (6\text{-}42)$$

This pattern continues up to the last term which has no cancellations due to Eq. (6-41); therefore, it must remain in the same form shown in Eq. (6-39).

After substituting all W_N terms, except the last, in their reduced forms, Eq. (6-37) becomes

$$A\left(n_{\gamma-1}, n_{\gamma-2}, \cdots, n_0\right)$$

$$= \sum_{m_0=0}^{1} \sum_{m_1=0}^{1} \cdots \sum_{m_{\gamma-1}=0}^{1} \left[P_0\left(m_{\gamma-1}, m_{\gamma-2}, \cdots, m_0\right) \times W_N^{2^{\gamma-1}(n_0 m_{\gamma-1})} \right.$$

$$\left. \times W_N^{2^{\gamma-2}(2n_1+n_0)m_{\gamma-2}} \times \cdots \times W_N^{(2^{\gamma-1}n_{\gamma-1}+2^{\gamma-2}n_{\gamma-2}+\cdots+n_0)m_0} \right] \qquad (6\text{-}43)$$

Carrying out all summations in this equation in succession gives

$$\sum_{m_{\gamma-1}=0}^{1} P_0(m_{\gamma-1}, m_{\gamma-2}, \cdots, m_0) W_N^{2^{\gamma-1}(n_0 m_{\gamma-1})} \equiv P_1(n_0, m_{\gamma-2}, \cdots, m_0)$$

$$\sum_{m_{\gamma-2}=0}^{1} P_1(n_0, m_{\gamma-2}, \cdots, m_0) W_N^{2^{\gamma-2}(2n_1+n_0)m_{\gamma-2}} \equiv P_2(n_0, n_1, m_{\gamma-3}, \cdots, m_0)$$

$$\vdots$$

$$\sum_{m_0=0}^{1} P_{\gamma-1}(n_0, n_1, \cdots, n_{\gamma-2}, m_0) W_N^{(2^{\gamma-1}n_{\gamma-1}+2^{\gamma-2}n_{\gamma-2}+\cdots+n_0)m_0}$$

$$= A(n_{\gamma-1}, n_{\gamma-2}, \cdots, n_0) \qquad (6\text{-}44)$$

These recursive equations, leading to the desired result $A(n_{\gamma-1}, n_{\gamma-2}, \cdots, n_0)$, represent the Cooley-Tukey algorithm used in modern FFT analysis. They are extremely efficient due to the fact that each summation is used immediately in the next summation. The fact that the exponential has unit value in the first term of each summation and that $W_N^{nm} = -W_N^{nm+N/2}$ adds to the efficiency. The reduction in computational effort which results from the use of Eqs. (6-44) is enormous when the time duration T_p is divided into a large number of intervals. For example, when $N = 1,024$ ($\gamma = 10$), the computer time required by the FFT to obtain all N A_n-values is approximately 0.5 percent of the time required to obtain the same values by direct use of Eq. (6-35). Not only is the FFT extremely efficient but it is very accurate as well, thus making the frequency-domain approach to the dynamic-response analysis of structures very attractive indeed.

Example E6-4. To illustrate the use of the recursive Eqs. (6-44), consider the case $\gamma = 2$ which corresponds to $N = 2^\gamma = 4$, $n = 2n_1 + n_0$, and $m = 2m_1 + m_0$. The first equation is

$$P_1(n_0, m_0) = \sum_{m_1=0}^{1} P_0(m_1, m_0)\, W_4^{2\,n_0\,m_1} \tag{a}$$

Since n_0 and m_0 can each take on values $+1$ or 0, this equation actually represents four equations as follows:

$$P_1(0,0) = P_0(0,0) + P_0(1,0)$$

$$P_1(0,1) = P_0(0,1) + P_0(1,1)$$

$$P_1(1,0) = P_0(0,0) + P_0(1,0)\, W_4^2 \tag{b}$$

$$P_1(1,1) = P_0(0,1) + P_0(1,1)\, W_4^2$$

Writing these equations in matrix form gives

$$\begin{Bmatrix} P_1(0,0) \\ P_1(0,1) \\ P_1(1,0) \\ P_1(1,1) \end{Bmatrix} = \begin{bmatrix} 1 & 0 & 1 & 0 \\ 0 & 1 & 0 & 1 \\ 1 & 0 & W_4^2 & 0 \\ 0 & 1 & 0 & W_4^2 \end{bmatrix} \begin{Bmatrix} P_0(0,0) \\ P_0(0,1) \\ P_0(1,0) \\ P_0(1,1) \end{Bmatrix} \tag{c}$$

The second of Eqs. (6-44), which is also the last for this simple case $\gamma = 2$, is

$$A(n_1, n_0) = \sum_{m_0=0}^{1} P_1(n_0, m_0)\, W_4^{(2n_1+n_0)\,m_0} \tag{d}$$

Substituting separately the four possible combinations of n_1 and n_0 into this expression yields

$$
\begin{Bmatrix} A(0,0) \\ A(0,1) \\ A(1,0) \\ A(1,1) \end{Bmatrix} = \begin{bmatrix} 1 & 1 & 0 & 0 \\ 0 & 0 & 1 & W_4^1 \\ 1 & W_4^2 & 0 & 0 \\ 0 & 0 & 1 & W_4^3 \end{bmatrix} \begin{Bmatrix} P_1(0,0) \\ P_1(0,1) \\ P_1(1,0) \\ P_1(1,1) \end{Bmatrix} = \begin{Bmatrix} A_0 \\ A_1 \\ A_2 \\ A_3 \end{Bmatrix} \qquad \text{(e)}
$$

Upon substitution of Eq. (c) into (e), the desired values of A_n ($n = 0, 1, 2, 3$) are obtained.

While the γ-value used in Example E6-4 is much too small to demonstrate the efficiency of the FFT algorithm (only about 50 percent reduction in computational effort is realized over the direct approach), it is adequate to develop an understanding of the application of Eqs. (6-44). As demonstrated, each of the FFT recursive equations yields N separate equations when applied to a solution.

Evaluation of Dynamic Response

To evaluate the dynamic response of a linear SDOF system in the frequency domain, the inverse Fourier transform expressing response, Eq. (6-24), should be used in its discrete form

$$
v_m \doteq \sum_{n=0}^{N-1} V_n \, \exp\left(i\,\frac{2\pi n m}{N}\right) \qquad m = 0, 1, 2, \cdots, N-1 \qquad (6\text{-}45)
$$

in which $v_m = v(t = t_m)$ and $V_n \equiv H_n P_n$. Discrete harmonic amplitudes P_n for all n-values are obtained by the FFT procedure described in the previous section while H_n is the discrete form of the complex frequency response function $H(i\bar{\omega})$. For the linear viscously-damped system, it is of the form shown in Eq. (4-18) while for the complex-stiffness form of damping as represented by Eqs. (3-69) and (3-79), it is

$$
H_n = \frac{1}{k}\left[\frac{1}{(1-\beta_n^2)+i\,(2\xi)}\right] = \frac{1}{k}\left[\frac{(1-\beta_n^2)-i\,(2\xi)}{(1-\beta_n^2)^2+(2\xi)^2}\right] \qquad (6\text{-}46)
$$

Having the values of P_n and H_n for $n = 0, 1, 2, \cdots, N-1$, the FFT algorithm is applied to Eq. (6-45) in exactly the same way it was applied to Eq. (6-35) giving the time-history response $v(t)$ in terms of its ordinates v_m for $m = 0, 1, 2, \cdots, N-1$. Note that both H_n and P_n are complex; this causes no difficulty, however, in the evaluation process.

To illustrate evaluating response by the above procedure, consider two SDOF systems represented by the following equations of motion:

$$\ddot{v}(t) + 10\,\pi\,\xi\,\dot{v}(t) + 25\,\pi^2\,v(t) = -\ddot{v}_g(t) \tag{6-47}$$

and

$$\ddot{v}(t) + 25\,\pi^2\,(1 + 2\,i\,\xi)\,v(t) = -\ddot{v}_g(t) \tag{6-48}$$

These two systems are identical except for the type of damping used, the first having viscous damping while the latter has complex-stiffness damping. The natural circular frequency of each system is $\omega = 5\pi$ corresponding to $T = 0.4\ sec$ and $f = 2.5\ Hz$.

Assuming the input acceleration $\ddot{v}_g(t)$ to be oscillatory over 6 sec as shown in Fig. 6-4a, one must select appropriate values for Δt and γ. Considering the frequency content in this accelerogram, it is reasonable to select $\Delta t = 0.01\ sec$. The value of γ should be selected so that period T_p is considerably longer than the duration of excitation; thus resulting in zero excitation ordinates over an interval of time following each

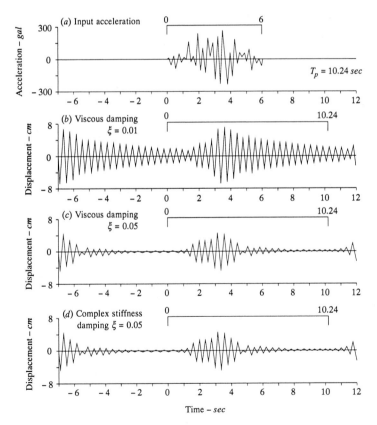

FIGURE 6-4
Evaluation of response through the frequency domain.

6 *sec* duration of excitation in the periodic FFT representation. This is a necessary requirement so that the free-vibration response during the intervals of zero excitation will damp out almost completely; otherwise, the assumed zero initial conditions at the start of the excitation will not be sufficiently satisfied. Suppose γ is set equal to 10 giving $N = 2^{10} = 1,024$ which corresponds to $T_p = 1,024\,\Delta t = 10.24\ sec$. Solving the viscously damped system of Eq. (6-47) in the frequency domain by the above described procedure for $\xi = 0.01$ and 0.05 gives the corresponding time-histories of response shown in Figs. 6-4b and 6-4c. Noting the periodic behavior in each case, it is clear that the response has sufficient time to damp out following the excitation for the case of $\xi = 0.05$ but not for the case of $\xi = 0.01$. Therefore, while $\gamma = 10$ is adequate for 5 percent critical damping, it is inadequate for 1 percent damping. In the latter case, one should use $\gamma = 11$ corresponding to $T_p = 2^{11}\,\Delta t = 20.48\ sec$. It is very apparent that response is sensitive to the amount of damping present in the system when the excitation is of the highly oscillatory type shown in Fig. 6-4a.

Solving the complex stiffness system defined by Eq. (6-48) in the frequency domain for $\xi = 0.05$ gives the displacement time-history shown in Fig. 6-4d. This periodic response is almost the same as that shown in Fig. 6-4c. This close comparison can be expected for such a low-damped system experiencing primary response near its natural frequency, since, as shown in Chapter 3, viscous and hysteretic damping produce identical results at the resonant condition ($\beta = 1$). If the input acceleration represented in Fig. 6-4a had been nearly void of frequency content in the neighborhood of the natural frequency but had high-intensity content in quite a different frequency range forcing the primary response to shift significantly away from the range near $\beta = 1$, the corresponding responses as represented in Figs. 6-4c and 6-4d would be quite different. If the damping energy loss per cycle E_D at fixed amplitude of response for the actual system being represented should be essentially independent of the excitation frequency, the response obtained using hysteretic damping would, of course, be much more realistic.

It should be noted that if parameters k and/or ξ in Eqs. (4-18) and Eq. (6-46) were frequency dependent, i.e., $k = k(\beta_n)$ and $\xi = \xi(\beta_n)$, the dynamic-response evaluation would proceed in exactly the same manner described above; however, in doing so, the frequency dependence of these parameters would enter into the calculation of the H_n-values. This compatibility of the frequency-domain analysis with frequency-dependent parameters is a valuable asset to be noted as such cases are common in engineering practice. One would have extreme difficulty in treating such cases by the time-domain approach described in the first part of this chapter.

Regarding efficiency of the frequency-domain method of dynamic response analysis, it is worthy of mention that the CPU time required to solve either Eq. (6-47) or Eq. (6-48) for one value of damping and for the excitation shown in Fig. 6-4a is about 1.2 *sec* on a Micro VAX-II computer.

6-3 RELATIONSHIP BETWEEN THE TIME- AND FREQUENCY-DOMAIN TRANSFER FUNCTIONS

As shown in the previous sections, response of the viscously-damped SDOF system to an arbitrary loading $p(t)$ can be obtained either through the time domain using the convolution integral,

$$v(t) = \int_{-\infty}^{t} p(\tau)\, h(t - \tau)\, d\tau \tag{6-49}$$

or through the frequency domain using the relation

$$v(t) = \frac{1}{2\pi} \int_{-\infty}^{\infty} H(i\overline{\omega})\, P(i\overline{\omega})\, \exp(i\overline{\omega}t)\, d\overline{\omega} \tag{6-50}$$

in which $h(t)$ and $H(i\overline{\omega})$ are the unit impulse and complex frequency response functions, respectively, given by

$$h(t) = \frac{1}{m\,\omega_D} \sin \omega_D t\, \exp(-\xi\omega t) \qquad 0 < \xi < 1 \tag{6-51}$$

and

$$H(i\overline{\omega}) = \frac{1}{k} \left[\frac{1}{(1 - \beta^2) + i\,(2\xi\beta)} \right] \qquad \xi \geq 0 \tag{6-52}$$

It is of interest to know at this point that these time- and frequency-domain transfer functions are related through the Fourier transform pair

$$H(i\overline{\omega}) = \int_{-\infty}^{\infty} h(t)\, \exp(-i\overline{\omega}t)\, dt \tag{6-53}$$

$$h(t) = \frac{1}{2\pi} \int_{-\infty}^{\infty} H(i\overline{\omega})\, \exp(i\overline{\omega}t)\, d\overline{\omega} \tag{6-54}$$

Proof of these relationships, along with an example solution, is presented in Chapter 12 (Section 12-7).

PROBLEMS

6-1. The undamped SDOF system of Fig. P6-1a is subjected to the half sine-wave loading of Fig. P6-1b. Calculate the spring force history $f_s(t)$ for the time $0 < t < 0.6\ sec$ by numerical evaluation of the Duhamel integral with $\Delta\tau = 0.1\ sec$ using:

    (a) Simple summation

    (b) Trapezoidal rule

    (c) Simpson's rule

Compare these results with those obtained with Eq. (5-1) evaluated at the same 0.1 *sec* time increments.

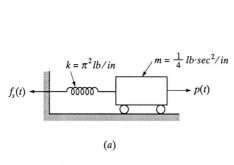

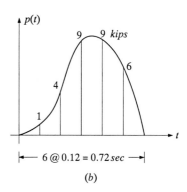

(a)

(b)

FIGURE P6-1

6-2. Solve Example E6-1 using the trapezoidal rule.

6-3. Solve Example E6-2 using the trapezoidal rule.

6-4. The SDOF frame of Fig. P6-2a is subjected to the blast loading history shown in Fig. P6-2b. Compute the displacement history for the time $0 < t < 0.72$ *sec* by numerical evaluation of the Duhamel integral using Simpson's rule with $\triangle\tau = 0.12$ *sec*.

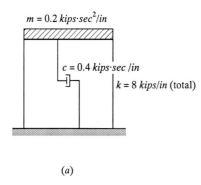

(a)

(b)

FIGURE P6-2

CHAPTER

7

RESPONSE TO GENERAL DYNAMIC LOADING: STEP-BY-STEP METHODS

7-1 GENERAL CONCEPTS

It is important to note that the response analysis procedures described in Chapter 6, whether formulated in the time domain or in the frequency domain, involve evaluation of many independent response contributions that are combined to obtain the total response. In the time-domain procedure (Duhamel integral), the loading $p(t)$ is considered to be a succession of short-duration impulses, and the free-vibration response to each impulse becomes a separate contribution to the total response at any subsequent time. In the frequency-domain method, it is assumed that the loading $p(t)$ is periodic and has been resolved into its discrete harmonic components P_n by Fourier transformation. The corresponding harmonic response components of the structure V_n are then obtained by multiplying these loading components by the frequency response coefficient of the structure H_n, and finally the total response of the structure is obtained by combining the harmonic response components (inverse Fourier transformation). Because superposition is applied to obtain the final result in both procedures, neither of these methods is suited for use in analysis of nonlinear response; therefore judgment must be used in applying them in earthquake engineering where it is expected that a severe earthquake will induce inelastic deformation in a code-designed structure.

The step-by-step procedure is a second general approach to dynamic response analysis, and it is well suited to analysis of nonlinear response because it avoids any use of superposition. There are many different step-by-step methods, but in all of them the loading and the response history are divided into a sequence of time intervals or "steps." The response during each step then is calculated from the initial conditions (displacement and velocity) existing at the beginning of the step and from the history of loading during the step. Thus the response for each step is an independent analysis problem, and there is no need to combine response contributions within the step. Nonlinear behavior may be considered easily by this approach merely by assuming that the structural properties remain constant during each step, and causing them to change in accordance with any specified form of behavior from one step to the next; hence the nonlinear analysis actually is a sequence of linear analyses of a changing system. Any desired degree of refinement in the nonlinear behavior may be achieved in this procedure by making the time steps short enough; also it can be applied to any type of nonlinearity, including changes of mass and damping properties as well as the more common nonlinearities due to changes of stiffness.

Step-by-step methods provide the only completely general approach to analysis of nonlinear response; however, the methods are equally valuable in the analysis of linear response because the same algorithms can be applied regardless of whether the structure is behaving linearly or not. Moreover, the procedures used in solving single-degree-of-freedom structures can easily be extended to deal with multidegree systems merely by replacing scalar quantities by matrices. In fact, these methods are so effective and convenient that time-domain analyses almost always are done by some form of step-by-step analysis regardless of whether or not the response behavior is linear; the Duhamel integral method seldom is used in practice.

7-2 PIECEWISE EXACT METHOD

The simplest step-by-step method for analysis of SDOF systems is the so-called "piecewise exact" method, which is based on the exact solution of the equation of motion for response of a linear structure to a loading that varies linearly during a discrete time interval. In using this method, the loading history is divided into time intervals, usually defined by significant changes of slope in the actual loading history; between these points, it is assumed that the slope of the load curve remains constant. Although the response expression derived for these linearly varying load steps is exact, it must be recognized that the actual loading history is only approximated by the constant slope steps. Thus the calculated response generally is not an exact representation of the true response to the real loading; however, the error can be reduced to any acceptable value merely by reducing the length of the time steps and thus better approximating the loading. If desired, the length of the time steps can be varied from one interval to the next in order to achieve the best possible fit of the

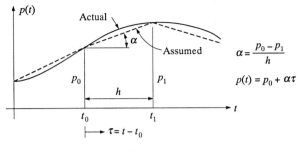

$$\alpha = \frac{p_0 - p_1}{h}$$

$$p(t) = p_0 + \alpha \tau$$

(a) Loading history

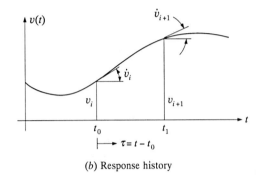

(b) Response history

FIGURE 7-1
Notation for piecewise exact analysis.

loading history by the sequence of straight line segments; however, for reasons of computational efficiency it is customary to use a constant time step.

The notation used in formulating this method of analysis is defined for one step of the loading history as shown in Fig. 7-1a; the duration of the step is denoted by h and it spans from t_0 to t_1. The assumed linearly varying loading during the time step is given by

$$p(\tau) = p_0 + \alpha \, \tau \tag{7-1}$$

where α is the constant slope, τ is the time variable during the step, and p_0 is the initial loading. Introducing this load expression in the equation of motion for a SDOF system with viscous damping leads to

$$m \, \ddot{v} + c \, \dot{v} + k \, v = p_0 + \alpha \, \tau \tag{7-2}$$

The response $v(\tau)$ during any time step (shown in Fig. 7-1b) consists of a free-vibration term $v_h(\tau)$ plus the particular solution to the specified linear load variation $v_p(\tau)$, thus

$$v(\tau) = v_h(\tau) + v_p(\tau) \tag{7-3}$$

where the damped free-vibration response, as shown by Eq. (2-48), is given by

$$v_h(\tau) = \exp(-\xi\omega\tau)\left[A\,\cos\omega_D\tau + B\,\sin\omega_D\tau\right]$$

and it is easy to verify that the linearly varying particular solution is

$$v_p(\tau) = \frac{1}{k}\left(p_0 + \alpha\tau\right) - \frac{\alpha c}{k^2} \tag{7-4}$$

Combining these expressions and evaluating the constants A and B by consideration of the initial conditions at time $\tau = 0$ leads finally to the following expression for the displacement during the time step:

$$v(\tau) = A_0 + A_1\tau + A_2\,\exp(-\xi\omega\tau)\,\cos\omega_D\tau + A_3\,\exp(-\xi\omega\tau)\,\sin\omega_D\tau \tag{7-5}$$

in which

$$A_0 = \frac{v_0}{\omega^2} - \frac{2\xi\alpha}{\omega^3}$$

$$A_1 = \frac{\alpha}{\omega^2}$$

$$A_2 = v_0 - A_0$$

$$A_3 = \frac{1}{\omega_D}\left[\dot{v}_0 + \xi\omega A_2 - \frac{\alpha}{\omega^2}\right]$$

Similarly, the velocity during the time step is found to be

$$\dot{v}(\tau) = A_1 + (\omega_D A_3 - \xi\omega A_2)\,\exp(-\xi\omega\tau)\,\cos\omega_D\tau$$

$$- (\omega_D A_2 + \xi\omega A_3)\,\exp(-\xi\omega\tau)\,\sin\omega_D\tau \tag{7-6}$$

Of course, the velocity and displacement at the end of this time step become the initial conditions for the next time step, and then the equivalent equations can be used to step forward to the end of that step, etc.

For situations where the applied loading may be approximated well by a series of straight line segments, this piecewise exact method undoubtedly is the most efficient means of calculating the response of a SDOF system. However, it must always be remembered that the loading being considered is only an approximation of the true loading history, which usually is a smoothly varying curve, and that the step lengths must be chosen so as to achieve an acceptable approximation of the true response history.

Example E7-1.   To demonstrate the effect of approximating a smoothly varying dynamic loading as a series of straight line segments, the response of a SDOF structure to various approximations of a single sine-wave loading has been calculated by the piecewise exact method. The properties of the structure are presented in Fig. E7-1a; three straight line approximations of the one and one-half cycle loading as sketched in Fig. E7-1b are defined by discrete values spaced at time intervals of (a) 0.0075, (b) 0.0225, and (c) 0.045 *sec*, respectively, (1/12, 1/4, and 1/2 of the 0.09 *sec* half cycle period).

The piecewise exact calculated responses to these three loadings are plotted in Fig. E7-2, together with the "static" ($\frac{p(t)}{k}$) response to loading (a); the response was evaluated at 0.0075 *sec* intervals in all cases. It is evident from these results that the applied loading is significantly diminished by the straight line assumption for the coarsest (case c) approximation; the corresponding reduction of the input work also is apparent in the plotted loading history for case c. However, the error is reduced greatly by taking twice as many straight line segments as shown for case b, and it may be concluded that the results for case a, using 0.0075 *sec* load segments, are quite close to the correct response for the theoretical sine-wave loading; certainly this solution is adequate for most engineering purposes.

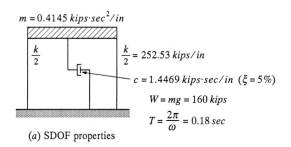

$m = 0.4145 \ kips \cdot sec^2/in$

$\dfrac{k}{2}$

$\dfrac{k}{2} = 252.53 \ kips/in$

$c = 1.4469 \ kips \cdot sec/in \ (\xi = 5\%)$

$W = mg = 160 \ kips$

$T = \dfrac{2\pi}{\omega} = 0.18 \ sec$

(*a*) SDOF properties

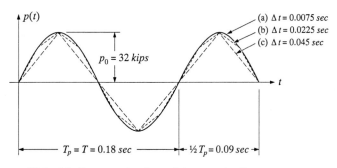

$p(t)$

(a) $\Delta t = 0.0075 \ sec$
(b) $\Delta t = 0.0225 \ sec$
(c) $\Delta t = 0.045 \ sec$

$p_0 = 32 \ kips$

t

$T_p = T = 0.18 \ sec$ $\frac{1}{2} T_p = 0.09 \ sec$

(*b*) Straight line approximations of sine-wave loading

FIGURE E7-1
Piecewise exact example – SDOF structure and loading.

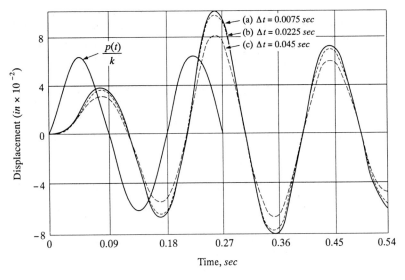

FIGURE E7-2
Piecewise exact calculated response.

7-3 NUMERICAL APPROXIMATION PROCEDURES — GENERAL COMMENTS

The other step-by-step methods employ numerical procedures to approximately satisfy the equations of motion during each time step — using either numerical differentiation or numerical integration. A vast body of literature has been written on these subjects, dealing with a range of applications as broad as the field of applied mechanics. However, only a very brief summary can be presented here, intended to give a general idea of how these techniques may be applied in the solution of structural dynamics problems, and to provide tools that will suffice for many practical uses. On the other hand, if a large amount of analytical work must be done, it probably will be worth while to study the literature and choose the solution procedure that is best adapted to the job at hand.

Before the details of any of the procedures are described, it will be useful to summarize a few basic facts about the numerical approximation step-by-step methods in general, as follows:

(1) The methods may be classified as either explicit or implicit. An explicit (or "open") method is defined as one in which the new response values calculated in each step depend only on quantities obtained in the preceding step, so that the analysis proceeds directly from one step to the next. In an implicit method, on the other hand, the expressions giving the new values for a given step include

one or more values pertaining to that same step, so that trial values of the necessary quantities must be assumed and then these are refined by successive iterations. Unless the calculations required for each step are very simple, the cost of iteration within a step may be prohibitive; thus it often is desirable to convert an implicit method to an explicit form by a procedure such as that described in Section 7-5.

(2) The primary factor to be considered in selecting a step-by-step method is efficiency, which concerns the computational effort required to achieve the desired level of accuracy over the range of time for which the response is needed. Accuracy alone cannot be a criterion for method selection because, in general, any desired degree of accuracy can be obtained by any method if the time step is made short enough (but with obvious corresponding increases of costs). In any case, the time steps must be made short enough to provide adequate definition of the loading and the response history — a high-frequency input or response cannot be described by long time steps.

(3) Factors that may contribute to errors in the results obtained from well-defined loadings include:

(a) Roundoff — resulting from calculations being done using numbers expressed by too few digits,

(b) Instability — caused by amplification of the errors from one step during the calculations in subsequent steps. Stability of any method is improved by reducing the length of the time step.

(c) Truncation — using too few terms in series expressions of quantities.

(4) Errors resulting from any causes may be manifested by either or both of the following effects:

(a) Phase shift or apparent change of frequency in cyclic results,

(b) Artificial damping, in which the numerical procedure removes or adds energy to the dynamically responding system.

All of the above mentioned topics are worthy of further discussion, but in this presentation it will be possible only to include a few additional remarks in the following descriptions of selected numerical methods.

7-4   SECOND CENTRAL DIFFERENCE FORMULATION

The basic concept in most finite-difference formulations is to write the equation of dynamic equilibrium for the beginning of the time step $t = t_0$, thus

$$m\,\ddot{v}_0 + c\,\dot{v}_0 + k\,v_0 = p_0$$

and then to solve this for the initial acceleration

$$\ddot{v}_0 = \frac{1}{m}\,[p_0 - c\,\dot{v}_0 - k\,v_0] \tag{7-7}$$

However, to formulate the numerical step-by-step procedure, the initial velocity and acceleration terms are approximated by finite-difference expressions. Thus, the variations between the various finite-difference formulations are embodied in the level of refinement adopted in writing these finite-difference expressions.

In this discussion, only one very simple method will be described — the second central difference method; this name relates to the finite-difference approximation used to express the acceleration at time $t = t_0$. To express the acceleration, the velocity is first approximated at the middle of the time steps before and after time t_0

$$\dot{v}_{-1/2} \doteq \frac{v_0 - v_{-1}}{h} \qquad\qquad \dot{v}_{1/2} \doteq \frac{v_1 - v_0}{h} \tag{7-8}$$

in which h denotes the duration of the time step, as shown in Fig. 7-2. Then the acceleration midway between these times is given by the equivalent velocity expression

$$\ddot{v}_0 \doteq \frac{\dot{v}_{1/2} - \dot{v}_{-1/2}}{h} \doteq \frac{1}{h^2}\,(v_1 - v_0) - \frac{1}{h^2}\,(v_0 - v_{-1})$$

from which

$$\ddot{v}_0 = \frac{1}{h^2}\,(v_1 - 2\,v_0 + v_{-1}) \tag{7-9}$$

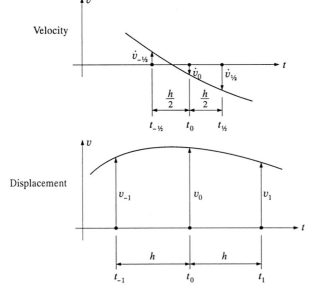

FIGURE 7-2
Second central difference notation.

Substituting this expression into Eq. (7-7) then leads to

$$v_1 - 2 v_0 + v_{-1} = \frac{h^2}{m} (p_0 - c \dot{v}_0 - k v_0)$$

and solving this for the displacement at the end of the time step results in

$$v_1 = \frac{h^2}{m} (p_0 - c \dot{v}_0 - k v_0) + 2 v_0 - v_{-1} \qquad (7\text{-}10)$$

But the value of the displacement at time t_{-1} is needed to evaluate this expression, so the required quantity is derived from the following finite-difference velocity expression

$$\dot{v}_0 = \frac{v_1 - v_{-1}}{2h}$$

from which

$$v_{-1} = v_1 - 2h \, \dot{v}_0$$

Introducing this into Eq. (7-10) and simplifying gives finally

$$v_1 = v_0 + h \, \dot{v}_0 + \frac{h^2}{2m} (p_0 - c \dot{v}_0 - k v_0) \qquad (7\text{-}11)$$

The velocity also must be stepped forward to time t_1 and for this purpose an expression is derived by assuming that the average of the velocities at times t_1 and t_0 is equal to the finite-difference expression for velocity within the time step, thus

$$\frac{1}{2} (\dot{v}_0 + \dot{v}_1) = \frac{v_1 - v_0}{h}$$

from which

$$\dot{v}_1 = \frac{2 (v_1 - v_0)}{h} - \dot{v}_0 \qquad (7\text{-}12)$$

Thus, the second central difference method merely uses Eqs. (7-11) and (7-12) to proceed from one step to the next throughout the time span of interest. It is a very simple explicit step-by-step method; however, it is only conditionally stable and will "blow up" if the time step is not made short enough. The specific condition for stability is

$$\frac{h}{T} \leq \frac{1}{\pi} = 0.318$$

however, this will not be a limiting requirement for SDOF systems. Clearly, the analysis requires more than three time steps per vibration period of the structure if the response is to be defined adequately, and in most earthquake response analyses a considerably shorter time step also must be adopted to permit effective definition of the earthquake input.

7-5 INTEGRATION METHODS

The other general numerical approach to step-by-step dynamic response analysis makes use of integration to step forward from the initial to the final conditions for each time step. The essential concept is represented by the following equations:

$$\dot{v}_1 = \dot{v}_0 + \int_0^h \ddot{v}(\tau) \, d\tau \tag{7-13a}$$

$$v_1 = v_0 + \int_0^h \dot{v}(\tau) \, d\tau \tag{7-13b}$$

which express the final velocity and displacement in terms of the initial values of these quantities plus an integral expression. The change of velocity depends on the integral of the acceleration history, and the change of displacement depends on the corresponding velocity integral. In order to carry out this type of analysis, it is necessary first to assume how the acceleration varies during the time step; this acceleration assumption controls the variation of the velocity as well and thus makes it possible to step forward to the next time step.

Euler-Gauss Procedure

The simplest integration method, known as the Euler-Gauss method, is based on assuming that the acceleration has a fixed constant value during the time step. The consequence of this assumption is that the velocity must vary linearly and the displacement as a quadratic curve during the time step. Figure 7-3 illustrates this type of behavior for a formulation where it is assumed that the constant acceleration is the average of the initial and the final values attained during the step. Also shown on this figure are expressions for acceleration, velocity, and displacement at any time τ during the step obtained by successive integration, and for the final velocity and displacement obtained by putting $\tau = h$ into these expressions.

To initiate this analysis for any step, it is necessary first to evaluate the initial acceleration $\ddot{v}_0$, and this may be obtained by solving the dynamic equilibrium expression at time $t = t_0$, as shown by Eq. (7-7). In addition, the final acceleration $\ddot{v}_1$ is needed to apply this implicit formulation, and this value may be obtained by iteration. Starting with an arbitrary assumption for $\ddot{v}_1$, values of $\dot{v}_1$ and v_1 are obtained from Eqs. (a) and (b) listed in Fig. 7-3. Then an improved value of $\ddot{v}_1$ is calculated from the dynamic equilibrium condition at time t_1 using an expression equivalent to Eq. (7-7), and this leads to improved values of velocity $\dot{v}_1$ and displacement v_1. Eventually, the iteration converges to a fixed value of the final acceleration for this time step and the procedure may be stepped forward to the next time step. A great advantage of this constant average acceleration method is that it is unconditionally stable; that is, the errors are not amplified from one step to the next no matter how long a time step is

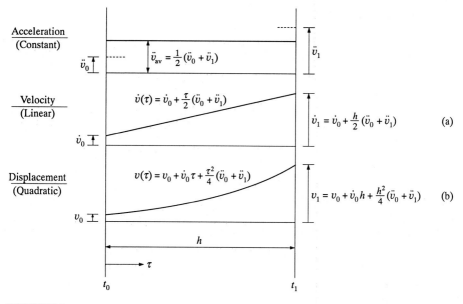

FIGURE 7-3
Motion based on constant average acceleration.

chosen. Consequently, the time step may be selected considering only the need for properly defining the dynamic excitation and the vibratory response characteristics of the structure.

Newmark Beta Methods

A more general step-by-step formulation was proposed by Newmark, which includes the preceding method as a special case, but also may be applied in several other versions. In the Newmark formulation, the basic integration equations [Eqs. (7-13)] for the final velocity and displacement are expressed as follows:

$$\dot{v}_1 = \dot{v}_0 + (1 - \gamma)\, h\, \ddot{v}_0 + \gamma\, h\, \ddot{v}_1 \qquad (7\text{-}14a)$$

$$v_1 = v_0 + h\, \dot{v}_0 + (\frac{1}{2} - \beta)\, h^2\, \ddot{v}_0 + \beta\, h^2\, \ddot{v}_1 \qquad (7\text{-}14b)$$

It is evident in Eq. (7-14a) that the factor γ provides a linearly varying weighting between the influence of the initial and the final accelerations on the change of velocity; the factor β similarly provides for weighting the contributions of these initial and final accelerations to the change of displacement.

From study of the performance of this formulation, it was noted that the factor γ controlled the amount of artificial damping induced by this step-by-step procedure;

there is no artificial damping if $\gamma = 1/2$, so it is recommended that this value be use for standard SDOF analyses. Adopting this factor $\gamma = 1/2$ and setting $\beta = 1/4$ in Eqs. (7-14a) and (7-14b), it may be seen that this Newmark formulation reduces directly to the expressions shown for the final velocity and displacement in Fig. 7-3. Thus, the Newmark $\beta = 1/4$ method may also be referred to as the constant average acceleration method.

On the other hand, if β is taken to be 1/6 (with $\gamma = 1/2$), the expressions for the final velocity and displacement become

$$\dot{v}_1 = \dot{v}_0 + \frac{h}{2}\,(\ddot{v}_0 + \ddot{v}_1) \tag{7-15a}$$

$$v_1 = v_0 + \dot{v}_0\,h + \frac{h^2}{3}\,\ddot{v}_0 + \frac{h^2}{6}\,\ddot{v}_1 \tag{7-15b}$$

These results also may be derived by assuming that the acceleration varies linearly during the time step between the initial and final values of $\ddot{v}_0$ and $\ddot{v}_1$, as shown in Fig. 7-4; thus the Newmark $\beta = 1/6$ method is also known as the linear acceleration method. Like the constant average acceleration procedure, this method is widely used in practice, but in contrast to the $\beta = 1/4$ procedure, the linear acceleration method is only conditionally stable; it will be unstable unless $h/T \le \sqrt{3}/\pi = 0.55$. However, as in the case of the second central difference method, this restriction has

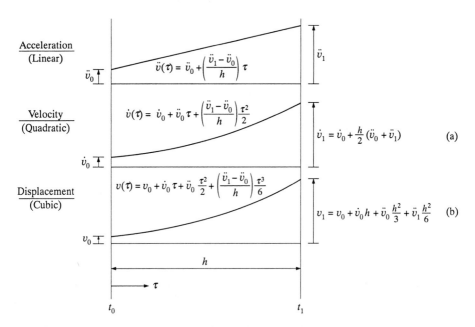

FIGURE 7-4
Motion based on linearly varying acceleration.

little significance in the analysis of SDOF systems because a shorter time step than this must be used to obtain a satisfactory representation of the dynamic input and response.

Conversion to Explicit Formulation

In general, the implicit formulations of the Beta methods are inconvenient to use because iteration is required at each time step to determine the acceleration at the end of the step. Accordingly, they are usually converted to an explicit form, and the conversion procedure will be explained here for the constant average acceleration ($\beta = 1/4$) method. The objective of the conversion is to express the final acceleration in terms of the other response quantities; accordingly Eq. (b) of Fig. 7-3 is solved for the final acceleration to obtain

$$\ddot{v}_1 = \frac{4}{h^2} \left(v_1 - v_0 \right) - \frac{4}{h} \dot{v}_0 - \ddot{v}_0 \tag{7-16a}$$

and this then is substituted into Eq. (a) of Fig. 7-3 to obtain an expression for the final velocity:

$$\dot{v}_1 = \frac{2}{h} \left(v_1 - v_0 \right) - \dot{v}_0 \tag{7-16b}$$

Writing the equations of dynamic equilibrium at time t_1

$$m \, \ddot{v}_1 + c \, \dot{v}_1 + k \, v_1 = p_1$$

and substituting Eqs. (7-16a) and (7-16b) leads to an expression in which the only unknown is the displacement at the end of the time step, v_1. With appropriate gathering of terms this may be written

$$\tilde{k}_c \, v_1 = \tilde{p}_{1c} \tag{7-17}$$

which has the form of a static equilibrium equation involving the effective stiffness

$$\tilde{k}_c = k + \frac{2c}{h} + \frac{4m}{h^2} \tag{7-17a}$$

and the effective loading

$$\tilde{p}_{1c} = p_1 + c \left(\frac{2v_0}{h} + \dot{v}_0 \right) + m \left(\frac{4v_0}{h^2} + \frac{4}{h} \dot{v}_0 + \ddot{v}_0 \right) \tag{7-17b}$$

In Eqs. (7-17) the subscript c is used to denote the constant average acceleration method.

Using this explicit formulation, the displacement at the end of the time step, v_1, can be calculated directly by solving Eq. (7-17), using only data that was available at the beginning of the time step. Then, the velocity at that time, $\dot{v}_1$, may be calculated from Eq. (7-16b). Finally, the acceleration at the end of the step, $\ddot{v}_1$, is derived by solving the dynamic equilibrium equation at that time

$$\ddot{v}_1 = \frac{1}{m}\left(p_1 - c\,\dot{v}_1 - k\,v_1\right)$$

[rather than from Eq. (7-16a)] thus preserving the equilibrium condition.

It will be noted that the linear acceleration method can be converted to explicit form similarly by using Eqs. (a) and (b) of Fig. 7-4 in exactly the same way. The only differences in the formulations, then, are in the expressions for the effective stiffness and effective loading and for the final velocity. Expressing the effective static equilibrium equation for the linear acceleration analysis by

$$\widetilde{k}_d\, v_1 = \widetilde{p}_{1d} \tag{7-18}$$

(in which the subscript d denotes the linear acceleration method) the effective stiffness and loading are given, respectively, by

$$\widetilde{k}_d = k + \frac{3c}{h} + \frac{6m}{h^2} \tag{7-18a}$$

$$\widetilde{p}_{1d} = p_1 + m\left(\frac{6v_0}{h^2} + \frac{6}{h}\,\dot{v}_0 + 2\,\ddot{v}_0\right) + c\left(\frac{3v_0}{h} + 2\,\dot{v}_0 + \frac{h}{2}\,\ddot{v}_0\right) \tag{7-18b}$$

When the displacement v_1 has been calculated from Eq. (7-18), the velocity at the same time is given by the following expression [equivalent to Eq. (7-16b)]:

$$\dot{v}_1 = \frac{3}{h}\left(v_1 - v_0\right) - 2\,\dot{v}_0 - \frac{h}{2}\,\ddot{v}_0 \tag{7-18c}$$

It is important to remember that the linear acceleration method is only conditionally stable, but this factor is seldom important in analysis of SDOF systems as was mentioned before. On the other hand, it is apparent that assuming a linear variation of acceleration during each step will give a better approximation of the true behavior than will a sequence of constant acceleration steps. In fact, numerical experiments have demonstrated the superiority of the linear acceleration method results compared with those obtained using constant acceleration steps, and for this reason the linear acceleration ($\beta = 1/6$) method is recommended for analysis of SDOF systems.

7-6 INCREMENTAL FORMULATION FOR NONLINEAR ANALYSIS

The step-by-step procedures described above are suitable for the analysis of linear systems in which the resisting forces are expressed in terms of the entire values of velocity and displacement that have been developed in the structure up to that time. However, for nonlinear analyses it is assumed that the physical properties remain constant only for short increments of time or deformation; accordingly it is convenient to reformulate the response in terms of the incremental equation of motion, as follows.

The structure to be considered in this discussion is the SDOF system shown in Fig. 7-5a. The properties of the system, m, c, k, and $p(t)$, may represent generalized quantities, as described in Chapter 8, instead of the simple localized properties implied in the sketch; thus the nonlinear step-by-step analysis discussed here is applicable to a generalized system exactly as it is applied to the simple system of Fig. 7-5a. The forces acting on the mass are indicated in Fig. 7-5b and the general nonlinear properties of the damping and spring mechanisms are depicted in Figs. 7-5c and d, respectively. The arbitrary external load history is shown in Fig. 7-5e.

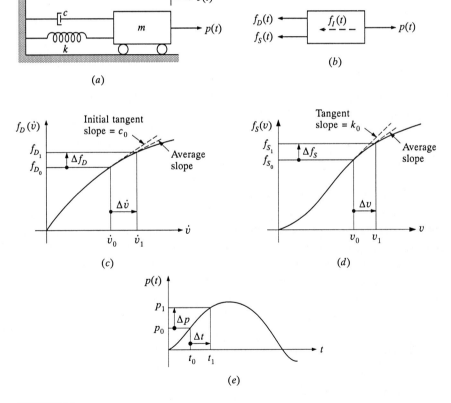

FIGURE 7-5
Definition of a nonlinear dynamic system: (a) basic SDOF structure; (b) force equilibrium; (c) nonlinear damping; (d) nonlinear stiffness; (e) applied load.

The equilibrium of forces acting on the mass at the time $t = t_0$ may be written

$$f_{I_0} + f_{D_0} + f_{S_0} = p_0 \tag{7-19a}$$

and a short time $h = t_1 - t_0$ later the equilibrium requirement is

$$f_{I_1} + f_{D_1} + f_{S_1} = p_1 \tag{7-19b}$$

Subtracting Eq. (7-19a) from Eq. (7-19b) then yields the incremental equation of motion

$$\Delta f_I + \Delta f_D + \Delta f_S = \Delta p \tag{7-20}$$

in which the incremental forces may be expressed as follows:

$$\Delta f_I = f_{I_1} - f_{I_0} = m\,\Delta\ddot{v} \tag{7-21a}$$

$$\Delta f_D = f_{D_1} - f_{D_0} = c(t)\,\Delta\dot{v} \tag{7-21b}$$

$$\Delta f_S = f_{S_1} - f_{S_0} = k(t)\,\Delta v \tag{7-21c}$$

$$\Delta p = p_1 - p_0 \tag{7-21d}$$

In Eqs. (7-21b) and (7-21c), the terms $c(t)$ and $k(t)$ represent average values of damping and stiffness properties that may vary during the time increment, as indicated by the average slopes in Figs. 7-5c and d, respectively. In practice, these average slopes could be evaluated only by iteration because the calculated velocity and displacement at the end of the time increment depend on these properties. To avoid this iteration, it is common practice to use the initial tangent slopes instead:

$$c(t) \doteq \left(\frac{d\,f_D}{d\dot{v}}\right)_0 \equiv c_0 \qquad k(t) \doteq \left(\frac{d\,f_S}{d v}\right)_0 \equiv k_0 \tag{7-22}$$

even though this approximation is not as good, in principle. Substituting the force expressions of Eqs. (7-21) into Eq. (7-20) leads to the final form of the incremental equilibrium equation for time t:

$$m\,\Delta\ddot{v} + c_0\,\Delta\dot{v} + k_0\,\Delta v = \Delta p \tag{7-23}$$

The step-by-step integration procedures discussed previously now may be easily modified into an incremental form. Considering, for example, the linear acceleration assumption presented in Fig. 7-4, the corresponding incremental equations are shown in Fig. 7-6. This implicit formulation may be transformed to explicit form by the same procedure described previously by Eq. (7-18). The resulting incremental effective static equilibrium equation may be stated as

$$\tilde{k}_d\,\Delta v = \Delta\tilde{p}_d \tag{7-24}$$

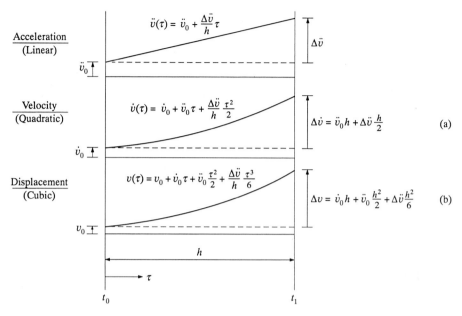

Acceleration
(Linear)

$$\ddot{v}(\tau) = \ddot{v}_0 + \frac{\Delta\ddot{v}}{h}\tau$$

$\Delta\ddot{v}$

$\ddot{v}_0$

Velocity
(Quadratic)

$$\dot{v}(\tau) = \dot{v}_0 + \ddot{v}_0\tau + \frac{\Delta\ddot{v}}{h}\frac{\tau^2}{2}$$

$$\Delta\dot{v} = \ddot{v}_0 h + \Delta\ddot{v}\frac{h}{2} \qquad (a)$$

$\dot{v}_0$

Displacement
(Cubic)

$$v(\tau) = v_0 + \dot{v}_0\tau + \ddot{v}_0\frac{\tau^2}{2} + \frac{\Delta\ddot{v}}{h}\frac{\tau^3}{6}$$

$$\Delta v = \dot{v}_0 h + \ddot{v}_0\frac{h^2}{2} + \Delta\ddot{v}\frac{h^2}{6} \qquad (b)$$

v_0

h

τ

$t_0 \qquad\qquad t_1$

FIGURE 7-6
Incremental motion based on linearly varying acceleration.

in which the effective stiffness expression [equivalent to Eq. (7-18a)] is

$$\tilde{k}_d = k_0 + \frac{3c_0}{h} + \frac{6m}{h^2} \tag{7-24a}$$

and the effective loading increment now becomes

$$\Delta\tilde{p}_d = \Delta p + m\left(\frac{6}{h}\dot{v}_0 + 3\ddot{v}_0\right) + c_0\left(3\dot{v}_0 + \frac{h}{2}\ddot{v}_0\right) \tag{7-24b}$$

When the incremental displacement has been evaluated from Eq. (7-24), the incremental velocity may be calculated from the following expression [derived from Eq. (7-18c)]:

$$\Delta\dot{v} = \frac{3}{h}\Delta v - 3\dot{v}_0 - \frac{h}{2}\ddot{v}_0 \tag{7-24c}$$

7-7 SUMMARY OF THE LINEAR ACCELERATION PROCEDURE

For any given time increment, the above described explicit linear acceleration analysis procedure consists of the following operations which must be carried out consecutively in the order given:

(1) Using the initial velocity and displacement values $\dot{v}_0$ and v_0, which are known either from the values at the end of the preceding time increment or as initial conditions of the response at time $t = 0$, and the specified nonlinear properties of the system as illustrated in Figs. 7-5c and d, establish the current values of the damping and spring forces, f_{D_0} and f_{S_0}, and the damping and spring coefficients, c_0 and k_0, to be used over the interval.

(2) Generate the initial acceleration for the interval using the equation of motion in the form

$$\ddot{v}_0 = \frac{1}{m} \left[p_0 - f_{D_0} - f_{S_0} \right] \tag{7-25}$$

(3) Compute the effective stiffness $\tilde{k}_d$ and effective load increment $\Delta \tilde{p}_d$ using Eqs. (7-24a) and (7-24b), respectively.

(4) Determine the displacement and velocity increments using Eqs. (7-24) and (7-24c), respectively.

(5) Finally, evaluate the velocity and displacement at the end of the time increment using

$$\dot{v}_1 = \dot{v}_0 + \Delta \dot{v} \tag{7-26a}$$

$$v_1 = v_0 + \Delta v \tag{7-26b}$$

When Step 5 has been completed, the calculation for this time increment is finished and the analysis may be stepped forward to the next time interval. Carrying out Steps 1 through 5 in this manner for consecutive time increments starting at $t = 0$ and ending after any desired number of intervals, one can obtain the complete time-history of response of any nonlinear SDOF system for which the varying stiffness and damping properties, k_0 and c_0, can be defined, when it is subjected to any arbitrary dynamic loading. Linear systems can also be treated by this same procedure, which becomes simplified due to the physical properties remaining constant over their entire time-histories of response.

As with any numerical-integration procedure the accuracy of this step-by-step method will depend on the length of the time increment h. Three factors must be considered in the selection of this interval: (1) the rate of variation of the applied loading $p(t)$, (2) the complexity of the nonlinear damping and stiffness properties, and (3) the period T of vibration of the structure. The time increment must be short enough to permit the reliable representation of all these factors, the last one being associated with the free-vibration behavior of the system. In general, the material-property variation is not a critical factor; however, if a significant sudden change takes place, as in the yielding of an elastoplastic spring, a special subdivided time increment may need to be introduced to treat this change accurately.

If the load history is relatively simple, the choice of the time interval will depend essentially on the period of vibration of the structure. While this linear acceleration method will usually give a convergent solution if the time increment is less than about one-half the vibration period, it must be considerably shorter than this to provide reasonable accuracy and to insure that numerical instability will not occur. In general, using an increment-period ratio $h/T \le 1/10$ will give reliable results. If there is any doubt about the adequacy of a given solution, a second analysis can be made after reducing the time increment by one-half; if the response is not changed appreciably in the second analysis, it may be assumed that the errors introduced by the numerical integration procedure are negligible.

Example E7-2. To demonstrate a hand-solution technique for applying the linear acceleration step-by-step method described above, the response of the elastoplastic SDOF frame shown in Fig. E7-3 to the loading history indicated has been calculated. A time step of 0.1 *sec* has been used for this analysis, which is longer than desirable for good accuracy but will be adequate for the present purpose.

In this structure, the damping coefficient has been assumed to remain constant; hence the only nonlinearity in the system results from the change of

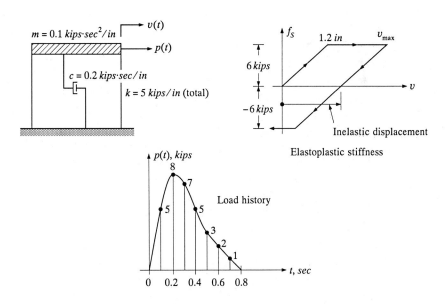

FIGURE E7-3
Elastoplastic frame and dynamic loading.

stiffness as yielding takes place. The effective stiffness thus may be expressed [see Eq. (7-24a)] as

$$\tilde{k}_d = k_0 + \frac{6}{(0.1)^2} m + \frac{3}{0.1} c = 66 + k_0$$

where k_0 is either 5 $kips/in$ or zero depending upon whether the frame is elastic or yielding. Also the effective incremental loading is given by [see Eq. (7-24b)]

$$\Delta \tilde{p}_d = \Delta p + \left(\frac{6\,m}{0.1} + 3\,c_0\right) \dot{v} + \left(3\,m + \frac{0.1}{2}\,c_0\right) \ddot{v} = \Delta p(t) + 6.6\,\dot{v} + 0.31\,\ddot{v}$$

The velocity increment given by Eq. (7-24c) becomes

$$\Delta \dot{v} = 30\,\Delta v - 3\,\dot{v} - 0.05\,\ddot{v}$$

A convenient tabular arrangement for the hand calculation of this response is shown in Table E7-1.

For this elastoplastic system, the response behavior changes drastically as the yielding starts and stops, and to obtain best accuracy it would be desirable to divide each time increment involving such a change of phase into two subincrements. The properties then would be constant during the sub-increments, and the analysis would be quite precise; however, an iterative procedure would be required to establish the lengths of the subincrements. In the present analysis, this refinement has not been used; the initial stiffness has been assumed to act during the entire increment, and thus significant errors may have arisen during the phase transitions.

The dynamic elastoplastic response calculated in Table E7-1 is plotted in Fig. E7-4, with the response during the yielding phase shown as a dashed line. Also plotted for comparison is the linear elastic response obtained by a similar step-by-step analysis but with $\tilde{k}_d = 71$ and $f_S = 5\,v$ throughout the calculations. The effect of the plastic yielding shows up clearly in this comparison; the permanent set (the position about which the subsequent free vibrations of the nonlinear system occur) amounts to about 1.49 in. Also shown to indicate the character of the loading is the static displacement p/k, that is, the deflection which would have occurred in the elastic structure if there had been no damping and inertia effects.

TABLE E7-1
Nonlinear response analysis: linear acceleration step-by-step method
Structure and loading in Fig. E7-3

t	p	v	$\dot{v}$	f_S	f_D	f_I	$\tilde{v}$	Δp	$6.6\dot{v}$	$0.31\ddot{v}$	$\Delta\tilde{p}\,d$	k	$\tilde{k}$	Δv	$30\,\Delta v$	$3\dot{v}$	$0.05\,\ddot{v}$	$\Delta\dot{v}$
				$5\,\bar{v}*$	$0.2\,\dot{v}$	$(2)-(5)-(6)$	$10\times(7)$				$(9)+(10)+(11)$		$66+(13)$	$(12)\div(14)$			$\bar{v}$	$(16)-(17)-(18)$
sec	kips	in	in/sec															
(1)	(2)	(3)	(4)	(5)	(6)	(7)	(8)	(9)	(10)	(11)	(12)	(13)	(14)	(15)	(16)	(17)	(18)	(19)
0.0	0	0	0	0	0	0	0	5	0	0	0	5	71	0.070	2.11	0	0	2.11
0.1	5	0.070	2.11	0.35	0.42	4.23	42.3	3	13.92	13.12	30.04	5	71	0.423	12.68	6.33	2.11	4.24
0.2	8	0.493	6.35	2.46	1.27	4.27	42.7	−1	41.90	13.25	54.15	5	71	0.763	22.88	19.06	2.14	1.68
0.3	7	1.256	8.03	6	1.61	−0.61	−6.1	−2	53.02	−1.89	49.13	0**	66	0.744	22.33	24.08	−0.30	−1.45
0.4	5	2.000	6.58	6	1.32	−2.32	−23.2	−2	43.43	−7.19	34.24	0	66	0.519	15.57	19.74	−1.16	−3.01
0.5	3	2.519	3.57	6	0.71	−3.71	−37.1	−1	23.56	−11.50	11.06	0	66	0.168	5.02	10.72	−1.85	−3.85
0.6	2	2.687	−0.28	6	−0.06	−3.94	−39.4	−1	−1.85	−12.22	−15.07	5	71	−0.212	−6.36	−0.84	−1.97	−3.55
0.7	1	2.475	−3.83	4.94	−0.77	−3.17	−31.7	−1	−25.28	−9.82	−36.10	5	71	−0.508	−15.24	−11.49	−1.58	−2.17
0.8	0	1.967	−6.00	2.40	−1.20	−1.20	−12.0	0	−39.60	−3.72	−43.32	5	71	−0.610	−18.30	−18.00	−0.60	0.30
0.9	0	1.357	−5.70	−0.65	−1.14	1.79	17.9	0	−37.62	5.55	−32.07	5	71	−0.452	−13.56	−17.10	0.90	2.64
1.0	0	0.905	−3.06															

* $\bar{v} = v - v_i$, where v_i = inelastic displacement = $v_{max} - 1.2\ in$;

** $k = 0$ while frame is yielding.

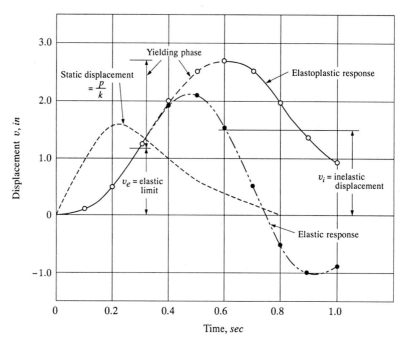

FIGURE E7-4
Comparison of elastoplastic with elastic response (frame of Fig. E8-1).

PROBLEMS

7-1. Solve the linear elastic response of Prob. 6-4 by step-by-step integration, using the linear acceleration method.

7-2. Solve Prob. 7-1, assuming an elastoplastic force-displacement relation for the columns and a yield force level of 8 *kips*, as shown in Fig. P7-1*a*.

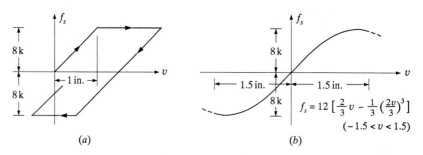

FIGURE P7-1

7-3. Solve Prob. 7-1, assuming the nonlinear elastic force-displacement relation, $f_S = 12[\frac{2}{3}v - \frac{1}{3}(2v/3)^3]$, which is sketched in Fig. P7-1*b* (f_S is in *kips*, v in *in*).

CHAPTER
8

GENERALIZED SINGLE-DEGREE-OF-FREEDOM SYSTEMS

8-1 GENERAL COMMENTS ON SDOF SYSTEMS

In formulating the SDOF equations of motion and response analysis procedures in the preceding chapters, it has been tacitly assumed that the structure under consideration has a single lumped mass that is constrained so that it can move only in a single fixed direction. In this case it is obvious that the system has only a single degree of freedom and that the response may be expressed in terms of this single displacement quantity.

However, the analysis of most real systems requires the use of more complicated idealizations, even when they can be included in the generalized single-degree-of-freedom category. In this chapter we will discuss these generalized SDOF systems, and in formulating their equations of motion it is convenient to divide them into two categories: (1) assemblages of rigid bodies in which elastic deformations are limited to localized weightless spring elements and (2) systems having distributed flexibility in which the deformations can be continuous throughout the structure, or within some of its components. In both categories, the structure is forced to behave like a SDOF system by the fact that displacements of only a single form or shape are permitted, and the assumed single degree of freedom expresses the amplitude of this permissible displacement configuration.

For structures in the category of rigid-body assemblages, discussed in Section 8-2, the limitation to a single displacement shape is a consequence of the assemblage configuration; i.e., the rigid bodies are constrained by supports and hinges arranged so that only one form of displacement is possible. The essential step in the analysis of such assemblages is the evaluation of the generalized elastic, damping, and inertial forces in terms of this single form of motion.

In the case of structures having distributed elasticity, considered in Section 8-3, the SDOF shape restriction is merely an assumption because the distributed elasticity actually permits an infinite variety of displacement patterns to occur. However, when the system motion is limited to a single form of deformation, it has only a single degree of freedom in a mathematical sense. Therefore, when the generalized mass, damping, and stiffness properties associated with this degree of freedom have been evaluated, the structure may be analyzed in exactly the same way as a true SDOF system.

From these comments it should be evident that the material on analysis of SDOF systems, presented in the preceding chapters, is equally applicable to generalized SDOF systems even though it was presented with reference to simple systems having only a single lumped mass.

8-2 GENERALIZED PROPERTIES: ASSEMBLAGES OF RIGID BODIES

In formulating the equations of motion of a rigid-body assemblage, the elastic forces developed during the SDOF displacements can be expressed easily in terms of the displacement amplitude because each elastic element is a discrete spring subjected to a specified deformation. Similarly the damping forces can be expressed in terms of the specified velocities of the attachment points of the discrete dampers. On the other hand, the mass of the rigid bodies need not be localized, and distributed inertial forces generally will result from the assumed accelerations. However, for the purposes of dynamic analysis, it usually is most effective to treat the rigid-body inertial forces as though the mass and the mass moment of inertia were concentrated at the center of mass. The inertial-force resultants which are obtained thereby are entirely equivalent to the distributed inertial forces insofar as the assemblage behavior is concerned. Similarly it is desirable to represent any distributed external loads acting on the rigid bodies by their force resultants. The total mass m and the centroidal mass moment of inertia j of a uniform rod and of uniform plates of various shapes are summarized in Fig. 8-1 for convenient reference.

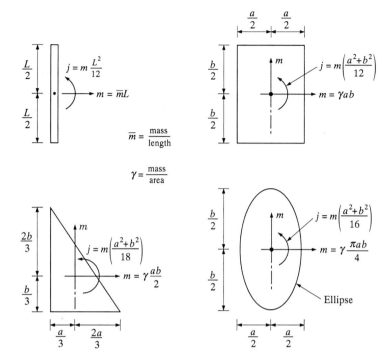

FIGURE 8-1
Rigid-body mass and centroidal mass moment of inertia for uniform rod and uniform
plates of unit thickness.

Example E8-1.   A representative example of a rigid-body assemblage,
shown in Fig. E8-1, consists of two rigid bars connected by a hinge at E and
supported by a pivot at A and a roller at H. Dynamic excitation is provided by a
transverse load $p(x,t)$ varying linearly along the length of bar AB. In addition,
a constant axial force N acts through the system, and the motion is constrained
by discrete springs and dampers located as shown along the lengths of the bars.

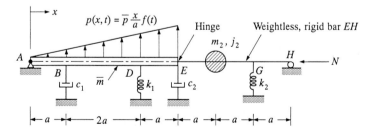

FIGURE E8-1
Example of a rigid-body-assemblage SDOF system.

The mass is distributed uniformly through bar AB, and the weightless bar BC supports a lumped mass m_2 having a centroidal mass moment of inertia j_2.

Because the two bars are assumed rigid, this system has only a single degree of freedom, and its dynamic response can be expressed with a single equation of motion. This equation could be formulated by direct equilibration (the reader may find this a worthwhile exercise), but because of the complexity of the system, it is more convenient to use a work or energy formulation. A virtual-work analysis will be employed here; although using Hamilton's principle, as described in Chapter 16, would be equally effective.

For the form of displacement which may take place in this SDOF structure (Fig. E8-2), the hinge motion $Z(t)$ may be taken as the basic quantity and all other displacements expressed in terms of it; for example, $BB'(t) = Z(t)/4$, $DD'(t) = 3\,Z(t)/4$, $FF'(t) = 2\,Z(t)/3$, etc. The force components acting on the system (exclusive of the axial applied force N, which will be discussed later) are also shown in this figure. Each resisting force component can be expressed in terms of $Z(t)$ or its time derivatives, as follows:

$$f_{I_1}(t) = m_1 \frac{1}{2}\ddot{Z}(t) = \overline{m}\,L\,\frac{1}{2}\ddot{Z}(t) = 2\,a\,\overline{m}\,\ddot{Z}(t)$$

$$M_{j_1}(t) = J_1 \frac{1}{4a}\ddot{Z}(t) = \frac{\overline{m}L}{4a}\frac{L^2}{12}\ddot{Z}(t) = \frac{4}{3}\,a^2\,\overline{m}\,\ddot{Z}(t)$$

$$f_{I_2}(t) = m_2 \frac{2}{3}\ddot{Z}(t)$$

$$M_{j_2}(t) = -J_2 \frac{1}{3a}\ddot{Z}(t)$$

$$f_{D_1}(t) = c_1 \left[\frac{d}{dt}DD'(t)\right] = c_1 \frac{1}{4}\dot{Z}(t)$$

$$f_{D_2}(t) = c_2 \dot{Z}(t)$$

$$f_{S_1}(t) = k_1 \left[DD'(t)\right] = k_1 \frac{3}{4}Z(t)$$

$$f_{S_2}(t) = k_2 \left[GG'(t)\right] = k_2 \frac{1}{3}Z(t)$$

The externally applied lateral load resultant is

$$p_1(t) = 8\,\overline{p}\,a\,f(t)$$

In these expressions, $\overline{m}$ and $\overline{p}$ denote reference values of mass and force, respectively, per unit length and $f(t)$ is a dimensionless time-dependent function which represents the dynamic load variation.

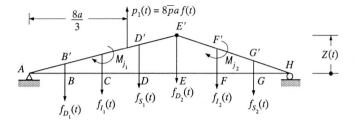

FIGURE E8-2
SDOF displacements and resultant forces.

The equation of motion of this system may be established by equating to zero all work done by these force components during an arbitrary virtual displacement δZ. The virtual displacements through which the force components move are proportional to $Z(t)$, as indicated in Fig. E8-2. Thus the total virtual work may be written

$$\delta W(t) = -2a\,\overline{m}\,\ddot{Z}(t)\,\frac{\delta Z}{2} - \frac{4}{3}a^2\,\overline{m}\,\ddot{Z}(t)\,\frac{\delta Z}{4a} - m_2\,\frac{2\ddot{Z}(t)}{3}\,\frac{2}{3}\,\delta Z$$

$$- j_2\,\frac{\ddot{Z}(t)}{3a} - c_1\,\frac{\dot{Z}(t)}{4}\,\frac{\delta Z}{4} - c_2\,\dot{Z}(t)\,\delta Z - k_1\,\frac{3}{4}\,Z(t)\,\frac{3}{4}\,\delta Z$$

$$- k_2\,\frac{Z(t)}{3}\,\frac{\delta Z}{3} + 8\overline{p}\,a\,f(t)\,\frac{2}{3}\,\delta Z = 0 \qquad \text{(a)}$$

which when simplified becomes

$$\left[\left(a\,\overline{m} + \frac{a\,\overline{m}}{3} + \frac{4}{9}\,m_2 + \frac{j_2}{9a^2}\right)\ddot{Z}(t) + \left(\frac{c_1}{16} + c_2\right)\dot{Z}(t)\right.$$

$$\left. + \left(\frac{9}{16}\,k_1 + \frac{k_2}{9}\right)Z(t) - \frac{16}{3}\,\overline{p}\,a\,f(t)\right]\delta Z = 0 \qquad \text{(b)}$$

Because the virtual displacement δZ is arbitrary, the term in square brackets must vanish; thus the final equation of motion becomes

$$\left(\frac{4}{3}\,\overline{m}\,a + \frac{4}{9}\,m_2 + \frac{j_2}{9a^2}\right)\ddot{Z}(t) + \left(\frac{c_1}{16} + c_2\right)\dot{Z}(t)$$

$$+ \left(\frac{9}{16}\,k_1 + \frac{k_2}{9}\right)Z(t) = \frac{16}{3}\,\overline{p}\,a\,f(t) \qquad \text{(c)}$$

This may be written in the simplified form

$$m^*\,\ddot{Z}(t) + c^*\,\dot{Z}(t) + k^*\,Z(t) = p^*(t) \qquad \text{(8-1)}$$

if the new symbols are defined as follows:

$$m^* = \frac{4}{3} \bar{m} a + \frac{4}{9} m_2 + \frac{j_2}{9a^2} \qquad c^* = \frac{1}{16} c_1 + c_2$$

$$k^* = \frac{9}{16} k_1 + \frac{1}{9} k_2 \qquad p^*(t) = \frac{16}{3} \bar{p} a f(t)$$

These quantities are termed, respectively, the *generalized mass, generalized damping, generalized stiffness,* and *generalized load* for this system; they have been evaluated with reference to the generalized coordinate $Z(t)$, which has been used here to define the displacements of the system.

Consider now the externally applied axial force N of Fig. E8-1. As may be seen in Fig. E8-3, the virtual work done by this force during the virtual displacement δZ is $N\delta e$. The displacement δe is made up of two parts, δe_1 and δe_2, associated with the rotations of the two bars. Considering the influence of bar AE only, it is clear from similar triangles (assuming small deflections) that $\delta e_1 = (Z(t)/4a)\,\delta Z$. Similarly $\delta e_2 = (Z(t)/3a)\,\delta Z$, thus the total displacement is

$$\delta e = \delta e_1 + \delta e_2 = \frac{7}{12} \frac{Z(t)}{a} \delta Z$$

and the virtual work done by the axial force N is

$$\delta W_P = \frac{7}{12} \frac{N Z(t)}{a} \delta Z \qquad (d)$$

Adding Eq. (d) and Eq. (a) and carrying out simplifying operations similar to those which led to Eq. (c) shows that only one term in the equation of motion is influenced by the axial force, the generalized stiffness. When the effect of

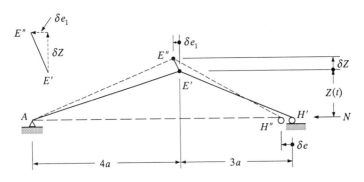

FIGURE E8-3
Displacement components in the direction of axial force.

the axial force in this system is included, the combined generalized stiffness $\bar{k}^*$ is

$$\bar{k}^* = k^* - \frac{7}{12}\frac{P}{a} = \frac{9}{16}k_1 + \frac{1}{9}k_2 - \frac{7}{12}\frac{N}{a} \qquad (e)$$

With this modified generalized-stiffness term, the equation of motion of the complete system of Fig. E8-1, including axial force, is given by an equation similar to Eq. (8-1). The last term in Eq. (e), which is directly proportional to the axial force N, often is given the name "geometric stiffness."

It is of interest to note that the condition of zero generalized stiffness represents a neutral stability or critical buckling condition in the system. The value of axial force N_{cr} which would cause buckling of this structure can be found by equating $\bar{k}^*$ of Eq. (e) to zero:

$$0 = \frac{9}{16}k_1 + \frac{1}{9}k_2 - \frac{7}{12}\frac{N_{cr}}{a}$$

Thus

$$N_{cr} = \left(\frac{27}{28}k_1 + \frac{4}{21}k_2\right)a \qquad (f)$$

In general, compressive axial forces tend to reduce the stiffness of a structural system, while tensile axial forces cause a corresponding increase of stiffness. Such loads can have a significant effect on the response of the structure to dynamic loads, and the resulting change of stiffness should always be evaluated to determine its importance in the given problem. It should be noted that *axial force* in this and in subsequent discussions refers to a force which acts parallel to the initial *undistorted* axis of the member; such a force is assumed not to change the direction of its line of action or its magnitude with the motion of the structure.

Example E8-2.   As a second example of the formulation of the equations of motion for a rigid-body assemblage, the system shown in Fig. E8-4 will be considered. The small-amplitude motion of this system can be characterized by

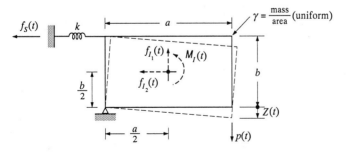

FIGURE E8-4
SDOF plate with dynamic forces.

the downward displacement of the load point $Z(t)$, and all the system forces resisting this motion can be expressed in terms of it:

$$f_S(t) = k\,\frac{b}{a}\,Z(t) \qquad\qquad f_{I_1}(t) = \gamma\,a\,b\,\frac{1}{2}\,\ddot{Z}(t)$$

$$f_{I_2}(t) = \gamma\,a\,b\,\frac{b}{2a}\,\ddot{Z}(t) \qquad\qquad M_I(t) = \gamma\,a\,b\,\frac{a^2+b^2}{12}\,\frac{1}{a}\,\ddot{Z}(t)$$

The equation of motion for this simple system can be written directly by expressing the equilibrium of moments about the plate hinge:

$$f_S(t)\,b + f_{I_1}(t)\,\frac{a}{2} + f_{I_2}(t)\,\frac{b}{2} + M_I(t) = p(t)\,a$$

Dividing by the length a and substituting the above expressions for the forces, this equation becomes

$$\gamma\,a\,b\left[\frac{1}{12}\left(\frac{b^2}{a^2}+1\right) + \frac{1}{4} + \frac{b^2}{4a^2}\right]\ddot{Z}(t) + k\,\frac{b^2}{a^2}\,Z(t) = p(t)$$

Finally, it may be written

$$m^*\,\ddot{Z}(t) + k^*\,Z(t) = p^*(t)$$

in which

$$m^* = \frac{\gamma\,a\,b}{3}\left(1 + \frac{b^2}{a^2}\right) \qquad k^* = k\,\frac{b^2}{a^2} \qquad p^*(t) = p(t)$$

8-3 GENERALIZED PROPERTIES: DISTRIBUTED FLEXIBILITY

The example of Fig. E8-1 is a true SDOF system in spite of the complex interrelationships of its various components because the two rigid bars are supported so that only one type of displacement pattern is possible. If the bars could deform in flexure, the system would have an infinite number of degrees of freedom. A simple SDOF analysis could still be made, however, if it were assumed that only a single flexural deflection pattern could be developed.

As an illustration of this method of approximating SDOF behavior in a flexure system actually having infinite degrees of freedom, consider the formulation of an equation of motion for the cantilever tower of Fig. 8-2a. The essential properties of the tower (excluding damping) are its flexural stiffness $EI(x)$ and its mass per unit of length $m(x)$. It is assumed to be subjected to horizontal earthquake ground-motion excitation $v_g(t)$, and it supports a constant vertical load N applied at the top.

To approximate the motion of this system with a single degree of freedom, it is necessary to assume that it will deform only in a single shape. The shape function will be designated $\psi(x)$, and the amplitude of the motion relative to the moving base will be represented by the generalized coordinate $Z(t)$; thus,

$$v(x,t) = \psi(x)Z(t) \tag{8-2}$$

Typically the generalized coordinate is selected as the displacement of some convenient reference point in the system, such as the tip displacement in this tower. Then the shape function is the dimensionless ratio of the local displacement to this reference displacement:

$$\psi(x) = \frac{v(x,t)}{Z(t)} \tag{8-3}$$

The equation of motion of this generalized SDOF system can be formulated conveniently only by work or energy principles, and the principle of virtual work will be used in this case.

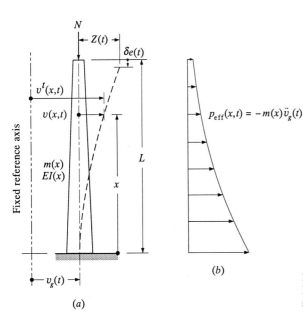

(a)

(b)

FIGURE 8-2
Flexure structure treated as a SDOF system.

Since the structure in this example is flexible in flexure, internal virtual work δW_I is performed by the real internal moments $M(x,t)$ acting through their corresponding virtual changes in curvature $\delta\left[\frac{\partial^2 v(x)}{\partial x^2}\right]$. The virtual-work principle requires that the external virtual work, $\delta W_E(t)$, performed by the external loadings acting through their corresponding virtual displacements be equated to the internal virtual work, i.e.,

$$\delta W_E = \delta W_I \tag{8-4}$$

To develop the equation of motion in terms of relative displacement $v(x,t)$, the base of the structure can be treated as fixed while an effective loading $p_{\text{eff}}(x,t)$ is applied as shown in Fig. 8-2b. The inertial loading is then given by

$$f_I(x,t) = m(x)\,\ddot{v}(x,t) \tag{8-5}$$

Using the full set of external forces, the external virtual work is given by

$$\delta W_E = -\int_0^L f_I(x)\,\delta v(x)\,dx + \int_0^L p_{\text{eff}}(x,t)\,\delta v(x)\,dx + N\delta e \tag{8-6}$$

and consistent with the above statement regarding internal virtual work,

$$\delta W_I(t) = \int_0^L M(x,t)\,\delta v''(x)\,dx \tag{8-7}$$

where $v''(x) = \partial^2 v(x)/\partial x^2$.

If it is assumed that damping stresses are developed in proportion to the strain velocity, a uniaxial stress-strain relation of the form

$$\sigma = E\left[\epsilon + a_1\,\dot{\epsilon}\right] \tag{8-8}$$

may be adopted, where E is Young's modulus and a_1 is a damping constant. Then the Euler-Bernouli hypothesis that plane sections remain plane leads to the relation

$$M(x,t) = EI(x)\left[v''(x,t) + a_1\,\dot{v}''(x,t)\right] \tag{8-9}$$

Using this equation, the basic relations may be expressed as follows:

$$
\begin{aligned}
v(x,t) &= \psi(x)\,Z(t) & \dot{v}''(x,t) &= \psi''(x)\,\dot{Z}(t)\\[4pt]
v'(x,t) &= \psi'(x)\,Z(t) & \delta v(x,t) &= \psi(x)\,\delta Z\\[4pt]
v''(x,t) &= \psi''(x)\,Z(t) & \delta v'(x,t) &= \psi'(x)\,\delta Z\\[4pt]
\ddot{v}(x,t) &= \psi(x)\,\ddot{Z}(t) & \delta v''(x,t) &= \psi''(x)\,\delta Z
\end{aligned}
\tag{8-10}
$$

Also, by analogy with the development of Eq. (d) of Example E8-1, the expressions for axial displacement take the form

$$e(t) = \frac{1}{2} \int_0^L [v'(x,t)]^2 \, dx \qquad \delta e = \int_0^L v'(x,t) \, \delta v'(x) \, dx \qquad (8\text{-}11)$$

Finally, expressions for the external and internal virtual work may be formulated by using Eqs. (8-10) and (8-11):

$$\delta W_E = \left[-\ddot{Z}(t) \int_0^L m(x) \, \psi(x)^2 \, dx \right.$$

$$\left. - \ddot{v}_g(t) \int_0^L m(x) \, \psi(x) \, dx + N Z(t) \int_0^L \psi'(x)^2 \, dx \right] \delta Z \qquad (8\text{-}12)$$

$$\delta W_I = \left[Z(t) \int_0^L EI(x) \, \psi''(x)^2 \, dx + a_1 \, \dot{Z}(t) \int_0^L EI(x) \, \psi''(x)^2 \, dx \right] \delta Z$$

Equating Eqs. (8-12) in accordance with Eq. (8-4) yields the generalized equation of motion

$$m^* \, \ddot{Z}(t) + c^* \, \dot{Z}(t) + k^* \, Z(t) - k_G^* \, Z(t) = p_{\text{eff}}^*(t) \qquad (8\text{-}13)$$

where

$$m^* = \int_0^L m(x) \, \psi(x)^2 \, dx = \text{generalized mass}$$

$$c^* = a_1 \int_0^L EI(x) \, \psi''(x)^2 \, dx = \text{generalized damping}$$

$$k^* = \int_0^L EI(x) \, \psi''(x)^2 \, dx = \text{generalized flexural stiffness} \qquad (8\text{-}14)$$

$$k_G^* = N \int_0^L \psi'(x)^2 \, dx = \text{generalized geometric stiffness}$$

$$p_{\text{eff}}^*(t) = -\ddot{v}_g(t) \int_0^L m(x) \, \psi(x) \, dx = \text{generalized effective load}$$

Combining the two stiffness terms, Eq. (8-13) can be written as

$$m^* \, \ddot{Z}(t) + c^* \, \dot{Z}(t) + \overline{k}^* \, Z(t) = p_{\text{eff}}^*(t) \qquad (8\text{-}15)$$

in which

$$\overline{k}^* = k^* - k_G^* \qquad (8\text{-}16)$$

is the combined generalized stiffness.

The critical buckling load can be calculated for this system by the same method used in Example E8-1, i.e., by equating to zero the combined generalized stiffness and solving for N_{cr}; thus, one obtains

$$N_{cr} = \frac{\int_0^L EI(x)\,\psi''(x)^2\,dx}{\int_0^L \psi'(x)^2\,dx} \tag{8-17}$$

This SDOF approximate analysis of the critical buckling load is called Rayleigh's method, which is discussed in the context of vibration analysis in Section 8-5. The value determined for the critical load depends, of course, upon the assumed shape function $\psi(x)$, but a very good approximation will be given by any shape that is consistent with the geometric boundary conditions.

Example E8-3. To provide a numerical example of the formulation of the equation of motion for a SDOF system with distributed flexibility, it will be assumed that the tower of Fig. 8-2 has constant flexural stiffness EI and constant mass distribution $\overline{m}$ along its length and damping in accordance with Eq. (8-8). Also, its deflected shape in free vibrations will be assumed as

$$\psi(x) = 1 - \cos\frac{\pi x}{2L} \tag{a}$$

which satisfies the geometric boundary conditions $\psi(0) = \psi'(0) = 0$. When Eqs. (8-14) are applied, one obtains

$$m^* = \overline{m}\int_0^L \left(1 - \cos\frac{\pi x}{2L}\right)^2 dx = 0.228\,\overline{m}\,L$$

$$c^* = a_1 EI \int_0^L \left(\frac{\pi^2}{4L^2}\,\cos\frac{\pi x}{2L}\right)^2 dx = \frac{a_1\,\pi^4 EI}{32\,L^3}$$

$$k^* = EI \int_0^L \left(\frac{\pi^2}{4L^2}\,\cos\frac{\pi x}{2L}\right)^2 dx = \frac{\pi^4\,EI}{32\,L^3} \tag{b}$$

$$k_G^* = N \int_0^L \left(\frac{\pi}{2L}\,\sin\frac{\pi x}{2L}\right)^2 dx = \frac{N\pi^2}{8L}$$

$$p_{\text{eff}}(t) = -\overline{m}\,\ddot{v}_g(t) \int_0^L \left(1 - \cos\frac{\pi x}{2L}\right) dx = 0.364\,\overline{m}\,L\,\ddot{v}_g(t)$$

which upon substitution into Eq. (8-13) gives the SDOF equation of motion:

$$\left(0.228\overline{m}L\right)\ddot{Z}(t) + \left(\frac{a_1\,\pi^4\,EI}{32\,L^3}\right)\dot{Z}(t) + \left(\frac{\pi^4\,EI}{32\,L^3} - \frac{N\pi^2}{8L}\right)Z(t)$$

$$= -0.364\,\overline{m}\,L\,\ddot{v}_g(t) \tag{c}$$

In addition, the buckling load for this column subjected to tip load will be evaluated by setting the combined stiffness equal to zero and solving for N_{cr}, with the following result:

$$N_{cr} = \frac{\pi^2}{4} \frac{EI}{L^2} \tag{d}$$

This is the true buckling load for an end-loaded uniform cantilever column because the assumed shape function of Eq. (a) is the true buckled shape.

Of course, one could select a different shape function $\psi(x)$ as long as it satisfies the geometric boundary conditions $\psi(0) = \psi'(0) = 0$. For example, if this function were assumed to be of the parabolic form

$$\psi(x) = \frac{x^2}{L^2} \tag{e}$$

the equation of motion obtained by the above procedure would be

$$\left(0.200\,\overline{m}\,L\right) \ddot{Z}(t) + \left(\frac{4\,a_1\,EI}{L^3}\right) \dot{Z}(t) + \left(\frac{4EI}{L^3} - \frac{4N}{3L}\right) Z(t) = -\frac{\overline{m}\,L}{3}\, \ddot{v}_g(t) \tag{f}$$

Setting the combined stiffness equal to zero, the critical load is given as

$$N_{cr} = \frac{3EI}{L^2} \tag{g}$$

which is about 22 percent higher than the true value given by Eq. (d).

When using the Rayleigh method of buckling analysis as given by Eq. (8-17), it should be recognized that assuming any shape other than the true buckled shape will require additional external constraints acting on the system to maintain its equilibrium. These additional external constraints represent a stiffening influence on the system; therefore the critical load computed by a Rayleigh analysis using any shape other than the true one must always be greater than the true critical load. In the above example, it is apparent that the parabolic shape is not a good assumption for this structure, even though it satisfies the geometric boundary conditions, because the constant curvature of this shape implies that the moment is constant along its length. It is obvious here that the moment must vanish at the top of the column, and any assumed shape that satisfies this force boundary condition (i.e., one having zero curvature at the top) will give much better results.

8-4 EXPRESSIONS FOR GENERALIZED SYSTEM PROPERTIES

As implied by the preceding examples, the equation of motion for any SDOF system, no matter how complex, can always be reduced to the form

$$m^* \ddot{Z}(t) + c^* \dot{Z}(t) + \overline{k^*}\, Z(t) = p^*(t)$$

in which $Z(t)$ is the single generalized coordinate expressing the motion of the system and the symbols with asterisks represent generalized physical properties corresponding to this coordinate. In general, the values of these properties can be determined by application of either the principle of virtual work, as illustrated by the previous examples, or Hamilton's principle as illustrated in Chapter 16. However, standardized forms of these expressions can be derived easily which are very useful in practice.

Consider an arbitrary one-dimensional system, as illustrated by the example in Fig. 8-3, assumed to displace only in a single shape $\psi(x)$ with displacements

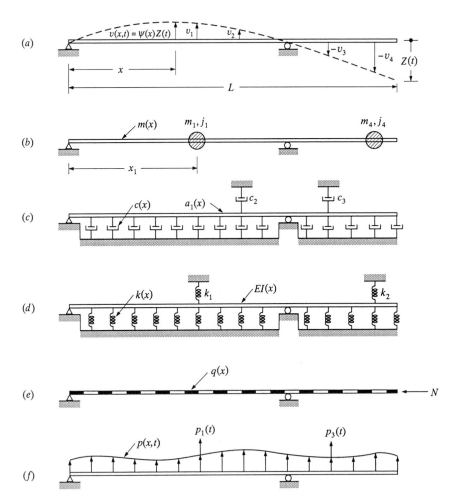

FIGURE 8-3
Properties of generalized SDOF system: (a) assumed shape; (b) mass properties; (c) damping properties; (d) elastic properties; (e) applied axial loading; (f) applied lateral loading.

expressed in terms of the generalized coordinate $Z(t)$ as given by

$$v(x, t) = \psi(x)\, Z(t)$$

Part of the total mass of the system is distributed in accordance with $m(x)$ and the remainder is lumped at discrete locations i ($i = 1, 2, \ldots$) as denoted by m_i. External damping is provided by distributed dashpots varying in accordance with $c(x)$ and by discrete dashpots as denoted by the c_i values, and internal damping is assumed to be present in flexure as controlled by the uniaxial stress-strain relation of Eq. (8-8). The elastic properties of the system result from distributed external springs varying in accordance with $k(x)$, from discrete springs as denoted by the k_i values, and from distributed flexural stiffness given by $EI(x)$. External loadings are applied to the system in both discrete and distributed forms as indicated by the time-independent axial forces $q(x)$ and N and the time-dependent lateral forces $p(x, t)$ and $p_i(t)$. These loadings produce internal axial force and moment distributions $N(x)$ and $M(x, t)$, respectively.

Applying the procedure of virtual work to this general SDOF system in the same manner as it was applied to the previous example solutions, one obtains the following useful expressions for the contributions to the generalized properties:

$$m^* = \int_0^L m(x)\, \psi(x)^2 \, dx + \sum m_i \, \psi_i^2 + \sum j_i \, \psi_i'^2$$

$$c^* = \int_0^L c(x)\, \psi(x)^2 \, dx + a_1 \int_0^L EI(x)\, \psi''(x)^2 \, dx + \sum c_i \, \psi_i^2$$

$$\overline{k^*} = \int_0^L k(x)\, \psi(x)^2 \, dx + \int_0^L EI(x)\, \psi''(x)^2 \, dx + \sum k_i \psi_i^2 \qquad (8\text{-}18)$$

$$\quad - \int_0^L N(x)\, \psi'(x)^2 \, dx$$

$$p^*(t) = \int_0^L p(x, t)\, \psi(x) \, dx + \sum p_i(t)\, \psi_i(x)$$

The vectorial nature of the force and displacement quantities in the last of Eqs. (8-18) must be carefully noted. Only components of the forces in the directions of the corresponding assumed displacements can be included, and the positive sense of each force component must be assigned in accordance with the positive sense of the corresponding displacement.

The above generalized-coordinate concepts apply equally in the reduction of two-dimensional systems to a single degree of freedom. Consider, for example, the rectangular floor slab shown in Fig. 8-4 subjected to a distributed downward loading

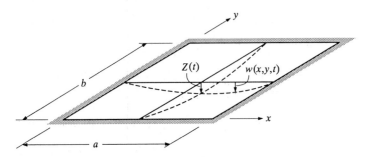

FIGURE 8-4
Simply supported two-dimensional slab treated as a SDOF system.

$p(x, y, t)$. If the deflections of this slab are assumed to have the shape $\psi(x, y)$ shown, and if the displacement amplitude at the middle is taken as the generalized coordinate, the displacements may be expressed

$$w(x, y, t) = \psi(x, y) \, Z(t) \tag{8-19}$$

For a uniform simply supported slab, the shape function might logically be of the form

$$\psi(x, y) = \sin \frac{\pi x}{a} \, \sin \frac{\pi y}{b} \tag{8-20}$$

but any other reasonable shape consistent with the support conditions could be used.

The generalized properties of this system can be calculated by expressions equivalent to those presented in Eqs. (8-18) for the one-dimensional case; however, the integrations must be carried out here in both the x and y directions. For this specific example, the generalized mass, stiffness, and loading would be given by

$$m^* = \int_0^a \int_0^b m(x, y) \, \psi(x, y)^2 \, dx \, dy$$

$$k^* = D \int_0^a \int_0^b \left\{ \left[\frac{\partial^2 \psi(x, y)}{\partial x^2} + \frac{\partial^2 \psi(x, y)}{\partial y^2} \right]^2 \right.$$

$$\left. - 2 \, (1 - \nu) \left[\frac{\partial^2 \psi(x, y)}{\partial x^2} \frac{\partial^2 \psi(x, y)}{\partial y^2} - \left(\frac{\partial^2 \psi(x, y)}{\partial x \, \partial y} \right)^2 \right] \right\} dx \, dy$$

$$p^*(t) = \int_0^a \int_0^b p(x, y) \, \psi(x, y) \, dx \, dy$$

where

$$D = Eh^3 / 12 \, (1 - \nu^2) = \text{flexural rigidity of the slab}$$

$$\nu = \text{Poisson's ratio}$$

$$h = \text{plate thickness}$$

It also should be evident that the same procedures can easily be extended to three-dimensional systems by assuming an appropriate displacement function in three dimensions. However, the difficulty of selecting a suitable shape increases rapidly with the number of dimensions of the system, and the reliability of the results so obtained is reduced accordingly.

8-5 VIBRATION ANALYSIS BY RAYLEIGH'S METHOD

It was pointed out in Section 8-2 that the critical buckling load for a flexural member can be calculated approximately from its generalized elastic and geometric stiffness properties, where these quantities are derived from an assumed buckling shape, and it was also noted that such an assumed shape formulation is generally called Rayleigh's method. Now Rayleigh's assumed shape concept will be extended further to develop an approximate method of evaluating the vibration frequency of the member. The essence of the concept is evident immediately from the fact that the SDOF frequency of vibrations is defined as

$$\omega = \sqrt{\frac{k}{m}} \tag{8-21}$$

where k and m are the system's mass and stiffness, respectively. The Rayleigh's method value of the vibration frequency is given directly by this expression if k^* and m^*, representing the generalized stiffness and mass associated with a given assumed shape, $\psi(x)$, are substituted.

Although this generalized-coordinate concept may be used to determine approximately the vibration frequency of any structure, it is instructive to examine the frequency analysis problem from another point of view, originated by Lord Rayleigh. The basic concept in the Rayleigh method is the principle of conservation of energy; the energy in a freely vibrating system must remain constant if no damping forces act to absorb it. Consider the free-vibration motion of the undamped spring-mass system shown in Fig. 8-5a. With an appropriate choice of time origin, the displacement can be expressed (Fig. 8-5b) by

$$v = v_0 \sin \omega t \tag{8-22a}$$

and the velocity (Fig. 8-5c) by

$$\dot{v} = v_0\, \omega \, \cos \omega t \tag{8-22b}$$

The potential energy of this system is represented entirely by the strain energy of the spring:

$$V = \frac{1}{2}\, k\, v^2 = \frac{1}{2}\, k\, v_0^2 \, \sin^2 \omega t \tag{8-23a}$$

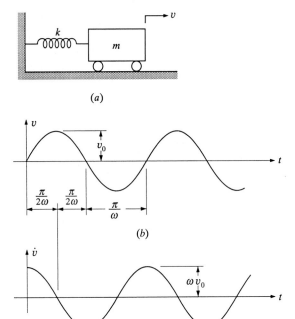

FIGURE 8-5
Free vibration of undamped SDOF
structure: (*a*) SDOF structure;
(*b*) displacement; (*c*) velocity.

while the kinetic energy of the mass is

$$T = \frac{1}{2} m \dot{v}^2 = \frac{1}{2} m v_0^2 \omega^2 \cos^2 \omega t \qquad (8\text{-}23b)$$

Now considering the time $t = \pi/2\omega$, it is clear from Fig. 8-5 [or from Eqs.(8-23)] that the kinetic energy is zero and that the potential energy has reached its maximum value:

$$V_{\max} = \frac{1}{2} k v_0^2 \qquad (8\text{-}24a)$$

Similarly, at the time $t = \pi/\omega$, the potential energy vanishes and the kinetic energy is maximum

$$T_{\max} = \frac{1}{2} m v_0^2 \omega^2 \qquad (8\text{-}24b)$$

Hence, if the total energy in the vibrating system remains constant (as it must in undamped free vibration), it is apparent that the maximum kinetic energy must equal the maximum potential energy, $V_{\max} = T_{\max}$; that is,

$$\frac{1}{2} k v_0^2 = \frac{1}{2} m v_0^2 \omega^2$$

from which

$$\omega^2 = \frac{k}{m}$$

This, of course, is the same frequency expression which was cited earlier; in this case it has been derived by the Rayleigh concept of equating expressions for the maximum strain energy and kinetic energy.

There is no advantage to be gained from the application of Rayleigh's method to vibration analysis of a spring-mass system as described above; its principal use is for the approximate frequency analysis of a system having many degrees of freedom. Consider, for example, the nonuniform simple beam shown in Fig. 8-6. This beam actually has an infinite number of degrees of freedom; that is, it can displace in an infinite variety of displacement patterns. To apply the Rayleigh procedure, it is necessary to assume the shape which the beam will take in its fundamental mode of vibration. As explained above, this assumption may be expressed by Eq. (8-2), or noting the harmonic variation of the generalized coordinate in free vibrations

$$v(x, t) = \psi(x)\, Z_0\, \sin \omega t \tag{8-25}$$

in which $\psi(x)$ is the shape function, which represents the ratio of the displacement at any point x to the reference displacement or generalized coordinate $Z(t)$. Equation (8-25) expresses the assumption that the shape of the vibrating beam does not change with time; only the amplitude of motion varies, and it varies harmonically in a free-vibration condition.

The assumption of the shape function $\psi(x)$ effectively reduces the beam to a SDOF system. Thus the frequency of vibration can be found by equating the maximum strain energy developed during the motion to the maximum kinetic energy. The strain energy of this flexural system is given by

$$V = \frac{1}{2} \int_0^L EI(x) \left(\frac{\partial^2 v}{\partial x^2} \right)^2 dx \tag{8-26}$$

Thus, substituting the assumed shape function of Eq. (8-25) and letting the displacement amplitude take its maximum value leads to

$$V_{\max} = \frac{1}{2}\, Z_0^2 \int_0^L EI(x)\, [\psi''(x)]^2\, dx \tag{8-27}$$

The kinetic energy of the nonuniformly distributed mass is

$$T = \frac{1}{2} \int_0^L m(x)\, (\dot{v})^2\, dx \tag{8-28}$$

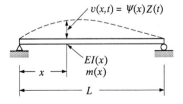

FIGURE 8-6
Vibration of a nonuniform beam.

Thus, when Eq.(8-25) is differentiated with respect to time to obtain the velocity and the amplitude is allowed to reach its maximum,

$$T_{\max} = \frac{1}{2} Z_0^2 \omega^2 \int_0^L m(x) \, [\psi(x)]^2 \, dx \tag{8-29}$$

Finally, after equating the maximum potential energy to the maximum kinetic energy, the squared frequency is found to be

$$\omega^2 = \frac{\int_0^L EI(x) \, [\psi''(x)]^2 \, dx}{\int_0^L m(x) \, [\psi(x)]^2 \, dx} \tag{8-30}$$

At this point, it may be noted that the numerator of Eq. (8-30) is merely the generalized stiffness of the beam k^* for this assumed displacement shape while the denominator is its generalized mass m^* [see Eqs. (8-18)]. Thus Rayleigh's method leads directly to the generalized form of Eq. (8-21), as is to be expected since it employs the same generalized-coordinate concept to reduce the system to a single degree of freedom.

8-6   SELECTION OF THE RAYLEIGH VIBRATION SHAPE

The accuracy of the vibration frequency obtained by Rayleigh's method depends entirely on the shape function $\psi(x)$ which is assumed to represent the vibration-mode shape. In principle, any shape may be selected which satisfies the geometric boundary conditions of the beam, that is, which is consistent with the specified support conditions. However, any shape other than the true vibration shape would require the action of additional external constraints to maintain equilibrium; these extra constraints would stiffen the system, adding to its strain energy, and thus would cause an increase in the computed frequency. Consequently, it may be recognized that the true vibration shape will yield the lowest frequency obtainable by Rayleigh's method, and in choosing between approximate results given by this method, the lowest frequency is always the best approximation.

Example   E8-4.   To illustrate this point, assume that the beam of Fig. 8-6 has uniform mass $\overline{m}$ and stiffness EI. As a first approximation for the frequency analysis, assume that the vibration shape is parabolic: $\psi(x) = (x/L)\,(x/L-1)$. Then, $\psi''(x) = 2/L^2$, and

$$V_{\max} = \frac{1}{2} Z_0^2 \, EI \int_0^L \left(\frac{2}{L^2}\right)^2 dx = \frac{1}{2} Z_0^2 \, \frac{4EI}{L^3}$$

while

$$T_{\max} = \frac{1}{2} Z_0^2 \omega^2 \, \overline{m} \int_0^L \left[\frac{x}{L}\left(\frac{x}{L} - 1\right)\right]^2 dx = \frac{1}{2} Z_0^2 \omega^2 \, \frac{\overline{m}L}{30}$$

from which

$$\omega^2 = \frac{V_{max}}{(1/\omega^2)\,T_{max}} = \frac{120\,EI}{\overline{m}L^4}$$

If the shape were assumed to be a sine curve, $\psi(x) = \sin(\pi x/L)$, the same type of analysis would lead to the result

$$\omega^2 = \frac{EI\,\pi^4/2L^3}{m\,L/2} = \pi^4\,\frac{EI}{\overline{m}\,L^4}$$

This second frequency is significantly less than the first (actually almost 20 percent less); thus it is a much better approximation. As a matter of fact, it is the exact answer because the assumed sine-curve shape is the true vibration shape of a uniform simple beam. The first assumption should not be expected to lead to very good results; the assumed parabolic shape implies a uniform bending moment along the span which does not correspond to the simple end-support conditions. It is a *valid* shape, since it satisfies the geometric requirements of zero end displacements, but is not a realistic assumption.

The question now arises of how a reasonable deflected shape can be selected in order to ensure good results with Rayleigh's method (or the equivalent generalized-coordinate approach described earlier). The concept to be used in selecting the vibration shape is that the displacements in free vibration result from the application of inertial forces and that the inertial forces (which are the product of mass and acceleration) are proportional to the mass distribution and to the displacement amplitude. Thus, the correct vibration shape $\psi_c(x)$ is that deflected shape resulting from a loading $p_c(x)$ proportional to $m(x)\,\psi_c(x)$. Of course, it is not possible to guess the exact shape $\psi_c(x)$, but the deflection shape computed from the loading $\overline{p}(x) = m(x)\,\overline{\psi}(x)$ [as shown in Fig. 8-7, where $\overline{\psi}(x)$ is any reasonable approximation of the true shape] will provide extremely good accuracy in the solution.

In general, the evaluation of the generalized coordinate shape on the basis of an assumed shape in this fashion involves more computational effort than is necessary in an approximate analysis. The Rayleigh procedure will give good accuracy with

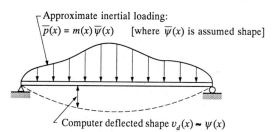

Approximate inertial loading:
$\overline{p}(x) = m(x)\,\overline{\psi}(x)$ [where $\overline{\psi}(x)$ is assumed shape]

Computer deflected shape $v_d(x) \sim \psi(x)$

FIGURE 8-7
Deflected shape resulting from inertial load of assumed shape.

a considerably less refined approach than this. One common assumption is that the inertial loading $\bar{p}(x)$ (see Fig. 8-7) is merely the weight of the beam, that is, $\bar{p}(x) = m(x)\,g$, where $m(x)$ is the mass distribution and g is the acceleration of gravity. The frequency then is evaluated on the basis of the deflected shape $v_d(x)$ resulting from this dead-weight load. The maximum strain energy can be found very simply in this case from the fact that the stored energy must be equal to the work done on the system by the applied loading:

$$V_{\max} = \frac{1}{2} \int_0^L \bar{p}(x)\,v_d(x)\,dx = \frac{1}{2}\,g\,Z_0 \int_0^L m(x)\,\psi(x)\,dx \qquad (8\text{-}31)$$

The kinetic energy is given still by Eq. (8-29), in which $\psi(x) = v_d(x)/Z_0$ is the shape function computed from the dead load. Thus the squared frequency found by equating the strain and kinetic-energy expressions is

$$\omega^2 = \frac{g}{Z_0} \frac{\int_0^L m(x)\,\psi(x)\,dx}{\int_0^L m(x)\,[\psi(x)]^2\,dx} = g\,\frac{\int_0^L m(x)\,v_d(x)\,dx}{\int_0^L m(x)\,[v_d(x)]^2\,dx} \qquad (8\text{-}32)$$

Equation (8-32) is commonly used for the approximate frequency analysis of any type of system. It should be noted that the reference amplitude Z_0 must be included in the expression if the shape is defined by the dimensionless shape function $\psi(x)$, but it is not involved if the actual dead-load deflections are used.

The loading $\bar{p}(x)$ used to calculate the dead-weight deflection $v_d(x)$ in Eq. (8-32) is actually a gravitational loading only in cases where the principal vibratory motion is in the vertical direction. For a structure like the vertical cantilever of Fig. 8-8a, in which the principal motion is horizontal, the loading must be applied laterally, as shown in this figure; in effect it is assumed that gravity acts horizontally

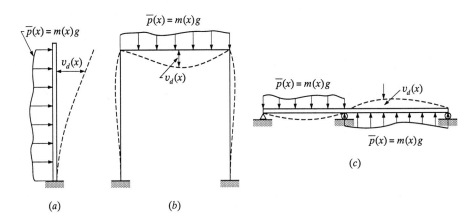

FIGURE 8-8
Assumed shapes resulting from dead loads.

for this purpose. An appropriate deflected shape to approximate the *symmetrical* vibration frequency of the frame of Fig. 8-8b could be obtained by applying a vertical gravity load, as shown. However, the fundamental vibrations of this type of structure will generally be in the horizontal direction; to obtain a shape $\psi(x)$ for approximating the lateral vibration frequency, the gravity forces should be applied *laterally*. Furthermore, in the fundamental mode of vibration of the two-span beam shown in Fig. 8-8c, the two spans will deflect in opposite directions. Thus, to obtain a deflected shape for this case, the gravitational forces should be applied in opposite directions in the adjacent spans. A considerably higher vibration frequency would be obtained from the deflected shape resulting from downward loads acting in both spans.

The reader must be cautioned, however, against spending too much time in computing deflected shapes which will yield extremely accurate results. The principal value of the Rayleigh method is in providing a simple and reliable approximation to the natural frequency. Almost any reasonable shape assumption will give useful results.

Example E8-5. The use of the Rayleigh method to compute the vibration frequency of a practical system will be illustrated by the analysis of the uniform cantilever beam supporting a weight at midspan, shown in Fig. E8-5. For this study, the vibration shape has been taken to be that produced by a load applied to the end of the cantilever, as shown in the lower sketch. The resulting deflected shape is

$$v(x) = \frac{pL^3}{3EI}\left[\frac{3x^2L - x^3}{2L^3}\right] \equiv Z_0\,\psi(x)$$

The maximum potential energy of the beam can be found in this case from

$$V_{\max} = \frac{1}{2}\,p\,Z_0 = \frac{1}{2}\frac{3EI}{L^3}\,Z_0^2$$

where Z_0 is the deflection under the load and p has been expressed in terms of this end deflection.

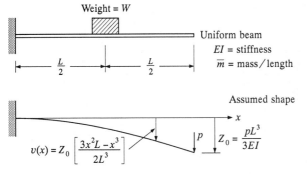

FIGURE E8-5
Rayleigh method analysis of beam vibration frequency.

The maximum kinetic energy of the beam can be calculated in two parts, considering separately the beam and the supported weight:

Beam:
$$T^B_{max} = \frac{\omega^2}{2} \int_0^L \overline{m}\, v^2 \, dx = \frac{\overline{m}}{2} \omega^2 Z_0^2 \int_0^L [\psi(x)]^2 \, dx$$

$$= \frac{33}{140} \frac{\overline{m}L}{2} \omega^2 Z_0^2$$

Weight:
$$T^W_{max} = \frac{W}{2g} \omega^2 \left[v\left(x = \frac{L}{2}\right) \right]^2 = \frac{W}{2g} \omega^2 \left(\frac{5}{16} Z_0\right)^2$$

$$= \frac{25}{256} \frac{W}{2g} \omega^2 Z_0^2$$

Hence the total kinetic energy is

$$T_{max} = \left(\frac{33}{140} + \frac{25}{256} \frac{W}{\overline{m}Lg} \right) \frac{\overline{m}L}{2} \omega^2 Z_0^2$$

and equating the maximum kinetic- and potential-energy expressions leads to the frequency equation

$$\omega^2 = \frac{3}{\left[\frac{33}{140} + \frac{25}{256} \frac{W}{\overline{m}Lg} \right]} \frac{EI}{\overline{m}L^4}$$

8-7   IMPROVED RAYLEIGH METHOD

The idea of using a deflected shape resulting from an inertial loading in a Rayleigh analysis, as described above, can be applied systematically to develop improved versions of the procedure. The standard analysis involves the arbitrary selection of a deflected shape which satisfies the geometric boundary conditions of the structure. For the purposes of this discussion, this initially selected shape will be identified with the superscript zero:

$$v^{(0)}(x, t) = \psi^{(0)}(x) Z_0^{(0)} \sin \omega t \qquad (8\text{-}33)$$

The maximum potential and kinetic energies associated with this shape are then given by

$$V_{max} = \frac{1}{2} \int_0^L EI(x) \left(\frac{\partial^2 v^{(0)}}{\partial x^2} \right)^2 dx = \frac{(Z_0^{(0)})^2}{2} \int_0^L EI(x)\, (\psi''^{(0)})^2 \, dx \qquad (8\text{-}34)$$

$$T_{max} = \frac{1}{2} \int_0^L m(x)\, (\dot{v}^{(0)})^2 \, dx = \frac{(Z_0^{(0)})^2}{2} \omega^2 \int_0^L m(x)\, (\psi^{(0)})^2 \, dx \qquad (8\text{-}35)$$

Method R_{00} — The standard Rayleigh frequency expression, designated as method R_{00}, is

$$\omega^2 = \frac{\int_0^L EI(x)\,(\psi''^{(0)})^2\,dx}{\int_0^L m(x)\,(\psi^{(0)})^2\,dx} \tag{8-36}$$

However, a better approximation of the frequency can be obtained by computing the potential energy from the work done in deflecting the structure by the inertial force associated with the assumed deflection. The distributed inertial force is (at the time of maximum displacement)

$$p^{(0)}(x) = \omega^2\,m(x)\,v^{(0)} = Z_0^{(0)}\,\omega^2\,m(x)\,\psi^{(0)} \tag{8-37}$$

The deflection produced by this loading may be written

$$v^{(1)} = \omega^2\,\frac{v^{(1)}}{\omega^2} = \omega^2\,\psi^{(1)}\,\frac{Z_0^{(1)}}{\omega^2} \equiv \omega^2\,\psi^{(1)}\,\overline{Z}_0^{(1)} \tag{8-38}$$

in which ω^2 is the unknown squared frequency. It may be looked upon as a proportionality factor in both Eqs. (8-37) and (8-38); it is not combined into the expression because its value is not known. The potential energy of the strain produced by this loading is given by

$$V_{\max} = \frac{1}{2}\int_0^L p^{(0)}\,v^{(1)}\,dx = \frac{Z_0^{(0)}\,\overline{Z}_0^{(1)}}{2}\,\omega^4\int_0^L m(x)\,\psi^{(0)}\,\psi^{(1)}\,dx \tag{8-39}$$

Method R_{01} — Equating this expression for the potential energy to the kinetic energy given by the originally assumed shape [Eq. (8-35)] leads to the improved Rayleigh frequency expression, here designated as method R_{01}:

$$\omega^2 = \frac{Z_0^{(0)}}{\overline{Z}_0^{(1)}}\,\frac{\int_0^L m(x)\,(\psi^{(0)})^2\,dx}{\int_0^L m(x)\,\psi^{(0)}\,\psi^{(1)}\,dx} \tag{8-40}$$

This expression often is recommended in preference to Eq. (8-36) because it avoids the differentiation operation required in the standard formula. In general, curvatures $\psi''(x)$ associated with an assumed deflected shape will be much less accurate than the shape function $\psi(x)$, and thus Eq. (8-40), which involves no derivatives, will give improved accuracy.

However, a still better approximation can be obtained with relatively little additional effort by computing the kinetic energy from the calculated shape $v^{(1)}$ rather than from the initial shape $v^{(0)}$. In this case the result is

$$T_{\max} = \frac{1}{2}\int_0^L m(x)\,(\dot{v}^{(1)})^2\,dx = \frac{1}{2}\,\omega^6\,(\overline{Z}^{(1)})^2\int_0^L m(x)\,(\psi^{(1)})^2\,dx \tag{8-41}$$

Method R_{11} — Equating this to the strain energy of Eq. (8-39) leads to the further improved result (here designated as method R_{11}):

$$\omega^2 = \frac{Z_0^{(0)}}{\overline{Z}_0^{(1)}} \frac{\int_0^L m(x)\,\psi^{(0)}\,\psi^{(1)}\,dx}{\int_0^L m(x)\,(\psi^{(1)})^2\,dx} \tag{8-42}$$

Further improvement could be made by continuing the process another step, that is, by using the inertial loading associated with $\psi^{(1)}$ to calculate a new shape $\psi^{(2)}$. In fact, as will be shown later, the process will eventually converge to the exact vibration shape if it is carried through enough cycles and therefore will yield the exact frequency. However, for practical use of the Rayleigh method there is no need to go beyond the improved procedure represented by Eq. (8-42). Also, it should be noted that the generalized-coordinate amplitudes $Z_0^{(0)}$ and $\overline{Z}_0^{(1)}$ in Eqs. (8-40) and (8-42) are arbitrary and can be set to unity if the shape functions $\psi^{(0)}$ and $\psi^{(1)}$ are defined appropriately. However, it is advisable to leave the generalized coordinates in the equations to show that the relative amplitude of $v^{(0)}$ and $v^{(1)}$ is a factor in computing the frequency.

Example E8-6. The two improved versions of the Rayleigh method will be demonstrated and compared with the standard method in carrying out the frequency analysis of the three-story frame shown in Fig. E8-6a. The mass of this frame is lumped in the girders, with values as shown, and the columns are assumed to be weightless. Also, the girders are assumed to be rigid, so that the columns in each story act as simple lateral springs with stiffness coefficients as indicated.

Method R_{00} — In order to demonstrate the effectiveness of the improvement procedures, a poor choice will be deliberately assumed for the initial vibration shape for the frame. This shape consists of equal displacements for the three stories, as shown in Fig. E8-6b; thus

$$v_1^{(0)} = v_2^{(0)} = v_3^{(0)} = 1.0 = Z_0^{(0)}\,\psi_i^{(0)} \qquad \text{where} \qquad \psi_i^{(0)} = Z_0^{(0)} = 1.0$$

From this shape, the maximum kinetic energy is given by

$$T_{\max}^{(0)} = \frac{1}{2}\sum m_i\,(v_i^{(0)})^2 = \frac{1}{2}\omega^2\,(Z_0^{(0)})^2 \sum m_i\,(\psi_i^{(0)})^2 = \frac{1}{2}\omega^2(4.5)$$

The maximum potential energy depends on the relative story-to-story deformations Δv_i and is given by

$$V_{\max}^{(0)} = \frac{1}{2}\sum k_i(\Delta v_i^{(0)})^2 = \frac{1}{2}(Z_0^{(0)})^2 \sum k_i(\Delta\psi_i^{(0)})^2 = \frac{1}{2}(1,800)$$

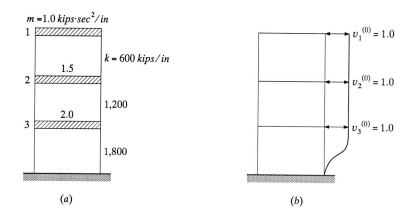

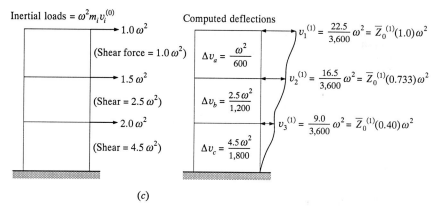

FIGURE E8-6
Frame for Rayleigh method frequency analysis: (a) mass and stiffness values; (b) initial assumed shape; (c) deflections resulting from initial inertial forces.

Hence, when the potential and kinetic energies are equated, the squared frequency is

$$\omega^2 = \frac{1,800}{4.5} = 400 \qquad \omega = 20 \ rad/sec$$

Method R_{01} — The assumption that the structure behaves as though the columns were rigid above the first story clearly is not reasonable for this frame and can be expected to give a gross overestimate of the frequency. Using the inertial forces associated with these initial deflections to calculate an improved shape, in accordance with the improved method R_{01}, leads to much better results.

The inertial loads of the initial shape and the deflections they produce are shown in Fig. E8-6c. The deflections can easily be calculated because the deformation Δv_i in each story is given by the story shear divided by the story stiffness. The maximum potential energy of this new shape $v_i^{(1)}$ may be found as follows:

$$V_{\max}^{(1)} = \frac{1}{2} \sum p_i^{(0)} v_i^{(1)} = \frac{\omega^4}{2} \overline{Z}_0^{(1)} \sum m_i \psi_i^{(0)} \psi_i^{(1)} = \frac{\omega^4}{2} \overline{Z}_0^{(1)} \quad (2.90)$$

When this is equated to the kinetic energy found previously, the frequency is

$$\omega^2 = \frac{1}{\overline{Z}_0^{(1)}} \frac{4.50}{2.90} = \frac{1}{22.5/3,600} \frac{4.5}{2.9} = 248 \qquad \omega = 15.73 \ rad/sec$$

It is apparent that this much smaller frequency represents a great improvement over the result obtained by the standard method R_{00}.

Method R_{11} — Still better results can be obtained by using the improved shape $\psi_1^{(1)}$ in calculating the kinetic as well as the potential energy. Thus the maximum kinetic energy becomes

$$T_{\max}^{(1)} = \frac{\omega^2}{2} (\overline{Z}_0^{(1)})^2 \sum m_i (\psi_i^{(1)})^2 = \frac{\omega^6}{2} \left(\frac{22.5}{3,600} \right)^2 \quad (2.124)$$

Hence, equating this to the improved potential-energy expression leads to the squared frequency value

$$\omega^2 = \frac{1}{\overline{Z}_0^{(1)}} \frac{2.90}{2.124} = \frac{3,600}{22.5} \frac{2.90}{2.124} = 218 \qquad \omega = 14.76 \ rad/sec$$

This is quite close to the exact first-mode frequency for this structure, $\omega_1 = 14.5 \ rad/sec$, as will be derived in Chapter 11.

It is interesting to note that method R_{11} gives the same result here as would be given by Eq. (8-32), where the deflections due to a lateral gravity acceleration are the basis of the analysis. This is because the inertial forces associated with *equal* story displacements are equivalent to the lateral gravity forces. However, if a more reasonable estimate had been made of the initial shape (rather than equal story deflections), the improved method R_{11} would have given a better result than Eq. (8-32).

PROBLEMS

8-1. For the uniform cantilever tower of Example E8-3, the following expressions for the generalized mass and stiffness were determined:

$$m^* = 0.228 \, \overline{m} \, L$$

$$k^* = \frac{\pi^4}{32} \frac{EI}{L^3}$$

Based on these expressions, compute the period of vibration for a concrete tower 200 ft high, with an outside diameter of 12 ft and wall thickness of 8 in, for which the following properties may be assumed:

$$\overline{m} = 110 \ lb \cdot sec^2/ft^2$$

$$EI = 165 \times 10^9 \ lb \cdot ft^2$$

8-2. Assuming that the tower of Prob. 8-1 supports an additional point weight of 400 *kips* at the top, determine the period of vibration (neglecting the geometric stiffness effect).

8-3. For the system shown in Fig. P8-1, determine the generalized physical properties m^*, c^*, k^*, and the generalized loading $p^*(t)$, all defined with respect to the displacement coordinate $Z(t)$. Express the results in terms of the given physical properties and dimensions.

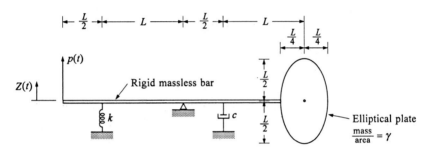

FIGURE P8-1

8-4. Repeat Prob. 8-3 for the structure shown in Fig. P8-2.

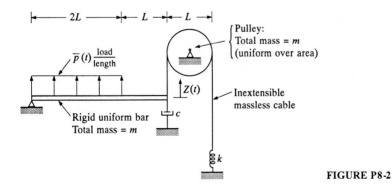

FIGURE P8-2

8-5. Repeat Prob. 8-3 for the structure shown in Fig. P8-3. (*Hint*: this system has only one dynamic degree of freedom; this is associated with the rotational inertia of the rigid bar of mass m.)

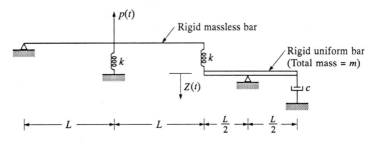

FIGURE P8-3

8-6. The column of Fig. P8-4 is to be treated as a SDOF system by defining its displaced shape as

$$\psi(x) = \frac{v(x,t)}{Z(t)} = \left(\frac{x}{L}\right)^2 \left(\frac{3}{2} - \frac{x}{2L}\right)$$

Denoting the uniformly distributed mass per unit length by $\overline{m}$, the uniform stiffness by EI, and the uniformly distributed load per unit length by $\overline{p}(t)$, evaluate the generalized physical properties m^* and k^* and the generalized loading $p^*(t)$.

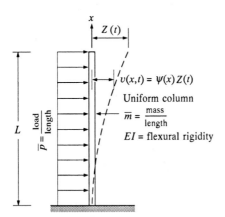

FIGURE P8-4

8-7. (*a*) If a downward load N is applied at the top of the column of Prob. 8-6, evaluate its combined generalized stiffness $\overline{k}^*$ using the same shape function $\psi(x)$.

(*b*) Repeat part *a* assuming that the axial force in the column varies linearly along its length as $N(x) = N(1 - x/L)$.

8-8. Assume that the uniform slab of Fig. 8-4 is square, with side length a, and is simply supported on all four edges.

(a) If its mass per unit area is γ and its flexural rigidity is D, determine its generalized properties m^* and k^* in terms of the central displacement coordinate $Z(t)$. Assume the displacement function is

$$\psi(x,y) = \sin \frac{\pi x}{a} \sin \frac{\pi y}{a}$$

(b) The uniformly distributed external loading per unit of area is $\bar{p}(t)$. Determine the generalized loading $p^*(t)$ based on the displacement function of part a.

8-9. The outer diameters, height, and material properties of a conical concrete smokestack are shown in Fig. P8-5. Assuming a uniform wall thickness of 8 in and that the deflected shape is given by

$$\psi(x) = 1 - \frac{\cos \pi x}{2L}$$

compute the generalized mass m^* and stiffness k^* of the structure. Dividing the height into two equal segments, use Simpson's rule to evaluate the integrals, including in the summations the integrand values for the bottom, middle, and top sections. For example

$$m^* \doteq \frac{\Delta x}{3} \left(y_0 + 4y_1 + y_2 \right)$$

where $y_i = m_i \psi_i^2$ evaluated at level "i."

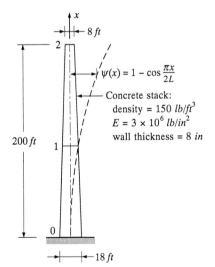

x

8 ft

2

$\psi(x) = 1 - \cos \frac{\pi x}{2L}$

Concrete stack:
density = 150 lb/ft^3
$E = 3 \times 10^6$ lb/in^2
wall thickness = 8 in

200 ft

1

0

18 ft

FIGURE P8-5

8-10. By Rayleigh's method, compute the period of vibration of the uniform beam supporting a central mass m_1 shown in Fig. P8-6. For the assumed shape, use the deflection produced by a central load p; i.e., $v(x) = px(3L^2 - 4x^2)/48EI$ for $0 \leq x \leq L/2$, symmetric with respect to $x = L/2$. Consider the cases: (a) $m_1 = 0$, and (b) $m_1 = 3\overline{m}L$.

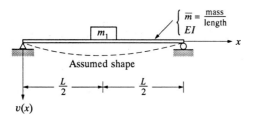

Assumed shape

FIGURE P8-6

8-11. (a) Determine the period of vibration of the frame shown in Fig. P8-7, assuming the girder to be rigid and the deflected shape of the columns to be that due to a lateral load p acting on the girder $v(x) = p\,(3L^2x + x^3)/12EI$;

(b) What fraction of the total column weight assumed lumped with the girder weight will give the same period of vibration as was found in part a?

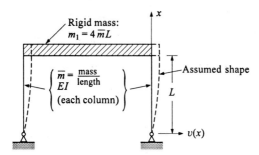

FIGURE P8-7

8-12. The shear building of Fig. P8-8 has its entire mass lumped in the rigid girders. For the given mass and stiffness properties, and assuming a linear initial shape (as shown), evaluate the period of vibration by:

(a) Rayleigh method R_{00}

(b) Rayleigh method R_{01}

(c) Rayleigh method R_{11}

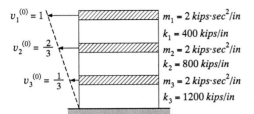

$v_1^{(0)} = 1$

$m_1 = 2 \ kips\cdot sec^2/in$

$k_1 = 400 \ kips/in$

$v_2^{(0)} = \dfrac{2}{3}$

$m_2 = 2 \ kips\cdot sec^2/in$

$k_2 = 800 \ kips/in$

$v_3^{(0)} = \dfrac{1}{3}$

$m_3 = 2 \ kips\cdot sec^2/in$

$k_3 = 1200 \ kips/in$

FIGURE P8-8

8-13. Repeat Prob. 8-12 if the building properties are $m_1 = 1$, $m_2 = 2$, $m_3 = 3 \ kips \cdot sec^2/in$ and $k_1 = k_2 = k_3 = 800 \ kips/in$.

PART

II

MULTI-
DEGREE-
OF-
FREEDOM
SYSTEMS

CHAPTER

9

FORMULATION OF THE MDOF EQUATIONS OF MOTION

9-1 SELECTION OF THE DEGREES OF FREEDOM

The discussion presented in Chapter 8 has demonstrated how a structure can be represented as a SDOF system the dynamic response of which can be evaluated by the solution of a single differential equation of motion. If the physical properties of the system are such that its motion can be described by a single coordinate and no other motion is possible, then it actually is a SDOF system and the solution of the equation provides the exact dynamic response. On the other hand, if the structure actually has more than one possible mode of displacement and it is reduced mathematically to a SDOF approximation by assuming its deformed shape, the solution of the equation of motion is only an approximation of the true dynamic behavior.

The quality of the result obtained with a SDOF approximation depends on many factors, principally the spatial distribution and time variation of the loading and the stiffness and mass properties of the structure. If the physical properties of the system constrain it to move most easily with the assumed shape, and if the loading is such as to excite a significant response in this shape, the SDOF solution will probably be a good approximation; otherwise, the true behavior may bear little resemblance to the computed response. One of the greatest disadvantages of the SDOF approximation is that it is difficult to assess the reliability of the results obtained from it.

169

In general, the dynamic response of a structure cannot be described adequately by a SDOF model; usually the response includes time variations of the displacement shape as well as its amplitude. Such behavior can be described only in terms of more than one displacement coordinate; that is, the motion must be represented by more than one degree of freedom. As noted in Chapter 1, the degrees of freedom in a discrete-parameter system may be taken as the displacement amplitudes of certain selected points in the structure, or they may be generalized coordinates representing the amplitudes of a specified set of displacement patterns. In the present discussion, the former approach will be adopted; this includes both the finite-element and the lumped-mass type of idealization. The generalized-coordinate procedure will be discussed in Chapter 16.

In this development of the equations of motion of a general MDOF system, it will be convenient to refer to the general simple beam shown in Fig. 9-1 as a typical example. The discussion applies equally to any type of structure, but the visualization of the physical factors involved in evaluating all the forces acting is simplified for this type of structure.

The motion of this structure will be assumed to be defined by the displacements of a set of discrete points on the beam: $v_1(t)$, $v_2(t)$, ..., $v_i(t)$, ..., $v_N(t)$. In principle, these points may be located arbitrarily on the structure; in practice, they should be associated with specific features of the physical properties which may be significant and should be distributed so as to provide a good definition of the deflected shape. The number of degrees of freedom (displacement components) to be considered is left to the discretion of the analyst; greater numbers provide better approximations of the true dynamic behavior, but in many cases excellent results can be obtained with only two or three degrees of freedom. In the beam of Fig. 9-1 only one displacement component has been associated with each nodal point on the beam. It should be noted, however, that several displacement components could be identified with each point; e.g., the rotation $\partial v / \partial x$ and longitudinal motions might be used as additional degrees of freedom at each point.

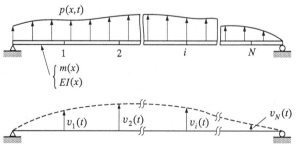

FIGURE 9-1
Discretization of a general beam-type structure.

9-2 DYNAMIC-EQUILIBRIUM CONDITION

The equation of motion of the system of Fig. 9-1 can be formulated by expressing the equilibrium of the effective forces associated with each of its degrees of freedom. In general four types of forces will be involved at any point i: the externally applied load $p_i(t)$ and the forces resulting from the motion, that is, inertia f_{Ii}, damping f_{Di}, and elastic f_{Si}. Thus for each of the several degrees of freedom the dynamic equilibrium may be expressed as

$$f_{I1} + f_{D1} + f_{S1} = p_1(t)$$
$$f_{I2} + f_{D2} + f_{S2} = p_2(t)$$
$$f_{I3} + f_{D3} + f_{S3} = p_3(t)$$

$$\cdots\cdots\cdots\cdots\cdots\cdots$$

(9-1)

or when the force vectors are represented in matrix form,

$$\mathbf{f}_I + \mathbf{f}_D + \mathbf{f}_S = \mathbf{p}(t) \tag{9-2}$$

which is the MDOF equivalent of the SDOF equation (2-1).

Each of the resisting forces is expressed most conveniently by means of an appropriate set of influence coefficients. Consider, for example, the elastic-force component developed at point 1; this depends in general upon the displacement components developed at all points of the structure:

$$f_{S1} = k_{11}v_1 + k_{12}v_2 + k_{13}v_3 + \cdots + k_{1N}v_N \tag{9-3a}$$

Similarly, the elastic force corresponding to the degree of freedom v_2 is

$$f_{S2} = k_{21}v_1 + k_{22}v_2 + k_{23}v_3 + \cdots + k_{2N}v_N \tag{9-3b}$$

and, in general,

$$f_{Si} = k_{i1}v_1 + k_{i2}v_2 + k_{i3}v_3 + \cdots + k_{iN}v_N \tag{9-3c}$$

In these expressions it has been tacitly assumed that the structural behavior is linear, so that the principle of superposition applies. The coefficients k_{ij} are called *stiffness influence coefficients*, defined as follows:

$$k_{ij} = \text{force corresponding to coordinate } i \text{ due to}$$
$$\text{a unit } \textit{displacement} \text{ of coordinate } j \tag{9-4}$$

In matrix form, the complete set of elastic-force relationships may be written

$$
\begin{Bmatrix} f_{S1} \\ f_{S2} \\ \cdot \\ f_{Si} \\ \cdot \end{Bmatrix}
=
\begin{bmatrix}
k_{11} & k_{12} & k_{13} & \cdots & k_{1i} & \cdots & k_{1N} \\
k_{21} & k_{22} & k_{23} & \cdots & k_{2i} & \cdots & k_{2N} \\
\cdots\cdots\cdots\cdots\cdots\cdots\cdots\cdots\cdots \\
k_{i1} & k_{i2} & k_{i3} & \cdots & k_{ii} & \cdots & k_{iN} \\
\cdots\cdots\cdots\cdots\cdots\cdots\cdots\cdots\cdots
\end{bmatrix}
\begin{Bmatrix} v_1 \\ v_2 \\ \cdot \\ v_i \\ \cdot \end{Bmatrix}
\tag{9-5}
$$

or, symbolically,

$$\mathbf{f}_S = \mathbf{k}\,\mathbf{v} \tag{9-6}$$

in which the matrix of stiffness coefficients $\mathbf{k}$ is called the *stiffness matrix* of the structure (for the specified set of displacement coordinates) and $\mathbf{v}$ is the displacement vector representing the displaced shape of the structure.

If it is assumed that the damping depends on the velocity, that is, the viscous type, the damping forces corresponding to the selected degrees of freedom may be expressed by means of damping influence coefficients in similar fashion. By analogy with Eq. (9-5), the complete set of damping forces is given by

$$
\left\{ \begin{array}{c} f_{D1} \\ f_{D2} \\ \cdot \\ f_{Di} \\ \cdot \end{array} \right\}
=
\left[\begin{array}{ccccccc}
c_{11} & c_{12} & c_{13} & \cdots & c_{1i} & \cdots & c_{1N} \\
c_{21} & c_{22} & c_{23} & \cdots & c_{2i} & \cdots & c_{2N} \\
\multicolumn{7}{c}{\dotfill} \\
c_{i1} & c_{i2} & c_{i3} & \cdots & c_{ii} & \cdots & c_{iN} \\
\multicolumn{7}{c}{\dotfill}
\end{array} \right]
\left\{ \begin{array}{c} \dot{v}_1 \\ \dot{v}_2 \\ \cdot \\ \dot{v}_i \\ \cdot \end{array} \right\} \tag{9-7}
$$

in which $\dot{v}_i$ represents the time rate of change (velocity) of the i displacement coordinate and the coefficients c_{ij} are called *damping influence coefficients*. The definition of these coefficients is exactly parallel to Eq. (9-4):

$$c_{ij} = \text{force corresponding to coordinate } i \text{ due to unit}$$
$$\textit{velocity} \text{ of coordinate } j \tag{9-8}$$

Symbolically, Eq. (9-7) may be written

$$\mathbf{f}_D = \mathbf{c}\,\dot{\mathbf{v}} \tag{9-9}$$

in which the matrix of damping coefficients $\mathbf{c}$ is called the *damping matrix* of the structure (for the specified degrees of freedom) and $\dot{\mathbf{v}}$ is the velocity vector.

The inertial forces may be expressed similarly by a set of influence coefficients called the *mass coefficients*. These represent the relationship between the accelerations of the degrees of freedom and the resulting inertial forces; by analogy with Eq. (9-5), the inertial forces may be expressed as

$$
\left\{ \begin{array}{c} f_{I1} \\ f_{I2} \\ \cdot \\ f_{Ii} \\ \cdot \end{array} \right\}
=
\left[\begin{array}{ccccccc}
m_{11} & m_{12} & m_{13} & \cdots & m_{1i} & \cdots & m_{1N} \\
m_{21} & m_{22} & m_{23} & \cdots & m_{2i} & \cdots & m_{2N} \\
\multicolumn{7}{c}{\dotfill} \\
m_{i1} & m_{i2} & m_{i3} & \cdots & m_{ii} & \cdots & m_{iN} \\
\multicolumn{7}{c}{\dotfill}
\end{array} \right]
\left\{ \begin{array}{c} \ddot{v}_1 \\ \ddot{v}_2 \\ \cdot \\ \ddot{v}_i \\ \cdot \end{array} \right\}
= \left\{ \begin{array}{c} \cdot \\ \cdot \\ \cdot \\ \cdot \\ \cdot \end{array} \right\} \tag{9-10}
$$

where $\ddot{v}_i$ is the acceleration of the i displacement coordinate and the coefficients m_{ij} are the *mass influence coefficients*, defined as follows:

$$m_{ij} = \text{force corresponding to coordinate } i \text{ due to}$$
$$\text{unit } acceleration \text{ of coordinate } j \qquad (9\text{-}11)$$

Symbolically, Eq. (9-10) may be written

$$\mathbf{f}_I = \mathbf{m}\,\ddot{\mathbf{v}} \qquad (9\text{-}12)$$

in which the matrix of mass coefficients $\mathbf{m}$ is called the *mass matrix* of the structure and $\ddot{\mathbf{v}}$ is its acceleration vector, both defined for the specified set of displacement coordinates.

Substituting Eqs. (9-6), (9-9), and (9-12) into Eq. (9-2) gives the complete dynamic equilibrium of the structure, considering all degrees of freedom:

$$\mathbf{m}\,\ddot{\mathbf{v}}(t) + \mathbf{c}\,\dot{\mathbf{v}}(t) + \mathbf{k}\,\mathbf{v}(t) = \mathbf{p}(t) \qquad (9\text{-}13)$$

This equation is the MDOF equivalent of Eq. (2-3); each term of the SDOF equation is represented by a matrix in Eq. (9-13), the order of the matrix corresponding to the number of degrees of freedom used in describing the displacements of the structure. Thus, Eq. (9-13) expresses the N equations of motion which serve to define the response of the MDOF system.

9-3 AXIAL-FORCE EFFECTS

It was observed in the discussion of SDOF systems that axial forces or any load which may tend to cause buckling of a structure may have a significant effect on the stiffness of the structure. Similar effects may be observed in MDOF systems; the force component acting parallel to the original axis of the members leads to additional load components which act in the direction (and sense) of the nodal displacements and which will be denoted by $\mathbf{f}_G$. When these forces are included, the dynamic-equilibrium expression, Eq. (9-2), becomes

$$\mathbf{f}_I + \mathbf{f}_D + \mathbf{f}_S - \mathbf{f}_G = \mathbf{p}(t) \qquad (9\text{-}14)$$

in which the negative sign results from the fact that the forces $\mathbf{f}_G$ are assumed to contribute to the deflection rather than oppose it.

These forces resulting from axial loads depend on the displacements of the structure and may be expressed by influence coefficients, called the *geometric-stiffness coefficients*, as follows:

$$
\begin{Bmatrix} f_{G1} \\ f_{G2} \\ \cdot \\ f_{Gi} \\ \cdot \end{Bmatrix} = \begin{bmatrix} k_{G_{11}} & k_{G_{12}} & k_{G_{13}} & \cdots & k_{G_{1i}} & \cdots & k_{G_{1N}} \\ k_{G_{21}} & k_{G_{22}} & k_{G_{23}} & \cdots & k_{G_{2i}} & \cdots & k_{G_{2N}} \\ \multicolumn{7}{c}{\cdots\cdots\cdots\cdots\cdots\cdots\cdots\cdots\cdots} \\ k_{G_{i1}} & k_{G_{i2}} & k_{G_{i3}} & \cdots & k_{G_{ii}} & \cdots & k_{G_{iN}} \\ \multicolumn{7}{c}{\cdots\cdots\cdots\cdots\cdots\cdots\cdots\cdots\cdots} \end{bmatrix} \begin{Bmatrix} v_1 \\ v_2 \\ \cdot \\ v_i \\ \cdot \end{Bmatrix} \tag{9-15}
$$

in which the geometric-stiffness influence coefficients $k_{G_{ij}}$ have the following definition:

$$
\begin{aligned}
k_{G_{ij}} = {}&\text{force corresponding to coordinate } i \text{ due to unit} \\
&\text{displacement of coordinate } j \text{ and resulting from} \\
&\text{axial-force components in the structure}
\end{aligned} \tag{9-16}
$$

Symbolically Eq. (9-15) may be written

$$
\mathbf{f}_G = \mathbf{k}_G \, \mathbf{v} \tag{9-17}
$$

where $\mathbf{k}_G$ is called the *geometric-stiffness matrix* of the structure.

When this expression is introduced, the equation of dynamic equilibrium of the structure [given by Eq. (9-13) without axial-force effects] becomes

$$
\mathbf{m}\,\ddot{\mathbf{v}}(t) + \mathbf{c}\,\dot{\mathbf{v}}(t) + \mathbf{k}\mathbf{v}(t) - \mathbf{k}_G\,\mathbf{v}(t) = \mathbf{p}(t) \tag{9-18}
$$

or when it is noted that both the elastic stiffness and the geometric stiffness are multiplied by the displacement vector, the combined stiffness effect can be expressed by a single symbol and Eq. (9-18) written

$$
\mathbf{m}\,\ddot{\mathbf{v}}(t) + \mathbf{c}\,\dot{\mathbf{v}}(t) + \overline{\mathbf{k}}\mathbf{v}(t) = \mathbf{p}(t) \tag{9-19}
$$

in which

$$
\overline{\mathbf{k}} = \mathbf{k} - \mathbf{k}_G \tag{9-20}
$$

is called the combined stiffness matrix, which includes both elastic and geometric effects. The dynamic properties of the structure are expressed completely by the four influence-coefficient matrices of Eq. (9-18), while the dynamic loading is fully defined by the load vector. The evaluation of these physical-property matrices and the evaluation of the load vector resulting from externally applied forces will be discussed in detail in the following chapter. The effective-load vector resulting from support excitation will be discussed in connection with earthquake-response analysis in Chapter 26.

CHAPTER 10

EVALUATION OF STRUCTURAL-PROPERTY MATRICES

10-1 ELASTIC PROPERTIES

Flexibility

Before discussing the elastic-stiffness matrix expressed in Eq. (9-5), it will be useful to define the inverse flexibility relationship. The definition of a flexibility influence coefficient $\tilde{f}_{ij}$ is

$$\tilde{f}_{ij} = \quad \text{deflection of coordinate } i \text{ due to unit load} \\ \text{applied to coordinate } j \tag{10-1}$$

For the simple beam shown in Fig. 10-1, the physical significance of some of the flexibility influence coefficients associated with a set of vertical-displacement degrees of freedom is illustrated. Horizontal or rotational degrees of freedom might also have

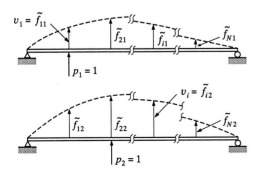

FIGURE 10-1
Definition of flexibility influence coefficients.

been considered, in which case it would have been necessary to use the corresponding horizontal or rotational unit loads in defining the complete set of influence coefficients; however, it will be convenient to restrict the present discussion to the vertical motions.

The evaluation of the flexibility influence coefficients for any given system is a standard problem of static structural analysis; any desired method of analysis may be used to compute these deflections resulting from the applied unit loads. When the complete set of influence coefficients has been determined, they are used to calculate the displacement vector resulting from any combination of the applied loads. For example, the deflection at point 1 due to any combination of loads may be expressed

$$v_1 = \tilde{f}_{11} p_1 + \tilde{f}_{12} p_2 + \tilde{f}_{13} p_3 + \ldots + \tilde{f}_{1N} p_N \tag{10-2}$$

Since similar expressions can be written for each displacement component, the complete set of displacements is expressed

$$\begin{Bmatrix} v_1 \\ v_2 \\ \cdot \\ v_i \\ \cdot \end{Bmatrix} = \begin{bmatrix} \tilde{f}_{11} & \tilde{f}_{12} & \tilde{f}_{13} & \cdots & \tilde{f}_{1i} & \cdots & \tilde{f}_{1N} \\ \tilde{f}_{21} & \tilde{f}_{22} & \tilde{f}_{23} & \cdots & \tilde{f}_{2i} & \cdots & \tilde{f}_{2N} \\ & & \cdots\cdots\cdots\cdots\cdots\cdots & & \\ \tilde{f}_{i1} & \tilde{f}_{i2} & \tilde{f}_{i3} & \cdots & \tilde{f}_{ii} & \cdots & \tilde{f}_{iN} \\ & & \cdots\cdots\cdots\cdots\cdots\cdots & & \end{bmatrix} \begin{Bmatrix} p_1 \\ p_2 \\ \cdot \\ p_i \\ \cdot \end{Bmatrix} \tag{10-3}$$

or symbolically

$$\mathbf{v} = \tilde{\mathbf{f}} \mathbf{p} \tag{10-4}$$

in which the matrix of flexibility influence coefficients $\tilde{\mathbf{f}}$ is called the *flexibility matrix* of the structure.

In Eq. (10-4) the deflections are expressed in terms of the vector of externally applied loads $\mathbf{p}$, which are considered positive when acting in the same sense as the positive displacements. The deflection may also be expressed in terms of the elastic forces $\mathbf{f}_S$ which *resist* the deflections and which are considered positive when acting opposite to the positive displacements. Obviously by statics $\mathbf{f}_S = \mathbf{p}$, and Eq. (10-4) may be revised to read

$$\mathbf{v} = \tilde{\mathbf{f}} \mathbf{f}_S \tag{10-5}$$

Stiffness

The physical meaning of the stiffness influence coefficients defined in Eq. (9-4) is illustrated for a few degrees of freedom in Fig. 10-2; they represent the forces developed in the structure when a unit displacement corresponding to one degree of freedom is introduced and no other nodal displacements are permitted. It should be

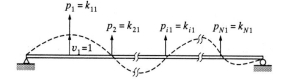

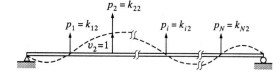

FIGURE 10-2
Definition of stiffness influence coefficients.

noted that the stiffness influence coefficients in Fig. 10-2 are numerically equal to the applied forces required to maintain the specified displacement condition. They are positive when the sense of the applied force corresponds to a positive displacement and negative otherwise.

Basic Structural Concepts

Strain energy — The strain energy stored in any structure may be expressed conveniently in terms of either the flexibility or the stiffness matrix. The strain energy U is equal to the work done in distorting the system; thus

$$U = \frac{1}{2} \sum_{i=1}^{N} p_i v_i = \frac{1}{2} \mathbf{p}^T \mathbf{v} \tag{10-6}$$

where the $\frac{1}{2}$ factor results from the forces which increase linearly with the displacements, and $\mathbf{p}^T$ represents the transpose of $\mathbf{p}$. By substituting Eq. (10-4) this becomes

$$U = \frac{1}{2} \mathbf{p}^T \tilde{\mathbf{f}} \mathbf{p} \tag{10-7}$$

Alternatively, transposing Eq. (10-6) and substituting Eq. (9-6) leads to the second strain-energy expression (note that $\mathbf{p} = \mathbf{f}_S$):

$$U = \frac{1}{2} \mathbf{v}^T \mathbf{k} \mathbf{v} \tag{10-8}$$

Finally, when it is noted that the strain energy stored in a stable structure during any distortion must always be positive, it is evident that

$$\mathbf{v}^T \mathbf{k} \mathbf{v} > 0 \qquad \text{and} \qquad \mathbf{p}^T \tilde{\mathbf{f}} \mathbf{p} > 0 \tag{10-9}$$

Matrices which satisfy this condition, where $\mathbf{v}$ or $\mathbf{p}$ is any arbitrary nonzero vector, are said to be *positive definite*; positive definite matrices (and consequently the flexibility and stiffness matrices of a stable structure) are nonsingular and can be inverted.

Inverting the stiffness matrix and premultiplying both sides of Eq. (9-6) by the inverse leads to

$$\mathbf{k}^{-1}\mathbf{f}_S = \mathbf{v}$$

which upon comparison with Eq. (10-5) demonstrates that the flexibility matrix is the inverse of the stiffness matrix:

$$\mathbf{k}^{-1} = \tilde{\mathbf{f}} \tag{10-10}$$

In practice, the evaluation of stiffness coefficients by direct application of the definition, as implied in Fig. 10-2, may be a tedious computational problem. In many cases, the most convenient procedure for obtaining the stiffness matrix is direct evaluation of the flexibility coefficients and inversion of the flexibility matrix.

Betti's law — A property which is very important in structural-dynamics analysis can be derived by applying two sets of loads to a structure in reverse sequence and comparing expressions for the work done in the two cases. Consider, for example, the two different load systems and their resulting displacements shown in Fig. 10-3. If the loads a are applied first followed by loads b, the work done will be as follows:

Case 1:

Loads a:
$$W_{aa} = \tfrac{1}{2}\sum p_{ia}v_{ia} = \tfrac{1}{2}\mathbf{p}_a{}^T\mathbf{v}_a$$

Loads b:
$$W_{bb} + W_{ab} = \tfrac{1}{2}\mathbf{p}_b{}^T\mathbf{v}_b + \mathbf{p}_a{}^T\mathbf{v}_b$$

Total:
$$W_1 = W_{aa} + W_{bb} + W_{ab} = \tfrac{1}{2}\mathbf{p}_a{}^T\mathbf{v}_a + \tfrac{1}{2}\mathbf{p}_b{}^T\mathbf{v}_b + \mathbf{p}_a{}^T\mathbf{v}_b \tag{10-11}$$

Note that the work done by loads a during the application of loads b is not multiplied by $\tfrac{1}{2}$; they act at their full value during the entire displacement $\mathbf{v}_b$. Now if the loads are applied in reverse sequence, the work done is:

Load system a:

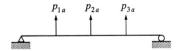

Load system b:

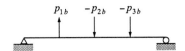

Deflections a:

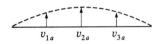

Deflections b:

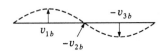

FIGURE 10-3
Two independent load systems and resulting deflections.

Case 2:

Loads b:

$$W_{bb} = \tfrac{1}{2}\,\mathbf{p}_b{}^T\mathbf{v}_b$$

Loads a:

$$W_{aa} + W_{ba} = \tfrac{1}{2}\,\mathbf{p}_a{}^T\mathbf{v}_a + \mathbf{p}_b{}^T\mathbf{v}_a$$

Total:

$$W_2 = W_{bb} + W_{aa} + W_{ba} = \tfrac{1}{2}\,\mathbf{p}_b{}^T\mathbf{v}_b + \tfrac{1}{2}\,\mathbf{p}_a{}^T\mathbf{v}_a + \mathbf{p}_b{}^T\mathbf{v}_a \tag{10-12}$$

The deformation of the structure is independent of the loading sequence, however; therefore the strain energy and hence also the work done by the loads is the same in both these cases; that is, $W_1 = W_2$. From a comparison of Eqs. (10-11) and (10-12) it may be concluded that $W_{ab} = W_{ba}$; thus

$$\mathbf{p}_a{}^T\mathbf{v}_b = \mathbf{p}_b{}^T\mathbf{v}_a \tag{10-13}$$

Equation (10-13) is an expression of *Betti's law*; it states that the work done by one set of loads on the deflections due to a second set of loads is equal to the work of the second set of loads acting on the deflections due to the first.

If Eq. (10-4) is written for the two sets of forces and displacements and substituted into both sides of Eq. (10-13):

$$\mathbf{p}_a{}^T\,\tilde{\mathbf{f}}\,\mathbf{p}_b = \mathbf{p}_b{}^T\,\tilde{\mathbf{f}}\,\mathbf{p}_a$$

it is evident that

$$\tilde{\mathbf{f}} = \tilde{\mathbf{f}}^T \tag{10-14}$$

Thus the flexibility matrix must be symmetric; that is, $\tilde{f}_{ij} = \tilde{f}_{ji}$. This is an expression of Maxwell's law of reciprocal deflections. Substituting similarly with Eq. (9-6) (and noting that $\mathbf{p} = \mathbf{f}_S$) leads to

$$\mathbf{k} = \mathbf{k}^T \tag{10-15}$$

That is, the stiffness matrix also is symmetric.

Finite-Element Stiffness

In principle, the flexibility or stiffness coefficients associated with any prescribed set of nodal displacements can be obtained by direct application of their definitions. In practice, however, the finite-element concept, described in Chapter 1, frequently provides the most convenient means for evaluating the elastic properties. By this approach the structure is assumed to be divided into a system of discrete elements which are interconnected only at a finite number of nodal points. The properties of the complete structure are then found by evaluating the properties of the individual finite elements and superposing them appropriately.

The problem of defining the stiffness properties of any structure is thus reduced basically to the evaluation of the stiffness of a typical element. Consider, for example,

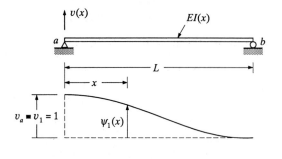

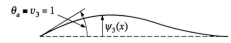

FIGURE 10-4
Beam deflections due to unit nodal displacements at left end.

the nonuniform straight-beam segment shown in Fig. 10-4. The two nodal points by which this type of element can be assembled into a structure are located at its ends, and if only transverse plane displacements are considered, it has two degrees of freedom at each node, vertical translation and rotation. The deflected shapes resulting from applying a unit displacement of each type at the left end of the element while constraining the other three nodal displacements are shown in Fig. 10-4. These displacement functions could be taken as any arbitrary shapes which satisfy nodal and internal continuity requirements, but they generally are assumed to be the shapes developed in a *uniform* beam subjected to these nodal displacements. These are cubic hermitian polynomials which may be expressed as

$$\psi_1(x) = 1 - 3\left(\frac{x}{L}\right)^2 + 2\left(\frac{x}{L}\right)^3 \tag{10-16a}$$

$$\psi_3(x) = x\left(1 - \frac{x}{L}\right)^2 \tag{10-16b}$$

The equivalent shape functions for displacements applied at the right end are

$$\psi_2(x) = 3\left(\frac{x}{L}\right)^2 - 2\left(\frac{x}{L}\right)^3 \tag{10-16c}$$

$$\psi_4(x) = \frac{x^2}{L}\left(\frac{x}{L} - 1\right) \tag{10-16d}$$

With these four interpolation functions, the deflected shape of the element can now be expressed in terms of its nodal displacements:

$$v(x) = \psi_1(x)\,v_1 + \psi_2(x)\,v_2 + \psi_3(x)\,v_3 + \psi_4(x)\,v_4 \tag{10-17}$$

where the numbered degrees of freedom are related to those shown in Fig. 10-4 as follows:

$$\begin{Bmatrix} v_1 \\ v_2 \\ v_3 \\ v_4 \end{Bmatrix} \equiv \begin{Bmatrix} v_a \\ v_b \\ \theta_a \\ \theta_b \end{Bmatrix} \tag{10-17a}$$

It should be noted that both rotations and translations are represented as basic nodal degrees of freedom v_i.

By definition, the stiffness coefficients of the element represent the nodal forces due to unit nodal displacements. The nodal forces associated with any nodal-displacement component can be determined by the principle of virtual displacements, as described in Section 2-5. Consider, for example, the stiffness coefficient k_{13} for the beam element of Fig. 10-4, that is, the vertical force developed at end a due to a unit rotation applied at that point.

This force component can be evaluated by introducing a virtual vertical displacement of end a, as shown in Fig. 10-5, while the unit rotation is applied as shown, and equating the work done by the external forces to the work done on the internal forces: $W_E = W_I$. In this case, the external work is done only by the vertical-force component at a because the virtual displacements of all other nodal components vanish; thus

$$W_E = \delta v_a\, P_a = \delta v_1\, k_{13} \tag{10-18}$$

The internal virtual work is done by the internal moments associated with $\theta_a = 1$ acting on the virtual curvatures, which are $\partial^2/\partial x^2[\delta v(x)] = \psi_1''(x)\,\delta v_1$ (neglecting the effects of shear distortion). However, the internal moments due to $\theta_a = 1$ may be expressed as

$$M(x) = EI(x)\,\psi_3''(x)$$

Thus the internal work is given by

$$W_I = \delta v_1 \int_0^L EI(x)\,\psi_1''(x)\,\psi_3''(x)\,dx \tag{10-19}$$

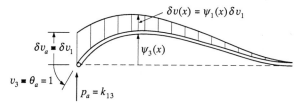

FIGURE 10-5
Beam subjected to real rotation and virtual translation of node.

When the work expressions of Eqs. (10-18) and (10-19) are equated, the expression for this stiffness coefficient is

$$k_{13} = \int_0^L EI(x)\, \psi_1''(x)\, \psi_3''(x)\, dx \tag{10-20}$$

Any stiffness coefficient associated with beam flexure therefore may be written equivalently as

$$k_{ij} = \int_0^L EI(x)\, \psi_i''(x)\, \psi_j''(x)\, dx \tag{10-21}$$

From the form of this expression, the symmetry of the stiffness matrix is evident; that is, $k_{ij} = k_{ji}$. Its equivalence to the corresponding term in the third of Eqs. (8-18) for the case where $i = j$ should be noted.

For the special case of a uniform beam segment, the stiffness matrix resulting from Eq. (10-21) when the interpolation functions of Eqs. (10-16) are used may be expressed by

$$\begin{Bmatrix} f_{S1} \\ f_{S2} \\ f_{S3} \\ f_{S4} \end{Bmatrix} = \frac{2EI}{L^3} \begin{bmatrix} 6 & -6 & 3L & 3L \\ -6 & 6 & -3L & -3L \\ 3L & -3L & 2L^2 & L^2 \\ 3L & -3L & L^2 & 2L^2 \end{bmatrix} \begin{Bmatrix} v_1 \\ v_2 \\ v_3 \\ v_4 \end{Bmatrix} \tag{10-22}$$

where the nodal displacements $\mathbf{v}$ are defined by Eq. (10-17a) and $\mathbf{f}_S$ is the corresponding vector of nodal forces. These stiffness coefficients are the exact values for a uniform beam without shear distortion because the interpolation functions used in Eq. (10-21) are the true shapes for this case. If the stiffness of the beam is not uniform, applying these shape functions in Eq. (10-21) will provide only an approximation to the true stiffness, but the final result for the complete beam will be very good if it is divided into a sufficient number of finite elements.

As mentioned earlier, when the stiffness coefficients of all the finite elements in a structure have been evaluated, the stiffness of the complete structure can be obtained by merely adding the element stiffness coefficients appropriately; this is called the *direct stiffness method*. In effect, any stiffness coefficient k_{ij} of the complete structure can be obtained by adding together the corresponding stiffness coefficients of the elements associated with those nodal points. Thus if elements m, n, and p were all attached to nodal point i of the complete structure, the structure stiffness coefficient for this point would be

$$\hat{k}_{ii} = \hat{k}_{ii}^{(m)} + \hat{k}_{ii}^{(n)} + \hat{k}_{ii}^{(p)} \tag{10-23}$$

in which the superscripts identify the individual elements. Before the element stiff-nesses can be superposed in this fashion, they must be expressed in a common global-coordinate system which is applied to the entire structure. The double hats are placed over each element stiffness symbol in Eq. (10-23) to indicate that they have been transformed from their local-coordinate form [for example, Eq. (10-22)] to the global coordinates.

Example E10-1. The evaluation of the structural stiffness matrix is a basic operation of the matrix-displacement method of static structural analysis; although a general discussion of this subject is beyond the scope of this structural-dynamics text, it may be useful to apply the procedure to a simple frame structure in order to demonstrate how the element stiffness coefficients of Eq. (10-22) may be used.

Consider the structure of Fig. E10-1a. If it is assumed that the members do not distort axially, this frame has the three joint degrees of freedom shown. The

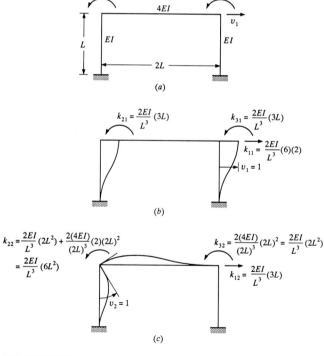

FIGURE E10-1
Analysis of frame stiffness coefficients: (a) frame properties and degrees of freedom; (b) forces due to displacement $v_1 = 1$; (c) forces due to rotation $v_2 = 1$.

corresponding stiffness coefficients can be evaluated by successively applying a unit displacement to each degree of freedom while constraining the other two and determining the forces developed in each member by the coefficients of Eq. (10-22).

When the sidesway displacement shown in Fig. E10-1b is applied, it is clear that only the vertical members are deformed; their end forces are given by elements 1, 3, and 4 in the first column of the stiffness matrix of Eq. (10-22). It will be noted that the structure coefficient k_{11} receives a contribution from each column.

Considering the joint rotation shown in Fig. E10-1c, both the girder and the left vertical contribute to the structure coefficient k_{22}, the contributions being given by element 3 of column 3 in the stiffness matrix of Eq. (10-22) (taking proper account of the girder properties, of course). Only the left vertical contributes to k_{12} and only the girder to k_{32}. The structure stiffness coefficients due to the right-joint rotation are analogous to these.

The structure stiffness matrix finally obtained by assembling all these coefficients is

$$
\begin{Bmatrix} f_{S1} \\ f_{S2} \\ f_{S3} \end{Bmatrix} = \frac{2EI}{L^3} \begin{bmatrix} 12 & 3L & 3L \\ 3L & 6L^2 & 2L^2 \\ 3L & 2L^2 & 6L^2 \end{bmatrix} \begin{Bmatrix} v_1 \\ v_2 \\ v_3 \end{Bmatrix}
$$

10-2 MASS PROPERTIES

Lumped-Mass Matrix

The simplest procedure for defining the mass properties of any structure is to assume that the entire mass is concentrated at the points at which the translational displacements are defined. The usual procedure for defining the point mass to be located at each node is to assume that the structure is divided into segments, the nodes serving as connection points. Figure 10-6 illustrates the procedure for a beam-type structure. The mass of each segment is assumed to be concentrated in point masses at each of its nodes, the distribution of the segment mass to these points being determined by statics. The total mass concentrated at any node of the complete structure then is the sum of the nodal contributions from all the segments attached to that node. In the beam system of Fig. 10-6 there are two segments contributing to each node; for example, $m_1 = m_{1a} + m_{1b}$.

For a system in which only translational degrees of freedom are defined, the lumped-mass matrix has a diagonal form; for the system of Fig. 10-6 it would be

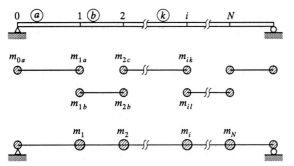

FIGURE 10-6
Lumping of mass at beam nodes.

written

$$\mathbf{m} = \begin{bmatrix} m_1 & 0 & 0 & \cdots & 0 & \cdots & 0 \\ 0 & m_2 & 0 & \cdots & 0 & \cdots & 0 \\ 0 & 0 & m_3 & \cdots & 0 & \cdots & 0 \\ & & \cdots\cdots\cdots\cdots\cdots\cdots\cdots & & \\ 0 & 0 & 0 & \cdots & m_i & \cdots & 0 \\ & & \cdots\cdots\cdots\cdots\cdots\cdots\cdots & & \\ 0 & 0 & 0 & \cdots & 0 & \cdots & m_N \end{bmatrix} \qquad (10\text{-}24)$$

in which there are as many terms as there are degrees of freedom. The off-diagonal terms m_{ij} of this matrix vanish because an acceleration of any mass point produces an inertial force at that point *only*. The inertial force at i due to a unit acceleration of point i is obviously equal to the mass concentrated at that point; thus the mass influence coefficient $m_{ii} = m_i$ in a lumped-mass system.

If more than one translational degree of freedom is specified at any nodal point, the same point mass will be associated with each degree of freedom. On the other hand, the mass associated with any rotational degree of freedom will be zero because of the assumption that the mass is lumped in points which have no rotational inertia. (Of course, if a rigid mass having a finite rotational inertia is associated with a rotational degree of freedom, the diagonal mass coefficient for that degree of freedom would be the rotational inertia of the mass.) Thus the lumped-mass matrix is a diagonal matrix which will include zero diagonal elements for the rotational degrees of freedom, in general.

Consistent-Mass Matrix

Making use of the finite-element concept, it is possible to evaluate mass influence coefficients for each element of a structure by a procedure similar to the analysis

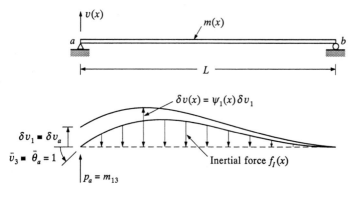

FIGURE 10-7
Node subjected to real angular acceleration and virtual translation.

of element stiffness coefficients. Consider, for example, the nonuniform beam segment shown in Fig. 10-7, which may be assumed to be the same as that of Fig. 10-4.
The degrees of freedom of the segment are the translation and rotation at each end,
and it will be assumed that the displacements within the span are defined by the same
interpolation functions $\psi_i(x)$ used in deriving the element stiffness.

If the beam were subjected to a unit angular acceleration of the left end, $\ddot{v}_3 =
\ddot{\theta}_a = 1$, accelerations would be developed along its length, as follows:

$$\ddot{v}(x) = \psi_3(x)\,\ddot{v}_3 \tag{10-25}$$

which can be obtained by taking the second derivative of Eq. (10-17). By
d'Alembert's principle, the inertial force resisting this acceleration is

$$f_I(x) = m(x)\,\ddot{v}(x) = m(x)\,\psi_3(x)\,\ddot{v}_3 \tag{10-26}$$

Now the mass influence coefficients associated with this acceleration are defined as the nodal inertial forces which it produces; these can be evaluated from the
distributed inertial force of Eq. (10-26) by the principle of virtual displacements. For
example, the vertical force at the left end can be evaluated by introducing a vertical
virtual displacement and equating the work done by the external nodal force p_a to the
work done on the distributed inertial forces $f_I(x)$. Thus

$$p_a\,\delta v_a = \int_0^L f_I(x)\,\delta v(x)\,dx$$

Expressing the vertical virtual displacement in terms of the interpolation function and
substituting Eq. (10-26) lead finally to

$$m_{13} = \int_0^L m(x)\,\psi_1(x)\,\psi_3(x)\,dx \tag{10-27}$$

It should be noted in Fig. 10-7 that the mass influence coefficient represents the inertial force opposing the acceleration, but that it is numerically equal to the external force producing the acceleration.

From Eq. (10-27) it is evident that any mass influence coefficient m_{ij} of an arbitrary beam segment can be evaluated by the equivalent expression

$$m_{ij} = \int_0^L m(x)\,\psi_i(x)\,\psi_j(x)\,dx \tag{10-28}$$

The symmetric form of this equation shows that the mass matrix (like the stiffness matrix) is symmetric; that is, $m_{ij} = m_{ji}$; also it may be noted that this expression is equivalent to the corresponding term in the first of Eqs. (8-18) in the case where $i = j$. When the mass coefficients are computed in this way, using the same interpolation functions which are used for calculating the stiffness coefficients, the result is called the *consistent-mass matrix*. In general, the cubic hermitian polynomials of Eqs. (10-16) are used for evaluating the mass coefficients of any straight beam segment. In the special case of a beam with uniformly distributed mass the results are

$$
\begin{Bmatrix} f_{I1} \\ f_{I2} \\ f_{I3} \\ f_{I4} \end{Bmatrix} = \frac{\overline{m}L}{420}
\begin{bmatrix}
156 & 54 & 22L & -13L \\
54 & 156 & 13L & -22L \\
22L & 13L & 4L^2 & -3L^2 \\
-13L & -22L & -3L^2 & 4L^2
\end{bmatrix}
\begin{Bmatrix} \ddot{v}_1 \\ \ddot{v}_2 \\ \ddot{v}_3 \\ \ddot{v}_4 \end{Bmatrix}
\tag{10-29}
$$

When the mass coefficients of the elements of a structure have been evaluated, the mass matrix of the complete element assemblage can be developed by exactly the same type of superposition procedure as that described for developing the stiffness matrix from the element stiffness [Eq. (10-23)]. The resulting mass matrix in general will have the same configuration, that is, arrangement of nonzero terms, as the stiffness matrix.

The dynamic analysis of a consistent-mass system generally requires considerably more computational effort than a lumped-mass system does, for two reasons: (1) the lumped-mass matrix is diagonal, while the consistent-mass matrix has many off-diagonal terms (leading to what is called *mass coupling*); (2) the rotational degrees of freedom can be eliminated from a lumped-mass analysis (by static condensation, explained later), whereas all rotational and translational degrees of freedom must be included in a consistent-mass analysis.

Example E10-2.   The structure of Example E10-1, shown again in Fig. E10-2a, will be used to illustrate the evaluation of the structural mass

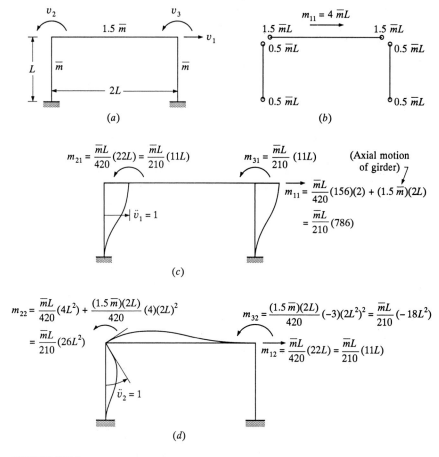

FIGURE E10-2

Analysis of lumped- and consistent-mass matrices: (*a*) uniform mass in members; (*b*) lumping of mass at member ends; (*c*) forces due to acceleration $\ddot{v}_1 = 1$ (consistent); (*d*) forces due to acceleration $\ddot{v}_2 = 1$ (consistent).

matrix. First the lumped-mass procedure is used: half the mass of each member is lumped at the ends of the members, as shown in Fig. E10-2*b*. The sum of the four contributions at the girder level then acts in the sidesway degree of freedom m_{11}; no mass coefficients are associated with the other degrees of freedom because these point masses have no rotational inertia.

The consistent-mass matrix is obtained by applying unit accelerations to each degree of freedom in succession while constraining the others and determining the resulting inertial forces from the coefficients of Eq. (10-29). Considering first the sidesway acceleration, as shown in Fig. E10-2*c*, it must be noted that the coefficients of Eq. (10-29) account only for the transverse inertia of the columns. The inertia of the girder due to the acceleration parallel to its axis must be added as a rigid-body mass $(3\overline{m}L)$, as shown.

The joint rotational acceleration induces only accelerations transverse to the members, and the resulting girder and column contributions are given by Eq. (10-29), as shown in Fig. E10-2d. The final mass matrices, from the lumped- and consistent-mass formulations, are

$$
\mathbf{m} = \frac{\overline{m}L}{210}
\begin{bmatrix}
840 & 0 & 0 \\
0 & 0 & 0 \\
0 & 0 & 0
\end{bmatrix}
\qquad
\mathbf{m} = \frac{\overline{m}L}{210}
\begin{bmatrix}
786 & 11L & 11L \\
11L & 26L^2 & -18L^2 \\
11L & -18L^2 & 26L^2
\end{bmatrix}
$$

$$
\text{Lumped} \qquad\qquad\qquad\qquad \text{Consistent}
$$

10-3 DAMPING PROPERTIES

If the various damping forces acting on a structure could be determined quantitatively, the finite-element concept could be used again to define the damping coefficients of the system. For example, the coefficient for any element might be of the form [compare with the corresponding term in the second of Eqs. (8-10) for the case where $i = j$]

$$
c_{ij} = \int_0^L c(x)\,\psi_i(x)\,\psi_j(x)\,dx \tag{10-30}
$$

in which $c(x)$ represents a distributed viscous-damping property. After the element damping influence coefficients were determined, the damping matrix of the complete structure could be obtained by a superposition process equivalent to the direct stiffness method. In practice, however, evaluation of the damping property $c(x)$ (or any other specific damping property) is impracticable. For this reason, the damping is generally expressed in terms of damping ratios established from experiments on similar structures rather than by means of an explicit damping matrix **c**. If an explicit expression of the damping matrix is needed, it generally will be computed from the specified damping ratios, as described in Chapter 12.

10-4 EXTERNAL LOADING

If the dynamic loading acting on a structure consists of concentrated forces corresponding with the displacement coordinates, the load vector of Eq. (9-2) can be written directly. In general, however, the load is applied at other points as well as the nodes and may include distributed loadings. In this case, the load terms in Eq. (9-2) are generalized forces associated with the corresponding displacement components.

Two procedures which can be applied in the evaluation of these generalized forces are described in the following paragraphs.

Static Resultants

The most direct means of determining the effective nodal forces generated by loads distributed between the nodes is by application of the principles of simple statics; in other words, the nodal forces are defined as a set of concentrated loads which are statically equivalent to the distributed loading. In effect, the analysis is made as though the actual loading were applied to the structure through a series of simple beams supported at the nodal points. The reactive forces developed at the supports then become the concentrated nodal forces acting on the structure. In this type of analysis it is evident that generalized forces will be developed corresponding only to the translational degrees of freedom; the rotational nodal forces will be zero unless external moments are applied directly to the joints.

Consistent Nodal Loads

A second procedure which can be used to evaluate nodal forces corresponding to all nodal degrees of freedom can be developed from the finite-element concept. This procedure employs the principle of virtual displacements in the same way as in evaluating the consistent-mass matrix, and the generalized nodal forces which are derived are called the *consistent* nodal loads. Consider the same beam segment as in the consistent-mass analysis but subjected to the externally applied dynamic loading shown in Fig. 10-8. When a virtual displacement δv_1 is applied, as shown in the sketch, and external and internal work are equated, the generalized force corresponding to v_1 is

$$p_1(t) = \int_0^L p(x,t)\,\psi_1(x)\,dx \qquad (10\text{-}31)$$

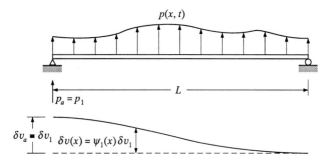

FIGURE 10-8
Virtual nodal translation of a laterally loaded beam.

Thus, the element generalized loads can be expressed in general as

$$p_i(t) = \int_0^L p(x,t)\,\psi_i(x)\,dx \tag{10-32}$$

The generalized load p_3 corresponding to $v_3 = \theta_a$ is an external moment applied at point a. The positive sense of the generalized loads corresponds to the positive coordinate axes. The equivalence of Eq. (10-32) to the corresponding term in the fourth of Eqs. (8-18) should be noted.

For the loads to be properly called consistent, the interpolation functions $\psi_i(x)$ used in Eq. (10-32) must be the same as those used to define the element stiffness coefficients. If linear interpolation functions

$$\psi_1(x) = 1 - \frac{x}{L} \qquad \psi_2(x) = \frac{x}{L} \tag{10-33}$$

were used instead, Eq. (10-32) would provide the static nodal resultants; in general this is the easiest way to compute the statically equivalent loads.

In some cases, the applied loading may have the special form

$$p(x,t) = \chi(x)\,f(t) \tag{10-34}$$

that is, the form of load distribution $\chi(x)$ does not change with time; only its amplitude changes. In this case the generalized force becomes

$$p_i(t) = f(t) \int_0^L \chi(x)\,\psi_i(x)\,dx \tag{10-34a}$$

which shows that the generalized force has the same time variation as the applied loading; the integral indicates the extent to which the load participates in developing the generalized force.

When the generalized forces acting on each element have been evaluated by Eq. (10-32), the total effective load acting at the nodes of the assembled structure can be obtained by a superposition procedure equivalent to the direct stiffness process.

10-5 GEOMETRIC STIFFNESS

Linear Approximation

The geometric-stiffness property represents the tendency toward buckling induced in a structure by axially directed load components; thus it depends not only on the configuration of the structure but also on its condition of loading. In this discussion, it is assumed that the forces tending to cause buckling are constant during the

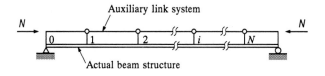

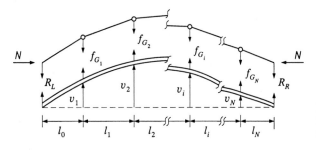

FIGURE 10-9
Idealization of axial-load
mechanism in beam.

dynamic loading; thus they are assumed to result from an independent static loading and are not significantly affected by the dynamic response of the structure. (When these forces do vary significantly with time, they result in a time-varying stiffness property, and analysis procedures based on superposition are not valid for such a nonlinear system.)

In general, two different levels of approximation can be established for the evaluation of geometric-stiffness properties, more or less in parallel with the preceding discussions for mass matrices and load vectors. The simplest approximation is conveniently derived from the physical model illustrated in Fig. 10-9, in which it is assumed that all axial forces are acting in an auxiliary structure consisting of rigid bar segments connected by hinges. The hinges are located at points where the transverse-displacement degrees of freedom of the actual beam are identified, and they are attached to the main beam by links which transmit transverse forces but no axial-force components.

When the actual beam is deflected by any form of loading, the auxiliary link system is forced to deflect equally, as shown in the sketch. As a result of these deflections and the axial forces in the auxiliary system, forces will be developed in the links coupling it to the main beam. In other words, the resistance of the main beam will be required to stabilize the auxiliary system.

The forces required for equilibrium in a typical segment i of the auxiliary system are shown in Fig. 10-10. The transverse force components f_{Gi} and f_{Gj} depend on the value of the axial-force component in the segment N_i and on the slope of the segment. They are assumed to be positive when they act in the positive-displacement

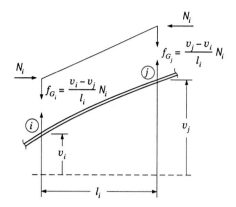

FIGURE 10-10
Equilibrium forces due to axial load in auxiliary link.

sense on the main beam. In matrix form, these forces may be expressed

$$\begin{Bmatrix} f_{Gi} \\ f_{Gj} \end{Bmatrix} = \frac{N_i}{l_i} \begin{bmatrix} 1 & -1 \\ -1 & 1 \end{bmatrix} \begin{Bmatrix} v_i \\ v_j \end{Bmatrix} \qquad (10\text{-}35)$$

By combining expressions of this type for all segments, the transverse forces due to axial loads can be written for the beam structure of Fig. 10-9 as follows:

$$\begin{Bmatrix} f_{G1} \\ f_{G2} \\ \cdot \\ f_{Gi} \\ \cdot \\ f_{GN} \end{Bmatrix} = \begin{bmatrix} \frac{N_0}{l_0} + \frac{N_1}{l_1} & -\frac{N_1}{l_1} & 0 & \cdots & 0 & \cdots \\ -\frac{N_1}{l_1} & \frac{N_1}{l_1} + \frac{N_2}{l_2} & -\frac{N_2}{l_2} & \cdots & 0 & \cdots \\ & & \cdots\cdots\cdots\cdots\cdots\cdots\cdots\cdots & & \\ 0 & 0 & 0 & \cdots & \frac{N_{i-1}}{l_{i-1}} + \frac{N_i}{l_i} & \cdots \\ & & \cdots\cdots\cdots\cdots\cdots\cdots\cdots\cdots & & \\ 0 & 0 & 0 & \cdots & 0 & \cdots \end{bmatrix} \begin{Bmatrix} v_1 \\ v_2 \\ \cdot \\ v_i \\ \cdot \\ v_N \end{Bmatrix}$$

$$(10\text{-}36)$$

in which it will be noted that magnitude of the axial force may change from segment to segment; for the loading shown in Fig. 10-9 all axial forces would be the same, and the term N could be factored from the matrix.

Symbolically, Eq. (10-36) may be expressed

$$\mathbf{f}_G = \mathbf{k}_G \mathbf{v} \qquad (10\text{-}37)$$

where the square symmetric matrix $\mathbf{k}_G$ is called the *geometric-stiffness matrix* of the structure. For this linear approximation of a beam system, the matrix has a tridiagonal form, as may be seen in Eq. (10-36), with contributions from two adjacent elements making up the diagonal terms and a single element providing each off-diagonal, or coupling, term.

Consistent Geometric Stiffness

The finite-element concept can be used to obtain a higher-order approximation of the geometric stiffness, as demonstrated for the other physical properties. Consider the same beam element used previously but now subjected to distributed axial loads which result in an arbitrary variation of axial force $N(x)$, as shown in Fig. 10-11. In the lower sketch, the beam is shown subjected to a unit rotation of the left end $v_3 = 1$. By definition, the nodal forces associated with this displacement component are the corresponding geometric-stiffness influence coefficients; for example, k_{G13} is the vertical force developed at the left end.

These coefficients may be evaluated by application of virtual displacements and equating the internal and external work components. The virtual displacement δv_1 required to determine k_{G13} is shown in the sketch. The external virtual work in this case is

$$W_E = f_{Ga} \, \delta v_a = k_{G13} \, \delta v_1 \qquad (10\text{-}38)$$

in which it will be noted that the positive sense of the geometric-stiffness coefficient corresponds with the positive displacements. To develop an expression for the internal virtual work, it is necessary to consider a differential segment of length dx, taken from the system of Fig. 10-11 and shown enlarged in Fig. 10-12. The work done in this segment by the axial force $N(x)$ during the virtual displacement is

$$dW_I = N(x) \, d(\delta e) \qquad (10\text{-}39)$$

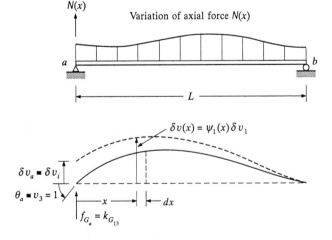

FIGURE 10-11
Axially loaded beam with real rotation and virtual translation of node.

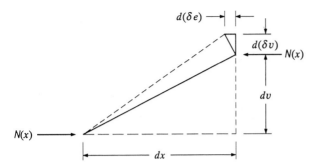

FIGURE 10-12
Differential segment of
deformed beam of Fig. 10-11.

where $d(\delta e)$ represents the distance the forces acting on this differential segment move toward each other. By similar triangles it may be seen in the sketch that

$$d(\delta e) = \frac{dv}{dx}\, d(\delta v)$$

Interchanging the differentiation and variation symbols on the right side gives

$$d(\delta e) = \frac{dv}{dx}\, \delta\!\left(\frac{dv}{dx}\, dx\right)$$

and hence introducing this into Eq. (10-39) leads to

$$dW_I = N(x)\,\frac{dv}{dx}\, \delta\!\left(\frac{dv}{dx}\right) dx$$

Expressing the lateral displacements in terms of interpolation functions and integrating finally gives

$$W_I = \delta v_1 \int_0^L N(x)\,\frac{d\psi_3(x)}{dx}\,\frac{d\psi_1(x)}{dx}\, dx \qquad (10\text{-}40)$$

Hence, by equating internal to external work, this geometric-stiffness coefficient is found to be

$$k_{G13} = \int_0^L N(x)\,\psi_3'(x)\,\psi_1'(x)\, dx \qquad (10\text{-}41)$$

or in general the element geometric-stiffness influence coefficients are

$$k_{Gij} = \int_0^L N(x)\,\psi_i'(x)\,\psi_j'(x)\, dx \qquad (10\text{-}42)$$

The equivalence of this equation to the last term in the third of Eqs. (8-18) should be noted; also its symmetry is apparent, that is, $k_{Gij} = k_{Gji}$.

If the hermitian interpolation functions [Eqs. (10-16)] are used in deriving the geometric-stiffness coefficients, the result is called the *consistent geometric-stiffness matrix.* In the special case where the axial force is constant through the length of the element, the consistent geometric-stiffness matrix is

$$
\begin{Bmatrix} f_{G1} \\ f_{G2} \\ f_{G3} \\ f_{G4} \end{Bmatrix} = \frac{N}{30L} \begin{bmatrix} 36 & -36 & 3L & 3L \\ -36 & 36 & -3L & -3L \\ 3L & -3L & 4L^2 & -L^2 \\ 3L & -3L & -L^2 & 4L^2 \end{bmatrix} \begin{Bmatrix} v_1 \\ v_2 \\ v_3 \\ v_4 \end{Bmatrix} \tag{10-43}
$$

On the other hand, if linear-interpolation functions [Eq. (10-33)] are used in Eq. (10-42), and if the axial force is constant through the element, its geometric stiffness will be as derived earlier in Eq. (10-35).

The assembly of the element geometric-stiffness coefficients to obtain the structure geometric-stiffness matrix can be carried out exactly as for the elastic-stiffness matrix, and the result will have a similar configuration (positions of the nonzero terms). Thus the consistent geometric-stiffness matrix represents rotational as well as translational degrees of freedom, whereas the linear approximation [Eq. (10-35)] is concerned only with the translational displacements. However, either type of relationship may be represented symbolically by Eq. (10-37).

10-6   CHOICE OF PROPERTY FORMULATION

In the preceding discussion, two different levels of approximation have been considered for the evaluation of the mass, elastic-stiffness, geometric-stiffness, and external-load properties: (1) an elementary approach taking account only of the translational degrees of freedom of the structure and (2) a "consistent" approach, which accounts for the rotational as well as translational displacements. The elementary approach is considerably easier to apply; not only are the element properties defined more simply but the number of coordinates to be considered in the analysis is much less for a given structural assemblage. In principle, the consistent approach should lead to greater accuracy in the results, but in practice the improvement is often slight. Apparently the rotational degrees of freedom are much less significant in the analysis than the translational terms. The principal advantage of the consistent approach is that all the energy contributions to the response of the structure are evaluated in a consistent manner, which makes it possible to draw certain conclusions regarding bounds on the vibration frequency; however, this advantage seldom outweighs the additional effort required.

The elementary lumped-mass approach presents a special problem when the elastic-stiffness matrix has been formulated by the finite-element approach or by any other procedure which includes the rotational degrees of freedom in the matrix. If the evaluation of all the other properties has excluded these degrees of freedom, it is necessary to exclude them also from the stiffness matrix before the equations of motion can be written.

The process of eliminating these unwanted degrees of freedom from the stiffness matrix is called *static condensation*. For the purpose of this discussion, assume that the rotational and translational degrees of freedom have been segregated, so that Eq. (9-5) can be written in partitioned form

$$
\begin{bmatrix} \mathbf{k}_{tt} & \mathbf{k}_{t\theta} \\ \mathbf{k}_{\theta t} & \mathbf{k}_{\theta\theta} \end{bmatrix} \begin{Bmatrix} \mathbf{v}_t \\ \mathbf{v}_\theta \end{Bmatrix} = \begin{Bmatrix} \mathbf{f}_{St} \\ \mathbf{f}_{S\theta} \end{Bmatrix} = \begin{Bmatrix} \mathbf{f}_{St} \\ \mathbf{0} \end{Bmatrix}
\tag{10-44}
$$

where $\mathbf{v}_t$ represents the translations and $\mathbf{v}_\theta$ the rotations, with corresponding subscripts to identify the submatrices of stiffness coefficients. Now, if none of the other force vectors acting in the structure include any rotational components, it is evident that the elastic rotational forces also must vanish, that is, $\mathbf{f}_{S\theta} = \mathbf{0}$. When this static constraint is introduced into Eq. (10-44), it is possible to express the rotational displacements in terms of the translations by means of the second submatrix equation, with the result

$$
\mathbf{v}_\theta = -\mathbf{k}_{\theta\theta}^{-1}\,\mathbf{k}_{\theta t}\,\mathbf{v}_t
\tag{10-45}
$$

Substituting this into the first of the submatrix equations of Eq. (10-44) leads to

$$
(\mathbf{k}_{tt} - \mathbf{k}_{t\theta}\,\mathbf{k}_{\theta\theta}^{-1}\,\mathbf{k}_{\theta t})\,\mathbf{v}_t = \mathbf{f}_{St}
$$

or

$$
\mathbf{k}_t\,\mathbf{v}_t = \mathbf{f}_{St}
\tag{10-46}
$$

where

$$
\mathbf{k}_t = \mathbf{k}_{tt} - \mathbf{k}_{t\theta}\,\mathbf{k}_{\theta\theta}^{-1}\,\mathbf{k}_{\theta t}
\tag{10-47}
$$

is the translational elastic stiffness. This stiffness matrix is suitable for use with the other elementary property expressions; in other words, it is the type of stiffness matrix implied in Fig. 10-2.

Example E10-3. To demonstrate the use of the static-condensation procedure, the two rotational degrees of freedom will be eliminated from the stiffness matrix evaluated in Example E10-1. The resulting condensed stiffness matrix will retain only the translational degree of freedom of the frame and thus will be compatible with the lumped-mass matrix derived in Example E10-2.

The stiffness submatrix associated with the rotational degrees of freedom of Example E10-1 is

$$\mathbf{k}_{\theta\theta} = \frac{2EI}{L^3}\begin{bmatrix} 6L^2 & 2L^2 \\ 2L^2 & 6L^2 \end{bmatrix} = \frac{4EI}{L}\begin{bmatrix} 3 & 1 \\ 1 & 3 \end{bmatrix}$$

and its inverse is

$$\mathbf{k}_{\theta\theta}{}^{-1} = \frac{L}{32EI}\begin{bmatrix} 3 & -1 \\ -1 & 3 \end{bmatrix}$$

When this is used in Eq. (10-45), the rotational degrees of freedom can be expressed in terms of the translation:

$$\begin{Bmatrix} v_2 \\ v_3 \end{Bmatrix} = -\frac{L}{32EI}\begin{bmatrix} 3 & -1 \\ -1 & 3 \end{bmatrix}\frac{2EI}{L^3}\begin{Bmatrix} 3L \\ 3L \end{Bmatrix} v_1 = -\frac{3}{8L}\begin{Bmatrix} 1 \\ 1 \end{Bmatrix} v_1$$

The condensed stiffness given by Eq. (10-47) then is

$$\mathbf{k}_t = \frac{2EI}{L^3}\left(12 - < 3L \quad 3L >\begin{Bmatrix} \frac{3}{8L} \\ \frac{3}{8L} \end{Bmatrix}\right) = \frac{2EI}{L^3}\frac{39}{4}$$

PROBLEMS

10-1. Using the hermitian polynomials, Eq. (10-16), as shape functions $\psi_i(x)$, evaluate by means of Eq. (10-21) the finite-element stiffness coefficient k_{23} for a beam having the following variation of flexural rigidity: $EI(x) = EI_0(1 + x/L)$.

10-2. Making use of Eq. (10-28), compute the consistent mass coefficient m_{23} for a beam with the following nonuniform mass distribution: $m(x) = \overline{m}(1 + x/L)$. Assume the shape functions of Eq. (10-16) and evaluate the integral by Simpson's rule, dividing the beam into four segments of equal length.

10-3. The distributed load applied to a certain beam may be expressed as

$$p(x,t) = \overline{p}\left(2 + \frac{x}{L}\right)\sin \overline{\omega}t$$

Making use of Eq. (10-34a), write an expression for the time variation of the consistent load component $p_2(t)$ based on the shape function of Eq. (10-16).

10-4. Using Eq. (10-42), evaluate the consistent geometric stiffness coefficient k_{G24} for a beam having the following distribution of axial force: $N(x) = N_0(2 - x/L)$. Make use of the shape functions of Eq. (10-16) and evaluate the integral by Simpson's rule using $\Delta x = L/4$.

10-5. The plane frame of Fig. P10-1 is formed of uniform members, with the properties of each as shown. Assemble the stiffness matrix defined for the three DOFs indicated, evaluating the member stiffness coefficients by means of Eq. (10-22).

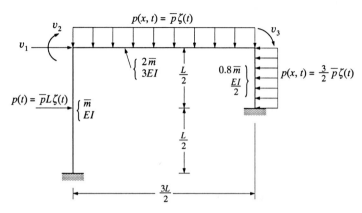

FIGURE P10-1

10-6. Assemble the mass matrix for the structure of Prob. 10-5, evaluating the individual member mass coefficients by means of Eq. (10-29).

10-7. Assemble the load vector for the structure of Prob. 10-5, evaluating the individual member nodal loads by means of Eq. (10-32).

10-8. For a plane frame of the same general form as that of Prob. 10-5, but having different member lengths and physical properties, the stiffness and lumped mass matrices are as follows:

$$\mathbf{k} = \frac{EI}{L^3} \begin{bmatrix} 20 & -10L & -5L \\ -10L & 15L^2 & -8L^2 \\ -5L & -8L^2 & 12L^2 \end{bmatrix} \qquad \mathbf{m} = \overline{m}L \begin{bmatrix} 30 & 0 & 0 \\ 0 & 0 & 0 \\ 0 & 0 & 0 \end{bmatrix}$$

(*a*) Using static condensation, eliminate the two rotational degrees of freedom from the stiffness matrix.

(*b*) Using the condensed stiffness matrix, write the SDOF equation of motion for undamped free vibrations.

CHAPTER 11

UNDAMPED
FREE
VIBRATIONS

11-1 ANALYSIS OF VIBRATION FREQUENCIES

The equations of motion for a freely vibrating undamped system can be obtained by omitting the damping matrix and applied-loads vector from Eq. (9-13):

$$\mathbf{m}\,\ddot{\mathbf{v}} + \mathbf{k}\,\mathbf{v} = \mathbf{0} \tag{11-1}$$

in which $\mathbf{0}$ is a zero vector. The problem of vibration analysis consists of determining the conditions under which the equilibrium condition expressed by Eq. (11-1) will be satisfied. By analogy with the behavior of SDOF systems, it will be assumed that the free-vibration motion is simple harmonic, which may be expressed for a MDOF system as

$$\mathbf{v}(t) = \hat{\mathbf{v}}\,\sin(\omega t + \theta) \tag{11-2}$$

In this expression $\hat{\mathbf{v}}$ represents the shape of the system (which does not change with time; only the amplitude varies) and θ is a phase angle. When the second time derivative of Eq. (11-2) is taken, the accelerations in free vibration are

$$\ddot{\mathbf{v}} = -\omega^2\,\hat{\mathbf{v}}\,\sin(\omega t + \theta) = -\omega^2\,\mathbf{v} \tag{11-3}$$

Substituting Eqs. (11-2) and (11-3) into Eq. (11-1) gives

$$-\omega^2\,\mathbf{m}\,\hat{\mathbf{v}}\,\sin(\omega t + \theta) + \mathbf{k}\,\hat{\mathbf{v}}\,\sin(\omega t + \theta) = \mathbf{0}$$

which (since the sine term is arbitrary and may be omitted) may be written

$$[\mathbf{k} - \omega^2\,\mathbf{m}]\,\hat{\mathbf{v}} = \mathbf{0} \tag{11-4}$$

Equation (11-4) is one way of expressing what is called an *eigenvalue* or *character-istic value* problem. The quantities ω^2 are the eigenvalues or characteristic values indicating the square of the free-vibration frequencies, while the corresponding displacement vectors $\hat{v}$ express the corresponding shapes of the vibrating system — known as the eigenvectors or mode shapes. Now it can be shown by Cramer's rule that the solution of this set of simultaneous equations is of the form

$$\hat{v} = \frac{0}{\|k - \omega^2\, m\|} \tag{11-5}$$

Hence a nontrivial solution is possible only when the denominator determinant vanishes. In other words, finite-amplitude free vibrations are possible only when

$$\|k - \omega^2\, m\| = 0 \tag{11-6}$$

Equation (11-6) is called the *frequency equation* of the system. Expanding the determinant will give an algebraic equation of the Nth degree in the frequency parameter ω^2 for a system having N degrees of freedom. The N roots of this equation $(\omega_1^2, \omega_2^2, \omega_3^2, \ldots, \omega_N^2)$ represent the frequencies of the N modes of vibration which are possible in the system. The mode having the lowest frequency is called the first mode, the next higher frequency is the second mode, etc. The vector made up of the entire set of modal frequencies, arranged in sequence, will be called the *frequency vector ω*:

$$\omega = \left\{ \begin{array}{c} \omega_1 \\ \omega_2 \\ \omega_3 \\ \vdots \\ \omega_N \end{array} \right\} \tag{11-7}$$

It can be shown that for the real, symmetric, positive definite mass and stiffness matrices which pertain to stable structural systems, all roots of the frequency equation will be real and positive.

Example E11-1. The analysis of vibration frequencies by the solution of the determinantal equation (11-6) will be demonstrated with reference to the structure of Fig. E11-1, the same frame for which an approximation of the fundamental frequency was obtained by the Rayleigh method in Example E8-3. The stiffness matrix for this frame can be determined by applying a unit displacement to each story in succession and evaluating the resulting story forces, as shown in the figure. Because the girders are assumed to be rigid, the story

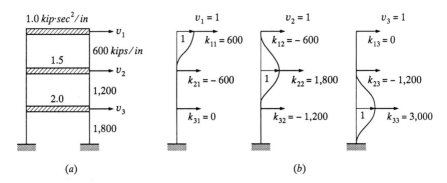

FIGURE E11-1
Frame used in example of vibration analysis: (a) structural system; (b) stiffness influence coefficients.

forces can easily be determined here by merely adding the sidesway stiffnesses of the appropriate stories.

The mass and stiffness matrices for this frame thus are

$$\mathbf{m} = (1 \ kip \cdot sec^2/in) \begin{bmatrix} 1.0 & 0 & 0 \\ 0 & 1.5 & 0 \\ 0 & 0 & 2.0 \end{bmatrix}$$

$$\mathbf{k} = (600 \ kips/in) \begin{bmatrix} 1 & -1 & 0 \\ -1 & 3 & -2 \\ 0 & -2 & 5 \end{bmatrix}$$

from which

$$\mathbf{k} - \omega^2\mathbf{m} = (600 \ kips/in) \begin{bmatrix} 1 - B & -1 & 0 \\ -1 & 3 - 1.5B & -2 \\ 0 & -2 & 5 - 2B \end{bmatrix} \quad (a)$$

where

$$B \equiv \frac{\omega^2}{600}$$

The frequencies of the frame are given by the condition that $\triangle = 0$, where $\triangle$ is the determinant of the square matrix in Eq. (a). Evaluating this determinant, simplifying, and equating to zero leads to the cubic equation

$$B^3 - 5.5 \, B^2 + 7.5 \, B - 2 = 0$$

The three roots of this equation may be solved directly or obtained by trial and error; their values are $B_1 = 0.3515$, $B_2 = 1.6066$, $B_3 = 3.5420$. Hence the frequencies are

$$
\begin{Bmatrix} \omega_1^2 \\ \omega_2^2 \\ \omega_3^2 \end{Bmatrix} = \begin{Bmatrix} 210.88 \\ 963.96 \\ 2,125.20 \end{Bmatrix} \qquad \begin{Bmatrix} \omega_1 \\ \omega_2 \\ \omega_3 \end{Bmatrix} = \begin{Bmatrix} 14.522 \\ 31.048 \\ 46.100 \end{Bmatrix} \; rad/sec
$$

11-2   ANALYSIS OF VIBRATION MODE SHAPES

When the frequencies of vibration have been determined from Eq. (11-6), the equations of motion [Eq. (11-4)] may be expressed as

$$
\widetilde{\mathbf{E}}^{(n)} \, \hat{\mathbf{v}}_n = \mathbf{0} \tag{11-8}
$$

in which

$$
\widetilde{\mathbf{E}}^{(n)} = \mathbf{k} - \omega_n^2 \, \mathbf{m} \tag{11-9}
$$

Thus $\widetilde{\mathbf{E}}^{(n)}$ represents the matrix obtained by subtracting $\omega_n^2 \, \mathbf{m}$ from the stiffness matrix; since it depends on the frequency, it is different for each mode. Equation (11-8) is satisfied identically because the frequencies were evaluated from this condition; therefore the *amplitude* of the vibrations is indeterminate. However, the *shape* of the vibrating system can be determined by solving for all the displacements in terms of any one coordinate.

For this purpose it will be assumed that the first element of the displacement vector has a unit amplitude; that is,

$$
\begin{Bmatrix} \hat{v}_{1n} \\ \hat{v}_{2n} \\ \hat{v}_{3n} \\ \vdots \\ \hat{v}_{Nn} \end{Bmatrix} = \begin{Bmatrix} 1 \\ \hat{v}_{2n} \\ \hat{v}_{3n} \\ \vdots \\ \hat{v}_{Nn} \end{Bmatrix} \tag{11-10}
$$

In expanded form, Eq. (11-8) may then be written

$$
\begin{bmatrix} e_{11}^{(n)} & | & e_{12}^{(n)} & e_{13}^{(n)} & \cdots & e_{1N}^{(n)} \\ \hline e_{21}^{(n)} & | & e_{22}^{(n)} & e_{23}^{(n)} & \cdots & e_{2N}^{(n)} \\ e_{31}^{(n)} & | & e_{32}^{(n)} & e_{33}^{(n)} & \cdots & e_{3N}^{(n)} \\ \cdots & & \cdots & \cdots & & \cdots \\ e_{N1}^{(n)} & | & e_{N2}^{(n)} & e_{N3}^{(n)} & \cdots & e_{NN}^{(n)} \end{bmatrix} \begin{Bmatrix} 1 \\ \hline \hat{v}_{2n} \\ \hat{v}_{3n} \\ \cdots \\ \hat{v}_{Nn} \end{Bmatrix} = \begin{Bmatrix} 0 \\ \hline 0 \\ 0 \\ \cdots \\ 0 \end{Bmatrix} \tag{11-11}
$$

in which partitioning is indicated to correspond with the as yet unknown displacement amplitudes. For convenience, Eq. (11-11) will be expressed symbolically as

$$
\begin{bmatrix} e_{11}^{(n)} & \widetilde{\mathbf{E}}_{10}^{(n)} \\ \widetilde{\mathbf{E}}_{01}^{(n)} & \widetilde{\mathbf{E}}_{00}^{(n)} \end{bmatrix} \begin{Bmatrix} 1 \\ \hat{\mathbf{v}}_{0n} \end{Bmatrix} = \begin{Bmatrix} 0 \\ \mathbf{0} \end{Bmatrix} \tag{11-11a}
$$

from which

$$
\widetilde{\mathbf{E}}_{01}^{(n)} + \widetilde{\mathbf{E}}_{00}^{(n)} \, \hat{\mathbf{v}}_{0n} = \mathbf{0} \tag{11-12}
$$

as well as

$$
e_{11}^{(n)} + \widetilde{\mathbf{E}}_{10}^{(n)} \, \hat{\mathbf{v}}_{0n} = 0 \tag{11-13}
$$

Equation (11-12) can be solved simultaneously for the displacement amplitudes

$$
\hat{\mathbf{v}}_{0n} = -\big(\widetilde{\mathbf{E}}_{00}^{(n)}\big)^{-1} \, \widetilde{\mathbf{E}}_{01}^{(n)} \tag{11-14}
$$

but Eq. (11-13) is redundant; the redundancy corresponds to the fact that Eq. (11-4) is satisfied identically. The displacement vector obtained in Eq. (11-14) must satisfy Eq. (11-13), however, and this condition provides a useful check on the accuracy of the solution. It should be noted that it is not always wise to let the first element of the displacement vector be unity; numerical accuracy will be improved if the unit element is associated with one of the larger displacement amplitudes. The same solution process can be employed in any case, however, by merely rearranging the order of the rows and columns of $\widetilde{\mathbf{E}}^{(n)}$ appropriately.

The displacement amplitudes obtained from Eq. (11-14) together with the unit amplitude of the first component constitute the displacement vector associated with the nth mode of vibration. For convenience the vector is usually expressed in dimensionless form by dividing all the components by one reference component (usually the largest). The resulting vector is called the nth mode shape $\boldsymbol{\phi}_n$; thus

$$
\boldsymbol{\phi}_n = \begin{Bmatrix} \phi_{1n} \\ \phi_{2n} \\ \phi_{3n} \\ \vdots \\ \phi_{Nn} \end{Bmatrix} \equiv \frac{1}{\hat{v}_{kn}} \begin{Bmatrix} 1 \\ \hat{v}_{2n} \\ \hat{v}_{3n} \\ \vdots \\ \hat{v}_{Nn} \end{Bmatrix} \tag{11-15}
$$

in which $\hat{v}_{kn}$ is the reference component, taken as the first component here.

The shape of each of the N modes of vibration can be found by this same process; the square matrix made up of the N mode shapes will be represented by $\boldsymbol{\Phi}$;

thus

$$
\boldsymbol{\Phi} = [\boldsymbol{\phi}_1 \quad \boldsymbol{\phi}_2 \quad \boldsymbol{\phi}_3 \quad \cdots \quad \boldsymbol{\phi}_N] =
\begin{bmatrix}
\phi_{11} & \phi_{12} & \cdots & \phi_{1N} \\
\phi_{21} & \phi_{22} & \cdots & \phi_{2N} \\
\phi_{31} & \phi_{32} & \cdots & \phi_{3N} \\
\phi_{41} & \phi_{42} & \cdots & \phi_{4N} \\
\cdots & \cdots & \cdots & \cdots \\
\phi_{N1} & \phi_{N2} & \cdots & \phi_{NN}
\end{bmatrix}
\tag{11-16}
$$

As was noted above, the vibration analysis of a structural system is a form of characteristic-value, or eigenvalue, problem of matrix-algebra theory. A brief discussion of the reduction of the equation of motion in free vibrations to standard eigenproblem form is presented in Chapter 14.

Example E11-2.   The analysis of vibration mode shapes by means of Eq. (11-14) will be demonstrated by applying it to the structure of Fig. E11-1. The vibration matrix for this structure was derived in Example E11-1, and when the second and third rows of this matrix are used, Eq. (11-14) may be expressed as

$$
\begin{Bmatrix} \phi_{2n} \\ \phi_{3n} \end{Bmatrix} = -
\begin{bmatrix} 3 - 1.5B_n & -2 \\ -2 & 5 - 2B_n \end{bmatrix}^{-1}
\begin{Bmatrix} -1 \\ 0 \end{Bmatrix}
$$

Thus the mode shapes can be found by introducing the values of B_n computed in Example E11-1, inverting, and multiplying as indicated. The calculations for the three mode shapes of this system follow.

Mode 1:

$$B_1 = 0.35$$

$$
\widetilde{\mathbf{E}}_{00}^{(1)} =
\begin{bmatrix} 2.4728 & -2 \\ -2 & 4.2971 \end{bmatrix}
\qquad
(\widetilde{\mathbf{E}}_{00}^{(1)})^{-1} = \frac{1}{6.6259}
\begin{bmatrix} 4.2971 & 2 \\ 2 & 2.4728 \end{bmatrix}
$$

$$
\begin{Bmatrix} \phi_{21} \\ \phi_{31} \end{Bmatrix} = \frac{1}{6.6259}
\begin{Bmatrix} 4.2971 \\ 2.000 \end{Bmatrix} =
\begin{Bmatrix} 0.64853 \\ 0.30185 \end{Bmatrix}
$$

Mode 2:

$B_2 = 1.61$

$$\widetilde{\mathbf{E}}_{00}^{(2)} = \begin{bmatrix} 0.5901 & -2 \\ -2 & 1.7868 \end{bmatrix} \quad (\widetilde{\mathbf{E}}_{00}^{(2)})^{-1} = -\frac{1}{2.9456} \begin{bmatrix} 1.7868 & 2 \\ 2 & 0.5901 \end{bmatrix}$$

$$\begin{Bmatrix} \phi_{22} \\ \phi_{32} \end{Bmatrix} = -\frac{1}{2.9456} \begin{Bmatrix} 1.7868 \\ 2.000 \end{Bmatrix} = -\begin{Bmatrix} 0.6066 \\ 0.6790 \end{Bmatrix}$$

Mode 3:

$B_3 = 3.54$

$$\widetilde{\mathbf{E}}_{00}^{(3)} = \begin{bmatrix} -2.3130 & -2 \\ -2 & -2.0840 \end{bmatrix} \quad (\widetilde{\mathbf{E}}_{00}^{(3)})^{-1} = \frac{1}{0.8203} \begin{bmatrix} -2.0840 & 2 \\ 2 & -2.3130 \end{bmatrix}$$

$$\begin{Bmatrix} \phi_{23} \\ \phi_{33} \end{Bmatrix} = \frac{1}{0.8203} \begin{Bmatrix} -2.0840 \\ 2.00 \end{Bmatrix} = -\begin{Bmatrix} -2.5405 \\ 2.4382 \end{Bmatrix}$$

Of course, the displacement of mass a in each mode has been assumed to be unity. The three mode shapes for this structure are sketched in Fig. E11-2.

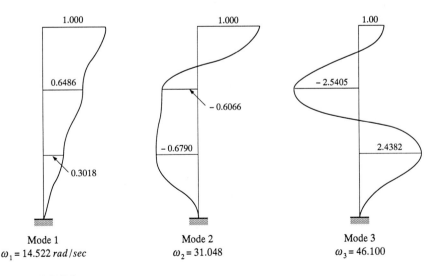

FIGURE E11-2
Vibration properties for the frame of Fig. E12-1.

11-3 FLEXIBILITY FORMULATION OF VIBRATION ANALYSIS

The preceding discussion of vibration analysis was based on a stiffness-matrix formulation of the equations of motion. In many cases it may be more convenient to express the elastic properties of the structure by means of the flexibility matrix rather than the stiffness matrix. Equation (11-4) can be converted readily into the flexibility form by multiplying by $(1/\omega^2)\tilde{\mathbf{f}}$, where the flexibility matrix $\tilde{\mathbf{f}}$ is the inverse of the stiffness matrix $\mathbf{k}$. The result is

$$\left[\frac{1}{\omega^2}\mathbf{I} - \tilde{\mathbf{f}}\mathbf{m}\right]\hat{\mathbf{v}} = \mathbf{0} \tag{11-17}$$

in which $\mathbf{I}$ represents an identity matrix of order N. As before, this set of homogeneous equations can have a nonzero solution only if the determinant of the square matrix vanishes; thus the frequency equation in this case is

$$\left\|\frac{1}{\omega^2}\mathbf{I} - \tilde{\mathbf{f}}\mathbf{m}\right\| = 0 \tag{11-18}$$

Evaluation of the roots of this equation can be carried out as for Eq. (11-6); similarly the mode shape corresponding to each frequency can be evaluated as before. The only basic difference between the solutions is that the roots of Eq. (11-18) represent the reciprocals of the frequency-squared values rather than the frequency squared.

It should be noted that the matrix product $\tilde{\mathbf{f}}\mathbf{m}$ in Eq. (11-18) is not symmetrical, in general, even though the mass and flexibility matrices are both symmetric. In digital-computer analyses of the eigenvalue problem it may be desirable to retain the symmetry of the matrix being solved; techniques for obtaining a symmetric form of the flexibility eigenvalue equation are presented in Chapter 13 (Section 13-6).

11-4 INFLUENCE OF AXIAL FORCES

Free Vibrations

The vibration mode shapes and frequencies of a structure which is subjected to a constant axial-force loading can be evaluated in exactly the same way as for a system without axial-force effects. In this case the geometric stiffness must be included in the equations of motion; thus Eq. (11-1) takes the form

$$\mathbf{m}\ddot{\mathbf{v}} + \mathbf{k}\mathbf{v} - \mathbf{k}_G\mathbf{v} = \mathbf{m}\ddot{\mathbf{v}} + \bar{\mathbf{k}}\mathbf{v} = \mathbf{0} \tag{11-19}$$

and the frequency equation becomes

$$\|\bar{\mathbf{k}} - \omega^2\mathbf{m}\| = 0 \tag{11-20}$$

In the mode-shape and frequency analysis, it is necessary only to substitute the combined stiffness $\bar{\mathbf{k}}$ for the elastic stiffness $\mathbf{k}$; otherwise the analysis is as described before. For any given condition of axial loading, the geometric-stiffness matrix (and therefore the combined stiffness) can be evaluated numerically. The effect of a compressive axial-force system is to reduce the effective stiffness of the structure, thus the frequencies of vibration are reduced; in addition the mode shapes generally are modified by the axial loads.

Buckling Load

If the frequency of vibration is zero, the inertial forces in Eq. (11-19) vanish and the equations of equilibrium become

$$\mathbf{k}\,\mathbf{v} - \mathbf{k}_G\,\mathbf{v} = \mathbf{0} \qquad (11\text{-}21)$$

The conditions under which a nonzero displacement vector is possible in this case constitute the static buckling condition; in other words, a useful definition of buckling is the condition in which the vibration frequency becomes zero. In order to evaluate the critical buckling loading of the structure, it is convenient to express the geometric stiffness in terms of a reference loading multiplied by a load factor λ_G. Thus

$$\mathbf{k}_G = \lambda_G\,\mathbf{k}_{G0} \qquad (11\text{-}22)$$

in which the element geometric-stiffness coefficients from which $\mathbf{k}_{G0}$ is formed are given by

$$k_{G_{ij}} = \int_0^L N_0(x)\,\psi_i'(x)\,\psi_j'(x)\,dx \qquad (11\text{-}23)$$

In this expression $N_0(x)$ is the reference axial loading in the element. The loading of the structure therefore is proportional to the parameter λ_G; its relative distribution, however, is constant. Substituting Eq. (11-22) into Eq. (11-21) leads to the eigenvalue equation

$$[\mathbf{k} - \lambda_G\,\mathbf{k}_{G0}]\,\hat{\mathbf{v}} = \mathbf{0} \qquad (11\text{-}24)$$

A nontrivial solution of this set of equations can be obtained only under the condition

$$\|\mathbf{k} - \lambda_G\,\mathbf{k}_{G0}\| = 0 \qquad (11\text{-}25)$$

which represents the buckling condition for the structure. The roots of this equation represent the values of the axial-load factor λ_G at which buckling will occur. The buckling mode shapes can be evaluated exactly like the vibration mode shapes. In practice, only the first buckling load and mode shape have any real significance; buckling in the higher modes generally is of little practical importance because the system will have failed when the load exceeds the lowest critical load.

Buckling with Harmonic Excitation

Although the concept has found little application in practice, it is at least of academic interest to note that a range of different "buckling" loads can be defined for a harmonically excited structure, just as a range of different vibration frequencies exists in an axially loaded structure. Suppose the structure is subjected to a harmonic excitation at the frequency $\bar{\omega}$; that is, assume that an applied-load vector of the following form is acting:

$$\mathbf{p}(t) = \mathbf{p}_0 \, \sin \bar{\omega} t \tag{11-26}$$

where $\bar{\omega}$ is the applied-load frequency. The undamped equation of equilibrium in this case becomes [from Eq. (9-18)]

$$\mathbf{m}\,\ddot{\mathbf{v}} + \mathbf{k}\,\mathbf{v} - \mathbf{k}_G\,\mathbf{v} = \mathbf{p}_0 \, \sin \bar{\omega} t \tag{11-27}$$

The steady-state response will then take place at the applied-load frequency,

$$\mathbf{v}(t) = \hat{\mathbf{v}} \, \sin \bar{\omega} t \tag{11-28a}$$

and the accelerations become

$$\ddot{\mathbf{v}}(t) = -\bar{\omega}^2 \, \hat{\mathbf{v}} \, \sin \bar{\omega} t \tag{11-28b}$$

Introducing Eqs. (11-28) into Eq. (11-27) gives (after dividing by $\sin \bar{\omega} t$):

$$-\bar{\omega}^2 \, \mathbf{m}\,\hat{\mathbf{v}} + \mathbf{k}\,\hat{\mathbf{v}} - \mathbf{k}_G\,\hat{\mathbf{v}} = \mathbf{p}_0 \tag{11-29}$$

The symbol $\bar{\bar{\mathbf{k}}}$ will be used to represent the *dynamic stiffness* of the system, where $\bar{\bar{\mathbf{k}}}$ is defined as

$$\bar{\bar{\mathbf{k}}} \equiv \mathbf{k} - \bar{\omega}^2 \, \mathbf{m} \tag{11-30a}$$

Substituting this into Eq. (11-29) and expressing the geometric stiffness in terms of the load factor λ_G leads to

$$[\bar{\bar{\mathbf{k}}} - \lambda_G \, \mathbf{k}_{G0}] \, \hat{\mathbf{v}} = \mathbf{p}_0 \tag{11-30b}$$

If the amplitude of the applied-load vector in this equation is allowed to approach zero, it is apparent by comparison with Eq. (11-5) that a nonzero response is still possible if the determinant of the square matrix is zero. Thus the condition

$$\|\bar{\bar{\mathbf{k}}} - \lambda_G \, \mathbf{k}_{G0}\| = 0 \tag{11-31}$$

defines the buckling condition for the harmonically excited structure.

When the applied load is allowed to vanish, Eq. (11-30b) may be written

$$[\mathbf{k} - \omega^2 \, \mathbf{m} - \lambda_G \, \mathbf{k}_{G0}] \, \hat{\mathbf{v}} = 0 \tag{11-32}$$

Now it is apparent that an infinite variety of combinations of buckling loads λ_G and frequencies ω^2 will satisfy this eigenvalue equation. For any given "buckling" load specified by a prescribed λ_G, the corresponding frequency of vibration can be found from Eq. (11-20). Similarly, for any given frequency of vibration ω^2, the corresponding buckling loading is defined by Eq. (11-31). It is interesting to note that a zero-axial-load condition causes "buckling" at the unstressed natural-vibration frequency according to this definition.

11-5 ORTHOGONALITY CONDITIONS

Basic Conditions

The free-vibration mode shapes ϕ_n have certain special properties which are very useful in structural-dynamics analyses. These properties, which are called *orthogonality relationships*, can be demonstrated by application of Betti's law. Consider, for example, two different modes of vibration of a structural system, as shown in Fig. 11-1. For convenience, the structure has been shown as a lumped-mass system, but the following analysis applies equally well to a consistent-mass idealization.

The equations of motion for a system in free vibration, Eq. (11-4), may be rewritten

$$\mathbf{k}\,\hat{\mathbf{v}}_n = \omega_n^2\,\mathbf{m}\,\hat{\mathbf{v}}_n \tag{11-33}$$

in which the right-hand side represents the applied-inertia-load vector $-\mathbf{f}_I$ and the left-hand side is the elastic-resisting-force vector $\mathbf{f}_S$. Thus the free-vibration motion may be considered to involve deflections produced by inertial forces acting as applied loads, as shown in Fig. 11-1. On this basis, the two vibration modes shown in the figure represent two different applied-load systems and their resulting displacements; consequently Betti's law may be applied as follows:

$$-\mathbf{f}_{Im}^T\,\hat{\mathbf{v}}_n = -\mathbf{f}_{In}^T\,\hat{\mathbf{v}}_m$$

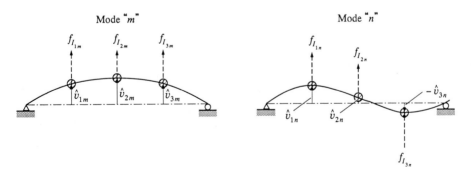

FIGURE 11-1
Vibration mode shapes and resulting inertial forces.

Introducing the inertial-force expression used in Eq. (11-33) gives

$$\omega_m^2 \, \hat{\mathbf{v}}_m^T \, \mathbf{m} \, \hat{\mathbf{v}}_n = \omega_n^2 \, \hat{\mathbf{v}}_n^T \, \mathbf{m} \, \hat{\mathbf{v}}_m \qquad (11\text{-}34)$$

where the rules of transposing matrix products have been observed, taking account of the symmetry of $\mathbf{m}$. When it is noted that the matrix products in Eq. (11-34) are scalars and can be transposed arbitrarily, it is evident that the equation may be written

$$(\omega_m^2 - \omega_n^2) \, \hat{\mathbf{v}}_m^T \, \mathbf{m} \, \hat{\mathbf{v}}_n = 0 \qquad (11\text{-}35)$$

Subject to the condition that the two mode frequencies are not the same, this gives the first orthogonality condition

$$\hat{\mathbf{v}}_m^T \, \mathbf{m} \, \hat{\mathbf{v}}_n = 0 \qquad \omega_m \neq \omega_n \qquad (11\text{-}36)$$

A second orthogonality condition can be derived directly from this by premultiplying Eq. (11-33) by $\hat{\mathbf{v}}_m^T$; thus

$$\hat{\mathbf{v}}_m^T \, k \, \hat{\mathbf{v}}_n = \omega_n^2 \, \hat{\mathbf{v}}_m^T \, \mathbf{m} \, \mathbf{v}_n$$

When Eq. (11-36) is applied to the right-hand side, it is clear that

$$\hat{\mathbf{v}}_m^T \, k \, \hat{\mathbf{v}}_n = \mathbf{0} \qquad \omega_m \neq \omega_n \qquad (11\text{-}37)$$

which shows that the vibrating shapes are orthogonal with respect to the stiffness matrix as well as with respect to the mass.

In general, it is convenient to express the orthogonality conditions in terms of the dimensionless mode-shape vectors $\boldsymbol{\phi}_n$ rather than for the arbitrary amplitudes $\hat{\mathbf{v}}_n$. Equations (11-36) and (11-37) are obviously equally valid when divided by any reference displacement value; thus the orthogonality conditions become

$$\boldsymbol{\phi}_m^T \, \mathbf{m} \, \boldsymbol{\phi}_n = 0 \qquad m \neq n \qquad (11\text{-}38a)$$

$$\boldsymbol{\phi}_m^T \, \mathbf{k} \, \boldsymbol{\phi}_n = 0 \qquad m \neq n \qquad (11\text{-}38b)$$

For systems in which no two modes have the same frequency, the orthogonality conditions apply to any two different modes, as indicated in Eqs. (11-38); they do *not* apply to two modes having the same frequency.

Additional Relationships

A complete family of additional orthogonality relationships can be derived directly from Eq. (11-33) by successive multiplications. In order to obtain the results in terms of mode-shape vectors, it is convenient to divide both sides of Eq. (11-33) by a reference amplitude, which gives the equivalent expression

$$\mathbf{k}\, \boldsymbol{\phi}_n = \omega_n^2\, \mathbf{m}\, \boldsymbol{\phi}_n \tag{11-39}$$

Premultiplying this by $\boldsymbol{\phi}_m^T\, \mathbf{k}\, \mathbf{m}^{-1}$ leads to

$$\boldsymbol{\phi}_m^T\, \mathbf{k}\, \mathbf{m}^{-1}\, \mathbf{k}\, \boldsymbol{\phi}_n = \omega_n^2\, \boldsymbol{\phi}_m^T\, \mathbf{k}\, \boldsymbol{\phi}_n$$

from which [using Eq. (11-38b)]

$$\boldsymbol{\phi}_m^T\, \mathbf{k}\, \mathbf{m}^{-1}\, \mathbf{k}\, \boldsymbol{\phi}_n = 0 \tag{11-40}$$

Premultiplying Eq. (11-39) by $\boldsymbol{\phi}_m^T \mathbf{k}\mathbf{m}^{-1}\mathbf{k}\mathbf{m}^{-1}$ leads to

$$\boldsymbol{\phi}_m^T \mathbf{k}\mathbf{m}^{-1}\mathbf{k}\mathbf{m}^{-1}\mathbf{k}\boldsymbol{\phi}_n = \omega_n^2 \boldsymbol{\phi}_m^T \mathbf{k}\mathbf{m}^{-1}\mathbf{k}\boldsymbol{\phi}_n$$

from which [using Eq. (11-40)]

$$\boldsymbol{\phi}_m^T \mathbf{k}\mathbf{m}^{-1}\mathbf{k}\mathbf{m}^{-1}\mathbf{k}\boldsymbol{\phi}_n = 0 \tag{11-41}$$

By proceeding similarly any number of orthogonality relationships of this type can be developed.

The first of a second series of relationships can be derived by premultiplying Eq. (11-39) by $(1/\omega^2)\boldsymbol{\phi}_m^T\mathbf{m}\widetilde{\mathbf{f}}$, with the result

$$\frac{1}{\omega_n^2}\, \boldsymbol{\phi}_m^T\, \mathbf{m}\, \boldsymbol{\phi}_n = \boldsymbol{\phi}_m^T\, \mathbf{m}\, \widetilde{\mathbf{f}}\, \mathbf{m}\, \boldsymbol{\phi}_n$$

from which [using Eq. (11-38a)]

$$\boldsymbol{\phi}_m^T\, \mathbf{m}\, \widetilde{\mathbf{f}}\, \mathbf{m}\, \boldsymbol{\phi}_n = 0 \tag{11-42}$$

Premultiplying Eq. (11-39) by $(1/\omega_n^2)\boldsymbol{\phi}_m^T\mathbf{m}\widetilde{\mathbf{f}}\mathbf{m}\widetilde{\mathbf{f}}$ then gives

$$\frac{1}{\omega_n^2}\, \boldsymbol{\phi}_m^T\, \mathbf{m}\widetilde{\mathbf{f}}\mathbf{m}\boldsymbol{\phi}_n = \boldsymbol{\phi}_m^T\, \mathbf{m}\widetilde{\mathbf{f}}\mathbf{m}\widetilde{\mathbf{f}}\mathbf{m}\, \boldsymbol{\phi}_n = 0 \tag{11-43}$$

Again the series can be extended indefinitely by similar operations.

Both complete families of orthogonality relationships, including the two basic relationships, can be compactly expressed as

$$\boldsymbol{\phi}_m^T\, \mathbf{m}\, [\mathbf{m}^{-1}\quad \mathbf{k}]^b\, \boldsymbol{\phi}_n = 0 \qquad -\infty < b < \infty \tag{11-44}$$

The two basic relationships, Eqs. (11-38a) and (11-38b), are given by exponents $b = 0$ and $b = +1$ in Eq. (11-44), respectively.

Normalizing

It was noted earlier that the vibration mode amplitudes obtained from the eigenproblem solution are arbitrary; any amplitude will satisfy the basic frequency equation (11-4), and only the resulting shapes are uniquely defined. In the analysis process described above, the amplitude of one degree of freedom (the first, actually) has been set to unity, and the other displacements have been determined relative to this reference value. This is called *normalizing* the mode shapes with respect to the specified reference coordinate.

Other normalizing procedures also are frequently used; e.g., in many computer programs the shapes are normalized relative to the maximum displacement value in each mode rather than with respect to any particular coordinate. Thus, the maximum value in each modal vector is unity, which provides convenient numbers for use in subsequent calculations. The normalizing procedure most often used in computer programs for structural-vibration analysis, however, involves adjusting each modal amplitude to the amplitude $\hat{\phi}_n$ which satisfies the condition

$$\hat{\phi}_n^T \, \mathbf{m} \, \hat{\phi}_n = 1 \tag{11-45}$$

This can be accomplished by computing the scalar factor

$$\hat{\mathbf{v}}_n^T \, \mathbf{m} \, \hat{\mathbf{v}}_m = \hat{M}_n \tag{11-46}$$

where $\hat{\mathbf{v}}_n$ represents an arbitrarily determined modal amplitude, and then computing the normalized mode shapes as follows:

$$\hat{\phi}_n = \hat{\mathbf{v}}_n \, \hat{M}_n^{-1/2} \tag{11-47}$$

By simple substitution it is easy to show that this gives the required result.

A consequence of this type of normalizing, together with the modal orthogonality relationships relative to the mass matrix [Eq. (11-38a)], is that

$$\hat{\mathbf{\Phi}}^T \, \mathbf{m}\hat{\mathbf{\Phi}} = \mathbf{I} \tag{11-48}$$

where $\hat{\mathbf{\Phi}}$ is the complete set of N normalized mode shapes and $\mathbf{I}$ is an $N \times N$ identity matrix. The mode shapes normalized in this fashion are said to be *orthonormal* relative to the mass matrix. Although the use of the orthonormalized mode shapes is convenient in the development of digital-computer programs for structural-dynamic analyses, it has no particular merit when the calculations are to be done by hand. For that reason, no specific normalizing procedure is assumed in the discussions which follow.

Example E11-3. The modal orthogonality properties and the orthonormalizing procedure will be demonstrated with the mode shapes calculated in Example E11-2. The normalizing factors obtained by applying Eq. (11-46) to these shapes are given in the lumped-mass case by

$$\hat{M}_n = \sum_{i=1}^{3} \phi_{in}^2 \, m_i$$

Their values are

$$\hat{M}_1 = 1.8131 \qquad \hat{M}_2 = 2.4740 \qquad \hat{M}_3 = 22.596$$

Dividing the respective mode shapes by the square root of these factors then leads to the orthonormalized mode-shape matrix

$$\hat{\Phi} = \begin{bmatrix} 0.74265 & 0.63577 & 0.21037 \\ 0.48164 & -0.38566 & -0.53475 \\ 0.22417 & -0.43168 & 0.51323 \end{bmatrix}$$

Finally, performing the multiplication of Eq. (11-48) gives the identity matrix, which serves as a check on the calculations:

$$\hat{\Phi}^T \, \mathbf{m} \, \hat{\Phi} = \mathbf{I}$$

PROBLEMS

11-1. The properties of a three-story shear building in which it is assumed that the entire mass is lumped in the rigid girders are shown in Fig. P8-8.

(a) By solving the determinantal equation, evaluate the undamped vibration frequencies of this structure.

(b) On the basis of the computed frequencies, evaluate the corresponding vibration mode shapes, normalizing them to unity at the top story.

(c) Demonstrate numerically that the computed mode shapes satisfy the orthogonality conditions with respect to mass and stiffness.

11-2. Repeat Prob. 11-1 for the mass and stiffness properties given in Prob. 8-13.

11-3. Two identical uniform beams are arranged, as shown in isometric view in Fig. P11-1, to support a piece of equipment weighing 3 *kips*. The flexural rigidity and weight per foot of the beams are shown. Assuming the distributed mass of each beam to be lumped half at its center and 1/4 at each end, compute the two frequencies and mode shapes in terms of the coordinates v_1 and v_2. [*Hint*: Note that the central deflection of a uniform beam with central load is $PL^3/48EI$. Use the flexibility formulation of the determinantal solution method, Eq. (11-18).]

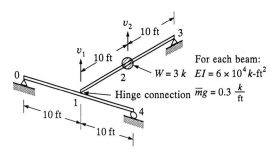

For each beam:
$W = 3\,k$ $EI = 6 \times 10^4 \, k\text{-ft}^2$
$\overline{m}g = 0.3 \dfrac{k}{\text{ft}}$

FIGURE P11-1

11-4. A rigid rectangular slab is supported by three columns rigidly attached to the slab and at the base (as shown in Fig. P11-2).

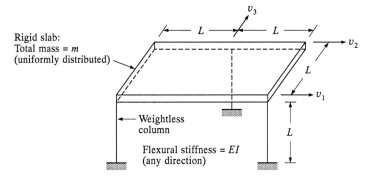

Rigid slab:
Total mass = m
(uniformly distributed)

Weightless column

Flexural stiffness = EI
(any direction)

FIGURE P11-2

   (*a*) Evaluate the mass and stiffness matrices for this system (in terms of m, EI, and L), considering the three displacement coordinates shown. (*Hint:* Apply a unit displacement or acceleration corresponding to each coordinate and evaluate the forces acting in each coordinate required for equilibrium.)
   (*b*) Compute the frequencies and mode shapes of this sytem, normalizing the mode shapes so that either v_2 or v_3 is unity.

11-5. Repeat Prob. 11-4 using the rotation and translation (parallel and perpendicular to the symmetry axis) of the center of mass as the displacement coordinates.

11-6. A rigid bar is supported by a wieghtless column as shown in Fig. P11-3.

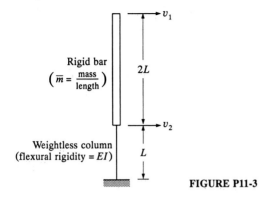

FIGURE P11-3

(*a*) Evaluate the mass and flexibility matrices of this system defined for the two coordinates shown.

(*b*) Compute the two mode shapes and frequencies of the system. Normalize the mode shapes so that the generalized mass for each mode is unity, i.e., so that $M_1 = M_2 = 1$.

CHAPTER
12

ANALYSIS OF DYNAMIC RESPONSE — USING SUPERPOSITION

12-1 NORMAL COORDINATES

In the preceding discussion of an arbitrary N-DOF linear system, the displaced position was defined by the N components in the vector $\mathbf{v}$. However, for the purpose of dynamic-response analysis, it is often advantageous to express this position in terms of the free-vibration mode shapes. These shapes constitute N independent displacement patterns, the amplitudes of which may serve as generalized coordinates to express any set of displacements. The mode shapes thus serve the same purpose as the trigonometric functions in a Fourier series, and they are used for the same reasons; because: (1) they possess orthogonality properties and (2) they are efficient in the sense that they usually can describe all N displacements with sufficient accuracy employing only a few shapes.

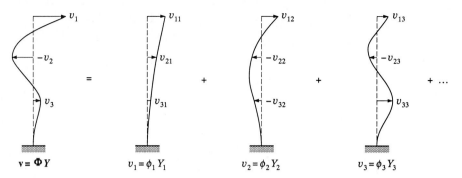

FIGURE 12-1
Representing deflections as sum of modal components.

Consider, for example, the cantilever column shown in Fig. 12-1, for which the deflected shape is expressed in terms of translational displacements at three levels. Any displacement vector $\mathbf{v}$ (static or dynamic) for this structure can be developed by superposing suitable amplitudes of the normal modes as shown. For any modal component $\mathbf{v}_n$, the displacements are given by the product of the mode-shape vector $\boldsymbol{\phi}_n$ and the modal amplitude Y_n; thus

$$\mathbf{v}_n = \boldsymbol{\phi}_n\, Y_n \tag{12-1}$$

The total displacement vector $\mathbf{v}$ is then obtained by summing the modal vectors as expressed by

$$\mathbf{v} = \boldsymbol{\phi}_1\, Y_1 + \boldsymbol{\phi}_2\, Y_2 + \cdots + \boldsymbol{\phi}_N\, Y_N = \sum_{n=1}^{N} \boldsymbol{\phi}_n\, Y_n \tag{12-2}$$

or, in matrix notation,

$$\mathbf{v} = \boldsymbol{\Phi}\, Y \tag{12-3}$$

In this equation, it is apparent that the $N \times N$ mode-shape matrix $\boldsymbol{\Phi}$ serves to transform the generalized coordinate vector Y to the geometric coordinate vector $\mathbf{v}$. The generalized components in vector Y are called the normal coordinates of the structure.

Because the mode-shape matrix consists of N independent modal vectors, $\boldsymbol{\Phi} = [\boldsymbol{\phi}_1 \ \boldsymbol{\phi}_2 \ \cdots \ \boldsymbol{\phi}_N]$, it is nonsingular and can be inverted. Thus, it is always possible to solve Eq. (12-3) directly for the normal-coordinate amplitudes in Y which are associated with any given displacement vector $\mathbf{v}$. In doing so, however, it is unnecessary to solve a set of simultaneous equations, due to the orthogonality property of the mode shapes. To evaluate any arbitrary normal coordinate, Y_n for example, premultiply Eq. (12-2) by $\boldsymbol{\phi}_n^T \mathbf{m}$ to obtain

$$\boldsymbol{\phi}_n^T \mathbf{m}\, \mathbf{v} = \boldsymbol{\phi}_n^T \mathbf{m}\, \boldsymbol{\phi}_1\, Y_1 + \boldsymbol{\phi}_n^T \mathbf{m}\, \boldsymbol{\phi}_2\, Y_2 + \cdots + \boldsymbol{\phi}_n^T \mathbf{m}\, \boldsymbol{\phi}_N\, Y_N \tag{12-4}$$

Because of the orthogonality property with respect to mass, i.e., $\boldsymbol{\phi}_n^T \mathbf{m} \boldsymbol{\phi}_m = 0$ for $m \neq n$, all terms on the right hand side of this equation vanish, except for the term containing $\boldsymbol{\phi}_n^T \mathbf{m} \boldsymbol{\phi}_n$, leaving

$$\boldsymbol{\phi}_n^T \mathbf{m} \mathbf{v} = \boldsymbol{\phi}_n^T \mathbf{m} \boldsymbol{\phi}_n Y_n \tag{12-5}$$

from which

$$Y_n = \frac{\boldsymbol{\phi}_n^T \mathbf{m} \mathbf{v}}{\boldsymbol{\phi}_n^T \mathbf{m} \boldsymbol{\phi}_n} \qquad n = 1, 2, \cdots, N \tag{12-6}$$

If vector $\mathbf{v}$ is time dependent, the Y_n coordinates will also be time dependent; in this case, taking the time derivative of Eq. (12-6) yields

$$\dot{Y}_n(t) = \frac{\boldsymbol{\phi}_n^T \mathbf{m} \dot{\mathbf{v}}(t)}{\boldsymbol{\phi}_n^T \mathbf{m} \boldsymbol{\phi}_n} \tag{12-7}$$

Note that the above procedure is equivalent to that used to evaluate the coefficients in the Fourier series given by Eqs. (4-3).

12-2   UNCOUPLED EQUATIONS OF MOTION: UNDAMPED

The orthogonality properties of the normal modes will now be used to simplify the equations of motion of the MDOF system. In general form these equations are given by Eq. (9-13) [or its equivalent Eq. (9-19) if axial forces are present]; for the undamped system they become

$$\mathbf{m} \, \ddot{\mathbf{v}}(t) + \mathbf{k} \, \mathbf{v}(t) = \mathbf{p}(t) \tag{12-8}$$

Introducing Eq. (12-3) and its second time derivative $\ddot{\mathbf{v}} = \boldsymbol{\Phi} \, \ddot{\mathbf{Y}}$ (noting that the mode shapes do not change with time) leads to

$$\mathbf{m} \, \boldsymbol{\Phi} \, \ddot{\mathbf{Y}}(t) + \mathbf{k} \, \boldsymbol{\Phi} \, \mathbf{Y}(t) = \mathbf{p}(t) \tag{12-9}$$

If Eq. (12-9) is premultiplied by the transpose of the nth mode-shape vector $\boldsymbol{\phi}_n^T$, it becomes

$$\boldsymbol{\phi}_n^T \mathbf{m} \, \boldsymbol{\Phi} \, \ddot{\mathbf{Y}}(t) + \boldsymbol{\phi}_n^T \mathbf{k} \, \boldsymbol{\Phi} \, \mathbf{Y}(t) = \boldsymbol{\phi}_n^T \mathbf{p}(t) \tag{12-10}$$

but if the two terms on the left hand side are expanded as shown in Eq. (12-4), all terms except the nth will vanish because of the mode-shape orthogonality properties; hence the result is

$$\boldsymbol{\phi}_n^T \mathbf{m} \boldsymbol{\phi}_n \, \ddot{Y}_n(t) + \boldsymbol{\phi}_n^T \mathbf{k} \boldsymbol{\phi}_n \, Y_n(t) = \boldsymbol{\phi}_n^T \mathbf{p}(t) \tag{12-11}$$

Now new symbols will be defined as follows:

$$M_n \equiv \boldsymbol{\phi}_n^T \mathbf{m} \boldsymbol{\phi}_n \tag{12-12a}$$

$$K_n \equiv \boldsymbol{\phi}_n^T \mathbf{k} \boldsymbol{\phi}_n \tag{12-12b}$$

$$P_n(t) \equiv \boldsymbol{\phi}_n^T \mathbf{p}(t) \tag{12-12c}$$

which are called the normal-coordinate generalized mass, generalized stiffness, and generalized load for mode n, respectively. With them Eq. (12-11) can be written

$$M_n \ddot{Y}_n(t) + K_n Y_n(t) = P_n(t) \tag{12-13}$$

which is a SDOF equation of motion for mode n. If Eq. (11-39), $\mathbf{k}\boldsymbol{\phi}_n = \omega_n^2 \mathbf{m}\boldsymbol{\phi}_n$, is multiplied on both sides by $\boldsymbol{\phi}_n^T$, the generalized stiffness for mode m is related to the generalized mass by the frequency of vibration

$$\boldsymbol{\phi}_n^T \mathbf{k} \boldsymbol{\phi}_n = \omega_n^2 \boldsymbol{\phi}_n^T \mathbf{m} \boldsymbol{\phi}_n$$

or

$$K_n = \omega_n^2 M_n \tag{12-12d}$$

(Capital letters are used to denote all normal-coordinate properties.)

The procedure described above can be used to obtain an independent SDOF equation for each mode of vibration of the undamped structure. Thus the use of the normal coordinates serves to transform the equations of motion from a set of N simultaneous differential equations, which are coupled by the off-diagonal terms in the mass and stiffness matrices, to a set of N independent normal-coordinate equations. The dynamic response therefore can be obtained by solving separately for the response of each normal (modal) coordinate and then superposing these by Eq. (12-3) to obtain the response in the original geometric coordinates. This procedure is called the *mode-superposition method*, or more precisely the *mode displacement superposition method*.

12-3   UNCOUPLED EQUATIONS OF MOTION: VISCOUS DAMPING

Now it is of interest to examine the conditions under which this normal-coordinate transformation will also serve to uncouple the damped equations of motion. These equations [Eq. (9-13)] are

$$\mathbf{m} \ddot{\mathbf{v}}(t) + \mathbf{c} \dot{\mathbf{v}}(t) + \mathbf{k} \mathbf{v}(t) = \mathbf{p}(t)$$

Introducing the normal-coordinate expression of Eq. (12-3) and its time derivatives and premultiplying by the transpose of the nth mode-shape vector $\boldsymbol{\phi}_n^T$ leads to

$$\boldsymbol{\phi}_n^T \mathbf{m}\boldsymbol{\Phi}\ddot{Y}(t) + \boldsymbol{\phi}_n^T \mathbf{c}\boldsymbol{\Phi}\dot{Y}(t) + \boldsymbol{\phi}_n^T \mathbf{k}\boldsymbol{\Phi}Y(t) = \boldsymbol{\phi}_n^T \mathbf{p}(t) \tag{12-14}$$

It was noted above that the orthogonality conditions

$$\boldsymbol{\phi}_m^T \mathbf{m} \boldsymbol{\phi}_n = 0$$
$$\boldsymbol{\phi}_m^T \mathbf{k} \boldsymbol{\phi}_n = 0$$
$$m \neq n$$

cause all components except the nth-mode term in the mass and stiffness expressions of Eq. (12-14) to vanish. A similar reduction will apply to the damping expression if it is *assumed* that the corresponding orthogonality condition applies to the damping matrix; that is, assume that

$$\phi_m^T \, \mathbf{c} \, \phi_n = 0 \qquad m \neq n \tag{12-15}$$

In this case Eq. (12-14) may be written

$$M_n \, \ddot{Y}_n(t) + C_n \, \dot{Y}_n(t) + K_n \, Y_n(t) = P_n(t) \tag{12-14a}$$

where the definitions of modal coordinate mass, stiffness, and load have been introduced from Eq. (12-12) and where the modal coordinate viscous damping coefficient has been defined similarly

$$C_n = \phi_n^T \, \mathbf{c} \, \phi_n \tag{12-15a}$$

If Eq. (12-14a) is divided by the generalized mass, this modal equation of motion may be expressed in alternative form:

$$\ddot{Y}_n(t) + 2 \, \xi_n \, \omega_n \, \dot{Y}_n(t) + \omega_n^2 \, Y_n(t) = \frac{P_n(t)}{M_n} \tag{12-14b}$$

where Eq. (12-12d) has been used to rewrite the stiffness term and where the second term on the left hand side represents a definition of the modal viscous damping ratio

$$\xi_n = \frac{C_n}{2 \, \omega_n \, M_n} \tag{12-15b}$$

As was noted earlier, it generally is more convenient and physically reasonable to define the damping of a MDOF system using the damping ratio for each mode in this way rather than to evaluate the coefficients of the damping matrix $\mathbf{c}$ because the modal damping ratios ξ_n can be determined experimentally or estimated with adequate precision in many cases.

12-4 RESPONSE ANALYSIS BY MODE DISPLACEMENT SUPERPOSITION

Viscous Damping

The normal coordinate transformation was used in Section 12-3 to convert the N coupled linear damped equations of motion

$$\mathbf{m} \, \ddot{\mathbf{v}}(t) + \mathbf{c} \, \dot{\mathbf{v}}(t) + \mathbf{k} \, \mathbf{v}(t) = \mathbf{p}(t) \tag{12-16}$$

to a set of N uncoupled equations given by

$$\ddot{Y}_n(t) + 2 \, \xi_n \, \omega_n \, \dot{Y}_n(t) + \omega_n^2 \, Y_n(t) = \frac{P_n(t)}{M_n} \qquad n = 1, 2, \cdots, N \tag{12-17}$$

in which

$$M_n = \boldsymbol{\phi}_n^T \, \mathbf{m} \, \boldsymbol{\phi}_n \qquad\qquad P_n(t) = \boldsymbol{\phi}_n^T \, \mathbf{p}(t) \qquad\qquad (12\text{-}18)$$

To proceed with the solution of these uncoupled equations of motion, one must first solve the eigenvalue problem

$$\left[\mathbf{k} - \omega^2 \, \mathbf{m}\right] \hat{\mathbf{v}} = \mathbf{0} \qquad\qquad (12\text{-}19)$$

to obtain the required mode shapes ϕ_n $(n = 1, 2, \cdots)$ and corresponding frequencies ω_n. The modal damping ratios ξ_n are usually assumed based on experimental evidence.

The total response of the MDOF system now can be obtained by solving the N uncoupled modal equations and superposing their effects, as indicated by Eq. (12-3). Each of Eqs. (12-17) is a standard SDOF equation of motion and can be solved in either the time domain or the frequency domain by the procedures described in Chapter 6. The time-domain solution is expressed by the Duhamel integral [see Eq. (6-7)]

$$Y_n(t) = \frac{1}{M_n \omega_n} \int_0^t P_n(\tau) \, \exp\left[-\xi_n \omega_n (t - \tau)\right] \, \sin \omega_{Dn}(t - \tau) \, d\tau \qquad (12\text{-}20)$$

which also may be written in standard convolution integral form

$$Y_n(t) = \int_0^t P_n(\tau) \, h_n(t - \tau) \, d\tau \qquad\qquad (12\text{-}21)$$

in which

$$h_n(t - \tau) = \frac{1}{M_n \omega_{Dn}} \, \sin \omega_{Dn}(t - \tau) \, \exp\left[-\xi_n \omega_n (t - \tau)\right] \quad 0 < \xi_n < 1 \quad (12\text{-}22)$$

is the unit-impulse response function, similar to Eq. (6-8).

In the frequency domain, the response is obtained similarly to Eq. (6-24) from

$$Y_n(t) = \frac{1}{2\pi} \int_{-\infty}^{\infty} H_n(i\overline{\omega}) \, P_n(i\overline{\omega}) \, \exp i\overline{\omega}t \, d\overline{\omega} \qquad\qquad (12\text{-}23)$$

In this equation, the complex load function $P_n(i\overline{\omega})$ is the Fourier transform of the modal loading $P_n(t)$, and similar to Eq. (6-23) it is given by

$$P_n(i\overline{\omega}) = \int_{-\infty}^{\infty} P_n(t) \, \exp(-i\overline{\omega}t) \, dt \qquad\qquad (12\text{-}24)$$

Also in Eq. (12-23), the complex frequency response function, $H_n(i\overline{\omega})$, may be expressed similarly to Eq. (6-25) as follows:

$$\begin{aligned} H_n(i\overline{\omega}) &= \frac{1}{\omega_n^2 M_n} \left[\frac{1}{(1 - \beta_n^2) + i(2\xi_n \beta_n)}\right] \\ &= \frac{1}{\omega_n^2 M_n} \left[\frac{(1 - \beta_n^2) - i(2\xi_n \beta_n)}{(1 - \beta_n^2)^2 + (2\xi_n \beta_n)^2}\right] \qquad \xi_n \geq 0 \qquad (12\text{-}25) \end{aligned}$$

In these functions, $\beta_n \equiv \overline{\omega}/\omega_n$ and $\omega_{Dn} = \omega_n\sqrt{1 - \xi_n^2}$. As indicated previously by Eqs. (6-53) and (6-54), $h_n(t)$ and $H_n(i\overline{\omega})$ are Fourier transform pairs. Solving Eq. (12-20) or (12-23) for any general modal loading yields the modal response $Y_n(t)$ for $t \geq 0$, assuming zero initial conditions, i.e., $Y_n(0) = \dot{Y}_n(0) = 0$. Should the initial conditions not equal zero, the damped free-vibration response [Eq. (3-31)]

$$Y_n(t) = \left[Y_n(0)\,\cos\omega_{Dn}t + \left(\frac{\dot{Y}_n(0) + Y_n(0)\xi_n\omega_n}{\omega_{Dn}} \right)\,\sin\omega_{Dn}t \right]\,\exp(-\xi_n\omega_n t)$$

$$(12\text{-}26)$$

must be added to the forced-vibration response given by Eqs. (12-20) or (12-23). The initial conditions $Y_n(0)$ and $\dot{Y}_n(0)$ in this equation are determined from $\mathbf{v}(0)$ and $\dot{\mathbf{v}}(0)$ using Eqs. (12-6) and (12-7) in the forms

$$Y_n(0) = \frac{\boldsymbol{\phi}_n^T\,\mathbf{m}\,\mathbf{v}(0)}{\boldsymbol{\phi}_n^T\,\mathbf{m}\,\boldsymbol{\phi}_n} \qquad (12\text{-}27)$$

$$\dot{Y}_n(0) = \frac{\boldsymbol{\phi}_n^T\,\mathbf{m}\,\dot{\mathbf{v}}(0)}{\boldsymbol{\phi}_n^T\,\mathbf{m}\,\boldsymbol{\phi}_n} \qquad (12\text{-}28)$$

Having generated the total response for each mode $Y_n(t)$ using either Eq. (12-20) or Eq. (12-23) and Eq. (12-26), the displacements expressed in the geometric coordinates can be obtained using Eq. (12-2), i.e.,

$$\mathbf{v}(t) = \boldsymbol{\phi}_1\,Y_1(t) + \boldsymbol{\phi}_2\,Y_2(t) + \cdots + \boldsymbol{\phi}_N\,Y_N(t) \qquad (12\text{-}29)$$

which superposes the separate modal displacement contributions; hence, the commonly referred to name mode superposition method. It should be noted that for most types of loadings the displacement contributions generally are greatest for the lower modes and tend to decrease for the higher modes. Consequently, it usually is not necessary to include all the higher modes of vibration in the superposition process; the series can be truncated when the response has been obtained to any desired degree of accuracy. Moreover, it should be kept in mind that the mathematical idealization of any complex structural system also tends to be less reliable in predicting the higher modes of vibration; for this reason, too, it is well to limit the number of modes considered in a dynamic-response analysis.

The displacement time-histories in vector $\mathbf{v}(t)$ may be considered to be the basic measure of a structure's overall response to dynamic loading. In general, other response parameters such as stresses or forces developed in various structural components can be evaluated directly from the displacements. For example, the elastic forces $\mathbf{f}_S$ which resist the deformation of the structure are given directly by

$$\mathbf{f}_S(t) = \mathbf{k}\,\mathbf{v}(t) = \mathbf{k}\,\boldsymbol{\Phi}\,Y(t) \qquad (12\text{-}30)$$

An alternative expression for the elastic forces may be useful in cases where the frequencies and mode shapes have been determined from the flexibility form of the eigenvalue equation [Eq. (11-17)]. Writing Eq. (12-30) in terms of the modal contributions

$$\mathbf{f}_S(t) = \mathbf{k}\,\boldsymbol{\phi}_1\,Y_1(t) + \mathbf{k}\,\boldsymbol{\phi}_2\,Y_2(t) + \mathbf{k}\,\boldsymbol{\phi}_3\,Y_3(t) + \cdots \qquad (12\text{-}31)$$

and substituting Eq. (11-39) leads to

$$\mathbf{f}_S(t) = \omega_1^2\,\mathbf{m}\,\boldsymbol{\phi}_1\,Y_1(t) + \omega_2^2\,\mathbf{m}\,\boldsymbol{\phi}_2\,Y_2(t) + \omega_3^2\,\mathbf{m}\,\boldsymbol{\phi}_3\,Y_3(t) + \cdots \qquad (12\text{-}32)$$

Writing this series in matrix form gives

$$\mathbf{f}_S(t) = \mathbf{m}\,\boldsymbol{\Phi}\,\left\{\omega_n^2\,Y_n(t)\right\} \qquad (12\text{-}33)$$

where $\left\{\omega_n^2\,Y_n(t)\right\}$ represents a vector of modal amplitudes each multiplied by the square of its modal frequency.

In Eq. (12-33), the elastic force associated with each modal component has been replaced by an equivalent modal inertial-force expression. The equivalence of these expressions was demonstrated from the equations of free-vibration equilibrium [Eq. (11-39)]; however, it should be noted that this substitution is valid at any time, even for a static analysis.

Because each modal contribution is multiplied by the square of the modal frequency in Eq. (12-33), it is evident that the higher modes are of greater significance in defining the forces in the structure than they are in the displacements. Consequently, it will be necessary to include more modal components to define the forces to any desired degree of accuracy than to define the displacements.

Example E12-1.   Various aspects of the mode-superposition procedure will be illustrated by reference to the three-story frame structure of Example E11-1 (Fig. E11-1). For convenience, the physical and vibration properties of the structure are summarized here:

$$\mathbf{m} = \begin{bmatrix} 1.0 & 0 & 0 \\ 0 & 1.5 & 0 \\ 0 & 0 & 2.0 \end{bmatrix} \; kips \cdot sec^2/in$$

$$\mathbf{k} = 600 \begin{bmatrix} 1 & -1 & 0 \\ -1 & 3 & -2 \\ 0 & -2 & 5 \end{bmatrix} \; kips/in$$

(a)

$$\boldsymbol{\omega} = \begin{Bmatrix} 14.522 \\ 31.048 \\ 46.100 \end{Bmatrix} \; rad/sec$$

$$\boldsymbol{\Phi} = \begin{bmatrix} 1.0000 & 1.0000 & 1.0000 \\ 0.6486 & -0.6066 & -2.5405 \\ 0.3018 & -0.6790 & 2.4382 \end{bmatrix}$$

Now the free vibrations which would result from the following arbitrary initial conditions will be evaluated, assuming the structure is undamped:

$$\mathbf{v}(t = 0) = \begin{Bmatrix} 0.5 \\ 0.4 \\ 0.3 \end{Bmatrix} \; in \qquad \dot{\mathbf{v}}(t = 0) = \begin{Bmatrix} 0 \\ 9 \\ 0 \end{Bmatrix} \; in/sec \qquad (b)$$

The modal coordinate amplitudes associated with the initial displacements are given by equations of the form of Eq. (12-5); writing the complete set of equations in matrix form leads to

$$\mathbf{Y}(t = 0) = \mathbf{M}^{-1} \, \boldsymbol{\Phi}^T \, \mathbf{m} \, \mathbf{v}(t = 0) \tag{c}$$

[which also could be derived by combining Eqs. (12-31) and (12-2)]. From the mass and mode-shape data given above, the generalized-mass matrix is

$$\mathbf{M} = \begin{bmatrix} 1.8131 & 0 & 0 \\ 0 & 2.4740 & 0 \\ 0 & 0 & 22.596 \end{bmatrix} \tag{d}$$

(where it will be noted that these terms are the same as the normalizing factors computed in Example E11-3). Multiplying the reciprocals of these terms by the mode-shape transpose and the mass matrix then gives

$$\boldsymbol{M}^{-1}\,\boldsymbol{\Phi}^T\,\mathbf{m} = \begin{bmatrix} 0.5515 & 0.5365 & 0.3330 \\ 0.4042 & -0.3678 & -0.5489 \\ 0.0443 & -0.1687 & 0.2159 \end{bmatrix}$$

Hence the initial modal coordinate amplitudes are given as the product of this matrix and the specified initial displacements

$$\boldsymbol{Y}(t=0) = \boldsymbol{M}^{-1}\,\boldsymbol{\Phi}^T\,\mathbf{m}\begin{Bmatrix} 0.5 \\ 0.4 \\ 0.3 \end{Bmatrix} = \begin{Bmatrix} 0.5903 \\ -0.1097 \\ 0.0194 \end{Bmatrix} in \qquad (e)$$

and the modal coordinate velocities result from multiplying this by the given initial velocities

$$\dot{\boldsymbol{Y}}(t=0) = \boldsymbol{M}^{-1}\,\boldsymbol{\Phi}^T\,\mathbf{m}\begin{Bmatrix} 0 \\ 9 \\ 0 \end{Bmatrix} = \begin{Bmatrix} 4.829 \\ -3.310 \\ -1.519 \end{Bmatrix} in/sec \qquad (f)$$

The free-vibration response of each modal coordinate of this undamped structure is of the form

$$Y_n(t) = \frac{\dot{Y}_n(t=0)}{\omega_n}\,\sin\omega_n t + Y_n(t=0)\,\cos\omega_n t \qquad (g)$$

Hence using the modal-coordinate initial conditions computed above, together with the modal frequencies, gives

$$\begin{Bmatrix} Y_1(t) \\ Y_2(t) \\ Y_3(t) \end{Bmatrix} = \begin{Bmatrix} 0.3325\,\sin\omega_1 t \\ -0.1066\,\sin\omega_2 t \\ -0.0329\,\sin\omega_3 t \end{Bmatrix} + \begin{Bmatrix} 0.5903\,\cos\omega_1 t \\ -0.1097\,\cos\omega_2 t \\ 0.0194\,\cos\omega_3 t \end{Bmatrix} \qquad (h)$$

From these modal results the free-vibration motion of each story could be obtained finally from the superposition relationship $\mathbf{v}(t) = \boldsymbol{\Phi}\boldsymbol{Y}(t)$. It is evident that the motion of each story includes contributions at each of the natural frequencies of the structure.

Example E12-2. As another demonstration of mode superposition, the response of the structure of Fig. E11-1 to a sine-pulse blast-pressure load is calculated. For this purpose, the load may be expressed as

$$
\begin{Bmatrix} p_1(t) \\ p_2(t) \\ p_3(t) \end{Bmatrix} = \begin{Bmatrix} 1 \\ 2 \\ 2 \end{Bmatrix} (500 \ kips) \cos \frac{\pi}{t_1} t \quad \text{where} \quad \begin{array}{c} t_1 = 0.02 \ sec \\ -\dfrac{t_1}{2} < t < \dfrac{t_1}{2} \end{array}
$$

With this short-duration loading, it may be assumed that the response in each mode is a free vibration with its amplitude defined by the sine-pulse displacement-response spectrum of Fig. 5-6. Thus during the early response era, when the effect of damping may be neglected, the modal response may be expressed as

$$
Y_n(t) = D_n \frac{P_{0n}}{K_n} \sin \omega_n t \tag{a}
$$

in which

$$
K_n = M_n \omega_n^2 \qquad P_{0n} = \phi_n^T \begin{Bmatrix} 1 \\ 2 \\ 2 \end{Bmatrix} 500 \ kips
$$

Using the data summarized in Example E11-1 gives

$$
\begin{Bmatrix} K_1 \\ K_2 \\ K_3 \end{Bmatrix} = \begin{bmatrix} 1.8131 \, \omega_1^2 \\ 2.4740 \, \omega_2^2 \\ 22.596 \, \omega_3^2 \end{bmatrix} = \begin{Bmatrix} 382.36 \\ 2,384.9 \\ 48,019 \end{Bmatrix} \ kips/in \tag{b}
$$

$$
\begin{Bmatrix} P_1 \\ P_2 \\ P_3 \end{Bmatrix} = \Phi^T \begin{Bmatrix} 500 \\ 1,000 \\ 1,000 \end{Bmatrix} = \begin{Bmatrix} 1,450.40 \\ -785.59 \\ 397.80 \end{Bmatrix} \ kips \tag{c}
$$

Also, the impulse length-period ratios for the modes of this structure are

$$
\begin{Bmatrix} \frac{t_1}{T_1} \\ \frac{t_1}{T_2} \\ \frac{t_1}{T_3} \end{Bmatrix} = \frac{0.02}{2\pi} \begin{Bmatrix} \omega_1 \\ \omega_2 \\ \omega_3 \end{Bmatrix} = \begin{Bmatrix} 0.046 \\ 0.099 \\ 0.147 \end{Bmatrix}
$$

and from Fig. 5-6, these give the following modal dynamic magnification factors:

$$
\begin{Bmatrix} D_1 \\ D_2 \\ D_3 \end{Bmatrix} = \begin{Bmatrix} 0.1865 \\ 0.4114 \\ 0.6423 \end{Bmatrix} \tag{d}
$$

Hence, using the results given in Eqs. (b) to (d) in Eq. (a) leads to

$$\begin{Bmatrix} Y_1(t) \\ Y_2(t) \\ Y_3(t) \end{Bmatrix} = \begin{Bmatrix} 0.7074 \sin w_1 t \\ -0.1355 \sin w_2 t \\ 0.0053 \sin w_3 t \end{Bmatrix} \; in \qquad (e)$$

It will be noted that the motion of the top story is merely the sum of the modal expressions of Eq. (e), because for each mode the modal shape has a unit amplitude at the top. However, for story 2, for example, the relative modal displacement at this level must be considered, that is, the mode-superposition expression becomes

$$v_2(t) = \sum \phi_{2n} Y_n(t)$$

$$= (0.4588 \; in) \; \sin w_1 t + (0.0822 \; in) \; \sin w_2 t$$

$$- (0.0135 \; in) \; \sin w_3 t \qquad (f)$$

Similarly, the elastic forces developed in this structure by the blast loading are given by Eq. (12-32), which for this lumped-mass system may be evaluated at story 2 as follows:

$$f_{S2}(t) = \sum m_2 w_n^2 \, Y_n(t) \, \phi_{2n}$$

$$= (145.13 \; kips) \, \sin w_1 t + (118.87 \; kips) \, \sin w_2 t$$

$$- (43.11 \; kips) \, \sin w_3 t \qquad (g)$$

That the higher-mode contributions are more significant with respect to the force response than for the displacements is quite evident from a comparison of expressions (f) and (g).

Complex-Stiffness Damping

As pointed out in Section 4-7, damping of the linear viscous form represented in Eqs. (12-17) has a serious deficiency because the energy loss per cycle at a fixed displacement amplitude is dependent upon the response frequency; see Eq. (3-61). Since this dependency is at variance with a great deal of test evidence which indicates that the energy loss per cycle is essentially independent of the frequency, it would be better to solve the uncoupled normal mode equations of motion in the frequency domain using complex-stiffness damping rather than viscous damping; in that case the energy loss per cycle would be independent of frequency; see Eq. (3-84).

Making this change in type of damping by using a complex-generalized-stiffness of the form given by Eq. (3-79), that is, using

$$\hat{K}_n = K_n \left[1 + i\,2\,\xi_n\right] \tag{12-34}$$

in which

$$K_n = \omega_n^2\,M_n \tag{12-35}$$

the response will be given by Eq. (12-23) using the complex-frequency-response transfer function

$$H_n(i\bar{\omega}) = \frac{1}{\omega_n^2\,M_n}\left[\frac{1}{(1-\beta_n^2)+i\,(2\xi_n)}\right] = \frac{1}{\omega_n^2\,M_n}\left[\frac{(1-\beta_n^2)-i\,(2\xi_n)}{(1-\beta_n^2)^2+(2\xi_n)^2}\right] \tag{12-36}$$

rather than the corresponding transfer function given by Eq. (12-25) for viscous damping; see Eq. (6-46). All quantities in this transfer function are defined the same as those in the transfer function of Eq. (12-25).

Having obtained the forced-vibration response $Y_n(t)$ for each normal mode of interest (a limited number of the lower modes) using Eqs. (12-23), (12-24), and (12-36), the free-vibration response of Eq. (12-26) can be added to it giving the total response. One can then proceed to obtain the displacement vector $\mathbf{v}(t)$ by superposition using Eq. (12-29) and the elastic force vector $\mathbf{f}_S(t)$ using either Eq. (12-30) or Eq. (12-33).

Example E12-3. A mechanical exciter placed on the top mass of the frame shown in Fig. E11-1 subjects the structure to a harmonic lateral loading of amplitude p_0 at frequency $\bar{\omega}$, i.e., it produces the force

$$p_1(t) = p_0\,\sin(\bar{\omega}t) \tag{a}$$

Calculate the steady-state amplitudes of acceleration produced at levels 1, 2, and 3 and the amplitude of the total shear force in the lowest story when the exciter is operating at $p_0 = 3\ kips$ and $\bar{\omega} = 4\pi\ rad/sec$. Assume 5 percent of critical damping in each of two separate forms: (1) viscous and (2) complex stiffness.

Making use of Eq. (a) and the second of Eqs. (2-23b) and recognizing that components $p_2(t)$ and $p_3(t)$ in force vector $\mathbf{p}(t)$ equal zero, one can state

$$\mathbf{p}(t) = -\frac{p_0}{2}\,i\left[\exp(i\bar{\omega}t) - \exp(-i\bar{\omega}t)\right] \cdot \begin{Bmatrix} 1 \\ 0 \\ 0 \end{Bmatrix} \tag{b}$$

Substituting Eq. (b) and separately the three mode-shape vectors ϕ_n of Fig. E11-2 into Eq. (12-12c) gives the same generalized load expression for each mode

$$P_n(t) = -\frac{p_o}{2} i \left[\exp(i\bar{\omega}t) - \exp(-i\bar{\omega}t) \right] \qquad n = 1, 2, 3 \qquad \text{(c)}$$

Since each term on the right hand side of this equation represents a discrete harmonic loading, the steady-state response of each normal mode coordinate $Y_n(t)$ is obtained by multiplying each discrete harmonic by its corresponding complex frequency response transfer function given by Eq. (12-25) for viscous damping and by Eq. (12-36) for complex-stiffness damping. Completing this step, one obtains

$$Y_n(t) = -\frac{p_0}{2} \frac{i}{K_n} \left\{ \left[\frac{(1 - \beta_n^2) - i(2\xi_n\beta_n)}{(1 - \beta_n^2)^2 + (2\xi_n\beta_n)^2} \right] \exp(i\bar{\omega}t) \right.$$

$$\left. - \left[\frac{(1 - \beta_n^2) + i(2\xi_n\beta_n)}{(1 - \beta_n^2)^2 + (2\xi_n\beta_n)^2} \right] \exp(-i\bar{\omega}t) \right\} \qquad n = 1, 2, 3 \qquad \text{(d)}$$

for the case of viscous damping and

$$Y_n(t) = -\frac{p_0}{2} \frac{i}{K_n} \left\{ \left[\frac{(1 - \beta_n^2) - i(2\xi_n)}{(1 - \beta_n^2)^2 + (2\xi_n)^2} \right] \exp(i\bar{\omega}t) \right.$$

$$\left. - \left[\frac{(1 - \beta_n^2) + i(2\xi_n)}{(1 - \beta_n^2)^2 + (2\xi_n)^2} \right] \exp(-i\bar{\omega}t) \right\} \qquad n = 1, 2, 3 \qquad \text{(e)}$$

for the case of complex-stiffness damping. Using $p_0 = 3$ $kips$, $\bar{\omega} = 4\pi$, the values of K_n given by Eq. (b) in Example E12-2, the relation $\beta_n = \bar{\omega}/\omega_n$ with corresponding values of ω_n as given in Eqs. (a) of Example E12-1, $\xi_n = 0.05$ for all values of n ($n = 1, 2, 3$), and changing exponential expressions to trigonometric form using Eqs. (2-23a), Eqs. (d) and (e) above yield

$$Y_1(t) = \begin{cases} [-9.879 \ \cos 4\pi t + 28.381 \ \sin 4\pi t] \ 10^{-3} \ in & \text{(d)} \\ [-11.005 \ \cos 4\pi t + 27.399 \ \sin 4\pi t] \ 10^{-3} \ in & \text{(e)} \end{cases}$$

$$Y_2(t) = \begin{cases} [-0.073 \ \cos 4\pi t + 1.506 \ \sin 4\pi t] \ 10^{-3} \ in & \text{(d)} \\ [-0.178 \ \cos 4\pi t + 1.488 \ \sin 4\pi t] \ 10^{-3} \ in & \text{(e)} \end{cases}$$

$$Y_3(t) = \begin{cases} [-0.002 \ \cos 4\pi t + 0.066 \ \sin 4\pi t] \ 10^{-3} \ in & \text{(d)} \\ [-0.007 \ \cos 4\pi t + 0.065 \ \sin 4\pi t] \ 10^{-3} \ in & \text{(e)} \end{cases}$$

Substituting the above Eqs. (d) and Eqs. (e) separately into Eq. (12-29) for $N = 3$ gives

$$v_1(t) = [-9.954 \ \cos \bar{\omega} t + 29.953 \ \sin \bar{\omega} t] \ 10^{-3} \ in$$

$$v_2(t) = [-6.313 \ \cos \bar{\omega} t + 17.203 \ \sin \bar{\omega} t] \ 10^{-3} \ in \qquad \text{(f)}$$

$$v_3(t) = [-2.920 \ \cos \bar{\omega} t + 7.659 \ \sin \bar{\omega} t] \ 10^{-3} \ in$$

for the case of viscous damping and

$$v_1(t) = [-11.190 \ \cos \bar{\omega} t + 28.952 \ \sin \bar{\omega} t] \ 10^{-3} \ in$$

$$v_2(t) = [-6.963 \ \cos \bar{\omega} t + 16.584 \ \sin \bar{\omega} t] \ 10^{-3} \ in \qquad \text{(g)}$$

$$v_3(t) = [-3.199 \ \cos \bar{\omega} t + 7.375 \ \sin \bar{\omega} t] \ 10^{-3} \ in$$

for the case of complex-stiffness damping. Taking the square root of the sum of the squares of the two coefficients in each of Eqs. (f) and (g) gives the displacement amplitude in each case which, when multiplied by $16\pi^2$ $(\bar{\omega}^2)$, yields the desired acceleration amplitudes

$$\bar{\bar{v}}_1 = 4.98 \quad \bar{\bar{v}}_2 = 2.89 \quad \bar{\bar{v}}_3 = 1.29 \ in/sec \quad \text{viscous damping}$$

$$\bar{\bar{v}}_1 = 4.90 \quad \bar{\bar{v}}_2 = 2.84 \quad \bar{\bar{v}}_3 = 1.27 \ in/sec^2 \quad \text{complex-stiffness damping} \quad \text{(h)}$$

The amplitude of the total shear force $\bar{V}$ in the lowest story is the product of the displacement amplitude $\bar{v}_3$ and the lowest story spring constant $1,800 \ kips/in$ (see Fig. E11-1); thus, one obtains

$$\bar{V} = 14.75 \ kips \qquad \text{viscous damping}$$

$$\bar{V} = 14.47 \ kips \qquad \text{complex-stiffness damping} \qquad \text{(i)}$$

In the above Example E12-3, the loading was of a simple harmonic form which allowed an easy solution using the appropriate transfer functions in the frequency domain. However, if each component in vector $\mathbf{p}(t)$ had been nonperiodic of arbitrary form giving corresponding nonperiodic normal-coordinate generalized loads $P_n(t)$ $(n = 1, 2, 3)$ in accordance with Eq. (12-12c), it would be necessary to Fourier transform each of these generalized loads as indicated by Eq. (12-24) using the FFT procedure described in Chapter 6, thus obtaining $N - 1$ discrete harmonics in accordance with $N = 2^\gamma$ where γ is an integer selected appropriately; see discussion of solutions in Fig. 6-4. Assuming zero initial conditions on each normal coordinate $Y_n(t)$ $(n = 1, 2, 3)$, its time-history of response following $t = 0$ would be obtained upon multiplying each discrete harmonic in $P_n(t)$ by the corresponding complex frequency response transfer function as illustrated in Example E12-3. The $N-1$ products would then be summed giving $Y_n(t)$. Carrying out this procedure for all values of $n = 1, 2, 3$, the time-histories of response $\mathbf{v}(t)$ would be obtained by superposition as in Example E12-3.

12-5 CONSTRUCTION OF PROPORTIONAL VISCOUS DAMPING MATRICES

Rayleigh Damping

As was stated above, generally there is no need to express the damping of a typical viscously damped MDOF system by means of the damping matrix because it is represented more conveniently in terms of the modal damping ratios ξ_n ($n = 1, 2, \cdots, N$). However, in at least two dynamic analysis situations the response is not obtained by superposition of the uncoupled modal responses, so the damping cannot be expressed by the damping ratios — instead an explicit damping matrix is needed. These two situations are: (1) nonlinear responses, for which the mode shapes are not fixed but are changing with changes of stiffness, and (2) analysis of a linear system having nonproportional damping. In both of these circumstances, the most effective way to determine the required damping matrix is to first evaluate one or more proportional damping matrices. In performing a nonlinear analysis, it is appropriate to define the proportional damping matrix for the initial elastic state of the system (before nonlinear deformations have occurred) and to assume that this damping property remains constant during the response even though the stiffness may be changing and causing hysteretic energy losses in addition to the viscous damping losses. In cases where the damping is considered to be nonproportional, an appropriate damping matrix can be constructed by assembling a set of suitably derived proportional damping matrices, as explained later in this section. Thus for these two situations, it is necessary to be able to derive appropriate proportional damping matrices.

Clearly the simplest way to formulate a proportional damping matrix is to make it proportional to either the mass or the stiffness matrix because the undamped mode shapes are orthogonal with respect to each of these. Thus the damping matrix might be given by

$$\mathbf{c} = a_0\,\mathbf{m} \qquad \text{or} \qquad \mathbf{c} = a_1\,\mathbf{k} \qquad (12\text{-}37a)$$

in which the proportionality constants a_0 and a_1 have units of sec^{-1} and sec, respectively. These are called mass proportional and stiffness proportional damping, and the damping behavior associated with them may be recognized by evaluating the generalized modal damping value for each [see Eq. (12-15a)],

$$C_n = \boldsymbol{\phi}_n^T\,c\,\boldsymbol{\phi}_n = a_0\,\boldsymbol{\phi}_n^T\,\mathbf{m}\,\boldsymbol{\phi}_n \qquad \text{or} \qquad a_1\,\boldsymbol{\phi}_n^T\,\mathbf{k}\,\boldsymbol{\phi}_n \qquad (12\text{-}37b)$$

or combining with Eq. (12-15b)

$$2\omega_n M_n \xi_n = a_0 M_n \quad \text{or} \quad a_1 K_n \quad (\text{where} \quad K_n = \omega_n^2 M_n) \qquad (12\text{-}37c)$$

from which

$$\xi_n = \frac{a_0}{2\omega_n} \quad \text{or} \quad \xi_n = \frac{a_1 \omega_n}{2} \qquad (12\text{-}37d)$$

These expressions show that for mass proportional damping, the damping ratio is inversely proportional to the frequency while for stiffness proportional damping it is directly in proportion with the frequency. In this regard it is important to note that the dynamic response generally will include contributions from all N modes even though only a limited number of modes are included in the uncoupled equations of motion. Thus, neither of these types of damping matrix is suitable for use with an MDOF system in which the frequencies of the significant modes span a wide range because the relative amplitudes of the different modes will be seriously distorted by inappropriate damping ratios.

An obvious improvement results if the damping is assumed to be proportional to a combination of the mass and the stiffness matrices as given by the sum of the two alternative expressions shown in Eq. (12-37a):

$$\mathbf{c} = a_0 \, \mathbf{m} + a_1 \, \mathbf{k} \qquad (12\text{-}38a)$$

This is called Rayleigh damping, after Lord Rayleigh, who first suggested its use. By analogy with the development in Eqs. (12-37b) to (12-37d), it is evident that Rayleigh damping leads to the following relation between damping ratio and frequency

$$\xi_n = \frac{a_0}{2\omega_n} + \frac{a_1 \omega_n}{2} \qquad (12\text{-}38b)$$

The relationships between damping ratio and frequency expressed by Eqs. (12-37d) and (12-38b) are shown graphically in Fig. 12-2.

Now it is apparent that the two Rayleigh damping factors, a_0 and a_1, can be evaluated by the solution of a pair of simultaneous equations if the damping ratios ξ_m and ξ_n associated with two specific frequencies (modes) ω_m, ω_n are known. Writing Eq. (12-38b) for each of these two cases and expressing the two equations in matrix

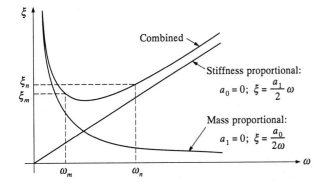

Combined

Stiffness proportional:
$$a_0 = 0; \quad \xi = \frac{a_1}{2}\,\omega$$

Mass proportional:
$$a_1 = 0; \quad \xi = \frac{a_0}{2\omega}$$

FIGURE 12-2
Relationship between damping ratio and frequency (for Rayleigh damping).

form leads to

$$\left\{ \begin{array}{c} \xi_m \\ \xi_n \end{array} \right\} = \frac{1}{2} \left[\begin{array}{cc} 1/\omega_m & \omega_m \\ 1/\omega_n & \omega_n \end{array} \right] \left\{ \begin{array}{c} a_0 \\ a_1 \end{array} \right\} \tag{12-39}$$

and the factors resulting from the simultaneous solution are

$$\left\{ \begin{array}{c} a_0 \\ a_1 \end{array} \right\} = 2 \frac{\omega_m \omega_n}{\omega_n^2 - \omega_m^2} \left[\begin{array}{cc} \omega_n & -\omega_m \\ -1/\omega_n & 1/\omega_m \end{array} \right] \left\{ \begin{array}{c} \xi_m \\ \xi_n \end{array} \right\} \tag{12-40}$$

When these factors have been evaluated, the proportional damping matrix that will give the required values of damping ratio at the specified frequencies is given by the Rayleigh damping expression, Eq. (12-38a), as shown by Fig. 12-2.

Because detailed information about the variation of damping ratio with frequency seldom is available, it usually is assumed that the same damping ratio applies to both control frequencies; i.e., $\xi_m = \xi_n \equiv \xi$. In this case, the proportionality factors are given by a simplified version of Eq. (12-40):

$$\left\{ \begin{array}{c} a_0 \\ a_1 \end{array} \right\} = \frac{2\xi}{\omega_m + \omega_n} \left\{ \begin{array}{c} \omega_m \omega_n \\ 1 \end{array} \right\} \tag{12-41}$$

In applying this proportional damping matrix derivation procedure in practice, it is recommended that ω_m generally be taken as the fundamental frequency of the MDOF system and that ω_n be set among the higher frequencies of the modes that contribute significantly to the dynamic response. The derivation ensures that the desired damping ratio is obtained for these two modes (i.e., $\xi_1 = \xi_n = \xi$); then as shown by Fig. 12-2, modes with frequencies between these two specified frequencies will have somewhat lower values of damping ratio, while all modes with frequencies greater than ω_n will have damping ratios that increases above ξ_n monotonically with frequency. The end result of this situation is that the responses of very high frequency modes are effectively eliminated by their high damping ratios.

Example E12-4. For the structure of Example E11-1, an explicit damping matrix is to be defined such that the damping ratio in the first and third modes will be 5 percent of critical. Assuming Rayleigh damping, the proportionality factors a_0 and a_1 can be evaluated from Eq. (12-39), using the frequency data listed in Example E12-1, as follows:

$$\left\{ \begin{array}{c} \xi_1 \\ \xi_3 \end{array} \right\} = \left\{ \begin{array}{c} 0.05 \\ 0.05 \end{array} \right\} = \frac{1}{2} \left[\begin{array}{cc} \frac{1}{14.522} & 14.522 \\ \frac{1}{46.100} & 46.100 \end{array} \right] \left\{ \begin{array}{c} a_0 \\ a_1 \end{array} \right\}$$

from which

$$\left\{ \begin{array}{c} a_0 \\ a_1 \end{array} \right\} = \left\{ \begin{array}{c} 1.1042 \\ 0.00165 \end{array} \right\}$$

Hence $c = 1.1042\,m + 0.00165\,k$ or, using the matrices listed in Example E12-1,

$$c = \begin{bmatrix} 2.094 & -0.990 & 0 \\ -0.990 & 4.626 & -1.980 \\ 0 & -1.980 & 7.157 \end{bmatrix} kip \cdot sec/in$$

Now it is of interest to determine what damping ratio this matrix will yield in the second mode. Introducing the second mode frequency in Eq. (12-38b) and putting it in matrix form gives

$$\xi_2 = \frac{1}{2} \left[\frac{1}{31.048} \quad 31.048 \right] \begin{Bmatrix} a_0 \\ a_1 \end{Bmatrix}$$

Then, introducing the values of a_0 and a_1 found above leads to

$$\xi_2 = 0.0434 = 4.34\%$$

Hence, even though only the first and third damping ratios were specified, the resulting damping ratio for the second mode is a reasonable value.

Extended Rayleigh Damping

The mass and stiffness matrices used to formulate Rayleigh damping are not the only matrices to which the free-vibration mode-shape orthogonality conditions apply; in fact, it was shown earlier in Eq. (11-44) that an infinite number of matrices have this property. Therefore a proportional damping matrix can be made up of any combination of these matrices, as follows:

$$c = m \sum_b a_b [m^{-1} k]^b \equiv \sum_b c_b \qquad (12\text{-}42)$$

in which the coefficients a_b are arbitrary. It is evident that Rayleigh damping is given by Eq. (12-42) if only the terms $b = 0$ and $b = 1$ are retained in the series. By retaining additional terms of the series a proportional damping matrix can be constructed that gives any desired damping ratio ξ_n at a specified frequency ω_n for as many frequencies as there are terms in the series of Eq. (12-42).

To understand the procedure, consider the generalized damping value C_n for any normal mode "n" [see Eq. (12-37b)]:

$$C_n = \phi_n^T c \, \phi_n = 2 \xi_n \omega_n M_n \qquad (12\text{-}43)$$

If **c** in this expression is given by Eq. (12-42), the contribution of term b to the generalized damping value is

$$C_{nb} = \boldsymbol{\phi}_n^T \mathbf{c}_b \boldsymbol{\phi}_n = a_b \, \mathbf{m} \, [\mathbf{m}^{-1} \mathbf{k}]^b \, \boldsymbol{\phi}_n \qquad (12\text{-}44a)$$

Now if Eq. (11-39) ($\mathbf{k} \, \boldsymbol{\phi}_n = \omega_n^2 \, m \, \boldsymbol{\phi}_n$) is premultiplied on both sides by $\boldsymbol{\phi}_n^T \mathbf{k} \mathbf{m}^{-1}$, the result is

$$\boldsymbol{\phi}_n^T \, \mathbf{k} \mathbf{m}^{-1} \, \mathbf{k} \boldsymbol{\phi}_n = \omega_n^2 \, \boldsymbol{\phi}_n^T \, \mathbf{k} \, \boldsymbol{\phi}_n \equiv \omega_n^4 \, M_n$$

By operations equivalent to this it can be shown that

$$\boldsymbol{\phi}_n^T \, \mathbf{m} \, [\mathbf{m}^{-1} \mathbf{k}]^b \, \boldsymbol{\phi}_n = \omega_n^{2b} \, M_n \qquad (12\text{-}45)$$

and consequently

$$C_{nb} = a_b \, \omega_n^{2b} \, M_n \qquad (12\text{-}44b)$$

On this basis, the generalized damping value associated with any mode n is

$$C_n = \sum_b C_{nb} = \sum_b a_b \, \omega_n^{2b} \, M_n = 2 \xi_n \, \omega_n \, M_n$$

from which

$$\xi_n = \frac{1}{2\omega_n} \sum_b a_b \, \omega_n^{2b} \qquad (12\text{-}46)$$

Equation (12-46) provides the means for evaluating the constants a_b to give the desired damping ratios at any specified number of modal frequencies. As many terms must be included in the series as there are specified modal damping ratios; then the constants are given by the solution of the set of equations, one written for each damping ratio. In principle, the values of b can lie anywhere in the range $-\infty < b < \infty$, but in practice it is desirable to select values of these exponents as close to zero as possible. For example, to evaluate the coefficients that will provide specified damping ratios in any four modes having the frequencies $\omega_m, \omega_n, \omega_o, \omega_p$, the equations resulting from Eq. (12-46) using the terms for $b = -1, 0, +1,$ and $+2$ are

$$\begin{Bmatrix} \xi_m \\ \xi_n \\ \xi_o \\ \xi_p \end{Bmatrix} = \frac{1}{2} \begin{bmatrix} 1/\omega_m^2 & 1/\omega_m & \omega_m & \omega_m^3 \\ 1/\omega_n^2 & 1/\omega_n & \omega_n & \omega_n^3 \\ 1/\omega_o^2 & 1/\omega_o & \omega_o & \omega_o^3 \\ 1/\omega_p^2 & 1/\omega_p & \omega_p & \omega_p^3 \end{bmatrix} \begin{Bmatrix} a_{-1} \\ a_0 \\ a_1 \\ a_2 \end{Bmatrix} \qquad (12\text{-}47)$$

When the coefficients $a_{-1}, a_0, a_1,$ and a_2 have been evaluated by the simultaneous solution of Eq. (12-47), the viscous damping matrix that provides the four required damping ratios at the four specified frequencies is obtained by superposing four matrices (one for each value of b) in accordance with Eq. (12-42). Figure 12-3a illustrates

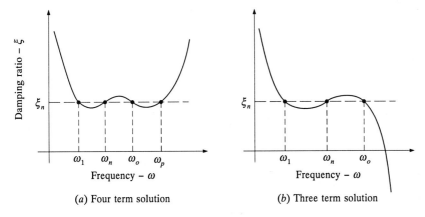

FIGURE 12-3
Extended Rayleigh damping (damping ratio vs. frequency).

the relation between damping ratio and frequency that would result from this matrix. To simplify the figure it has been assumed here that the same damping ratio, ξ_x, was specified for all four frequencies; however, each of the damping ratios could have been specified arbitrarily. Also, ω_m has been taken as the fundamental mode frequency, ω_1, and ω_p is intended to approximate the frequency of the highest mode that contributes significantly to the response, while ω_n and ω_0 are spaced about equally within the frequency range. It is evident in Fig. 12-3a that the damping ratio remains close to the desired value ξ_x throughout the frequency range, being exact at the four specified frequencies and ranging slightly above or below at other frequencies in the range. It is important to note, however, that the damping increases monotonically with frequency for frequencies increasing above ω_p. This has the effect of excluding any significant contribution from any modes with frequencies much greater than ω_p, thus such modes need not be included in the response superposition.

An even more important point to note is the consequence of including only three terms in the derivation of the viscous damping matrix using Eq. (12-42). In that case a set of three simultaneous equations equivalent to Eq. (12-47) would be obtained and solved for the coefficients a_{-1}, a_0, a_1, and if these were substituted into Eq. (12-42), the resulting damping ratio-frequency relation would be as shown in Fig. 12-3b. As required by the simultaneous equation solution, the desired damping ratio is obtained exactly at the three specified frequencies, and is approximated well at intermediate frequencies. However, the serious defect of this result is that the damping decreases monotonically with frequencies increasing above ω_o and negative damping is indicated for all the highest modal frequencies. This is an unacceptable result because the contribution of the negatively damped modes would tend to increase without limit in the analysis but certainly would not do so in actuality.

The general implication of this observation is that extended Rayleigh damping may be used effectively only if an even number of terms is included in the series expression, Eq. (12-42). In such cases, modes with frequencies greater than the range considered in evaluating the coefficients will be effectively excluded from the mode superposition response. However, if an odd number of terms (greater than one) were included in Eq. (12-42), the modes with frequencies much greater than the controlled range would be negatively damped and would invalidate the results of the analysis.

Alternative Formulation

A second method is available for evaluating the damping matrix associated with any given set of modal damping ratios. In principle, the procedure can be explained by considering the complete diagonal matrix of generalized damping coefficients, given by pre- and postmultiplying the damping matrix by the mode-shape matrix:

$$C = \Phi^T c \Phi = 2 \begin{bmatrix} \xi_1 \omega_1 M_1 & 0 & 0 & \cdots \\ 0 & \xi_2 \omega_2 M_2 & 0 & \cdots \\ 0 & 0 & \xi_3 \omega_3 M_3 & \vdots \\ \vdots & \vdots & \vdots & \vdots \end{bmatrix} \tag{12-48}$$

It is evident from this equation that the damping matrix can be obtained by pre- and postmultiplying matrix C by the inverse of the transposed mode-shape matrix and the inverse of the mode-shape matrix, respectively, yielding

$$\left[\Phi^T\right]^{-1} C \Phi^{-1} = \left[\Phi^T\right]^{-1} \Phi^T c \Phi \Phi^{-1} = c \tag{12-49}$$

Since for any specified set of modal damping ratios ξ_n, the generalized damping coefficients in matrix C can be evaluated, as indicated in Eq. (12-43), the damping matrix c can be evaluated using Eq. (12-49).

In practice, however, this is not a convenient procedure because inversion of the mode-shape matrix requires a large computational effort. Instead, it is useful to take advantage of the orthogonality properties of the mode shapes relative to the mass matrix. The diagonal generalized-mass matrix of the system is obtained using the relation

$$M = \Phi^T m \Phi \tag{12-50}$$

Premultiplying this equation by the inverse of the generalized-mass matrix then gives

$$I = M^{-1} M = \left[M^{-1} \Phi^T m\right] \Phi = \Phi^{-1} \Phi \tag{12-51}$$

from which it is evident that the mode-shape-matrix inverse is

$$\Phi^{-1} = M^{-1} \Phi^T m \tag{12-52}$$

Operating on this expression, one can obtain

$$\left[\mathbf{\Phi}^T\right]^{-1} = \mathbf{m} \, \mathbf{\Phi} \, M^{-1} \tag{12-53}$$

Substituting Eqs. (12-52) and (12-53) into Eq. (12-49) yields

$$\mathbf{c} = \left[\mathbf{m} \, \mathbf{\Phi} \, M^{-1}\right] C \left[M^{-1} \, \mathbf{\Phi}^T \, \mathbf{m}\right] \tag{12-54}$$

Since matrix C is a diagonal matrix containing elements $C_n = 2\,\xi_n\,\omega_n\,M_n$, the elements of the diagonal matrix obtained as the product of the three central diagonal matrices in this equation are

$$d_n \equiv \frac{2\,\xi_n\,\omega_n}{M_n} \tag{12-55}$$

so that Eq. (12-54) may be written

$$\mathbf{c} = \mathbf{m} \, \mathbf{\Phi} \, \mathbf{d} \, \mathbf{\Phi}^T \, \mathbf{m}$$

where $\mathbf{d}$ is the diagonal matrix containing elements d_n. In the analysis, however, it is more convenient to note that each modal damping ratio provides an independent contribution to the damping matrix, as follows:

$$\mathbf{c}_n = \mathbf{m} \, \boldsymbol{\phi}_n \, d_n \, \boldsymbol{\phi}_n^T \, \mathbf{m} \tag{12-56a}$$

Thus the total damping matrix is obtained as the sum of the modal contributions

$$\mathbf{c} = \sum_{n=1}^{N} \mathbf{c}_n = \mathbf{m} \left[\sum_{n=1}^{N} \boldsymbol{\phi}_n \, d_n \, \boldsymbol{\phi}_n^T \right] \mathbf{m} \tag{12-56b}$$

By substituting from Eq. (12-55), this equation may be written

$$\mathbf{c} = \mathbf{m} \left[\sum_{n=1}^{N} \frac{2\,\xi_n\omega_n}{M_n} \, \boldsymbol{\phi}_n \, \boldsymbol{\phi}_n^T \right] \mathbf{m} \tag{12-56c}$$

In this equation, the contribution to the damping matrix from each mode is proportional to the modal damping ratio; thus any undamped mode will contribute nothing to the damping matrix. In other words, only those modes specifically included in the formation of the damping matrix will have any damping and all other modes will be undamped.

In order to avoid undesirable amplification of undamped modal responses, damping of the type provided by Eq. (12-56c) should be used only as a supplement to a stiffness proportional damping matrix, for which the damping ratio increases in proportion with the modal frequencies as shown by the right hand expression of Eq. (12-37d); i.e., $\xi = \frac{a_1\omega}{2}$. The coefficient a_1 of this stiffness proportional damping matrix should be calculated to provide the damping ratio ξ_c required at the frequency ω_c of the highest mode for which damping is specified; thus from Eq. (12-37d),

$$a_1 = \frac{2\xi_c}{\omega_c} \tag{12-57a}$$

The stiffness proportional damping ratios at other frequencies then are given by

$$\hat{\xi}_n = \frac{a_1 \omega_n}{2} = \xi_c \left(\frac{\omega_n}{\omega_c}\right) \tag{12-57b}$$

Hence if the total damping ratio desired in any mode n is ξ_n, it is evident that the damping of the type of Eq. (12-56c), designated $\overline{\xi}_n$, required to supplement the stiffness proportional damping must be

$$\overline{\xi}_n = \xi_n - \xi_c \left(\frac{\omega_n}{\omega_c}\right) \tag{12-57c}$$

The final result of this development is a proportional damping matrix **c** given by

$$\mathbf{c} = a_1 \mathbf{k} + \mathbf{m} \left[\sum_{n=1}^{c-1} \frac{2\overline{\xi}_n \omega_n}{M_n} \phi_n \phi_n^T\right] \mathbf{m} \tag{12-57d}$$

which provides the desired modal damping ratios for frequencies less than or equal to ω_c and which has linearly increasing damping for higher frequencies.

Construction of Nonproportional Damping Matrices

The proportional damping matrices described in the preceding paragraphs are suitable for modeling the behavior of most structural systems, in which the damping mechanism is distributed rather uniformly throughout the structure. However, for structures made up of more than a single type of material, where the different materials provide drastically differing energy-loss mechanisms in various parts of the structure, the distribution of damping forces will not be similar to the distribution of the inertial and elastic forces; in other words, the resulting damping will be nonproportional.

A nonproportional damping matrix that will represent this situation may be constructed by applying procedures similar to those discussed above in developing proportional damping matrices, with a proportional matrix being developed for each distinct part of the structure and then the combined system matrix being formed by direct assembly. The procedure is explained with reference to Fig. 12-4, which portrays a five-story steel building frame erected on top of a five-story reinforced concrete building frame. As shown, it is assumed that the modal damping of the steel

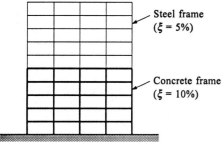

FIGURE 12-4
Combined frame: steel and concrete.

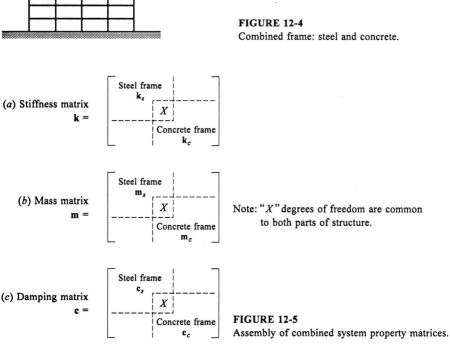

(a) Stiffness matrix
$\mathbf{k} =$

(b) Mass matrix
$\mathbf{m} =$

Note: "X" degrees of freedom are common to both parts of structure.

(c) Damping matrix
$\mathbf{c} =$

FIGURE 12-5
Assembly of combined system property matrices.

frame alone would be 5 percent of critical while that of the concrete frame alone would be 10 percent of critical.

The stiffness and mass matrices of the combined system are shown qualitatively in Fig. 12-5, with the contributions from the steel frame located in the upper left corner of the combined matrices and the concrete frame contributions in the lower right corner. The contributions associated with the common degrees of freedom at the interface between the two substructures (designated areas "X" in the figure) include contributions from both the steel and the concrete frames. The damping matrix for the combined frame may be developed by a similar assembly procedure as shown in Fig. 12-5c after the damping submatrices for the steel and concrete substructures have been derived. In principle these could be evaluated by any of the procedures for developing proportional damping matrices described above, but for most cases the recommended procedure is to assume Rayleigh damping. Thus the steel and the concrete submatrices will be given respectively by [see Eq. (12-38a)]

$$\mathbf{c}_s = a_{0s}\,\mathbf{m}_s + a_{1s}\,\mathbf{k}_s$$

$$\mathbf{c}_c = a_{0c}\,\mathbf{m}_c + a_{1c}\,\mathbf{k}_c$$

in which the constants for the steel frame are evaluated as shown by Eq. (12-41):

$$\left\{ \begin{matrix} a_{0s} \\ a_{1s} \end{matrix} \right\} = \frac{2\xi_s}{\omega_m + \omega_n} \left\{ \begin{matrix} \omega_m\omega_n \\ 1 \end{matrix} \right\}$$

and the corresponding values for the concrete frame, a_{0c} and a_{1c}, are twice as great because $\xi_c = 10\%$ is twice as great as $\xi_s = 5\%$.

These values depend on the frequencies ω_m and ω_n, and the frequencies to be used must be determined by solving the eigenproblem of the *combined* system (i.e., using the combined stiffness and mass matrices $\mathbf{k}$ and $\mathbf{m}$. As was mentioned above, it is recommended that ω_m be taken as the first mode frequency of the combined system, while for this 10-story frame it would be appropriate to use the seventh or eighth mode frequency as ω_n. The nonproportional damping matrix for the combined system is obtained finally by assembly as shown in Fig. 12-5c.

Using this damping matrix in the equations of motion [Eq. (9-13)] and transforming to normal coordinates by pre- and postmultiplying by the mode-shape matrix $\boldsymbol{\Phi}$ for the combined system leads to the modal coordinate equations of motion

$$M\,\ddot{Y} + C\,\dot{Y} + K\,Y = P(t) \qquad (12\text{-}58)$$

where M and K are the diagonal modal coordinate mass and stiffness matrices and $P(t)$ is the standard modal coordinate load vector. However, the modal coordinate damping matrix

$$C = \boldsymbol{\Phi}^T \mathbf{c}\, \boldsymbol{\Phi} = \begin{bmatrix} C_{11} & C_{12} & C_{13} & \cdots \\ C_{21} & C_{22} & C_{23} & \cdots \\ C_{31} & C_{32} & C_{33} & \cdots \\ \vdots & \vdots & \vdots & \vdots \end{bmatrix} \qquad (12\text{-}59)$$

is not diagonal but includes modal coupling coefficients C_{ij} $(i \neq j)$ because the matrix $\mathbf{c}$ is nonproportional.

An effective method of solving for the dynamic response using this coupled modal equation set is to merely use direct step-by-step integration, as is explained by means of an example in Chapter 15. An approximate solution may be obtained by ignoring the off-diagonal coupling coefficients of the modal damping matrix and then solving the resulting uncoupled equations as a typical mode superposition analysis.

The errors resulting from this approximation are indicated in the example presented in Chapter 15; however, it must be remembered that the errors resulting in other cases from this assumed uncoupling may be larger or smaller than those found in this example.

12-6   RESPONSE ANALYSIS USING COUPLED EQUATIONS OF MOTION

Mode superposition is a very effective means of evaluating the dynamic response of structures having many degrees of freedom because the response analysis is performed only for a series of SDOF systems. However, the computational cost in this type of calculation is transferred from the MDOF dynamic analysis to the solution of the N degree of freedom undamped eigenproblem followed by the modal coordinate transformation, which must be done before the individual modal responses can be evaluated. Certainly the eigenproblem solution represents the major part of the cost of a typical mode superposition analysis, but also it must be recalled that the equations of motion will be uncoupled by the resulting undamped mode shapes only if the damping is represented by a proportional damping matrix.

For these reasons it is useful to examine the possibility of avoiding the modal coordinate transformation by carrying out the dynamic response analysis directly in the original geometric coordinate equations of motion; these were stated previously by Eq. (9-13) and are renumbered here for convenience:

$$\mathbf{m} \, \ddot{\mathbf{v}}(t) + \mathbf{c} \, \dot{\mathbf{v}}(t) + \mathbf{k} \, \mathbf{v}(t) = \mathbf{p}(t) \tag{12-60}$$

One approach to the solution of this set of coupled equations that often may be worth consideration is the step-by-step procedure, as is described in Chapter 15. However, for linear systems to which superposition is applicable, a more convenient solution may be obtained by Fourier transform (frequency-domain) procedures, as well as — at least in principle — by applying convolution integral (time-domain) methods; these MDOF procedures are analogous to the corresponding methods described previously for SDOF systems. A brief conceptual description of these techniques follows; however, the convolution integral approach is not generally suitable for practical use, and it is not discussed further after this brief description.

Time Domain

First is considered the case where the MDOF structure is subjected to a unit-impulse loading in the jth degree of freedom, while no other loads are applied. Thus the force vector $\mathbf{p}(t)$ consists only of zero components except for the jth term, and that term is expressed by $p_j(t) = \delta(t)$, where $\delta(t)$ is the Dirac delta function defined as

$$\delta(t) = \begin{cases} 0 & t \neq 0 \\ \infty & t = 0 \end{cases} \qquad \int_{-\infty}^{\infty} \delta(t) \, dt = 1 \tag{12-61}$$

Assuming now that Eq. (12-60) can be solved for the displacements caused by this loading, the ith component in the resulting displacement vector $\mathbf{r}(t)$ will then be the free-vibration response in that degree of freedom caused by a unit-impulse loading in coordinate j; therefore by definition this i component motion is a unit-impulse transfer function, which will be denoted herein by $h_{ij}(t)$.

If the loading in coordinate j were a general time varying load $p_j(t)$ rather than a unit-impulse loading, the dynamic response in coordinate i could be obtained by superposing the effects of a succession of impulses in the manner of the Duhamel integral, assuming zero initial conditions. The generalized expression for the response in coordinate i to the load at j is the convolution integral, as follows:

$$v_{ij}(t) = \int_0^t p_j(\tau)\, h_{ij}(t-\tau)\, d\tau \qquad i = 1, 2, \cdots, N \qquad (12\text{-}62)$$

and the total response in coordinate i produced by a general loading involving all components of the load vector $p(t)$ is obtained by summing the contributions from all load components:

$$v_i(t) = \sum_{j=1}^N \left[\int_0^t p_j(\tau)\, h_{ij}(t-\tau)\, d\tau \right] \qquad i = 1, 2, \cdots, N \qquad (12\text{-}63)$$

Frequency Domain

The frequency-domain analysis is similar to the time-domain procedure in that it involves superposition of the effects in coordinate i of a unit load applied in coordinate j; however, in this case both the load and the response are harmonic. Thus the loading is an applied force vector $\mathbf{p}(t)$ having all zero components except for the jth term which is a unit harmonic loading, $p_j(t) = 1 \exp(i\overline{\omega}t)$. Assuming now that the steady-state solution of Eq. (12-60) to this loading can be obtained, the resulting steady-state response in the ith component of the displacement vector $\mathbf{v}(t)$ will be $H_{ij}(i\overline{\omega}) \exp(i\overline{\omega}t)$ in which $H_{ij}(i\overline{\omega})$ is defined as the complex-frequency-response transfer function.

If the loading in coordinate j were a general time-varying load $p_j(t)$ rather than a unit-harmonic loading, the forced-vibration response in coordinate i could be obtained by superposing the effects of all the harmonics contained in $p_j(t)$. For this purpose the time-domain expression of the loading is Fourier transformed to obtain

$$P_j(i\overline{\omega}) = \int_{-\infty}^{\infty} p_j(t)\, \exp(-i\overline{\omega}t)\, dt \qquad (12\text{-}64)$$

and then by inverse Fourier transformation the responses to all of these harmonics are combined to obtain the total forced-vibration response in coordinate i as follows (assuming zero initial conditions):

$$v_{ij}(t) = \frac{1}{2\pi} \int_{-\infty}^{\infty} H_{ij}(i\bar{\omega}) \, P_j(i\bar{\omega}) \, \exp(i\bar{\omega}t) \, d\bar{\omega} \qquad (12\text{-}65)$$

Finally, the total response in coordinate i produced by a general loading involving all components of the load vector $\mathbf{p}(t)$ could be obtained by superposing the contributions from all the load components:

$$v_i(t) = \frac{1}{2\pi} \sum_{j=1}^{N} \left[\int_{-\infty}^{\infty} H_{ij}(i\bar{\omega}) \, P_j(i\bar{\omega}) \, \exp(i\bar{\omega}t) \, d\bar{\omega} \right] \qquad i = 1, 2, \cdots, N \quad (12\text{-}66)$$

Equations (12-63) and (12-66) consistute general solutions to the coupled equations of motion (12-60), assuming zero initial conditions. Their successful implementation depends on being able to generate the transfer functions $h_{ij}(t)$ and $H_{ij}(i\bar{\omega})$ efficiently, and it was suggested above that this is not practical for the time-domain functions, in general. However, procedures for implementing the frequency-domain formulation will be developed in this chapter after the following section.

12-7 RELATIONSHIP BETWEEN TIME- AND FREQUENCY-DOMAIN TRANSFER FUNCTIONS

To develop the interrelationships between transfer functions $h_{ij}(t)$ and $H_{ij}(i\bar{\omega})$, it is necessary to define a complex function $V_{ij}(i\bar{\omega})$ as the Fourier transform of function $v_{ij}(t)$ given by Eq. (12-62); thus,

$$V_{ij}(i\bar{\omega}) \equiv \int_{-\infty}^{\infty} \left[\int_{-\infty}^{t} p_j(\tau) \, h_{ij}(t-\tau) \, d\tau \right] \exp(-i\bar{\omega}t) \, dt \qquad (12\text{-}67)$$

Note that because Eq. (12-62) assumes zero initial conditions, which is equivalent to assuming $p_j(t) = 0$ for $t < 0$, one can change the lower limit of the integral in that equation from zero to $-\infty$ as shown in Eq. (12-67) without affecting the results of the integral. It will be assumed here that damping is present in the system so that the integral

$$I_1 \equiv \int_{-\infty}^{\infty} \left| v_{ij}(t) \right| \, dt$$

is finite. This is a necessary condition for the Fourier transform given by Eq. (12-67) to exist.

Since the function $h_{ij}(t-\tau)$ equals zero for $\tau > t$, the upper limit of the second integral in Eq. (12-67) can be changed from t to ∞ without influencing the final result. Therefore, Eq. (12-67) can be expressed in the equivalent form

$$V_{ij}(i\bar{\omega}) = \lim_{s \to \infty} \int_{-s}^{s} \int_{-s}^{s} p_j(\tau) \, h_{ij}(t-\tau) \, \exp(-i\bar{\omega}t) \, dt \, d\tau \qquad (12\text{-}68)$$

When a new variable $\theta \equiv t - \tau$ is introduced, this equation becomes

$$V_{ij}(i\bar{\omega}) = \lim_{s\to\infty} \int_{-s}^{s} p_j(\tau) \exp(-i\bar{\omega}t)\, d\tau \int_{-s-\tau}^{s-\tau} h_{ij}(\theta) \exp(-i\bar{\omega}\theta)\, d\theta \quad (12\text{-}69)$$

The expanding domain of integration given by this equation is shown in Fig. 12-6a. Since the function $V_{ij}(i\bar{\omega})$ exists only when the integrals

$$I_2 \equiv \int_{-\infty}^{\infty} |p_j(\tau)|\, d\tau \qquad I_3 \equiv \int_{-\infty}^{\infty} |h_{ij}(\theta)|\, d\theta$$

are finite, which is always the case in practice due to the loadings being of finite duration and the unit-impulse-response function being a decayed function, it is valid to drop τ from the limits of the second integral in Eq. (12-69), resulting in

$$V_{ij}(i\bar{\omega}) = \left[\lim_{s\to\infty} \int_{-s}^{s} p_j(\tau) \exp(-i\bar{\omega}t)\, d\tau \right] \left[\lim_{s\to\infty} \int_{-s}^{s} h_{ij}(\theta) \exp(-i\bar{\omega}\theta)\, d\theta \right]$$
$$(12\text{-}70)$$

which changes the expanding domain of integration to that shown in Fig. 12-6b. Variable θ can now be changed to t since it is serving only as a dummy time variable. Equation (12-70) then becomes

$$V_{ij}(i\bar{\omega}) = P_j(i\bar{\omega}) \int_{-\infty}^{\infty} h_{ij}(t) \exp(-i\bar{\omega}t)\, dt \quad (12\text{-}71)$$

When it is noted that Eq. (12-65) in its inverse form gives

$$V_{ij}(i\bar{\omega}) = H_{ij}(i\bar{\omega})\, P_j(i\bar{\omega}) \quad (12\text{-}72)$$

a comparison of Eqs. (12-71) and (12-72) makes it apparent that

$$H_{ij}(i\bar{\omega}) = \int_{-\infty}^{\infty} h_{ij}(t) \exp(-i\bar{\omega}t)\, dt$$

and $(12\text{-}73)$

$$h_{ij}(t) = \frac{1}{2\pi} \int_{-\infty}^{\infty} H_{ij}(i\bar{\omega}) \exp(i\bar{\omega}t)\, d\bar{\omega}$$

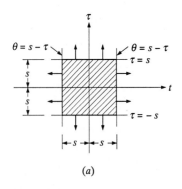
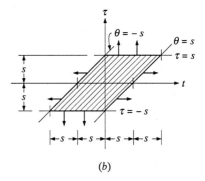

(a)   (b)

FIGURE 12-6
Expanding domains of integration.

This derivation shows that any unit-impulse-response transfer function $h_{ij}(t)$ and the corresponding complex-frequency-response transfer function $H_{ij}(i\overline{\omega})$ are Fourier transform pairs, provided damping is present in the system. This is a requirement for mathematical stability to exist.

Example E12-5.   Show that the complex-frequency-response function given by Eq. (6-52) and the unit-impulse-response function given by Eq. (6-51) are Fourier transform pairs in accordance with Eqs. (6-53) and (6-54) which correspond to Eqs. (12-73).

Substituting Eq. (6-52) into Eq. (6-54) gives

$$h(t) = \frac{-1}{2\pi \, m \, \omega} \int_{-\infty}^{\infty} \frac{\exp(i\omega\beta t)}{(\beta - r_1)(\beta - r_2)} \, d\beta \qquad (a)$$

after introducing

$$\beta = \frac{\overline{\omega}}{\omega} \qquad\qquad k = m\,\omega^2 \qquad (b)$$

$$r_1 = i\xi + \sqrt{1 - \xi^2} \qquad r_2 = i\xi - \sqrt{1 - \xi^2} \qquad (c)$$

The integration of Eq. (a) is best carried out using the complex β plane and contour integration as indicated in Fig. E12-1. The integrand in the integral is an analytic function everywhere in the β plane except at $\beta = r_1$ and $\beta = r_2$. At these two points, poles of order 1 exist, for damping in the ranges $0 < \xi < 1$

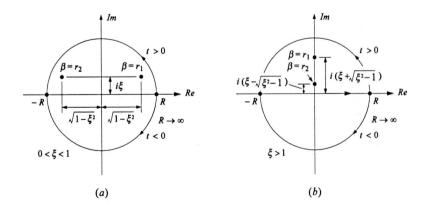

FIGURE E12-1
Poles for the integrand function of Eq. (a).

and $\xi > 1$. Note that for $\xi = 1$, points $\beta = r_1$ and $\beta = r_2$ coincide at location $(0, i)$, thus forming a single pole of order 2 in this case. The arrows along the closed paths in Fig. E12-1 indicate the directions of contour integration for the ranges of time shown. The poles mentioned above have residues as follows:

$$\text{Res}(\beta = r_1) = \frac{\exp[i\omega(i\xi + \sqrt{1 - \xi^2})\,t]}{2\sqrt{1 - \xi^2}} \qquad 0 < \xi < 1; \xi > 1$$

$$\text{Res}(\beta = r_2) = \frac{\exp[i\omega(i\xi - \sqrt{1 - \xi^2})\,t]}{-2\sqrt{2 - \xi^2}} \qquad 0 < \xi < 1; \xi > 1 \qquad \text{(d)}$$

$$\text{Res}(\beta = r_1 = r_2) = i\omega t \, \exp(-\omega t) \qquad \xi = 1$$

According to Cauchy's residue theorem, the integral in Eq. (a) equals $-2\pi i \sum \text{Res}$ and $+2\pi i \sum \text{Res}$ when integration is clockwise and counterclockwise, respectively, around a closed path and when the integral is analytic along the entire path, as in the case treated here. Thus one obtains the results

$$
h(t) = \begin{cases}
\begin{aligned}
&\frac{-2\pi i}{2\pi m\omega} \times \left\{ \frac{\exp[i\omega(i\xi + \sqrt{1 - \xi^2})t]}{2\sqrt{1 - \xi^2}} \right. \\
&\qquad \left. + \frac{\exp[i\omega(i\xi - \sqrt{1 - \xi^2})t]}{-2\sqrt{1 - \xi^2}} \right\} && t > 0 \\
&0 && t < 0
\end{aligned} & \Big\} \ 0 < \xi < 1 \\[2em]
\begin{aligned}
&\frac{-2\pi i}{2\pi m\omega} \times \left(\frac{\exp\{i\omega[i(\xi + \sqrt{\xi^2 - 1})]t\}}{2i\sqrt{\xi^2 - 1}} \right. \\
&\qquad \left. + \frac{\exp\{i\omega[i(\xi - \sqrt{\xi^2 - 1})]t\}}{-2i\sqrt{\xi^2 - 1}} \right) && t > 0 \\
&0 && t < 0
\end{aligned} & \Big\} \ \xi > 1 \\[2em]
\begin{aligned}
&\frac{-2\pi i}{2\pi m\omega} [i\omega t \, \exp(-\omega t)] && t > 0 \\
&0 && t < 0
\end{aligned} & \Big\} \ \xi = 1
\end{cases}
$$

(e)

It is easily shown that Eqs. (e) reduce to

$$
h(t) = \begin{cases}
\begin{cases}
\dfrac{1}{\omega_D m} \sin \omega_D t \, \exp(-\omega \xi t) & t > 0 \\
0 & t < 0
\end{cases} \quad \Big\} \quad 0 < \xi < 1 \\[3em]
\begin{cases}
\dfrac{1}{2\omega m \sqrt{\xi^2 - 1}} \exp(-\omega \xi t) & \\
\quad \times [\exp(\omega \sqrt{\xi^2 - 1}\, t) - \exp(-\omega \sqrt{\xi^2 - 1}\, t)] & t > 0 \\
0 & t < 0
\end{cases} \Big\} \quad \xi > 1 \\[3em]
\begin{cases}
\dfrac{t}{m} \exp(-\omega t) & t > 0 \\
0 & t < 0
\end{cases} \quad \Big\} \quad \xi = 1
\end{cases}
\tag{f}
$$

in which $\omega_D = \omega \sqrt{1 - \xi^2}$. Note that the first of Eqs. (f) does indeed agree with Eq. (6-51), thus showing the validity of Eqs. (6-53) and (6-54) in this case. Note also that the inverse Fourier transform of $H(i\overline{\omega})$ yields the unit-impulse response functions for all values of damping, i.e., for $0 < \xi < 1$, $\xi > 1$, and $\xi = 1$.

12-8 PRACTICAL PROCEDURE FOR SOLVING COUPLED EQUATIONS OF MOTION

The solution of coupled sets of equations of motion is carried out most easily in the frequency domain; therefore, this section will be devoted to developing procedures for this approach only. In doing so, consideration will be given to three different sets of equations as expressed in the frequency domain by

$$
\left[(\mathbf{k} - \overline{\omega}^2 \, \mathbf{m}) + i\, \hat{\mathbf{k}}\right] \mathbf{V}(i\overline{\omega}) = \mathbf{P}(i\overline{\omega})
\tag{12-74}
$$

$$
\left[(\mathbf{k} - \overline{\omega}^2 \, \mathbf{m}) + i\, (\overline{\omega}\, \mathbf{c})\right] \mathbf{V}(i\overline{\omega}) = \mathbf{P}(i\overline{\omega})
\tag{12-75}
$$

$$
\left[(K - \overline{\omega}^2 \, M) + i\, (\overline{\omega}\, C)\right] \mathbf{Y}(i\overline{\omega}) = \overline{\mathbf{P}}(i\overline{\omega})
\tag{12-76}
$$

in which the complex matrix in the bracket term on the left hand side of each equation is the impedance (or dynamic stiffness) matrix for the complete structural system being represented.

Equation (12-74) represents a complete N-DOF system using the complex-stiffness form of damping equivalent to Eq. (3-79) for the SDOF system. Matrix $\hat{\mathbf{k}}$ in this equation is a stiffness matrix for the entire system obtained by assembling individual finite-element stiffness matrices $\hat{\mathbf{k}}^{(m)}$ [superscript (m) denotes element m] of the form

$$
\hat{\mathbf{k}}^{(m)} = 2\, \xi^{(m)}\, \mathbf{k}^{(m)}
\tag{12-77}
$$

in which $\mathbf{k}^{(m)}$ denotes the individual elastic stiffness matrix for finite element m as used in the assembly process to obtain matrix $\mathbf{k}$ for the entire system; and $\xi^{(m)}$ is a damping ratio selected to be appropriate for the material used in finite element m. If the material is the same throughout the system so that the same damping ratio is used for each element, i.e., $\xi^{(1)} = \xi^{(2)} = \cdots = \xi$, then the overall system matrix $\hat{\mathbf{k}}$ would be proportional to $\mathbf{k}$ as given by $\hat{\mathbf{k}} = 2\xi\,\mathbf{k}$. Matrix $\hat{\mathbf{k}}$ would then possess the same orthogonality property as $\mathbf{k}$. However, when different materials are included in the system, e.g., soil and steel, the finite elements consisting of these materials would be assigned different values of $\xi^{(m)}$. In this case, the assembled matrix $\hat{\mathbf{k}}$ would not satisfy the orthogonality condition, and modal coupling would be present. Vectors $\mathbf{V}(i\overline{\omega})$ and $\mathbf{P}(i\overline{\omega})$ in Eq. (12-74) are the Fourier transforms of vectors $\mathbf{v}(t)$ and $\mathbf{p}(t)$, respectively, and all other quantities are the same as previously defined.

Equation (12-75) is the Fourier transform of Eq. (12-60) which represents an N-DOF system having the viscous form of damping. Using the solution procedure developed subsequently in this section, it is not necessary for matrix $\mathbf{c}$ to satisfy the orthogonality condition. Therefore, the case of modal coupling through damping can be treated, whether it is of the viscous form or of the complex-stiffness form described above.

Equation (12-76) gives the normal mode equations of motion [Eq. (12-58)] in the frequency domain, in which $\overline{\mathbf{P}}(i\overline{\omega})$ is the Fourier transform of the generalized (modal) loading vector $P(t)$ which contains components $P_1(t)$, $P_2(t)$, $\cdots$, $P_n(t)$ as defined by Eq. (12-12c), $\mathbf{Y}(i\overline{\omega})$ is the Fourier transform of the normal coordinate vector $Y(t)$, K and M are the diagonal normal mode stiffness and mass matrices containing elements in accordance with Eqs. (12-12b) and (12-12a), respectively, and C is the normal mode damping matrix having elements as given by Eq. (12-15a). As noted earlier, if the damping matrix $\mathbf{c}$ possesses the orthogonality property, matrix C will be of diagonal form; however, if matrix $\mathbf{c}$ does not possess the orthogonality property, the modal damping matrix will be full. The analysis procedure developed subsequently can treat this coupled form of matrix without difficulty, however. Note that Eqs. (12-76) may contain all N normal mode equations or only a smaller specified number representing the lower modes according to the degree of approximation considered acceptable. Reducing the number of equations to be solved does not change the analysis procedure but it does reduce the computational effort involved.

To develop the analysis procedure, let us consider only Eq. (12-74) since the procedure is applied to the other cases [Eqs. (12-75) and (12-76)] in exactly the same way. Equation (12-74) may be written in the abbreviated form:

$$\mathbf{I}(i\overline{\omega})\,\mathbf{V}(i\overline{\omega}) = \mathbf{P}(i\overline{\omega}) \tag{12-78}$$

in which the impedance matrix $\mathbf{I}(i\overline{\omega})$ is given by the entire bracket matrix on the left hand side. Premultiplying both sides of this equation by the inverse of the impedance matrix, response vector $\mathbf{V}(i\overline{\omega})$ can be expressed in the form

$$\mathbf{V}(i\overline{\omega}) = \mathbf{I}(i\overline{\omega})^{-1} \, \mathbf{P}(i\overline{\omega}) \tag{12-79}$$

which implies that multiplying a complex matrix by its inverse results in the identity matrix, similar to the case involving a real matrix. The inversion procedure is the same as that involving a real matrix with the only difference being that the coefficients involved are complex rather than real. Although computer programs are readily available for carrying out this type of inversion solution, it is impractical for direct use as it involves inverting the $N \times N$ complex impedance matrix for each of the closely-spaced discrete values of $\overline{\omega}$ as required in performing the fast Fourier transform (FFT) of loading vector $\mathbf{p}(t)$ to obtain the vector $\mathbf{P}(i\overline{\omega})$; this approach requires an excessive amount of computer time. The required time can be reduced to a practical level, however, by first solving for the complex-frequency-response transfer functions $H_{ij}(i\overline{\omega})$ at a set of widely-spaced discrete values of $\overline{\omega}$, and then using an effective and efficient interpolation procedure to obtain the transfer functions at the intermediate closely-spaced discrete values of $\overline{\omega}$ required by the FFT procedure.

The complex-frequency-response transfer functions $H_{ij}(i\overline{\omega})$ are obtained for the widely spaced discrete values of $\overline{\omega}$ using Eq. (12-79) consistent with the definition of these functions given previously; that is, using

$$< H_{1j}(i\overline{\omega}) \quad H_{2j}(i\overline{\omega}) \quad \cdots \quad H_{Nj}(i\overline{\omega}) >^{T} = \mathbf{I}(i\overline{\omega})^{-1} \, \mathbf{I}_j \qquad j = 1, 2, \cdots, N \tag{12-80}$$

in which $\mathbf{I}_j$ denotes an N-component vector containing all zeros except for the jth component which equals unity. Because these transfer functions are smooth, as indicated in Fig. 12-7, even though they peak at the natural frequencies of the system, interpolation can be used effectively to obtained their complex values at the intermediate closely-spaced discrete values of $\overline{\omega}$. Note that natural frequencies can be obtained, corresponding to the frequencies at the peaks in the transfer functions, without solving

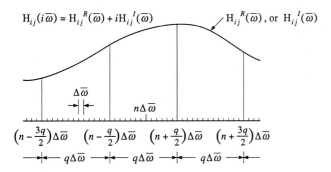

FIGURE 12-7
Interpolation of transfer function.

the eigenvalue problem. The effective interpolation procedure required to carry out the analysis in this way will be developed in the following Section 12-9.

Having obtained all transfer functions $H_{ij}(i\overline{\omega})$ using Eq. (12-80) and the interpolation procedure of Section 12-9, the response vector $\mathbf{V}(i\overline{\omega})$ is easily obtained by superposition using

$$\mathbf{V}(i\overline{\omega}) = \mathbf{H}(i\overline{\omega})\,\mathbf{P}(i\overline{\omega}) \qquad (12\text{-}81)$$

in which $\mathbf{H}(i\overline{\omega})$ is the $N \times N$ complex-frequency-response transfer matrix

$$\mathbf{H}(i\overline{\omega}) = \begin{bmatrix} H_{11}(i\overline{\omega}) & H_{12}(i\overline{\omega}) & \cdots & H_{1N}(i\overline{\omega}) \\ H_{21}(i\overline{\omega}) & H_{22}(i\overline{\omega}) & \cdots & H_{2N}(i\overline{\omega}) \\ \vdots & \vdots & \vdots & \vdots \\ H_{N1}(i\overline{\omega}) & H_{N2}(i\overline{\omega}) & \cdots & H_{NN}(i\overline{\omega}) \end{bmatrix} \qquad (12\text{-}82)$$

obtained for each frequency required in the response analysis. Note that once this transfer matrix has been obtained, the responses of the system to multiple sets of loadings can be obtained very easily by simply Fourier transforming each set by the FFT procedure and then multiplying the resulting vector set in each case by the transfer matrix in accordance with Eq. (12-81). Having vector $\mathbf{V}(i\overline{\omega})$ for each set, it can be inverse transformed by the FFT procedure to obtain the corresponding set of displacements in vector $\mathbf{v}(t)$.

It is evident that by Fourier transforming each element $H_{ij}(i\overline{\omega})$ in Eq. (12-82), one could easily obtain the corresponding unit-impulse-response function $h_{ij}(t)$ as shown by the second of Eqs. (12-73). This is of academic interest only, however, as one would not use the convolution integral formulation given by Eq. (12-63) to evaluate the response of a complicated structural system.

12-9 INTERPOLATION PROCEDURE FOR GENERATION OF TRANSFER FUNCTIONS

Because both the real and imaginary parts of a complex-frequency-response transfer function are smooth functions of $\overline{\omega}$, interpolation of their values at equal intervals $\Delta\overline{\omega}$ over relatively wide frequency bands can be done effectively using an interpolation function corresponding to the forms of the complex-frequency-response transfer functions for a 2-DOF system having the complex-stiffness uncoupled-type of damping. The frequency-domain normal mode equations of motion for such a system are

$$\left[(K_1 - \overline{\omega}^2 M_1) + i\,(2\xi\,K_1)\right]\,Y_1(i\overline{\omega}) = \phi_1^T\,\mathbf{P}(i\overline{\omega}) \qquad (12\text{-}83)$$

$$\left[(K_2 - \overline{\omega}^2 M_2) + i\,(2\xi\,K_2)\right]\,Y_2(i\overline{\omega}) = \phi_2^T\,\mathbf{P}(i\overline{\omega}) \qquad (12\text{-}84)$$

in which vector $\mathbf{P}(i\bar{\omega})$ is the Fourier transform of loading vector $\mathbf{p}(t)$.

Let us now generate a single complex-frequency-response transfer function, e.g., $H_{11}(i\bar{\omega})$, which is the transfer function between loading $p_1(t)$ and displacement $v_1(t)$. In the frequency domain $v_1(t)$ is given in terms of the normal mode coordinates by

$$V_1(i\bar{\omega}) = \phi_{11} \, Y_1(i\bar{\omega}) + \phi_{12} Y_2(i\bar{\omega}) \qquad (12\text{-}85)$$

To generate H_{11}, let $\mathbf{P}(i\bar{\omega}) = <1 \ 0>^T$ giving

$$\boldsymbol{\phi}_1^T \, \mathbf{P}(i\bar{\omega}) = <\phi_{11} \quad \phi_{21}> \ <1 \ 0>^T = \phi_{11}$$

and

$$\boldsymbol{\phi}_2^T \, \mathbf{P}(i\bar{\omega}) = <\phi_{12} \quad \phi_{22}> \ <1 \ 0>^T = \phi_{12}$$

in which case, substituting the resulting values of $Y_1(i\bar{\omega})$ and $Y_2(i\bar{\omega})$ given by Eqs. (12-83) and (12-84), respectively, into Eq. (12-85) gives $V_1(i\bar{\omega}) = H_{11}(i\bar{\omega})$. Taking this action, one obtains

$$H_{11}(i\bar{\omega}) = \frac{\phi_{11}^2}{[(K_1 - \bar{\omega}^2 M_1) + i\,(2\xi_1 K_1)]} + \frac{\phi_{12}^2}{[(K_2 - \bar{\omega}^2 M_2) + i\,(2\xi_2 K_2)]} \qquad (12\text{-}86)$$

By operating on this equation, it can be put in the equivalent single-fraction form

$$H_{11}(i\bar{\omega}) = \frac{A\,\bar{\omega}^2 + B}{\bar{\omega}^4 + C\,\bar{\omega}^2 + D} \qquad (12\text{-}87)$$

in which A is a real constant and B, C, and D are complex constants, all expressed in terms of the known quantities in Eq. (12-86). The forms of these expressions are of no interest, however, as only the functional form of $H_{11}(i\bar{\omega})$ with respect to $\bar{\omega}$ is needed. Repeating the above development, one finds that each of the other three transfer functions $H_{12}(i\bar{\omega})$, $H_{21}(i\bar{\omega})$, and $H_{22}(i\bar{\omega})$ has the same form as that given by Eq. (12-87).

To use Eq. (12-87) purely as an interpolation function for any transfer function $H_{ij}(i\bar{\omega})$ of a complex N-DOF system, express it in the discrete form

$$\mathbf{H}_{ij}(i\bar{\omega}_m) = \frac{A_{mn}\,\bar{\omega}_m^2 + B_{mn}}{\bar{\omega}_m^4 + C_{mn}\,\bar{\omega}_m^2 + D_{mn}} \qquad \left(n - \frac{3}{2}\,q\right) < m < \left(n + \frac{3}{2}\,q\right) \qquad (12\text{-}88)$$

in which $\bar{\omega}_m = m\,\Delta\bar{\omega}$, $\Delta\bar{\omega}$ being the constant frequency interval of the narrowly spaced discrete frequencies required by the FFT procedure in generating loading vector $\mathbf{P}(i\bar{\omega})$, and A_{mn}, B_{mn}, C_{mn}, and D_{mn} are all treated as complex constants, even though coefficient A in Eq. (12-87) for a 2-DOF system is real. These four constants are evaluated by applying Eq. (12-88) separately to four consecutive widely-spaced discrete values of $\bar{\omega}$, as given by $m = \left(n - \frac{3}{2}\,q\right)$, $m = \left(n - \frac{1}{2}\,q\right)$, $m = \left(n + \frac{1}{2}\,q\right)$, and $m = \left(n + \frac{3}{2}q\right)$, as shown in Fig. 12-7, in which q represents the number of closely-spaced frequency intervals within one of the widely-spaced intervals. Knowing $H_{ij}(i\bar{w})$ for the above four values of m as obtained using Eq. (12-80), separate applications of Eq. (12-88) to the corresponding four values of $\bar{w}_m$ yields four simultaneous complex algebraic equations involving unknowns A_{mn}, B_{mn}, C_{mn}, and D_{mn}. Solving for these constants and entering their numerical values back into Eq. (12-88), this equation can be used to calculate the intermediate values of $H_{ij}(i\bar{w}_m)$ at the closely-spaced discrete frequencies in the range $\left(n - \frac{3}{2}q\right) < m < \left(n + \frac{3}{2}q\right)$. This same procedure is then repeated for $n = \frac{3}{2}q$, $\frac{9}{2}q$, $\frac{15}{2}q$, $\frac{21}{2}q$, $\cdots$ so as to cover the entire range of frequencies of interest. Better accuracy can be obtained by this interpolation if it is applied over only the central frequency interval, i.e., over the range $\left(n - \frac{1}{2}q\right) < m < \left(n + \frac{1}{2}q\right)$; this is a greater computational task, however, because the set of constants in Eq. (12-88) then must be evaluated for $n = \frac{3}{2}q$, $\frac{5}{2}q$, $\frac{7}{2}q$, $\cdots$.

To set the optimum value of q, considering computational effort and accuracy, requires considerable experience with the procedure. While it is difficult to provide guidelines for this purpose, one should at least be aware that the frequency interval of $3\,q\,\Delta\bar{w}$ should never include more than two natural frequencies because the form of the interpolation function is that of a transfer function for a 2-DOF system, for which only two peaks can be represented.

PROBLEMS

12-1. A cantilever beam supporting three equal lumped masses is shown in Fig. P12-1; also listed there are its undamped mode shapes $\boldsymbol{\Phi}$ and frequencies of vibration $\boldsymbol{\omega}$. Write an expression for the dynamic response of mass 3 of this system after an 8-*kips* step function load is applied at mass 2 (i.e., 8 *kips* is suddenly applied at time $t = 0$ and remains on the structure permanently), including all three modes and neglecting damping. Plot the history of response $v_3(t)$ for the time interval $0 < t < T_1$ where $T_1 = 2\pi/\omega_1 = 2\pi/3.61$.

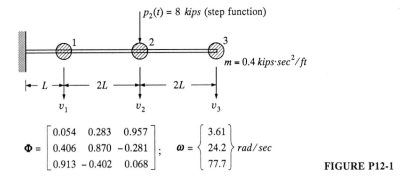

$$\boldsymbol{\Phi} = \begin{bmatrix} 0.054 & 0.283 & 0.957 \\ 0.406 & 0.870 & -0.281 \\ 0.913 & -0.402 & 0.068 \end{bmatrix}; \qquad \boldsymbol{\omega} = \begin{Bmatrix} 3.61 \\ 24.2 \\ 77.7 \end{Bmatrix} rad/sec$$

FIGURE P12-1

12-2. Consider the beam of Prob. 12-1, but assume that a harmonic load is applied to mass 2, $p_2(t) = 3\,k\,\sin\bar{\omega}t$, where $\bar{\omega} = \frac{3}{4}\omega_1$.

(a) Write an expression for the steady-state response of mass 1, assuming that the structure is undamped.

(b) Evaluate the displacements of all masses at the time of maximum steady-state response and plot the deflected shape at that time.

12-3. Repeat part (a) of Prob. 12-2, assuming that the structure has 10 percent critical damping in each mode.

12-4. The mass and stiffness properties of a three-story shear building, together with its undamped vibration mode shapes and frequencies, are shown in Fig. P12-2. The structure is set into free vibration by displacing the floors as follows: $v_1 = 0.3$ in, $v_2 = -0.8$ in, and $v_3 = 0.3$ in, and then releasing them suddenly at time $t = 0$. Determine the displaced shape at time $t = 2\pi/\omega_1$:

(a) Assuming no damping.

(b) Assuming $\xi = 10\%$ in each mode.

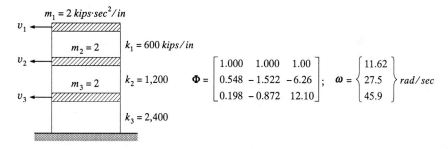

$$m_1 = 2\ kips\cdot sec^2/in$$

$$k_1 = 600\ kips/in$$
$$m_2 = 2$$
$$k_2 = 1{,}200$$
$$m_3 = 2$$
$$k_3 = 2{,}400$$

$$\Phi = \begin{bmatrix} 1.000 & 1.000 & 1.00 \\ 0.548 & -1.522 & -6.26 \\ 0.198 & -0.872 & 12.10 \end{bmatrix}; \quad \omega = \begin{Bmatrix} 11.62 \\ 27.5 \\ 45.9 \end{Bmatrix}\ rad/sec$$

FIGURE P12-2

12-5. The building of Prob. 12-4b is subjected to a harmonic loading applied at the top floor: $p_1(t) = 5k \sin \bar{\omega}t$, where $\bar{\omega} = 1.1\,\omega_1$. Evaluate the steady-state amplitude of motion at the three floor levels and the phase angle θ between the applied load vector and the displacement response vector at each floor.

12-6. Assuming that the building of Prob. 12-4 has Rayleigh damping, by using Eqs. (12-40) and (12-38a) evaluate a damping matrix for the structure which will provide 5 percent and 15 percent damping ratios in the first and third modes, respectively. What damping ratio will this matrix give in the second mode?

12-7. For the building of Prob. 12-4, evaluate a viscous damping matrix that will provide 8 percent, 10 percent, and 12 percent critical damping in the first, second, and third modes, respectively. Use Eq. (12-57a) to obtain the coefficient a_1 that corresponds to the third mode frequency and the required third mode damping. Then form the damping matrix by combining the resulting stiffness proportional contribution ($c_s = a_1k$) with the contributions from the first two modes given by Eq. (12-56c) using the required supplementary damping ratios given by Eq. (12-57c).

CHAPTER

13

VIBRATION ANALYSIS BY MATRIX ITERATION

13-1 PRELIMINARY COMMENTS

It is evident from the preceding discussion that the mode displacement superposition method provides an efficient means of evaluating the dynamic response of most structures — those for which the undamped mode shapes serve to uncouple the equations of motion. The response analysis for the individual modal equations requires very little computational effort, and in most cases only a relatively small number of the lowest modes of vibration need be included in the superposition. In this regard, it is important to realize that the physical properties of the structure and the characteristics of the dynamic loading generally are known only approximately; hence the structural idealization and the solution procedure should be formulated to provide only a corresponding level of accuracy. Nevertheless, the mathematical models developed to solve practical problems in structural dynamics range from very simplified systems having only a few degrees of freedom to highly sophisticated finite-element models including hundreds or even thousands of degrees of freedom in which as many as 50 to 100 modes may contribute significantly to the response. To deal effectively with these practical problems, much more efficient means of vibration analysis are needed than the determinantal solution procedure described earlier, and this chapter describes the matrix iteration approach which is the basis of many of the vibration or "eigenproblem" solution techniques that are used in practice.

The basic concept is explained first with reference to the simplest application, the evaluation of the fundamental (or first-) mode shape and frequency of an N-degree-of-freedom system. This is followed by a proof of the fact that the iteration will converge to the first-mode properties; the essential concept of the proof is then used as a means for evaluating the higher modes of vibration, one mode at a time in sequence. Because this procedure involves increasing computational costs as more modes are calculated, an alternative method that employs "shifting" of the eigenvalues (frequencies) is described. Also included is a brief discussion of elastic buckling, noting that both the vibrations and buckling are represented by equivalent eigenproblem equations.

13-2 FUNDAMENTAL MODE ANALYSIS

The use of iteration to evaluate the fundamental vibration mode of a structure is a very old concept that originally was called the Stodola method after its originator. Now it is recognized to be part of a broad segment of structural mechanics in which iteration procedures are used. The starting point of this formulation is the statement of the undamped free-vibration equations of motion given by Eq. (11-33):

$$\mathbf{k}\,\hat{\mathbf{v}}_n = \omega_n^2\,\mathbf{m}\,\hat{\mathbf{v}}_n$$

This equation expresses the fact that in undamped free vibrations, the inertial forces induced by the motion of the masses must be equilibrated by the elastic forces resulting from the system deformations. This equilibrium will be satisfied only if the displacements $\hat{\mathbf{v}}_n$ are in the shape of the nth mode of vibration and are varying harmonically at the nth-mode frequency ω_n. Expressing the inertial forces on the right hand side of Eq. (11-33) as

$$\mathbf{f}_{I_n} = \omega_n^2\,\mathbf{m}\,\hat{\mathbf{v}}_n \tag{13-1}$$

the displacements resulting from these forces may be calculated by solving the static deflection problem

$$\hat{\mathbf{v}}_n = \mathbf{k}^{-1}\,\mathbf{f}_{I_n} \tag{13-2}$$

or using Eq. (13-1),

$$\hat{\mathbf{v}}_n = \omega_n^2\,\mathbf{k}^{-1}\,\mathbf{m}\,\hat{\mathbf{v}}_n \tag{13-3}$$

The matrix product in this expression summarizes the dynamic properties of the structure. It is called the *dynamic* matrix, denoted as

$$\mathbf{D} \equiv \mathbf{k}^{-1}\,\mathbf{m} \tag{13-4}$$

and when this is introduced, Eq. (13-3) becomes

$$\hat{\mathbf{v}}_n = \omega_n^2\,\mathbf{D}\,\hat{\mathbf{v}}_n \tag{13-5}$$

To initiate the iteration procedure for evaluating the first-mode shape, a trial displacement vector $\mathbf{v}_1^{(0)}$ is assumed that is a reasonable estimate of this shape. The zero superscript indicates that this is the initial shape used in the iteration sequence; for convenience the vector is normalized so that a selected reference element is unity. Introducing this on the right side of Eq. (13-1) gives an expression for the inertial forces induced by the system masses moving harmonically in this shape at the as yet unknown vibration frequency

$$\mathbf{f}_{I_1}^{(0)} = \omega_1^2 \, \mathbf{m} \, \mathbf{v}_1^{(0)} \tag{13-6}$$

The displacement vector resulting from applying these forces in Eq. (13-2) is a better approximation of the first-mode shape than was the initial vector, and it may be expressed in a form equivalent to Eq. (13-5) as follows:

$$\mathbf{v}_1^{(1)} = \omega_1^2 \, \mathbf{D} \, \mathbf{v}_1^{(0)} \tag{13-5a}$$

where the "one" superscript indicates that this is the result of the first cycle of iteration.

It is evident that the amplitude of this vector depends on the unknown frequency, but only the shape is needed in the iteration process so the frequency is dropped from the expression and the resulting improved shape is denoted by a bar over the vector symbol:

$$\bar{\mathbf{v}}_1^{(1)} \equiv \mathbf{D} \, \mathbf{v}_1^{(0)} \tag{13-7}$$

Then the improved iteration vector is obtained finally by normalizing this shape, dividing it by an arbitrary reference element of the vector, $\text{ref}(\bar{\mathbf{v}}_1^{(1)})$; thus,

$$\mathbf{v}_1^{(1)} = \frac{\bar{\mathbf{v}}_1^{(1)}}{\text{ref}(\bar{\mathbf{v}}_1^{(1)})} \tag{13-8}$$

which has the effect of scaling the reference element of the vector to unity. In principle any element of the improved shape vector $\bar{\mathbf{v}}_1^{(1)}$ (except for zero elements) could be used as the reference or normalizing factor in Eq. (13-8), but the best results generally are obtained by normalizing with the largest element of the vector, designated $\max(\bar{\mathbf{v}}_1^{(1)})$; thus $\max(\bar{\mathbf{v}}_1^{(1)}) \equiv \text{ref}(\bar{\mathbf{v}}_1^{(1)})$ is used as the denominator in the standard iteration procedure.

Now if it is assumed that the computed displacement vector is the same as the initially assumed vector (as it would be if it were the true mode shape), Eq. (13-5a) can be used to obtain an approximate value of the vibration frequency. Introducing Eq. (13-7) on the right side of Eq. (13-5a) and then assuming the new vector is approximately equal to the initial vector lead to

$$\mathbf{v}_1^{(1)} = \omega_1^2 \, \bar{\mathbf{v}}_1^{(1)} \doteq \mathbf{v}_1^{(0)}$$

Considering any arbitrary degree of freedom, k, in the vector then provides an expression that may be solved to obtain an approximation of the frequency

$$\omega_1^2 \doteq \frac{v_{k1}^{(0)}}{\overline{v}_{k1}^{(1)}} \tag{13-9}$$

If the assumed shape were a true mode shape, then the same frequency would be obtained by taking the ratio expressed in Eq. (13-9) for any degree of freedom of the structure. In general, however, the derived shape $\mathbf{v}_1^{(1)}$ will differ from $\mathbf{v}_1^{(0)}$ and a different frequency will be obtained for each displacement coordinate. In this case, the true first-mode frequency lies between the maximum and minimum values obtainable from Eq. (13-9):

$$\left(\frac{v_{k1}^{(0)}}{\overline{v}_{k1}^{(1)}}\right)_{\min} < \omega_1^2 < \left(\frac{v_{k1}^{(0)}}{\overline{v}_{k1}^{(1)}}\right)_{\max} \tag{13-10}$$

Because of this fact, it is evident that a better approximation of the frequency can be obtained by an averaging process. Often the best averaging procedure involves including the mass distribution as a weighting factor. Thus writing the vector equivalent of Eq. (13-9) and premultiplying numerator and denominator by $\left(\overline{\mathbf{v}}_1^{(1)}\right)^T \mathbf{m}$ give

$$\omega_1^2 \doteq \frac{(\overline{\mathbf{v}}_1^{(1)})^T \mathbf{m} \mathbf{v}_1^{(0)}}{(\overline{\mathbf{v}}_1^{(1)})^T \mathbf{m} \overline{\mathbf{v}}_1^{(1)}} \tag{13-11}$$

Equation (13-11) represents the best frequency approximation obtainable by a single iteration step, in general, from any assumed shape $\mathbf{v}_1^{(0)}$. [Its equivalence to the improved Rayleigh expression of Eq. (8-42) should be noted]. However, the derived shape $\overline{\mathbf{v}}_1^{(1)}$ is a better approximation of the first-mode shape than was the original assumption $\mathbf{v}_1^{(0)}$. Thus if $\mathbf{v}_1^{(1)}$ and its derived shape $\overline{\mathbf{v}}_1^{(2)}$ were used in Eq. (13-9) or (13-11), the resulting frequency approximations would be better than those computed from the initial assumption. By repeating the process sufficiently, the mode-shape approximation can be improved to any desired level of accuracy. In other words, after s cycles

$$\overline{\mathbf{v}}_1^{(s)} = \frac{1}{\omega_1^2} \mathbf{v}_1^{(s-1)} \doteq \frac{1}{\omega_1^2} \boldsymbol{\phi}_1 \tag{13-12}$$

in which the proportionality between $\overline{\mathbf{v}}_1^{(s)}$ and $\mathbf{v}_1^{(s-1)}$ can be achieved to any specified number of decimal places; the resulting shape is accepted as the first-mode shape. When the desired degree of convergence has been achieved, the frequency may be obtained by equating the displacements of any selected degree of freedom before and after the improvement calculation. However, the most accurate results are obtained by selecting the degree of freedom having the maximum displacement, and this also is a convenient choice because the normalizing procedure that has been adopted gives this displacement a unit value. Thus the frequency is expressed by

$$\omega_1^2 = \frac{\max\left(\mathbf{v}_1^{(s-1)}\right)}{\max\left(\bar{\mathbf{v}}_1^{(s)}\right)} = \frac{1}{\max\left(\bar{\mathbf{v}}_1^{(s)}\right)} \tag{13-13}$$

or in other words it is equal to the reciprocal of the normalizing factor used in the final iteration cycle. When the iteration has converged completely, there is no need to apply the averaging process of Eq. (13-11) to improve the result.

Example E13-1. The matrix iteration method will be demonstrated by calculating the first-mode shape and frequency of the three-story building frame of Fig. E11-1 (shown again in Fig. E13-1). Although the flexibility matrix of this structure could be obtained easily by inversion of the stiffness matrix derived in Example E11-1, it will be derived here for demonstration purposes by applying a unit load to each degree of freedom successively. By definition, the deflections resulting from these unit loads, shown in Fig. E13-1, represent the flexibility influence coefficients.

Thus the flexibility matrix of this structure is

$$\tilde{\mathbf{f}} = \mathbf{k}^{-1} = \frac{1}{3,600} \begin{bmatrix} 11 & 5 & 2 \\ 5 & 5 & 2 \\ 2 & 2 & 2 \end{bmatrix} \; in/kip$$

Multiplying this by the mass matrix gives the dynamic matrix

$$\mathbf{D} = \tilde{\mathbf{f}}\mathbf{m} = \frac{1}{3,600} \begin{bmatrix} 11 & 7.5 & 4 \\ 5 & 7.5 & 4 \\ 2 & 3 & 4 \end{bmatrix} \; sec^2$$

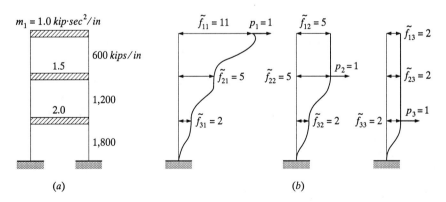

FIGURE E13-1
Frame used in example Stodola analysis: (*a*) structural system; (*b*) flexibility influence coefficients ($\times$ 3,600).

The iteration process indicated by Eq. (13-7) can conveniently be carried out in the tabular form shown below. A relatively poor trial vector $\mathbf{v}_1^{(0)}$ has been used in this example to demonstrate the good convergence of the procedure

$$
\frac{1}{3,600}
\begin{array}{c} \mathbf{D} \\ \begin{bmatrix} 11 & 7.5 & 4 \\ 5 & 7.5 & 4 \\ 2 & 3 & 4 \end{bmatrix} \end{array}
\begin{array}{c} \mathbf{v}_1^{(0)} \\ \begin{Bmatrix} 1 \\ 1 \\ 1 \end{Bmatrix} \end{array}
=
\begin{array}{c} \overline{\mathbf{v}}_1^{(1)} \\ \begin{Bmatrix} 22.50 \\ 16.50 \\ 9.00 \end{Bmatrix} \end{array}
$$

$\mathbf{v}_1^{(1)}$	$\overline{\mathbf{v}}_1^{(2)}$	$\mathbf{v}_1^{(2)}$	$\overline{\mathbf{v}}_1^{(3)}$
1.000	18.10	1.000	17.296
0.733	12.10	0.669	11.296
0.400	5.80	0.320	5.287

$\mathbf{v}_1^{(3)}$	$\overline{\mathbf{v}}_1^{(4)}$	$\mathbf{v}_1^{(4)}$	$\overline{\mathbf{v}}_1^{(5)}$
1.000	17.121	1.000	17.082
0.653	11.121	0.650	11.082
0.306	5.182	0.303	5.159

Final shape

Note that the factor 1/3,600 has not been considered in this phase of the analysis because only the relative shape is important. The shapes have been normalized by dividing by the largest displacement component [as suggested following Eq. (13-8)]. After four cycles, the shape has converged to adequate accuracy and agrees with that obtained by the determinantal approach (Example E11-2).

From Eq. (13-13) using the largest displacement component, the first-mode frequency is found to be

$$
\omega_1^2 = \frac{v_{11}^{(4)}}{\overline{v}_{11}^{(5)}} = \frac{1.000}{(1/3,600)(17.082)} = 210.77 \qquad \omega_1 = 14.52 \; rad/sec
$$

in which it will be noted that the factor 1/3,600 has now been included with the value of $\overline{v}_{11}^{(5)}$.

It also is of interest to determine the range of frequencies obtained after one cycle, as shown by Eq. (13-10):

$$
(\omega_1^2)_{min} = \frac{v_{21}^{(0)}}{\overline{v}_{21}^{(1)}} = \frac{3,600}{22.5} = 160 \qquad (\omega_1^2)_{max} = \frac{v_{31}^{(0)}}{\overline{v}_{31}^{(1)}} = \frac{3,600}{9} = 400
$$

Hence the frequency is not well established in this case after one cycle (due to the poor trial vector). However, a very good approximation can be achieved after this first cycle by applying the averaging process of Eq. (13-11):

$$\omega_1^2 = \frac{< 22.5 \quad 24.75 \quad 18.00 > \begin{Bmatrix} 1 \\ 1 \\ 1 \end{Bmatrix} (3,600)}{< 22.5 \quad 24.75 \quad 18.00 > \begin{Bmatrix} 22.5 \\ 16.5 \\ 9.0 \end{Bmatrix}} = \frac{65.25(3,600)}{1,077} = 218$$

This first-cycle approximation is identical to the improved Rayleigh method (R_{11}) demonstrated in Example E9-3.

13-3 PROOF OF CONVERGENCE

That the Stodola iteration process must converge to the first-mode shape, in general, can be demonstrated by recognizing that it essentially involves computing the inertial forces corresponding to any assumed shape, then computing the deflections resulting from those forces, then computing the inertial forces due to the computed deflections, etc. The concept is illustrated in Fig. 13-1 and explained mathematically in the following paragraph.

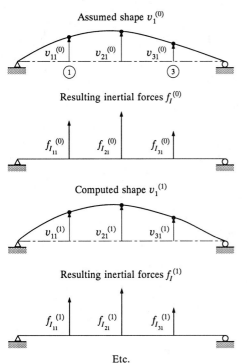

Assumed shape $v_1^{(0)}$

$v_{11}^{(0)}$ $v_{21}^{(0)}$ $v_{31}^{(0)}$

Resulting inertial forces $f_I^{(0)}$

$f_{I_{11}}^{(0)}$ $f_{I_{21}}^{(0)}$ $f_{I_{31}}^{(0)}$

Computed shape $v_1^{(1)}$

$v_{11}^{(1)}$ $v_{21}^{(1)}$ $v_{31}^{(1)}$

Resulting inertial forces $f_I^{(1)}$

$f_{I_{11}}^{(1)}$ $f_{I_{21}}^{(1)}$ $f_{I_{31}}^{(1)}$

Etc.

FIGURE 13-1
Physical interpretation of Stodola iteration sequence.

The initially assumed shape is expressed in normal coordinates [see Eq. (12-2)] as

$$\mathbf{v}_1^{(0)} = \mathbf{\Phi}\,Y^{(0)} = \boldsymbol{\phi}_1 Y_1^{(0)} + \boldsymbol{\phi}_2 Y_2^{(0)} + \boldsymbol{\phi}_3 Y_3^{(0)} + \cdots \tag{13-14}$$

in which $Y_1^{(0)}$ will be relatively large if a good guess has been made of the trial shape. The inertial forces associated with this shape vibrating at the first-mode frequency will be [see Eq. (11-33)]

$$\mathbf{f}_I{}^{(0)} = \omega_1^2\,\mathbf{m}\,\mathbf{v}_1^{(0)} = \omega_1^2\,\mathbf{m}\,\mathbf{\Phi}\,Y^{(0)} \tag{13-15}$$

Expanding $\mathbf{v}_1^{(0)}$ as in Eq. (13-14) and writing $\omega_1^2 = \omega_n^2(\omega_1/\omega_n)^2$ gives

$$\mathbf{f}_I{}^{(0)} = \mathbf{m}\left[\boldsymbol{\phi}_1\,\omega_1^2\,Y_1^{(0)} + \boldsymbol{\phi}_2\,\omega_2^2\,Y_2^{(0)}\left(\frac{\omega_1}{\omega_2}\right)^2 + \boldsymbol{\phi}_3\,\omega_3^2\,Y_3^{(0)}\left(\frac{\omega_1}{\omega_3}\right)^2 + \cdots\right] \tag{13-16}$$

The deflections derived from these inertial forces are

$$\overline{\mathbf{v}}_1^{(1)} = \mathbf{k}^{-1}\mathbf{f}_I{}^{(0)} = \mathbf{k}^{-1}\mathbf{m}\left[\boldsymbol{\phi}_1\,\omega_1^2\,Y_1^{(0)} + \boldsymbol{\phi}_2\,\omega_2^2\,Y_2^{(0)}\left(\frac{\omega_1}{\omega_2}\right)^2 + \cdots\right]$$

or

$$\overline{\mathbf{v}}_1^{(1)} = \sum_{n=1}^{N}\mathbf{D}\,\boldsymbol{\phi}_n\,\omega_n^2\,Y_n^{(0)}\left(\frac{\omega_1}{\omega_n}\right)^2 \tag{13-17}$$

Now multiplying Eq. (11-39) by $\mathbf{k}^{-1}$ shows that

$$\boldsymbol{\phi}_n = \omega_n^2\,\mathbf{D}\,\boldsymbol{\phi}_n \tag{13-18}$$

and using this in Eq. (13-17) leads to

$$\overline{\mathbf{v}}_1^{(1)} = \sum_{n=1}^{N}\boldsymbol{\phi}_n\,Y_n^{(0)}\left(\frac{\omega_1}{\omega_n}\right)^2 \tag{13-19}$$

The final improved first cycle shape $\mathbf{v}_1^{(1)}$ then is obtained by normalizing this vector, dividing it by its largest element, $\max(\overline{\mathbf{v}}_1^{(1)})$; thus

$$\mathbf{v}_1^{(1)} = \frac{\overline{\mathbf{v}}_1^{(1)}}{\max(\overline{\mathbf{v}}_1^{(1)})} = \frac{\sum_{n=1}^{N}\boldsymbol{\phi}_n\,Y_n^{(0)}\left(\frac{\omega_1}{\omega_n}\right)^2}{\max(\overline{\mathbf{v}}_1^{(1)})} \tag{13-20}$$

Following the same procedure for another cycle of iteration then leads to

$$\mathbf{v}_1^{(2)} = \frac{\overline{\mathbf{v}}_1^{(2)}}{\max(\overline{\mathbf{v}}_1^{(2)})} = \frac{\sum_{n=1}^{N}\boldsymbol{\phi}_n\,Y_n^{(0)}\left(\frac{\omega_1}{\omega_n}\right)^4}{\max(\overline{\mathbf{v}}_1^{(2)})} \tag{13-21}$$

so continuing in this fashion for s cycles gives the result

$$\mathbf{v}_1^{(s)} = \frac{\overline{\mathbf{v}}_1^{(s)}}{\max(\overline{\mathbf{v}}_1^{(s)})} = \frac{1}{\max(\overline{\mathbf{v}}_1^{(s)})}\left[\boldsymbol{\phi}_1 Y_1^{(0)} + \boldsymbol{\phi}_2 Y_2^{(0)}\left(\frac{\omega_1}{\omega_2}\right)^{2s} + \cdots\right] \qquad (13\text{-}22)$$

and noting that

$$1 \gg \left(\frac{\omega_1}{\omega_2}\right)^{2s} \gg \left(\frac{\omega_1}{\omega_3}\right)^{2s} \gg \cdots \qquad (13\text{-}23)$$

the result finally is seen to be

$$\mathbf{v}_1^{(s)} \doteq \frac{\boldsymbol{\phi}_1 Y_1^{(0)}}{\max(\boldsymbol{\phi}_1 Y_1^{(0)})} \equiv \boldsymbol{\phi}_1 \qquad (13\text{-}24)$$

It is obvious from Eq. (13-23) that the contributions of the higher modes to the vector $\mathbf{v}_1^{(s)}$ can be made as small as desired by iterating for a sufficient number of cycles; thus the procedure converges to the first-mode shape $\boldsymbol{\phi}_1$ in which the normalizing procedure produces a maximum element of unity. This convergence is contingent on having a nonzero first-mode contribution $Y_1^{(0)}$ in the initially assumed shape $\mathbf{v}_1^{(0)}$.

13-4 ANALYSIS OF HIGHER MODES

Second-Mode Analysis

The above proof of the convergence of the matrix iteration procedure to the first mode of vibration also suggests the manner in which matrix iteration can be used to evaluate higher modes as well. From Eq. (13-22) it is apparent that if the first-mode contribution in the assumed shape is zero ($Y_1^{(0)} = 0$), then the dominant contribution will be the second-mode shape; similarly, if both $Y_1^{(0)}$ and $Y_2^{(0)}$ are zero, the iteration will converge to the third-mode shape, etc. Thus to calculate the second mode it is necessary merely to assume a trial shape $\widetilde{\mathbf{v}}_2^{(0)}$ which contains no first-mode component. The tilde over the symbol designates a shape which has been purified of any first-mode contribution.

The means of eliminating the first-mode component from any assumed second-mode shape is provided by the orthogonality condition. Consider any arbitrary assumption of the second-mode shape, expressed in terms of its modal components, as follows:

$$\mathbf{v}_2^{(0)} = \boldsymbol{\Phi} \, Y^{(0)} \qquad (13\text{-}25)$$

Premultiplying both sides by $\boldsymbol{\phi}_1^T \mathbf{m}$ leads to

$$\boldsymbol{\phi}_1^T \mathbf{m} \, \mathbf{v}_2^{(0)} = \boldsymbol{\phi}_1^T \mathbf{m} \boldsymbol{\phi}_1 Y_1^{(0)} + \boldsymbol{\phi}_1^T \mathbf{m} \boldsymbol{\phi}_2 Y_2^{(0)} + \cdots \qquad (13\text{-}26)$$

in which the right hand side is reduced to a first-mode term only because of the modal orthogonality properties. Hence, Eq. (13-26) can be solved for the amplitude of the first-mode component in $\mathbf{v}_2^{(0)}$:

$$Y_1^{(0)} = \frac{\boldsymbol{\phi}_1^T \mathbf{m} \mathbf{v}_2^{(0)}}{M_1} \qquad (13\text{-}27)$$

Thus, if this component is removed from the assumed shape, the vector which remains may be said to be *purified*:

$$\widetilde{\mathbf{v}}_2^{(0)} = \mathbf{v}_2^{(0)} - \boldsymbol{\phi}_1 Y_1^{(0)} \qquad (13\text{-}28)$$

This purified trial vector will now converge toward the second-mode shape in the iteration process. However, round-off errors are introduced in the numerical operations which permit first-mode components to reappear in the trial vector; therefore it is necessary to repeat this purification operation during each cycle of the iterative solution to ensure its convergence to the second mode.

A convenient means of purifying the trial vector of the first-mode component is provided by a *sweeping* matrix, which can be derived by substituting the value of $Y_1^{(0)}$ from Eq. (13-27) into Eq. (13-28), that is,

$$\widetilde{\mathbf{v}}_2^{(0)} = \mathbf{v}_2^{(0)} - \frac{1}{M_1} \boldsymbol{\phi}_1 \boldsymbol{\phi}_1^T \mathbf{m} \mathbf{v}_2^{(0)} \equiv \mathbf{S}_1 \mathbf{v}_2^{(0)} \qquad (13\text{-}29)$$

where the first-mode sweeping matrix $\mathbf{S}_1$ is given by

$$\mathbf{S}_1 \equiv \mathbf{I} - \frac{1}{M_1} \boldsymbol{\phi}_1 \boldsymbol{\phi}_1^T \mathbf{m} \qquad (13\text{-}30)$$

As is shown by Eq. (13-29), this matrix has the property of removing the first-mode component from any trial vector to which it is premultiplied, leaving only the purified shape.

The matrix iteration procedure can now be formulated with this sweeping matrix so that it converges toward the second mode of vibration. In this case, Eq. (13-7) can be written

$$\frac{1}{\omega_2^2} \widetilde{\mathbf{v}}_2^{(1)} = \mathbf{D} \widetilde{\mathbf{v}}_2^{(0)} \qquad (13\text{-}31)$$

which states that a second-mode trial shape which contains no first-mode component will converge toward the second mode. Substituting Eq. (13-29) into Eq. (13-31) gives

$$\frac{1}{\omega_2^2} \mathbf{v}_2^{(1)} = \mathbf{D} \ \mathbf{S}_1 \mathbf{v}_2^{(0)} \equiv \mathbf{D}_2 \mathbf{v}_2^{(0)} \qquad (13\text{-}32)$$

where

$$\mathbf{D}_2 \equiv \mathbf{D} \mathbf{S}_1 \qquad (13\text{-}33)$$

is a new dynamic matrix which eliminates the first-mode component from any trial shape $\mathbf{v}_2^{(0)}$ and thus automatically converges toward the second mode. When $\mathbf{D}_2$ is used, the second-mode analysis is entirely equivalent to the first-mode analysis discussed above. Thus the frequency can be approximated by the equivalent of Eq. (13-11):

$$\omega_2^2 \doteq \frac{(\bar{\mathbf{v}}_2^{(1)})^T \mathbf{m} \, \mathbf{v}_2^{(0)}}{\bar{\mathbf{v}}_2^{(1)} \mathbf{m} \, \bar{\mathbf{v}}_2^{(1)}} \tag{13-34}$$

in which

$$\bar{\mathbf{v}}_2^{(1)} = \mathbf{D}_2 \mathbf{v}_2^{(0)}$$

or the analysis may be carried to any desired level of convergence. It is obvious that the first mode must be evaluated before the second mode can be determined by this method. Also, the first-mode shape $\boldsymbol{\phi}_1$ must be determined with considerable accuracy in evaluating the sweeping matrix $\mathbf{S}_1$ if satisfactory results are to be obtained in the second-mode analysis. In general, the second-mode-shape ordinates will have about one less significant figure than the first-mode values.

Example E13-2.   To demonstrate the matrix iteration analysis of a higher vibration mode, the second mode of the building of Example E13-1 will be calculated. The sweeping matrix to eliminate any first-mode displacement contribution is given by Eq. (13-30), which is repeated here for convenience:

$$\mathbf{S}_1 = \left[\mathbf{I} - \frac{1}{M_1}\boldsymbol{\phi}_1\boldsymbol{\phi}_1^T\mathbf{m}\right]$$

Using the following data from Example E13-1,

$$\boldsymbol{\phi}_1^T = [1.000 \quad 0.6485 \quad 0.3018] \qquad \mathbf{m} = \begin{bmatrix} 1 & 0 & 0 \\ 0 & 1.5 & 0 \\ 0 & 0 & 2.0 \end{bmatrix} \; kips/in \cdot sec^2$$

and noting that the first-mode generalized mass is $M_1 = 1.8174 \; kips/in \cdot sec^2$ the second term in the sweeping matrix is found to be

$$\frac{1}{M_1}\boldsymbol{\phi}_1\boldsymbol{\phi}_1^T\mathbf{m} = \begin{bmatrix} 0.55157 & 0.53654 & 0.33293 \\ 0.35770 & 0.34795 & 0.21590 \\ 0.16646 & 0.16193 & 0.10048 \end{bmatrix}$$

Introducing this in Eq. (13-30), the sweeping matrix becomes

$$\mathbf{S}_1 = \begin{bmatrix} 0.44843 & -0.53654 & -0.33293 \\ -0.35770 & 0.65205 & -0.21590 \\ -0.16646 & -0.16193 & 0.89952 \end{bmatrix}$$

and then using the dynamic matrix $\mathbf{D}$ from Example E13-1, the second-mode dynamic matrix is found to be

$$\mathbf{D}_2 = \mathbf{D}\,\mathbf{S}_1 = \begin{bmatrix} 0.44003\text{E} - 03 & -0.46092\text{E} - 03 & -0.46762\text{E} - 03 \\ -0.30375\text{E} - 03 & 0.43332\text{E} - 03 & 0.87264\text{E} - 04 \\ -0.23391\text{E} - 03 & 0.65376\text{E} - 04 & 0.63459\text{E} - 03 \end{bmatrix}$$

Using this dynamic matrix, the iteration solution for the second-mode shape and frequency is carried out below following the same format used in Example E13-1:

$$[\mathbf{D}_2]\begin{Bmatrix} \mathbf{v}_2^{(0)} \\ 1.0000 \\ 0.0000 \\ -1.0000 \end{Bmatrix} = \begin{Bmatrix} \overline{\mathbf{v}}_2^{(1)} \\ 0.90765\text{E} - 03 \\ -0.39461\text{E} - 03 \\ -0.86850\text{E} - 03 \end{Bmatrix} ; \quad \begin{vmatrix} \mathbf{v}_2^{(1)} & \overline{\mathbf{v}}_2^{(2)} \\ 1.0000 & 0.10879\text{E} - 02 \\ -0.43476 & -0.57924\text{E} - 03 \\ -0.95687 & -0.86955\text{E} - 03 \end{vmatrix}$$

$$\begin{vmatrix} \mathbf{v}_2^{(2)} & \overline{\mathbf{v}}_2^{(3)} \\ 1.0000 & 0.10592\text{E} - 02 \\ -0.53245 & -0.60782\text{E} - 03 \\ -0.79932 & -0.77596\text{E} - 03 \end{vmatrix} \quad \begin{vmatrix} \mathbf{v}_2^{(3)} & \overline{\mathbf{v}}_2^{(4)} \\ 1.0000 & 0.10472\text{E} - 02 \\ -0.57383 & -0.61993\text{E} - 03 \\ -0.73258 & -0.73631\text{E} - 03 \end{vmatrix} \cdots$$

The relatively slow rate of convergence of this second-mode iteration compared with the first-mode solution in Example E13-1 is quite apparent. Continuing the process for twelve cycles led to the the following estimate of the second-mode shape:

$$\boldsymbol{\phi}_2^T = \begin{bmatrix} 1.0000 & -0.6069 & -0.6793 \end{bmatrix}$$

which compares well with the results obtained by the determinantal analysis in Example E11-1.

The frequency of the second-mode vibration derived from the top story displacement after the first cycle of iteration is given by

$$\left[\omega_2^{(1)}\right]^2 = \frac{1.0000}{0.00090675} = 1,102 \qquad \omega_2^{(1)} = 33.19 \; rad/sec$$

On the other hand, the Rayleigh quotient expression of Eq. (13-34) applied after one cycle of iteration gives the frequency $\omega_2 = 32.10$. For comparison, after four cycles, the frequency based on the top story displacement is given by

$$\left[\omega_2^{(4)}\right]^2 = \frac{1.0000}{0.0010471} = 955 \qquad \omega_2^{(4)} = 30.90 \; rad/sec$$

while Eq. (13-34) gives $\omega_2 = 31.06$ which agrees well with the value given in Example E11-1. This example demonstrates that many cycles of iteration are required to obtain the second-mode shape with good accuracy whereas Eq. (13-34) gives a good approximation of the frequency after only a few cycles. To be specific in this example, the correct frequency ($\omega_1 = 31.048$) was given to five-figure accuracy after only 6 cycles of iteration, whereas the top story displacement was still changing in the fifth significant figure after 12 cycles of iteration.

Analysis of Third and Higher Modes

It should now be evident that the same sweeping process can be extended to purify a trial vector of both the first- and second-mode components, with the result that the iteration procedure will converge toward the third mode. Expressing the purified trial third-mode shape [by analogy with Eq. (13-28)] as

$$\tilde{v}_3^{(0)} = v_3^{(0)} - \phi_1 Y_1^{(0)} - \phi_2 Y_2^{(0)} \tag{13-35}$$

and applying the conditions that $\tilde{v}_3^{(0)}$ be orthogonal to both ϕ_1 and ϕ_2,

$$\phi_1^T m \tilde{v}_3^{(0)} = 0 = \phi_1^T m v_3^{(0)} - M_1 Y_1^{(0)}$$

$$\phi_2^T m \tilde{v}_3^{(0)} = 0 = \phi_2^T m v_3^{(0)} - M_2 Y_2^{(0)}$$

lead to expressions for the first- and second-mode amplitudes in the trial vector $v_3^{(0)}$

$$Y_1^{(0)} = \frac{1}{M_1} \phi_1^T m v_3^{(0)} \tag{13-36a}$$

$$Y_2^{(0)} = \frac{1}{M_2} \phi_2^T m v_3^{(0)} \tag{13-36b}$$

which are equivalent to Eq. (13-27). Substituting these into Eq. (13-35) leads to

$$\tilde{v}_3^{(0)} = v_3^{(0)} - \frac{1}{M_1} \phi_1 \phi_1^T m \, v_3^{(0)} - \frac{1}{M_2} \phi_2 \phi_2^T m \, v_3^{(0)}$$

or

$$\tilde{v}_3^{(0)} = \left[I - \frac{1}{M_1} \phi_1 \phi_1^T m - \frac{1}{M_2} \phi_2 \phi_2^T m \right] v_3^{(0)} \tag{13-37}$$

Equation (13-37) shows that the sweeping matrix S_2 which eliminates both first- and second-mode components from $v_3^{(0)}$ can be obtained by merely subtracting a second-mode term from the first-mode sweeping matrix given by Eq. (13-30), that is,

$$S_2 = S_1 - \frac{1}{M_2} \phi_2 \phi_2^T m \tag{13-38}$$

where the sweeping-matrix operation is expressed by

$$\widetilde{\mathbf{v}}_3^{(0)} = \mathbf{S}_2 \mathbf{v}_3^{(0)} \tag{13-39}$$

The matrix iteration relationship for analysis of the third mode can now be written by analogy with Eq. (13-32):

$$\frac{1}{\omega_3^2} \mathbf{v}_3^{(1)} = \mathbf{D}\widetilde{\mathbf{v}}_3^{(0)} = \mathbf{D}\,\mathbf{S}_2 \mathbf{v}_3^{(0)} \equiv \mathbf{D}_3 \mathbf{v}_3^{(0)} \tag{13-40}$$

Hence this modified dynamic matrix $\mathbf{D}_3$ performs the function of sweeping out first- and second-mode components from the trial vector $\mathbf{v}_3^{(0)}$ and thus produces convergence toward the third-mode shape.

This same process obviously can be extended successively to analysis of higher and higher modes of the system. For example, to evaluate the fourth mode, the sweeping matrix $\mathbf{S}_3$ would be calculated as follows:

$$\mathbf{S}_3 = \mathbf{S}_2 - \frac{1}{M_3}\boldsymbol{\phi}_3 \boldsymbol{\phi}_3^T \mathbf{m} \tag{13-41}$$

where it would perform the function

$$\widetilde{\mathbf{v}}_4^{(0)} = \mathbf{S}_3 \mathbf{v}_4^{(0)} \tag{13-42}$$

The corresponding dynamic matrix would be

$$\mathbf{D}_4 = \mathbf{D}\,\mathbf{S}_3$$

The matrices suitable for calculating any mode can be obtained easily by analogy from these; that is,

$$\mathbf{S}_n = \mathbf{S}_{n-1} - \frac{1}{M_n}\boldsymbol{\phi}_n \boldsymbol{\phi}_n^T \mathbf{m} \qquad \mathbf{D}_{n+1} = \mathbf{D}\,\mathbf{S}_n \tag{13-43}$$

Clearly the most important limitation of this procedure is that all the lower-mode shapes must be calculated before any given higher mode can be evaluated. Also, it is essential to evaluate these lower modes with great precision if the sweeping matrix for the higher modes is to perform effectively. Generally this process is used directly for the calculation of no more than four or five modes.

Analysis of Highest Mode

It is of at least academic interest to note that the matrix iteration method can also be applied for the analysis of the highest mode of vibration of any structure. If Eq. (13-3) is premultiplied by $\mathbf{m}^{-1}\mathbf{k}$, the result can be written

$$\omega_n^2 \hat{\mathbf{v}}_n = \mathbf{E}\,\hat{\mathbf{v}}_n \tag{13-44}$$

in which the dynamic properties of the system are now contained in the matrix

$$\mathbf{E} \equiv \mathbf{m}^{-1}\mathbf{k} \equiv \mathbf{D}^{-1} \tag{13-45}$$

If a trial shape for the highest (Nth) mode of vibration is introduced, Eq. (13-44) becomes

$$\omega_N^2 \mathbf{v}_N^{(1)} = \mathbf{E}\mathbf{v}_N^{(0)} \tag{13-46}$$

which is equivalent to Eq. (13-5a). By analogy with Eqs. (13-9) and (13-11), approximations of the Nth-mode frequency are given by

$$\omega_N^2 \doteq \frac{\overline{v}_{kN}^{(1)}}{v_{kN}^{(0)}} \tag{13-47a}$$

or

$$\omega_N^2 \doteq \frac{(\overline{\mathbf{v}}_N^{(1)})^T \mathbf{m} \overline{\mathbf{v}}_N^{(1)}}{(\overline{\mathbf{v}}_N^{(1)})^T \mathbf{m}\, \mathbf{v}_N^{(0)}} \tag{13-47b}$$

in which $\overline{\mathbf{v}}_N^{(1)} = \mathbf{E}\,\mathbf{v}_N^{(0)}$.

Moreover, the computed shape $\overline{\mathbf{v}}_N^{(1)}$ is a better approximation of the highest-mode shape than the original assumption was; thus if it is used as a new trial shape and the process repeated a sufficient number of times, the highest-mode shape can be determined to any desired degree of approximation.

The proof of the convergence of this process to the highest mode can be carried out exactly as for the lowest mode. The essential difference in the proof is that the term ω_N^2 is in the numerator rather than in the denominator, with the result that the equivalent of Eq. (13-23) takes the form

$$1 \gg \left(\frac{\omega_{N-1}}{\omega_N}\right)^{2s} \gg \left(\frac{\omega_{N-2}}{\omega_N}\right)^{2s} \gg \left(\frac{\omega_{N-3}}{\omega_N}\right)^{2s} \gg \cdots \tag{13-48}$$

which emphasizes the highest rather than the lowest mode.

Analysis of the next highest mode can be accomplished by developing a highest-mode-shape sweeping matrix from the orthogonality principle, and in principle, the entire analysis could proceed from the top downward. However, since the convergence of the iteration process is much less rapid when applied with Eq. (13-46) than for the normal iteration analysis of the lower modes, this method is seldom used except to obtain an estimate of the highest frequency of vibration which can be expected in the structure.

Example E13-3.   The analysis of the third vibration mode for the three-story structure of Example E13-1 could be carried out by evaluating the second-mode sweeping matrix and using that to obtain a dynamic matrix which would converge directly to the third mode. However, it generally is easier and more accurate to evaluate the highest mode of a structure by iterating with the stiffness form of the dynamic matrix; that approach is demonstrated here.

The stiffness matrix and the inverse of the mass matrix for the structure of Fig. E13-1 are (see Example E11-1)

$$\mathbf{k} = 600 \begin{bmatrix} 1 & -1 & -0 \\ -1 & 3 & -2 \\ 0 & -2 & 5 \end{bmatrix} kips/in \qquad \mathbf{m}^{-1} = \frac{1}{6} \begin{bmatrix} 6 & 0 & 0 \\ 0 & 4 & 0 \\ 0 & 0 & 3 \end{bmatrix} in/kip \cdot sec^2$$

Hence the stiffness form of the dynamic matrix is

$$\mathbf{E} = \mathbf{m}^{-1}\mathbf{k} = 100 \begin{bmatrix} 6 & -6 & 0 \\ -4 & 12 & -8 \\ 0 & -6 & 15 \end{bmatrix} sec^{-2}$$

Using an initial shape which is a reasonable guess of the third mode, the iteration is carried out below, following the format of Example E13-1.

$$\begin{array}{cccc} \mathbf{E} & \mathbf{v}_3^{(0)} & \bar{\mathbf{v}}_3^{(1)} \\ 100 \begin{bmatrix} 6 & -6 & 0 \\ -4 & 12 & -8 \\ 0 & -6 & 15 \end{bmatrix} \begin{Bmatrix} 1 \\ -1 \\ 1 \end{Bmatrix} = \begin{Bmatrix} -12 \\ 24 \\ -21 \end{Bmatrix} \end{array}$$

$\mathbf{v}_3^{(1)}$	$\bar{\mathbf{v}}_3^{(2)}$	$\mathbf{v}_3^{(2)}$	$\bar{\mathbf{v}}_3^{(3)}$
0.5714	10.286	0.4706	9.412
−1.1429	−24.000	−1.0980	−23.059
1.0000	21.857	1.0000	21.588

$\mathbf{v}_3^{(3)}$	$\bar{\mathbf{v}}_3^{(4)}$	$\mathbf{v}_3^{(6)}$	$\bar{\mathbf{v}}_3^{(7)}$
0.4360	9.024	0.4123	8.740
−1.0681	−22.561	−1.0444	−22.182
1.0000	21.409	1.0000	21.266

Final shape

It is evident that this iteration process converges toward the highest-mode shape much more slowly than the convergence toward the lowest mode in Example

E13-1; this is characteristic of matrix iteration in general. However, the final shape agrees well with that obtained from the determinantal solution (Example E11-1) showing it has essentially converged. The frequency obtained from the last iteration cycle [see Eq. (13-47a)] is

$$\omega_3^2 = \frac{\bar{v}_{23}^{(7)}}{v_{23}^{(6)}} = \frac{21.266(100)}{1} = 2,127$$

which also agrees well with the value obtained in Example E11-1. The factor of 100 in this expression is the multiplier which has been factored out of the dynamic matrix $\mathbf{E}$.

13-5 BUCKLING ANALYSIS BY MATRIX ITERATION

The matrix iteration procedure for evaluating eigenvalues and eigenvectors is applicable also when axial forces act in the members of the structure, if the axial forces do not vary with the vibratory motion of the structure. For any specified condition of axial loading, an equation equivalent to Eq. (13-5a) may be written

$$\mathbf{v}_1^{(1)} = \omega_1^2 \, \overline{\mathbf{D}} \, \mathbf{v}_1^{(0)} \tag{13-49a}$$

in which

$$\overline{\mathbf{D}} = \overline{\mathbf{k}}^{-1} \mathbf{m} \tag{13-49b}$$

where $\overline{\mathbf{k}} = \mathbf{k} - \mathbf{k}_{G0}$ is the combined stiffness matrix, taking account of the geometric-stiffness effect [see Eq. (9-20)]. The vibration mode shapes and frequencies can be determined from Eq. (13-49a) by iteration, just as they are without axial loads.

The effect of compressive axial forces is to reduce the stiffnesses of the members of the structure, thus tending to reduce the frequencies of vibration. In the limiting (buckling) case, the vibration frequency goes to zero, and the static eigenvalue equation takes the form

$$(\mathbf{k} - \lambda_G \mathbf{k}_{G0}) \, \hat{\mathbf{v}} = \mathbf{0} \tag{11-24}$$

Premultiplying this equation by $(1/\lambda_G)\widetilde{\mathbf{f}}$ gives

$$\frac{1}{\lambda_G} \hat{\mathbf{v}} = \mathbf{G} \, \hat{\mathbf{v}} \tag{13-50a}$$

in which

$$\mathbf{G} = \widetilde{\mathbf{f}} \, \mathbf{k}_{G0} \tag{13-50b}$$

Equation (13-50a) has the same form as the vibration eigenvalue equations and may be solved by the same type of iterative procedure. The eigenvalues which permit nonzero values of $\hat{\mathbf{v}}$ to be developed are the buckling loads, which are represented by the values of the load parameter λ_G. Thus, if a trial shape for the first buckling mode is designated $\mathbf{v}_1^{(0)}$, the iterative process is indicated by

$$\frac{1}{\lambda_{G1}}\mathbf{v}_1^{(1)} = \mathbf{G}\,\mathbf{v}_1^{(0)} \tag{13-51}$$

When the iterative procedure is used to evaluate buckling modes in this way, it has been called the *Vianello method*, after the man who first used it for this purpose.

The matrix iteration analysis of buckling is identical in principle and technique to the iteration analysis of vibration and need not be discussed further except to mention that the orthogonality condition used in evaluating the higher buckling modes is

$$\boldsymbol{\phi}_m^T \mathbf{k}_{G0}\boldsymbol{\phi}_n = 0 \qquad m \neq n \tag{13-52}$$

However, generally only the lowest mode of buckling is of interest, and there is little need to consider procedures for evaluating higher buckling modes.

Example E13-4. The matrix iteration analysis of buckling will be demonstrated by the evaluation of the critical buckling load of a uniform cantilever column loaded by its own weight (Fig. E13-2). The structure has been discretized by dividing it into three equal segments and using the lateral displacement of each node as the degrees of freedom. It is assumed that the uniformly

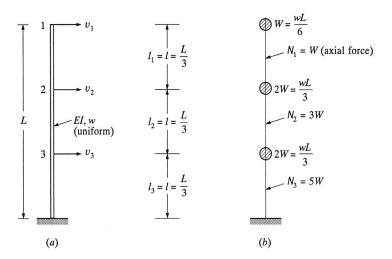

(a)   *(b)*

FIGURE E13-2
Analysis of column buckling due to its own weight: (*a*) uniform column;
(*b*) discretized model.

distributed weight of the column is lumped at the ends of the segments; hence one-sixth of its total weight is concentrated at the top and one-third at each of the two interior nodes. The axial forces in the three segments of the column due to these concentrated weights are shown in the figure.

When the linear-displacement approximation [Eq. (10-36)] is used, the geometric stiffness of this column is given by

$$
\mathbf{k}_G = \begin{bmatrix} \frac{N_1}{l_1} & \frac{-N_1}{l_1} & 0 \\ -\frac{N_1}{l_1} & \frac{N_1}{l_1} + \frac{N_2}{l_2} & -\frac{N_2}{l_2} \\ 0 & -\frac{N_2}{l_2} & \frac{N_2}{l_2} + \frac{N_3}{l_3} \end{bmatrix} = \frac{W}{l} \begin{bmatrix} 1 & -1 & 0 \\ -1 & 4 & -3 \\ 0 & -3 & 8 \end{bmatrix}
$$

and this will be taken as the reference geometric stiffness $\mathbf{k}_{G0}$. By applying unit loads successively at the three nodes and calculating the resulting deflections by standard static-analysis procedures, the flexibility matrix of the column is found to be

$$
\tilde{\mathbf{f}} = \frac{l^3}{6EI} \begin{bmatrix} 54 & 28 & 8 \\ 28 & 16 & 5 \\ 8 & 5 & 2 \end{bmatrix}
$$

Hence the stability matrix $\mathbf{G}$ is given by

$$
\mathbf{G} = \tilde{\mathbf{f}} \, \mathbf{k}_{G0} = \frac{Wl^2}{6EI} \begin{bmatrix} 26 & 34 & -20 \\ 12 & 21 & -8 \\ 3 & 6 & 1 \end{bmatrix}
$$

A parabola is taken as a reasonable guess for the first-mode buckled shape, and the matrix iteration is carried out below, following the same format as the vibration examples.

$$
\begin{array}{cccc}
& \mathbf{G} & \mathbf{v}_1^{(0)} & \bar{\mathbf{v}}_1^{(1)} \\[4pt]
\frac{Wl^2}{6EI} & \begin{bmatrix} 26 & 34 & -20 \\ 12 & 21 & -8 \\ 3 & 6 & 1 \end{bmatrix} & \begin{Bmatrix} 1.00 \\ 0.44 \\ 0.11 \end{Bmatrix} = & \begin{Bmatrix} 38.76 \\ 20.36 \\ 5.75 \end{Bmatrix}
\end{array}
$$

$$
\begin{array}{cccccc}
\mathbf{v}_1^{(1)} & \bar{\mathbf{v}}_1^{(2)} & \mathbf{v}_1^{(2)} & \bar{\mathbf{v}}_1^{(3)} & \mathbf{v}_1^{(3)} & \bar{\mathbf{v}}_1^{(4)} \\[4pt]
\begin{vmatrix} 1.0000 & 40.89 \\ 0.5253 & 81.84 \\ 0.1484 & 6.30 \end{vmatrix} & & \begin{vmatrix} 1.0000 & 41.08 \\ 0.5342 & 21.99 \\ 0.1541 & 6.36 \end{vmatrix} & & \begin{vmatrix} 1.0000 & 41.10 \\ 0.5352 & 22.00 \\ 0.1548 & 6.37 \end{vmatrix} \\
& & & & & \text{True shape}
\end{array}
$$

This process converges as quickly as the first-mode iteration in vibration analysis. The critical buckling-load factor obtained from the final iteration cycle is

$$\lambda_{cr} = \frac{v_{11}^{(3)}}{\bar{v}_{11}^{(4)}} = \frac{1.000}{41.10(Wl^2/6EI)} = 1.3139 \frac{EI}{WL^2}$$

where the final result is expressed in terms of the total length L. From this, the critical weight per unit length is found to be

$$w_{cr} = \frac{\lambda_{cr}W}{L/6} = 1.3139(6)\frac{EI}{L^3} = 7.883\frac{EI}{L^3}$$

Since this compares very well with the exact result of $7.83EI/L^3$, it is evident that the geometric stiffness derived from the simple linear-displacement assumption is quite effective.

The influence of geometric stiffness on the vibration frequency of this column can also be calculated by matrix iteration. Of course, if its unit weight has the critical value calculated above, the vibration frequency will be zero. However, for any smaller value of unit weight, a corresponding frequency can be determined. Suppose, for example, that $W = (27/26)(EI/L^2)$, which is $(27/26)/1.3139 = 79$ percent of the critical value. Then the geometric stiffness is given by substituting this value into the expression for k_G above.

The elastic stiffness of the column, obtained by inverting the flexibility matrix, is

$$k = \frac{6}{26}\frac{EI}{l^3}\begin{bmatrix} 7 & -16 & 12 \\ -16 & 44 & -46 \\ 12 & -46 & 80 \end{bmatrix}$$

Hence the combined stiffness matrix which takes account of the axial-force effects is given by [Eq. (10-20)]

$$\bar{k} = k - k_G = \frac{6}{26}\frac{EI}{l^3}\begin{bmatrix} 7 & -16 & 12 \\ -16 & 44 & -46 \\ 12 & -46 & 80 \end{bmatrix} - \frac{27}{26}\frac{EI}{9l^3}\begin{bmatrix} 1 & -1 & 0 \\ -1 & 4 & -3 \\ 0 & -3 & 8 \end{bmatrix}$$

$$= \frac{3}{26}\frac{EI}{l^3}\begin{bmatrix} 13 & -31 & 24 \\ -31 & 84 & -89 \\ 24 & -89 & 152 \end{bmatrix}$$

Finally, the vibration analyses could be carried out by iterating with a modified dynamic matrix $\bar{D} = \bar{k}^{-1}m$, where $\bar{k}^{-1}$ is the inverse of the combined stiffness matrix shown above. The completion of this example is left to the reader.

13-6 INVERSE ITERATION — THE PREFERRED PROCEDURE

In all the discussions of matrix iteration presented in the foregoing sections of this chapter, the improvement in calculated shape achieved during each cycle of iteration is obtained by simply multiplying the vector for the preceding cycle by the dynamic matrix $\mathbf{D} \equiv \mathbf{k}^{-1}\mathbf{m}$; for this reason the procedure is called direct iteration. It is apparent in these descriptions that the method is easy to apply; also, because it is based on the flexibility version of the dynamic matrix, it converges toward the shape of the lowest vibration mode, as is necessary for the procedure to be used as a general tool for structural dynamics. The major disadvantage of this procedure is that the flexibility matrix is fully populated, and this leads to computational inefficiency in comparison with what can be achieved by operating with the narrowly-banded stiffness matrix. Of course direct iteration with the stiffness-based dynamic matrix $\mathbf{E} = \mathbf{m}^{-1}\mathbf{k}$ is not appropriate because it would converge to the highest-mode shape, as was discussed earlier. Also the dynamic matrix $\mathbf{E}$ is not narrowly banded even though both $\mathbf{k}$ and $\mathbf{m}$ are, so an alternative technique is needed.

Inverse iteration is the preferred method for taking advantage of the narrow banding of the stiffness matrix; because it is applied inversely, it converges toward the lowest-mode shape. In order to retain the narrow banding of $\mathbf{k}$, the dynamic matrix $\mathbf{E}$ is never formed. Instead, the mass matrix is combined with the assumed displacement vector to obtain an inertial load vector, and then the stiffness-based simultaneous equations of equilibrium are solved to obtain the improved displacement vector.

As in the above-described direct iteration method, the initially assumed displacement vector will be designated $\mathbf{v}_1^{(0)}$; then the inertial forces due to harmonic motions with this shape are given by an expression similar to Eq. (13-6). However, noting that the effect of the frequency will be removed subsequently by the normalization step, in this formulation the frequency is assumed to be unity ($\omega_1^2 = 1$) and the resulting inertial forces are denoted by

$$\mathbf{W}_1^{(0)} \equiv \mathbf{m}\mathbf{v}_1^{(0)} \tag{13-53}$$

Now the improved displacement vector $\overline{\mathbf{v}}_1^{(1)}$ resulting from the action of these forces is obtained by solving the equilibrium equations of the structure subjected to these forces,

$$\mathbf{k}\overline{\mathbf{v}}^{(1)} = \mathbf{W}_1^{(0)} \tag{13-54}$$

Of course, one way to solve these equations would be to calculate the flexibility matrix by inversion of the stiffness matrix ($\widetilde{\mathbf{f}} = \mathbf{k}^{-1}$) and to multiply the inertial forces by that flexibility,

$$\overline{\mathbf{v}}^{(1)} = \widetilde{\mathbf{f}}\,\mathbf{W}_1^{(0)}$$

This procedure actually would be entirely equivalent to the direct iteration analysis described before and would be inefficient because of the need to invert and then multiply by a fully populated flexibility matrix, as explained before.

In the inverse iteration procedure recommended here, the equilibrium equations, Eq. (13-54), are solved after first using Gauss elimination to decompose the stiffness matrix to the following form[1]

$$\mathbf{k} = \mathbf{L}\,\mathbf{d}\,\mathbf{L}^T \equiv \mathbf{L}\mathbf{U} \tag{13-55}$$

where $\mathbf{L}$ is called the lower triangular matrix, $\mathbf{L}^T$ is its transpose and $\mathbf{d}$ is a diagonal matrix defined such that

$$\mathbf{d}\,\mathbf{L}^T \equiv \mathbf{U}$$

is the upper triangular matrix. With the substitution of Eq. (13-55), Eq. (13-54) becomes

$$\mathbf{L}\mathbf{U}\,\bar{\mathbf{v}}_1^{(1)} = \mathbf{W}_1^{(0)} \tag{13-54a}$$

and the simultaneous solution then is carried out in two steps:

(1) Define

$$\mathbf{y}_1^{(1)} \equiv \mathbf{U}\,\bar{\mathbf{v}}_1^{(1)} \tag{13-56}$$

and solve for $\mathbf{y}_1^{(1)}$ from

$$\mathbf{L}\mathbf{y}_1^{(1)} = \mathbf{W}_1^{(0)} \tag{13-57}$$

(2) Solve for $\bar{\mathbf{v}}_1^{(1)}$ from

$$\mathbf{U}\,\bar{\mathbf{v}}_1^{(1)} = \mathbf{y}_1^{(1)} \tag{13-58}$$

As was described before, this derived vector then is normalized by dividing it by its largest element to obtain the improved first-mode shape that is the final result of the first iteration cycle:

$$\mathbf{v}_1^{(1)} = \frac{\bar{\mathbf{v}}_1^{(1)}}{\max(\bar{\mathbf{v}}_1^{(1)})} \tag{13-59}$$

It is important to note that the narrow banded character of the stiffness matrix $\mathbf{k}$ is retained in the triangular matrices $\mathbf{L}$ and $\mathbf{U}$, consequently the efficiency of this inverse displacement analysis is greatly enhanced relative to the flexibility matrix formulation used with direct iteration.

Because the only difference between this inverse iteration procedure and the previously described direct iteration lies in the more efficient Gauss decomposition

---

[1] K-J. Bathe and E. L. Wilson, *Numerical Methods in Finite Element Analysis*, Prentice-Hall, 1976, p. 248.

technique used to calculate the derived displacement vector $v_1^{(1)}$, the entire earlier description of direct matrix iteration is equally applicable to inverse iteration if Eq. (13-7) that was used previously to calculate $v_1^{(1)}$ is replaced by the simultaneous equation solution described above. However, even though this difference may appear to be minor, the tremendous computational advantage of inverse iteration based on Eqs. (13-54a) to (13-58) must not be overlooked, especially when the system being analyzed has a large number of degrees of freedom.

13-7 INVERSE ITERATION WITH SHIFTS

In principle, the inverse iteration procedure just described can be combined with sweeping matrix concepts to obtain a more efficient method for calculating the second and higher-modes of vibration. However, the calculation of the sweeping matrices becomes increasingly expensive and the sweeping operation becomes less and less effective as the mode number increases. For this reason, other methods have been developed for calculating the higher mode vibration properties, and one of these that has proven to be useful in practice is based on the concept of "shifting" the eigenvalues. Although shifting can be employed with either direct or inverse iteration, it is most effective with inverse iteration analyses, and it will be discussed here in that context.

For this explanation, it is convenient to express the eigenproblem equation as the inverse of the flexibility form of Eq. (13-5):

$$\mathbf{E}\,\boldsymbol{\phi}_n = \boldsymbol{\phi}_n\,\lambda_n \tag{13-60}$$

in which $\lambda_n \equiv \omega_n^2$ represents the eigenvalue or frequency, $\mathbf{E} = \mathbf{D}^{-1}$ is the stiffness form of the dynamic matrix, and $\boldsymbol{\phi}_n$ is the eigenvector (mode shape). When rewritten to express the full set of mode shapes and frequencies, Eq. (13-60) becomes

$$\mathbf{E}\,\boldsymbol{\Phi} = \boldsymbol{\Phi}\,\boldsymbol{\Lambda} \tag{13-61}$$

where $\boldsymbol{\Phi}$ is the array of all mode-shape vectors and $\boldsymbol{\Lambda}$ is the diagonal array of all frequencies, ω_n^2.

The essential concept of shifting is the representation of each eigenvalue λ_n as the sum of a shift μ plus a residual δ_n, thus

$$\lambda_n = \delta_n + \mu \tag{13-62}$$

or considering the entire diagonal matrix of eigenvalues

$$\boldsymbol{\Lambda} = \hat{\boldsymbol{\delta}} + \mu\,\mathbf{I} \tag{13-63}$$

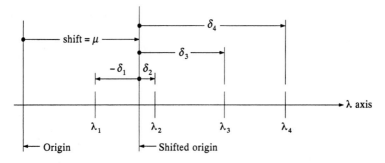

FIGURE 13-2
Demonstration of a shift on the eigenvalue axis.

in which $\hat{\delta}$ is the diagonal matrix of residuals and $\mu\mathbf{I}$ expresses the shift applied to each eigenvalue.

The shift can be visualized as a displacement of the origin in a plot of the eigenvalues, as shown in Fig. 13-2. Its effect is to transform the eigenvalue problem to the analysis of the residuals rather than the actual eigenvalues, as is evident if Eq. (13-63) is substituted into Eq. (13-61):

$$\mathbf{E}\,\boldsymbol{\Phi} = \boldsymbol{\Phi}\,[\hat{\delta} + \mu\mathbf{I}]$$

which can be rewritten as

$$[\mathbf{E} - \mu\mathbf{I}]\,\boldsymbol{\Phi} = \boldsymbol{\Phi}\hat{\delta} \tag{13-64}$$

Here the term in brackets represents a modified dynamic matrix to which the residual eigenvalues apply, and it will be denoted as $\hat{\mathbf{E}}$ for convenience; thus

$$\hat{\mathbf{E}}\,\boldsymbol{\Phi} = \boldsymbol{\Phi}\,\hat{\delta} \tag{13-65}$$

It is apparent that Eq. (13-65) is entirely equivalent to Eq. (13-61) and that the shifted matrix has the same eigenvectors as $\mathbf{E}$.

The solution of this new eigenproblem may be carried out by inverse iteration following a procedure analogous to that described above, and the result of the first cycle of iteration can be expressed as follows [by analogy with Eqs. (13-7) and (13-8)]:

$$\mathbf{v}_k^{(1)} = \frac{\hat{\mathbf{E}}^{-1}\mathbf{v}_k^{(0)}}{\max(\hat{\mathbf{E}}^{-1}\mathbf{v}_k^{(0)})}$$

where $\mathbf{v}_k^{(0)}$ is an initial approximation of the kth-mode shape. After s cycles the result becomes

$$\mathbf{v}_k^{(s)} = \frac{\hat{\mathbf{E}}^{-s}\mathbf{v}_k^{(0)}}{\max(\hat{\mathbf{E}}^{-s}\mathbf{v}_k^{(0)})} = \frac{\sum_{n=1}^{N}\delta_n^{-s}\phi_n Y_n^{(0)}}{\max(\hat{\mathbf{E}}^{-s}\mathbf{v}_k^{(0)})}$$

or by comparison with Eq. (13-22), this may be expressed as

$$\mathbf{v}_k^{(s)} = \frac{\delta_k^{-s}}{\max(\hat{\mathbf{E}}^{-s}\mathbf{v}_k^{(0)})} \left[\phi_k Y_k^{(0)} + \sum_{n=1}^{k-1} \left(\frac{\delta_k}{\delta_n}\right)^s \phi_n Y_n^{(0)} + \sum_{n=k+1}^{N} \left(\frac{\delta_k}{\delta_n}\right)^s \phi_n Y_n^{(0)} \right] \quad (13\text{-}66)$$

where δ_k represents the smallest residual eigenvalue, that is,

$$|\delta_k| < \delta_{k+1} < \delta_{k+2} \cdots \quad \text{and} \quad |\delta_k| < -\delta_{k-1} < -\delta_{k-2} \cdots$$

Thus it is evident that the two summations in Eq. (13-66) will become negligibly small after a sufficient number of iteration cycles, so the computed mode shape converges to

$$\mathbf{v}_k^{(s)} = \frac{\delta_k^{-s}\phi_k Y_k^{(0)}}{\max(\delta_k^{-s}\phi_k Y_k^{(0)})} = \frac{\phi_k}{\max(\phi_k)} \equiv \phi_k \quad (13\text{-}67)$$

This analysis therefore shows that the process of inverse iteration with eigenvalue shift converges to the mode shape for which the eigenvalue is closest to the shift position; e.g., it would converge to the second mode for the case illustrated in Fig. 13-2. By analogy with Eq. (13-13) it may be seen that the residual eigenvalue for this mode is given by the maximum term in the derived eigenvector (before normalization):

$$\delta_k = \frac{1}{\max(\overline{\mathbf{v}}_k^{(s)})}$$

Hence the actual eigenvalue is obtained by adding the shift to this residual value,

$$\lambda_k = \mu + \frac{1}{\max(\overline{\mathbf{v}}_k^{(s)})} \quad (13\text{-}68)$$

By appropriate selection of the shift points, this inverse iteration analysis can be caused to converge to any or all modes of the structural system. Moreover, because the speed of convergence can be accelerated by shifting to a value very close to the root that is sought, it is good practice to shift at intervals during iteration, as better approximations of the root are obtained. A useful formula for approximating the shift point can be derived from the averaging expression given by Eq. (13-11):

$$\mu_k = \frac{\overline{\mathbf{v}}_k^{(s)}\mathbf{m}\mathbf{v}_k^{(s-1)}}{\overline{\mathbf{v}}_k^{(s)}\mathbf{m}\overline{\mathbf{v}}_k^{(s)}} \quad (13\text{-}69)$$

It is evident that the shifting procedure would be less effective using the stiffness formulation with direct iteration because the convergence in that case is toward the largest root, and only the first or last residuals (δ_1 or δ_N) can be made largest by shifting.

This description of the shifting concept has been presented in the context of the dynamic matrix $\mathbf{E} = \mathbf{m}^{-1}\mathbf{k}$ for convenience in the explanation. However, this relatively inefficient displacement analysis based on inversion of the fully populated dynamic matrix should be replaced by a procedure that takes advantage of the narrow banding property of the stiffness matrix in the solution, as already has been noted. The recommended inverse iteration analysis procedure using shifting is carried out in much the same way that was described for the case without shifting in Section 13-6. Starting with the eigenproblem in the form

$$\mathbf{k}\,\boldsymbol{\phi}_n = \mathbf{m}\,\boldsymbol{\phi}_n\,\omega_n^2$$

and introducing the eigenvalue shift $\mu = \omega_n^2 - \delta_n$ lead to

$$\mathbf{k}\,\boldsymbol{\phi}_n = \mathbf{m}\,\boldsymbol{\phi}_n\,(\mu + \delta_n)$$

which may be expressed alternatively as

$$[\mathbf{k} - \mu\,\mathbf{m}]\,\boldsymbol{\phi}_n = \mathbf{m}\,\boldsymbol{\phi}_n\,\delta_n \tag{13-70}$$

The term in brackets in Eq. (13-70) is the shifted stiffness matrix of the structure; it represents the effective stiffness when the system is moving harmonically at the shift frequency, and it will be denoted here by

$$\hat{\mathbf{k}} \equiv \mathbf{k} - \mu\,\mathbf{m} \tag{13-71}$$

With this substituted into Eq. (13-70), the iterative solution for the displacements is initiated by assuming a trial vector for mode k, $\mathbf{v}_k^{(0)}$, and multiplying it by the mass matrix to obtain a trial inertial load vector, $\mathbf{W}_k^{(0)} = \mathbf{m}\,\mathbf{v}_k^{(0)}$. The resulting iteration form of Eq. (13-70) becomes

$$\hat{\mathbf{k}}\overline{\mathbf{v}}_k^{(1)} = \mathbf{W}_k^{(0)} \tag{13-72}$$

which is solved simultaneously to obtain the improved displacements $\overline{\mathbf{v}}_k^{(1)}$.

In order to take advantage of the narrow banding of $\hat{\mathbf{k}}$ (which usually has the same band width as $\mathbf{k}$) in the simultaneous solution, it is reduced by Gauss elimination to upper and lower triangular form,

$$\hat{\mathbf{k}} = \hat{\mathbf{L}}\,\hat{\mathbf{U}} \tag{13-73}$$

as was described earlier for $\mathbf{k}$ [see Eq. (13-55)]. Then the solution for the improved displacements is carried out in two steps equivalent to those mentioned before with Eqs. (13-56) to (13-58) and the final shape for the first iteration cycle, $\mathbf{v}_k^{(1)}$, is obtained by normalization as shown by Eq. (13-59). As was discussed earlier, this iteration will converge after sufficient cycles to the mode shape having its frequency closest to the shift frequency.

13-8 SPECIAL EIGENPROBLEM TOPICS

In the foregoing discussion it has become apparent that the eigenproblem equation used in the analysis of vibration mode shapes and frequencies may be stated in various ways, and these expressions may be used either with or without shifting of the eigenvalues. The undamped free-vibration equation that expresses the equilibrium between the vibration inertial forces and the elastic resisting forces will be adopted here as the basic eigenproblem and will be stated as follows for mode "n":

$$\mathbf{k}\,\boldsymbol{\phi}_n = \mathbf{m}\,\boldsymbol{\phi}_n\,\omega_n^2 \tag{13-74}$$

This is the form that was recommended above for analysis of structures having very many degrees of freedom because the narrow banding of $\mathbf{k}$ can be used to reduce the computational effort. The equivalent form given by Eq. (13-70) is recommended for analyses in which the stiffness is modified by shifting.

However, other forms of the eigenproblem equation have been used in the preceding sections of this chapter for convenience in the presentation. The first of these may be obtained by multiplying Eq. (13-74) by the flexibility matrix, $\mathbf{k}^{-1}$, leading to an expression equivalent to Eq. (13-3),

$$\boldsymbol{\phi}_n = \mathbf{k}^{-1}\,\mathbf{m}\,\boldsymbol{\phi}_n\,\omega_n^2$$

which may be stated as

$$\frac{1}{\omega_n^2}\,\boldsymbol{\phi}_n = \mathbf{D}\,\boldsymbol{\phi}_n \tag{13-75}$$

Because the dynamic matrix $\mathbf{D}$ contains both the flexibility and the mass properties of the structure, each cycle of the iteration solution for the mode shapes involves merely multiplication by $\mathbf{D}$ followed by normalizing (scaling) which is accomplished by dividing the improved displacement vector by its largest element. This direct iteration procedure converges toward the lowest-mode shape because the eigenvalue is in the denominator of the eigenproblem equation, Eq. (13-75).

The other major alternative formulation of the eigenproblem is derived by multiplying Eq. (13-74) by the inverse of the mass matrix, leading to a result equivalent to Eq. (13-60),

$$\mathbf{m}^{-1}\,\mathbf{k}\,\boldsymbol{\phi}_n = \boldsymbol{\phi}_n\,\omega_n^2$$

but which will be expressed here as

$$\boldsymbol{\phi}_n\,\omega_n^2 = \mathbf{E}\,\boldsymbol{\phi}_n \tag{13-76}$$

Direct iteration with this equation, performed by multiplying the displacement vector by the dynamic matrix $\mathbf{E}$ followed by normalizing for each cycle would converge toward the highest-mode shape because the eigenvalue is in the numerator; therefore Eq. (13-76) is used with inverse iteration as has been described above.

The discussion of special eigenproblem topics that follows in this section is presented with reference to the eigenproblem equation given in Eq. (13-76), as a matter of convenience. However, it is emphasized here again that practical matrix iteration solutions should be based on the form of the eigenproblem equation given by Eq. (13-74).

Eigenproperty Expansion

An eigenproblem concept that is worth mentioning here is the expansion of a matrix in terms of its eigenvalues and eigenvectors. This discussion is based on the following form of the eigenproblem equation rewritten from Eq. (13-76):

$$\mathbf{E}\,\boldsymbol{\phi}_n = \boldsymbol{\phi}_n\,\lambda_n \tag{13-77}$$

where $\lambda_n \equiv \omega_n^2$. It is evident from study of the determinantal-equation approach to evaluating eigenvalues that the eigenvalues of the transposed matrix are the same as those of the original matrix. However, the eigenvectors of the transpose of an unsymmetrical matrix like $\mathbf{E}$ are different from those of the original. Hence for the transpose $\mathbf{E}^T$, the eigenproblem can be written

$$\mathbf{E}^T\,\boldsymbol{\phi}_{Ln} = \boldsymbol{\phi}_{Ln}\,\lambda_n$$

where $\boldsymbol{\phi}_{Ln}$ is the nth eigenvector of $\mathbf{E}^T$. Transposing this relationship gives

$$\boldsymbol{\phi}_{Ln}^T\,\mathbf{E} = \lambda_n\,\boldsymbol{\phi}_{Ln}^T \tag{13-78}$$

which shows why the eigenvectors $\boldsymbol{\phi}_{Ln}$ often are called the *left hand eigenvectors* of $\mathbf{E}$ while $\boldsymbol{\phi}_n$ are the *right hand eigenvectors*.

The orthogonality property of the left and right hand eigenvectors can be demonstrated if Eq. (13-77) is premultiplied by the eigenvector $\boldsymbol{\phi}_{Lm}^T$,

$$\boldsymbol{\phi}_{Lm}^T\,\mathbf{E}\,\boldsymbol{\phi}_n = \boldsymbol{\phi}_{Lm}^T\boldsymbol{\phi}_n\lambda_n \tag{13-79}$$

while Eq. (13-78) is written for mode m and postmultiplied by $\boldsymbol{\phi}_n$,

$$\boldsymbol{\phi}_{Lm}^T\,\mathbf{E}\,\boldsymbol{\phi}_n = \lambda_m\,\boldsymbol{\phi}_{Lm}^T\,\boldsymbol{\phi}_n \tag{13-80}$$

Subtracting Eq. (13-80) from Eq. (13-79) then gives

$$0 = (\lambda_n - \lambda_m)\,\boldsymbol{\phi}_{Lm}^T\,\boldsymbol{\phi}_n$$

which represents the orthogonality property

$$\phi_{Lm}^T \phi_n = 0 \qquad (\lambda_m \neq \lambda_n) \qquad \text{(13-81)}$$

If the eigenvectors are normalized to satisfy the condition $\phi_{Ln}^T \phi_n = 1$ (note that this does not fix the amplitude of ϕ_{Ln} or ϕ_n individually, only their product), and if the square matrices of the sets of all right and left hand eigenvectors are designated $\mathbf{\Phi}$ and $\mathbf{\Phi}_L$, respectively, it is evident from the normalizing and orthogonality conditions that

$$\mathbf{\Phi}_L^T \, \mathbf{\Phi} = \mathbf{I} \qquad \text{(13-82a)}$$

Hence the transpose of the left hand eigenvectors is the inverse of the right-hand eigenvectors:

$$\mathbf{\Phi}_L^T = \mathbf{\Phi}^{-1} \qquad \text{(13-82b)}$$

The expansion of $\mathbf{E}$ can now be demonstrated by writing the eigenproblem expression of Eq. (13-77) for the full set of eigenvectors and eigenvalues:

$$\mathbf{E}\mathbf{\Phi} = \mathbf{\Phi}\mathbf{\Lambda} \qquad \text{(13-83)}$$

in which $\mathbf{\Lambda}$ is the diagonal matrix of eigenvalues. Premultiplying Eq. (13-83) by $\mathbf{\Phi}_L^T$ and invoking Eq. (13-82b) lead to an expression for the eigenvalues:

$$\mathbf{\Phi}_L^T \, \mathbf{E} \, \mathbf{\Phi} = \mathbf{\Lambda} \qquad \text{(13-84)}$$

Alternatively, $\mathbf{E}$ can be expressed in terms of the eigenvalues and eigenvectors by premultiplying Eq. (13-84) by $\mathbf{\Phi}$, postmultiplying it by $\mathbf{\Phi}_L^T$, and invoking Eq. (13-82b) which leads to

$$\mathbf{E} = \mathbf{\Phi} \, \mathbf{\Lambda} \, \mathbf{\Phi}_L^T \qquad \text{(13-85)}$$

This result also can be expressed as the sum of the modal contributions:

$$\mathbf{E} = \sum_{n=1}^{N} \lambda_n \phi_n \phi_{Ln}^T \qquad \text{(13-85a)}$$

Furthermore, the square of matrix $\mathbf{E}$ is

$$\mathbf{E}^2 = \mathbf{\Phi}\mathbf{\Lambda}\mathbf{\Phi}_L^T\mathbf{\Phi}\mathbf{\Lambda}\mathbf{\Phi}_L^T = \mathbf{\Phi}\mathbf{\Lambda}^2\mathbf{\Phi}_L^T \qquad \text{(13-86)}$$

and by continued multiplication, the sth power of $\mathbf{E}$ is

$$\mathbf{E}^s = \mathbf{\Phi}\mathbf{\Lambda}^s\mathbf{\Phi}_L^T \qquad \text{(13-87)}$$

It must be remembered that the expansion of Eq. (13-87) is based on the type of eigenvector normalizing which has been used ($\mathbf{\Phi}_L^T\mathbf{\Phi} = \mathbf{I}$). A specific expression for the left hand eigenvectors can be obtained if an additional normalizing condition is introduced. For example, if Eq. (13-83) is premultiplied by $\mathbf{\Phi}^T\mathbf{m}$ (note that $\mathbf{E} = \mathbf{m}^{-1}\mathbf{k}$), it becomes

$$\mathbf{\Phi}^T\mathbf{k}\mathbf{\Phi} = \mathbf{\Phi}^T\mathbf{m}\mathbf{\Phi}\mathbf{\Lambda} \tag{13-88}$$

Now if the right hand eigenvectors are normalized so that

$$\mathbf{\Phi}^T\mathbf{m}\mathbf{\Phi} = \mathbf{I} \tag{13-89}$$

it is apparent from comparison of the transpose of Eq. (13-82a) with Eqs. (13-89) and (13-88) that

$$\mathbf{\Phi}_L = \mathbf{m}\mathbf{\Phi} = \mathbf{k}\mathbf{\Phi}\mathbf{\Lambda}^{-1} \tag{13-90}$$

Symmetric Form of Dynamic Matrix

It was noted above that the eigenproblem equation could be expressed in terms of the stiffness form of the dynamic matrix as follows [see Eq. (13-77)]:

$$\mathbf{E}\boldsymbol{\phi}_n = \boldsymbol{\phi}_n\omega_n^2$$

and a number of efficient techniques for solving this eigenproblem have been discussed. However, it should be noted that the matrix $\mathbf{E} = \mathbf{m}^{-1}\mathbf{k}$ is unsymmetric, even though both $\mathbf{m}$ and $\mathbf{k}$ are symmetric, and thus this problem cannot be solved by many efficient standard solution procedures developed to take advantage of symmetry in the eigenproblem, e.g., the Householder method. For this reason it is useful to be able to transform the general vibration eigenproblem [Eq. (13-74)] to the standard symmetric form

$$\mathbf{B}\mathbf{y}_n = \mathbf{y}_n\lambda_n \tag{13-91}$$

The transformation from general to standard form can be accomplished by manipulation of the mass matrix, and the type of transformation required depends on the form of the mass matrix. Two cases will be considered here: (1) a diagonal mass matrix representing a lumped-mass system and (2) a general (nondiagonal) mass matrix which might result from a consistent finite-element formulation. In both cases, the transformation matrix which converts Eq. (13-74) into Eq. (13-91) is obtained by decomposing the mass matrix into the product of a matrix and its transpose.

Diagonal Mass Matrix — In this case the transformation matrix is obtained very simply as the square root of the mass matrix,

$$\mathbf{m} = \mathbf{m}^{1/2}\mathbf{m}^{1/2}$$

and the square-root matrix is obtained by merely taking the square root of the diagonal terms (of course the diagonal matrix is unchanged in transposition). The transformation of Eq. (13-74) is performed by expressing its eigenvectors as

$$\boldsymbol{\phi}_n = \mathbf{m}^{-1/2}\mathbf{y}_n \tag{13-92}$$

where the inverse is formed with the reciprocals of the diagonal terms in $\mathbf{m}^{1/2}$. Substituting Eq. (13-92) into Eq. (13-74) and premultiplying by $\mathbf{m}^{-1/2}$ lead to

$$\mathbf{m}^{-1/2}\,\mathbf{km}^{-1/2}\mathbf{y}_n = \mathbf{y}_n\lambda_n$$

which is of the form of Eq. (13-91) with $\mathbf{B} = \mathbf{m}^{-1/2}\mathbf{km}^{-1/2}$. Solving this symmetric eigenvalue problem leads directly to the frequencies of the original equation (13-74); but the eigenvectors $\mathbf{y}_n$ of this new eigenproblem must be transformed to obtain the desired vibration mode shapes $\boldsymbol{\phi}_n$, using Eq. (13-92).

It is evident that this transformation procedure cannot be applied if any of the diagonal mass elements is zero. Therefore, it is necessary to eliminate these degrees of freedom from the analysis by the method of static condensation as described subsequently in Chapter 14 before performing the transformation to standard symmetric form.

Consistent Mass Matrix — Two methods are available for the transformation when the mass matrix is banded, as it would be in a consistent mass formulation, rather than diagonal. The more reliable of these is based on evaluating the eigenvalues v_n and eigenvectors $\mathbf{t}_n$ of the mass matrix from the equation

$$\mathbf{mt}_n = \mathbf{t}_n v_n$$

The eigenvectors of this symmetric matrix satisfy the orthogonality condition $\mathbf{t}_m^T\mathbf{t}_n = 0$ (if $m \neq n$); hence if they also are normalized, so that $\mathbf{t}_n^T\mathbf{t}_n = 1$, the complete set of eigenvectors $\mathbf{T}$ is orthonormal:

$$\mathbf{T}^T\mathbf{T} = \mathbf{I} \qquad \mathbf{T}^T = \mathbf{T}^{-1}$$

[This expression corresponds with Eq. (13-82a); note that the left hand and right hand eigenvectors are the same for a symmetric matrix.] Finally, the mass matrix can be expressed in terms of these eigenvectors and the set of eigenvalues $\hat{\mathbf{v}}$ as follows [by analogy with Eq. (13-85)]:

$$\mathbf{m} = \mathbf{T}\hat{\mathbf{v}}\mathbf{T}^T \tag{13-93}$$

From Eq. (13-93) it is apparent that the transformation matrix is $\mathbf{T}\hat{\mathbf{v}}^{1/2}$, and thus the transformation of Eq. (13-74) is performed by expressing its eigenvectors as

$$\boldsymbol{\phi}_n = \mathbf{T}\hat{\mathbf{v}}^{-1/2}\mathbf{y}_n \tag{13-94}$$

Substituting this into Eq. (13-74) and premultiplying by $\hat{\mathbf{v}}^{-1/2}\mathbf{T}^T$ then lead to

$$(\hat{\mathbf{v}}^{-1/2}\mathbf{T}^T\mathbf{k}\mathbf{T}\hat{\mathbf{v}}^{-1/2})\,\mathbf{y}_n = \mathbf{y}_n\lambda_n \tag{13-95}$$

where Eq. (13-93) has been used to simplify the right hand side. Equation (13-95) is of the form of Eq. (13-91) with $\mathbf{B} = \hat{\mathbf{v}}^{-1/2}\mathbf{T}^T\mathbf{k}\mathbf{T}\hat{\mathbf{v}}^{-1/2}$; hence the solution of this symmetric eigenproblem gives the desired vibration frequencies directly, and the vibration mode shapes are obtained from the eigenvectors $\mathbf{y}_n$ by Eq. (13-94).

Inasmuch as this transformation requires the solution of a preliminary eigenproblem of the order of the original eigenproblem, it is apparent that the use of a vibration-analysis procedure based on the solution of the symmetric eigenproblem Eq. (13-91) will be relatively expensive if the mass matrix is not diagonal. It is possible to obtain a simpler transformation by performing a Choleski decomposition of the mass matrix, that is,

$$\mathbf{m} = \mathbf{L}\mathbf{L}^T$$

where $\mathbf{L}$ is the lower triangular component [equivalent to the decomposition of $\mathbf{k}$ in Eq. (13-55)]. The transformation can then be performed as described above with $(\mathbf{L}^T)^{-1}$ taking the place of $\mathbf{T}\hat{\mathbf{v}}^{-1/2}$ [in Eq. (13-94)] or of $\mathbf{m}^{-1/2}$ [in Eq. (13-92)]. However, it has been found in many practical cases that the eigenproblem resulting from the Choleski transformation may be quite sensitive and difficult to solve accurately. For this reason, the eigenvector decomposition of Eq. (13-93) is preferable for general use, even though it is more expensive.

Analysis of Unconstrained Structures

Structures which are unconstrained or only partially constrained against rigid-body displacements by their external support system present a special problem in eigenproblem analysis because their stiffness matrices are singular and the vibration frequencies corresponding to the rigid-body motions are zero. Although the determinantal equation method (and some other formal mathematical procedures) can deal directly with a dynamic system having a singular stiffness matrix, it is evident that inverse iteration (or any other method making direct use of the stiffness inverse) cannot be applied without modification. Two simple methods of avoiding difficulty with a singular stiffness matrix are described here.

Spring Constraints — The most direct way to deal with an unconstrained structure is to modify it by adding small spring constraints to the unconstrained degrees of freedom. First a minimum set of constraints sufficient to prevent any rigid-body motions must be identified. Then if a spring is connected between the structure and the ground in each of these degrees of freedom, the singularity of the stiffness matrix will be removed. Analytically these springs are represented by adding terms to the diagonal elements of the stiffness matrix for these degrees of freedom. If the added spring stiffnesses are very small relative to the original stiffness-matrix coefficients, they will have negligible effect on the vibration mode shapes and frequencies associated with deformations of the structure, but an additional set of rigid-body modes will be defined having frequencies much smaller than the deformation modes. These constraint springs can be introduced automatically by the computer program when the eigenproblem is solved by inverse iteration. If the stiffness equations are solved by Choleski or Gauss decomposition, any singularity leads to a zero in the diagonal position and would prevent continuation of the decomposition. However, the program can be written so that each diagonal zero is replaced by a small number which physically represents the spring constraint; in this way the singularities are overcome and the decomposition process can be carried to completion.

Eigenvalue Shift — A similar effect can be achieved mathematically by means of an eigenvalue shift. From Eq. (13-70),

$$[\mathbf{k} - \mu\mathbf{m}]\ \boldsymbol{\phi}_n = \delta_n \mathbf{m} \boldsymbol{\phi}_n$$

it is evident that the shifted stiffness matrix $\hat{\mathbf{k}} = \mathbf{k} - \mu\mathbf{m}$ will be nonsingular in general even if $\mathbf{k}$ is singular. If the mass matrix is diagonal, introducing a negative shift causes a positive quantity to be added to the diagonal elements of the stiffness matrix; hence this is equivalent to connecting a spring to each degree of freedom. The essential difference between this procedure and the physical approach mentioned first is that a "spring" is added corresponding to each mass coefficient, rather than just a minimum set. The shift approach has the advantage that the mode shapes are not changed and the frequency effect is accounted for exactly by the shift.

PROBLEMS

13-1. Evaluate the fundamental vibration-mode shape and frequency for the building of Prob. 8-12 using the matrix iteration method. Note that the flexibility matrix may be obtained from the given story shear stiffness either by inverting the stiffness matrix or by applying a unit load successively at each story, and evaluating the resulting displacements at each story.

13-2. Evaluate the highest mode shape and frequency for the building of Prob. 13-1 by matrix iteration, using the stiffness form of the dynamic matrix Eq. (13-45).

13-3. Repeat Prob. 13-1 for the building properties of Prob. 8-13.

13-4. Evaluate the second mode shape and frequency for the shear building of Prob. 12-4 by matrix iteration. To form the first mode sweeping matrix S_1, use the given first mode shape ϕ_1 and Eq. (13-30).

13-5. Repeat Prob. 13-4 using inverse iteration with shifts, as indicated by Eq. (13-65) and the discussion that follows it. For this demonstration problem, use a shift $\mu = 98\%(\omega_2)^2$, where ω_2 is as given in Prob. 12-4.

13-6. A beam with three lumped masses is shown in Fig. P13-1; also shown are its flexibility and stiffness matrices. By matrix iteration, determine the axial force N_{cr} that will cause this beam to buckle. In this analysis, use the linear approximation, Eq. (10-36), to express the geometric stiffness of the beam.

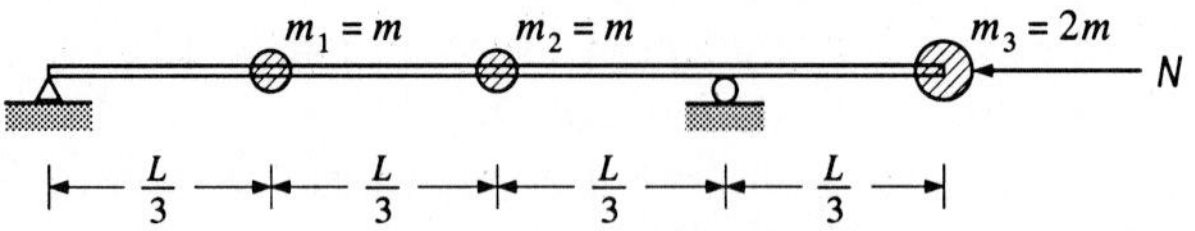

$$\tilde{f} = \frac{L^3}{243EI} \begin{bmatrix} 8 & 7 & -8 \\ 7 & 8 & -10 \\ -8 & -10 & 24 \end{bmatrix}; \quad k = \begin{bmatrix} 92 & -88 & -6 \\ -88 & 128 & 24 \\ -6 & 24 & 15 \end{bmatrix} \frac{243}{168} \frac{EI}{L^3}$$

FIGURE P13-1

13-7. By matrix iteration, compute the frequency of vibration of the beam of Prob. 13-6 if the axial force has the value $N = 2(EI/L^2)$.

CHAPTER

14

SELECTION
OF DYNAMIC
DEGREES
OF FREEDOM

14-1 FINITE-ELEMENT DEGREES OF FREEDOM

Several procedures for calculating the linear response to arbitrary dynamic load-ings of a system with multiple degrees of freedom were described in Chapter 13; some of these are considered suitable only for systems with very few degrees of freedom while others are well adapted to use with mathematical models having large numbers of degrees of freedom. However, little was said about the selection of the degrees of freedom to be used in analysis — that is, about the number that may be needed to obtain satisfactory results — and it is the purpose of this chapter to discuss many aspects of that question.

It was stated previously that the formulation of the mathematical model is the most critical step in any dynamic analysis, because the validity of the calculated results depends directly on how well the mathematical description can represent the behavior of the real physical system, and a few comments will be made here on the model definition. For the purpose of this discussion, it will be assumed that the mathematical model is an assemblage of finite elements and that the displacements of the interconnected nodes are the degrees of freedom of the model. Only framed structures, i.e., assemblages of one-dimensional elements, are considered in detail in this text. However, the same analysis procedures may be applied to the analysis of any MDOF system regardless of the types of finite elements employed; for this reason some comments also will be made here about structures made up of two- or three-dimensional elements.

One-Dimensional Elements

A finite-element model of a framed structure typically is formed by assembling a set of one-dimensional elements which are in one-to-one correspondence with the beams, struts, girders, etc., that make up the actual structure. The number of degrees of freedom in the model, therefore, is fixed by the physical arrangement of the structure, and in general all of the degrees of freedom would be involved in the analysis of stresses and displacements resulting from application of a general static load distribution. On the other hand, not all of the degrees of freedom need be considered as independent variables in analysis of the response to an arbitrary dynamic loading. Depending on both the time variation as well as the spatial distribution of the load, the dynamic analysis often may be performed effectively with a much smaller number of independent degrees of freedom using procedures to be explained later in this chapter.

Two- and Three-Dimensional Elements

Many structures can be treated as two- or three-dimensional continua or as combinations of such continuum components, and appropriate two- or three-dimensional elements are most effective in modeling such structures. In formulating models of this type, the number of degrees of freedom to be used is not dictated just by the configuration of the structure; in addition the degree of mesh refinement that is required to obtain a reasonable approximation of the actual strain distribution is an important consideration. The basic factor that controls the stiffness properties of the individual finite elements is the variation of displacements within the elements as expressed by the assumed displacement interpolation functions. For the one-dimensional flexural elements described in Chapter 10, the variation of displacements with position along the element's length was assumed to be expressed by cubic Hermitian polynomials. For two- and three-dimensional elements, the displacement variations must be assumed similarly with respect to position axes in two or three directions. The strain distributions that may be developed within the elements clearly depend directly on the displacement functions that are assumed: constant strains result from linear displacement variations, linear strains from quadratic displacements, etc.

Thus in order for any required strain variation to be developed by a finite-element mesh, such as might be associated with stress concentrations in a plane stress system, for example, it may be necessary to provide a very fine finite-element mesh with many degrees of freedom to achieve the necessary variations of strain gradients. Fortunately, however, the nodal displacements that control the inertial forces in a dynamic analysis are not as sensitive to local strain variations as is the stress distribution. Consequently fewer degrees of freedom are needed to perform an adequate analysis of the dynamic displacements, and the resulting stress distributions can then be determined from these displacements by a static analysis using a more refined finite-element mesh as necessary. Various techniques that may be used for reducing a mathematical model from the number of degrees of freedom suitable for stress analysis to a less refined system that is adequate and efficient for analysis of dynamic nodal displacements are described in the following sections.

14-2 KINEMATIC CONSTRAINTS

Probably the simplest means of reducing the number of degrees of freedom in a mathematical model is by assuming kinematic constraints which express the displacements of many degrees of freedom in terms of a much smaller set of primary displacement variables. In principle, the displacement interpolation concept that was introduced in evaluating the stiffness properties of beam elements (Fig. 10-4) exerts a form of kinematic constraint on the displacements within the span of the element. However, in the present context, the constraints will be expressed in terms of the displacements imposed on a group of degrees of freedom by the displacements specified at one (or more) degrees of freedom.

One of the most widely used applications of this type of constraint is introduced in the modeling of multistory building frames. For example, consider the 20-story rectangular building frame shown in Fig. 14-1, which includes six frames parallel to the Y-Z plane and four frames parallel to the X-Z plane. The Y-Z frames contain a total of $20 \times 6 \times 3 = 360$ girders while there are $20 \times 4 \times 5 = 400$ girders in the X-Z frames. The model also includes $20 \times 4 \times 6 = 480$ column elements which are common to both the Y-Z and the X-Z frames; thus there is a total of 1240 one-

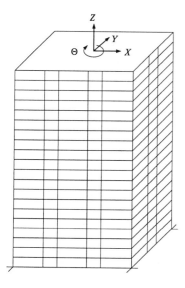

FIGURE 14-1
Twenty-story building frame (2880 degrees of freedom).

dimensional elements in the model. The number of joints interconnecting the elements is the same as the number of column elements, 480, so considering 3 translation and 3 rotation displacements per joint, the building frame includes a total of 2880 degrees of freedom.

If the constraining effect of the floor slabs is considered, however, this number can be reduced almost by half. It usually is assumed that each floor diaphragm is rigid in its own plane but is flexible in the vertical direction, which is a reasonable representation of the true floor system behavior. Introducing this assumption reduces the independent degrees of freedom of each joint from 6 to 3 (Z displacement plus rotation about the X and Y axes). In addition the diaphragm at each floor level has 3 rigid-body degrees of freedom in its own plane: X and Y translation plus rotation about the Z axis, as indicated in the figure. Consequently, after introducing the diaphragm constraint, the total number of degrees of freedom that would be considered in a static analysis is $1440 + 60 = 1500$.

A further reduction in the number of degrees of freedom that need be considered in a dynamic analysis of the building frame may be made by the method of static condensation. This concept was introduced previously in Section 10-6, and it will be described more fully in Section 14-3 of this chapter. It is sufficient here to note that static condensation can reduce the dynamic degrees of freedom of this frame to only the three rigid-body motions of each floor slab in its own plane. Thus the final result of this reduction is a total of 60 dynamic degrees of freedom, only about 2 percent of the 2880 included in the original finite-element model.

Additional kinematic constraints sometimes have been assumed in both the static and the dynamic analysis of building frames, such as that the columns are inextensible and/or that the floor slabs are rigid-out-of-plane as well as in-plane. However, these assumptions seldom are justified by the actual stiffness properties of the components of which the building is assembled and they should not be employed except in special circumstances. It is important to recognize that all members are free to distort in flexure and that all columns have axial flexibility in the type of model described above.

14-3   STATIC CONDENSATION

In contrast to the kinematic constraint idea described above, the concept of static condensation is based on static equilibrium constraints — hence the name of the procedure. To apply this principle, the degrees of freedom of the structural system are divided into two categories: those in which no mass participates so that inertial forces are not developed and those having mass that induces inertial forces. As the procedure was described in Section 10-6, the degrees of freedom were classified as either rotational or translational because it was assumed that the mass was concentrated in point lumps which had no inertial resistance to rotation. However, the fundamental concept involves merely the recognition of those degrees of freedom that can develop inertial forces as distinguished from those that cannot.

Consider, for example, the equations of motion in free vibration [Eq. (11-33)] written in the form

$$\mathbf{k}\,\hat{\mathbf{v}} = \omega^2\,\mathbf{m}\,\hat{\mathbf{v}} \tag{14-1}$$

in which the vector $\hat{\mathbf{v}}$ represents the vibration displacements of all degrees of freedom. If these displacements are partitioned into a subvector $\hat{\mathbf{v}}_0$ for which no inertial forces are developed and a subvector $\hat{\mathbf{v}}_t$ which is associated with the nonzero mass coefficients, and if the mass and stiffness matrices are partitioned to correspond, Eq. (14-1) can be written

$$\begin{bmatrix} k_{00} & k_{0t} \\ k_{t0} & k_{tt} \end{bmatrix} \begin{bmatrix} \hat{v}_0 \\ \hat{v}_t \end{bmatrix} = \omega^2 \begin{bmatrix} 0 & 0 \\ 0 & m_t \end{bmatrix} \begin{bmatrix} \hat{v}_0 \\ \hat{v}_t \end{bmatrix} \tag{14-2}$$

in which it is assumed that the mass matrix is diagonal as would result from a lumped-mass idealization.

The first of this pair of submatrix equations provides the static restraint relation between the two types of degrees of freedom, i.e.,

$$\mathbf{k}_{00}\,\hat{\mathbf{v}}_0 + \mathbf{k}_{0t}\,\hat{\mathbf{v}}_t = 0$$

from which

$$\hat{\mathbf{v}}_0 = -\mathbf{k}_{00}^{-1}\,\mathbf{k}_{0t}\,\hat{\mathbf{v}}_t \tag{14-3}$$

Using this expression to eliminate $\hat{\mathbf{v}}_0$ from the second submatrix equation of Eq. (14-2) then leads to the reduced free-vibration equation

$$\mathbf{k}_t\,\hat{\mathbf{v}}_t = \omega^2\,\mathbf{m}_t\,\hat{\mathbf{v}}_t \tag{14-4a}$$

in which $\mathbf{k}_t$ is the reduced stiffness matrix expressed by

$$\mathbf{k}_t = \mathbf{k}_{tt} - \mathbf{k}_{t0}\,\mathbf{k}_{00}^{-1}\,\mathbf{k}_{0t} \tag{14-4b}$$

This static condensation procedure can be used to effect a very considerable reduction in the number of degrees of freedom to be used in a dynamic analysis, such as the reduction from 1500 to only 60 in the building frame example discussed above; however, the reduction in actual computational effort may be much less significant than these data suggest. This is because the narrow banding of the stiffness matrix $\mathbf{k}$ in Eq. (14-1) makes possible a very efficient solution procedure when the analysis is performed in the original coordinates, whereas the analysis using Eq. (14-4a) is much more expensive per degree of freedom because the reduced stiffness $\mathbf{k}_t$ becomes fully populated as a result of the condensation procedure. For this reason, the advisability of using static condensation should be evaluated carefully on a case-by-case basis.

It is of interest to note that equilibrium constraints such as are used in static condensation often are utilized even when the structure has no massless degrees of freedom. However, in such cases the constraints are used merely as a basis for defining patterns of nodal displacements for use in a Rayleigh or Rayleigh-Ritz type of analysis as is described in the following sections of this chapter.

14-4   RAYLEIGH METHOD IN DISCRETE COORDINATES

The Rayleigh method of vibration analysis described in Section 8-5 may be recognized as a demonstration of the fact that a useful dynamic analysis often may be performed using fewer degrees of freedom than are required for a static analysis. In the example presented there, a continuum definition of a beam having an infinite number of degrees of freedom was reduced to a system in which a single degree of freedom served to express the amplitude of displacement. However, the Rayleigh concept applies equally to systems for which the displacements are expressed in discrete coordinates. To apply the method, it is necessary to express the displacement of the structure in terms of an assumed shape and a generalized-coordinate amplitude. In matrix notation, the assumed free-vibration displacements may be expressed [compare with Eq. (8-25)]

$$\mathbf{v}(t) = \boldsymbol{\psi}\, Z(t) = \boldsymbol{\psi}\, Z_0 \,\sin \omega t \tag{14-5a}$$

in which $\boldsymbol{\psi}$ is the assumed shape vector and $Z(t)$ is the generalized coordinate expressing its amplitude. The velocity vector in free vibrations then is

$$\dot{\mathbf{v}}(t) = \boldsymbol{\psi}\, \omega\, Z_0 \,\cos \omega t \tag{14-5b}$$

In matrix form, the maximum kinetic energy of the structure is given by

$$T_{\max} = \frac{1}{2}\, \dot{\mathbf{v}}_{\max}^T \,\mathbf{m}\, \dot{\mathbf{v}}_{\max} \tag{14-6a}$$

and the maximum potential energy by

$$V_{\max} = \frac{1}{2}\, \mathbf{v}_{\max}^T \,\mathbf{k}\, \mathbf{v}_{\max} \tag{14-6b}$$

When the maximum displacement and velocity, obtained from Eqs. (14-5), are substituted, these are written

$$T_{\max} = \frac{1}{2}\, Z_0^2\, \omega^2 \,\boldsymbol{\psi}^T \,\mathbf{m}\, \boldsymbol{\psi} \tag{14-7a}$$

$$V_{\max} = \frac{1}{2}\, Z_0^2 \,\boldsymbol{\psi}^T \,\mathbf{k}\, \boldsymbol{\psi} \tag{14-7b}$$

Then the frequency can be obtained by equating the maximum potential- and kinetic-energy expressions, according to the Rayleigh principle, so that

$$\omega^2 = \frac{\psi^T k \psi}{\psi^T m \psi} \equiv \frac{k^*}{m^*} \tag{14-8}$$

in which the asterisks denote generalized-coordinate properties. It should be noted that Eq. (14-8) is merely the matrix equivalent of Eq. (8-30).

The improved Rayleigh method of Eqs. (8-40) or (8-42) can also be developed in matrix form. If the initial displacement assumption is designated

$$\mathbf{v}^{(0)} = \psi \, Z \tag{14-9}$$

then the inertial forces developed in free vibrations will be [from Eq. (11-33)]

$$\mathbf{f}_I = \omega^2 \, \mathbf{m} \, \mathbf{v}^{(0)} = \omega^2 \, \mathbf{m} \, \psi \, Z \tag{14-10}$$

and the deflections produced by these inertial forces are

$$\mathbf{v}^{(1)} = \widetilde{\mathbf{f}} \mathbf{f}_I = \omega^2 \widetilde{\mathbf{f}} \mathbf{m} \, \psi \, Z \tag{14-11}$$

which is a better approximation of the first-mode shape, as noted in the discussion of the matrix iteration method. Thus if this derived shape is used in the Rayleigh method, it will produce a better result than the initial assumption would. The result of introducing Eq. (14-11) into Eqs. (14-6) and equating them is

$$\omega^2 = \frac{\psi^T \mathbf{m} \widetilde{\mathbf{f}} \mathbf{m} \psi}{\psi^T \mathbf{m} \widetilde{\mathbf{f}} \mathbf{m} \widetilde{\mathbf{f}} \mathbf{m} \psi} \tag{14-12}$$

which is the improved Rayleigh method expression (method R_{11}). By comparing Eq. (14-12) with Eq. (13-11) it can be seen that the frequency obtained from the improved Rayleigh procedure is identical to that given by a single-step matrix iteration analysis using the mass as a weighting factor in the averaging process.

14-5 RAYLEIGH-RITZ METHOD

Although the Rayleigh method can provide a satisfactory approximation of the first mode of vibration in many structures, it frequently is necessary to include more than one mode in a dynamic analysis to give adequate accuracy in the results. The Ritz extension of the Rayleigh method is one of the most convenient procedures for evaluating the first several modes of vibration. The basic assumption of the Ritz method is that the displacement vector can be expressed in terms of a set of assumed shapes Ψ of amplitude Z as follows:

$$\mathbf{v} = \psi_1 Z_1 + \psi_2 Z_2 + \psi_3 Z_3 + \cdots$$

or

$$\mathbf{v} = \boldsymbol{\Psi}\,\mathbf{Z} \tag{14-13}$$

in which the generalized-coordinate amplitudes $\mathbf{Z}$ are as yet unknown. To obtain the best results from the least possible number of coordinates, each of the vectors ψ_n should be taken as an approximation of the corresponding true vibration mode shape ϕ_n, although many other schemes have been proposed for selecting the trial vectors. For example, the static-condensation process can be looked upon as a means for defining a set of Ritz shapes as was mentioned above. The fact that a specified set of elastic forces is to be set to zero constitutes a constraint which makes it possible to express the corresponding set of displacements in terms of all the others. This type of relationship is given by Eq. (10-45) or by Eq. (14-3) in the notation used here. Hence, the complete displacement vector can also be expressed in terms of the non-zero-force degrees of freedom merely by incorporating an identity matrix of appropriate dimensions into the transformation:

$$\hat{\mathbf{v}} = \begin{bmatrix} \hat{\mathbf{v}}_0 \\ \hat{\mathbf{v}}_t \end{bmatrix} = \begin{bmatrix} -\mathbf{k}_{00}^{-1}\mathbf{k}_{0t} \\ \mathbf{I} \end{bmatrix} \hat{\mathbf{v}}_t \tag{14-14}$$

Here the second matrix in square brackets clearly is equivalent to the assumed shapes $\boldsymbol{\Psi}$ of Eq. (14-13), and the vector $\hat{\mathbf{v}}_t$ represents the generalized coordinates $\mathbf{Z}$. As many trial vectors as desired may be used in the Ritz analysis. In general, it may be advisable to use as many as s assumed shapes $\boldsymbol{\Psi}$ if it is desired to obtain $s/2$ vibration mode shapes and frequencies with good accuracy.

Expressions for the maximum kinetic and potential energy in the system can be obtained by introducing Eq. (14-13) into Eqs. (14-6), giving

$$T_{\max} = \frac{1}{2}\,\omega^2 \mathbf{Z}^T \boldsymbol{\Psi}^T \mathbf{m} \boldsymbol{\Psi} \mathbf{Z} \tag{14-15a}$$

$$V_{\max} = \frac{1}{2}\,\mathbf{Z}^T \boldsymbol{\Psi}^T \mathbf{k} \boldsymbol{\Psi} \mathbf{Z} \tag{14-15b}$$

Equating these then leads to the frequency expression

$$\omega^2 = \frac{\mathbf{Z}^T \boldsymbol{\Psi}^T \mathbf{k} \boldsymbol{\Psi} \mathbf{Z}}{\mathbf{Z}^T \boldsymbol{\Psi}^T \mathbf{m} \boldsymbol{\Psi} \mathbf{Z}} \equiv \frac{\tilde{\mathbf{k}}(\mathbf{Z})}{\tilde{\mathbf{m}}(\mathbf{Z})} \tag{14-16}$$

Equation (14-16) is not an explicit expression for the frequency of vibration, of course; both the numerator and denominator are functions of the generalized-coordinate amplitudes $\mathbf{Z}$, which are not yet known. To evaluate these, the fact that the Rayleigh analysis provides an upper bound to the vibration frequency will be utilized; in other words, any assumed shape leads to a calculated frequency which is higher than the true frequency, and so the best approximation of the shape, that is, the best choice of Z, will minimize the frequency.

Thus differentiating the frequency expression with respect to any one of the generalized coordinates Z_n and equating to zero gives

$$\frac{\partial \omega^2}{\partial Z_n} = \frac{\tilde{m}(\partial \tilde{k}/\partial Z_n) - \tilde{k}(\partial \tilde{m}/\partial Z_n)}{\tilde{m}^2} = 0 \qquad (14\text{-}17)$$

But from Eq. (14-16), $\tilde{k} = \omega^2 \tilde{m}$; thus Eq. (14-17) leads to

$$\frac{\partial \tilde{k}}{\partial Z_n} - \omega^2 \frac{\partial \tilde{m}}{\partial Z_n} = 0 \qquad (14\text{-}18)$$

Now from the definitions given in Eq. (14-16)

$$\frac{\partial \tilde{k}}{\partial Z_n} = 2Z^T \psi^T k\psi \frac{\partial}{\partial Z_n}(Z) = 2Z^T \Psi^T k\psi_n \qquad (14\text{-}19a)$$

and similarly

$$\frac{\partial \tilde{m}}{\partial Z_n} = 2Z^T \Psi^T m\psi_n \qquad (14\text{-}19b)$$

Substituting Eqs. (14-19) into Eq. (14-18) and transposing gives

$$\psi_n^T k\Psi Z - \omega^2 \psi_n^T m\Psi Z = 0 \qquad (14\text{-}20)$$

Minimizing the frequency successively with respect to each of the generalized coordinates leads to an equation like Eq. (14-20) for each of the shape vectors ψ_n; thus the entire set of equations may be expressed as

$$\Psi^T k\Psi Z - \omega^2 \Psi^T m\Psi Z = 0$$

With the notation

$$k^* = \Psi^T k\Psi \qquad (14\text{-}21a)$$

$$m^* = \Psi^T m\Psi \qquad (14\text{-}21b)$$

this becomes

$$(k^* - \omega^2 m^*)\,\hat{Z} = 0 \qquad (14\text{-}22)$$

where $\hat{Z}$ represents each of the eigenvectors (relative values of Z) which satisfies this eigenvalue equation.

Comparing Eq. (14-22) with Eq.(11-4) shows that the Rayleigh-Ritz analysis has the effect of reducing the system from N degrees of freedom, as represented by the geometric coordinates $\mathbf{v}$, to s degrees of freedom representing the number of generalized coordinates Z and the corresponding assumed shapes. Equation (14-13) is the coordinate transformation, and Eqs. (14-21) are the generalized-mass and stiffness matrices (of dimensions $s \times s$). Each element of these matrices is a generalized-mass or stiffness term; thus

$$k^*_{mn} = \boldsymbol{\psi}^T_m \mathbf{k} \boldsymbol{\psi}_n \qquad (14\text{-}23a)$$

$$m^*_{mn} = \boldsymbol{\psi}^T_m \mathbf{m} \boldsymbol{\psi}_n \qquad (14\text{-}23b)$$

In general, the assumed shapes $\boldsymbol{\psi}_n$ do not have the orthogonality properties of the true mode shapes, thus the off-diagonal terms do not vanish from these generalized-mass and stiffness matrices; however, a good choice of assumed shapes will tend to make the off-diagonal terms relatively small. In any case, it is much easier to obtain the dynamic response for the reduced number of coordinates s than for the original N equations.

Equation (14-22) can be solved by any standard eigenvalue-equation solution procedure, including the determinantal equation approach discussed earlier for systems having only a few generalized coordinates Z. The frequency vector $\boldsymbol{\omega}$ so obtained represents approximations to the true frequencies of the lower modes of vibration, the accuracy generally being excellent for the lowest modes ($1 < n < s/2$) and relatively poor in the highest modes. When the mode-shape vectors Z_n are normalized by dividing by some reference coordinate, they will be designated $\boldsymbol{\phi}_{Zn}$, where the subscript Z indicates that they represent the mode shapes expressed in generalized coordinates. The complete set of generalized-coordinate mode shapes can then be denoted $\boldsymbol{\Phi}_Z$, representing a square $s \times s$ matrix.

The generalized coordinates Z expressed in terms of the modal amplitudes [by analogy with Eq. (12-3)] are

$$\mathbf{Z} = \boldsymbol{\Phi}_Z \mathbf{Y} \qquad (14\text{-}24)$$

It is of interest that these mode shapes are orthogonal with respect to the generalized-mass and stiffness matrices:

$$\boldsymbol{\phi}_{Zm} \mathbf{m}^* \boldsymbol{\phi}_{Zn} = 0$$
$$\qquad\qquad\qquad\qquad m \neq n \qquad (14\text{-}25)$$
$$\boldsymbol{\phi}_{Zm} \mathbf{k}^* \boldsymbol{\phi}_{Zn} = 0$$

By introducing Eq. (14-24) into Eq. (14-13) the geometric coordinates can be expressed in terms of the normal modal coordinates

$$\mathbf{v} = \boldsymbol{\Psi} \boldsymbol{\Phi}_Z \mathbf{Y} \qquad (14\text{-}26)$$

Thus it is seen that the approximate mode shapes in geometric coordinates are given by the product of the assumed shapes and the generalized-coordinate mode shapes

$$\boldsymbol{\Phi} = \boldsymbol{\Psi} \boldsymbol{\Phi}_Z \qquad (14\text{-}27)$$

which is of dimensions $N \times s$. Substituting Eqs. (14-21) into Eqs. (14-25) and applying Eqs. (14-26) demonstrates that these approximate geometric mode shapes are orthogonal with respect to the mass and stiffness expressed in geometric coordinates. They can therefore be used in the standard mode-superposition dynamic-analysis procedure.

It is important to note that the same type of improvement described above for the Rayleigh method is applicable to the Rayleigh-Ritz procedure. Thus, by analogy with Eq. (14-21), the improved generalized-coordinate stiffness and mass matrices are given by

$$\mathbf{k}^* = \mathbf{\Psi}^T \mathbf{m}\tilde{\mathbf{f}}\mathbf{m}\mathbf{\Psi} \tag{14-28a}$$

$$\mathbf{m}^* = \mathbf{\Psi}^T \mathbf{m}\tilde{\mathbf{f}}\mathbf{m}\tilde{\mathbf{f}}\mathbf{m}\mathbf{\Psi} \tag{14-28b}$$

in place of Eqs. (14-21). The principal advantage of these equations is that the inertial-force deflections on which they are based provide reasonable assumed shapes from very crude initial assumptions. In large, complex structures it is very difficult to make detailed estimates of the shapes, and it is possible with this improved procedure merely to indicate the general character of each shape. Another major advantage in many analyses is that it avoids use of the stiffness matrix. In fact, if the initial assumed shapes are designated $\mathbf{\Psi}^{(0)}$ and the deflections resulting from inertial forces associated with those shapes are called $\mathbf{\Psi}^{(1)}$, that is,

$$\mathbf{\Psi}^{(1)} = \tilde{\mathbf{f}}\mathbf{m}\mathbf{\Psi}^{(0)} \tag{14-29}$$

then Eqs. (14-28) may be written

$$\mathbf{k}^* = (\mathbf{\Psi}^{(1)})^T \mathbf{m}\mathbf{\Psi}^{(0)} \tag{14-30a}$$

$$\mathbf{m}^* = (\mathbf{\Psi}^{(1)})^T \mathbf{m}\mathbf{\Psi}^{(1)} \tag{14-30b}$$

Consequently it is not necessary to have an explicit expression for the flexibility either; it is necessary only to be able to compute the deflections resulting from a given loading (which in this case is $\mathbf{m}\mathbf{\Psi}^{(0)}$).

This improvement process in the Rayleigh-Ritz method may be looked upon as the first cycle of an iterative solution, just as the improved Rayleigh method is equivalent to a single cycle of the basic matrix iteration method. However, that type of analysis results in only a single mode shape and frequency, whereas the continuation of the Ritz improvement process evaluates simultaneously the entire reduced set of mode shapes and frequencies. This method, called *simultaneous* or *subspace iteration*, is described in the following section.

14-6   SUBSPACE ITERATION

Because subspace iteration is essentially a continuation of the Rayleigh-Ritz improvement procedure in which the improvement is continued iteratively, it is convenient to use the Ritz analysis notation in this presentation. In order to obtain a set of p mode shapes and frequencies that are established with adequate accuracy, it is desirable to start with a somewhat larger number q of trial vectors. Denoting these trial vectors by the superscript (0), the displacements of the structure can be expressed as combinations of these shapes [see Eq. (14-13)] as follows:

$$\mathbf{v}^{(0)} = \mathbf{\Psi}^{(0)}\mathbf{Z}^{(0)} = \mathbf{\Psi}^{(0)} \tag{14-31}$$

in which the initial generalized-coordinate matrix $\mathbf{Z}^{(0)}$ is merely an identity matrix (indicating that the trial vectors are the assumed Ritz shapes $\mathbf{\Psi}^{(0)}$).

For the large systems to which this method is usually applied, it is important to take advantage of the banding properties of the mass and stiffness matrices; hence the free-vibration equation [Eq. (14-1)] is written for the set of p eigenvalues and eigenvectors as

$$\mathbf{k\Phi} = \mathbf{m\Phi\Lambda} \tag{14-32}$$

in which Λ is the diagonal matrix of the eigenvalues. Introducing the q trial vectors on the right side of this equation leads to

$$\mathbf{k}\overline{\mathbf{\Psi}}^{(1)} = \mathbf{m}\mathbf{\Psi}^{(0)} \equiv \mathbf{w}^{(0)} \tag{14-33}$$

which is equivalent to Eq. (13-72) written for multiple vectors and with no shift. The unscaled improved shapes are obtained by solving Eq. (14-33); thus

$$\overline{\mathbf{\Psi}}^{(1)} = \mathbf{k}^{-1}\mathbf{w}^{(0)} \tag{14-34}$$

and, as explained above, it will be more efficient to use the Choleski decomposition of $\mathbf{k}$ [Eq. (13-72)] rather than its inverse in obtaining the solution.

Before the improved shapes of Eq. (14-34) can be used in a new iteration cycle, they must be modified in two ways: normalized to maintain reasonable number sizes in the calculations and orthogonalized so that each vector will converge toward a different mode (rather than all toward the lowest mode). These operations can be performed in many different ways, but it is convenient to accomplish both at once by carrying out a Ritz eigenproblem analysis. Thus the first-cycle generalized-coordinate stiffness and mass matrices are computed [see Eqs. (14-21)] as follows:

$$\mathbf{k}_1^* = \overline{\mathbf{\Psi}}^{(1)T}\mathbf{k}\overline{\mathbf{\Psi}}^{(1)} \equiv \overline{\mathbf{\Psi}}^{(1)T}\mathbf{m}\mathbf{\Psi}^{(0)}$$
$$\mathbf{m}_1^* = \overline{\mathbf{\Psi}}^{(1)T}\mathbf{m}\overline{\mathbf{\Psi}}^{(1)} \tag{14-35}$$

in which the subscripts identify the first-cycle values, and then the corresponding eigenproblem

$$\mathbf{k}_1^* \hat{\mathbf{Z}}^{(1)} = \mathbf{m}_1^* \hat{\mathbf{Z}}^{(1)} \Omega_1^2 \tag{14-36}$$

is solved for the first-cycle generalized-coordinate mode shapes $\hat{\mathbf{Z}}^{(1)}$ and frequencies Ω_1^2. Any suitable eigenproblem-analysis procedure may be used in the solution of Eq. (14-36), but since it is a much smaller equation system than the original eigenproblem, that is, $q \ll N$, it can often be done by a standard computer-center library program. Usually it is convenient to normalize the generalized-coordinate modal vector so that the generalized masses have unit values:

$$\hat{\mathbf{Z}}^{(1)T} \mathbf{m}_1^* \hat{\mathbf{Z}}^{(1)} = \mathbf{I}$$

When the normalized generalized-coordinate vectors are used, the improved trial vectors are given by

$$\mathbf{v}^{(1)} = \mathbf{\Psi}^{(1)} = \overline{\mathbf{\Psi}}^{(1)} \hat{\mathbf{Z}}^{(1)} \tag{14-37}$$

The entire process can now be repeated iteratively, solving for the unscaled improved shapes $\overline{\mathbf{\Psi}}^{(2)}$, as indicated by Eq. (14-34), and then solving the corresponding Ritz eigenproblem [Eq. (14-36)] to provide for scaling and orthogonalization:

$$\mathbf{\Psi}^{(2)} = \overline{\mathbf{\Psi}}^{(2)} \hat{\mathbf{Z}}^{(2)}$$

and so on. Eventually the process will converge to the true mode shapes and frequencies, that is,

$$\begin{aligned}\mathbf{\Psi}^{(s)} &\to \mathbf{\Phi} \\[4pt] \Omega_s^2 &\to \Lambda \end{aligned} \qquad \text{as} \quad s \to \infty \tag{14-38}$$

In general, the lower modes converge most quickly, and the process is continued only until the desired p modes are obtained with the necessary accuracy. The additional $q - p$ trial vectors are included because they accelerate the convergence process, but obviously they require additional computational effort in each cycle, so that a reasonable balance must be maintained between the number of vectors used and the number of cycles required for convergence. By experience it has been found that a suitable choice is given by the smaller of $q = 2p$ and $q = p + 8$.

This subspace, or simultaneous-iteration, procedure has proved to be one of the most efficient methods for solving large-scale structural-vibration problems where probably no more than 40 modes are required for the dynamic analysis of systems having many hundreds to a few thousand degrees of freedom. Although this may be considered as a Rayleigh-Ritz coordinate-reduction scheme, it has the great advantage that the resulting modal coordinates can be obtained to any desired degree of precision. Other coordinate-reduction procedures involve approximations which make the accuracy of the final results uncertain; hence subspace iteration is strongly recommended for practical applications.

14-7 REDUCTION OF MODAL TRUNCATION ERRORS

General Comments on Coordinate Reduction

Based on the preceding discussions, it is evident that the Rayleigh-Ritz method is an excellent procedure for reducing a model of a structural system from the set of finite-element degrees of freedom chosen to define the static stress distribution to the smaller number of coordinates needed to evaluate the system vibration properties, and also that subspace iteration is an efficient method for solving the vibration eigenproblem. Furthermore, it is apparent that the calculated mode shapes are extremely efficient in depicting the dynamic response of the system, so a severely truncated set of modal coordinates can produce results with satisfactory precision.

Two final questions remain to be answered in establishing a recommended dynamic analysis method:

(1) How should the trial vectors $\overline{\Psi}^{(0)}$ be selected for use in the subspace iteration analysis?

(2) How many modal coordinates are needed to avoid significant modal truncation errors?

For convenience, the modal truncation error is considered first in this section of the chapter; then the selection of the Ritz displacement patterns used in the vibration eigenproblem is discussed in the following Section 14-8.

In beginning this examination of the modal truncation error, it must be recalled that the entire dynamic analysis procedure involves a succession of approximations. First is the selection of a finite-element mesh that approximates the true strain distribution only in a virtual work sense, and next is the transformation to Ritz coordinates that only approximate the displacements of the larger number of finite-element coordinates.

A final transformation then is made expressing the Ritz coordinates in terms of the undamped vibration mode shapes. If the full set of modal coordinates is used, this transformation involves no approximation; in other words, a mode-superposition analysis will give exactly the same results as a step-by-step solution of the coupled Ritz coordinate equations if all modes are included in the superposition. However, in view of the approximations accepted in the other coordinate transformations, there is no need to try to obtain an exact mode-superposition analysis by including all of the modal coordinates. Undoubtedly significant discrepancies exist between the individual modal coordinate responses and the corresponding modal contributions to the response of the real structure, especially in the higher modes; for this reason the additional error that may result from truncation of some higher modes need not be a major concern.

Modal Contributions

In order to evaluate the errors that may result from modal truncation, it is necessary to consider the independent dynamic response contributions associated with the individual modes. For any arbitrary mode n, Eq. (12-17) expresses the equation of motion:

$$\ddot{Y}_n(t) + 2\xi_n\omega_n\dot{Y}_n(t) + \omega_n^2 Y_n(t) = \frac{P_n(t)}{M_n}$$

in which the modal mass and modal load, respectively, are given by Eqs. (12-18):

$$M_n = \boldsymbol{\phi}_n^T \mathbf{m} \boldsymbol{\phi}_n \qquad P_n = \boldsymbol{\phi}_n^T \mathbf{p}(t)$$

The load vector $\mathbf{p}(t)$ in Eq. (12-18) may be caused by any external loading mechanism, and in general it may vary with time both in amplitude and in spatial distribution. However, for the purpose of the present discussion it is assumed that the distribution does not vary with time so that only the amplitude is time-varying. Thus the load vector may be expressed as the product of a load distribution vector $\mathbf{R}$ and an amplitude function $f(t)$:

$$\mathbf{p}(t) = \mathbf{R} \, f(t) \tag{14-39}$$

This type of external loading expression applies to many practical situations, including earthquake excitation. The effective earthquake loading vector generally is most conveniently expressed as

$$\mathbf{p}_{\text{eff}}(t) = \mathbf{m} \, \mathbf{r} \, \ddot{v}_g(t) \tag{14-40a}$$

in which $\mathbf{m}$ is the structure mass matrix, $\ddot{v}_g(t)$ is the earthquake acceleration history applied at the structure's supports, and $\mathbf{r}$ is a displacement transformation vector that expresses the displacement of each structure degree of freedom due to static application of a unit support displacement. Equation (14-40a) may be put in the form of Eq. (14-39) if the seismic input is expressed as a fraction of the acceleration of gravity, g,

$$f(t) = \frac{1}{g} \, \ddot{v}_g(t) \tag{14-40b}$$

Then the corresponding load distribution vector is given by

$$\mathbf{R} = \mathbf{m} \, \mathbf{r} \, g \tag{14-40c}$$

Introducing Eq. (14-39), the equation of motion [Eq. (12-17)] becomes for this special class of loading:

$$\ddot{Y}_n(t) + 2\xi_n\omega_n\dot{Y}_n(t) + \omega_n^2 Y_n(t) = \frac{\boldsymbol{\phi}_n^T \mathbf{R}}{\boldsymbol{\phi}_n^T \mathbf{m} \boldsymbol{\phi}_n} \, f(t) \tag{14-41a}$$

and for the particular case of earthquake loading, using Eqs. (14-40) it becomes

$$\ddot{Y}_n(t) + 2\xi_n\omega_n Y_n(t) + \omega_n^2 Y_n(t) = \frac{\phi_n^T \mathbf{m} \mathbf{r}}{\phi_n^T \mathbf{m} \phi_n} \ddot{v}_g(t) \qquad (14\text{-}41b)$$

The dynamic response given by this equation of motion may be calculated in either the time domain or in the frequency domain, as explained in Section 12-4, using either form of Eq. (14-41). For a frequency-domain analysis, the damping might be expressed in complex stiffness form rather than by the modal viscous damping ratio indicated here, but that distinction is not pertinent to the present discussion. However, it is important to note that two factors control the relative importance of any mode in the total dynamic response obtained by use of these equations: (1) the modal participation factor (MPF) which depends on the interaction of the mode shape with the spatial distribution of the external load and (2) the dynamic magnification factor that depends on the ratios of the applied loading harmonic frequencies to the modal frequency. These two factors are discussed in the following paragraphs.

Modal Participation Factor — The ratios shown on the right side of Eqs. (14-41) define the modal participation factor as follows:

$$\text{MPF}_n = \frac{\phi_n^T \mathbf{R}}{\phi_n^T \mathbf{m} \phi_n} \qquad \text{or} \qquad \frac{\phi_n^T \mathbf{m} \mathbf{r}}{\phi_n^T \mathbf{m} \phi_n} \qquad (14\text{-}42)$$

where the second expression applies to the case of earthquake loading. The denominator in these expressions is the modal mass, a constant that depends on the mode shape and the mass distribution. As was mentioned previously, the mode shape often is normalized to produce a unit value for this quantity; however, the complete expression is retained here for generality.

It is apparent from Eq. (14-42) that the amplitude of the response due to any given mode depends on how the applied load distribution interacts with the mode shape. For a typical multistory building, subjected to horizontal ground motion, the earthquake motion transformation vector $\mathbf{r}$ is a unit column, so a lumped-mass model load distribution vector is merely the story mass vector $\mathbf{m}_s$. Considering the mass distribution of a typical building as sketched in Fig. 14-2a, as well as the mode shapes depicted in Fig. 14-2b, it is evident that the vector product $\phi_n^T \mathbf{m}_s$ will be relatively large for the first mode because the first-mode shape is all positive. However, for the second and third modes the product will be much smaller because these mode shapes include both positive and negative zones. It is for this reason that an earthquake tends to excite response of a structure mainly in its first mode.

On the other hand, an arbitrary external load distribution that might be applied to a building could be of any shape, in principle, and thus it might accentuate response in any of the modes. For example, the first-mode response would be excited only

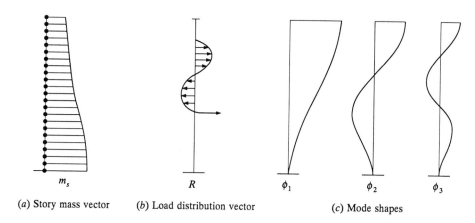

(a) Story mass vector (b) Load distribution vector (c) Mode shapes

FIGURE 14-2
Mass and load distribution and vibration mode shapes for typical building.

slightly by the load distribution vector **R** shown in Fig. 14-2b, because it has positive and negative portions acting on parts of the mode-shape pattern having rather similar displacements. In contrast, this loading would be very effective in exciting the second-mode response because the reversal in direction of the loading tends to match the direction reversal of the mode shape. Considering instead the concentrated load also shown in Fig. 14-2b, it is evident that this would tend to excite response in the first and second modes (as well as in most of the other modes that are not shown here); however it would not excite any third-mode response because this load is applied at a node in the third-mode shape.

Dynamic Magnification Factor — Because the individual modes respond to the applied loading independently in a mode superposition analysis, the dynamic magnification effects of the applied loading can be evaluated for each mode in the same way as for any single-degree-of-freedom system; thus the modal amplification may be represented by frequency response curves such as those shown in Fig. 3-3. To simplify this discussion, only the undamped case is considered here as shown by the solid line in Fig. 14-3; this depicts the response in terms of the ratio of the modal elastic resistance F_{s_n} to the harmonic modal applied force P_n. The abscissa of this plot is the frequency ratio, β_n, that is, the ratio of the excitation frequency $\bar{\omega}$ to the modal frequency ω_n. For values of β_n less than one, this curve is identical to the undamped response curve of Fig. 3-3; for values greater than one the plot shows the negative of the curve in Fig. 3-3 — the reversal of sign showing that the response is 180° out of phase with the applied load for these larger frequency ratios. For the static load case ($\beta = 0$), the response ratio is unity, indicating that the applied load is balanced directly by the elastic resistance ($F_{s_n} \equiv P_n$).

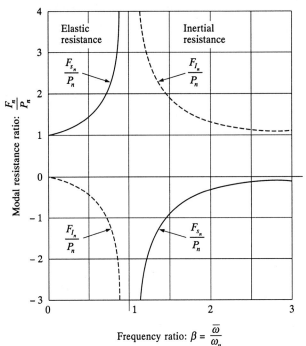

Frequency ratio: $\beta = \dfrac{\bar{\omega}}{\omega_n}$

FIGURE 14-3
Resistance ratio response curves.

The dashed curve in Fig. 14-3 shows the ratio of the modal inertial resistance F_{I_n} to the harmonic modal load P_n. This resistance decreases from zero for the static case to negative infinity at resonance ($\beta_n = 1$), the negative sign merely indicating that the inertial resistance acts in the direction opposing the elastic resistance. For input frequencies greater than the resonance condition, the inertial resistance undergoes a phase reversal equivalent to that shown for the elastic resistance, and as the excitation frequencies continue to increase, the inertial resistance ratio approaches unity asymptotically. Study of this response graph reveals that the inertial force ratio and the elastic force ratio always are of opposite sign, and that they change together with frequency in such a way that their combined effect is equal to unity for all frequency ratios thus

$$\frac{F_{s_n}(\beta)}{P_n} + \frac{F_{I_n}(\beta)}{P_n} = 1 \tag{14-43}$$

For applied frequencies exceeding the reasonance condition, the signs of both contributions are reversed but they still combine so as to equilibrate the applied load.

For a given harmonic of the input excitation, $\bar{\omega}$, it is apparent that the frequency ratio β_n tends toward zero as higher modal frequencies, ω_n, are considered. Thus, as shown by Fig. 14-3, for the higher modes of the system the resistance tends toward purely static behavior and inertial effects are negligible. On the other hand, for the lower modes of the system the frequency ratio β_n is greater and in the limit the elastic resistance is negligible; that is, at these higher excitation frequencies the resistance becomes entirely inertial.

Static Correction Procedure[1]

In order to take advantage of the fact that the response of the higher frequency modes can be calculated by static analysis because their inertial effects are negligible, the standard mode displacement superposition equation given by Eq. (12-2):

$$r(t) = \sum_{n=1}^{N} \phi_n \, Y_n(t)$$

is divided into two terms, the first being the sum of the lower mode contributions and the other being the sum of the remaining higher modes for which dynamic amplification effects may be neglected. Thus Eq. (12-2) becomes

$$\mathbf{v}(t) = \mathbf{v}_d(t) + \mathbf{v}_s(t) = \sum_{n=1}^{d} \phi_n Y_n(t) + \sum_{n=d+1}^{N} \phi_n Y_n(t) \qquad (14\text{-}44)$$

in which the subscript "d" identifies the response from those modes that are subject to dynamic amplification effects while the subscript "s" denotes the response that can be approximated by static analysis.

The response $Y_n(t)$ given by each of the first "d" modes may be calculated by any standard SDOF dynamic analysis procedure such as the Duhamel integral, step-by-step integration, or in the case of a simple form of dynamic loading by direct solution of the differential equation. For each of the remaining $N - d$ modes, the response $Y_{sn}(t)$ at any time "t" may be obtained by ordinary static analysis, dividing the modal load $P_n(t)$ by the modal stiffness, thus

$$Y_{sn}(t) = \frac{P_n(t)}{K_n} = \frac{\phi_n^T \mathbf{p}(t)}{\phi_n^T \mathbf{k} \phi_n} \qquad (14\text{-}45)$$

Hence the "static" contribution to the displacement due to that mode is given by

$$\mathbf{v}_{sn}(t) = \phi_n \, Y_{sn}(t) = \frac{\phi_n \phi_n^T}{K_n} \, \mathbf{p}(t)$$

and for convenience this is written as

$$\mathbf{v}_{sn}(t) = \mathbf{F}_n \, \mathbf{p}(t) \qquad (14\text{-}46)$$

---

[1] O. E. Hansteen and K. Bell "On the Accuracy of Mode Superposition Analysis in Structural Dynamics," *Earthquake Engineering and Structural Dynamics*, Vol. 7, No. 5, 1979.

in which

$$\mathbf{F}_n \equiv \frac{\phi_n \phi_n^T}{K_n} \tag{14-47}$$

is the modal flexibility matrix that gives the nth-mode static deflection resulting from the applied load vector $\mathbf{p}(t)$.

Using such a modal flexibility matrix for each of the "static" response modes and incorporating the load distribution vector from Eq. (14-39) the total static response could be expressed as

$$\mathbf{v}_s(t) = \sum_{n=d+1}^{N} \mathbf{F}_n \mathbf{R} f(t) \tag{14-48}$$

and the combined "static" plus dynamic response then would be

$$\mathbf{v}(t) = \sum_{n=1}^{d} \phi_n Y_n(t) + \sum_{n=d+1}^{N} \mathbf{F}_n \mathbf{R} f(t) \tag{14-49}$$

In this formulation, although only the first "d" modes are solved dynamically, it still is necessary to solve for all "N" mode shapes so that the static contribution from each of the higher modes may be calculated. However, the evaluation of the higher mode shapes may be avoided by calculating the total static response given by all modes and then subtracting the static response developed in the first "d" modes. Therefore, this more convenient form of the static response analysis can be expressed as

$$v_s(t) = \mathbf{k}^{-1}\mathbf{R} f(t) - \sum_{n=1}^{d} \mathbf{F}_n \mathbf{R} f(t) \tag{14-50}$$

in which the first term on the right side constitutes a standard static displacement analysis (expressed here with the flexibility matrix $\mathbf{k}^{-1}$) and the summation includes the static response of the first "d" modes calculated with the modal flexibility matrices.

The total response equation including this static correction now is obtained by substituting Eq. (14-50) in Eq. (14-44) with the following final result:

$$\mathbf{v}(t) = \sum_{n=1}^{d} \phi_n Y_n(t) + \left[\mathbf{k}^{-1} - \sum_{n=1}^{d} \mathbf{F}_n\right] \mathbf{R} f(t) \tag{14-51}$$

in which the first term represents a mode displacement superposition analysis using "d" modes and the other term is the corresponding static correction for the higher $(N - d)$ modes. A computer solution using this formulation requires only adding the correction term, which is given as the product of a constant matrix and the load amplitude factor $f(t)$, to the standard mode displacement solution for "d" modes.

From the rationale behind its development, this static correction method may be expected to be effective in analyses where many higher modes must be included to account for the spatial distribution of the applied load, but where the time variation function subjects only a few of the lower-mode responses to significant amplification. In these circumstances, the dynamic superposition of a few modes together with the static correction will give results comparable to a standard mode superposition analysis using many more modes.

Mode Acceleration Method

Although the static correction method was a modern development in structural dynamics, another method intended to serve the purpose of avoiding certain higher mode errors had been formulated several decades earlier following a different line of reasoning.[2] This procedure, usually known as the Mode Acceleration Method, may be derived by making minor changes in the modal coordinate equation of motion, Eq. (12-14a):

$$M_n \ddot{Y}_n(t) + C_n \dot{Y}_n(t) + K_n Y_n(t) = P_n(t)$$

Dividing this by K_n and rearranging gives the following expression for the modal response:

$$Y_n(t) = \frac{P_n(t)}{K_n} - \frac{1}{\omega_n^2} \ddot{Y}_n(t) - \frac{2\xi_n}{\omega_n} \dot{Y}_n(t) \tag{14-52}$$

Therefore the total response may be obtained in the usual way by superposition of these modal responses:

$$\mathbf{v}(t) = \sum_{n=1}^{N} \phi_n Y_n(t) = \sum_{n=1}^{N} \phi_n \frac{P_n(t)}{K_n} - \sum_{n=1}^{N} \phi_n \left[\frac{1}{\omega_n^2} \ddot{Y}_n(t) + \frac{2\xi_n}{\omega_n} \dot{Y}_n(t) \right] \tag{14-53}$$

However, the first summation on the right hand side of Eq. (14-53) may be written as

$$\sum_{n=1}^{N} \phi_n \frac{\phi_n^T \mathbf{p}(t)}{K_n} = \sum_{n=1}^{N} \mathbf{F}_n \, \mathbf{p}(t) \equiv \mathbf{k}^{-1} \mathbf{R} \, f(t) \tag{14-54}$$

where it is apparent that the sum of all the modal flexibilities must be the total flexibility of the structure $\mathbf{k}^{-1}$. On the other hand, the second summation in Eq. (14-53) represents the dynamic amplification effects of the applied loading, which have negligible influence in the response of the higher modes, hence the upper limit of

---

[2] R. E. Cornwell, R. R. Craig, and C. P. Johnston "On the Application of the Mode Acceleration Method to Structural Dynamics Problems," *Earthquake Engineering and Structural Dynamics*, Vol. 11, No. 6, 1983, pp. 679–688.

this summation may be changed to "d." On this basis, the final form of the Mode Acceleration response equation is

$$\mathbf{v}(t) = \mathbf{k}^{-1}\mathbf{R}\,f(t) - \sum_{n=1}^{d} \boldsymbol{\phi}_n \left[\frac{1}{\omega_n^2}\ddot{Y}_n(t) + \frac{2\xi_n}{\omega_n}\dot{Y}_n(t)\right] \tag{14-55}$$

Now for comparison purposes, the static correction method Eq. (14-51) will be rewritten as

$$\mathbf{v}(t) = \mathbf{k}^{-1}\mathbf{R}\,f(t) + \sum_{n=1}^{d} \left[\boldsymbol{\phi}_n Y_n(t) - \boldsymbol{\phi}_n \frac{\boldsymbol{\phi}_n^T}{K_n}\mathbf{R}\,f(t)\right]$$

$$= \mathbf{k}^{-1}\mathbf{R}\,f(t) + \sum_{n=1}^{d} \boldsymbol{\phi}_n \left[Y_n(t) - \frac{P_n(t)}{K_n}\right] \tag{14-56}$$

But Eq. (14-52) shows that the term in brackets can be expressed in terms of the modal acceleration and modal velocity, with the results

$$\mathbf{v}(t) = \mathbf{k}^{-1}\mathbf{R}\,f(t) - \sum_{n=1}^{d} \boldsymbol{\phi}_n \left[\frac{1}{\omega_n^2}\ddot{Y}_n(t) + \frac{2\xi_n}{\omega_n}\dot{Y}_n(t)\right]$$

which is identical to Eq. (14-55), the Mode Acceleration Method equation. Thus it makes no difference which of these two procedures is used, but the static correction method has an advantage in that it provides a more direct indication of the reason for its superiority over standard mode displacement superposition.

14-8 DERIVED RITZ VECTORS

Preliminary Comments

The introduction of Rayleigh-Ritz coordinates in the dynamic analysis of a structural system may be viewed as the second stage of a three-stage discretization procedure in which the finite-element idealization constitutes the first stage and the transformation to uncoupled modal coordinates is the third stage. The discussion in Section 14-5 shows that the Ritz coordinates provide a very effective means of reducing the number of degrees of freedom that must be considered in the analysis of the system vibration properties. A truncated set of the resulting eigenvectors (undamped mode shapes) may then be used to obtain the uncoupled set of equations of motion which is solved in the mode superposition analysis.

The critical step in this analysis sequence is the choice of the Ritz coordinates, which must be efficient in the sense that a relatively small number of assumed shapes will yield vibration properties accurate enough for reliable analysis of the dynamic response. A very effective set of Ritz vectors, often called Lanczos coordinates,[3] may be derived in a routine way by a procedure that is similar in many respects to the matrix iteration analysis of the fundamental vibration mode. The basic difference in the derivation of the Lanczos coordinates is that each step of the iteration sequence yields one Lanczos shape, whereas the standard matrix iteration procedure gives only the fundamental mode shape and the iteration serves only to improve the approximation to the true vibration shape.

The significant advantage that the Ritz vectors derived by the procedure described here have over the Lanczos coordinates as originally proposed is that the initial vector of this coordinate sequence is the deflected shape resulting from static application of the dynamic load distribution. For this reason, the first vector serves as a static correction and the subsequent vectors need only account for inertial effects on the dynamic response.

In this discussion these special Lanczos coordinates will be called derived Ritz vectors (DRV). The essential operations in the derivation of each vector are: (1) solution of a set of simultaneous equilibrium equations to determine the deflected shape resulting from the inertial load associated with the preceding derived vector, (2) application of the Gram-Schmidt procedure to make this new shape "mass orthogonal" to the DRV derived in preceding steps, and (3) normalization to give the new vector a unit generalized mass. (It is apparent that the orthogonalization step is not applicable to the derivation of the first vector.) Furthermore, it has been shown that the properties of these shapes are such that when a vector is made orthogonal to the two preceding shapes it automatically is orthogonal to *all* preceding shapes to within the accuracy allowed by roundoff errors. However, in order to avoid accumulation of roundoff errors, it is necessary to reestablish Gram-Schmidt orthogonality with *all* preceding DRV at intervals in the derivation sequence after several new vectors have been made orthogonal to only the two preceding vectors.

The operations followed in deriving the first and subsequent DRV are detailed in the following section. It is assumed that the external loading causing the dynamic response is of the form given by Eq. (14-39), i.e., $\mathbf{p}(t) = \mathbf{R} f(t)$, where the external load distribution, $\mathbf{R}$, may have any form and be due to any cause. The preliminary deflected shape calculated as the first step in the derivation of each vector is denoted by the symbol $\mathbf{q}_i$ where the subscript is the number of the derived vector; after "purification" (i.e., orthogonalizing with respect to the preceding vectors) the vector is distinguished by a tilde over the symbol; and after normalization the final form of the derived vector is designated by the standard Ritz vector symbol $\boldsymbol{\psi}_i$ [see Eq. (14-5a)].

---

[3] B. Nour-Omid and R. W. Clough, "Dynamics Analysis of Structures Using Lanczos Coordinates," *Earthquake Engineering and Structural Dynamics*, Vol. 12, 1984, pp. 565–577.

Derivation Details

First Vector — As was noted above, the first step in the derivation is the solution of the static equilibrium equations

$$\mathbf{k}\,\mathbf{q}_1 = \mathbf{R}$$

to obtain the deflected shape $\mathbf{q}_1$ due to application of the applied load distribution $\mathbf{R}$. In this solution, advantage should be taken of the banded form of the stiffness matrix $\mathbf{k}$, as discussed in Section 13-6. The normalizing factor β_1 is then calculated from the relationship

$$\beta_1^2 = \mathbf{q}_1^T\,\mathbf{m}\,\mathbf{q}_1 \tag{14-57}$$

which scales the first DRV, given by

$$\boldsymbol{\psi}_1 = \frac{1}{\beta_1}\,\mathbf{q}_1 \tag{14-58}$$

so that it provides a unit generalized mass; that is,

$$\boldsymbol{\psi}_1^T\,\mathbf{m}\,\boldsymbol{\psi}_1 = 1$$

Second Vector — The equivalent calculation for the second vector starts with solution of the equilibrium equations

$$\mathbf{k}\,\mathbf{q}_2 = \mathbf{m}\,\boldsymbol{\psi}_1 \tag{14-59}$$

to obtain the deflected shape $\mathbf{q}_2$ resulting from the inertial load $\mathbf{m}\boldsymbol{\psi}_1$ induced when the system is vibrating in the first vector shape $\boldsymbol{\psi}_1$. Then this shape is purified by the Gram-Schmidt procedure, making it mass orthogonal to the first vector as follows:

$$\tilde{\mathbf{q}}_2 = \mathbf{q}_2 - \alpha_1\,\boldsymbol{\psi}_1 \tag{14-60}$$

where the factor α_1 is given by

$$\alpha_1 = \boldsymbol{\psi}_1^T\,\mathbf{m}\,\mathbf{q}_2 \tag{14-61}$$

Finally this shape is normalized to obtain the second DRV,

$$\boldsymbol{\psi}_2 = \frac{1}{\beta_2}\,\tilde{\mathbf{q}}_2$$

where the normalizing factor, β_2, given by

$$\beta_2 = \sqrt{\tilde{\mathbf{q}}_2^T\,\mathbf{m}\,\tilde{\mathbf{q}}_2} \tag{14-62}$$

scales the shape so it has the desired unit generalized mass:

$$\boldsymbol{\psi}_2^T \, \mathbf{m} \, \boldsymbol{\psi}_2 = 1$$

Third Vector — Derivation of the third DRV proceeds in essentially the same way, starting with calculation of the preliminary shape $\mathbf{q}_3$ by solution of the static equilibrium equations formulated with the inertial load associated with the second DRV:

$$\mathbf{k} \, \mathbf{q}_3 = \mathbf{m} \, \boldsymbol{\psi}_2$$

In this case, however, the preliminary shape has to be purified by elimination of the displacement components associated with each of the two preceding vectors, i.e.,

$$\tilde{\mathbf{q}}_3 = \mathbf{q}_3 - \alpha_2 \boldsymbol{\psi}_2 - \beta_2 \boldsymbol{\psi}_1 \tag{14-63}$$

in which by analogy with the preceding operations

$$\alpha_2 = \boldsymbol{\psi}_2^T \, \mathbf{m} \, \mathbf{q}_3 \tag{14-64}$$

$$\beta_2 = \boldsymbol{\psi}_1^T \, \mathbf{m} \, \mathbf{q}_3 \tag{14-65}$$

It may be shown by simple algebra that this value of β_2 is identical to the normalizing factor given by Eq. (14-62). Furthermore, by analogy with Eq. (14-62), the normalizing factor β_3 for the third DRV is given by

$$\beta_3 = \sqrt{\tilde{\mathbf{q}}_3^T \, \mathbf{m} \, \tilde{\mathbf{q}}_3} \tag{14-66}$$

When scaled by this factor, the third DRV

$$\boldsymbol{\psi}_3 = \frac{1}{\beta_3} \, \tilde{\mathbf{q}}_3 \tag{14-67}$$

has the desired unit generalized mass.

Fourth Vector — Continuing similarly, the fourth preliminary shape is made orthogonal to the preceding derived vectors by eliminating components of those vectors, as follows:

$$\tilde{\mathbf{q}}_4 = \mathbf{q}_4 - \alpha_3 \, \boldsymbol{\psi}_3 - \beta_3 \, \boldsymbol{\psi}_2 - \gamma_3 \, \boldsymbol{\psi}_1 \tag{14-68}$$

where

$$\mathbf{q}_4 = \mathbf{k}^{-1} \, \mathbf{m} \, \boldsymbol{\psi}_3$$

$$\alpha_3 = \boldsymbol{\psi}_3^T \, \mathbf{m} \, \mathbf{q}_4$$

$$\beta_3 = \boldsymbol{\psi}_2^T \, \mathbf{m} \, \mathbf{q}_4$$

$$\gamma_3 = \boldsymbol{\psi}_1^T \, \mathbf{m} \, \mathbf{q}_4$$

However, as shown in the previously mentioned reference, the factor γ_3 that is obtained by this procedure is identically equal to zero, so it is necessary only to make the new vector orthogonal to the two preceding vectors. In addition, by analogy with the discussion concerning Eq. (14-63) it may be shown that the above defined factor β_3 is identical to the normalizing factor for the third vector, given by Eq. (14-67). Thus it is now necessary only to evaluate the fourth mode normalizing factor

$$\beta_4 = \sqrt{\tilde{\mathbf{q}}_4^T \, \mathbf{m} \, \tilde{\mathbf{q}}_4}$$

from which the fourth DRV is obtained.

General Vector — The foregoing discussion makes it apparent that any DRV ψ_{i+1} can be evaluated when the two preceding vectors are known, by applying the following algorithm:

(1) Solve $\quad \mathbf{k}\, \mathbf{q}_{i+1} = \mathbf{m}\, \psi_i \quad$ to obtain $\quad \mathbf{q}_{i+1}$.

(2) Orthogonalize with respect to the two preceding vectors:

$$\tilde{\mathbf{q}}_{i+1} = \mathbf{q}_{i+1} - \alpha_i \, \psi_i - \beta_i \, \psi_{i-1} \qquad (14\text{-}68a)$$

where

$$\alpha_i = \psi_i^T \, \mathbf{m} \, \mathbf{q}_{i+1}$$

$$\beta_i = \psi_{i-1}^T \, \mathbf{m} \, \mathbf{q}_{i+1}$$

$$= \sqrt{\tilde{\mathbf{q}}_i^T \, \mathbf{m} \, \tilde{\mathbf{q}}_i} \qquad \text{(preceding normalizing factor)}$$

(3) Normalize

$$\psi_{i+1} = \frac{1}{\beta_{i+1}} \, \tilde{\mathbf{q}}_{i+1}$$

where

$$\beta_{i+1} = \sqrt{\tilde{\mathbf{q}}_{i+1}^T \, \mathbf{m} \, \tilde{\mathbf{q}}_{i+1}}$$

in order to obtain the desired unit generalized mass:

$$\psi_{i+1}^T \, \mathbf{m} \, \psi_{i+1} = 1$$

This procedure may be followed to obtain any desired number of DRV, except that at intervals in the sequence it will be necessary to force orthogonality with respect to *all* preceding vectors when the loss of orthogonality due to roundoff is found to be excessive. A convenient test for the loss of orthogonality is described later in this section.

Tridiagonal Equations of Motion

Orthogonality Condition — When the desired number of DRV has been obtained using the algorithm stated above, they could be used to perform a dynamic analysis in the same way as any other set of Ritz vectors, as described in Section 14-5. However, the unique orthogonality properties of these Lanczos vectors make it possible to organize the equations of motion in a special tridiagonal form that facilitates the dynamic analysis.

To formulate these special equations, the mass orthogonality conditions for the DRV are arranged in tridiagonal form after first writing the Gram-Schmidt equation for each Ritz vector as follows:

$$\tilde{\mathbf{q}}_1 \equiv \beta_1 \, \boldsymbol{\psi}_1 = \mathbf{k}^{-1} \, \mathbf{R}$$

$$\tilde{\mathbf{q}}_2 \equiv \beta_2 \, \boldsymbol{\psi}_2 = \mathbf{k}^{-1} \, \mathbf{m} \, \boldsymbol{\psi}_1 - \boldsymbol{\psi}_1 \, \alpha_1$$

$$\tilde{\mathbf{q}}_3 \equiv \beta_3 \, \boldsymbol{\psi}_3 = \mathbf{k}^{-1} \, \mathbf{m} \, \boldsymbol{\psi}_2 - \boldsymbol{\psi}_2 \, \alpha_2 - \boldsymbol{\psi}_1 \, \beta_2$$

$$\tilde{\mathbf{q}}_4 \equiv \beta_4 \, \boldsymbol{\psi}_4 = \mathbf{k}^{-1} \, \mathbf{m} \, \boldsymbol{\psi}_3 - \boldsymbol{\psi}_3 \, \alpha_3 - \boldsymbol{\psi}_2 \, \beta_3 \qquad (14\text{-}69)$$

$$\tilde{\mathbf{q}}_5 \equiv \beta_5 \, \boldsymbol{\psi}_5 = \mathbf{k}^{-1} \, \mathbf{m} \, \boldsymbol{\psi}_4 - \boldsymbol{\psi}_4 \, \alpha_4 - \boldsymbol{\psi}_3 \, \beta_4$$

etc.

where it will be noted again that only the two preceding vector components need be eliminated in the purification process for any given vector. Omitting the first equation, the remaining set of equations is rearranged to the following form:

$$\mathbf{k}^{-1} \, \mathbf{m} \, \boldsymbol{\psi}_1 - \boldsymbol{\psi}_1 \, \alpha_1 - \boldsymbol{\psi}_2 \, \beta_2 = 0$$

$$\mathbf{k}^{-1} \, \mathbf{m} \, \boldsymbol{\psi}_2 - \boldsymbol{\psi}_1 \, \beta_2 - \boldsymbol{\psi}_2 \, \alpha_2 - \boldsymbol{\psi}_3 \, \beta_3 = 0$$

$$\mathbf{k}^{-1} \, \mathbf{m} \, \boldsymbol{\psi}_3 - \boldsymbol{\psi}_2 \, \beta_3 - \boldsymbol{\psi}_3 \, \alpha_3 - \boldsymbol{\psi}_4 \, \beta_4 = 0$$

$$\mathbf{k}^{-1} \, \mathbf{m} \, \boldsymbol{\psi}_4 - \boldsymbol{\psi}_3 \, \beta_4 - \boldsymbol{\psi}_4 \, \alpha_4 - \boldsymbol{\psi}_5 \, \beta_5 = 0$$

etc.

which may be expressed in matrix form as

$$\mathbf{k}^{-1}\mathbf{m}\begin{bmatrix} \boldsymbol{\psi}_1 & \boldsymbol{\psi}_2 & \cdots & \boldsymbol{\psi}_{i-1} & \boldsymbol{\psi}_i \end{bmatrix}$$

$$- \begin{bmatrix} \boldsymbol{\psi}_1 & \boldsymbol{\psi}_2 & \cdots & \boldsymbol{\psi}_{i+1} & \boldsymbol{\psi}_i \end{bmatrix} \begin{bmatrix} \alpha_1 & \beta_2 & 0 & \cdots & 0 & 0 \\ \beta_2 & \alpha_2 & \beta_3 & \cdots & 0 & 0 \\ 0 & \beta_3 & \alpha_3 & \cdots & 0 & 0 \\ \vdots & \vdots & \vdots & \ddots & \vdots & \vdots \\ 0 & 0 & 0 & \cdots & \alpha_{i-1} & \beta_i \\ 0 & 0 & 0 & \cdots & \beta_i & \alpha_i \end{bmatrix} = \mathbf{0}$$

Now this equation set is abbreviated as follows:

$$\mathbf{k}^{-1}\,\mathbf{m}\,\mathbf{\Psi}_i = \mathbf{\Psi}_i\,\mathbf{T}_i \qquad (14\text{-}70)$$

in which the set of all "i" DRV is denoted by

$$\mathbf{\Psi}_i \equiv \begin{bmatrix} \psi_3 & \psi_2 & \cdots & \psi_{i-1} & \psi_i \end{bmatrix}$$

and the corresponding tridiagonal set of coefficients is designated

$$\mathbf{T}_i = \begin{bmatrix} \alpha_1 & \beta_2 & 0 & \cdots & 0 & 0 \\ \beta_2 & \alpha_2 & \beta_3 & \cdots & 0 & 0 \\ 0 & \beta_3 & \alpha_3 & \cdots & 0 & 0 \\ \vdots & \vdots & \vdots & \ddots & \vdots & \vdots \\ 0 & 0 & 0 & \cdots & \alpha_{i-1} & \beta_i \\ 0 & 0 & 0 & \cdots & \beta_i & \alpha_i \end{bmatrix} \qquad (14\text{-}71)$$

Finally multiplying Eq. (14-70) by $\mathbf{\Psi}_i^T\,\mathbf{m}$ to invoke the mass orthogonality condition, a simple expression is derived for $\mathbf{T}_i$ as follows:

$$\mathbf{\Psi}_i^T\,\mathbf{m}\,\mathbf{k}^{-1}\,\mathbf{m}\,\mathbf{\Psi}_i = \mathbf{\Psi}_i^T\,\mathbf{m}\,\mathbf{\Psi}_i\,\mathbf{T}_i = \mathbf{I}\,\mathbf{T}_i$$

in which the orthonormal property of the DRV has been noted, which leads to the following formulation:

$$\mathbf{T}_i = \mathbf{\Psi}_i^T\,\mathbf{m}\,\mathbf{k}^{-1}\,\mathbf{m}\,\mathbf{\Psi}_i \qquad (14\text{-}72)$$

Transformed Equations of Motion — Now in order to take advantage of the tridiagonal form of the coefficient matrix $\mathbf{T}_i$, the standard finite-element equations of motion,

$$\mathbf{m}\,\ddot{\mathbf{v}}(t) + \mathbf{c}\,\dot{\mathbf{v}}(t) + \mathbf{k}\,\mathbf{v}(t) = \mathbf{p}(t) = \mathbf{R}\,f(t)$$

are transformed to the DRV coordinates using the transformation

$$\mathbf{v}(t) = \mathbf{\Psi}_i\,\mathbf{Z}_i(t)$$

which leads to

$$\mathbf{m}\,\mathbf{\Psi}_i\,\ddot{\mathbf{Z}}_i(t) + \mathbf{c}\,\mathbf{\Psi}_i\,\dot{\mathbf{Z}}_i(t) + \mathbf{k}\,\mathbf{\Psi}_i\,\mathbf{Z}_i(t) = \mathbf{R}\,f(t)$$

But assuming the damping is of the Rayleigh form, $\mathbf{c} = a_0\,\mathbf{m} + a_1\,\mathbf{k}$, and premultiplying by $\mathbf{\Psi}_i^T\mathbf{m}\mathbf{k}^{-1}$, the equation becomes

$$\mathbf{\Psi}_i^T\mathbf{m}\mathbf{k}^{-1}\mathbf{m}\mathbf{\Psi}_i\ddot{\mathbf{Z}}_i(t) + a_0\mathbf{\Psi}_i^T\mathbf{m}\mathbf{k}^{-1}\mathbf{m}\mathbf{\Psi}_i\dot{\mathbf{Z}}_i(t)$$
$$+ a_1\mathbf{\Psi}_i^T\mathbf{m}\mathbf{\Psi}_i\dot{\mathbf{Z}}_i(t) + \mathbf{\Psi}_i^T\mathbf{m}\mathbf{\Psi}_i\mathbf{Z}_i(t) = \mathbf{\Psi}_i^T\mathbf{m}\mathbf{k}^{-1}\mathbf{R}\,f(t)$$

Now using the definition of $\mathbf{T}_i$ given by Eq. (14-72) and noting again that $\mathbf{\Psi}_i^T \mathbf{m} \mathbf{\Psi}_i = \mathbf{I}$, this is reduced to the following simple form:

$$\mathbf{T}_i \ddot{\mathbf{Z}}_i(t) + [a_0 \mathbf{T}_i + a_1 \mathbf{I}] \dot{\mathbf{Z}}_i(t) + \mathbf{Z}_i(t) = \left\{ \begin{array}{c} \beta_1 \\ 0 \\ 0 \\ \vdots \end{array} \right\} f(t) \qquad (14\text{-}73)$$

It is interesting to observe here that only the first Ritz coordinate equation is subjected directly to the excitation; the orthogonality conditions eliminate any direct excitation effects in the other coordinates so they are put into motion only by their tridiagonal coupling to the adjacent Ritz vectors.

Solution of the Transformed Equations — Throughout this discussion of the DRV it is assumed that the ultimate objective of the formulation is to perform a dynamic response analysis of a structural system having many degrees of freedom — several dozens to several hundreds in number. Mode superposition, including appropriate correction for static effects if necessary, usually provides the most efficient linear response analyses for such systems; a possible exception to this conclusion may occur if the structure is subjected only to a very short duration impulsive load as will be explained later. For the mode superposition analysis, an important question is the amount of computational effort required to evaluate the modal coordinates used in the analysis. In the past, the mode shapes typically have been evaluated by subspace iteration of the eigenproblem associated with the original finite-element coordinates, but it is evident here that the tridiagonal eigenproblem

$$\mathbf{T}_i \ddot{\mathbf{Z}}_i(t) + \mathbf{Z}_i(t) = 0 \qquad (14\text{-}74)$$

offers a much more efficient solution.

The great advantage given by the DRV coordinates in dynamic response analysis has been demonstrated by many research studies; in one of these[4] it was shown that subspace iteration analysis of a specified number of mode shapes required about nine times the computational effort expended in solution of the DRV eigenproblem, Eq. (14-74). Moreover, it must be recalled that the "mode shapes" obtained from Eq. (14-74) include a static correction effect as mentioned earlier, so significantly fewer of these coordinates may be required to express the dynamic response with a given degree of precision than if the true vibration mode shapes obtained by subspace iteration were used.

Another potential advantage of the DRV formulation is that the tridiagonal equations of motion, Eq. (14-73), can be solved directly by step-by-step procedures using

---

[4] P. Leger, "Numerical Techniques for the Dynamic Analysis of Large Structural Systems," Ph.D. Dissertation, University of California, Berkeley, March 1986.

only about 40 percent more computational effort per time step compared with solution of the uncoupled SDOF equations. Thus the cost of the tridiagonal eigenproblem solution [Eq. (14-74)] will not be justified if the response is to be determined for only a very short duration impulsive load. On the other hand, the time savings resulting from a mode superposition analysis of the uncoupled equations for a system subjected to a long duration load such as an earthquake will easily compensate for the cost of the eigenproblem solution. In this regard it should be noted that the mode shapes and frequencies obtained by solving the DRV eigenproblem are only approximations of the true values if they are obtained from a truncated set of Ritz vectors. However, these shapes are sufficiently accurate to aid in understanding the dynamic response behavior of the structure and they are significantly more efficient in calculating the dynamic response, as was mentioned before.

Loss of Orthogonality

The fact that roundoff errors gradually will lead to loss of orthogonality if each new DRV is made orthogonal to only the two preceding vectors (as in the above-described algorithm) has been stated previously. To guard against this eventuality, an orthogonality test vector denoted as

$$\mathbf{W}_i = \begin{bmatrix} W_1 & W_2 & W_3 & \cdots & W_{i-1} & W_i \end{bmatrix}$$

should be calculated as soon as DRV $\boldsymbol{\psi}_{i+1}$ is derived. This test vector may be expressed by

$$\mathbf{W}_i = \boldsymbol{\psi}_{i+1}\, \mathbf{m}\, \boldsymbol{\Psi}_i \tag{14-75}$$

where $\boldsymbol{\Psi}_i$ was defined earlier as the set of DRV including $\boldsymbol{\psi}_i$. However, the test vector is evaluated most conveniently in a step-by-step sequence as follows:

$$\mathbf{W}_i = \frac{1}{\beta_i}\left[\mathbf{W}_{i-1} - \alpha_{i-1}\,\mathbf{W}_{i-2} - \beta_{i-1}\,\mathbf{W}_{i-2}\right] \tag{14-76}$$

in which the initiating scalars required to calculate $\mathbf{W}_2$ are $W_0 = 0$ and $W_1 = \boldsymbol{\psi}_2^T \mathbf{m} \boldsymbol{\psi}_1$; the coefficients α_{i-1}, β_{i-1}, etc., are the same as those included in Eq. (14-71).

From Eq. (14-75) it is apparent that the elements of $\mathbf{W}_i$ are coefficients that express the mass coupling between the new DRV $\boldsymbol{\psi}_{i+1}$ and each preceding derived vector. Of course, the values of $\mathbf{W}_i$ and $\mathbf{W}_{i-1}$ are forced to be zero by the Gram-Schmidt procedure used in the DRV algorithm, but the values of $\mathbf{W}_{i-2}$, $\mathbf{W}_{i-3}$, etc., demonstrate the extent to which the new vector fails to achieve orthogonality with the preceding vectors. When any element of $\mathbf{W}_i$ is found to be excessive, the Gram-Schmidt procedure should be applied to make $\boldsymbol{\psi}_{i+1}$ orthogonal to all preceding vectors. Then the simple algorithm involving only two-term orthogonality can be continued until a new test vector $\mathbf{W}_i$ again indicates the need for full orthogonalization. In a test case involving 100 degrees of freedom, it was found that full orthogonalization was required for about every fifth derived vector.

Required Number of Vectors

Because the first DRV is the static displacement shape caused by the applied load distribution $\mathbf{R}$, the function of the subsequent DRV used in the response analysis is to represent the dynamic effect of this loading. The contribution to $\mathbf{R}$ associated with vector "i" is indicated by the Ritz participation factor, RPF_i; this is entirely analogous to the modal participation factor given by Eq. (14-42) which represents the load contribution of mode "n" when the response is expressed in modal coordinates. Thus by analogy with Eq. (14-42), the Ritz participation factor for vector $\boldsymbol{\psi}_i$ is given by

$$\mathrm{RPF}_i = \frac{\boldsymbol{\psi}_i^T \mathbf{R}}{\boldsymbol{\psi}_i^T \mathbf{m}\, \boldsymbol{\psi}_i} = \boldsymbol{\psi}_i^T \mathbf{R} \qquad (14\text{-}77)$$

where advantage is taken of the fact that the DRV algorithm has normalized the generalized mass in the denominator to unity.

From Eq. (14-75) it is evident that a vector listing all participation factors up to and including that associated with DRV $\boldsymbol{\psi}_i$ could be obtained by forming the matrix product $\boldsymbol{\psi}_i^T \mathbf{R}$; then these sucessive values could be judged as a basis for termination of the DRV algorithm. However, rather than using this matrix multiplication to calculate the participation factors, it is preferable to calculate each factor successively as the final step of the derivation algorithm. A convenient formula to serve this purpose may be derived by multiplying Eq. (14-68a) by $\mathbf{R}^T$, leading to

$$\mathbf{R}^T \tilde{\mathbf{q}}_{i+1} = \mathbf{R}^T \mathbf{q}_{i+1} - \alpha_i \mathbf{R}^T \boldsymbol{\psi}_i - \beta_i \mathbf{R}^T \boldsymbol{\psi}_{i-1}$$

Then noting that $\mathbf{q}_{i+1} = \mathbf{k}^{-1}\mathbf{m}\boldsymbol{\psi}_i$, that $\mathbf{R}^T\mathbf{k}^{-1} = \boldsymbol{\psi}_1^T \beta_1$, and that $\mathbf{R}^T \Psi_{i+1} = \mathrm{RPF}_{i+1}$, etc., this may be reduced to

$$\mathrm{RPF}_{i+1} = -\left[\frac{\alpha_i\, \mathrm{RPF}_i + \beta_i\, \mathrm{RPF}_{i-1}}{\beta_{i+1}} \right] \qquad (14\text{-}78)$$

Adding this simple scalar calculation at the end of the DRV algorithm provides a basis for termination of the derivation when RPF_{i+1} drops below a specified value.

PROBLEMS

14-1. The four-story shear frame of Fig. P14-1 has the same mass m lumped in each rigid girder and the same story-to-story stiffness k in the columns of each story. Using the indicated linear and quadratic shape functions; ψ_1 and ψ_2, as generalized coordinates, obtain the approximate shapes and the frequencies of the first two modes of vibration by the Rayleigh-Ritz method, Eqs. (14-21) and (14-22).

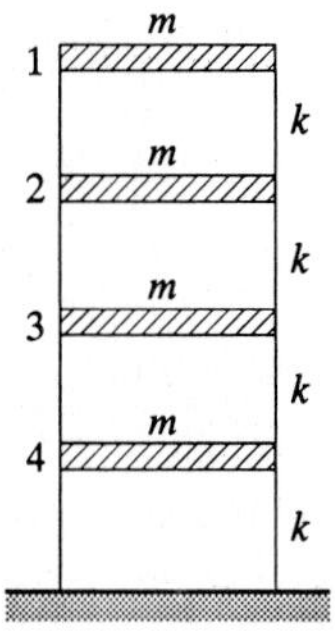

$$[\psi_1, \psi_2] = \begin{bmatrix} 1.00 & 1.00 \\ 0.75 & 0.56 \\ 0.50 & 0.25 \\ 0.25 & 0.06 \end{bmatrix}$$

FIGURE P14-1

14-2. Repeat Prob. 14-1 usign the "improved" expressions of Eq. (14-28) to define the generalized coordinate mass and stiffness properties.

CHAPTER
15

ANALYSIS OF MDOF DYNAMIC RESPONSE: STEP-BY-STEP METHODS

15-1 Preliminary Comments

In the presentation of SDOF dynamic analysis procedures in Part I of this text, two classes of approach were considered: those making use of superposition which consequently are limited to the analysis of linear system and step-by-step methods which may be applied to either linear or nonlinear systems. In the treatment of MDOF dynamic analysis given in Chapter 12, the principle of superposition was employed so all of that discussion concerns only linear systems, but two different categories of superposition were utilized in those analyses. One involves superposition with regard to time, employing either convolution or Fourier integrals as was explained in the SDOF discussion. The other category is spatial superposition, in which the MDOF response is represented as the combination of a set of independent SDOF modal coordinate responses. The great advantage of this spatial or modal coordinate superposition, as was explained in Chapter 12, is that an adequate approximation of the dynamic response often can be obtained from only a few modes of vibration or derived Ritz vectors even when the system may have dozens or even hundreds of degrees of freedom.

Both of these types of superposition require that the system remain linear during the response; any nonlinearity indicated by a change in the coefficients of the structure property matrices would invalidate the analytical results. However, in many practical situations the coefficients cannot be assumed to remain constant; for example, the stiffness influence coefficients may be altered by yielding of the structural materials — a very likely situation during response to a severe earthquake. Other possibilities are that changes in the member axial forces may cause appreciable changes in their geometric stiffness and that the mass or damping coefficients will undergo changes during the dynamic response; each of those mechanisms may have an important effect on the uncoupling of the modal coordinate equations of motion. In addition it must be recalled that although linearity is a necessary condition for modal coordinate uncoupling, this result will be achieved only if the system is proportionally damped; for any other type of damping the modal coordinate equations of motion will be coupled by modal damping coefficients.

The only generally applicable procedure for analysis of an arbitrary set of non-linear response equations, and also an effective means of dealing with coupled linear modal equations, is by numerical step-by-step integration. The analysis can be carried out as the exact MDOF equivalent of the SDOF step-by-step analyses described in Chapter 7. The response history is divided into a sequence of short, equal time steps, and during each step the response is calculated for a linear system having the physical properties existing at the beginning of the interval. At the end of the interval, the properties are modified to conform to the state of deformation and stress at that time for use during the subsequent time step. Thus the nonlinear MDOF analysis is approximated as a sequence of MDOF analyses of successively changing linear systems.

When step-by-step integration is applied to linear structures, the computation is greatly simplified because the structural properties need not be modified at each step. In some cases it may be advantageous to use direct integration rather than mode superposition in order to avoid the great computational effort required for the eigen-problem solution of a system with very many degrees of freedom. This possibility was discussed briefly in Chapter 14 with regard to the tridiagonal equations of motion obtained from the derived Ritz vector transformation.

One potential difficulty in the step-by-step response integration of MDOF systems is that the damping matrix c must be defined explicitly rather than in terms of modal damping ratios. It is very difficult to estimate the magnitudes of the damping influence coefficients of a complete damping matrix. In general, the most effective means for deriving a suitable damping matrix is to assume appropriate values of modal damping ratios for all the modes which are considered to be important and then to compute an orthogonal damping matrix which has those properties, as described in Chapter 12.

On the other hand, the fact that the damping matrix is defined explicitly rather than by modal damping ratios may be advantageous in that it increases the generality of the step-by-step method over mode superposition. There is no need for uncoupling the modal response; therefore the damping matrix need not be selected to satisfy modal orthogonality conditions. Any desired set of damping-matrix coefficients can be employed in the analysis, and they may represent entirely different levels of damping in different parts of the structure, as explained in Chapter 12. The modal coordinate transformation still may be used to reduce the number of coordinates used in the analysis, followed by step-by-step solution of the resulting coupled modal equations.

Finally, it is worth noting that the transformation to normal coordinates may be useful even in the analysis of nonlinear systems. Of course, the undamped free-vibration mode shapes will serve to uncouple the equations of motion only so long as the stiffness matrix remains unchanged from the state for which the vibration analysis was made. As soon as the stiffness changes, due to yielding or other damage, the normal-coordinate transformation will introduce off-diagonal terms in the generalized stiffness matrix which cause coupling of the modal response equations. However, if the nonlinear deformation mechanisms in the structure do not cause major changes in its deflection patterns, the dynamic response still may be expressed efficiently in terms of the original undamped mode shapes. Thus it often will be worthwhile to evaluate the response of a complex structure by direct step-by-step integration of a limited set of normal-coordinate equations of motion, even though the equations will become coupled as soon as any significant nonlinearity develops in the response. This treatment of a system with stiffness coupling of the normal-coordinate equations is equivalent to the approach suggested above for the analysis of systems in which the damping matrix is such as to introduce normal-coordinate coupling.

15-2 Incremental Equations of Motion

In the step-by-step analysis of MDOF systems it is convenient to use an incremental formulation equivalent to that described for SDOF systems in Section 7-6 because the procedure then is equally applicable to either linear or nonlinear analyses. Thus taking the difference between vector equilibrium relationships defined for times t_0 and $t_1 = t_0 + h$ gives the incremental equilibrium equation

$$\Delta \mathbf{f}_I + \Delta \mathbf{f}_D + \Delta \mathbf{f}_S = \Delta \mathbf{p} \tag{15-1}$$

The force vector increments in this equation, by analogy with the SDOF expressions [Eqs. (7-20) and (7-21)], can be written as follows:

$$\Delta \mathbf{f}_I = \mathbf{f}_{I_1} - \mathbf{f}_{I_0} = \mathbf{m} \, \Delta \ddot{\mathbf{v}}$$

$$\Delta \mathbf{f}_D = \mathbf{f}_{D_1} - \mathbf{f}_{D_0} = \mathbf{c}_0 \, \Delta \dot{\mathbf{v}}$$

$$\Delta \mathbf{f}_S = \mathbf{f}_{S_1} - \mathbf{f}_{S_0} = \mathbf{k}_0 \, \Delta \mathbf{v} \tag{15-2}$$

$$\Delta \mathbf{p} = \mathbf{p}_1 - \mathbf{p}_0$$

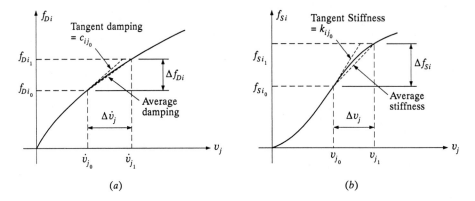

FIGURE 15-1
Definition of nonlinear influence coefficients: (a) nonlinear viscous damping c_{ij}; (b) nonlinear stiffness k_{ij}.

where it has been assumed that the mass does not change with time.

The elements of the incremental damping and stiffness matrices c_0 and k_0 are influence coefficients c_{ij_0} and k_{ij_0} defined for the time increment; typical representations of these coefficients are shown in Fig. 15-1. As was explained with regard to the SDOF coefficients, it is convenient to use the initial tangent rather than the average slope as a measure of the damping or stiffness property in order to avoid the need for iteration at each step of the solution. Hence the influence coefficients are given by

$$c_{ij_0} = \left(\frac{d\mathbf{f}_{D_i}}{d\mathbf{v}_j}\right)_0 \qquad k_{ij_0} = \left(\frac{d\mathbf{f}_{S_i}}{d\mathbf{v}_j}\right)_0 \qquad (15\text{-}3)$$

for the time increment h starting at time t_0. When Eqs. (15-3) are substituted into Eq. (15-1), the incremental equation of motion becomes

$$\mathbf{m}\,\triangle\ddot{\mathbf{v}} + \mathbf{c}_0\,\triangle\dot{\mathbf{v}} + \mathbf{k}_0\,\triangle\mathbf{v} = \triangle\mathbf{p} \qquad (15\text{-}4)$$

The incremental force expressions on the left side of Eq. (15-4) are only approximations because of the use of initial tangent values for $\mathbf{c}_0$ and $\mathbf{k}_0$. However, accumulation of errors due to this factor will be avoided if the acceleration at the beginning of each time step is calculated from the total equilibrium of forces at that time, as was mentioned in discussing the SDOF case.

15-3   Step-by-Step Integration: Constant Average Acceleration Method

The step-by-step solution of the incremental equations of motion [Eq. (15-4)] is formulated by specifying a simple relationship among the displacement, velocity, and acceleration which is assumed to be valid for the short time step h. On this basis, the incremental changes of velocity and displacement can be expressed in terms of the changes in acceleration, or the changes in velocity and acceleration, alternatively, can be expressed in terms of the incremental displacements. In either case, only one unknown vector remains in the incremental equilibrium equations [Eq. (15-4)], and this may be evaluated by any standard procedure for solving simultaneous equations.

As described in Chapter 7, the relationship among displacement, velocity, and acceleration can be established conveniently by assuming the manner of variation of the acceleration vector with time. For the analysis of general MDOF systems, the constant average acceleration assumption has the very important advantage that it provides an unconditionally stable integration procedure. Any method that is only conditionally stable may require use of extremely short time steps to avoid instability in the higher mode responses, and such instability will cause the analysis to "blow up" even if the unstable modes make no significant contribution to the actual dynamic response behavior.

The constant average acceleration assumption leads to a linear variation of the velocity vector and a quadratic variation of the displacement vector, as was described for the SDOF case, and the explicit MDOF incremental analysis procedure may be derived by complete analogy to the SDOF formulation presented in Chapter 7. The final results of the derivation are analgous to the expressions given by Eq. (7-24) for the linear acceleration assumption. In this case the incremental pseudostatic equilibrium equation will be stated as

$$\widetilde{\mathbf{k}}_c \, \triangle\mathbf{v} = \triangle\widetilde{\mathbf{p}}_c \tag{15-5}$$

where the subscript c denotes the constant average acceleration assumption. The effective stiffness matrix in this case is given by

$$\widetilde{\mathbf{k}}_c = \mathbf{k}_0 + \frac{2}{h}\,\mathbf{c}_0 + \frac{4}{h^2}\,\mathbf{m} \tag{15-6a}$$

while the incremental effective load vector is as follows:

$$\triangle\widetilde{\mathbf{p}}_c = \triangle\mathbf{p} + 2\,\mathbf{c}_0\,\dot{\mathbf{v}}_0 + \mathbf{m}\left[\frac{4}{h}\,\dot{\mathbf{v}}_0 + 2\,\ddot{\mathbf{v}}_0\right] \tag{15-6b}$$

The step-by-step analysis is carried out using Eq. (15-5) by first evaluating $\widetilde{\mathbf{k}}_c$ from the mass, damping, and stiffness properties determined from the conditions at the beginning of the time step and also evaluating $\triangle\widetilde{\mathbf{p}}_c$ from the damping property as well as the velocity and acceleration vectors at the beginning of the time step combined with the load increment specified for the step. Then the simultaneous equations [Eq. (15-5)] are solved for the displacement increment $\triangle\mathbf{v}$, usually using Gauss or Choleski decomposition; it should be noted that the changing values of $\mathbf{k}_0$ and $\mathbf{c}_0$ in a nonlinear analysis require that the decomposition be performed for each time step, and this is a major computational effort for a system with very many degrees of freedom.

When the displacement increment has been calculated, the velocity increment is given by the following expression, which is analogous to Eq. (7-24c) but is based on the constant average acceleration assumption

$$\triangle \dot{\mathbf{v}} = \frac{2}{h} \triangle \mathbf{v} - 2\dot{\mathbf{v}}_0 \tag{15-7}$$

Thus using the incremental Eqs. (15-5), (15-6), and (15-7) the analysis may be carried out for any MDOF system for which the varying properties $\mathbf{k}_0$ and $\mathbf{c}_0$ can be defined for each step. The response vectors calculated for time $t_1 = t_0 + h$ at the end of one step serve as the initial vectors for the next step. However, to avoid accumulation of errors, as noted before, the initial acceleration vector is calculated directly from the condition of equilibrium at the beginning of the step; thus,

$$\ddot{\mathbf{v}}_0 = \mathbf{m}^{-1} \left[\mathbf{p}_0 - \mathbf{f}_{D_0} - \mathbf{f}_{S_0} \right] \tag{15-8}$$

where $\mathbf{f}_{D_0}$ and $\mathbf{f}_{S_0}$ depend on the velocity and displacement vectors at the beginning of the step. Because the inverse of the mass matrix, $\mathbf{m}^{-1}$, is used at each step of the analysis, it should be calculated at the beginning and stored by the computer program.

15-4 Step-by-Step Integration: Linear Acceleration Method

The constant average acceleration method described in the preceding section is a convenient and relatively efficient procedure for nonlinear structural analysis; however, as was mentioned in Part I, comparative numerical tests have demonstrated that the linear acceleration method gives better results using any specified step length that does not approach the integration stability limit. In certain types of structures, notably multistory buildings that are modelled with one degree of freedom per story for planar response or with three degrees of freedom for general three-dimensional response, there is little difficulty in adopting a time step that ensures stability in the response of even the highest modes. In such situations, the linear acceleration version of the above-described procedure is recommended, replacing Eqs. (15-5), (15-6a, b), and (15-7) by their linear acceleration equivalents as follows:

$$\tilde{\mathbf{k}}_d \triangle \mathbf{v} = \triangle \tilde{\mathbf{p}}_d \tag{15-9}$$

$$\tilde{\mathbf{k}}_d = \mathbf{k}_0 + \frac{3}{h} \mathbf{c}_0 + \frac{6}{h^2} \mathbf{m} \tag{15-10a}$$

$$\triangle \tilde{\mathbf{p}}_d = \triangle \mathbf{p} + \mathbf{c}_0 \left[3\dot{\mathbf{v}}_0 + \frac{h}{2} \ddot{\mathbf{v}}_0 \right] + \mathbf{m} \left[\frac{6}{h} \dot{\mathbf{v}}_0 + 3\ddot{\mathbf{v}}_0 \right] \tag{15-10b}$$

$$\triangle \dot{\mathbf{v}} = \frac{3}{h} \triangle \mathbf{v} - 3\dot{\mathbf{v}}_0 - \frac{h}{2} \ddot{\mathbf{v}}_0 \tag{15-11}$$

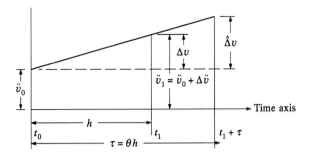

FIGURE 15-2
Linear acceleration; normal and extended time steps.

In order to avoid instability in the response calculated by these equations the length of the time step must be limited by the condition

$$h \le \frac{1}{1.8} T_N \tag{15-12}$$

where T_N is the vibration period of the highest mode (i.e., the shortest period) associated with the system eigenproblem.

For a more general type of structure that is modelled by finite elements, the period of the highest mode is related to the properties of the individual elements, with the result that Eq. (15-12) may require the use of an extremely short time step in the response analyses. In such cases the analysis of response to an actual earthquake loading or even a relatively short duration impulsive load may involve a prohibitive computational effort, making it necessary to adopt an unconditionally stable integration procedure instead of the linear acceleration algorithm. Of course the constant acceleration method could be used, but a preferable alternative may be an unconditionally stable modification of the linear acceleration method called the Wilson θ-method.[1]

This modification is based on the assumption that the acceleration varies linearly over an extended computation interval $\tau = \theta h$. The parameters associated with this assumption are depicted in Fig. 15-2. The acceleration increment $\widetilde{\Delta} \ddot{v}$ is calculated by the standard linear acceleration procedure applied to the extended time step τ; from this the increment $\Delta \ddot{v}$ for the normal time step h is obtained by interpolation. For a value of $\theta = 1$, the procedure reverts to the standard linear acceleration method, but for $\theta > 1.37$ it becomes unconditionally stable.

The analysis procedure can be derived merely by rewriting the basic relationships of the linear acceleration method for the extended time step τ. Thus, by analogy

[1] E. L. Wilson at the University of California, Berkeley.

with Eqs. (a) and (b) of Fig. 7-6b

$$\hat{\triangle}\dot{\mathbf{v}} = \ddot{\mathbf{v}}_0\,\tau + \hat{\triangle}\ddot{\mathbf{v}}\,\frac{\tau}{2} \tag{15-13a}$$

$$\hat{\triangle}\mathbf{v} = \dot{\mathbf{v}}_0\,\tau + \ddot{\mathbf{v}}_0\,\frac{\tau^2}{2} + \hat{\triangle}\ddot{\mathbf{v}}\,\frac{\tau^2}{6} \tag{15-13b}$$

in which the "^" symbol denotes an increment associated with the extended time step. Solving these to express $\hat{\triangle}\ddot{\mathbf{v}}$ and $\hat{\triangle}\dot{\mathbf{v}}$ in terms of $\hat{\triangle}\mathbf{v}$ and substituting into the equation of motion lead to expressions equivalent to Eqs. (15-9) and (15-10) but written for the extended time step:

$$\hat{\mathbf{k}}\,\hat{\triangle}\mathbf{v} = \hat{\triangle}\mathbf{p} \tag{15-14}$$

where

$$\hat{\mathbf{k}} = \mathbf{k}_0 + \frac{3}{\tau}\,\mathbf{c}_0 + \frac{6}{\tau^2}\,\mathbf{m} \tag{15-15a}$$

$$\hat{\triangle}\mathbf{p} = \triangle\mathbf{p} + \mathbf{c}_0\left[3\dot{\mathbf{v}}_0 + \frac{\tau}{2}\ddot{\mathbf{v}}_0\right] + \mathbf{m}\left[\frac{6}{\tau}\dot{\mathbf{v}}_0 + 3\ddot{\mathbf{v}}_0\right] \tag{15-15b}$$

Finally the pseudostatic relationship Eq. (15-14) can be solved for $\hat{\triangle}\mathbf{v}$ and substituted into the following equation [obtained by solving Eq. (15-13b)]:

$$\hat{\triangle}\ddot{\mathbf{v}} = \hat{\triangle}\mathbf{v}\,\frac{6}{\tau^2} - \dot{\mathbf{v}}_0\,\frac{6}{\tau} - 3\ddot{\mathbf{v}}_0 \tag{15-16}$$

to obtain the increment of acceleration during the extended time step. From this, the acceleration increment for the normal time step h is obtained by linear interpolation:

$$\triangle\ddot{\mathbf{v}} = \frac{1}{\Theta}\,\hat{\triangle}\ddot{\mathbf{v}} \tag{15-17}$$

and then the corresponding incremental velocity and displacement vectors are obtained from expressions like Eqs. (15-13) but written for the normal time step h. Using these results, the time stepping analysis proceeds exactly as was described above for the constant average acceleration method.

15-5   Strategies for Analysis of Coupled MDOF Systems

Localized Nonlinearity[2]

The major advantage of the step-by-step methods, that they permit direct analysis of the coupled equations of motion resulting from the finite-element idealization,

---

[2] R. W. Clough and E. L. Wilson "Dynamic Analysis of Large Structural Systems with Localized Nonlinearities," *Computer Methods in Applied Mechanics and Engineering*, North Holland Publ. Co., Vol. 17/18, 1979, pp. 107–129.

was pointed out in the preceding sections of this chapter. With such direct analysis there is no need to reduce the system to a set of SDOF equations before calculating the dynamic response, even if the equations are linear so that the eigenproblem analysis is applicable. However, although the direct solution avoids the potentially very large cost of the modal coordinate evaluation, the step-by-step analysis that is used as an alternative often requires even greater computational effort; consequently this approach should be adopted only after the number of degrees of freedom treated in the analysis has been reduced to a minimum.

Coordinate reduction procedures can be especially effective if the structural forces for most of the system degrees of freedom are linearly related to the displacements, and the nonlinear response is associated with relatively few degrees of freedom. Many examples of systems having such localized nonlinearity may be recognized in practice, including elastic structures mounted on yielding supports, as well as bridge piers or tall liquid storage tanks that are not anchored to their foundations and thus may be expected to tip or uplift during strong earthquake excitation. An important feature of such systems is that the locations where nonlinear displacements may occur are known in advance. This situation makes it convenient to eliminate the purely linear response degrees of freedom by static condensation before performing the dynamic analysis; to implement this reduction scheme, the structure usually is idealized as an assemblage of substructures. For the simple case described here it will be assumed that there are only two substructures: a nonlinear zone which includes all parts of the structure that may exhibit any nonlinear behavior and a second zone which accounts for the remainder of the system and is completely linear in its dynamic response.

Such a two-component idealization of a structural system is depicted conceptually in Fig. 15-3. The degrees of freedom associated with the linear substructure, denoted v_0, include only nodes within the linear elastic region and exclude any nodes at the substructure boundary. The nonlinear substructure degrees of freedom, designated v_i, include all degrees of freedom that serve to interconnect the boundaries of

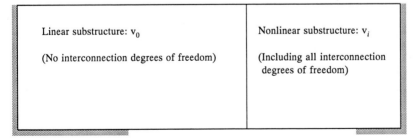

FIGURE 15-3
Definition of degrees of freedom for localized nonlinearity.

the two substructures, as well as all degrees of freedom in the interior of the nonlinear substructure. The coordinate reduction performed for this idealization eliminates the internal elastic degrees of freedom; clearly the reduction is most effective if the number of linear degrees of freedom, v_0, greatly exceeds the number of retained degrees of freedom, v_i.

The coordinate reduction procedure applied here is derived from the equations of motion expressed as the incremental pseudostatic equilibrium relationship, either Eq. (15-5) or (15-9) depending on the type of integration to be employed. For present purposes, either equation will be expressed as

$$\tilde{\mathbf{k}} \, \Delta \mathbf{v} = \Delta \tilde{\mathbf{p}} \tag{15-18}$$

in which the effective stiffness and incremental load matrices are given by either Eq. (15-6) or (15-10). To formulate the static condensation expressions, the incremental displacements are partitioned into linear and nonlinear sets, $\Delta \mathbf{v}_0$ and $\Delta \mathbf{v}_i$, and the pseudostatic equilibrium relation [Eq. (15-18)] is partitioned correspondingly thus

$$\begin{bmatrix} \tilde{k}_{00} & \tilde{k}_{0i} \\ \tilde{k}_{i0} & \tilde{k}_{ii} \end{bmatrix} \begin{bmatrix} \Delta v_0 \\ \Delta v_i \end{bmatrix} = \begin{bmatrix} \tilde{\Delta p_0} \\ \tilde{\Delta p_i} \end{bmatrix} \tag{15-19}$$

The static constraint relation then is obtained by solving the first of these submatrix equations for $\Delta \mathbf{v}_0$, that is,

$$\Delta \mathbf{v}_0 = \tilde{\mathbf{k}}_{00}^{-1} \left[\tilde{\Delta}\mathbf{p}_0 - \tilde{\mathbf{k}}_{0i} \, \Delta \mathbf{v}_i \right] \tag{15-20a}$$

Introducing this in the second submatrix equation leads to

$$\tilde{\mathbf{k}}_{i0} \tilde{\mathbf{k}}_{00}^{-1} \left[\tilde{\Delta}\mathbf{p}_0 - \tilde{\mathbf{k}}_{0i} \, \Delta \mathbf{v}_i \right] + \tilde{\mathbf{k}}_{ii} \, \Delta \mathbf{v}_i = \tilde{\Delta}\mathbf{p}_i \tag{15-20b}$$

which may be simplified to provide the reduced pseudostatic equation of equilibrium

$$\tilde{\tilde{\mathbf{k}}}_i \, \Delta \mathbf{v}_i = \Delta \tilde{\tilde{\mathbf{p}}}_i \tag{15-21}$$

in which the reduced pseudostatic stiffness is

$$\tilde{\tilde{\mathbf{k}}}_i = \tilde{\mathbf{k}}_{ii} - \tilde{\mathbf{k}}_{i0} \tilde{\mathbf{k}}_{00}^{-1} \tilde{\mathbf{k}}_{0i} \tag{15-21a}$$

and the reduced pseudo load increment is

$$\Delta \tilde{\tilde{\mathbf{p}}}_i = \Delta \tilde{\mathbf{p}}_i - \tilde{\mathbf{k}}_{i0} \tilde{\mathbf{k}}_{00}^{-1} \Delta \tilde{\mathbf{p}}_0 \tag{15-21b}$$

It is worth noting here that full advantage may be taken of the fact that the nonlinearity is localized when calculating the changing stiffness properties during the incremental analysis; specifically, the stiffness coefficients of the linear substructure are constants.

This static condensation reduction of coordinates may be looked upon as a specialized Rayleigh-Ritz coordinate transformation as was explained in Chapter 14 (Section 14-5). Thus by analogy with Eq. (14-14) the reduction transformation may be written

$$\triangle \mathbf{v} = \begin{bmatrix} \triangle v_0 \\ \triangle v_i \end{bmatrix} = \mathbf{T}_s \, \triangle \mathbf{v}_i \tag{15-22a}$$

where

$$\mathbf{T}_s \equiv \begin{bmatrix} -k_{00}^{-1} k_{0i} \\ I \end{bmatrix} \tag{15-22b}$$

and it is easily demonstrated that the reduced pseudostatic stiffness and reduced incremental pseudo load expressions shown above in Eq. (15-21) may be obtained by applying the standard coordinate transformation to Eq. (15-18) leading to

$$\widetilde{\widetilde{\mathbf{k}}}_i = \mathbf{T}_s^T \, \widetilde{\mathbf{k}} \, \mathbf{T}_s \tag{15-23a}$$

$$\triangle \widetilde{\widetilde{\mathbf{p}}}_i = \mathbf{T}_s^T \, \widetilde{\triangle \mathbf{p}} \tag{15-23b}$$

However, it must be realized that the assumed Ritz shapes given by the static condensation procedure [matrix $\mathbf{T}_s$] are not very effective in structural dynamics analysis because they do not account for any inertial effects in the dynamic response. Better results will be obtained if the coordinate transformation matrix $\mathbf{T}_s$ is supplemented by some shapes which represent such inertial forces, and a convenient approach is to express the internal displacements by a few derived Ritz vectors. These may be calculated by the procedure of Section 14-8 for the linear substructure with all of its boundary degrees of freedom restrained. Expressing the displacements given by a chosen number "n" of these vectors as

$$\triangle \mathbf{v}_0 = \mathbf{\Psi}_n \, \triangle \mathbf{Z}_n \tag{15-24a}$$

the expanded coordinate transformation may be written

$$\triangle \mathbf{v} = \begin{bmatrix} \triangle v_0 \\ \triangle v_i \end{bmatrix} = \mathbf{T}_{SZ} \, \triangle \mathbf{v}_{iZ} \tag{15-24b}$$

where

$$\mathbf{T}_{SZ} \equiv \begin{bmatrix} \mathbf{\Psi}_n & [-\widetilde{\mathbf{k}}_{00}^{-1} \widetilde{\mathbf{k}}_{0i}] \\ 0 & I \end{bmatrix} \tag{15-24c}$$

and

$$\triangle \mathbf{v}_{iZ} \equiv \begin{bmatrix} \triangle Z_n \\ \triangle v_i \end{bmatrix} \tag{15-24d}$$

It is evident here that the inclusion of the derived Ritz vector coordinates increases the number of degrees of freedom to be considered in the dynamic response analysis. However, the pseudostatic equilibrium equation derived with this transformation

$$\widetilde{\widetilde{\mathbf{k}}}_{iZ}\,\mathbf{v}_{iZ} = \triangle\widetilde{\widetilde{\mathbf{p}}}_{iZ} \tag{15-25}$$

in which

$$\widetilde{\widetilde{\mathbf{k}}}_{iZ} \equiv \mathbf{T}_{SZ}^T\,\widetilde{\mathbf{k}}\,{}^{\text{'}}\mathbf{T}_{SZ} \tag{15-25a}$$

and

$$\widetilde{\widetilde{\mathbf{p}}}_{iZ} \equiv \mathbf{T}_{SZ}^T\,\triangle\widetilde{\mathbf{p}} \tag{15-25b}$$

generally provides a great improvement in the step-by-step analysis results compared with those obtained from Eqs. (15-21), even if no more than two or three derived Ritz vectors are included in the supplemental degrees of freedom expressed by Eq. (15-24a).

Coupled Effects Treated as Pseudo-Forces

A different strategy for dealing with coupling of the modal response coordinates may be applied in cases where the coupling terms do not dominate the dynamic behavior. In this approach, the property coefficients that contribute to the coordinate coupling are transferred to the right hand side of the equation of motion, and they then serve to define a system of pseudo-forces acting on the structure. The objective in transforming these coefficients is to leave on the left side of the equation a set of property matrices that may be used in a standard mode superposition analysis, thus making it possible to exploit the modal uncoupling and coordinate truncation advantages of that approach.

Thus the eigenproblem associated with the mass and stiffness matrices remaining on the left side is solved, and the resulting mode shapes are used to transform the equations of motion resulting from the coefficient transfer into a set of uncoupled modal equations. Of course this imposes the condition that only a damping matrix of proportional form and constant stiffness and mass matrices may be retained on the left side. Consequently any nonproportional part of the damping matrix as well as any stiffness or mass changes associated with nonlinear behavior must be transferred to the right hand side. The principal disadvantage of this procedure for eliminating modal coordinate coupling is that the pseudo-force terms which have been transferred to the right side are functions of the response quantities thus the response solution may be obtained only by iteration. However, if the pseudo-force terms are relatively small, they will have only a secondary effect on the response, and a satisfactory equilibrium state generally can be achieved with only a few iterations.

It is convenient here to describe this pseudo-force procedure separately for cases where the modal coordinate coupling results from changes of stiffness (nonlinearity) and nonproportional damping; other situations might be handled similarly but will not be included in this discussion. In these analyses that involve "right-hand-siding," it is desirable to perform the dynamic response analysis by the piece-wise exact method, to avoid introducing errors except in the iterative solution; then the errors may be controlled by adopting an appropriate tolerance limit in the iteration cycles. Accordingly, the time step used in the analysis is selected to provide a reasonable approximation of the applied load history by a sequence of linearly varying load segments. The analysis for each step involves a standard linear evaluation of the displacements in each mode resulting from the linearly varying applied loads and pseudo-forces, followed by iteration until the modal forces on the left and right sides of the equations are balanced to within the specified tolerance.

Changes of Stiffness — The analysis of a system with nonlinear structural resistance may be formulated by expressing the changes of stiffness as deviations from the original linear elastic stiffness, i.e., from the stiffness matrix that was used in the eigenproblem solution for the mode shapes. These shapes do not change, so the analysis accounts for all of the nonlinearity effects in terms of pseudo-forces. The concept is depicted in a qualitative sense in Fig. 15-4 which shows the force-displacement relation for a single degree of freedom. Similar relationships might be assumed for each of the structure's stiffness coefficients; however, subscripts that might identify the force and displacement components (as shown in Fig. 15-1, for example) are omitted here to avoid confusion with the subscripts that identify various constituents of the response, such as "o" and "n" for linear and nonlinear, respectively.

In Fig. 15-4, the linear elastic stiffness is represented by the initial tangent slope, k_0, while the nonlinear stiffness associated with the displacement **v** is indicated by the average slope, k_n. Using this notation, the nonlinear structural resistance in this degree of freedom may be expressed as

$$f_{sn}(t) = K_n\, v(t) \tag{15-26}$$

However, to use the pseudo-force procedure, it is necessary to express this force in terms of the change from the linear elastic force, thus it is given as

$$f_{sn} = f_{se} - f_{sd}$$

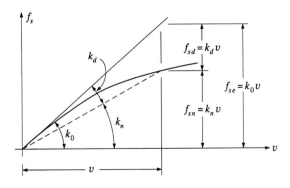

FIGURE 15-4
Definitions of stiffnesses and structural forces.

where f_{se} and f_{sd} are derived from the linear elastic stiffness and the change from that stiffness, respectively (i.e., $f_{se} = k_0\, v$ and $f_{sd} = k_d\, v$), so that the nonlinear structural resistance is given by

$$f_{sn}(t) = (k_0 - k_d)\, v(t) \tag{15-27}$$

in preference to the expression of Eq. (15-26).

Extending this concept to a system with multiple degrees of freedom, the nonlinear structural resistance vector is given by the equivalent matrix expression

$$\mathbf{f}_{sn}(t) = \begin{bmatrix} \mathbf{k}_0 - \mathbf{k}_d \end{bmatrix} \mathbf{v}(t) \tag{15-28}$$

and using this expression the nonlinear equations of motion may be stated as

$$\mathbf{m}\,\ddot{\mathbf{v}}(t) + \mathbf{c}\,\dot{\mathbf{v}}(t) + \begin{bmatrix} \mathbf{k}_0 - \mathbf{k}_d \end{bmatrix} \mathbf{v}(t) = \mathbf{p}(t)$$

However if the nonlinear change of resistance is transferred to the right hand side, the equation of motion becomes

$$\mathbf{m}\,\ddot{\mathbf{v}}(t) + \mathbf{c}\,\dot{\mathbf{v}}(t) + \mathbf{k}_0\,\mathbf{v}(t) = \mathbf{p}(t) + \mathbf{k}_d\,\mathbf{v}(t) \tag{15-29}$$

where the left hand side is the standard linear response expression and the nonlinear change of resistance acts as a pseudo-force on the right side. The change of stiffness matrix $\mathbf{k}_d$ typically is a function of the displacements $\mathbf{v}$. In practice it usually is evaluated from the stress-strain relationships specified for the materials incorporated in the structure, making use of the strain-displacement transformations developed in the formulation of the finite-element model.

It would be possible to use Eq. (15-29) in a direct nonlinear step-by-step analysis of the response; however, it is more efficient to transform the equations first to a truncated set of modal coordinates in order to reduce the number of quantities that must be evaluated by iteration. Thus, using the standard modal coordinate transformation, $\mathbf{v}(t) = \mathbf{\Phi} Y(t)$, Eq. (15-29), becomes a set of modal equations:

$$M\,\ddot{Y}(t) + C\,\dot{Y}(t) + K\,Y(t) = P(t) + F_{sd}(t) \tag{15-30}$$

These equations are uncoupled on the left hand side if the system is proportionally damped. Furthermore, if the modal coordinates are normalized in the usual way, M becomes an identity matrix, C is a diagonal array of terms $2\xi_n \omega_n$, and K is a diagonal array of the squared modal frequencies. The force vectors on the right hand side of Eq. (15-30) include the usual modal forces, $P_n(t) = \boldsymbol{\phi}_n^T \mathbf{p}(t)$, and the corresponding modal pseudo-forces:

$$F_{sd_n}(t) = \boldsymbol{\phi}_n^T\, \mathbf{k}_d\, \boldsymbol{\phi}\, Y(t) \tag{15-31}$$

The mode shapes Φ contained in this expression were derived using the original elastic stiffness matrix, $\mathbf{k}_0$, so they are not orthogonal with respect to the change of stiffness matrix, $\mathbf{k}_d$. Consequently, the modal change of stiffness matrix, $K_d = \Phi^T \mathbf{k}_d \Phi$, contains modal coupling coefficients $K_{d_{np}} = \phi_n^T \mathbf{k}_d \phi_p$, and the modal pseudo-forces therefore may be functions of all the modal displacements; thus

$$F_{sd_n}(t) = \sum_{p=1}^{m} K_{d_{np}} Y_p(t) \tag{15-32}$$

Introducing all of the above-named quantities in Eq. (15-30) the equation of motion for each mode finally takes the form

$$\ddot{Y}_n(t) + 2\xi_n\omega_n\dot{Y}_n(t) + \omega_n^2 Y_n(t) = \phi_n^T \mathbf{p}(t) + \sum_{p=1}^{m} K_{d_{np}} Y_p(t) \tag{15-33}$$

in which iteration must be used to obtain balance between the two sides of the equation. The equilibrium expression for any time step for any mode "n" during the "k" cycle of iteration may be stated as

$$\ddot{Y}_n^{(k)} + 2\xi_n\omega_n\dot{Y}_n^{(k)} + \omega_n^2 Y_n^{(k)} = \phi_n^T \mathbf{p} + \sum_{p=1}^{m} K_{d_{np}} Y_p^{(k-1)} \tag{15-34}$$

in which the functional relationship to time has not been shown to avoid confusion with the iteration cycle indicator "k."

During each iteration cycle, the equation is integrated by the piecewise exact method to determine the modal displacement and velocity at the end of the time step. For this purpose, it is necessary that the modal applied loading, $P_n(t)$, as well as the modal pseudo-force $F_{sd_n}(t)$ be expressed as a linear variation during the time step, as described in Section 7-2; of course, this implies that the modal displacements $Y_n(t)$ are assumed to vary linearly during the time step. The iteration for each time step is terminated when the force balance indicated by Eq. (15-33) has converged to the desired tolerance level.

Nonproportional Damping[3] — The pseudo-force concept can be applied in a very similar way to account for modal coupling effects in the analysis of systems with nonproportional damping. In this case, the essential step is to separate the damping matrix into a proportional component that is uncoupled by the modal coordinate

---

[3] A. Ibrahimbegovic and E. L. Wilson, "Simple Numerical Algorithm for the Mode Superposition Analysis of Linear Structural Systems with Nonproportional Damping," *Computers and Structures*, Vol. 33, 1989, pp. 523–533.

transformation, plus a nonproportional component that is transferred to the right hand side of the equations of motion where its effects are represented as pseudo-forces.

For the purpose of this discussion, it will be assumed that an appropriate non-proportional viscous damping matrix $\mathbf{C}$ has been constructed to represent the system's actual damping mechanism, by the method described in Section 12-5 or otherwise. Applying the modal coordinate transformation to this damping matrix then leads to a modal damping matrix, $\mathbf{C} = \Phi^T \mathbf{c} \Phi$, in which the diagonal elements

$$C_{nn} \equiv \phi_n^T \mathbf{c} \phi_n \equiv 2\xi_n \omega_n$$

represent the proportional damping contribution, while the off-diagonal elements

$$C_{np} \equiv \phi_n^T \mathbf{c} \phi_p \qquad (= 0, \quad n = p)$$

represent the nonproportional damping effects. These off-diagonal coefficients express the damping coupling between the modes and are treated as pseudo-forces applied on the right side of the equation of motion. It must be noted that the diagonal modal damping coefficients, C_{nn}, make no contribution to these pseudo-forces.

By analogy with the development of Eq. (15-33), it is evident that the modal equation of motion for the case of nonproportional damping may be expressed as follows:

$$\ddot{Y}_n(t) + 2\xi_n \omega_n \dot{Y}_n(t) + \omega_n^2 Y_n(t) = \phi_n^T \mathbf{p}(t) + \sum_{p=1}^{m} C_{np} \dot{Y}_p(t) \qquad (15\text{-}35)$$

in which it is noted that $C_{nn} = 0$. Following the procedure described above for the analysis of nonlinear response, this equation must be solved mode by mode, iterating to achieve force balance for each time step. To use the piecewise exact analysis, the modal damping pseudo-force, $F_{dn}(t) = \sum_{p=1}^{m} C_{np} \dot{Y}_n(t)$, must be assumed to vary linearly during the time step together with the applied modal loads, $P_n(t)$; this implies that the modal velocities, $\dot{Y}_n(t)$, are assumed to vary linearly.

CHAPTER
16

VARIATIONAL
FORMULATION
OF THE
EQUATIONS
OF MOTION

16-1 GENERALIZED COORDINATES

The significant advantages of describing the response of dynamic systems by means of generalized coordinates, rather than by merely expressing the displacements of discrete points on the structure, have been emphasized many times in this text, and various types of generalized coordinates have been considered for this purpose. It has also been pointed out that different approaches may be used to advantage in establishing the equations of motion for a structure, depending on its geometric form and complexity as well as the type of coordinates used. Up to this point, only the direct equilibration and the virtual-work approaches have been employed. The purpose of this chapter is to describe and demonstrate by examples the formulation of the equations of motion for MDOF systems by the variational approach.

In formulating the variational MDOF technique, extensive use will be made of generalized coordinates, and in this development a precise definition of the concept is needed rather than the somewhat loose terminology that has sufficed until now. Thus, generalized coordinates for a system with N degrees of freedom are defined here as any set of N independent quantities which completely specify the position of every point within the system. Being completely independent, generalized coordinates must not be related in any way through geometric constraints imposed on the system.

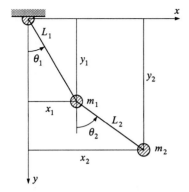

FIGURE 16-1
Double pendulum with hinge support.

In the classical double pendulum shown in Fig. 16-1, the position of the two masses m_1 and m_2 could be specified using the coordinates x_1, y_1, x_2, y_2; however, two geometric constraint conditions must be imposed on these coordinates, namely,

$$x_1^2 + y_1^2 - L_1^2 = 0$$
$$(x_2 - x_1)^2 + (y_2 - y_1)^2 - L_2^2 = 0 \qquad (16\text{-}1)$$

Because of these constraint relations, x_1, y_1, x_2, and y_2 are not independent and therefore cannot be considered as generalized coordinates.

Suppose, on the other hand, the angles θ_1 and θ_2 were specified as the coordinates to be used in defining the positions of masses m_1 and m_2. Clearly either of these coordinates can be changed while holding the other constant; thus, they are seen to be completely independent and therefore a suitable set of generalized coordinates.

16-2   HAMILTON'S PRINCIPLE

To establish a variational statement of dynamics, consider mass particle m shown in Fig. 16-2 which moves in response to the applied force vector $\mathbf{F}(t)$ along the real path indicated, leaving point 1 at time t_1 and arriving at point 2 at time t_2. It should be noted that this force includes the combined effects of the externally applied load $p(t)$, the structural resistance $f_S(t)$, and the damping resistance $f_D(t)$; by d'Alembert's principle, it is equilibrated by the inertial resistance $f_I(t)$. If, at time t, the mass particle is subjected to the resultant virtual displacement $\delta r(t)$, the virtual work of all forces, including the inertial force, must equal zero as expressed by

$$\left[F_x(t) - m\,\ddot{x}(t) \right] \delta x(t) + \left[F_y(t) - m\,\ddot{y}(t) \right] \delta y(t)$$

$$+ \left[F_z(t) - m\,\ddot{z}(t) \right] \delta z(t) = 0 \qquad (16\text{-}2)$$

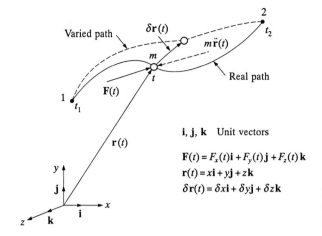

i, j, k Unit vectors

$\mathbf{F}(t) = F_x(t)\mathbf{i} + F_y(t)\mathbf{j} + F_z(t)\mathbf{k}$
$\mathbf{r}(t) = x\mathbf{i} + y\mathbf{j} + z\mathbf{k}$
$\delta\mathbf{r}(t) = \delta x\mathbf{i} + \delta y\mathbf{j} + \delta z\mathbf{k}$

FIGURE 16-2
Real and varied motions of mass
particle m.

Rearranging terms and integrating this equation from time t_1 to time t_2 give

$$\int_{t_1}^{t_2} -m\left[\ddot{x}(t)\,\delta x(t) + \ddot{y}(t)\,\delta y(t) + \ddot{z}(t)\,\delta z(t)\right]dt$$

$$+ \int_{t_1}^{t_2}\left[F_x(t)\,\delta x(t) + F_y(t)\,\delta y(t) + F_z(t)\,\delta z(t)\right]dt = 0 \qquad (16\text{-}3)$$

Integrating the first integral (I_1) by parts and recognizing that the virtual displacement must vanish at the beginning and the end of this varied path, i.e., that $\delta\mathbf{r}(t_1)$ and $\delta\mathbf{r}(t_2)$ equal zero, one obtains

$$I_1 = \int_{t_1}^{t_2} m\left[\dot{x}(t)\,\delta\dot{x}(t) + \dot{y}(t)\,\delta\dot{y}(t) + \dot{z}(t)\,\delta\dot{z}(t)\right]dt$$

$$= \int_{t_1}^{t_2} \delta T(t)\,dt = \delta\int_{t_1}^{t_2} T(t)\,dt \qquad (16\text{-}4)$$

in which $T(t)$ is the kinetic energy of the particle given by

$$T(t) = \frac{1}{2}m\left[\dot{x}(t)^2 + \dot{y}(t)^2 + \dot{z}(t)^2\right] \qquad (16\text{-}5)$$

In this discussion, it is helpful to separate the force vector $\mathbf{F}(t)$ into its conservative and nonconservative components as represented by

$$\mathbf{F}(t) = \mathbf{F}_c(t) + \mathbf{F}_{nc}(t) \qquad (16\text{-}6)$$

A potential energy function, $V(x, y, z, t)$, is then defined such that the conservative force vector $\mathbf{F}_c(t)$, by definition, must satisfy the component relations

$$\frac{\partial V(x, y, z, t)}{\partial x} = -F_{x,c}(t) \quad \frac{\partial V(x, y, z, t)}{\partial y} = -F_{y,c}(t) \quad \frac{\partial V(x, y, z, t)}{\partial z} = -F_{z,c}(t)$$
$$(16\text{-}7)$$

Making use of Eqs. (16-6) and (16-7), the second integral (I_2) in Eq. (16-3) becomes

$$I_2 = \int_{t_1}^{t_2} -\delta V(x, y, z, t) \, dt + \int_{t_1}^{t_2} \delta W_{nc}(t) \, dt \qquad (16\text{-}8)$$

in which $W_{nc}(t)$ equals the virtual work done by the nonconservative forces in vector $\mathbf{F}_{nc}(t)$. Making use of Eqs. (16-4) and (16-8), Eq. (16-3) can be expressed in the form

$$\int_{t_1}^{t_2} \delta[T(t) - V(t)] \, dt + \int_{t_1}^{t_2} \delta W_{nc}(t) \, dt = 0 \qquad (16\text{-}9)$$

Upon considering a summation of equations of this type for all mass particles, it becomes apparent that Eq. (16-9) is also valid for any complicated system, linear or nonlinear, provided quantities $T(t)$, $V(t)$, and $W_{nc}(t)$ represent the summation of such quantities for the entire system.

Equation (16-9), which is generally known as Hamilton's variational statement of dynamics, shows that the sum of the time-variations of the difference in kinetic and potential energies and the work done by the nonconservative forces over any time interval t_1 to t_2 equals zero. The application of this principle leads directly to the equations of motion for any given system.

The above variational procedure differs from the virtual-work procedure used previously in that the external load as well as the inertial and elastic forces are not explicitly involved; the variations of the kinetic- and potential-energy terms, respectively, are utilized instead. It therefore has the advantage of dealing only with purely scalar energy quantities, whereas the forces and displacements used to represent corresponding effects in the virtual-work procedure are all vectorial in character even though the work terms themselves are scalar.

It is of interest to note that Hamilton's equation can also be applied to statics problems. In this case, the kinetic-energy term T vanishes, and the remaining terms in the integrands of Eq. (16-8) are invariant with time; thus, the equation reduces to

$$\delta(V - W_{nc}) = 0 \qquad (16\text{-}10)$$

which is the well-known principle of minimum potential energy, so widely used in static analyses.

16-3   LAGRANGE'S EQUATIONS OF MOTION

The equations of motion for an N-DOF system can be derived directly from Hamilton's equation [Eq. (16-9)] by simply expressing the total kinetic energy T, the total potential energy V, and the total virtual work δW_{nc} in terms of a set of generalized coordinates, $q_1, q_2, \cdots, q_N$.

For most mechanical or structural systems, the kinetic energy can be expressed in terms of the generalized coordinates and their first time derivatives, and the potential energy can be expressed in terms of the generalized coordinates alone. In addition, the virtual work which is performed by the nonconservative forces as they act through the virtual displacements caused by an arbitrary set of variations in the generalized coordinates can be expressed as a linear function of those variations. In mathematical terms the above three statements are expressed in the form

$$T = T(q_1, q_2, \cdots, q_N, \dot{q}_1, \dot{q}_2, \cdots, \dot{q}_N) \tag{16-11a}$$

$$T = V(q_1, q_2, \cdots, q_N) \tag{16-11b}$$

$$\delta W_{nc} = Q_1 \, \delta q_1 + Q_2 \, \delta q_2 + \cdots + Q_N \, \delta q_N \tag{16-11c}$$

where the coefficients Q_1, Q_2, $\cdots$, Q_N are the generalized forcing functions corresponding to the coordinates q_1, q_2, $\cdots$, q_N, respectively.

Introducing Eqs. (16-11) into Eq. (16-9) and completing the variation of the first term give

$$\int_{t_2}^{t_2} \left(\frac{\partial T}{\partial q_1} \delta q_1 + \frac{\partial T}{\partial q_2} \delta q_2 + \cdots + \frac{\partial T}{\partial q_N} \delta q_N + \frac{\partial T}{\partial \dot{q}_1} \delta \dot{q}_1 + \frac{\partial T}{\partial \dot{q}_2} \delta \dot{q}_2 + \cdots + \frac{\partial T}{\partial \dot{q}_N} \delta \dot{q}_N \right.$$
$$\left. - \frac{\partial V}{\partial q_1} \delta q_1 - \frac{\partial V}{\partial q_2} \delta q_2 - \cdots - \frac{\partial V}{\partial q_N} \delta q_N + Q_1 \, \delta q_1 + Q_2 \, \delta q_2 + \cdots + Q_N \, \delta q_N \right) dt = 0 \tag{16-12}$$

Integrating the velocity-dependent terms in Eq. (16-12) by parts leads to

$$\int_{t_1}^{t_2} \frac{\partial T}{\partial \dot{q}_i} \delta \dot{q}_i \, dt = \left[\frac{\partial T}{\partial \dot{q}_i} \delta q_i \right]_{t_1}^{t_2} - \int_{t_1}^{t_2} \frac{d}{dt} \left(\frac{\partial T}{\partial \dot{q}_i} \right) \delta q_i \, dt \tag{16-13}$$

The first term on the right hand side of Eq. (16-13) is equal to zero for each coordinate since $\delta q_i(t_1) = \delta q_i(t_2) = 0$ is the basic condition imposed upon the variations. Substituting Eq. (16-13) into Eq. (16-12) gives, after rearranging terms,

$$\int_{t_1}^{t_2} \left\{ \sum_{i=1}^{N} \left[-\frac{d}{dt} \left(\frac{\partial T}{\partial \dot{q}_i} \right) + \frac{\partial T}{\partial q_i} - \frac{\partial V}{\partial q_i} + Q_i \right] \delta q_i \right\} dt = 0 \tag{16-14}$$

Since all variations δq_i ($i = 1, 2, \cdots, N$) are arbitrary, Eq. (16-14) can be satisfied in general only when the term in brackets vanishes, i.e.,

$$\frac{d}{dt} \left(\frac{\partial T}{\partial \dot{q}_i} \right) - \frac{\partial T}{\partial q_i} + \frac{\partial V}{\partial q_i} = Q_i \tag{16-15}$$

Equations (16-15) are the well-known Lagrange's equations of motion, which have found widespread application in various fields of science and engineering.

The beginning student of structural dynamics should take special note of the fact that Lagrange's equations are a direct result of applying Hamilton's variational principle, under the specific condition that the energy and work terms can be expressed in terms of the generalized coordinates, and of their time derivatives and variations, as indicated in Eqs. (16-11). Thus Lagrange's equations are applicable to all systems which satisfy these restrictions, and they may be nonlinear as well as linear. The following examples should clarify the application of Lagrange's equations in structural-dynamics analysis.

Example E16-1.   Consider the double pendulum shown in Fig. 16-1 under free-vibration conditions. The x- and y-coordinate positions along with their first time derivatives can be expressed in terms of the set of generalized coordinates $q_1 \equiv \theta_1$ and $q_2 \equiv \theta_2$ as follows:

$$x_1 = L_1 \sin q_1 \qquad\qquad \dot{x}_1 = L_1 \dot{q}_1 \cos q_1$$

$$y_1 = L_1 \cos q_1 \qquad\qquad \dot{y}_1 = -L_1 \dot{q}_1 \sin q_1$$

$$x_2 = L_1 \sin q_1 + L_2 \sin q_2 \qquad \dot{x}_2 = L_1 \dot{q}_1 \cos q_1 + L_2 \dot{q}_2 \cos q_2$$

$$y_2 = L_1 \cos q_1 + L_2 \sin q_2 \qquad \dot{y}_2 = -L_1 \dot{q}_1 \sin q_1 - L_2 \dot{q}_2 \sin q_2 \qquad \text{(a)}$$

Substituting the above velocity expressions into the basic expression for kinetic energy, namely,

$$T = \frac{1}{2} m_1 \left(\dot{x}_1^2 + \dot{y}_1^2 \right) + \frac{1}{2} m_2 \left(\dot{x}_2^2 + \dot{y}_2^2 \right) \qquad \text{(b)}$$

gives

$$T = \frac{1}{2} m_1 L_1^2 \dot{q}_1^2 + \frac{1}{2} m_2 \left[L_1^2 \dot{q}_1^2 + L_2^2 \dot{q}_2^2 + 2 L_1 L_2 \dot{q}_1 \dot{q}_2 \cos(q_2 - q_1) \right] \qquad \text{(c)}$$

The only potential energy present in the double pendulum of Fig. 16-1 is that due to gravity. If zero potential energy is assumed when $q_1 = q_2 = 0$, the potential-energy relation is

$$V = (m_1 + m_2) g L_1 (1 - \cos q_1) + m_2 g L_2 (1 - \cos q_2) \qquad \text{(d)}$$

where g is the acceleration of gravity. There are, of course, no nonconservative forces acting on this system; therefore, the generalized forcing functions Q_1 and Q_2 are both equal to zero.

Substituting Eqs. (c) and (d) into Lagrange's Eqs. (16-15) for $i = 1$ and $i = 2$ separately gives the two equations of motion

$$(m_1 + m_2) L_1^2 \ddot{q}_1 + m_2 L_1 L_2 \ddot{q}_2 \cos(q_2 - q_1)$$

$$- m_2 L_1 L_2 \dot{q}_2^2 \sin(q_2 - q_1) + (m_1 + m_2) g L_1 \sin q_1 = 0$$

$$m_2 L_2^2 \ddot{q}_2 + m_2 L_1 L_2 \ddot{q}_1 \cos(q_2 - q_1) \qquad \text{(e)}$$

$$+ m_2 L_1 L_2 \dot{q}_1^2 \sin(q_2 - q_1) + m_2 g L_2 \sin q_2 = 0$$

These equations are highly nonlinear for large-amplitude oscillation; however, for small-amplitude oscillation Eqs. (e) can be reduced to their linear forms

$$(m_1 + m_2) L_1^2 \ddot{q}_1 + m_2 L_1 L_2 \ddot{q}_2 + (m_1 + m_2) g L_1 q_1 = 0$$

$$m_2 L_1 L_2 \ddot{q}_1 + m_2 L_2^2 \ddot{q}_2 + m_2 g L_2 q_2 = 0 \qquad \text{(f)}$$

The small-amplitude mode shapes and frequencies can easily be obtained from the linearized equations of motion by any of the standard eigenproblem analysis methods, e.g., the determinantal-solution procedure.

Example E16-2. Assume a uniform rigid bar of length L and total mass m to be supported by an elastic, massless flexure spring and subjected to a uniformly distributed time-varying external loading as shown in Fig. E16-1. If the downward vertical deflections of points 1 and 2 from their static-equilibrium positions are selected as the generalized coordinates q_1 and q_2, respectively, the governing equations of motion for small-displacement theory can be obtained from Lagrange's equations as follows.

The total kinetic energy of the rigid bar is the sum of its translational and rotational kinetic energies, that is,

$$T = \frac{1}{2} m \left(\frac{\dot{q}_1 + \dot{q}_2}{2} \right)^2 + \frac{1}{2} \frac{m L^2}{12} \left(\frac{\dot{q}_1 - \dot{q}_2}{L} \right)^2 \qquad \text{(a)}$$

or

$$T = \frac{m}{6} (\dot{q}_1^2 + \dot{q}_1 \dot{q}_2 + \dot{q}_2^2)$$

Since q_1 and q_2 are displacements from the static-equilibrium position, gravity forces can be ignored provided that the potential energy of the system is evaluated as only the strain energy stored in the flexure spring. Where this strain

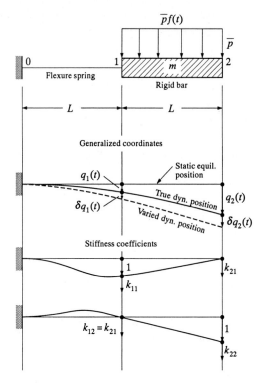

FIGURE E16-1
Rigid bar on massless flexure spring.

energy is expressed in terms of the stiffness influence coefficients (defined in Fig. E16-1) the potential-energy term becomes

$$V = \frac{1}{2}\left(k_{11}\, q_1^2 + 2\,k_{12}\, q_1\, q_2 + k_{22}\, q_2^2\right) \tag{b}$$

The virtual work performed by the nonconservative loading $\bar{p}\, f(t)$ as it acts through the virtual displacements produced by the arbitrary variations $\delta q_1(t)$ and $\delta q_2(t)$ is given by

$$\delta W_{nc} = \frac{\bar{p}\, L\, f(t)}{2}\left(\delta q_1 + \delta q_2\right) \tag{c}$$

From a comparison of Eq. (c) with Eq. (16-11c) it is clear that

$$Q_1(t) = Q_2(t) = \frac{\bar{p}\, L}{2}\, f(t) \tag{d}$$

Substituting Eqs. (a), (b), and (d) into Lagrange's Eqs. (16-15) gives the linear equations of motion for this structure:

$$\frac{m}{6}\left(2\,\ddot{q}_1 + \ddot{q}_2\right) + k_{11}\, q_1 + k_{12}\, q_2 = \frac{\bar{p}\, L}{2}\, f(t)$$

$$\frac{m}{6}\left(\ddot{q}_1 + 2\,\ddot{q}_2\right) + k_{12}\, q_1 + k_{22}\, q_2 = \frac{\bar{p}\, L}{2}\, f(t) \tag{e}$$

Example E16-3. Three uniform rigid bars of length L and mass m are hinged together at points 1 and 2, as shown in Fig. E16-2, and are supported by a roller at point 3 and a hinge at point 0. Concentrated moment-resisting elastic springs and viscous rotational dashpots are attached to adjoining bars at points 1 and 2, having property constants k_1, c_1, k_2, and c_2, respectively. A constant axial load N acts at point 3. If this system is excited by the applied lateral loading $p_1(t)$ and by a small vertical support motion $v_0(t)$ at end 0, the governing equations of motion based on small-deflection theory can be derived directly from Lagrange's equations as follows.

The kinetic energy of the three bars is

$$T = \frac{m}{6} \left(3\,\dot{v}_0^2 + 2\,\dot{q}_1^2 + 2\,\dot{q}_2^2 + 4\,\dot{v}_0\,\dot{q}_1 + 2\,\dot{v}_0\,\dot{q}_2 + \dot{q}_1\,\dot{q}_2 \right) \qquad (a)$$

The movement toward the left of end 3 due to the vertical joint displacement is

$$d = \frac{1}{L} \left(\frac{v_0^2}{6} + q_1^2 + q_2^2 - q_1\,q_2 \right) \qquad (b)$$

The relative rotations of the bars at joints 1 and 2 and their variations are given by

$$\theta_1 = \frac{1}{L}\,(2\,q_1 - q_2) \qquad \delta\theta_1 = \frac{1}{L}\,(2\,\delta q_1 - \delta q_2) \qquad (c)$$

$$\theta_2 = \frac{1}{L}\,(2\,q_2 - q_1) \qquad \delta\theta_2 = \frac{1}{L}\,(2\,\delta q_2 - \delta q_1) \qquad (d)$$

Hence the potential energy of the springs and of the axial force N is

$$V = \left[\frac{1}{2L^2}(4\,k_1 + k_2) - \frac{N}{L} \right] q_1^2 + \left[\frac{1}{2L^2}(k_1 + 4\,k_2) - \frac{N}{L} \right] q_2^2$$

$$+ \left[\frac{1}{2L^2}(-4\,k_1 - 4\,k_2) + \frac{N}{L} \right] q_1\,q_2 - \frac{N v_0^2}{6L} \qquad (e)$$

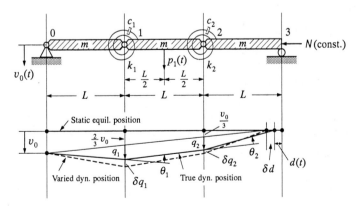

FIGURE E16-2
A 2-DOF rigid-bar assemblage with rotational springs and dashpots.

The virtual work done by the nonconservative forces is

$$\delta W_{nc} = \frac{1}{2} p_1(t) \left(\delta q_1 + \delta q_2 \right) - c_1 \dot{\theta}_1 \, \delta \theta_1 - c_2 \dot{\theta}_2 \, \delta \theta_2$$

or

$$\delta W_{nc} = \left[\frac{p_1}{2} - \frac{2 c_1}{L^2} (2 \dot{q}_1 - \dot{q}_2) + \frac{c_2}{L^2} (2 \dot{q}_2 - \dot{q}_1) \right] \delta q_1$$

$$+ \left[\frac{p_1}{2} + \frac{c_1}{L^2} (2 \dot{q}_1 - \dot{q}_2) - \frac{2 c_2}{L^2} (2 \dot{q}_2 - \dot{q}_1) \right] \delta q_2 \qquad \text{(f)}$$

from which the generalized forces are seen to be

$$Q_1 = \frac{p_1}{2} - \frac{2 c_1}{L^2} (2 \dot{q}_1 - \dot{q}_2) + \frac{c_2}{L^2} (2 \dot{q}_2 - \dot{q}_1)$$

$$Q_2 = \frac{p_1}{2} - \frac{c_1}{L^2} (2 \dot{q}_1 - \dot{q}_2) - \frac{2 c_2}{L^2} (2 \dot{q}_2 - \dot{q}_1)$$
$$\text{(g)}$$

Substituting Eqs. (a), (e), and (g) into Eqs. (16-15) gives the following two equations of motion, from which the dynamic response can be computed:

$$\frac{2}{3} m \ddot{q}_1 + \frac{m}{6} \ddot{q}_2 + \left(\frac{4 c_1}{L^2} + \frac{c_2}{L^2} \right) \dot{q}_1 + \left(- \frac{2 c_1}{L^2} - \frac{2 c_2}{L^2} \right) \dot{q}_2$$

$$+ \left[\frac{1}{L^2} (4 k_1 + k_2) - \frac{2N}{L} \right] q_1 + \left[\frac{1}{2L^2} (-4 k_1 - 4 k_2) + \frac{N}{L} \right] q_2 = \frac{p_1}{2} - \frac{m}{3} \ddot{v}_0 \quad \text{(h)}$$

$$\frac{m}{6} \ddot{q}_1 + \frac{2m}{3} \ddot{q}_2 + \left(- \frac{2 c_1}{L^2} - \frac{2 c_2}{L^2} \right) \dot{q}_1 + \left(\frac{c_1}{L^2} + \frac{4 c_2}{L^2} \right) \dot{q}_2$$

$$+ \left[\frac{1}{2L^2} (-4 k_1 - 4 k_2) + \frac{N}{L} \right] q_1 + \left[\frac{1}{L^2} (k_1 + 4 k_2) - \frac{2N}{L} \right] q_2 = \frac{p_1}{2} - \frac{m}{3} \ddot{v}_0 \quad \text{(i)}$$

By setting the accelerations and velocities to zero and removing the sources of excitation $p_1(t)$ and $v_0(t)$ from the system, Eqs. (h) and (i) reduce to the static-equilibrium conditions

$$\left[\frac{1}{L^2} (4 k_1 + k_2) - \frac{2N}{L} \right] q_1 + \left[\frac{1}{2L^2} (-4 k_1 - 4 k_2) + \frac{N}{L} \right] q_2 = 0$$

$$\left[\frac{1}{2L^2} (-4 k_1 - 4 k_2) + \frac{N}{L} \right] q_1 + \left[\frac{1}{L^2} (k_1 + 4 k_2) - \frac{2N}{L} \right] q_2 = 0$$
$$\text{(j)}$$

Now a nontrivial solution of Eqs. (j) is possible only when the structure buckles under the acting of the axial force N, and this is indicated when the determinant of the coefficient matrix equals zero, that is, when

$$\left\| \begin{matrix} \frac{1}{L^2} (4 k_1 + k_2) - \frac{2N}{L} & \frac{1}{2L^2} (-4 k_1 - 4 k_2) + \frac{N}{L} \\ \frac{1}{2L^2} (-4 k_1 - 4 k_2) + \frac{N}{L} & \frac{1}{L^2} (k_1 + 4 k_2) - \frac{2N}{L} \end{matrix} \right\| = 0 \qquad \text{(k)}$$

Expanding the determinant given by Eq. (k) and solving for N gives

$$N_{\text{cr}} = -\frac{3}{2L}(k_1 - k_2) \pm \sqrt{\frac{1}{12L^2}(13k_1^2 - 118k_1k_2 + 13k_2^2)} \qquad (\text{l})$$

Equation (l) gives two values for N_{cr} corresponding to the first and second buckling modes. The two mode shapes are found by substituting these two critical loads separately into Eq. (k) and solving for one of the generalized coordinates in terms of the other.

16-4 DERIVATION OF THE GENERAL EQUATIONS OF MOTION FOR LINEAR SYSTEMS

As is evident in the above three examples, the kinetic and potential energies of linear engineering systems subjected to small-amplitude oscillations can be expressed in the quadratic forms

$$T = \frac{1}{2}\sum_{j=1}^{N}\sum_{i=1}^{N} m_{ij}\,\dot{q}_i\,\dot{q}_j = \frac{1}{2}\dot{\mathbf{q}}^T\,\mathbf{m}\,\dot{\mathbf{q}} \qquad (16\text{-}16)$$

$$V = \frac{1}{2}\sum_{j=1}^{N}\sum_{i=1}^{N} k_{ij}\,q_i\,q_j = \frac{1}{2}\mathbf{q}^T\,\mathbf{k}\,\mathbf{q} \qquad (16\text{-}17)$$

where N is the number of degrees of freedom in the system. For such systems, the second term of Eqs. (16-15), namely, $\partial T/\partial q_i$ $(i = 1, 2, \cdots, N)$, equals zero, which reduces Lagrange's equations to the form

$$\frac{d}{dt}\left(\frac{\partial T}{\partial \dot{q}_i}\right) + \frac{\partial V}{\partial q_i} = Q_i \qquad i = 1, 2, \cdots, N \qquad (16\text{-}18)$$

When Eqs. (16-16) and (16-17) are substituted into Eqs. (16-18), Lagrange's equations of motion, when placed in matrix form, become

$$\mathbf{m}\,\ddot{\mathbf{q}} + \mathbf{k}\,\mathbf{q} = \mathbf{Q} \qquad (16\text{-}19)$$

which are similar to the discrete-coordinate equations formulated earlier by virtual work. It must be remembered, however, that *all* nonconservative forces, *including damping forces*, are contained here in the generalized forcing functions $Q_1, Q_2, \cdots, Q_N$.

Now the discretization problem will be considered, i.e., approximating infinite-DOF systems by a finite number of coordinates. For example, the lateral deflections $v(x, t)$ of a flexural member can be approximated by the relation

$$v(x, t) \doteq q_1(t)\,\psi_1(x) + q_2(t)\,\psi_2(x) + \cdots + q_N(t)\,\psi_N(x) \qquad (16\text{-}20)$$

where q_i $(i = 1, 2, \cdots, N)$ are generalized coordinates and ψ_i $(i = 1, 2, \cdots, N)$ are assumed dimensionless shape functions which satisfy the prescribed geometric boundary conditions for the member.

If $m(x)$ is the mass per unit length for the member, the kinetic energy (neglecting rotational inertial effects) can be expressed

$$T = \frac{1}{2} \int m(x) \, \dot{v}(x, t)^2 \, dx \tag{16-21}$$

Substituting Eq. (16-20) into Eq. (16-21) gives Eq. (16-16):

$$T = \frac{1}{2} \sum_{j=1}^{N} \sum_{i=1}^{N} m_{ij} \, \dot{q}_i \, \dot{q}_j$$

in which

$$m_{ij} = \int m(x) \, \psi_i(x) \, \psi_j(x) \, dx \tag{16-22}$$

The flexural strain energy is given by

$$V = \frac{1}{2} \int EI(x) \, [v''(x, t)]^2 \, dx \tag{16-23}$$

Substituting Eq. (16-20) into Eq. (16-23) gives

$$V = \frac{1}{2} \sum_{j=1}^{N} \sum_{i=1}^{N} k_{ij} \, q_i \, q_j \tag{16-17}$$

in which

$$k_{ij} = \int EI(x) \, \psi_i''(x) \, \psi_j''(x) \, dx \tag{16-24}$$

To obtain the generalized forcing functions $Q_1, Q_2, \cdots, Q_N$, the virtual work δW_{nc} must be evaluated. This is the work performed by *all* nonconservative forces acting on or within the flexural member while an arbitrary set of virtual displacements $\delta q_1, \delta q_2, \cdots, \delta q_N$ is applied to the system. To illustrate the principles involved in this evaluation, it will be assumed that the material of the flexure member obeys the uniaxial stress-strain relation

$$\sigma(t) = E \left[\epsilon(t) + a_1 \, \dot{\epsilon}(t) \right] \tag{16-25}$$

Using Eq. (16-25) and the Bernoulli-Euler hypothesis that the normal strains vary linearly over the member cross section leads to the moment-displacement relation

$$M(x, t) = EI(x) \left[v''(x, t) + a_1 \, \dot{v}''(x, t) \right] \tag{16-26}$$

The first term on the right hand side of Eq. (16-26) results from the internal conservative forces, which have already been accounted for in the potential-energy term V, while the second term results from the internal nonconservative forces. The virtual work performed by these nonconservative forces per unit length along the member equals the negative of the product of the nonconservative moment $a_1 EI(x) \dot{v}''(x,t)$ times the variation in the curvature $\delta v''(x,t)$. Therefore, the total virtual work performed by these internal nonconservative forces is

$$\delta W_{nc,\text{int}} = -a_1 \int EI(x)\, \dot{v}''(x,t)\, \delta v''(x,t)\, dx \qquad (16\text{-}27)$$

If the externally applied nonconservative forces are assumed in this case to be limited to a distributed transverse loading $p(x,t)$, the virtual work performed by these forces equals

$$\delta W_{nc,\text{ext}} = \int p(x,t)\, \delta v(x,t)\, dx \qquad (16\text{-}28)$$

Substituting Eq. (16-20) into Eqs. (16-27) and (16-28) and adding gives

$$\delta W_{nc,\text{total}} = \sum_{i=1}^{N} \left(p_i - \sum_{j=1}^{N} c_{ij}\, \dot{q}_j \right) \delta q_i \qquad (16\text{-}29)$$

where

$$p_i = \int p(x,t)\, \psi_i(x)\, dx \qquad (16\text{-}30)$$

$$c_{ij} = a_1 \int EI(x)\, \psi_i''(x)\, \psi_j''(x)\, dx \qquad (16\text{-}31)$$

When Eq. (16-29) is compared with Eq. (16-11c), it is evident that

$$Q_i = p_i - \sum_{j=1}^{N} c_{ij}\, \dot{q}_j \qquad (16\text{-}32)$$

Finally, substituting Eqs. (16-16), (16-17), and (16-32) into Lagrange's equations (16-15) gives the governing equations of motion in matrix form:

$$\mathbf{m}\,\ddot{\mathbf{q}} + \mathbf{c}\,\dot{\mathbf{q}} + \mathbf{k}\,\mathbf{q} = \mathbf{p} \qquad (16\text{-}33)$$

Note from the definitions of m_{ij}, c_{ij}, and k_{ij} as given by Eqs. (16-22), (16-31), and (16-24), respectively, that

$$m_{ij} = m_{ji} \qquad c_{ij} = c_{ji} \qquad k_{ij} = k_{ji} \qquad (16\text{-}34)$$

Therefore, the mass, damping, and stiffness coefficient matrices of Eq. (16-33) are symmetric in form.

Example E16-4. The formulation of the equations of motion by the general Lagrange's equation procedure described above will be illustrated for the rigid-bar assemblage shown in Fig. E16-3. The bars are interconnected by hinges, and their relative rotations are resisted by rotational springs and dashpots located at each hinge with values as indicated. The generalized coordinates of this system are taken to be the rotation angles q_i of the rigid bars, as shown in the sketch; it will be assumed that the displacements are small so that the small-deflection theory is valid.

With the kinetic energy of the rigid bars due to rotation about their individual centroids and due to translation of the centroids considered separately, the total kinetic energy is

$$T = \frac{1}{2}\frac{WL^2}{12g}(\dot{q}_1^2 + \dot{q}_2^2 + \dot{q}_3^2)$$

$$+ \frac{1}{2}\frac{W}{g}\left[\left(\frac{\dot{q}_1 L}{2}\right)^2 + \left(\dot{q}_1 L + \frac{\dot{q}_2 L}{2}\right)^2 + \left(\dot{q}_1 L + \dot{q}_2 L + \frac{\dot{q}_3 L}{2}\right)^2\right]$$

$$= \frac{WL^2}{6g}(2\dot{q}_1^2 + 4\dot{q}_2^2 + \dot{q}_3^2 + 9\dot{q}_1\dot{q}_2 + 3\dot{q}_2\dot{q}_3 + 3\dot{q}_1\dot{q}_3) \qquad (a)$$

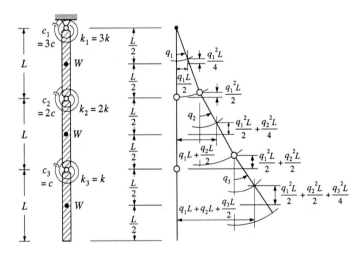

FIGURE E16-3
A 3-DOF rigid-body assemblage (including geometric-stiffness effect due to dead weight).

Also with the potential energy associated with deformation of the rotational springs and due to the raising of the bars above their vertical base position considered separately (the amounts of the vertical displacements of the centers of gravity are indicated on the sketch), the total potential energy of the system is given by

$$
V = W \left[\frac{q_1^2 L}{4} + \left(\frac{q_2^2 L}{2} + \frac{q_2^2 L}{4} \right) + \left(\frac{q_1^2 L}{2} + \frac{q_2^2 L}{2} + \frac{q_3^2 L}{4} \right) \right]
$$

$$
+ \frac{1}{2} [k_1 q_1^2 + k_2 (q_2 - q_1)^2 + k_3 (q_3 - q_2)^2]
$$

$$
= \frac{1}{4} [(5WL + 10k) q_1^2 + (3WL + 6k) q_2^2 + (WL + 4k) q_3^2 - 8 q_1 q_2 - 4 q_2 q_3]
$$

(b)

Finally, the virtual work done by the rotational dashpots during the virtual displacements of the structure is given by

$$
\delta W_{nc} = -c_1 \dot{q}_1 \, \delta q_1 - c_2 (\dot{q}_2 - \dot{q}_1)(\delta q_2 - \delta q_1) - c_3 (\dot{q}_3 - \dot{q}_2)(\delta q_3 - \delta q_2)
$$

$$
= c \left[(-5\dot{q}_1 + 2\dot{q}_2)\delta q_1 + (2\dot{q}_1 - 3\dot{q}_2 + \dot{q}_3)\delta q_2 + (\dot{q}_2 - \dot{q}_3)\delta q_3 \right]
$$

(c)

from which the nonconservative forces, which are due only to damping, become

$$
Q_1 = c(-5\dot{q}_1 + 2\dot{q}_2)
$$

$$
Q_2 = c(2\dot{q}_1 - 3\dot{q}_2 + \dot{q}_3)
$$

(d)

$$
Q_3 = c(\dot{q}_2 - \dot{q}_3)
$$

Substituting Eqs. (a) to (d) into

$$
\frac{d}{dt} \left(\frac{\partial T}{\partial \dot{q}_i} \right) + \frac{\partial v}{\partial q_i} = Q_i \qquad i = 1, 2, 3
$$

(e)

gives the three equations of motion of the system, which, arranged in matrix form, are

$$
\frac{WL^2}{6g}
\begin{bmatrix} 14 & 9 & 3 \\ 9 & 8 & 3 \\ 3 & 3 & 2 \end{bmatrix}
\begin{bmatrix} \ddot{q}_1 \\ \ddot{q}_2 \\ \ddot{q}_3 \end{bmatrix}
+ c
\begin{bmatrix} 5 & -2 & 0 \\ -2 & 3 & -1 \\ 0 & -1 & 1 \end{bmatrix}
\begin{bmatrix} \dot{q}_1 \\ \dot{q}_2 \\ \dot{q}_3 \end{bmatrix}
$$

$$
+ \frac{1}{2}
\begin{bmatrix} 5WL + 10k & -4k & 0 \\ -4k & 3WL + 6k & -2k \\ 0 & -2k & WL + 4k \end{bmatrix}
\begin{bmatrix} q_1 \\ q_2 \\ q_3 \end{bmatrix}
=
\begin{bmatrix} 0 \\ 0 \\ 0 \end{bmatrix}
$$

(f)

16-5   CONSTRAINTS AND LAGRANGE MULTIPLIERS

Usually when determining the dynamic response of an N-DOF system, the equations of motion are written in terms of a set of generalized coordinates q_1, q_2, $\cdots$, q_N; however, there are cases where in order to maintain symmetry in the equations of motion, it is preferable to select a set of coordinates q_1, q_2, $\cdots$, q_c, where $c > N$. These coordinates cannot be generalized coordinates since their number exceeds the number of degrees of freedom in the system. Therefore, one must impose m ($m = c - N$) equations of constraint on the system. For example, returning to the double pendulum shown in Fig. 16-1, it was pointed out earlier that the equations of motion could be expressed in terms of generalized coordinates θ_1 and θ_2 ($N = 2$) or in terms of coordinates x_1, y_1, x_2, y_2 ($c = 4$). If the latter coordinates are used, two equations of constraint, namely Eqs. (16-1), must be satisfied.

Suppose the m equations of constraint for a general case are expressed in the form

$$f_1(g_1, g_2, \cdots, g_c) = 0$$

$$f_2(g_1, g_2, \cdots, g_c) = 0$$

$$\cdots \cdots \cdots \cdots \cdots$$

$$f_m(g_1, g_2, \cdots, g_c) = 0$$

(16-35)

Taking the variations of Eqs. (16-35) results in

$$\delta f_1 = \frac{\partial f_1}{\partial g_1}\delta g_1 + \frac{\partial f_1}{\partial g_2}\delta g_2 + \cdots + \frac{\partial f_1}{\partial g_c}\delta g_c = 0$$

$$\delta f_2 = \frac{\partial f_2}{\partial g_1}\delta g_1 + \frac{\partial f_2}{\partial g_2}\delta g_2 + \cdots + \frac{\partial f_2}{\partial g_c}\delta g_c = 0$$

$$\cdots \cdots \cdots \cdots \cdots \cdots \cdots \cdots \cdots \cdots \cdots$$

$$\delta f_m = \frac{\partial f_m}{\partial g_1}\delta g_1 + \frac{\partial f_m}{\partial g_2}\delta g_2 + \cdots + \frac{\partial f_m}{\partial g_c}\delta g_c = 0$$

(16-36)

Now if each δf_i ($i = 1, 2, \cdots, m$) is multiplied by an unknown time function $\lambda_i(t)$ and the product is integrated over the time interval t_1 to t_2 [assuming Eqs. (16-11) to apply, when expressed in terms of coordinates q_1, q_2, $\cdots$, q_c], then if each of the above integrals is added to Hamilton's variational equation [Eq. (16-9)], the following equation is obtained after completing the variation:

$$\int_{t_1}^{t_2}\left\{\sum_{i=1}^{c}\left[-\frac{d}{dt}\left(\frac{\partial T}{\partial \dot{g}_i}\right) + \frac{\partial T}{\partial g_i} - \frac{\partial V}{\partial g_i} + Q_i \right.\right.$$

$$\left.\left. + \lambda_1\frac{\partial f_1}{\partial g_i} + \lambda_2\frac{\partial f_2}{\partial g_i} + \cdots + \lambda_m\frac{\partial f_m}{g_i}\right]\delta g_i\right\}dt = 0 \qquad (16\text{-}37)$$

Since the variations δg_i $(i = 1, 2, \cdots, c)$ are all arbitrary, it is necessary that each square-bracket term in Eq. (16-37) equal zero, i.e.,

$$\frac{d}{dt}\left(\frac{\partial T}{\partial \dot{g}_i}\right) - \frac{\partial T}{\partial g_i} + \frac{\partial V}{\partial g_i} = Q_i + \lambda_1 \frac{\partial f_1}{\partial g_i} + \lambda_2 \frac{\partial f_2}{\partial g_i} + \cdots + \lambda_m \frac{\partial f_m}{\partial g_i}$$

$$= 0 \qquad i = 1, 2, \cdots, c \qquad (16\text{-}38)$$

Equation (16-38) is a modified form of Lagrange's equations which will permit the use of coordinates $g_1, g_2, \cdots, g_c$. This procedure of developing Eqs. (16-38) may seem trivial at first because a number of integrals equating to zero have been added to Hamilton's equation; however, it should be noted that while each δf_i $(i = 1, 2, \cdots, m)$ equals zero, the individual terms given on the right hand side of Eqs. (16-36) are not equal to zero. The time-dependent functions λ_i $(i = 1, 2, \cdots, m)$ are known as *Lagrange multipliers.*

When a reduced potential-energy term $\overline{V}$ is defined as

$$\overline{V} = V(g_1, g_2, \cdots, g_c) - (\lambda_1 f_1 + \lambda_2 f_2 + \cdots + \lambda_m f_m) \qquad (16\text{-}39)$$

Eqs. (16-38) can be written

$$\frac{d}{dt}\left(\frac{\partial T}{\partial \dot{g}_i}\right) - \frac{\partial T}{\partial g_i} + \frac{\partial \overline{V}}{\partial g_i} = Q_i \qquad i = 1, 2, \cdots, c \qquad (16\text{-}40)$$

which contain the unknown time functions $g_1, g_2, \cdots, g_c, \lambda_1, \lambda_2, \cdots, \lambda_m$. Since there are $c + m$ unknown time functions, $c + m$ equations are required for their solution. These equations include the c modified Lagrange's equations [Eqs. (16-40)] and the m constraint equations [Eqs. (16-35)].

Example E16-5. The use of Lagrange multipliers in satisfying specified constraint conditions will be illustrated with reference to the end-supported cantilever beam of Fig. E16-4. This beam is subjected to a time-varying loading, $\overline{p} f(t)$, uniformly distributed along its length, as well as to a constant axial force N, as shown in the sketch; its stiffness is uniform along the length, and there

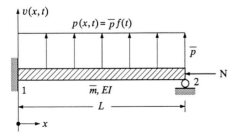

FIGURE E16-4
Uniform beam used to demonstrate
Lagrange multipliers.

is no damping. To obtain an approximate solution which is reasonably valid if the frequency components in the loading function are low enough, it will be assumed that the beam deflections can be expressed as

$$v(x, t) = g_1(t) \sin \frac{\pi x}{L} + g_2(t) \sin \frac{2\pi x}{L} \tag{a}$$

Expressing the kinetic and potential energies and the virtual work performed by the external loading in terms of the coordinates g_1 and g_2 leads to

$$T = \frac{1}{2} \int_0^L \overline{m} \left(\dot{g}_1^2 \sin^2 \frac{\pi x}{L} + 2\dot{g}_1 \dot{g}_2 \sin \frac{\pi x}{L} \sin \frac{2\pi x}{L} + \dot{g}_2^2 \sin^2 \frac{2\pi x}{L} \right) dx \tag{b}$$

$$V = \frac{1}{2} \int_0^L EI \left(g_1^2 \frac{\pi^4}{L^4} \sin^2 \frac{\pi x}{L} + \frac{8\pi^4}{L^4} g_1 g_2 \sin \frac{\pi x}{L} \sin \frac{2\pi x}{L} \right.$$
$$\left. + g_2^2 \frac{16\pi^4}{L^4} \sin^2 \frac{2\pi x}{L} \right) dx$$
$$- \frac{N}{2} \int_0^L \left(\frac{\pi^2}{L^2} g_1^2 \cos^2 \frac{\pi x}{L} + \frac{4\pi^2}{L^2} g_1 g_2 \cos \frac{\pi x}{L} \cos \frac{2\pi x}{L} \right.$$
$$\left. + \frac{4\pi^2}{L^2} g_2^2 \cos^2 \frac{2\pi x}{L} \right) dx \tag{c}$$

$$\delta W_{nc} = \delta g_1 \int_0^L p(x, t) \sin \frac{\pi x}{L} dx + \delta g_2 \int_0^L p(x, t) \sin \frac{2\pi x}{L} dx \tag{d}$$

Completing the integrals of Eqs. (b) to (d) gives

$$T = \frac{\overline{m} L}{4} (\dot{g}_1^2 + \dot{g}_2^2) \tag{e}$$

$$V = \frac{\pi^4 EI}{4L^3} (g_1^2 + 16 g_2^2) - \frac{N\pi^2}{4L} (g_1^2 + 4 g_2^2) \tag{f}$$

$$\delta W_{nc} = \frac{2L}{\pi} \overline{p} f(t) \delta g_1 \tag{g}$$

and comparing Eq. (g) with Eq. (16-11c) gives the external loads

$$Q_1 = \frac{2L\overline{p} f(t)}{\pi} \qquad Q_2 = 0 \tag{h}$$

When the fixed-support condition at the left end of the beam is considered, it is evident that the solution must satisfy the constraint condition

$$f_1(g_1, g_2) = g_1 + 2 g_2 = 0 \tag{i}$$

Substituting Eqs. (f) and (i) into Eqs. (16-39) thus leads to the reduced potential

$$\overline{V} = \frac{\pi^4 EI}{4L^3}(g_1^2 + 16g_2^2) - \frac{N\pi^2}{4L}(g_1^2 + 4g_2^2) - \lambda_1(g_1 + 2g_2) \qquad \text{(j)}$$

Substituting Eqs. (e), (h), and (j) into the Lagrange reduced equations of motion [Eqs. (16-40)] finally gives

$$\frac{\overline{m}L}{2}\ddot{g}_1 + \left(\frac{\pi^4 EI}{2L^3} - \frac{\pi^2 N}{2L}\right)g_1 - \lambda_1 = \frac{2L\overline{p}\,f(t)}{\pi}$$

$$\frac{\overline{m}L}{2}\ddot{g}_2 + \left(\frac{8\pi^4 EI}{L^3} - \frac{2\pi^2 N}{L}\right)g_2 - 2\lambda_1 = 0 \qquad \text{(k)}$$

From this point the complete solution of the problem can be obtained by solving Eq. (i) and Eqs. (k) for $g_1(t)$, $g_2(t)$, and $\lambda_1(t)$. The resulting solution shows that $\lambda_1(t)$ is proportional to the fixed-end moment at $x = 0$. This moment performs zero virtual work on the member because the constraint at that location does not permit a virtual rotation of the member cross section.

PROBLEMS

16-1. Applying Lagrange's equations, Eqs. (16-15), and permitting large displacements, determine the equation of motion for the system shown in Fig. E8-4. What is the linearized equation of motion for small amplitude oscillation?

16-2. Applying Lagrange's equations and permitting large displacements, determine the equations of motion for the system shown in Fig. P16-1. What are the linearized equations of motion for small-amplitude oscillations?

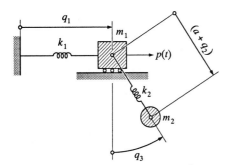

FIGURE P16-1

16-3. Repeat Prob. 16-1 for the system shown in Prob. 8-4.

16-4. Repeat Prob. 16-1 for the system shown in Prob. 8-5.

16-5. Obtain the equations of motion for the uniform cantilever beam shown in Fig. P16-2 when the deflected shape can be approximated by the relation

$$v(x,t) \doteq q_1(t)\left(\frac{x}{L}\right)^2 + q_2(t)\left(\frac{x}{L}\right)^3 + q_3(t)\left(\frac{x}{L}\right)^4$$

Assume small deflection theory.

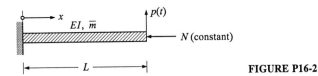

FIGURE P16-2

16-6. A ball of radius R_1 and mass m_1 is placed at rest on top of a fixed cylindrical surface of radius R_2. Assume a very slight disturbance that starts the ball rolling to the left, as shown in Fig. P16-3, under the influence of gravity. If the ball rolls without slippage and angles θ_1 and θ_2 are taken as displacement coordinates:

(a) Determine the equation of constraint between θ_1 and θ_2.

(b) Write the equation of motion in terms of one displacement coordinate by eliminating the other through the constraint equation.

(c) Write the equation of motion using both displacement coordinates and in addition using a Lagrange multiplier λ_1. (What does λ_1 represent physically in this case?)

(d) Determine the value of θ_2 when the ball leaves the surface of the cylinder.

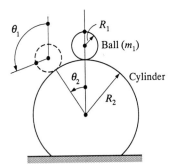

FIGURE P16-3

16-7. A uniform rigid bar of total mass m_1 and length L swings as a pendulum under the influence of gravity. A concertrated mass m_2 is constrained to slide along the axis of the bar and is attached to a massless spring, as shown in Fig. P16-4. Assuming a frictionless system and large amplitude displacements, determine the equations of motion in terms of generalized coordinates q_1 and q_2.

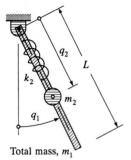

Total mass, m_1 **FIGURE P16-4**

16-8. Determine the linearized equations of motion for small-amplitude oscillations of the system defined in Prob. 16-7.

PART III

DISTRIBUTED-PARAMETER SYSTEMS

CHAPTER
17

PARTIAL
DIFFERENTIAL
EQUATIONS
OF MOTION

17-1 INTRODUCTION

The discrete-coordinate systems described in Part Two provide a convenient and practical approach to the dynamic-response analysis of arbitrary structures. However, the solutions obtained can only approximate their actual dynamic behavior because the motions are represented by a limited number of displacement coordinates. The precision of the results can be made as refined as desired by increasing the number of degrees of freedom considered in the analyses. In principle, however, an infinite number of coordinates would be required to converge to the exact results for any real structure having distributed properties; hence this approach to obtaining an exact solution is manifestly impossible.

The formal mathematical procedure for considering the behavior of an infinite number of connected points is by means of differential equations in which the position coordinates are taken as independent variables. Inasmuch as time is also an independent variable in a dynamic-response problem, the formulation of the equations of motion in this way leads to partial differential equations. Different classes of continuous systems can be identified in accordance with the number of independent variables required to describe the distribution of their physical properties. For example, the wave-propagation formulas used in seismology and geophysics are derived from the equations of motion expressed for general three-dimensional solids. Similarly, in studying the dynamic behavior of thin-plate or thin-shell structures, special equations of motion must be derived for these two-dimensional systems. In the present discussion, however, attention will be limited to one-dimensional structures, that is, beam- and rod-type systems which may have variable mass, damping, and stiffness properties along their elastic axes. The partial differential equations of these systems involve only two independent variables: time and distance along the elastic axis of each component member.

It is possible to derive the equations of motion for rather complex one-dimensional structures, including assemblages of many members in three-dimensional space. Moreover, the axes of the individual members might be arbitrarily curved in three-dimensional space, and the physical properties might vary as a complicated function of position along the axis. However, the solutions of the equations of motion for such complex systems generally can be obtained only by numerical means, and in most cases a discrete-coordinate formulation is preferable to a continuous-coordinate formulation. For this reason, the present treatment will be limited to simple systems involving members having straight elastic axes and assemblages of such members. In formulating the equations of motion, general variations of the physical properties along each axis will be permitted, although in subsequent solutions of these equations, the properties of each member will be assumed to be constant. Because of these severe limitations of the cases which may be considered, this presentation is intended mainly to demonstrate the general concepts of the partial-differential-equation formulation rather than to provide a tool for significant practical application to complex systems. Closed form solutions through this formulation can, however, be very useful when treating simple uniform systems.

17-2 BEAM FLEXURE: ELEMENTARY CASE

The first case to be considered in the formulation of partial differential equations of motion is the straight, nonuniform beam shown in Fig. 17-1a. The significant physical properties of this beam are assumed to be the flexural stiffness $EI(x)$ and the mass per unit length $m(x)$, both of which may vary arbitrarily with position x along the span L. The transverse loading $p(x,t)$ is assumed to vary arbitrarily with position and time, and the transverse-displacement response $v(x,t)$ also is a function of these variables. The end-support conditions for the beam are arbitrary, although they are pictured as simple supports for illustrative purposes.

The equation of motion of this simple system can readily be derived by considering the equilibrium of forces acting on the differential segment of beam shown in

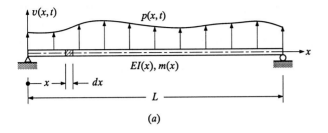

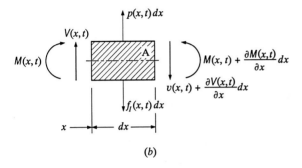

FIGURE 17-1
Basic beam subjected to dynamic loading: (*a*) beam properties and
coordinates; (*b*) resultant forces acting on differential element.

Fig. 17-1*b*, in much the same way that the equations were developed for a discrete-
parameter system. Summing all forces acting vertically leads to the first dynamic-
equilibrium relationship

$$V(x,t) + p(x,t)\, dx - \left[V(x,t) + \frac{\partial V(x,t)}{\partial x}\, dx \right] - f_I(x,t)\, dx = 0 \qquad (17\text{-}1)$$

in which $V(x,t)$ is the vertical force acting on the cut section and $f_I(x,t)\, dx$ is the
resultant transverse inertial force equal to the mass of the element multiplied by its
transverse acceleration, i.e.,

$$f_I(x,t)\, dx = m(x)\, dx\, \frac{\partial^2 v(x,t)}{\partial t^2} \qquad (17\text{-}2)$$

Substituting Eq. (17-2) into Eq. (17-1) and dividing the resulting equation by dx yield

$$\frac{\partial V(x,t)}{\partial x} = p(x,t) - m(x)\, \frac{\partial^2 v(x,t)}{\partial t^2} \qquad (17\text{-}3)$$

This equation is similar to the standard static relationship between shear force and
transverse loading but with the loading now being the resultant of the applied and
inertial-force loadings.

The second equilibrium relationship is obtained by summing moments about point A on the elastic axis. After dropping the two second-order moment terms involving the inertia and applied loadings, one gets

$$M(x,t) + V(x,t)\, dx - \left[M(x,t) + \frac{\partial M(x,t)}{\partial x}\, dx\right] = 0 \qquad (17\text{-}4)$$

Because rotational inertia is neglected, this equation simplifies directly to the standard static relationship between shear and moment

$$\frac{\partial M(x,t)}{\partial x} = V(x,t) \qquad (17\text{-}5)$$

Differentiating this equation with respect to x and substituting the result into Eq. (17-3) give

$$\frac{\partial^2 M(x,t)}{\partial x^2} + m(x)\,\frac{\partial^2 v(x,t)}{\partial t^2} = p(x,t) \qquad (17\text{-}6)$$

which, upon introducing the basic moment-curvature relationship $M = EI\frac{\partial^2 v}{\partial x^2}$, becomes

$$\frac{\partial^2}{\partial x^2}\left[EI(x)\,\frac{\partial^2 v(x,t)}{\partial x^2}\right] + m(x)\,\frac{\partial^2 v(x,t)}{\partial t^2} = p(x,t) \qquad (17\text{-}7)$$

This is the partial differential equation of motion for the elementary case of beam flexure. The solution of this equation must, of course, satisfy the prescribed boundary conditions at $x = 0$ and $x = L$.

17-3   BEAM FLEXURE: INCLUDING AXIAL-FORCE EFFECTS

If the beam considered in the above case is subjected to a time-invariant axial loading in the horizontal direction as shown in Fig. 17-2a in addition to the lateral loading shown in Fig. 17-1, the local equilibrium of forces is altered because the internal axial force $N(x)$ interacts with the lateral displacements to produce an additional term in the moment-equilibrium expression. It is apparent in Fig. 17-2b that transverse equilibrium is not affected by the axial force because its *direction* does not change with the beam deflection; hence Eq. (17-3) is still valid. However, the line of action of the axial force changes with the beam deflection so that the moment-equilibrium equation now becomes

$$M(x,t) + V(x,t)\, dx + N(x)\,\frac{\partial v(x,t)}{\partial x}\, dx - \left[M(x,t) + \frac{\partial M(x,t)}{\partial x}\, dx\right] = 0 \quad (17\text{-}8)$$

from which the vertical section force $V(x,t)$ is found to be

$$V(x,t) = -N(x)\,\frac{\partial v(x,t)}{\partial x} + \frac{\partial M(x,t)}{\partial x} \qquad (17\text{-}9)$$

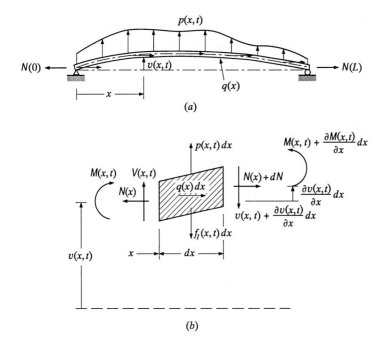

FIGURE 17-2
Beam with static axial loading and dynamic lateral loading: (*a*) beam deflected
due to loadings; (*b*) resultant forces acting on differential element.

Introducing this modified expression for $V(x,t)$ into Eq. (17-3) and proceeding
as before, one obtains the following partial differential equation of motion, including
axial-force effects:

$$\frac{\partial^2}{\partial x^2}\left[EI(x)\,\frac{\partial^2 v(x,t)}{\partial x^2}\right] - \frac{\partial}{\partial x}\left[N(x)\,\frac{\partial v(x,t)}{\partial x}\right] + m(x)\,\frac{\partial^2 v(x,t)}{\partial t^2} = p(x,t) \quad (17\text{-}10)$$

Comparing Eqs. (17-10) and (17-7), it is evident that the longitudinal loading producing the internal axial-force distribution $N(x)$ gives rise to an additional effective
transverse loading acting on the beam. Note that the vertical section force $V(x,t)$ is
not the section shear force in the usual sense because it is not acting normal to the
elastic axis.

17-4 BEAM FLEXURE: INCLUDING VISCOUS DAMPING

In the preceding formulations of the partial differential equations of motion for
beam-type members, no damping was included. Now distributed viscous damping
of two types will be included: (1) an external damping force per unit length as
represented by $c(x)$ in Fig. 8-3 and (2) internal resistance opposing the strain velocity

as represented by the second parts of Eqs. (8-8) and (8-9). The first of these requires that a transverse force opposing velocity

$$f_D(x,t)\, dx = c(x)\, \frac{\partial v(x,t)}{\partial t}\, dx \qquad (17\text{-}11)$$

be added to the element free body in Fig. 17-1b, the second requires that the section moment expression in Eq. (17-7) be changed to the form of Eq. (8-9), i.e.,

$$M(x,t) = EI(x)\left[\frac{\partial^2 v(x,t)}{\partial x^2} + a_1 \frac{\partial^3 v(x,t)}{\partial x^2\, \partial t}\right] \qquad (17\text{-}12)$$

in which a_1 is the stiffness proportionality factor defined previously for Rayleigh damping. Making these changes, the derivation procedure applied in Section 17-2 leads finally to

$$\frac{\partial^2}{\partial x^2}\left[EI(x)\left(\frac{\partial^2 v(x,t)}{\partial x^2} + a_1 \frac{\partial^3 v(x,t)}{\partial x^2\, \partial t}\right)\right] + m(x)\, \frac{\partial^2 v(x,t)}{\partial t^2}$$

$$+ c(x)\, \frac{\partial v(x,t)}{\partial t} = p(x,t) \qquad (17\text{-}13)$$

If in addition to the above two forms of viscous damping, one included at the same time axial-force effects, the left hand side of this equation would also contain the term $\frac{\partial}{\partial x}\left[N(x)\, \frac{\partial v(x,t)}{\partial x}\right]$ shown in Eq. (17-10).

17-5   BEAM FLEXURE: GENERALIZED SUPPORT EXCITATIONS

As discussed previously in Part One, SDOF Systems, and Part Two, MDOF Systems, structural and mechanical systems are often excited dynamically through support motions rather than by applied external loadings, e.g., piping systems in a nuclear power plant subjected to support motions at their connections to containment buildings and heavy equipment, which in turn are responding to earthquake ground motion inputs at their supports. While the practical analysis of such complex systems, as discussed in Chapter 26, must be carried out using discrete-parameter modeling, it is instructive here to formulate the partial differential equation governing the response of a distributed-parameter beam as shown in Fig. 17-1 when dynamically excited by support excitations.

Assume first that this beam is subjected to specified support motions (translations and rotations) at the two ends

$$v^t(0,t) = \delta_1(t) \qquad \left[\frac{\partial v^t(x,t)}{\partial x}\right]_{x=0} = \delta_3(t)$$

$$v^t(L,t) = \delta_2(t) \qquad \left[\frac{\partial v^t(x,t)}{\partial x}\right]_{x=L} = \delta_4(t) \qquad (17\text{-}14)$$

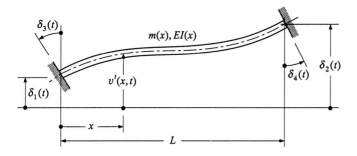

FIGURE 17-3
Basic beam subject to dynamic support displacements.

as shown in Fig. 17-3. The superscript "t" has been added to $v(x,t)$ in these equations to indicate total displacement of the beam's elastic axis from its original position. The governing partial differential equation for the viscously-damped case is given by Eq. (17-13) after removing the term $p(x,t)$ and substituting $v^t(x,t)$ for $v(x,t)$. This modified equation of motion must be solved so as to satisfy the specified geometric boundary conditions given by Eqs. (17-14). In doing so, it is convenient to express the beam's total displacement $v^t(x,t)$ as the sum of the displacement which would be induced by static application of the support motions $\delta_1(t)$, $\delta_2(t)$, $\delta_3(t)$, and $\delta_4(t)$, i.e., the so-called quasi-static displacement $v^s(x,t)$, plus the additional displacement $v(x,t)$ due to dynamic inertial and viscous force effects; thus

$$v^t(x,t) = v^s(x,t) + v(x,t) \tag{17-15}$$

Substituting this equation into the modified form of Eq. (17-13) and transferring all terms associated with the quasi-static displacement $v^s(x,t)$ and its derivatives to the right hand side lead to

$$\frac{\partial^2}{\partial x^2}\left[EI(x)\left(\frac{\partial^2 v(x,t)}{\partial x^2} + a_1\frac{\partial^3 v(x,t)}{\partial x^2\,\partial t}\right)\right] + m(x)\frac{\partial^2 v(x,t)}{\partial t^2}$$

$$+ c(x)\frac{\partial v(x,t)}{\partial t} = p_{\text{eff}}(x,t) \tag{17-16}$$

in which

$$p_{\text{eff}}(x,t) \equiv -\frac{\partial^2}{\partial x^2}\left[EI(x)\left(\frac{\partial^2 v^s(x,t)}{\partial x^2} + a_1\frac{\partial^3 v^s(x,t)}{\partial x^2\,\partial t}\right)\right]$$

$$- m(x)\frac{\partial^2 v^s(x,t)}{\partial t^2} - c(x)\frac{\partial v^s(x,t)}{\partial t} \tag{17-17}$$

represents the effective distributed dynamic loading caused by the prescribed support excitations. Note that because $v^s(x,t)$ is produced by static support displacements only, the first term on the right hand side of Eq. (17-17) equals zero; therefore, the effective loading can be simplified to the form

$$p_{\text{eff}}(x,t) = -\frac{\partial^2}{\partial x^2}\left[a_1\ EI(x)\ \frac{\partial^3 v^s(x,t)}{\partial x^2\ \partial t}\right] - m(x)\ \frac{\partial^2 v^s(x,t)}{\partial t^2}$$

$$- c(x)\ \frac{\partial v^s(x,t)}{\partial t} \tag{17-18}$$

The quantity $v^s(x,t)$, which is the source of the effective loading, was previously defined as the quasi-static displacement produced by the four specified support displacements $\delta_1(t)$, $\delta_2(t)$, $\delta_3(t)$, and $\delta_4(t)$; therefore, it can be expressed in the form

$$v^s(x,t) = \sum_{i=1}^{4}\psi_i(x)\ \delta_i(t) \tag{17-19}$$

where each static influence function $\psi_i(x)$ ($i = 1,2,3,4$) is the beam's deflection caused by a unit static displacement in the corresponding coordinate δ_i. If the beam is uniform, i.e., $EI(x) = $ constant, these influence functions are the cubic Hermitian polynomials given by Eqs. (10-16a) to (10-16d). Substituting Eq. (17-19) into Eq. (17-18) gives

$$p_{\text{eff}}(x,t) = -\sum_{i=1}^{4}\left\{m(x)\ \psi_i(x)\ \ddot{\delta}_i(t) + c(x)\ \psi_i(x)\ \dot{\delta}_i(t)\right.$$

$$\left. +\frac{\partial^2}{\partial x^2}\left[a_1\ EI(x)\ \dot{\delta}_i(t)\ \frac{\partial^2\psi_i(x)}{\partial x^2}\right]\right\} \tag{17-20}$$

In most practical cases, the damping contributions to the effective loading are small compared with the inertial contribution; thus, the last two terms in Eq. (17-20) are usually omitted. This allows the effective loading to be expressed in its approximate form

$$p_{\text{eff}}(x,t) \doteq -\sum_{i=1}^{4} m(x)\ \psi_i(x)\ \ddot{\delta}_i(t) \tag{17-21}$$

Substituting this equation into Eq. (17-16), one can solve for $v(x,t)$ which when added to $v^s(x,t)$ given by Eq. (17-19) yields the total displacement $v^t(x,t)$. Noting that $v^s(x,t)$ by itself satisfies the specified geometric boundary conditions of Eqs. (17-14), the end conditions $v(0,t)$, $v(L,t)$, $\left[\partial v(x,t)/\partial x\right]_{x=0}$, and $\left[\partial v(x,t)/\partial x\right]_{x=L}$ must all be set equal to zero when solving Eq. (17-16).

Next consider the beam in Fig. 17-1 when subjected to only two specified end motions as given by

$$v^t(0,t) = \delta_1(t) \qquad\qquad \left[\partial v^t(x,t)/\partial x\right]_{x=0} = \delta_3(t) \qquad (17\text{-}22)$$

Assuming its end at $x = L$ to be totally free, the quasi-static displacement of the resulting cantilever beam is

$$v^s(x,t) = 1\,\delta_1(t) + x\,\delta_3(t) \qquad (17\text{-}23)$$

which yields the approximate effective loading

$$p_{\text{eff}}(x,t) \doteq -m(x)\left[1\,\ddot{\delta}_1(t) + x\,\ddot{\delta}_3(t)\right] \qquad (17\text{-}24)$$

Since $v^s(x,t)$ as given by Eq. (17-23) totally satisfies both geometric boundary conditions at $x = 0$ as specified in Eqs. (17-22), one must impose the conditions $v(0,t) = 0$ and $\left[\partial v(x,t)/\partial x\right]_{x=0} = 0$ when solving Eq. (17-16) for $v(x,t)$. In addition, one must satisfy the zero moment and shear boundary conditions at the free end, i.e., the conditions

$$\left[\partial^2 v(x,t)/\partial x^2\right]_{x=L} = 0 \qquad\qquad \left[\partial^3 v(x,t)/\partial x^3\right]_{x=L} = 0 \qquad (17\text{-}25)$$

Since no flexural deformations are imposed in this cantilever beam due to the pseudo-static displacements of Eq. (17-23), stresses are produced only by the dynamic response $v(x,t)$.

Finally, consider the above cantilever beam subjected to only one specified support motion $v^t(0,t) = \delta_1(t)$. In this simple case, $v^s(x,t) = \delta_1(t)$, $p_{\text{eff}}(x,t) \doteq -m(x)\,\ddot{\delta}_1(t)$, and the boundary conditions which must be imposed on the solution of Eq. (17-16) are $v(0,t) = 0$, $\left[\partial v(x,t)/\partial x\right]_{x=0} = 0$, $\left[\partial^2 v(x,t)/\partial x^2\right]_{x=L} = 0$, and $\left[\partial^3 v(x,t)/\partial x^3\right]_{x=L} = 0$.

17-6 AXIAL DEFORMATIONS: UNDAMPED

The preceding discussions in Sections 17-2 through 17-5 have been concerned with beam flexure, in which case the dynamic displacements are in the direction transverse to the elastic axis. While this bending mechanism is the most common type of behavior encountered in the dynamic analysis of one-dimensional members, some important cases involve only axial displacements, e.g., a pile subjected to hammer blows during the driving process. The equations of motion governing such behavior can be derived by a procedure similar to that used in developing the equations of motion for flexure. However, derivation is simpler for the axial-deformation case, since equilibrium need be considered only in one direction rather than two. In this

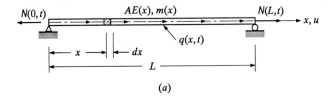

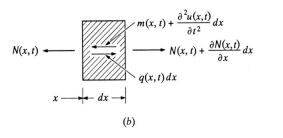

FIGURE 17-4
Bar subjected to dynamic axial deformations: (*a*) bar properties and
coordinates; (*b*) forces acting on differential element.

formulation, damping is neglected because it usually has little effect on the behavior
in axial deformation.

Consider a straight bar for which the axial stiffness EA and mass per unit
length m vary along its length as indicated in Fig. 17-4*a*. If it is subjected to an
arbitrary external distributed axial loading $q(x,t)$, an internal time-varying axial-force
distribution $N(x,t)$ will be produced as indicated on the differential element of the
beam in Fig. 17-4*b*. Note that $N(x,t)$ is used in this formulation to distinguish the
time-varying axial force from the time-invariant axial force $N(x)$ used previously in
Section 17-3. Summing the forces on this element in the x-direction, one obtains

$$N(x,t) + f_I(x,t)\,dx - \left[N(x,t) + \frac{\partial N(x,t)}{\partial x}\,dx\right] - q(x,t)\,dx = 0 \qquad (17\text{-}26)$$

in which $f_I(x,t)$ is the inertial force per unit length given by

$$f_I(x,t) = m(x)\,\frac{\partial^2 u(x,t)}{\partial t^2} \qquad (17\text{-}27)$$

where $u(x,t)$ is the displacement in the axial direction.

Substituting Eq. (17-27) and the axial force-deformation relationship

$$N(x,t) = \sigma(x,t)\,A(x) = \epsilon(x,t)\,EA(x) = \frac{\partial u(x,t)}{\partial x}\,EA(x) \qquad (17\text{-}28)$$

into Eq. (17-26), one obtains the partial differential equation of axial motion

$$m(x) \frac{\partial^2 u(x,t)}{\partial t^2} - \frac{\partial}{\partial x} \left[EA(x) \frac{\partial u(x,t)}{\partial x} \right] = q(x,t) \qquad (17\text{-}29)$$

Usually, the external axial loading consists only of end loads, in which case the right hand side of this equation would be zero. However, when solving Eq. (17-29) the boundary conditions imposed at $x = 0$ and $x = L$ must be satisfied.

PROBLEMS

17-1. Using Hamilton's principle, Eq. (16-9), determine the differential equation of motion and boundary conditions of the uniform cantilever beam loaded as shown in Fig. P17-1. Assume small deflection theory and neglect shear and rotary inertia effects.

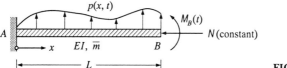

FIGURE P17-1

17-2. Using Hamilton's principle, determine the differential equation of motion and boundary conditions of the simply supported uniform pipe (shown in Fig. P17-2) through which fluid of density ρ and zero viscosity flows with constant velocity $\dot{v}_f$ relative to the pipe. Flexible moment connections are provided at each end of the pipe. Does the presence of the flowing fluid provide damping in the system? If the same pipe is supported as a cantilever member discharging the fluid at its free end, can fluid damping of the system be present (neglect material damping in the pipe)? Let A equal the inside cross-sectional area of the pipe. Assume small deflection theory and neglect shear and rotary inertia effects.

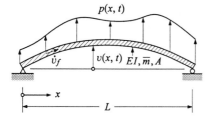

FIGURE P17-2

17-3. As shown in Fig. P17-3, a concertrated lumped mass m_1 traveling to the right with constant velocity $\dot{v}_{m_1}$ crosses a simply supported uniform beam during the time interval $0 < t < L/\dot{v}_{m_1}$. Determine the governing equations of motion for this system using Lagrange's equation of motion, Eqs. (16-15), and state the required boundary and initial conditions that must be imposed to obtain the vertical forced-vibration response of the simple beam. Assume small deflection theory and neglect shear and rotary inertia effects.

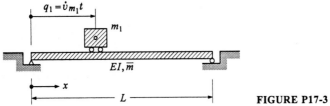

$$q_1 = \dot{v}_{m_1} t$$

m_1

$EI, \overline{m}$

x

L

FIGURE P17-3

ANALYSIS
OF
UNDAMPED
FREE
VIBRATION

18-1 BEAM FLEXURE: ELEMENTARY CASE

Following the same general approach employed with discrete-parameter systems, the first step in the dynamic-response analysis of a distributed-parameter system is to evaluate its undamped mode shapes and frequencies. Because of the mathematical complications of treating systems having variable properties, the following discussion will be limited to beams having uniform properties along their lengths and to frames assembled from such members. This is not a serious limitation, however, because it is more efficient to treat any variable-property systems using discrete-parameter modeling.

First, let us consider the elementary case presented in Section 17-2 with $EI(x)$ and $m(x)$ set equal to constants EI and $\overline{m}$, respectively. As shown by Eq. (17-7), the free-vibration equation of motion for this system is

$$EI \, \frac{\partial^4 v(x,t)}{\partial x^4} + \overline{m} \, \frac{\partial^2 v(x,t)}{\partial t^2} = 0 \qquad (18\text{-}1)$$

377

After dividing by EI and adopting the prime and dot notations to indicate partial derivatives with respect to x and t, respectively, this equation becomes

$$v^{iv}(x,t) + \frac{\overline{m}}{EI}\,\ddot{v}(x,t) = 0 \tag{18-2}$$

Since $\overline{m}/EI$ is a constant, one form of solution of this equation can be obtained easily by separation of variables using

$$v(x,t) = \phi(x)\,Y(t) \tag{18-3}$$

which indicates that the free-vibration motion is of a specific shape $\phi(x)$ having a time-dependent amplitude $Y(t)$. Substituting this equation into Eq. (18-2) gives

$$\phi^{iv}(x)\,Y(t) + \frac{\overline{m}}{EI}\,\phi(x)\,\ddot{Y}(t) = 0 \tag{18-4}$$

Dividing by $\phi(x)\,Y(t)$, the variables can be separated as follows:

$$\frac{\phi^{iv}(x)}{\phi(x)} + \frac{\overline{m}}{EI}\,\frac{\ddot{Y}(t)}{Y(t)} = 0 \tag{18-5}$$

Because the first term in this equation is a function of x only and the second term is a function of t only, the entire equation can be satisfied for arbitrary values of x and t only if each term is a constant in accordance with

$$\frac{\phi^{iv}(x)}{\phi(x)} = -\frac{\overline{m}}{EI}\,\frac{\ddot{Y}(t)}{Y(t)} = a^4 \tag{18-6}$$

where the single constant involved is designated in the form a^4 for later mathematical convenience. This equation yields two ordinary differential equations

$$\ddot{Y}(t) + \omega^2\,Y(t) = 0 \tag{18-7a}$$

$$\phi^{iv}(x) - a^4\,\phi(x) = 0 \tag{18-7b}$$

in which

$$\omega^2 \equiv \frac{a^4\,EI}{\overline{m}} \qquad (\text{i.e.,} \quad a^4 = \frac{\omega^2\,\overline{m}}{EI}) \tag{18-8}$$

The first of these [Eq. (18-7a)] is the familiar free-vibration expression for an undamped SDOF system having the solution [see Eq. (2-31)]

$$Y(t) = A\,\cos\omega t + B\,\sin\omega t \tag{18-9}$$

in which constants A and B depend upon the initial displacement and velocity conditions, i.e.,

$$Y(t) = Y(0)\,\cos\omega t + \frac{\dot{Y}(0)}{\omega}\,\sin\omega t \tag{18-10}$$

The second equation can be solved in the usual way by introducing a solution of the form

$$\phi(x) = G \, \exp(st) \tag{18-11}$$

leading to

$$(s^4 - a^4) \, G \, \exp(sx) = 0 \tag{18-12}$$

from which

$$s_{1,2} = \pm ia \qquad s_{3,4} = \pm a \tag{18-13}$$

Incorporating each of these roots into Eq. (18-11) separately and adding the resulting four terms, one obtains the complete solution

$$\phi(x) = G_1 \, \exp(iax) + G_2 \, \exp(-iax) + G_3 \, \exp(ax) + G_4 \, \exp(-ax) \tag{18-14}$$

in which G_1, G_2, G_3, and G_4 must be treated as complex constants. Expressing the exponential functions in terms of their trigonometric and hyperbolic equivalents and setting the entire imaginary part of the right hand side of this equation to zero lead to

$$\phi(x) = A_1 \, \cos ax + A_2 \, \sin ax + A_3 \, \cosh ax + A_4 \, \sinh ax \tag{18-15}$$

where A_1, A_2, A_3, and A_4 are real constants which can be expressed in terms of the components of G_1, G_2, G_3, and G_4. These real constants must be evaluated so as to satisfy the known boundary conditions (displacement, slope, moment, or shear) at the ends of the beam. Taking this action, any three of the four constants can be expressed in terms of the fourth and an expression (called the frequency equation) can be obtained from which the frequency parameter a is determined. The fourth constant cannot be evaluated directly in a free-vibration analysis because it represents an arbitrary amplitude of the shape function $\phi(x)$. However, having given it a numerical value, say unity, the values of $Y(0)$ and $\dot{Y}(0)$ in Eq. (18-10) must be set consistent with it so that the initial conditions on $v(x,t)$ given by Eq. (18-3) are satisfied, i.e., $Y(0) = v(x,0)/\phi(x)$ and $\dot{Y}(0) = \dot{v}(x,0)/\phi(x)$.

The above described free-vibration analysis procedure will now be illustrated through a series of examples as follows:

Example E18-1.   *Simple Beam*   Considering the uniform simple beam shown in Fig. E18-1a, its four known boundary conditions are

$$\phi(0) = 0 \qquad M(0) = EI \, \phi''(0) = 0 \tag{a}$$

$$\phi(L) = 0 \qquad M(L) = EI \, \phi''(L) = 0 \tag{b}$$

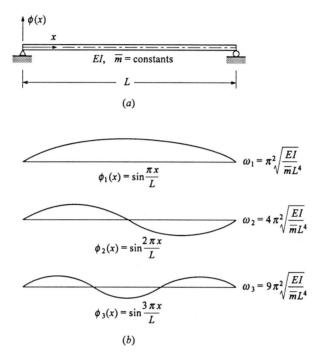

FIGURE E18-1
Simple beam-vibration analysis: (*a*) basic properties of simple beam;
(*b*) first three vibration modes.

Making use of Eq. (18-15) and its second partial derivative with respect to x,
Eqs. (a) can be written as

$$\phi(0) = A_1 \cos 0 + A_2 \sin 0 + A_3 \cosh 0 + A_4 \sinh 0 = 0$$

$$\phi''(0) = a^2 \left(-A_1 \cos 0 - A_2 \sin 0 + A_3 \cosh 0 + A_4 \sinh 0\right) = 0 \tag{c}$$

from which one obtains $(A_1 + A_3) = 0$ and $(-A_1 + A_3) = 0$; hence, $A_1 = A_3 = 0$. Similarly, Eqs. (b) can be written in the form

$$\phi(L) = A_2 \sin aL + A_4 \sinh aL = 0$$

$$\phi''(L) = a^2 \left(-A_2 \sin aL + A_4 \sinh aL\right) = 0 \tag{d}$$

after setting A_1 and A_3 equal to zero. Adding these two equations, after cancelling out a^2, gives

$$2 A_4 \sinh aL = 0 \tag{e}$$

thus $A_4 = 0$ since $\sinh aL \neq 0$. Only A_2 remains as a nonzero constant,
therefore

$$\phi(x) = A_2 \sin ax \tag{f}$$

Excluding the trivial solution $A_2 = 0$, boundary condition $\phi(L) = 0$ can be satisfied only when

$$\sin aL = 0 \qquad\qquad\text{(g)}$$

which is the system frequency equation; it requires that

$$a = n\pi/L \qquad\qquad n = 0, 1, 2, \cdots \qquad\qquad\text{(h)}$$

Substituting this expression into Eq. (18-8) and taking the square root of both sides yield the frequency expression

$$\omega_n = n^2\,\pi^2\,\sqrt{\frac{EI}{mL^4}} \qquad\qquad\text{(i)}$$

The corresponding vibration mode shapes are now given by Eq. (f) upon substitution of Eq. (h) for the frequency parameter a in the sine term; thus, ignoring the trivial case $n = 0$, one obtains

$$\phi_n(x) = A_2 \sin\frac{n\pi}{L}x \qquad\qquad n = 1, 2, \cdots \qquad\qquad\text{(j)}$$

The first three of these mode shapes are shown in Fig. E18-1b along with their circular frequencies.

Example E18-2.   *Cantilever Beam*   The free-vibration analysis of the simple beam in the previous example was not difficult because its mode shapes were defined by only one term in the shape-function expression of Eq. (18-15). To provide a more representative example of the analysis procedure requiring all four terms, consider the cantilever beam shown in Fig. E18-2a. Its four boundary conditions to be satisfied are

$$\phi(0) = 0 \qquad\qquad\qquad \phi'(0) = 0 \qquad\qquad\text{(a)}$$

$$M(L) = EI\,\phi''(L) = 0 \qquad\qquad V(L) = EI\,\phi'''(L) = 0 \qquad\text{(b)}$$

Substituting Eq. (18-15) and its derivative expressions into these equations gives

$$\phi(0) = (A_1\,\cos 0 + A_2\,\sin 0 + A_3\,\cosh 0 + A_4\,\sinh 0) = 0$$
$$\phi'(0) = a\,(-A_1\,\sin 0 + A_2\,\cos 0 + A_3\,\sinh 0 + A_4\,\cosh 0) = 0$$
$$\phi''(L) = a^2\,(-A_1\,\cos aL - A_2\,\sin aL + A_3\,\cosh aL + A_4\,\sinh aL) = 0$$
$$\phi'''(L) = a^3\,(A_1\,\sin aL - A_2\,\cos aL + A_3\,\sinh aL + A_4\,\cosh aL) = 0$$
$$\text{(c)}$$

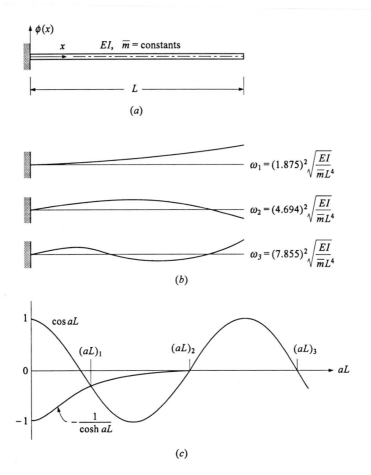

FIGURE E18-2
Cantilever-beam vibration analysis: (a) properties of cantilever beam;
(b) first three vibration modes; (c) frequency equation terms.

Making use of $\cos 0 = \cosh 0 = 1$ and $\sin 0 = \sinh 0 = 0$, the first two of these equations yield $A_3 = -A_1$ and $A_4 = -A_2$. Substituting these equalities into the last two equations, changing all signs, and placing the resulting expressions in matrix form, one obtains

$$\begin{bmatrix} (\cos aL + \cosh aL) & (\sin aL + \sinh aL) \\ (\sinh aL - \sin aL) & (\cos aL + \cosh aL) \end{bmatrix} \begin{Bmatrix} A_1 \\ A_2 \end{Bmatrix} = \begin{Bmatrix} 0 \\ 0 \end{Bmatrix} \qquad (d)$$

For coefficients A_1 and A_2 to be nonzero, the determinant of the square matrix in this equation must equal zero, thus giving the frequency equation

$$\sinh^2 aL - \sin^2 aL - \cos^2 aL - 2 \cosh aL \, \cos aL - \cosh^2 aL = 0 \qquad (e)$$

which reduces to the form

$$\cos aL = -(1/\cosh aL) \tag{f}$$

The solution of this transcendental equation provides the values of aL which represent the frequencies of vibration of the cantilever beam. Figure E18-2c shows a plot of functions $\cos aL$ and $-(1/\cosh aL)$; their crossing points give the values of aL which satisfy Eq. (f). Note that as the function $-(1/\cosh aL)$ approaches the axis asymptotically, the crossing points approach the values of aL given by $\cos aL = 0$; therefore, the roots of Eq. (f) higher than the third can be obtained within at least four-place accuracy using the approximate relation

$$(aL)_n \doteq \frac{\pi}{2}(2n - 1) \qquad n = 4, 5, 6, \cdots \tag{g}$$

Introducing the values of aL given by Eqs. (f) and (g) into Eq. (18-8) the corresponding circular frequencies can be obtained as shown by

$$\omega_n = (aL)_n^2 \sqrt{\frac{EI}{\overline{m}L^4}} \qquad n = 1, 2, 3, \cdots \tag{h}$$

Either of Eqs. (d) can now be employed to express coefficient A_2 in terms of A_1; the first gives

$$A_2 = -\frac{(\cos aL + \cosh aL)}{(\sin aL + \sinh aL)} A_1 \tag{i}$$

This result along with the previously obtained conditions that $A_3 = -A_1$ and $A_4 = -A_2$ allows the mode-shape expression of Eq. (18-15) to be written in the form

$$\phi(x) = A_1 \left[\cos ax - \cosh ax - \frac{(\cos aL + \cosh aL)}{(\sin aL + \sinh aL)}(\sin ax - \sinh ax) \right] \tag{j}$$

Substituting separately the frequency-equation roots for aL into this expression, one obtains the corresponding mode-shape functions. Plots of these functions for the first three modes are shown in Fig. E18-2b along with their corresponding circular frequencies.

Example E18-3. *Cantilever Beam with Rigid Mass at Free End* In this example the same uniform cantilever beam of Example E18-2 is used but a rigid lumped mass m_1 having a rotary mass moment of inertia j_1 is attached

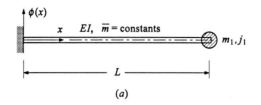

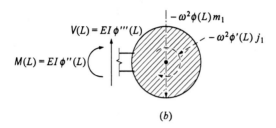

FIGURE E18-3
Beam with lumped mass at end: (a) beam properties; (b) forces acting on the end mass.

by fixed connection to its free end as shown in Fig. E18-3a. The boundary conditions at the beam's fixed end are the same as before; however, the moment and shear are no longer equal to zero at its other end due to the presence of the lumped mass. These internal force components are shown on the free-body diagram in Fig. E18-3b along with the translational and rotary inertial force components $m_1 \ddot{v}(L,t)$ and $j_1 \ddot{v}'(L,t)$, respectively. Noting that under free-vibration conditions, as given by Eqs. (18-3) and (18-9),

$$
\begin{aligned}
\ddot{v}(L,t) &= \phi(L)\ddot{Y}(t) = -\omega^2\,\phi(L)\,Y(t) \\
\ddot{v}'(L,t) &= \phi'(L)\ddot{Y}(t) = -\omega^2\,\phi'(x)\,Y(t)
\end{aligned}
\tag{a}
$$

force and moment equilibrium of the rigid mass requires that boundary conditions

$$
\begin{aligned}
EI\,\phi'''(L) &= -\omega^2\,\phi(L)\,m_1 \\
EI\,\phi''(L) &= \omega^2\,\phi'(L)\,j_1
\end{aligned}
\tag{b}
$$

be satisfied. Using these relations instead of the previous conditions $\phi''(L) = \phi'''(L) = 0$, the free-vibration analysis leading to mode shapes and frequencies can proceed exactly as outlined in Example E18-2.

Example   E18-4.   *Two-Member Frame*   To illustrate the free-vibration analysis procedure for multimember systems, consider the two-member frame

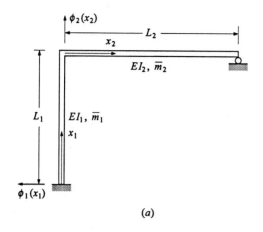

FIGURE E18-4
Simple two-member frame:
(a) arrangement of frame;
(b) forces acting on the beam.

shown in Fig. E18-4a. Each member has uniform properties as indicated; however, they may differ from one member to the other. Because of its fixed base, the vertical column must satisfy the two conditions: $\phi_1(0) = \phi_1'(0) = 0$. Neglecting axial distortions, the following three beam-support conditions must be satisfied for the upper horizontal member:

$$\phi_2(0) \qquad \phi_2(L_2) = 0 \qquad \phi_2''(L_2) = \frac{M(L_2)}{EI_2} = 0 \qquad \text{(a)}$$

Continuity of slope and moment equilibrium at the joint between the two members provides two additional conditions

$$\phi_1'(L_1) = \phi_2'(0) \qquad EI_1\,\phi_1''(L_1) = EI_2\,\phi_2''(0) \qquad \text{(b)}$$

Finally, equilibrium of the shear force at the top of the column with the inertial force developed in the upper member by sidesway motion Fig. E18-4b provides the eighth condition

$$EI_1\,\phi_1'''(L_1) + \bar{m}_2\,L_2\,\omega^2\,\phi_1(L_1) = 0 \qquad \text{(c)}$$

in which $\bar{m}_2$ represents the mass per unit length of the horizontal beam.

Expressing each of $\phi_1(x_1)$ and $\phi_2(x_2)$ in the form of Eq. (18-15), taking their partial derivatives as needed, and substituting the results into Eqs. (a), (b), and (c), one obtains eight equations containing the eight unknown mode-shape coefficients (four for each beam segment). Writing these equations in matrix form and setting the determinant of the resulting 8×8 square matrix to zero, one obtains the frequency equation of the system containing the single parameter a of Eq. (18-6). Finding the roots of this equation and substituting each one separately back into the matrix equation, any seven of the eight mode-shape coefficients for the corresponding mode can be evaluated in terms of the eighth. This final coefficient remains as an arbitrary measure of the mode amplitude. Finally, each root of the frequency equation is substituted separately into Eq. (18-8) to find the corresponding circular frequency.

The above example shows that a free-vibration analysis by the distributed-parameter procedure can lead to a sizable computational problem even for a simple two-member frame. While in the past it has been found useful to use such solutions for appropriate systems, discrete-parameter forms of solutions are now more convenient and more commonly used.

18-2 BEAM FLEXURE: INCLUDING AXIAL-FORCE EFFECTS

As was discussed previously in Chapter 8, axial forces acting in a flexural element may have a very significant influence on the vibration behavior of the member, resulting generally in modifications of both frequencies and mode shapes. When considering free vibrations of a prismatic member having uniform physical properties, the equation of motion, including the effect of a time-invariant uniform axial force throughout its length, is [from Eq. (17-10)]

$$EI\, \frac{\partial^4 v(x,t)}{\partial x^4} + N \frac{\partial^2 v(x,t)}{\partial x^2} + \overline{m}\, \frac{\partial^2 v(x,t)}{\partial t^2} = 0 \qquad (18\text{-}16)$$

Separating variables as before using the solution of Eq. (18-3) leads to

$$\frac{\phi^{iv}(x)}{\phi(x)} + \frac{N}{EI}\frac{\phi''(x)}{\phi(x)} = -\frac{\overline{m}}{EI}\frac{\ddot{Y}(t)}{Y(t)} = a^4 \qquad (18\text{-}17)$$

from which two independent equations are obtained as given by

$$\ddot{Y}(t) + \omega^2\, Y(t) = 0 \qquad (18\text{-}18a)$$

$$\phi^{iv}(x) + g^2\, \phi''(x) - a^4\, \phi(x) = 0 \qquad (18\text{-}18b)$$

in which ω^2 is again defined by Eq. (18-8) and g^2 is given by

$$g^2 \equiv \frac{N}{EI} \qquad (18\text{-}19)$$

Equation (18-18a) is the same time-dependent equation obtained before [Eq. (18-7a)], showing that a uniformly distributed axial force does not affect the simple-harmonic character of the free vibrations; however, it does affect the mode shapes and frequencies due to the presence of the term $g^2 \phi''(x)$ in Eq. (18-18b). This equation can be solved by the introduction of Eq. (18-11) giving

$$(s^4 + g^2 s^2 - a^4) G \exp(sx) = 0 \tag{18-20}$$

Cancelling the term $G \exp(sx)$ and solving the remaining equation for s yield the four roots

$$s_{1,2} = \pm i\delta \qquad\qquad s_{3,4} = \pm \epsilon \tag{18-21}$$

where

$$\delta \equiv \sqrt{\left(a^4 + \frac{g^4}{4}\right)^{1/2} + \frac{g^2}{2}} \qquad \epsilon \equiv \sqrt{\left(a^4 + \frac{g^4}{4}\right)^{1/2} - \frac{g^2}{2}} \tag{18-22}$$

Introducing separately each of the four roots given in Eq. (18-21) into Eq. (18-11), adding the resulting four terms to get the complete solution for $\phi(x)$ in exponential form, converting the exponential functions in terms of their trigonometric and hyperbolic equivalents, and setting the entire imaginary part of $\phi(x)$ equal to zero lead to

$$\phi(x) = D_1 \cos \delta x + D_2 \sin \delta x + D_3 \cosh \epsilon x + D_4 \sinh \epsilon x \tag{18-23}$$

This equation defines the shape of the vibrating beam segment for any value of axial force which might be specified. The coefficients D_1, D_2, D_3, and D_4 can be evaluated by exactly the same procedure presented in Section 18-1 for the system without axial force. In fact, it is evident that when the axial force P equals zero, so that $g = 0$, then $\delta = \epsilon = a$ in which case Eq. (18-23) is identical to Eq. (18-15).

Retaining the constant axial force N, Eq. (18-23) can be used to find the static buckling loads and corresponding shapes. For this nonvibrating case where $\omega = 0$ so that $a = 0$, $\delta = g$, and $\epsilon = 0$, the four roots of Eq. (18-21) are $s_{1,2} = \pm ig$ and $s_{3,4} = 0$ which lead to the complete solution

$$\phi(x) = D_1 \cos gx + D_2 \sin gx + D_3 x + D_4 \tag{18-24}$$

in which the last two terms correspond to the zero values of s_3 and s_4. Following the same procedure as for finding the frequency equation in the vibratory case, one obtains an equation containing the single unknown parameter g. The roots of this equation give the critical values of N, i.e., N_{cr}. In a static buckling-load analysis only the first buckling mode is important. The shape of this mode is obtained using the lowest critical value of N in exactly the same way the lowest value of the frequency parameter a was used in finding the first vibratory mode shape.

18-3   BEAM FLEXURE: WITH DISTRIBUTED ELASTIC SUPPORT

Consider the same uniform beam segment treated in Section 18-1, but in addition to having prescribed support conditions at its ends assume it to be supported transversely by distributed elastic springs of the type shown in Fig. 8-3d where $k(x) = \overline{k}$. The free-vibration equation of motion for this system is given by Eq. (18-1) with one term added to take care of the transverse force provided by the distributed elastic support, i.e., by

$$EI\,\frac{\partial^4\,v(x,t)}{\partial x^4} + \overline{m}\,\frac{\partial^2\,v(x,t)}{\partial t^2} + \overline{k}\,v(x,t) = 0 \tag{18-25}$$

Separating variables as before using the solution of Eq. (18-3) leads to

$$\frac{\phi^{iv}(x)}{\phi(x)} + \frac{\overline{k}}{EI} = -\frac{\overline{m}}{EI}\,\frac{\ddot{Y}(t)}{Y(t)} = a^4 \tag{18-26}$$

giving two independent equations

$$\ddot{Y}(t) + \omega^2\,Y(t) = 0 \tag{18-27a}$$

$$\phi^{iv}(x) - b^4\,\phi(x) = 0 \tag{18-27b}$$

where

$$\omega^2 \equiv \frac{a^4\,EI}{\overline{m}} \qquad b^4 \equiv a^4 - \frac{\overline{k}}{EI} \tag{18-28}$$

Since Eq. (18-27b) is in the identical form of Eq. (18-7b), it has the same type of solution, namely

$$\phi(x) = B_1\,\cos bx + B_2\,\sin bx + B_3\,\cosh bx + B_4\sinh bx \tag{18-29}$$

Following the procedure of Section 18-1, the same frequency equation would be obtained for a prescribed set of boundary conditions, but it would now contain the frequency parameter b, as defined by the second of Eqs. (18-28), rather than the parameter a. Therefore, the numerical values of b would be exactly the same as those obtained previously for parameter a. For example, considering the uniform cantilever beam of Fig. E18-3 having a uniform distributed elastic support over its entire length of spring constant $\overline{k}$ per unit length, its frequency equation would yield $(bL)_1 = 1.875$, $(bL)_2 = 4.694$, and $(bL)_3 = 7.855$ which are identical to the correpording values of aL in Example E18-2. The frequencies are now higher, however, in accordance with Eqs. (18-28), i.e.,

$$\omega_n = \left[(bL)^4_n\,\frac{EI}{\overline{m}L^4} + \frac{\overline{k}}{\overline{m}}\right]^{1/2} \tag{18-30}$$

Also, since Eqs. (18-15) and (18-29) are of identical form, the corresponding mode shapes are exactly the same.

18-4   BEAM FLEXURE: ORTHOGONALITY OF VIBRATION MODE SHAPES

The vibration mode shapes derived for beams with distributed properties have orthogonality relationships equivalent to those defined previously for the discrete-parameter systems, which can be demonstrated in essentially the same way — by application of Betti's law. Consider the beam shown in Fig. 18-1. For this discussion, the beam may have arbitrarily varying stiffness and mass along its length, and it could have arbitrary support conditions, although only simple supports are shown. Two different vibration modes, m and n, are shown for the beam. In each mode, the displaced shape and the inertial forces producing the displacements are indicated.

Betti's law applied to these two deflection patterns means that the work done by the inertial forces of mode n acting on the deflection of mode m is equal to the work of the forces of mode m acting on the displacement of mode n; that is,

$$\int_0^L v_m(x)\, f_{I_n}(x)\, dx = \int_0^L v_n(x)\, f_{I_m}(x)\, dx \qquad (18\text{-}31)$$

Expressing these in terms of the modal shape functions shown in Fig. 18-1 gives

$$Y_m(t)\, Y_n(t)\, \omega_n^2 \int_0^L \phi_m(x)\, m(x)\, \phi_n(x)\, dx$$

$$= Y_m(t)\, Y_n(t)\, \omega_m^2 \int_0^L \phi_n(x)\, m(x)\, \phi_m(x)\, dx \qquad (18\text{-}32)$$

which may be rewritten

$$(\omega_n^2 - \omega_m^2) \int_0^L \phi_m(x)\, \phi_n(x)\, m(x)\, dx = 0 \qquad (18\text{-}33)$$

Since the frequencies of these two modes are different, their mode shapes must satisfy

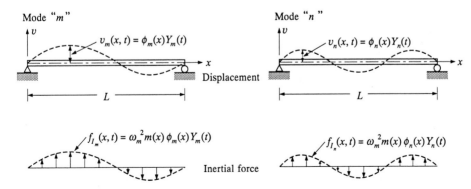

FIGURE 18-1

Two modes of vibration for the same beam.

the orthogonality condition

$$\int_0^L \phi_m(x)\,\phi_n(x)\,m(x)\,dx = 0 \qquad\qquad \omega_m \neq \omega_n \qquad\qquad (18\text{-}34)$$

which is clearly the distributed-parameter equivalent of the discrete-parameter orthogonality condition of Eq. (11-38a). If the two modes have the same frequency, the orthogonality condition does not apply, but this condition does not occur often in ordinary structural problems.

A second orthogonality condition, involving the stiffness property rather than the mass as a weighting parameter, can be derived for the distributed-parameter systems as it was earlier for the discrete-parameter case. For a nonuniform beam, the equation of motion in free vibrations [see Eq. (17-7)] is

$$\frac{\partial^2}{\partial x^2}\left[EI(x)\,\frac{\partial^2 v(x,t)}{\partial x^2}\right] + m(x)\,\frac{\partial^2 v(x,t)}{\partial t^2} = 0 \qquad\qquad (18\text{-}35)$$

In accordance with Eqs. (18-3) and (18-9), the harmonic motion in the nth mode can be written

$$v_n(x,t) = \phi_n(x)\,\rho_n\,\sin(\omega_n t + \phi_n) \qquad\qquad (18\text{-}36)$$

where $\rho_n = (A_n^2 + B_n^2)^{1/2}$ and ϕ_n is its phase angle. Substituting this expression into Eq. (18-35) and cancelling the common factor $\rho_n\,\sin(\omega_n t + \phi_n)$, one obtains

$$\omega_n^2\,m(x)\,\phi_n(x) = \frac{d^2}{dx^2}\left[EI(x)\,\frac{d^2\,\phi_n(x)}{dx^2}\right] \qquad\qquad (18\text{-}37)$$

Substituting this relation into both sides of Eq. (18-32) and cancelling the common term $Y_m(t)\,Y_n(t)$ give

$$(\omega_n^2 - \omega_m^2)\int_0^L \phi_m(x)\,\frac{d^2}{dx^2}\left[EI(x)\,\frac{d^2\,\phi_n(x)}{dx^2}\right]\,dx = 0 \qquad\qquad (18\text{-}38)$$

Since the frequencies are different, modes m and n must satisfy the orthogonality condition

$$\int_0^L \phi_m(x)\,\frac{d^2}{dx^2}\left[EI(x)\,\frac{d^2\,\phi_n(x)}{dx^2}\right]\,dx = 0 \qquad\qquad (18\text{-}39)$$

A more convenient symmetric form of this orthogonality relationship can be obtained by integrating twice by parts resulting in

$$\phi_m(x)\,V_n(x)\,\Big|_0^L - \phi_m'(x)\,M_n(x)\,\Big|_0^L + \int_0^L \phi_m''(x)\,\phi_n''(x)\,EI(x)\,dx = 0$$

$$\omega_m \neq \omega_n \qquad\qquad (18\text{-}40)$$

The first two terms in this equation represent the work done by the boundary vertical section forces of mode n acting on the end displacements of mode m and the work done by the end moments of mode n on the corresponding rotations of mode m. For the standard clamped-, hinged-, or free-end conditions, these terms will vanish. However, they contribute to the orthogonality relationship if the beam has elastic supports or if it has a lumped mass at its end; therefore they must be retained in the expression when considering such cases.

18-5 FREE VIBRATIONS IN AXIAL DEFORMATION

The analysis of free vibrations associated with axial motions of a one-dimensional member can be carried out in a manner similar to the case of flexural vibrations. Considering a prismatic member having uniform properties along its length, the free-vibration equation of motion is [see Eq. (17-29)]

$$EA \frac{\partial^2 u(x,t)}{\partial x^2} - \overline{m} \frac{\partial^2 u(x,t)}{\partial t^2} = 0 \tag{18-41}$$

Using the solution

$$u(x,t) = \overline{\phi}(x) Y(t) \tag{18-42}$$

and separating the variables, Eq. (18-41) can be written in the form

$$\frac{\overline{\phi}''(x)}{\overline{\phi}(x)} = \frac{\overline{m}}{EA} \frac{\ddot{Y}(t)}{Y(t)} = -c^2 \tag{18-43}$$

yielding two separate differential equations

$$\ddot{Y}(t) + \omega^2 Y(t) = 0 \tag{18-44a}$$

$$\overline{\phi}''(x) + c^2 \overline{\phi}(x) = 0 \tag{18-44b}$$

where

$$\omega^2 \equiv c^2 \frac{EA}{\overline{m}} \tag{18-45}$$

Equation (18-44a) is the same as Eq. (18-7a) and has the harmonic free-vibration solution shown by Eq. (18-9). Equation (18-44b) is of identically the same form as Eq. (18-44a) but with the independent variable being x rather than t. It therefore has the same type of solution as given by

$$\overline{\phi}(x) = C_1 \cos cx + C_2 \sin cx \tag{18-46}$$

in which the coefficients C_1 and C_2 determine the vibration mode shape. Considering the two known boundary conditions, one of these can be expressed in terms of the other and a frequency equation involving parameter c can be obtained.

Example   E18-5. *Cantilever Bar*   Consider the bar of Fig. E18-5a subjected to axial deformations under free-vibration conditions. The two boundary conditions to be satisfied are

$$\overline{\phi}(0) = 0 \qquad\qquad N(L) = EA\,\overline{\phi}'(L) = 0 \tag{a}$$

Substituting Eq. (18-46) into the first of these equation yields

$$C_1 \, \cos 0 + C_2 \, \sin 0 = 0 \tag{b}$$

showing that $C_1 = 0$. Taking the first derivative of Eq. (18-46) and substituting the result into the second of Eqs. (a) give

$$EA\,C_2\,c\,\cos cL = 0 \tag{c}$$

Excluding the trivial solution $C_2 = 0$, the frequency equation is seen to be

$$\cos cL = 0 \tag{d}$$

from which

$$c_n\,L = \frac{\pi}{2}\,(2n - 1) \tag{e}$$

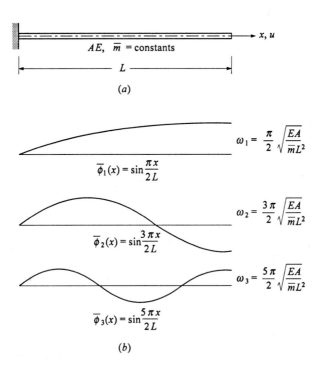

(a)

$$\overline{\phi}_1(x) = \sin\frac{\pi x}{2L} \qquad\qquad \omega_1 = \frac{\pi}{2}\sqrt{\frac{EA}{\overline{m}L^2}}$$

$$\overline{\phi}_2(x) = \sin\frac{3\pi x}{2L} \qquad\qquad \omega_2 = \frac{3\pi}{2}\sqrt{\frac{EA}{\overline{m}L^2}}$$

$$\overline{\phi}_3(x) = \sin\frac{5\pi x}{2L} \qquad\qquad \omega_3 = \frac{5\pi}{2}\sqrt{\frac{EA}{\overline{m}L^2}}$$

(b)

FIGURE E18-5
Axial vibrations of bar: (a) basic properties of cantilever bar; (b) first three vibration modes.

from which

$$c_n L = \frac{\pi}{2}(2n - 1) \tag{e}$$

The vibrating mode shapes of the rod are thus given by

$$\overline{\phi}_n = C_2 \sin\left[\frac{\pi}{2}(2n - 1)\frac{\pi}{L}\right] \qquad n = 1, 2, 3, \cdots \tag{f}$$

where C_2 is the arbitrary amplitude. Substituting Eq. (e) into Eq. (18-45), the corresponding circular frequencies are

$$\omega_n = c_n \sqrt{\frac{EA}{m}} = \frac{\pi}{2}(2n - 1)\sqrt{\frac{EA}{mL^2}} \qquad n = 1, 2, 3, \cdots \tag{g}$$

The first three mode shapes and corresponding frequencies are shown in Fig. E18-5b.

18-6 ORTHOGONALITY OF AXIAL VIBRATION MODES

The axial vibration mode shapes have orthogonality properties which are entirely equivalent to those demonstrated earlier for the flexural vibration modes. In fact, the orthogonality of the axial mode shapes with respect to the mass distribution can be derived using Betti's law in the same way as for the flexural modes with the equivalent result:

$$\int_0^L \overline{\phi}_m(x)\,\overline{\phi}_n(x)\,m(x)\,dx = 0 \tag{18-47}$$

The orthogonality relationship with respect to the axial stiffness property can be derived from the homogeneous form of the equation of motion [Eq. (17-29)] in which the harmonic time variation of free vibrations has been substituted. In other words, when the nth-mode displacements are expressed as

$$u_n(x, t) = \overline{\phi}_n(x)\,\rho_n\,\sin(\omega_n t + \phi_n) \tag{18-48}$$

and this displacement expression is substituted into the homogeneous form of Eq. (17-29), one obtains

$$\omega_n^2\,m(x)\,\overline{\phi}_n(x) = -\frac{d}{dx}\left[EA(x)\frac{d\overline{\phi}_n}{dx}\right] \tag{18-49}$$

Thus the inertial-force term in the orthogonality relationship of Eq. (18-47) can be replaced by the equivalent axial elastic-force term, with the result

$$\int_0^L \overline{\phi}_m(x)\frac{d}{dx}\left[EA(x)\frac{d\overline{\phi}_n(x)}{dx}\right] dx = 0 \tag{18-50}$$

Integrating by parts then leads to the more convenient symmetric form

$$\overline{\phi}_m(x)\,N_n(x)\,\bigg|_0^L - \int_0^L \overline{\phi}'_m(x)\,\overline{\phi}'_n(x)\,EA(x)\,dx = 0 \qquad \omega_m \neq \omega_n \tag{18-51}$$

The first term in this equation represents the work done by the boundary axial forces of mode n acting on the end displacements of mode m; this term will vanish if the bar has the standard free- or fixed-end conditions but may have to be included in more complex situations.

PROBLEMS

18-1. Evaluate the fundamental frequency for the cantilever beam with a mass at the end shown in Fig. E18-3, if the end lumped mass $m_1 = 2\overline{m}L$ and if its mass moment of inertia $j_1 = 0$. Plot the shape of this mode, evaluating at increments $L/5$ along the span.

18-2. Evaluate the fundamental frequency for the frame of Fig. E18-4 if the two members are identical, with properties L, EI, $\overline{m}$. Plot the shape of this mode, evaluating at increments $L/4$ along each span.

18-3. Evaluate the fundamental *flexural* frequency of the beam of Fig. P18-1 and plot its mode shape, evaluated at increments $L/5$ along its length. Note that the lowest frequency of this unstable structure is zero; the frequency of interest is the lowest nonzero value.

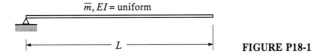

$\overline{m}, EI$ = uniform

L

FIGURE P18-1

18-4. The uniform beam of Fig. P18-2 is continuous over two spans as shown. Evaluate the fundamental flexural frequency of this structure and plot its mode shape at increments $L/2$ along the two spans.

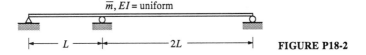

$\overline{m}, EI$ = uniform

L $2L$

FIGURE P18-2

18-5. A reinforced concrete beam having a cross section 8 *in* wide by 18 *in* deep is simply supported with a span of 28 *ft*. Assuming that the modulus of the material is 3×10^6 *lb/in²* and that its unit weight is 150 *lb/ft³*, evaluate the frequencies of its first five vibration modes neglecting shear distortion and rotary inertia.

18-6. Evaluate the fundamental frequency of *axial* vibration of the structure of Fig. E18-3 if the end lumped mass is $m_1 = 2\overline{m}L$ and if the cross-sectional area of the beam is A. Plot the shape of this mode, evaluating at increments $L/5$ along the span.

18-7. A column is assembled with two uniform bars, of the same length but having different properties, as shown in Fig. P18-3. For this structure:

(a) List the four boundary conditions required to evaluate the constants in deriving the axial vibration frequency equation.

(b) Write the transcendental axial frequency equation, and evaluate the first mode frequency and mode shape. Plot the mode shape evaluated at intervals $L/3$ along its length, normalized to unit amplitude at the free end.

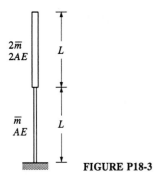

FIGURE P18-3

CHAPTER

19

ANALYSIS
OF DYNAMIC
RESPONSE

19-1 NORMAL COORDINATES

The mode-superposition analysis of a distributed-parameter system is entirely equivalent to that of a discrete-coordinate system once the mode shapes and frequencies have been determined, because in both cases the amplitudes of the modal-response components are used as generalized coordinates in defining the response of the structure. In principle an infinite number of these coordinates are available for a distributed-parameter system since it has an infinite number of modes of vibration, but in practice only those modal components need be considered which provide significant contributions to the response. Thus the problem is actually converted into a discrete-parameter form in which only a limited number of modal (normal) coordinates is used to describe the response.

The essential operation of the mode-superposition analysis is the transformation from the geometric displacement coordinates to the modal-amplitude or normal coordinates. For a one-dimensional system, this transformation is expressed as

$$v(x, t) = \sum_{i=1}^{\infty} \phi_i(x)\, Y_i(t) \tag{19-1}$$

which is simply a statement that any physically permissible displacement pattern can be made up by superposing appropriate amplitudes of the vibration mode shapes for

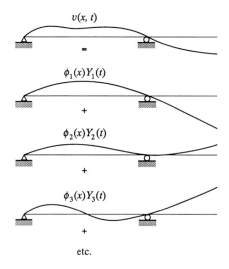

$v(x, t)$

$\phi_1(x)Y_1(t)$

$+$

$\phi_2(x)Y_2(t)$

$+$

$\phi_3(x)Y_3(t)$

$+$

etc.

FIGURE 19-1
Arbitrary beam displacements represented by
normal coordinates.

the structure. This principle is illustrated in Fig. 19-1, which shows an arbitrary displacement of a beam with an overhanging end developed as the sum of a set of modal components.

The modal components contained in any given shape, such as the top curve of Fig. 19-1, can be evaluated by applying the orthogonality conditions; usually it is most convenient to make use of the form involving the mass weighting parameter [Eq. (18-34)]. To evaluate the contribution of mode n in any arbitrary shape $v(x, t)$, Eq. (19-1) is multiplied by $\phi_n(x)\, m(x)$ on both sides and integrated, with the result

$$\int_0^L \phi_n(x)\, m(x)\, v(x,t)\, dx = \sum_{i=1}^{\infty} Y_i(t) \int_0^L \phi_i(x)\, m(x)\, \phi_n(x)\, dx$$

$$= Y_n(t) \int_0^L \phi_n(x)^2\, m(x)\, dx \qquad (19\text{-}2)$$

where only one term remains of the infinite series on the far right hand side by virtue of the orthogonality condition. Hence the expression can be solved directly for the one remaining amplitude term

$$Y_n(t) = \frac{\int_0^L \phi_n(x)\, m(x)\, v(x,t)\, dx}{\int_0^L \phi_n(x)^2\, m(x)\, dx} \qquad (19\text{-}3)$$

which is entirely equivalent to the discrete-parameter expression, Eq. (12-6). Given the initial beam displacement $v(x,0)$ and velocity $\dot{v}(x,0)$, the corresponding modal amplitude $Y_n(0)$ and its velocity $\dot{Y}_n(0)$ can be obtained directly from this equation and its velocity equivalent, so that the free-vibration response for each mode can be expressed by Eq. (18-10).

Example E19-1.   The uniform bar of length L shown in Fig. E19-1 is lifted from its right hand support as indicated and then dropped producing a rotation about its left hand pinned support. Assuming it rotates as a rigid body, the velocity distribution upon initial impact is

$$\dot{v}(x,0) = \frac{x}{L}\,\dot{v}_t \qquad\qquad (a)$$

where $\dot{v}_t$ represents the tip velocity. The displacement at the same time is $v(x,0) = 0$, corresponding to the rigid-body rotation concept.

The nth vibration mode shape for this simple beam is given by

$$\phi_n(x) = \sin\frac{n\,\pi\,x}{L} \qquad\qquad (b)$$

Hence the denominator integral of Eq. (19-3) that defines the modal amplitude is

$$\int_0^L \phi_n(x)^2\,m(x)\,dx = m\int_0^L \sin^2\left(\frac{n\,\pi\,x}{L}\right)dx = \frac{m\,L}{2} \qquad\qquad (c)$$

The numerator integral of Eq. (19-3) defining the modal amplitude at $t = 0$ obviously is zero because $v(x,0)$ is zero, hence $Y_n(0) = 0$ for all modes. However, taking the first time derivative of the equation, the numerator integral for the modal velocity is

$$\int_0^L \phi_n(x)\,m(x)\,\dot{v}(x,0)\,dx = m\,\dot{v}_t\int_0^L \frac{x}{L}\sin\frac{n\,\pi\,x}{L}\,dx$$

$$= \pm\frac{m\,L}{n\,\pi}\,\dot{v}_t \quad \begin{cases} + & n = \text{odd no.} \\ - & n = \text{even no.} \end{cases} \qquad\qquad (d)$$

Combining the numerator and denominator integrals, one obtains the initial normal-coordinate velocity

$$\dot{Y}_n(0) = \pm\frac{2\,\dot{v}_t}{n\,\pi} \qquad\qquad (e)$$

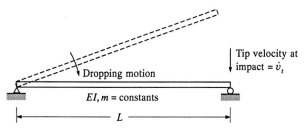

Dropping motion

$EI, m = $ constants

L

Tip velocity at impact $= \dot{v}_t$

FIGURE E19-1
Example of free-vibration amplitude analysis.

Making use of Eq. (18-9), the modal vibration is given by

$$Y_n(t) = \pm \frac{2\,\dot{v}_t}{n\,\pi\,\omega_n} \sin\omega_n t \qquad\qquad (f)$$

which when introduced into Eq. (19-1) yields

$$v(x,t) = \sum_{n=1}^{\infty} \phi_n(x) \left(\pm \frac{2\,\dot{v}_t}{n\,\pi\,\omega_n} \sin\omega_n t \right)$$

$$= \frac{2\,\dot{v}_t}{\pi} \left(\frac{1}{\omega_1} \sin\frac{\pi x}{L} \sin\omega_1 t - \frac{1}{2\omega_2} \sin\frac{2\pi x}{L} \sin\omega_2 t + \cdots \right) \quad (g)$$

Note that this analysis assumes that the right hand end of the beam is held in contact with the support so that $v(L,t) = 0$ at all times after the impact.

19-2   UNCOUPLED FLEXURAL EQUATIONS OF MOTION: UNDAMPED CASE

The two orthogonality conditions [Eqs. (18-34) and (18-39) or (18-40)] provide the means for decoupling the equations of motion for the distributed-parameter system in the same way that decoupling was accomplished for the discrete-parameter system. Introducing Eq. (19-1) into the equation of motion [Eq. (17-7)]

$$\frac{\partial^2}{\partial x^2} \left[EI(x)\, \frac{\partial^2 v(x,t)}{\partial x^2} \right] + m(x)\, \frac{\partial^2 v(x,t)}{\partial t^2} = p(x,t) \qquad (19\text{-}4)$$

this normal-coordinate expression leads to

$$\sum_{i=1}^{\infty} m(x)\, \phi_i(x)\, \ddot{Y}_i(t) + \sum_{i=1}^{\infty} \frac{d^2}{dx^2}\left[EI(x)\, \frac{d^2\phi_i(x)}{dx^2} \right] Y_i(t) = p(x,t) \qquad (19\text{-}5)$$

Multiplying each term by $\phi_n(x)$ and integrating gives

$$\sum_{i=1}^{\infty} \ddot{Y}_i(t) \int_0^L m(x)\, \phi_i(x)\, \phi_n(x)\, dx + \sum_{i=1}^{\infty} Y_i(t) \int_0^L \phi_n(x) \frac{d^2}{dx^2}\left[EI(x) \frac{d^2\phi_i(x)}{dx^2} \right] dx$$

$$= \int_0^L \phi_n(x)\, p(x,t)\, dx \qquad (19\text{-}6)$$

When the two orthogonality relationships are applied to the first two terms, it is evident that all terms in the series expansions, except the nth, vanish; thus,

$$\ddot{Y}_n(t) \int_0^L m(x)\, \phi_n(x)^2\, dx + Y_n(t) \int_0^L \phi_n(x) \frac{d^2}{dx^2}\left[EI(x) \frac{d^2\phi_n(x)}{dx^2} \right] dx$$

$$= \int_0^L \phi_n(x)\, p(x,t)\, dx \qquad (19\text{-}7)$$

Multiplying Eq. (18-37) by $\phi_n(x)$ and integrating yields

$$\int_0^L \phi_n(x) \frac{d^2}{dx^2} \left[EI(x) \frac{d^2\phi_n(x)}{dx^2} \right] dx = \omega_n^2 \int_0^L \phi_n(x)^2 \, m(x) \, dx \qquad (19\text{-}8)$$

Recognizing that the integral on the right hand side of this equation is the generalized mass of the nth mode [Eqs. (8-14)]

$$M_n = \int_0^L \phi_n(x)^2 \, m(x) \, dx \qquad (19\text{-}9)$$

Eq. (19-8) shows that the second term of Eq. (19-7) is $\omega_n^2 \, M_n Y_n(t)$; therefore, the latter equation can be written in the form

$$M_n \ddot{Y}_n(t) + \omega_n^2 \, M_n \, Y_n(t) = P_n(t) \qquad (19\text{-}10)$$

where

$$P_n(t) = \int_0^L \phi_n(x) \, p(x,t) \, dx \qquad (19\text{-}11)$$

is the generalized loading associated with mode shape $\phi_n(x)$.

An equation of the type of Eq. (19-10) can be established for each vibration mode of the structure, using Eqs. (19-9) and (19-11) to evaluate its generalized mass and loading, respectively. It should be noted that these expressions are the distributed-parameter equivalents of the matrix expressions previously derived for the discrete-parameter systems. Also, it should be noted that they are applicable to beams having nonuniform properties, if their mode shapes can be defined.

Example E19-2. To illustrate the above mode-superposition analysis procedure, the dynamic response of a uniform simple beam subjected to a central step-function loading as shown in Fig. E19-2 will be evaluated.

Determine Mode Shapes and Frequencies: This vibration analysis is accomplished by substituting into the controlling set of boundary-condition equations the modal-shape expression [Eq. (18-15)]. For this simple beam,

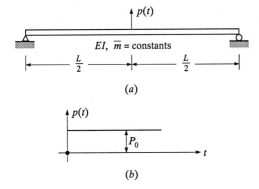

(a)

(b)

FIGURE E19-2
Example of dynamic-response analysis:
(a) arrangement of beam and load;
(b) applied step-function loading.

the results were found in Example E18-1 to be

$$\phi_n(x) = \sin \frac{n \pi x}{L} \qquad n = 1, 2, \cdots \qquad \text{(a)}$$

$$\omega_n = n^2 \pi^2 \sqrt{\frac{EI}{\overline{m} L^4}} \qquad n = 1, 2, \cdots \qquad \text{(b)}$$

Compute Generalized Mass and Loading: From Eqs. (19-9) and (19-11), these quantities are found to be

$$M_n = \int_0^L \phi_n(x)^2 \, m(x) \, dx = \overline{m} \int_0^L \sin^2 \left(\frac{n \pi x}{L} \right) dx = \frac{\overline{m} L}{2} \qquad \text{(c)}$$

$$P_n(t) = \int_0^L \phi_n(x) \, p(x,t) \, dx = P_0 \, \phi_n \left(\frac{L}{2} \right) = \alpha_n \, P_0 \qquad \text{(d)}$$

where

$$\alpha_n = \begin{cases} 1 & n = 1, 5, 9, \cdots \\ -1 & n = 3, 7, 11, \cdots \\ 0 & n = \text{even no.} \end{cases}$$

Solve the Normal-Coordinate Response Equation: This is exactly the same equation considered for the discrete-parameter case, i.e.,

$$M_n \, \ddot{Y}_n(t) + \omega_n^2 \, M_n \, Y_n(t) = P_n(t) \qquad \text{(e)}$$

The Duhamel solution of this equation is

$$Y_n(t) = \frac{1}{M_n \, \omega_n} \int_0^t P_n(\tau) \, \sin \omega_n \, (t - \tau) \, d\tau$$

$$= \frac{2 \alpha_n P_0}{\overline{m} L \omega_n} \int_0^t \sin \omega_n \, (t - \tau) \, d\tau = \frac{2 \alpha_n P_0}{\overline{m} L \omega_n^2} \, (1 - \cos \omega_n t) \qquad \text{(f)}$$

Evaluate Displacement, Moment and Shear Response: Substituting Eq. (f) into the normal-coordinate expression, Eq. (19-1), and letting $\omega_n^2 = n^4 \pi^4 EI / \overline{m} L^4$, one obtains

$$v(x,t) = \sum_{n=1}^{\infty} \phi_n(x) \, Y_n(t) = \frac{2 P_0 L^3}{\pi^4 EI} \sum_{n=1}^{\infty} \frac{\alpha_n}{n^4} \, (1 - \cos \omega_n t) \sin \frac{n \pi x}{L} \qquad \text{(g)}$$

which can be introduced into

$$M(x,t) = EI \frac{\partial^2 v(x,t)}{\partial x^2} \qquad V(x,t) = EI \frac{\partial^3 v(x,t)}{\partial x^2} \qquad \text{(h)}$$

giving

$$M(x,t) = -\frac{2 P_0 L}{\pi^2} \sum_{n=1}^{\infty} \frac{\alpha_n}{n^2} (1 - \cos \omega_n t) \sin \frac{n \pi x}{L} \qquad \text{(i)}$$

$$V(x,t) = -\frac{2 P_0}{\pi} \sum_{n=1}^{\infty} \frac{\alpha_n}{n} (1 - \cos \omega_n t) \cos \frac{n \pi x}{L} \qquad \text{(j)}$$

Note that the higher modes contribute an insignificant amount to displacement due to the position of n^4 in Eq. (g); however, their contributions become more significant for the moment response and even more significant for shear. In other words, the series in Eq. (j) converges much more slowly with mode number n than does the series in Eq. (i), which in turn converges much more slowly than the series in Eq. (g). Therefore, one should be careful when limiting the number of lower modes in estimating response because proper selection of that number depends upon the response quantity being evaluated.

When the dynamic response of a one-dimensional distributed-parameter system is caused by support motions, the effective loading on the structure is given by Eq. (17-21). The corresponding normal-coordinate load term resulting from each support acceleration $\ddot{\delta}_i(t)$ is therefore

$$P_{ni}(t) = \int_0^L \phi_n(x) \, p_{\text{eff},i}(x,t) \, dx = -\ddot{\delta}_i(t) \int_0^L m(x) \, \phi_n(x) \, \psi_i(x) \, dx \qquad \text{(19-12)}$$

and the total loading is the sum of the contributions from all support accelerations.

19-3 UNCOUPLED FLEXURAL EQUATIONS OF MOTION: DAMPED CASE

To determine the effect of the normal-coordinate transformation [Eq. (19-1)] on the damped equation of motion, substitute Eq. (19-1) into Eq. (17-13) to get

$$\sum_{i=1}^{\infty} m(x) \, \phi_i(x) \, \ddot{Y}_i(t) + \sum_{i=1}^{\infty} c(x) \, \phi_i(x) \, \dot{Y}_i(t) + \sum_{i=1}^{\infty} \frac{d^2}{dx^2} \left[a_1 \, EI(x) \, \frac{d^2 \phi_i(x)}{dx^2} \right] \dot{Y}_i(t)$$

$$+ \sum_{i=1}^{\infty} \frac{d^2}{dx^2} \left[EI(x) \, \frac{d^2 \phi_i(x)}{dx^2} \right] Y_i(t) = p(x,t) \qquad \text{(19-13)}$$

Multiplying by $\phi_n(x)$, integrating, and applying the two orthogonality relationships together with the definitions of generalized mass and generalized loading leads to

$$M_n \, \ddot{Y}_n(t) + \sum_{i=1}^{\infty} \dot{Y}_i(t) \int_0^L \phi_n(x) \left\{ c(x) \, \phi_i(x) + \frac{d^2}{dx^2} \left[a_1 \, EI(x) \, \frac{d^2 \phi_i(x)}{dx^2} \right] \right\} dx$$

$$+ \omega_n^2 \, M_n \, Y_n(t) = P_n(t) \qquad \text{(19-14)}$$

Because of the stiffness orthogonality condition [Eq. (18-39)], all terms in the series involving constant a_1 [having dimension of time as defined through Eq. (8-8)] will be zero, except for the term $i = n$. With only this term remaining, the modes are obviously uncoupled as far as the stiffness-proportional damping is concerned. Coupling will be present, however, due to $c(x)$, unless it takes on a form allowing only the term $i = n$ to remain in the first series of Eq. (19-14). This is indeed the case for mass-proportional damping; that is, if one lets

$$c(x) = a_0 \, m(x) \qquad (19\text{-}15)$$

in which the proportionality constant a_0 has the dimension time^{-1} and is the same factor defined in Eq. (12-37a). Substituting this relation into Eq. (19-14) and making use of of the mass and stiffness orthogonality conditions [Eqs. (18-34) and (18-39)], one obtains the uncoupled modal equation

$$M_n \, \ddot{Y}_n(t) + \left(a_0 \, M_n + a_1 \, \omega_n^2 \, M_n \right) \dot{Y}_n(t) + \omega_n^2 \, M_n \, Y_n(t) = P_n(t) \qquad (19\text{-}16)$$

Finally, introducing the damping ratio for the nth mode [see Eq. (12-38b)]

$$\xi_n = \frac{C_n}{2 \, M_n \, \omega_n} = \frac{a_0}{2 \, \omega_n} + \frac{a_1 \, \omega_n}{2} \qquad (19\text{-}17)$$

and dividing through by the generalized mass, Eq. (19-16) becomes the standard SDOF equation

$$\ddot{Y}_n(t) + 2 \, \xi_n \, \omega_n \, \dot{Y}_n(t) + \omega_n^2 \, Y_n(t) = \frac{P_n(t)}{M_n} \qquad n = 1, 2, \cdots \qquad (19\text{-}18)$$

Thus, it is clear that when the viscous damping is of the Rayleigh mass- and stiffness-proportional type, the distributed-parameter equations of motion can be uncoupled in the same way as for the discrete-parameter systems. From Eq. (19-17), it is seen that for mass-proportional damping, the damping ratio is inversely proportional to the frequency; while for stiffness-proportional damping, it is directly proportional. This is the same result presented earlier in Eq. (12-38b) for discrete-parameter systems.

Note that, similar to the discrete-parameter case, one can evaluate coefficients a_0 and a_1 by assigning appropriate numerical values for ξ_n for two values of n, say $n = 1$ and $n = m$, using the corresponding known values of ω_n, and then solving the resulting two simultaneous equations given by Eq. (19-17). While this procedure leads to values of a_0 and a_1 which give the required damping in the two selected modes, it provides different damping in the other modes in accordance with Eq. (19-17) and as shown by Fig. 12-2. While this observation is of instructional value, it is not too significant in a practical sense as Eqs. (19-18) are usually solved after assigning numerical values to all damping ratios consistent with experimental information and judgment.

Example E19-3. To further illustrate the mode-superposition analysis procedure, consider the simple beam shown in Fig. E19-2 when excited by a harmonic vertical displacement of its right hand support as given by

$$\delta_2 = \bar{\delta}_2 \, \sin \bar{\omega} t \tag{a}$$

where $\bar{\delta}_2$ is the single amplitude of the support motion. The total displacement of the beam from its original position can be expressed in the form

$$v^t(x,t) = v(x,t) + \frac{x}{L} \, \delta_2(t) \tag{b}$$

in which $v(x,t)$ is given by Eq. (19-1). The mode shapes, frequencies, and generalized masses to be used in this formulation are identical to those found previously for the simple beam, i.e., $\phi_n(x) = \sin(n\pi x/L)$, $\omega_n = n^2 \, \pi^2 \, \sqrt{EI/m L^4}$, and $M_n = m L/2$.

The viscously-damped normal-coordinate equations of motion are

$$\ddot{Y}_n(t) + 2 \, \xi_n \, \omega_n \, \dot{Y}_n(t) + \omega_n^2 \, Y_n(t) = \frac{P_n(t)}{M_n} \qquad n = 1, 2, \cdots \tag{c}$$

where $P_n(t)$ is obtained from Eq. (19-12) upon substitution of $\psi_2(x) = \frac{x}{L}$, thus giving

$$P_n(t) = -\ddot{\delta}_2(t) \int_0^L m(x) \, \phi_n(x) \, \psi_2(x) \, dx$$

$$= -\bar{\omega}^2 \, \bar{\delta}_2 \, \sin \bar{\omega} t \int_0^L \frac{m \, x}{L} \, \sin \frac{n \pi x}{L} \, dx \tag{d}$$

or

$$P_n(t) = \pm \frac{m \, L \bar{\omega}^2 \, \bar{\delta}_2}{n \, \pi} \, \sin \bar{\omega} t \qquad \begin{cases} + \quad n = \text{odd} \\ - \quad n = \text{even} \end{cases} \tag{e}$$

The steady-state solution of Eq. (c) takes the form of Eq. (4-20) which, upon substitution of

$$\frac{P_{0n}}{K_n} = \frac{P_{0n}}{\omega_n^2 \, M_n}$$

$$= \pm \left(\frac{m \, L \bar{\omega}^2 \, \bar{\delta}_2}{n \, \pi} \right) \left(\frac{m \, L^4}{n^4 \, \pi^4 \, EI} \right) \left(\frac{2}{m \, L} \right) = \pm \frac{2 \, m \, L^4 \bar{\omega}^2 \, \bar{\delta}_2}{n^5 \, \pi^5 \, EI} \tag{f}$$

for p_0/k, $\beta_n \equiv \bar{\omega}/\omega_n$ for β, and ξ_n for ξ, becomes

$$Y_n(t) = \pm \frac{2 \, m \, L^4 \bar{\omega}^2 \, \bar{\delta}_2}{n^5 \, \pi^5 \, EI} \times \left[\frac{1}{(1 - \beta_n^2)^2 + (2 \, \xi_n \, \beta_n)^2} \right]$$

$$\times \left[(1 - \beta_n^2) \, \sin \bar{\omega} t - 2 \, \xi_n \, \beta_n \, \cos \bar{\omega} t \right] \tag{g}$$

Making use of Eq. (19-1), the displacement $v(x, t)$ becomes

$$v(x, t) = \frac{2 \, m \, L^4 \, \bar{\omega}^2 \, \bar{\delta}_2}{\pi^5 \, EI}$$

$$\times \sum_{n=1}^{\infty} \pm \frac{1}{n^5} \left[\frac{1}{(1 - \beta_n^2)^2 + (2 \, \xi_n \, \beta_n)^2} \right]$$

$$\times \left[(1 - \beta_n^2) \, \sin \bar{\omega} t - 2 \xi_n \beta_n \, \cos \bar{\omega} t \right] \sin \frac{n \pi x}{L}$$

$$\begin{cases} + & n = \text{odd} \\ - & n = \text{even} \end{cases} \qquad \text{(h)}$$

Finally the internal moment and shear distribution expressions can be obtained therefrom using

$$M(x, t) = EI \, \frac{\partial^2 v(x, t)}{\partial x^2} \qquad V(x, t) = EI \, \frac{\partial^3 v(x, t)}{\partial x^3} \qquad \text{(i)}$$

After assigning numerical values to the damping ratios of a limited number of the lower modes, the distributed response quantity of interest can be evaluated. One must be careful in selecting the number of modes required for engineering accuracy in the solution as it depends upon the response quantity of interest. As in Example E19-2, the series in the expression for $v(x, t)$ converges much faster than does the series in $M(x, t)$, which in turn converges much faster than the series in $V(x, t)$.

The solutions of Eqs. (19-18) for the case of general excitation, either direct loading or support motion, can be obtained either through the time domain or the frequency domain by the procedures given in Chapter 7. Since the linear viscous damping used in these equations results in the energy absorption per cycle (at fixed response amplitude) being dependent upon the response frequency $\bar{\omega}$ as discussed in Section 4-7, one may find it more appropriate to use the complex-stiffness form of hysteretic damping; in this case, the uncoupled normal mode equations of motion in the frequency domain are

$$\left[(\omega_n^2 - \bar{\omega}^2) + 2i \, \xi_n \, \omega_n^2 \right] Y_n(i\bar{\omega}) = \frac{P_n(i\bar{\omega})}{M_n} \qquad n = 1, 2, \cdots \qquad \text{(19-19)}$$

After assigning numerical values to the damping ratios, these equations can be solved in the frequency domain using the FFT analysis procedure. The energy absorption per cycle at fixed amplitude is now independent of the response frequency $\bar{\omega}$.

19-4 UNCOUPLED AXIAL EQUATIONS OF MOTION: UNDAMPED CASE

The mode-shape (normal) coordinate transformation serves to uncouple the equations of motion of any dynamic system and therefore is applicable to the axial as well as the flexural equations of motion of a one-dimensional member. Introducing Eq. (19-1) into the equation of axial motion, Eq. (17-29), leads to

$$\sum_{i=1}^{\infty} m(x)\,\overline{\phi}(x)\,\ddot{Y}_i(t) - \sum_{i=1}^{\infty} \frac{d}{dx}\left[EA(x)\,\frac{d\overline{\phi}_i(x)}{dx}\right] Y_i(t) = q(x,t) \qquad (19\text{-}20)$$

Multiplying each term by $\overline{\phi}_n(x)$ and applying the orthogonality relationships [Eqs. (18-47) and (18-50)] leads to

$$\ddot{Y}_n(t) \int_0^L m(x)\,\overline{\phi}_n(x)^2\,dx - Y_n(t) \int_0^L \overline{\phi}_n(x)\,\frac{d}{dx}\left[EA(x)\,\frac{d\overline{\phi}_n(x)}{dx}\right] dx$$
$$= \int_0^L \overline{\phi}_n(x)\,q(x,t)\,dx \qquad (19\text{-}21)$$

Substituting the inertial force for the elastic-force term [from Eq. (18-49)] and introducing the standard expressions for generalized mass and load

$$M_n = \int_0^L m(x)\,\overline{\phi}_n^2(x)\,dx \qquad (19\text{-}22)$$

$$P_n = \int_0^L \overline{\phi}_n(x)\,q(x,t)\,dx \qquad (19\text{-}23)$$

results in the final uncoupled axial equation of motion

$$M_n\,\ddot{Y}_n(t) + \omega_n^2\,M_n\,Y_n(t) = P_n(t) \qquad (19\text{-}24)$$

which is exactly the same as the uncoupled equation for flexural motion [Eq. (19-10)]. From this discussion it is apparent that after the vibration mode shapes have been determined, the reduction to the normal-coordinate form involves exactly the same type of operations for all structures.

Example E19-4. Because the dynamic response of a prismatic bar to axial loading has special characteristics which will be the subject of later discussion, it will be instructive to perform an example analysis of this type. Consider a pile fixed rigidly at its base and subjected to a step-function compression loading P_0 at the upper end, as shown in Fig. E19-3. The mode-superposition analysis

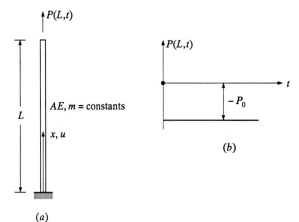

FIGURE E19-3
Pile subjected to end loading:
(*a*) geometric configuration;
(*b*) step-function loading.

of this system can be carried out by the same sequence of steps employed in the beam-response example of Fig. E19-2.

Determine Mode Shapes and Frequencies (see Example E18-5):

$$\overline{\phi}_n(x) = \sin\left[(2n-1)\frac{\pi x}{2L}\right] \qquad n = 1, 2, \cdots \tag{a}$$

$$\omega_n = (2n-1)\frac{\pi}{2}\sqrt{\frac{EA}{mL^2}} \qquad n = 1, 2, \cdots \tag{b}$$

Compute Generalized Mass and Loading:

$$M_n = \int_0^L m(x)\,\overline{\phi}_n(x)^2\,dx = m\int_0^L \sin^2\left[(2n-1)\frac{\pi x}{2L}\right]\,dx = \frac{mL}{2} \tag{c}$$

$$P_n = \int_0^L q(x,t)\,\overline{\phi}_n(x)\,dx = -P_0\,\overline{\phi}_n(L) = \pm P_0 \quad \begin{cases} + & n = \text{even no.} \\ - & n = \text{odd no.} \end{cases} \tag{d}$$

Solve the Generalized-Coordinate Response (see Example E19-2):

$$Y_n(t) = \pm\frac{2P_0}{mL\omega_n^2}\,(1 - \cos\omega_n t) \tag{e}$$

Evaluate Displacement and Axial-Force Response:

$$u(x,t) = \sum_{n=1}^{\infty} \overline{\phi}_n(x)\,Y_n(t)$$

$$= \frac{2P_0}{mL\omega_1^2}\left[-\left(\frac{1-\cos\omega_1 t}{1}\right)\sin\frac{\pi x}{2L}\right.$$

$$+ \left(\frac{1 - \cos \omega_2 t}{9} \right) \sin \frac{3\pi x}{2L} - \left(\frac{1 - \cos \omega_3 t}{25} \right) \sin \frac{5\pi x}{2L} + \cdots \Bigg]$$

$$= \frac{8 P_0}{\pi^2} \frac{L}{EA} \sum_{n=1}^{\infty} \left\{ \pm \left(\frac{1 - \cos \omega_n t}{(2n-1)^2} \right) \sin \left[\frac{(2n-1)}{2} \frac{\pi x}{L} \right] \right\} \qquad (f)$$

$$N(x,t) = EA \frac{\partial u(x,t)}{\partial x}$$

$$= \frac{8 P_0 L}{\pi^2} \sum_{n=1}^{\infty} \left\{ \pm \left(\frac{1 - \cos \omega_n t}{(2n-1)^2} \right) \left[\frac{(2n-1)}{2} \frac{\pi}{L} \right] \cos \left[\frac{(2n-1)}{2} \frac{\pi x}{L} \right] \right\}$$

$$= \frac{4 P_0}{\pi} \sum_{n=1}^{\infty} \left\{ \pm \left(\frac{1 - \cos \omega_n t}{2n-1} \right) \cos \left[\frac{(2n-1)}{2} \frac{\pi x}{L} \right] \right\} \qquad (g)$$

The response at any time t can be obtained by summing terms in the series expressions (f) and (g) representing the displacement and force distributions. For this purpose, it is convenient to express the time-variation parameter $\omega_n t$ in the form

$$\omega_n t = \left(\frac{2n-1}{2} \pi \right) \frac{ct}{L}$$

where $c = \sqrt{EA/\overline{m}}$ has the dimensions of velocity. Thus the product ct becomes a distance, and the time parameter may be considered as the ratio of this distance to the length of the pile. The displacement and force distribution in the pile at four different values of this time parameter have been obtained by evaluating the series expressions; the results obtained by summing the series are plotted in Fig. E19-4. The simple form of the response produced by the step-function loading is evident in these sketches. For any time $t_1 < L/c$, the pile has no load ahead of the distance ct_1 but is subject to the constant force P_0 behind this distance. Thus the response may be interpreted as a force wave of amplitude P_0 propagating ahead with the velocity c. The displacement is consistent with this load distribution, of course, showing a linear variation in the section of the pile in which there is constant load and no displacement in the zone ahead of the force wave. In the time interval $L/c < t_2 < 2L/c$ the force wave is doubled in the zone from the rigid support to a point $ct_2 - L$ from this support. This response behavior may be interpreted as a reflection of the force wave, resulting in a double amplitude as it propagates back along the pile. In the time interval $2L/c < t_3 < 3L/c$ the response may be interpreted as a negative reflection from the free end of the pile, causing a reduction of the force amplitude which propagates with the velocity c. During the fourth phase,

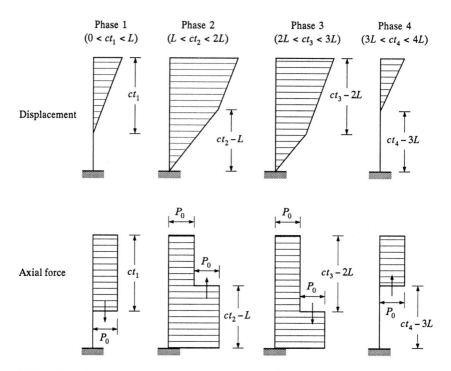

FIGURE E19-4
Response of pile to step-function loading.

$3L/c < t_4 < 4L/c$, the negative wave of phase 3 is reflected from the rigid base and causes a reduction of the axial force to zero value. At the end of the fourth phase, at time $t = 4L/c$, the pile is completely unstressed, as it was at time $t = 0$; a negative reflection of the negative wave then initiates a positive wave propagating down the pile in a form exactly equivalent to phase 1.

From the preceding discussion it is apparent that the free vibrations of the pile subjected to the step-function loading can be interpreted as an axial-force wave propagating along the pile and being subjected to positive and negative reflections at the fixed and free ends. This wave oscillation will continue indefinitely in the absence of damping or of any change in the loading. It is important to note that this response was evaluated in this example by the superposition of the axial vibration modes of the pile, each of which involves the entire extent of the pile. For example, in phase 1 the unstressed zone in the pile ahead of the advancing wave was obtained by the superposition of an infinite number of modes, each of which included stresses ahead of the wave-front. It is evident that the mode-superposition method is a rather cumbersome way to represent the very simple wave-propagation concept, and a more direct analysis is presented in the next section of this chapter. The fact that the mode-superposition process does account for the wave-propagation mechanism is most significant, however; this type of analysis provides the complete solution for any structure subjected to any type of dynamic loading.

19-5 WAVE-PROPAGATION ANALYSIS

Basic Axial-Wave-Propagation Equation

It was pointed out in the preceding section that the dynamic response of a uniform bar to a suddenly applied axial load can be interpreted as the propagation of a stress wave (and its associated deformation wave) along its length. This wave-propagation result has many practical applications in areas as diverse as earthquake engineering (as described in Chapters 25 and 26) and pile driving, as described later in this section. The analytical result may be obtained directly by using a different form of solution of the equation of axial motion—one not based on separation of variables as was used in the mode-superposition analysis presented before.

For this derivation, the equation of motion [Eq. (18-41)] is written

$$\ddot{u}(x,t) = V_p^2\, u''(x,t) = 0 \tag{19-25}$$

in which

$$V_p = \sqrt{\frac{EA}{m}} = \sqrt{\frac{E}{\rho}} \tag{19-26}$$

has the dimensions of velocity and where ρ is the mass density. It can be shown by simple substitution that

$$u(x,t) = f_1(x - V_p t) + f_2(x + V_p t) \tag{19-27}$$

is a solution of Eq. (19-25), f_1 and f_2 being arbitrary functional relationships of the parameters $x - V_p t$ and $x + V_p t$, respectively. This expression represents a pair of displacement waves propagating in the positive and negative directions, along the axis of the bar, as shown in Fig. 19-2. The instant of time represented in this figure

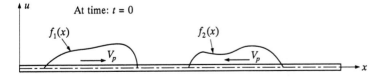

FIGURE 19-2
Axial displacement waves propagating along bar.

has been taken arbitrarily to be $t = 0$, so that the two waves are shown as specified functions of position only. The specific waveshapes f_1 and f_2 shown might be the result of specified displacement or force conditions applied earlier at the two ends of the bar.

The nature of the wave-propagation mechanism can easily be understood by considering the forward-propagating wave at two instants of time, $t = 0$ and $t = \Delta t$, as shown in Fig. 19-3. If a new position variable $x' = x - V_p \Delta t$ is considered, then $f_1(x - V_p \Delta t) \equiv f_1(x')$ and the shape of the wave relative to the variable x' in Fig. 19-3b is the same as the shape relative to x in Fig. 19-3a. Thus the wave has merely advanced a distance $V_p \Delta t$ during the time Δt, with no change of shape; the velocity of this wave propagation is V_p. By similar reasoning, it can be shown that the second term in Eq. (19-27) represents a waveform f_2 moving in the negative x direction.

The dynamic behavior of the bar can also be expressed in terms of its stress distribution rather than its displacements. With $\sigma = E\varepsilon$ and $\varepsilon = \partial u/\partial x$, the stress wave is given by

$$\sigma(x,t) = E \frac{\partial u}{\partial x} = E\frac{\partial f_1}{\partial x}(x - V_p t) + E\frac{\partial f_2}{\partial x}(x + V_p t) \qquad (19\text{-}28)$$

When the stress wave functions $E\,\partial f_1/\partial x$ and $E\,\partial f_2/\partial x$ are designated by g_1 and g_2, this may be written

$$\sigma(x,t) = g_1(x - V_p t) + g_2(x + V_p t) \qquad (19\text{-}29)$$

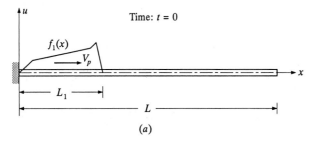

(a)

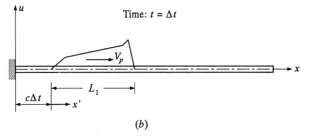

(b)

FIGURE 19-3
Propagating of wave during time interval Δt.

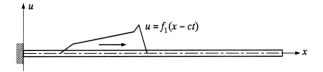

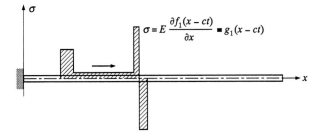

FIGURE 19-4
Relationship between
displacement and stress waves.

The relation between an arbitrary displacement waveform and the corresponding stress wave is illustrated in Fig. 19-4; obviously the stress wave also propagates with the velocity V_p and with unchanging shape.

Example   E19-5.   The general nature of the axial-wave-propagation mechanism will be demonstrated by studying the stress wave generated by the impact of a pile-driving hammer at the top of a pile, as shown in Fig. E19-5. For the purpose of this example, it will be assumed that the hammer generates a force pulse $P(t) = (600 \ kips) \sin(\pi t/0.005)$ and the stress distribution will be evaluated at the end of the pulse ($t_1 = 0.005 \ sec$) in both the steel and the concrete piles whose properties are shown in Fig. E19-5a.

To consider the steel pile first, the velocity of wave propagation given by Eq. (19-26) is

$$V_{ps} = \sqrt{\frac{E}{\rho}} = \sqrt{\frac{(30 \times 10^6)(1,728)(386)}{490}}$$
$$= 202,000 \ in/sec = 16,800 \ ft/sec$$

The stress at the origin generated by the hammer blow is

$$\sigma_0(t) = -\frac{P(t)}{A} = -(20 \ kips/in^2) \sin\frac{\pi t}{0.005}$$

but from Eq. (19-29) evaluated at the origin and considering only the forward-propagating wave,

$$\sigma_0(t) = g_1(-V_{ps}t)$$

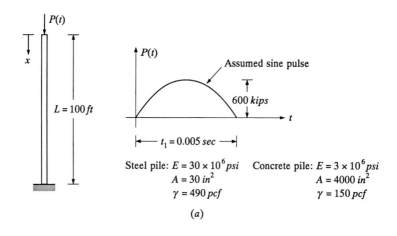

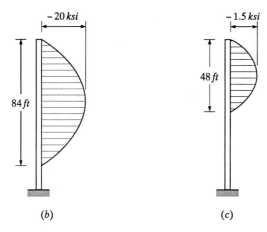

FIGURE E19-5
Propagation of applied stress wave: (*a*) properties of pile and loading; (*b*) stress in
steel pile at *t* = 0.005 *sec*; (*c*) stress in concrete pile at *t* = 0.005 *sec*.

Hence g_1 can be evaluated by equating these expressions, giving

$$\sigma_0(t) = (-20 \ kips/in^2) \ \sin\left(-\frac{\pi}{84}\right) V_{ps} t$$

Thus the general expression for the forward-propagating wave is

$$\sigma(x, t) = (-20 \ kips/in^2) \ \sin\frac{\pi}{84}(V_{ps} t - x)$$

Evaluating this at $t_1 = 0.005 \ sec$ leads to

$$\sigma(x, 0.005) = (-20 \ kips/in^2) \ \sin\pi\left(1 - \frac{x}{84}\right)$$

which is plotted in Fig. E19-5b.

Following the same procedure for the concrete pile gives

$$V_{pc} = \sqrt{\frac{(3 \times 10^6)(1,728)(386)}{150}} = 115,000 \ in/sec = 9,600 \ ft/sec$$

$$\sigma(x,t) = (-1.5 \ kips/in^2) \ \sin \frac{\pi}{48}(V_{pc}t - x)$$

$$\sigma(x,0.005) = (-1.5 \ kips/in^2) \ \sin \pi \left(1 - \frac{x}{48}\right)$$

and this last result is plotted in Fig. E19-5c.

Consideration of Boundary Conditions

The function defining the shape of any wave propagating through a uniform bar is controlled by the conditions imposed at the ends of the bar; that is, the waveform within the bar is generated by the requirements of equilibrium and compatibility at the boundaries. For example, the displacement waveform shown in Fig. 19-3a could have been initiated by introducing the displacement history at $x = 0 : u(0,t) = f_1(-V_pt)$, as shown in Fig. 19-5.

If the right end $(x = L)$ of the bar is free, as indicated in Fig. 19-3, the condition of zero stress must be maintained at all times at that end. This condition may be satisfied by a second stress wave propagating toward the left, which, when superposed on the incident wave, cancels the end-section stresses. Expressing this concept mathematically by means of Eq. (19-28) leads to

$$\sigma_{x=L} = 0 = E \frac{\partial f_1}{\partial x}(L - V_pt) + E \frac{\partial f_2}{\partial x}(L + V_pt)$$

from which

$$\frac{\partial f_1}{\partial x}(L - V_pt) = -\frac{\partial f_2}{\partial x}(L + V_pt) \tag{19-30}$$

Hence it is evident that the slope $\partial u/\partial x$ of the left-propagating wave must be the negative of the slope of the forward-propagating wave as each part of the waves

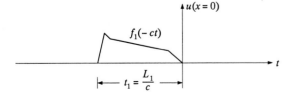

FIGURE 19-5
Displacement imposed at end $(x = 0)$ of bar of Fig. 19-3.

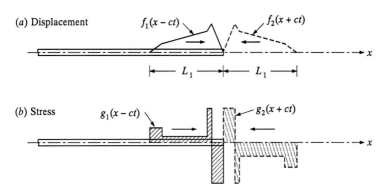

(a) Displacement $f_1(x - ct)$ $f_2(x + ct)$

L_1 L_1

(b) Stress $g_1(x - ct)$ $g_2(x + ct)$

FIGURE 19-6
Reflection of displacement and stress waves at free end.

passes the end of the rod. The displacement waves shown in Fig. 19-6a demonstrate this condition, and the corresponding stress waves in Fig. 19-6b show clearly how the stresses at the tip are canceled.

Although the concept of a left-moving wave coming from beyond the end of the bar makes it easier to visualize the mechanism by which the boundary condition is satisfied, it should be understood that this wave actually is created at the end of the bar as the forward-propagating wave reaches that point. In other words, the incident wave is reflected at the free end; the reflected wave has the *same deflections* as the incident wave, but the *stresses* are *reversed* because the direction of travel is reversed. It will be noted that the total deflection at the free end is doubled by the superposition of the incident and reflected waves, while the two stress components cancel each other.

To consider now the case where the right end of the bar is fixed rather than free, it is evident that the boundary condition imposed on the two propagating waves is

$$u_{x=L} = 0 = f_1(L - V_p t) + f_2(L + V_p t)$$

from which the reflected wave may be expressed in terms of the incident wave as

$$f_2(L + V_p t) = -f_1(L - V_p t) \tag{19-31}$$

Thus the displacement waves in this case are seen to have opposite signs, and by analogy with the preceding discussion it can be inferred that the incident and reflected stress waves have the same sign, as shown in Fig. 19-7. Hence, in satisfying the required zero-displacement condition, the reflected wave produces a doubling of stress at the fixed end of the bar.

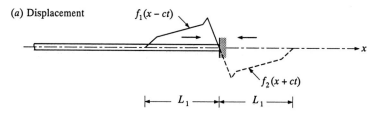

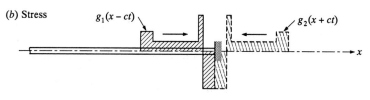

FIGURE 19-7
Reflection of displacement and stress waves at fixed end.

Example E19-6. To demonstrate these boundary-reflection phenomena, the stress wave produced by the driving hammer on the concrete pile in Example E19-5 will be considered further. The stress wave shown in Fig. E19-5c is traveling with a velocity of $9,600$ ft/sec, and so the forward end of this wave reaches the tip of the pile at a time

$$t_2 = \frac{100}{9,600} = 0.0104 \ sec$$

The subsequent behavior then depends on the nature of the tip support condition.

Assuming first that the pile rests on a rigid support, so that no displacement can take place at this point, the reflected stress wave must be compression, the same as the incident wave. The total stress at subsequent times is then given by the sum of the incident and reflected components. As a specific example, the distribution of stress at the time when the stress wave has traveled 128 ft,

$$t_3 = \frac{128}{9,600} = 0.0133 \ sec$$

is shown in Fig. E19-6a.

The other limiting case occurs if the end of the pile is resting on very soft mud, so that there is essentially no resistance to its displacement and the tip stress is required to be zero. In this case, the reflected stress wave must be tensile, and the total stress in the pile is given by the difference between the tensile and compressive components. Taking again the time t_3, when the stress wave has traveled 128 ft, the distribution of stress is as shown in Fig. E19-6b.

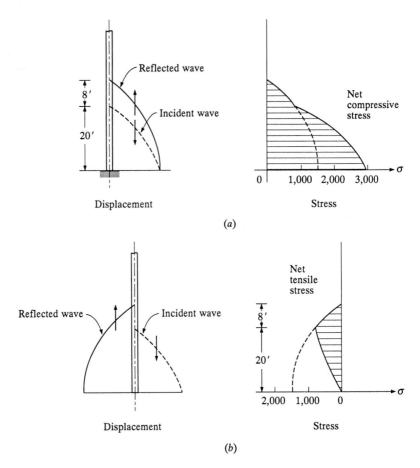

FIGURE E19-6
Stress distribution at $t_3 = 0.0133 \; sec$ (concrete pile of Fig. E21-1): (a) fixed end at $x = L$; (b) free end at $x = L$.

It is significant that the net stress is tensile over the lower 28 ft of the pile and that the greatest tensile stress occurs 20 ft from the tip. This illustrates how tensile stresses will be developed in a pile during the driving process if the material through which it is driven offers little resistance, and that fracture due to these tensile stresses might occur at a significant distance from the tip. Of course, the behavior in any specific case will depend on the tensile strength of the concrete and on the duration of the hammer-force impulse.

Discontinuity in Bar Properties

The wave reflections which take place at the fixed or free end of a uniform bar may be considered as special cases of the general reflection and refraction phenomena

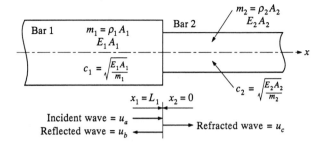

Incident wave = u_a
Reflected wave = u_b
Refracted wave = u_c

FIGURE 19-8
Wave reflection and refraction
at discontinuity.

occurring at any discontinuity in the bar properties. The conditions of equilibrium and compatibility which must be satisfied at all points along the bar require that additional reflected and refracted waves be generated at the juncture between bars of different properties in response to the action of any given incident wave.

Consider, for example, the juncture between bars 1 and 2 shown in Fig. 19-8. The properties of the bars on each side of the juncture are characterized by their mass per unit length $\overline{m}$ and axial stiffness EA. Also the wave-propagation velocity on each side is given by $V_p = \sqrt{AE/\overline{m}} = \sqrt{E/\rho}$. The forward-propagating wave u_a which arrives at the juncture in bar 1 generates a reflection u_b which travels in the negative direction in bar 1 and at the same time creates a refracted wave u_c which propagates forward in bar 2.

Two continuity conditions are imposed at the juncture:

Displacement: $\qquad\qquad u_1 = u_2 \qquad$ or $\qquad u_a + u_b = u_c \qquad\qquad$ (19-32a)

Force: $\qquad\qquad N_1 = N_2 \qquad$ or $\qquad N_a + N_b = N_c \qquad\qquad$ (19-32b)

where the fact that both incident and reflected waves act in bar 1 has been indicated. Because these continuity conditions must be satisfied at all times, the time derivative of the displacement condition also must be satisfied, that is,

$$\frac{\partial u_a}{\partial t} + \frac{\partial u_b}{\partial t} = \frac{\partial u_c}{\partial t} \qquad\qquad (19\text{-}33)$$

But the incident wave can be expressed in the form

$$u_a = f_a(x - V_p t) \equiv f_a(\zeta)$$

where the variable ζ has been introduced for convenience. Now the derivatives of u_a can be expressed as

$$\frac{\partial u_a}{\partial x} = \frac{\partial f_a}{\partial \zeta}\frac{\partial \zeta}{\partial x} = \frac{\partial f_a}{\partial \zeta} \qquad\qquad \frac{\partial u_a}{\partial t} = \frac{\partial f_a}{\partial \zeta}\frac{\partial \zeta}{\partial t} = -V_{p1}\frac{\partial f_a}{\partial \zeta}$$

from which it is evident that the time and position derivatives are related by the velocity of wave propagation

$$\frac{\partial u_a}{\partial t} = -V_{p1}\frac{\partial u_a}{\partial x} \qquad\qquad (19\text{-}34a)$$

Similar analyses for the reflected and refracted waves result in

$$\frac{\partial u_b}{\partial t} = +V_{p1} \frac{\partial u_b}{\partial x} \tag{19-34b}$$

$$\frac{\partial u_c}{\partial t} = -V_{p2} \frac{\partial u_c}{\partial x} \tag{19-34c}$$

where the positive sign in Eq. (19-34b) is due to the negative direction of the reflected-wave propagation.

Substituting Eqs. (19-34) into (19-33) yields

$$-V_{p1} \frac{\partial u_a}{\partial x} + V_{p1} \frac{\partial u_b}{\partial x} = -V_{p2} \frac{\partial u_c}{\partial x} \tag{19-35}$$

but the strains, $\partial u_a / \partial x = \varepsilon_a$, etc., can be expressed in terms of the forces acting in the bars: $\varepsilon_a = \sigma_a / E = N_a / A_1 E_1$, etc; hence, the compatibility condition of Eq. (19-35) can be expressed in terms of the force waves

$$-\frac{V_{p1}}{A_1 E_1} N_a + \frac{V_{p1}}{A_1 E_1} N_b = -\frac{V_{p2}}{A_2 E_2} N_c$$

or more simply

$$N_c = \alpha \, (N_a - N_b) \tag{19-36}$$

where

$$\alpha = \frac{V_{p1}}{V_{p2}} \frac{A_2 E_2}{A_1 E_1} = \sqrt{\frac{m_2 E_2 A_2}{m_1 E_1 A_1}} \tag{19-37}$$

Finally, this compatibility condition [Eq. (19-36)] can be introduced into the force-equilibrium condition [Eq. (19-32b)] to express the refracted and reflected waves in terms of the incident wave

$$N_a + N_b = \alpha \, (N_a - N_b)$$

from which

$$N_b = N_a \frac{\alpha - 1}{\alpha + 1} \tag{19-38}$$

and from Eq. (19-36)

$$N_c = N_a \frac{2\alpha}{\alpha + 1} \tag{19-39}$$

Equations (19-38) and (19-39) express the relationships between the incident, reflected, and refracted force waves at the bar discontinuity. Corresponding relationships can be obtained for the displacement waves by noting that

$$N = AE \frac{\partial u}{\partial x} = \pm \frac{AE}{V_p} \frac{\partial u}{\partial t}$$

Substituting this into Eq. (19-38) and integrating lead to

$$\frac{A_1 E_1}{V_{p1}} u_b = -\frac{A_1 E_1}{V_{p1}} u_a \frac{\alpha - 1}{\alpha + 1}$$

from which

$$u_b = -u_a \frac{\alpha - 1}{\alpha + 1} \tag{19-40}$$

Similarly, substituting into Eq. (19-39) and integrating give

$$-\frac{A_2 E_2}{V_{p2}} u_c = -\frac{A_1 E_1}{V_{p1}} u_a \frac{2\alpha}{\alpha + 1}$$

from which

$$u_c = u_a \frac{2}{\alpha + 1} \tag{19-41}$$

It is evident that the factor α defines the character of the discontinuity at the juncture between two bars and controls the relative amplitudes of the reflected and refracted waves. Where the properties of two adjoining bars are identical or related in any manner such that the value of α given by Eq. (19-37) is unity, there is no discontinuity and no reflected wave. For increasing stiffness in bar 2, the value of α increases and the reflected force wave is of the same sign as the incident wave; for decreasing stiffness in bar 2, the value of α becomes less than unity, and the reflected force wave is of opposite sign to the incident wave. In this context, the fixed- and free-end conditions discussed above can be considered as limiting cases of bar discontinuity and are defined by infinite and zero values of α, respectively. The relationships between incident, reflected, and refracted waves for various cases of discontinuity are listed in Table 19-1.

TABLE 19-1
Wave relationships for various discontinuities

Case	$\alpha = \sqrt{\dfrac{A_2 E_2 m_2}{A_1 E_1 m_1}}$	$\overrightarrow{N_a}$ +	$\overleftarrow{N_b}$ =	$\overrightarrow{N_c}$	$\overrightarrow{u_a}$ +	$\overleftarrow{u_b}$ =	$\overrightarrow{u_c}$
		Force waves			**Displacement waves**		
No discontinuity	1	1	0	1	1	0	1
Fixed end	∞	1	1	2	1	-1	0
Free end	0	1	-1	0	1	1	2
$\dfrac{A_2 E_2}{A_1 E_1} = \dfrac{m_2}{m_1} = 2$	2	1	$\dfrac{1}{3}$	$\dfrac{4}{3}$	1	$-\dfrac{1}{3}$	$\dfrac{2}{3}$
$\dfrac{A_2 E_2}{A_1 E_1} = \dfrac{m_2}{m_1} = \dfrac{1}{2}$	$\dfrac{1}{2}$	1	$-\dfrac{1}{3}$	$\dfrac{2}{3}$	1	$\dfrac{1}{3}$	$\dfrac{4}{3}$

Another relationship of considerable interest becomes immediately apparent from Eq. (19-34a) if the particle velocity on the left side is denoted by $\partial u_a / \partial t = \dot{u}_a$ and the strain on the right side by $\partial u_a / \partial x = \varepsilon_a = \sigma_a / E_1$. Making these substitutions, the relation becomes

$$\dot{u}_a = -\frac{V_{p1}}{E_1} \sigma_a \tag{19-42}$$

Expressing this in words, the positive particle velocity in propagation of normal stress waves is directly related to the wave compressive stress by the proportionality factor for the material, V_p / E, where V_p is the wave-propagation velocity.

Example E19-7. To illustrate the effects caused by discontinuities on the propagation of force waves through a multiple-segment bar, the stepped bar shown in Fig. E19-7a will be considered. Since the material is the same in each section, the discontinuities are due only to the changes of area. At each step

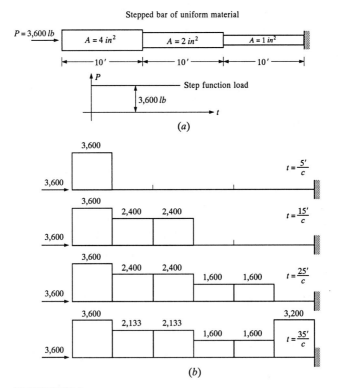

FIGURE E19-7
Force reflection and refraction at bar discontinuities: (a) definition of bar and load; (b) force distribution at various times.

$A_2/A_1 = 1/2$, and so $m_2/m_1 = 1/2$; thus $\alpha = 1/2$. This corresponds to the last case in Table 19-1 and, as indicated there, at each step

$$\frac{N_b}{N_a} = -\frac{1}{3} \qquad \frac{N_c}{N_a} = \frac{2}{3}$$

If the left end of the bar is subjected to a constant force of $3,600 \ lb$, the force distributions at the times required for the stress wave to propagate 5, 15, 25, and 35 ft will be as shown in Fig. E19-7b. Stress distributions can be derived from these sketches by dividing by the appropriate area of each segment. Because the one-dimensional wave equation has been used in this analysis, it must be assumed that the segments are interconnected by rigid disks which maintain the uniaxial stress state through the discontinuities.

PROBLEMS

19-1. Assume that the undamped uniform beam of Fig. E19-2 is subjected to a static central load p_0 and then set into free vibration by suddenly releasing the load at time $t = 0$. The initial deflected shape is given by

$$v(x) = \frac{p_0 x}{48EI}(3L^2 - 4x^2) \qquad 0 < x < \frac{L}{2}$$

(a) From this information, evaluate the amplitude of the midspan free-vibration displacement in each of the first three modes of vibration, expressing results as fractions of the static midspan displacement.

(b) Evaluate the amplitude of the midspan free-vibration moment in each of the first three modes of vibration, expressing the results as fractions of the static midspan moment.

19-2. Assuming that the step-function load of Fig. E19-2 is applied at the quarter span point ($x = L/4$), rather than at midspan, write expressions for the undamped displacement response and bending moment response at the load point. Plot this moment history, considering the first three modes, over the time interval $0 < t < T_1$.

19-3. Assume that the beam of Fig. E19-2 is subjected to a harmonic load applied at the quarter span point: $p(t) = p_0 \sin \bar{\omega} t$ where $\bar{\omega} = \frac{5}{4}\omega_1$. Considering the first three modes of vibration, plot the steady-state displacement response amplitude of the beam, evaluating it at increments $L/4$ along the span:

(a) Neglecting damping.

(b) Assuming the damping in each mode is 10 percent of critical.

19-4. A uniform simple beam having flexural rigidity $EI = 78 \times 10^8 \ lb \cdot in^2$ supports a total weight of $1,000 \ lb/ft$. When immersed in a viscous fluid and set into first-mode vibration with an amplitude of 1 in, it is observed that the motion is reduced to 0.1 in amplitude in 3 cycles.

    (a) Assuming that the damping resistance per unit velocity, $c(x)$, is uniform along the span, determine its numerical value.

    (b) Assuming that the same beam is set into second-mode vibration with a 1 in amplitude, determine how many cycles will be required to reduce this motion to 0.1 in.

19-5. Repeat Prob. 19-3 for the uniform rod shown in Fig. P19-1. Note that the harmonic load $p(t) = p_0 \sin \bar{\omega} t$ is applied in the axial direction at midlength, and that the axial displacement response is to be plotted by both neglecting and including the influence of modal damping.

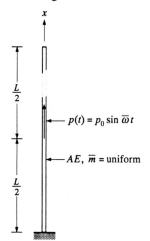

$$p(t) = p_0 \sin \bar{\omega} t$$

$$AE, \bar{m} = \text{uniform}$$

FIGURE P19-1

19-6. The uniform simple beam shown in Fig. P19-2 is subjected to a lateral loading $p(x, t) = \delta(x - a)\delta(t)$, where $\delta(x - a)$ and $\delta(t)$ are Dirac delta functions. (See Section 20-1 for a definition of the Dirac delta function.) Using elementary beam theory and the mode superposition method, determine the series expressions for lateral deflection $v(x, t)$, internal moment $M(x, t)$, and internal shear $V(x, t)$ caused by the loading $p(x, t)$ defined above. Discuss the relative rates of convergence of these three series expressions.

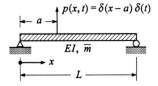

$$p(x, t) = \delta(x - a)\,\delta(t)$$

$$EI, \bar{m}$$

FIGURE P19-2

RANDOM VIBRATIONS

PROBABILITY
THEORY

20-1 SINGLE RANDOM VARIABLE

It is assumed that the reader has had some experience with various games of chance and has an intuitive grasp of simple probability theory even though he or she may never have studied this subject formally. Let us begin by formalizing the basic probability concepts for a simple experiment.

Consider the familiar rotating disk shown in Fig. 20-1a, which has 10 equally spaced pegs driven into its side with the intervals between pegs representing numbers 1 through 10 as shown. When the disk is spun, it will eventually come to rest with

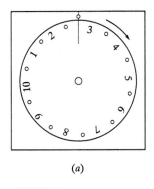
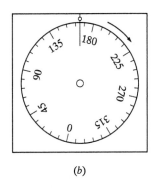

| (a) | (b) |

FIGURE 20-1
Single-random-variable experiment: (a) discrete variable N; (b) continuous variable θ.

427

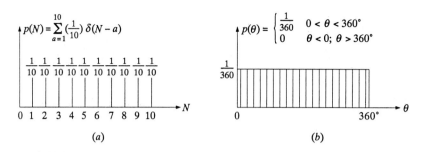

FIGURE 20-2
Probability density functions for single random variables N and θ : (a) discrete variable;
(b) continuous variable.

one of these numbers at the indicator. Assuming an unbiased disk, each number has a $1/10$ probability of occurrence; that is, after sampling n times, each number will have been sampled $n/10$ times in the limit as n approaches infinity. If N represents the value of the number sampled and $p(N)$ its probability of occurrence, the probability relationship for this experiment will be the bar diagram shown in Fig. 20-2a. N is said to be a discrete random variable in this case since only discrete values can be sampled.

Consider now an unbiased rotating disk as shown in Fig. 20-1b which has no pegs but is marked off in degrees similar to a full 360° compass. In this experiment, if the disk is spun and the angle θ to which the indicator points when it comes to rest is noted, values can be sampled throughout the range $0 \leq \theta \leq 360°$ with equal chance of occurrence; that is, its probability relation will be continuous and uniform, as shown in Fig. 20-2b. Both probability relations in Fig. 20-2 are called probability density functions.

To clarify further the definition of probability density, consider a general experiment involving a single random variable x which has the probability density function shown in Fig. 20-3a. This function is defined so that $p(x_1)\,dx$ equals the chances that a sampled value of x will be in the range $x_1 < x < x_1 + dx$. When unity represents a certainty of occurrence, the above definition requires that the probability density function be normalized so that the area between the x axis and the function itself, that is, $\int_{-\infty}^{\infty} p(x)\,dx$, equals unity.

From the above definition, it should be noted that a zero probability exists that a sampled value of x will be exactly equal to some preselected value in the continuous case. In other words, a finite probability can be associated only with x falling in a certain finite range. To illustrate this point, consider again the simple experiment shown in Fig. 20-1b, where a zero probability exists that the indicator will point exactly to, say, 256°; however, the probability that a sampled value of θ will be in

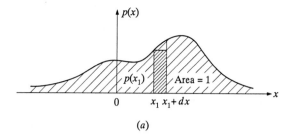

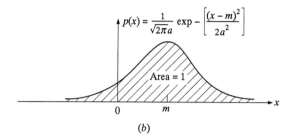

FIGURE 20-3
Probability density function for random variable x: (a) general probability density function; (b) normal, or gaussian, probability density function.

the range $256° < \theta < 257°$ is $1/360$.

Further, it should be noted that to satisfy the above definition of probability density in the discrete case like that shown in Fig. 20-2a, the probability density function must consist of a number of Dirac delta functions. The Dirac delta function $\delta(x - a)$ is simply any function which satisfies the conditions

$$\delta(x - a) = \begin{cases} 0 & x \neq a \\ \infty & x = a \end{cases}$$

$$\int_{-\infty}^{\infty} \delta(x - a)\, dx = 1$$

(20-1)

Example E20-1. Show by proper selection of the constant C in the function $f(x - a)$, defined below, that this function satisfies the conditions for a Dirac delta function given by Eqs. (20-1):

$$f(x - a) = C \lim_{\varepsilon \to 0} \frac{1}{\varepsilon} \exp\left[-\frac{(x - a)^2}{2\varepsilon^2}\right]$$

(a)

For $x \neq a$, it is quite apparent that the function equals zero since in the limit the exponential term approaches zero much more rapidly than ε itself. When $x = a$, the exponential term equals unity; therefore, the entire function approaches infinity in the limit at this point. Now, consider the integral

$$I \equiv \int_{-\infty}^{\infty} \frac{C}{\varepsilon} \exp\left[-\frac{(x-a)^2}{2\,\varepsilon^2}\right] dx = \frac{C}{\varepsilon} \int_{-\infty}^{\infty} \exp\left[-\frac{(x-a)^2}{2\,\varepsilon^2}\right] dx \qquad (b)$$

Substituting the change of variable

$$u = \frac{x-a}{\sqrt{2}\,\varepsilon} \qquad du = \frac{1}{\sqrt{2}\,\varepsilon} dx \qquad (c)$$

gives

$$I = \sqrt{2}\,C \int_{-\infty}^{\infty} \exp[-u^2]\, du = \sqrt{2\pi}\,C \qquad (d)$$

Note that the value of this integral is independent of ε; therefore

$$I = \int_{-\infty}^{\infty} f(x-a)\, dx = \sqrt{2\pi}\,C = 1 \qquad (e)$$

giving

$$C = \frac{1}{\sqrt{2\pi}} \qquad (f)$$

The most commonly used probability density function of a single random variable is the so-called *normal*, or *gaussian*, *distribution* shown in Fig. 20-3b, which is defined by the symmetric relation

$$p(x) = \frac{1}{\sqrt{2\pi}\,a} \exp[-(x-m)^2/2a^2] \qquad (20\text{-}2)$$

where a and m are constants. A plot of this relation shows that a is a measure of the spread of the function in the neighborhood of $x = m$. From the above example solution, it is seen that the integral of Eq. (20-2) between the limits $x = -\infty$ and $x = +\infty$ equals unity, as it should, regardless of the numerical values of a and m.

If a random variable x is transformed into a second random variable r, which is a known single-valued function of x as defined in general form by the relation

$$r \equiv r(x) \qquad (20\text{-}3)$$

the probability density function for r is easily obtained from the relation

$$p(r) = p(x) \left|\frac{dx}{dr}\right| \qquad (20\text{-}4)$$

provided the inverse relation $x = x(r)$ is also a single-valued function. The validity of Eq. (20-4) is obvious since (as shown in Fig. 20-4) all sampled values of x which fall in the range $x_1 < x < x_1 + dx$ correspond to values of r in the range $r_1 < r < r_1 + dr$. The absolute value of dx/dr is necessary since for some functions $r(x)$ a positive dx corresponds to a negative dr or vice versa.

Another probability function which is useful when treating single random variables is the probability distribution function defined by

$$P(x) \equiv \int_{-\infty}^{x} p(u)\, du \qquad (20\text{-}5)$$

In accordance with this definition, the function $P(x)$ either becomes or approaches zero and unity with increasing negative and positive values of x, respectively, as shown in Fig. 20-5. Equation (20-5) in its differential form

$$p(x) = \frac{d\,P(x)}{dx} \qquad (20\text{-}6)$$

is also very useful.

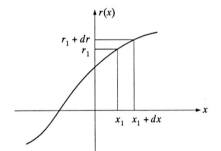

FIGURE 20-4
Relation between random variable x and random variable r.

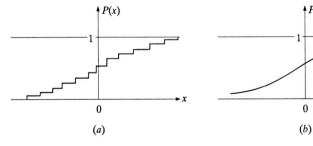

(a) (b)

FIGURE 20-5
Probability distribution function for random variable x: (a) discrete variable; (b) continuous variable.

Example   E20-2.   A random variable has the probability density function

$$p(x) = \begin{cases} 1/2 & -1 < x < +1 \\ 0 & x < 1 \; ; \; x > 1 \end{cases} \tag{a}$$

If random variable r is related to x through the relation

$$r(x) = x \, |x| \tag{b}$$

find the probability density function $p(r)$ and show that it satisfies the condition

$$\int_{-\infty}^{\infty} p(r) \, dr = 1 \tag{c}$$

Fist, taking the derivative of Eq. (b) gives $dr/dx = 2|x|$. Then using Eq. (20-4) gives

$$p(r) = \begin{cases} \dfrac{1}{4|x|} & -1 < x < +1 \\ 0 & x < 1 \; ; \; x > 1 \end{cases} \tag{d}$$

or

$$p(r) = \begin{cases} \dfrac{1}{4\sqrt{|r|}} & -1 < r < +1 \\ 0 & r < 1 \; ; \; r > 1 \end{cases} \tag{e}$$

Substituting Eq. (e) into Eq. (c) leads to

$$I \equiv \int_{-\infty}^{\infty} p(r) \, dr = \frac{1}{4} \int_{-1}^{1} \frac{dr}{\sqrt{|r|}} = 1 \tag{f}$$

thus showing that Eq. (c) is satisfied.

20-2   IMPORTANT AVERAGES OF A SINGLE RANDOM VARIABLE

If a certain random variable x is sampled n times and is each time used to evaluate a second random variable r defined by a single-valued function $r(x)$, the average of this second variable as n approaches infinity, that is,

$$\bar{r} \equiv \lim_{n \to \infty} \frac{1}{n} \sum_{i=1}^{n} r(x_i) \tag{20-7}$$

where x_i is the ith sampled value of x, can be determined using the relation

$$\bar{r} = \int_{-\infty}^{\infty} r(x) \, p(x) \, dx \tag{20-8}$$

A bar placed above any random variable is used to indicate average value.

Averages most commonly used in nondeterministic analyses are (1) mean value of x, (2) mean square value of x, (3) variance of x, and (4) standard deviation of x, defined as follows:

Mean value

$$\bar{x} = \int_{-\infty}^{\infty} x \, p(x) \, dx \tag{20-9}$$

Mean square value

$$\overline{x^2} = \int_{-\infty}^{\infty} x^2 \, p(x) \, dx \tag{20-10}$$

Variance

$$\sigma_x^2 = \overline{(x - \bar{x})^2} = \int_{-\infty}^{\infty} (x - \bar{x})^2 \, p(x) \, dx = \overline{x^2} - \bar{x}^2 \tag{20-11}$$

Standard deviation

$$\sigma_x = \sqrt{\text{variance}} \tag{20-12}$$

Example E20-3.   Find the mean, mean square value, and variance of a random variable x having the normal probability distribution given by Eq. (20-2).

From Eq. (20-9), the mean value can be written in the form

$$\bar{x} = \frac{1}{\sqrt{2\pi} \, a} \int_{-\infty}^{\infty} x \, \exp[-(x - m)^2 / 2 \, a^2] \, dx \tag{a}$$

Substituting the change of variable

$$u \equiv \frac{x - m}{\sqrt{2} \, a} \qquad du = \frac{1}{\sqrt{2} \, a} \, dx \tag{b}$$

gives

$$\bar{x} = \frac{\sqrt{2} \, a}{\sqrt{\pi}} \int_{-\infty}^{\infty} u \, \exp(-u^2) \, du + \frac{m}{\sqrt{\pi}} \int_{-\infty}^{\infty} \exp(-u^2) \, du \tag{c}$$

The first integral in Eq. (c) equals zero while the second equals $\sqrt{\pi}$, thus showing that

$$\bar{x} = m \tag{d}$$

The mean square value of x as given by Eq. (20-10) becomes

$$\overline{x^2} = \frac{1}{\sqrt{2\pi} \, a} \int_{-\infty}^{\infty} x^2 \, \exp[-(x - m)^2 / 2 \, a^2] \, dx \tag{e}$$

Using the same change of variable indicated above gives

$$\overline{x^2} = \frac{2\,a^2}{\sqrt{\pi}} \int_{-\infty}^{\infty} u^2 \exp(-u^2)\,du$$

$$+ \frac{2\sqrt{2}\,am}{\sqrt{\pi}} \int_{-\infty}^{\infty} u \exp(-u^2)\,du + \frac{m^2}{\sqrt{\pi}} \int_{-\infty}^{\infty} \exp(-u^2)\,du \quad \text{(f)}$$

Upon integrating by parts, the first integral is shown equal to $\sqrt{\pi}/2$, the second integral equals zero, and the third equals $\sqrt{\pi}$, thus yielding

$$\overline{x^2} = a^2 + m^2 \tag{g}$$

Substituting Eqs. (d) and (g) into Eq. (20-11) gives

$$\sigma_x^2 = a^2 \tag{h}$$

20-3   ONE-DIMENSIONAL RANDOM WALK

Assume in this experiment that n individuals are walking along a straight line without interference. If all individuals start walking from the same point $(x = 0)$ and each separate step length L is controlled by the probability density function

$$p(L) = \frac{1}{4}\delta(L + \Delta L) + \frac{3}{4}\delta(L - \Delta L) \tag{20-13}$$

that is, there exists a $1/4$ probability of taking a backward step of length ΔL and a $3/4$ probability of taking a forward step of the same length, the probability density function $p(x_i)$ for distance x_i as defined by $x_i \equiv \sum_{j=1}^{i} L_j$ will be as given in Fig. 20-6 for $i = 0, 1, 2, 3,$ and 4. (Vertical heavy arrows will be used herein to indicate Dirac delta functions.) Since all n individuals performing this experiment are at the origin before taking their first step, the probability density function $p(x_0)$ is a single Dirac delta function of unit intensity located at the origin. If n is considered to approach infinity, it follows directly from Eq. (20-13) that $3n/4$ individuals will be located at $x_1 = \Delta L$ after taking their first step and $n/4$ individuals will be located at $x_1 = -\Delta L$. Upon taking their second step three-fourths of those individuals located at $x_1 = \Delta L$, that is, $9n/16$, will move to $x_2 = 2\,\Delta L$ and the remaining one-fourth will step backward to the origin. Similarly, upon taking their second step, three-fourths of those individuals located at $x_1 = -\Delta L$, that is, $3n/16$, will step forward to the origin while the remaining one-fourth will step backward to $x_2 = -2\,\Delta L$. Such

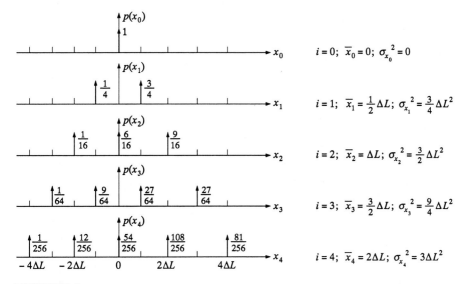

FIGURE 20-6
Example of one-dimensional random walk.

reasoning can be continued to establish each successive probability density function in the same way.

If the probability density function for length of step is given by the somewhat more general form

$$p(L) = g\,\delta(L + \Delta L) + h\,\delta(L - \Delta L) \tag{20-14}$$

where $g + h = 1$, and if the numerical values of g and h are known, it is possible to find the probability density functions $p(x_i)$ $(i = 1, 2, \cdots)$ by the same procedure used above for $g = 1/4$ and $h = 3/4$. While it will not be proved here, it can be easily shown that the probability density function $p(x_i)$ is given by the well-known binomial relation

$$p(x_i) = \sum_{k=-i,-i+2,\cdots}^{i} \frac{i!\,\delta(x_i - k\,\Delta L)}{[(i+k)/2]!\,[(i-k)/2]!}\,h^{(i+k)/2}\,(1-h)^{(i-k)/2}$$

$$i = 0, 1, 2, \cdots \tag{20-15}$$

and that the mean value and variance of x_i as defined by Eqs. (20-9) and (20-11) are, respectively,

$$\bar{x}_i = i\bar{L} = i(h - g)\,\Delta L \tag{20-16}$$

$$\sigma_{x_i}^2 = i\sigma_L^2 = i[1 - (h - g)^2]\Delta L^2 \tag{20-17}$$

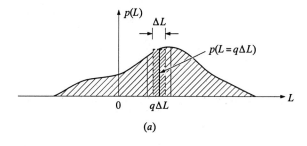

(a)

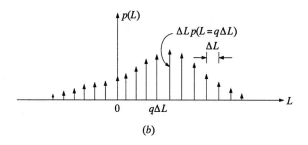

(b)

FIGURE 20-7
Arbitrary probability density
function for length of step.

The reader can easily check Eqs. (20-15) through (20-17) with the results previously obtained by straightforward means as given in Fig. 20-6.

Consider the one-dimensional random walk in its most general form, that is, one with an arbitrarily prescribed probability density function for length of step L, as shown in Fig. 20-7a. This function can be approximated by the discrete distribution shown in Fig. 20-7bb obtained by simply concentrating the area $\Delta L \, p(L = q \, \Delta L)$ in the form of a Dirac delta function. Of course, in the limit as ΔL approaches zero, this discrete representation becomes exact. Likewise, the continuous probability density functions for distance from the origin $x_i \equiv \sum_{j=1}^{i} L_j$ (Fig. 20-8a) can be approximated by a discrete distribution as shown in Fig. 20-8b. With Δx chosen equal to ΔL, it is possible to determine the probability density function $p(x_{i+1})$ in exactly the same way as for the simpler case shown in Fig. 20-6; that is, the contribution by the Dirac delta function of intensity $\Delta x \, p(x_i = r \, \Delta x)$ to the Dirac delta function of intensity $\Delta x \, p(x_{i+1} = s \, \Delta x)$ is the product $\Delta x \, p(x_i = r \, \Delta x) \Delta L \, p(L = q \, \Delta L)$ where $q \equiv s - r$. Therefore, the contribution of all delta functions in Fig. 20-8b to the intensity $p(x_{i+1} = s \, \Delta x)$ can be obtained by superposition, thus giving

$$p(x_{i+1} = s \, \Delta x) = \sum_{r=-\infty}^{\infty} p(x_i = r \, \Delta x) \, p(L = q \, \Delta L) \, \Delta L \qquad (20\text{-}18)$$

It will become apparent later in this development that it is advantageous to express the probability density function of Fig. 20-8b and c in terms of distances X_i and X_{i+1} measured from points $x_i = iA$ and $x_{i+1} = (i + 1)A$, respectively,

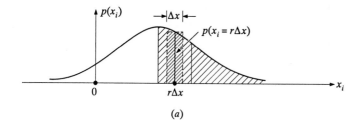

(a)

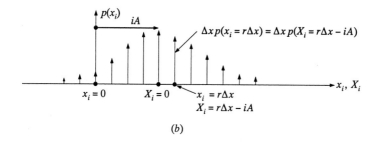

(b)

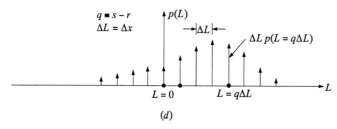

(c)

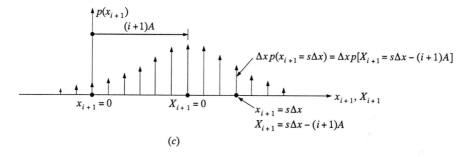

(d)

FIGURE 20-8
Probability density functions for the general one-dimensional random walk.

where A is some integer number of Δx. With this type of coordinate transformation, Eq. (20-18) becomes.

$$p[X_{i+1} = s\,\Delta x - (i+1)A] = \sum_{r=-\infty}^{\infty} p(X_i = r\,\Delta x - iA)\,p(L = q\,\Delta L)\,\Delta L \quad (20\text{-}19)$$

If in the above random walk, each individual is located at the origin $x = 0$ at time $t = 0$, and if each individual takes his ith step at the instant $t_i = i\,\Delta t$, Eq. (20-19) can be written

$$p(X;t_i + \Delta t) = \sum_{q=-\infty}^{\infty} p(X - q\,\Delta x + A;t_i)\ p(L = q\,\Delta L)\,\Delta L \qquad (20\text{-}20)$$

where

$$X_{i+1} \equiv X \qquad (20\text{-}21)$$

$$X_i = X - q\,\Delta x + A \qquad (20\text{-}22)$$

When a limiting process is now used by letting $\Delta x = \Delta L \to 0$ in such a manner that the quantity X remains finite, Eq. (20-20) converts to its continuous form with respect to distance, that is,

$$p(X;t_i + \Delta t) = \int_{-\infty}^{\infty} p(X - L + A;t_i)\ p(L)\ dL \qquad (20\text{-}23)$$

When the function $p(X - L + A;t_i)$ is expanded in a Taylor's series about point $X = 0$ and the integral is completed, Eq. (20-23) becomes

$$p(X;t_i + \Delta t) = p(X;t_i) + (A - \overline{L})\ p'(X;t_i)$$
$$+ \frac{A^2 - 2\,A\overline{L} + \overline{L^2}}{2}\ p''(X;t_i) + \cdots \qquad (20\text{-}24)$$

It now becomes apparent why (as previously noted) it is helpful to express the probability density functions in terms of X rather than x, since the second term on the right hand side of Eq. (20-24) can be eliminated by letting

$$A = \overline{L} \qquad (20\text{-}25)$$

Substituting Eq. (20-25) into Eq. (20-24) and dividing both sides of this equation by Δt and making use of Eq. (20-11) gives

$$\frac{p(X;t_i + \Delta t) - p(X;t_i)}{\Delta t} = \frac{\sigma_L^2}{2\,\Delta t}\,p''(X;t_i) + \cdots \qquad (20\text{-}26)$$

If during the limiting process mentioned above, the variance of the function $p(L)$, that is, σ_L^2, also approaches zero but in such a way that the ratio $\sigma_L^2/\Delta t$ equals a constant C, the terms on the right hand side of Eq. (20-26) beyond the first term will be of higher order and can be dropped. Also in the limit as $\Delta t \to 0$, the left hand side of Eq. (20-26) equals $\dot{p}(X,t)$; thus, Eq. (20-26) becomes the well-known one-dimensional diffusion equation

$$\frac{\partial p(X,t)}{\partial t} = \frac{C}{2}\frac{\partial^2 p(X,t)}{\partial X^2} \tag{20-27}$$

From the known initial condition

$$p(X,0) = \delta(X) \tag{20-28}$$

and the boundary conditions

$$\lim_{Q\to\infty}\left[\frac{\partial p(X,t)}{\partial X}\Big|_{X=Q}\right] = \lim_{Q\to\infty}\left[\frac{\partial p(X,t)}{\partial X}\Big|_{X=-Q}\right] = 0 \tag{20-29}$$

the solution of Eq. (20-27) is

$$p(X,t) = \frac{1}{\sqrt{2\pi Ct}}\exp\left[-\frac{X^2}{2Ct}\right] \tag{20-30}$$

The probability density function for random variable X_i, after i steps, is given by Eq. (20-30) after substitutuing $C = \sigma_L^2/\Delta t$ and $i = t/\Delta t$, thus giving

$$p(X_i) = \frac{1}{\sqrt{2\pi i\sigma_L^2}}\exp\left[-\frac{X_i^2}{2i\sigma_L^2}\right] \tag{20-31}$$

In this case, use of Eqs. (20-9) through (20-11) and (20-31) shows that

$$\overline{X}_i = 0 \qquad \sigma_{X_i}^2 = i\sigma_L^2 \tag{20-32}$$

From Eqs. (20-25) and (20-31) and the information provided in Fig. 20-8, the relation

$$p(x_i) = \frac{1}{\sqrt{2\pi}\sigma_{x_i}}\exp\left[-\frac{(x_i - \overline{x}_i)^2}{2\sigma_{x_i}^2}\right] \tag{20-33}$$

is obtained, where

$$\overline{x}_i = i\overline{L} \qquad \sigma_{xi}^2 = i\sigma_L^2 \tag{20-34}$$

This treatment of the general one-dimensional random walk, which follows the method used originally by Lord Rayleigh, has far-reaching significance since it shows that the probability density functions $p(x_i)$ for the algebraic sum of i random variables, namely,

$$x_i \equiv \sum_{j=1}^{i} L_j \tag{20-35}$$

where $L_j(j = 12, \cdots, i)$ are selected in accordance with an arbitrary probability density function $p(L)$ like that shown in Fig. 20-7, approach a gaussian distribution in the limit as $i \to \infty$. This fact is contained in the so-called central-limit theorem, which is found in most textbooks on probability theory. Fortunately, the probability density function $p(x_i)$ approaches a gaussian distribution rapidly as i increases (except for large values of x); therefore, the often assumed gaussian distribution in engineering applications of Eq. (20-35) is usually justified.

Example   E20-4.   Consider the one-dimensional random walk as defined by Eq. (20-35), where the probability density function for a single step length is given in the discrete form

$$p(L) = 0.05\,\delta(L + 2\triangle x) + 0.15\,\delta(L + \triangle x) + 0.30\,\delta(L)$$

$$+0.40\,\delta(L - \triangle x) + 0.10\,\delta(L - 2\triangle x) \qquad \text{(a)}$$

This function is also the probability density function for random variable x_1. By successively distributing the Dirac delta function intensities, as done for the simpler case in Fig. 20-6, probability density functions $p(x_2), p(x_3)$, etc., can be obtained, as shown in Fig. E20-1. To ensure a complete understanding of this method, it is suggested that the student check the numerical values given in the figure for distributions $p(x_2)$ and $p(x_3)$. For comparison continuous normal distributions are plotted in Fig. E20-1 by dashed lines. These distributions have the same mean values and variances as the corresponding discrete distributions. Note the very rapid rate at which the discrete distributions are approaching the normal distributions with increasing values of i.

Obviously for large values of i, the above distribution technique for obtaining $p(x_i)$ is extremely tedious and time-consuming. However, a good approximation of this function can be obtained by assuming a normal distribution having a mean value and variance as given by Eqs. (20-34). Thus for the case represented by Eq. (a), the continuous distribution is

$$p(x_i) \doteq \frac{1}{\sqrt{2\pi}\,\sigma_{x_i}} \exp\left[-\frac{(x_i - \overline{x}_i)^2}{2\sigma_{x_i}^2} \right] \qquad \text{(b)}$$

where

$$\overline{x}_i = 0.35\,i\,\triangle x \qquad \sigma_{x_i}^2 = 1.0275\,i\,\triangle x^2 \qquad \text{(c)}$$

For large values of i, this distribution when discretized will give a very good approximation of the true distribution. Significant differences will appear only in the extreme "tail regions" of the distributions.

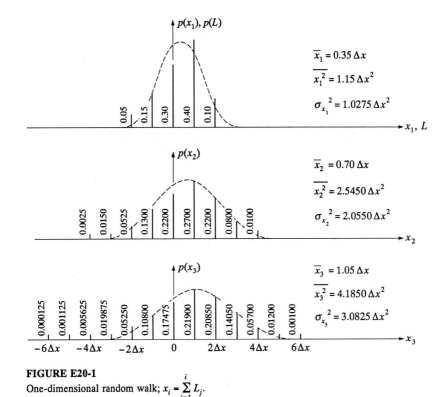

FIGURE E20-1
One-dimensional random walk; $x_i = \displaystyle\sum_{j=1}^{i} L_j$.

Let us now consider a random variable y_i defined as the product of i random variables, rather than the sum as given by Eq. (20-35), i.e., the variable defined by

$$y_i = L_1 L_2 \cdots L_j \cdots L_{i-1} L_i \qquad (20\text{-}36)$$

with the values of L_j $(j = 1, 2, \cdots, i)$ selected in accordance with an arbitrary probability density function $p(L)$ as represented in Fig. 20-7a. Taking the natural logarithm of this equation gives

$$z_i \equiv \ln y_i = \sum_{j=1}^{i} \ln L_j \qquad (20\text{-}37)$$

Based on the previous discussion regarding the probability density function for x_i, Eq. (20-35), it is clear that as i increases, the probability density function for z_i must approach the normal distribution

$$p(z_i) = \frac{1}{\sqrt{2\pi}\,\sigma_{z_i}} \exp\left[\frac{-(z_i - \bar{z}_i)^2}{2\,\sigma_{z_i}^2}\right] \qquad -\infty < z < \infty \qquad (20\text{-}38)$$

Using the relation

$$p(y_i) = p(z_i) \left| \frac{dz_i}{dy_i} \right| \tag{20-39}$$

corresponding to Eq. (20-4), one obtains

$$p(y_i) = \frac{1}{\sqrt{2\pi}\,\sigma_{\ln y_i}\,y_i} \exp\left[-\frac{(\ln y_i - \overline{\ln y_i})^2}{2\,\sigma_{\ln y_i}^2} \right] \qquad 0 \le y_i < \infty \tag{20-40}$$

which is called the lognormal distribution. It can be shown[1] that the mean and variance of $\ln y_i$ are given by

$$\overline{\ln y_i} = \ln y_m \qquad \sigma_{\ln y_i}^2 = \ln\left(1 + \frac{\sigma_{y_i}^2}{\overline{y_i}^2}\right) \tag{20-41}$$

respectively, in which y_m refers to the median (50 percentile) value of y_i, $\sigma_{y_i}^2$ is the variance of y_i, and $\overline{y_i}$ is the mean value of y_i; for small values of the ratio $\sigma_{y_i}/\overline{y_i}$, say less than 0.3, $\sigma_{\ln y_i} \doteq \sigma_{y_i}/\overline{y_i}$. The relationship between the median and mean values of y_i is

$$y_{im} = \frac{\overline{y_i}}{\sqrt{1 + (\sigma_{y_i}/\overline{y_i})^2}} \tag{20-42}$$

20-4   TWO RANDOM VARIABLES

This section is concerned with experiments involving two random variables. Suppose, for example, a discrete random variable N is obtained by spinning the disk shown in Fig. 20-1a while a second variable M is obtained by spinning a second disk of identical design. Obtaining n $(n \to \infty)$ pairs of numbers N and M in such a manner would in the limit give a discrete distribution of number pairs, as shown in Fig. 20-9. This distribution $p(N, M)$, which consists of 100 two-dimensional Dirac delta functions of intensity 1/100, is called the *joint probability density function* for random variables N and M. If instead of two disks of the type shown in Fig. 20-1a, two disks of the type shown in Fig. 20-1b are used to sample random variables θ and ϕ, sampling n $(n \to \infty)$ pairs would give the uniform distribution shown in Fig. 20-10. Note that the volume between the plane of the two random-variable axes and the surface of the joint probability function is normalized to unity in each case.

The joint probability density function $p(x, y)$ for a general experiment involving random variables x and y is shown in Fig. 20-11. This function is defined so that the element volume $p(x_1, y_1)\, dx\, dy$ as shown represents the probability that a pair of sampled values will be within the region $x_1 < x < x_1 + dx$ and $y_1 < y < y_1 + dy$.

---

[1] Alfredo H-S. Ang and Wilson H. Tang, *Probability Concepts in Engineering Planning and Design*, Volume I, Basic Principles, John Wiley & Sons, 1975.

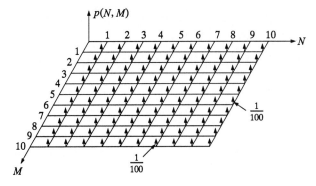

FIGURE 20-9
Joint probability density function for discrete random variables N and M.

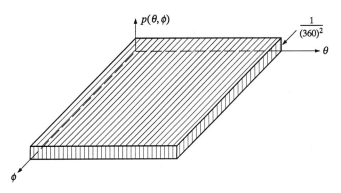

FIGURE 20-10
Joint probability density function for continuous random variables θ and ϕ.

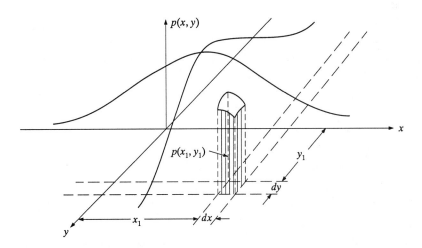

FIGURE 20-11
General joint probability density function for random variables x and y.

This definition requires that the total volume between the xy plane and the $p(x, y)$ surface equal unity, that is, $\int_{-\infty}^{\infty} \int_{-\infty}^{\infty} p(x, y)\, dx\, dy = 1$.

The most common joint probability density function (later used extensively) is the *normal*, or *gaussian, distribution* given by

$$p(x, y) = \frac{1}{2\pi ab\sqrt{1 - c^2}}$$

$$\times \exp\left\{ -\frac{1}{2(1 - c^2)} \left[\frac{(x - d)^2}{a^2} - \frac{2c(x - d)(x - e)}{ab} + \frac{(y - e)^2}{b^2} \right] \right\} \qquad (20\text{-}43)$$

where a, b, c, d, and e are constants.

Suppose the joint probability density function for a new set of random variables r and s as defined by the relations

$$r = r(x, y) \qquad s = s(x, y) \qquad (20\text{-}44)$$

is desired, where Eqs. (20-44) and their inverse relations

$$x = x(r, s) \qquad y = y(r, s) \qquad (20\text{-}45)$$

are single-valued functions. Because, as shown in Fig. 20-12, a square infinitesimal area $dr\, ds$ in the rs plane will map as a parallelogram of area

$$\left| \frac{\partial x}{\partial r} \frac{\partial y}{\partial s} - \frac{\partial x}{\partial s} \frac{\partial y}{\partial r} \right| dr\, ds \qquad (20\text{-}46)$$

in the xy plane, it is necessary that

$$p(r, s) = \left| \frac{\partial x}{\partial r} \frac{\partial y}{\partial s} - \frac{\partial x}{\partial s} \frac{\partial y}{\partial r} \right| p(x, y) \qquad (20\text{-}47)$$

since sampled values of x and y which fall within the parallelogram correspond to values of r and s within the square. The absolute value has been indicated in

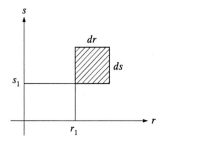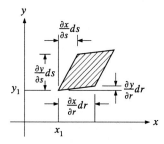

FIGURE 20-12
Jacobian transformation of two random variables.

Eq. (20-47), as the area of the parallelogram must always be a positive quantity. The transformation indicated by this equation is known as the *jacobian transformation*.

Next certain probability functions closely associated with the joint probability density function $p(x, y)$ are defined.

Marginal probability density function $p(x)$ is defined such that $p(x_1)\, dx$ equals the chances that a sampled value of x will be in the range $x_1 < x < x_1 + dx$ regardless of the value of y sampled. Likewise, marginal probability density function $p(y)$ is defined so that $p(y_1)\, dy$ equals the chances that a sampled value of y will be in the range $y_1 < y < y_1 + dy$ regardless of the value of x sampled. In accordance with the above definitions,

$$p(x_1)\, dx = \int_{-\infty}^{\infty} \int_{x_1}^{x_1+dx} p(x, y)\, dx\, dy = dx \int_{-\infty}^{\infty} p(x_1, y)\, dy \qquad (20\text{-}48)$$

Therefore, the marginal probability density functions in unrestricted form are given by the relations

$$p(x) = \int_{-\infty}^{\infty} p(x, y)\, dy \qquad p(y) = \int_{-\infty}^{\infty} p(x, y)\, dx \qquad (20\text{-}49)$$

Probability distribution function $P(X, Y)$ is defined such that $P(X_1, Y_1)$ equals the chances that sampled values of x and y will be within the ranges $-\infty < x < X_1$ and $-\infty < y < Y_1$, respectively; thus

$$P(X, Y) = \int_{-\infty}^{Y} \int_{-\infty}^{X} p(x, y)\, dx\, dy \qquad (20\text{-}50)$$

In differential form this becomes

$$p(X, Y) = \frac{\partial^2 P(X, Y)}{\partial X\, \partial Y} \qquad (20\text{-}51)$$

Conditional probability density function $p(x|y)$ is defined such that $p(x_1|y_1)dx$ equals the chances that x will be in the range $x_1 < x < x_1 + dx$ when considering *only* those sampled values of x and y which are in the ranges $-\infty < x < \infty$ and $y_1 < y < y_1 + dy$, respectively; that is, $p(x_1|y_1) = p(x_1, y_1)/p(y_1)$ or, in its unrestricted form,

$$p(x|y) = \frac{p(x, y)}{p(y)} \qquad (20\text{-}52)$$

Likewise

$$p(y|x) = \frac{p(x, y)}{p(x)} \qquad (20\text{-}53)$$

It should be noted that the conditional probability density functions are ratios of marginal and joint probability density functions.

The conditional probability density functions $p(x|y)$ and $p(y|x)$ are often functions of x and y, respectively. In such cases, Eqs. (20-52) and (20-53) require that

$$p(x|y) = p(x) \qquad \text{and} \qquad p(y|x) = p(y) \tag{20-54}$$

and that

$$p(x,y) = p(x)\,p(y) \tag{20-55}$$

Random variables which satisfy Eqs. (20-54) and (20-55) are said to be *statistically independent*. Physically this means that when values of x and y are sampled, the sampled values of x are not influenced by corresponding sampled values of y and vice versa. The random variables represented in Figs. 20-9 and 20-10 are examples of statistically independent variables.

Suppose someone involved with statistically independent random variables x and y wishes to obtain the probability density function for a random variable r defined as the sum of random variables x and y. This probability density function can easily be obtained by using the Jacobian transformation Eq. (20-47) and Eq. (20-48) as follows. Define a new set of random variables r and s in the form

$$r \equiv x + y \qquad s \equiv y \tag{20-56}$$

which, in inverse form, are

$$x = r - s \qquad y = s \tag{20-57}$$

Equations (20-47), (20-55), and (20-57) give

$$p(r,s) = p(x,y) = p_x(x)\,p_y(y) = p_x(r-s)\,p_y(s) \tag{20-58}$$

Subscripts have been added here to identify the random variables involved. The marginal probability density function $p(r)$ now becomes

$$p(r) = \int_{-\infty}^{\infty} p_x(r-s)\,p_y(s)\,ds \tag{20-59}$$

Example E20-5. Consider the one-dimensional random walk defined by Eq. (20-35), where the probability density function for a single step length is given in the continuous form

$$p(L) = \begin{cases} \dfrac{1}{\Delta x} & 0 < L < \Delta x \\ 0 & L < 0\,;\ L > \Delta x \end{cases} \tag{a}$$

This function is also the probability density function for random variable x_1. To obtain the probability density function for $x_2 = x_1 + L_2$, apply the convolution integral given by Eq. (20-59), writing

$$p(x_2) = \int_{-\infty}^{\infty} p_L(x_2 - s)\, p_{x_1}(s)\, ds \qquad \text{(b)}$$

where

$$p_{x_1}(s) = \begin{cases} \dfrac{1}{\Delta x} & 0 < s < \Delta x \\[2mm] 0 & s < 0 \; ; \; s > \Delta x \end{cases} \qquad \text{(c)}$$

$$p_L(x_2 - s) = \begin{cases} \dfrac{1}{\Delta x} & x_2 - \Delta x < s < x_2 \\[2mm] 0 & s < x_2 - \Delta x \; ; \; s > x_2 \end{cases} \qquad \text{(d)}$$

Substituting Eqs. (c) and (d) into Eq. (b) gives

$$p(x_2) = \begin{cases} \dfrac{1}{(\Delta x)^2} \displaystyle\int_0^{x_2} ds = \dfrac{1}{(\Delta x)^2} x_2 & 0 \le x_2 \le \Delta x \\[4mm] \dfrac{1}{(\Delta x)^2} \displaystyle\int_{x_2 - \Delta x}^{\Delta x} ds = \dfrac{1}{(\Delta x)^2}(2\Delta x - x_2) & \Delta x \le x_2 \le 2\Delta x \\[4mm] 0 & x_2 \le 0; x_2 \ge 2\Delta x \end{cases} \qquad \text{(e)}$$

When the probability density function for x_2 is known, the same convolution integral can be used once again to find the probability density function for random variable $x_3 = x_2 + L_3$, giving

$$p(x_3) = \int_{-\infty}^{\infty} p_L(x_3 - s)\, p_{x_2}(s)\, ds \qquad \text{(f)}$$

With Eqs. (a) and (e), the integrand terms of this integral can be written

$$p_{x_2}(s) = \begin{cases} \dfrac{1}{(\Delta x)^2} s & 0 < s < \Delta x \\[3mm] \dfrac{1}{(\Delta x)^2}(2\Delta x - s) & \Delta x < s < 2\Delta x \\[3mm] 0 & s < 0; s > 2\Delta x \end{cases} \qquad \text{(g)}$$

$$p_L(x_3 - s) = \begin{cases} \dfrac{1}{\Delta x} & x_3 - \Delta x < s < x_3 \\[2mm] 0 & s < x_3 - \Delta x; s > x_3 \end{cases} \qquad \text{(h)}$$

Substituting Eqs. (g) and (h) into Eq. (f) gives

$$
p(x_3) = \begin{cases}
\dfrac{1}{(\Delta x)^3} \displaystyle\int_0^{x_3} s\,ds = \dfrac{x_3^2}{2(\Delta x)^3} & 0 \le x_3 \le \Delta x \\[2ex]
\dfrac{1}{(\Delta x)^3}\left[\displaystyle\int_{x_3-\Delta x}^{\Delta x} s\,ds + \int_{\Delta x}^{x_3}(2\Delta x - s)\,ds\right] & \\[1ex]
\quad = \dfrac{1}{(\Delta x)^3}\left(-x_3^2 + 3\Delta x\,x_3 - \dfrac{3}{2}\Delta x^2\right) & \Delta x \le x_3 \le 2\Delta x \\[2ex]
\dfrac{1}{(\Delta x)^3}\displaystyle\int_{x_3-\Delta x}^{2\Delta x}(2\Delta x - s)\,ds & \\[1ex]
\quad = \dfrac{1}{(\Delta x)^3}\left(\dfrac{x_3^2}{2} - 3\Delta x\,x_3 - \dfrac{9}{2}\Delta x^2\right) & 2\Delta x \le x_3 \le 3\Delta x \\[2ex]
0 & x_3 \le 0 \,;\; x_3 \ge 3\Delta x
\end{cases}
$$

(i)

Probability density functions $p(x_1)$, $p(x_2)$, and $p(x_3)$ as given by Eqs. (a), (e), and (i), respectively, are plotted in Fig. E20-2.

For comparison, normal distributions are plotted in this figure by dashed lines. These distributions have the same mean values and variances as the corresponding exact distributions shown by the solid lines. Note the very rapid manner in which $p(x_i)$ approaches the normal distribution with increasing values of i. Although $p(x_4)$, $p(x_5)$, etc., could be obtained by repeated use of the convolution integral as above, this procedure would be very time-consuming. Therefore, as with the discrete case of Example E20-4, convergence toward the normal distribution as noted above allows one to assume a normal distribution having a mean value and variance given by Eq. (20-34). Thus for large values of i, one can use the normal form

$$
p(x_i) \doteq \frac{1}{\sqrt{2\pi}\,\sigma_{x_i}} \exp\left[-\frac{(x_i - \overline{x}_i)^2}{2\sigma_{x_i}^2}\right]
\tag{j}
$$

where

$$
\overline{x}_i = \frac{i\Delta x}{2} \qquad \sigma_{x_i}^2 = \frac{i\Delta x^2}{12}
\tag{k}
$$

Example E20-6.   Given the joint probability density function

$$
p(x, y) = C \exp\left[-\frac{2}{3}\left(\frac{x^2}{4} - \frac{xy}{6} + \frac{y^2}{9}\right)\right]
\tag{a}
$$

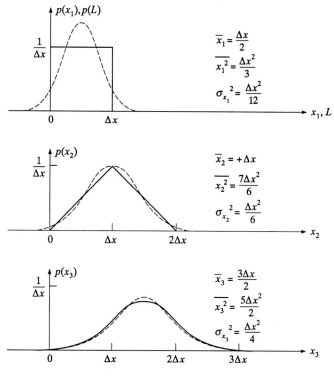

FIGURE E20-2
One-dimensional random walk; $x_i = \sum\limits_{j=1}^{i} L_j$.

find (1) the numerical value of C so that the function is normalized properly, (2) the marginal probability density functions $p(x)$ and $p(y)$, and (3) the conditional probability density functions $p(x|y)$ and $p(y|x)$. Show that the random variables x and y are statistically dependent.

The function $p(x, y)$ is properly normalized when its double integral over the infinite x and y domains equals unity, that is, when

$$C \int\limits_{-\infty}^{\infty} \int\limits_{-\infty}^{\infty} \exp\left[-\frac{2}{3}\left(\frac{x^2}{4} - \frac{xy}{6} + \frac{y^2}{9}\right)\right] dx\, dy = 1 \qquad (b)$$

Equation (b) can be separated and put into the equivalent form

$$C \int_{-\infty}^{\infty} \exp\left(-\frac{y^2}{18}\right)\left\{\int_{-\infty}^{\infty} \exp\left[-\frac{2}{3}\left(\frac{x}{2} - \frac{y}{6}\right)^2\right] dx\right\} dy = 1 \qquad (c)$$

By substituting the change of variable

$$u = \sqrt{\frac{2}{3}}\left(\frac{x}{2} - \frac{y}{6}\right) \qquad dx = \sqrt{6}\,du \tag{d}$$

Equation (c) becomes

$$\sqrt{6}\,C\int_{-\infty}^{\infty} \exp\left(-\frac{y^2}{18}\right)\left[\int_{-\infty}^{\infty} \exp(-u^2)\,du\right] dy = 1 \tag{e}$$

With another change of variable $v = y/3\sqrt{2}$ and the fact that the second integral in Eq. (e) equals $\sqrt{\pi}$ the result is

$$6\sqrt{3\pi}\,C\int_{-\infty}^{\infty} \exp(-v^2)\,dv = 1 \tag{f}$$

or

$$C = \frac{1}{6\sqrt{3}\,\pi} \tag{g}$$

With the first of Eqs. (20-49), the marginal probability density function $p(x)$ can be written

$$p(x) = \frac{1}{6\sqrt{3}\,\pi}\int_{-\infty}^{\infty} \exp\left[-\frac{2}{3}\left(\frac{x^2}{4} - \frac{xy}{6} + \frac{y^2}{9}\right)\right] dy \tag{h}$$

or

$$p(x) = \frac{1}{6\sqrt{3}\,\pi}\exp\left(-\frac{x^2}{8}\right)\int_{-\infty}^{\infty} \exp\left[-\frac{2}{3}\left(\frac{y}{3} - \frac{x}{4}\right)^2\right] dy \tag{i}$$

With the change of variable

$$u = \sqrt{\frac{2}{3}}\left(\frac{y}{3} - \frac{x}{4}\right) \qquad dy = 3\sqrt{\frac{3}{2}}\,du \tag{j}$$

Equation (i) becomes

$$p(x) = \frac{1}{2\sqrt{2}\,\pi}\exp\left(-\frac{x^2}{8}\right)\int_{-\infty}^{\infty} \exp(-u^2)\,du \tag{k}$$

or

$$p(x) = \frac{1}{2\sqrt{2\pi}}\exp\left(-\frac{x^2}{8}\right) \tag{l}$$

Similarly the second of Eqs. (20-49) gives

$$p(y) = \frac{1}{3\sqrt{2\pi}}\exp\left(-\frac{y^2}{18}\right) \tag{m}$$

Substituting Eqs. (a) and (m) into Eq. (20-52) and dividing as required gives

$$p(x|y) = \frac{1}{\sqrt{6\pi}} \exp\left[-\frac{2}{3}\left(\frac{x^2}{4} - \frac{xy}{6} + \frac{y^2}{36}\right)\right]$$ (n)

Likewise, substituting Eqs. (a) and (l) into Eq. (20-53) gives

$$p(y|x) = \frac{\sqrt{2}}{3\sqrt{3\pi}} \exp\left[-\frac{2}{3}\left(\frac{x^2}{16} - \frac{xy}{6} + \frac{y^2}{9}\right)\right]$$ (o)

Since the above marginal and conditional probability density functions do not satisfy Eqs. (20-54), random variables x and y are statistically dependent.

20-5 IMPORTANT AVERAGES OF TWO RANDOM VARIABLES

In an experiment involving random variables x and y, sampling in pairs is done n times, and each time a third random variable r is evaluated as defined by a single-valued function $r(x, y)$. The average of this third random variable as $n \to \infty$, that is,

$$\bar{r} = \lim_{n\to\infty} \frac{1}{n} \sum_{i=1}^{n} r(x_i, y_i)$$ (20-60)

where x_i and y_i are the ith sampled values of x and y, respectively, can be determined using the relation

$$\bar{r} = \int_{-\infty}^{\infty} \int_{-\infty}^{\infty} r(x, y)\, p(x, y)\, dx\, dy$$ (20-61)

The validity of Eq. (20-61) can easily be rationalized since $p(x, y)\, dx\, dy$ represents the fractional number of samples falling in the infinitesimal area $dx\, dy$ located at point (x, y).

Averages most commonly used when treating two random variables are the following: Mean values:

$$\bar{x} = \int_{-\infty}^{\infty} \int_{-\infty}^{\infty} x\, p(x, y)\, dx\, dy = \int_{-\infty}^{\infty} x\, p(x)\, dx$$

$$\bar{y} = \int_{-\infty}^{\infty} \int_{-\infty}^{\infty} y\, p(x, y)\, dx\, dy = \int_{-\infty}^{\infty} y\, p(y)\, dy$$ (20-62)

Mean square values:

$$\overline{x^2} = \int\limits_{-\infty}^{\infty} \int\limits_{-\infty}^{\infty} x^2 \, p(x,y) \, dx \, dy = \int_{-\infty}^{\infty} x^2 \, p(x) \, dx$$

$$\overline{y^2} = \int\limits_{-\infty}^{\infty} \int\limits_{-\infty}^{\infty} y^2 \, p(x,y) \, dx \, dy = \int_{-\infty}^{\infty} y^2 \, p(y) \, dy$$

(20-63)

Variances:

$$\sigma_x^2 = \overline{(x - \overline{x})^2} = \int\limits_{-\infty}^{\infty} \int\limits_{-\infty}^{\infty} (x - \overline{x})^2 \, p(x,y) \, dx \, dy = \overline{x^2} - \overline{x}^2$$

$$\sigma_y^2 = \overline{(y - \overline{y})^2} = \int\limits_{-\infty}^{\infty} \int\limits_{-\infty}^{\infty} (y - \overline{y})^2 \, p(x,y) \, dx \, dy = \overline{y^2} - \overline{y}^2$$

(20-64)

Standard deviations:

$$\sigma_x, \qquad \sigma_y \tag{20-65}$$

Covariance:

$$\mu_{xy} = \overline{(x - \overline{x})(y - \overline{y})} = \int\limits_{-\infty}^{\infty} \int\limits_{-\infty}^{\infty} (x - \overline{x})\,(y - \overline{y})\, p(x,y) \, dx \, dy = \overline{xy} - \overline{x}\,\overline{y} \tag{20-66}$$

Correlation coefficient:

$$\rho_{xy} \equiv \frac{\mu_{xy}}{\sigma_x \, \sigma_y} \tag{20-67}$$

Note that when x and y are statistically independent,

$$\overline{xy} = \int\limits_{-\infty}^{\infty} \int\limits_{-\infty}^{\infty} xy \, p(x)\,p(y) \, dx \, dy = \overline{x}\,\overline{y} \tag{20-68}$$

in which case both the covariance μ_{xy} and the correlation coefficient ρ_{xy} equal zero.

Substituting the normal, or gaussian, distribution as expressed by Eq. (20-43) into the above relations gives

$$\overline{x} = d \qquad \overline{y} = e \qquad \sigma_x = a \qquad \sigma_y = b \qquad \rho_{xy} = c \tag{20-69}$$

Therefore, the normal distribution can be, and usually is, expressed in the form

$$p(x,y) = \frac{1}{2\pi\sigma_x \, \sigma_y \sqrt{1 - \rho_{xy}^2}}$$

$$\times \exp\left\{ -\frac{1}{2(1 - \rho_{xy}^2)} \times \left[\frac{(x - \overline{x})^2}{\sigma_x^2} - \frac{2\rho_{xy}(x - \overline{x})(y - \overline{y})}{\sigma_x \, \sigma_y} + \frac{(y - \overline{y})^2}{\sigma_y^2} \right] \right\} \tag{20-70}$$

Example E20-7. Random variables x_1 and x_2 are statistically independent and are both uniformly distributed over the range 0 to 1. Two new random variables r_1 and r_2 are defined by

$$r_1 = (-2 \ln x_1)^{1/2} \cos 2\pi x_2 \qquad r_2 = (-2 \ln x_1)^{1/2} \sin 2\pi x_2 \qquad \text{(a)}$$

Find (1) the joint probability density function $p(r_1, r_2)$, (2) the marginal probability density functions $p(r_1)$ and $p(r_2)$, (3) the mean values of r_1 and r_2, (4) the variances of r_1 and r_2, and (5) the covariance of r_1 and r_2.

Inverting Eqs. (a) gives

$$x_1 = \exp\left[-\frac{1}{2}(r_1^2 + r_2^2)\right]$$

$$x_2 = \frac{1}{2\pi} \cos^{-1} \frac{r_1}{\sqrt{r_1^2 + r_2^2}} = \frac{1}{2\pi} \sin^{-1} \frac{r_2}{\sqrt{r_1^2 + r_2^2}} \qquad \text{(b)}$$

Thus,

$$\frac{\partial x_1}{\partial r_1} = -r_1 \exp\left[-\frac{1}{2}(r_1^2 + r_2^2)\right]$$

$$\frac{\partial x_1}{\partial r_2} = -r_2 \exp\left[-\frac{1}{2}(r_1^2 + r_2^2)\right]$$

$$\frac{\partial x_2}{\partial r_1} = -\frac{1}{2\pi} \frac{r_2}{r_1^2 + r_2^2} \qquad \text{(c)}$$

$$\frac{\partial x_2}{\partial r_2} = +\frac{1}{2\pi} \frac{r_1}{r_1^2 + r_2^2}$$

With the jacobian transformation, Eq. (20-47), the joint probability density function $p(r_1, r_2)$ can be expressed as

$$p(r_1, r_2) = \left| \frac{\partial x_1}{\partial r_1} \frac{\partial x_2}{\partial r_2} - \frac{\partial x_1}{\partial r_2} \frac{\partial x_2}{\partial r_1} \right| p(x_1, x_2) \qquad \text{(d)}$$

where

$$p(x_1, x_2) = p(x_1)p(x_2) = \begin{cases} 1 & \begin{cases} 0 < x_1 < 1 \\ 0 < x_2 < 1 \end{cases} \\ 0 & \begin{cases} x_1 < 0 \; ; \; x_1 > 1 \\ x_2 < 0 \; ; \; x_2 > 1 \end{cases} \end{cases} \qquad \text{(e)}$$

Substituting Eqs. (c) and (e) into Eq. (d) gives the normal distribution

$$p(r_1, r_2) = \frac{1}{2\pi} \exp\left[-\frac{1}{2}(r_1^2 + r_2^2)\right] \qquad \text{(f)}$$

Making use of Eqs. (20-49) results in the relations

$$p(r_1) = \frac{1}{2\pi} \exp\left(-\frac{r_1^2}{2}\right) \int_{-\infty}^{\infty} \exp\left(-\frac{r_2^2}{2}\right) dr_2 = \frac{1}{\sqrt{2\pi}} \exp\left(-\frac{r_1^2}{2}\right)$$

$$p(r_2) = \frac{1}{2\pi} \exp\left(-\frac{r_2^2}{2}\right) \int_{-\infty}^{\infty} \exp\left(-\frac{r_1^2}{2}\right) dr_1 = \frac{1}{\sqrt{2\pi}} \exp\left(-\frac{r_2^2}{2}\right)$$

(g)

Integrating in accordance with Eqs. (20-62) and (20-63) gives

$$\bar{r}_1 = \frac{1}{\sqrt{2\pi}} \int_{-\infty}^{\infty} r_1 \exp\left(-\frac{r_1^2}{2}\right) dr_1 = 0$$

$$\bar{r}_2 = \frac{1}{\sqrt{2\pi}} \int_{-\infty}^{\infty} r_2 \exp\left(-\frac{r_2^2}{2}\right) dr_2 = 0$$

(h)

$$\overline{r_1^2} = \frac{1}{\sqrt{2\pi}} \int_{-\infty}^{\infty} r_1^2 \exp\left(-\frac{r_1^2}{2}\right) dr_1 = 1$$

$$\overline{r_2^2} = \frac{1}{\sqrt{2\pi}} \int_{-\infty}^{\infty} r_2^2 \exp\left(-\frac{r_2^2}{2}\right) dr_2 = 1$$

Thus,

$$\sigma_{r_1}^2 = \overline{r_1^2} - \bar{r}_1^2 = 1 \qquad \sigma_{r_2}^2 = \overline{r_2^2} - \bar{r}_2^2 = 1 \tag{i}$$

Since r_1 and r_2 appear in an uncoupled form in Eq. (f), they are shown to be statistically independent. The mean value of $r_1 r_2$ is of the form given by Eq. (20-68), that is,

$$\overline{r_1 r_2} = \bar{r}_1 \bar{r}_2 \tag{j}$$

Therefore the covariance becomes

$$\mu_{r_1 r_2} = \overline{r_1 r_2} - \bar{r}_1 \bar{r}_2 = 0 \tag{k}$$

Example E20-8.   Given the joint probability density function used in Example E20-6, namely,

$$p(x,y) = \frac{1}{6\sqrt{3}\,\pi} \exp\left[-\frac{2}{3}\left(\frac{x^2}{4} - \frac{xy}{6} + \frac{y^2}{9}\right)\right] \tag{a}$$

find (1) the mean values, (2) the mean square values, (3) the variances, and (4) the covariance of random variables x and y.

These quantities could be obtained from the general relations given by Eqs. (20-62) to (20-64) and (20-66). However, comparison of this equation with the general form of the normal distribution given by Eq. (20-70) shows that it is obviously of similar form. Therefore these quantities can be obtained directly by setting the coefficients of terms in Eq. (a) equal to their corresponding coefficients in Eq. (20-70), giving

$$\frac{1}{6} = \frac{1}{2\sigma_x^2(1 - \rho_{xy}^2)} \qquad \frac{1}{9} = \frac{\rho_{xy}}{\sigma_x\sigma_y(1 - \rho_{xy}^2)} \qquad \frac{2}{27} = \frac{1}{2\sigma_y^2(1 - \rho_{xy}^2)} \qquad \text{(b)}$$

Solving Eqs. (b) for the three unknowns gives

$$\sigma_x = 2 \qquad \sigma_y = 3 \qquad \rho_{xy} = 1/2 \qquad \text{(c)}$$

The mean values $\bar{x}$ and $\bar{y}$ are obviously zero from the form of the equation; therefore,

$$\overline{x^2} = 4 \qquad \overline{y^2} = 9 \qquad \text{(d)}$$

The covariance is easily obtained since

$$\mu_{xy} = \sigma_x\sigma_y\,\rho_{xy} = 3 \qquad \text{(e)}$$

20-6 SCATTER DIAGRAM AND CORRELATION OF TWO RANDOM VARIABLES

The so-called scatter diagram can be helpful to the beginner in understanding the basic concepts and definitions of probability related to two random variables x and y. This diagram is obtained by sampling pairs of random variables and each time plotting them as a point on the xy plane, as shown in Fig. 20-13. Suppose n pairs are sampled and that (x_1, y_1), (x_2, y_2), $\cdots$, (x_n, y_n) represent their coordinates on the scatter diagram. If n_1, n_2, and n_3 represent the numbers of sampled pairs falling in regions $X < x < X + \Delta x$ and $Y < y < Y + \Delta y$, $X < x < X + \Delta x$ and $-\infty < y < +\infty$, and $-\infty < x < \infty$ and $Y < y < Y + \Delta y$, respectively, the joint,

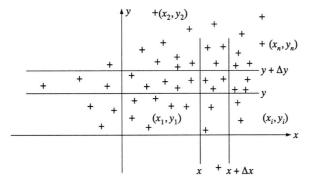

FIGURE 20-13
Scatter diagram for random variables x and y.

marginal, and conditional probability density functions as previously defined will be given by

$$p(x, y) = \lim_{\substack{\Delta x \to 0 \\ \Delta y \to 0 \\ n \to \infty}} \frac{n_1}{n \, \Delta x \, \Delta y} \tag{20-71}$$

$$p(x) = \lim_{\substack{\Delta x \to 0 \\ n \to \infty}} \frac{n_2}{n \, \Delta x} \qquad p(y) = \lim_{\substack{\Delta y \to 0 \\ n \to \infty}} \frac{n_3}{n \, \Delta y} \tag{20-72}$$

$$p(x|y) = \lim_{\substack{\Delta x \to 0 \\ \Delta y \to 0 \\ n \to \infty}} \frac{n_1}{n_3 \, \Delta x} \qquad p(y|x) = \lim_{\substack{\Delta x \to 0 \\ \Delta y \to 0 \\ n \to \infty}} \frac{n_1}{n_2 \, \Delta y} \tag{20-73}$$

Further, it is quite apparent that

$$\bar{x} = \lim_{n \to \infty} \frac{1}{n} \sum_{i=1}^{n} x_i \qquad \bar{y} = \lim_{n \to \infty} \frac{1}{n} \sum_{i=1}^{n} y_i \tag{20-74}$$

$$\overline{x^2} = \lim_{n \to \infty} \frac{1}{n} \sum_{i=1}^{n} x_i^2 \qquad \overline{y^2} = \lim_{n \to \infty} \frac{1}{n} \sum_{i=1}^{n} y_i^2 \tag{20-75}$$

$$\sigma_x^2 = \lim_{n \to \infty} \frac{1}{n} \sum_{i=1}^{n} (x_i - \bar{x})^2 \qquad \sigma_y^2 = \lim_{n \to \infty} \frac{1}{n} \sum_{i=1}^{n} (y_i - \bar{y})^2 \tag{20-76}$$

$$\mu_{xy} = \lim_{n \to \infty} \frac{1}{n} \sum_{i=1}^{n} (x_i - \bar{x})(y_i - \bar{y}) \tag{20-77}$$

Let us examine certain features of the correlation coefficient ρ_{xy} as defined by Eq. (20-67). First, to establish the range of possible numerical values which it may possess, consider two new random variables r and s as defined by the relations

$$r \equiv \frac{x - \bar{x}}{\sigma_x} \qquad s \equiv \frac{y - \bar{y}}{\sigma_y} \tag{20-78}$$

This transformation represents a translation of the coordinate axes and a scale-factor change along each axis, so that

$$\bar{r} = \bar{s} = 0 \qquad \overline{r^2} = \sigma_r^2 = \overline{s^2} = \sigma_s^2 = 1 \qquad \overline{rs} = \rho_{rs} = \rho_{xy} \tag{20-79}$$

Consider now the mean square value of $r \pm s$. Use of Eqs. (20-79) leads to

$$\overline{(r \pm s)^2} = 2 (1 \pm \rho_{rs}) \tag{20-80}$$

Since the mean square values given above must always be positive, the correlation coefficient must always be in the range

$$-1 < \rho_{rs} < +1 \tag{20-81}$$

From the normal distribution as given by Eq. (20-70), the joint probability density function for variables r and s as defined by Eq. (20-78) is easily obtained by using the Jacobian transformation, Eq. (20-47), thus yielding the relation

$$p(r,s) = \frac{1}{2\pi\sqrt{1-\rho_{rs}^2}} \exp\left[-\frac{1}{2(1-\rho_{rs}^2)}(r^2 - 2\rho_{rs}\,rs + s^2)\right] \tag{20-82}$$

Contour lines representing equal values of $p(r,s)$ are shown in Fig. 20-14 for one particular positive value of ρ_{rs}. To obtain the analytical expression for such contour lines, the natural logarithm of both sides of Eq. (20-82) is taken, giving

$$r^2 - 2\rho_{rs}\,rs + s^2 = C^2 \tag{20-83}$$

where C^2 is a constant which can be varied to correspond to a particular value of $p(r,s)$. When the correlation coefficient is positive, that is, in the range $0 < \rho_{rs} < 1$, Eq. (20-83) is the equation of an ellipse with its major and minor axes oriented as shown in Fig. 20-14. On the other hand, when the correlation coefficient is in the range $-1 < \rho_{rs} < 0$, this same equation represents an ellipse but with the directions of its major and minor principal axes reversed from those shown in Fig. 20-14. As the correlation coefficient approaches $+1$, profiles of $p(r,s)$ normal to the major principal axis at $+45°$ approach Dirac delta functions centered on this axis. Likewise, as the correlation coefficient approaches -1, profiles of $p(r,s)$ normal to the major principal axis at $-45°$ approach Dirac delta functions centered on this axis. When the correlation coefficient equals zero, Eq. (20-83) is the equation of a circle.

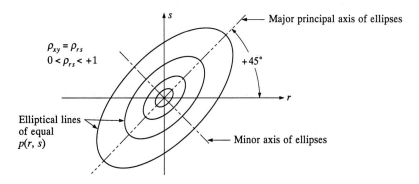

FIGURE 20-14
Contour lines of equal $p(r, s)$.

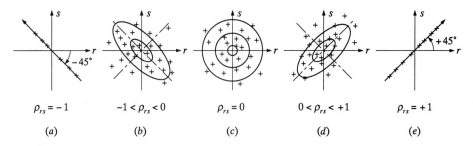

$$\rho_{rs} = -1 \qquad\quad -1 < \rho_{rs} < 0 \qquad\quad \rho_{rs} = 0 \qquad\quad 0 < \rho_{rs} < +1 \qquad\quad \rho_{rs} = +1$$

$$(a) \qquad\qquad\quad (b) \qquad\qquad\quad (c) \qquad\qquad\quad (d) \qquad\qquad\quad (e)$$

FIGURE 20-15
Contour lines of equal probability $p(r, s)$ with limited number of scatter points ($\rho_{rs} = \rho_{xy}$).

Contour lines of equal probability $p(r, s)$ as given by Eq. (20-82) along with a limited number of scatter points are shown in Fig. 20-15 for each of the above five cases. It is clear from the diagrams of this figure that random variables r and s (or x and y) are completely dependent upon each other when the correlation coefficient is either $+1$ or -1. In other words, only one random variable really exists in these cases, as one of the random variables can be determined directly from the other. However, when the correlation coefficient equals zero, as in Fig. 20-15c, the random variables are completely independent of each other. The cases in Fig. 20-15b and d are intermediate examples, representing partial statistical dependence of one random variable upon the other.

20-7 PRINCIPAL AXES OF JOINT PROBABILITY DENSITY FUNCTION

Consider random variables x and y having an arbitrary distribution as indicated in Fig. 20-16 which leads to nonzero mean values, variances, and covariance. It is often desirable to transform these random variables to a new set u and v having zero mean values and covariance, i.e., having nonzero variances only.

This conversion is easily accomplished in two steps using the coordinate transformations indicated in Fig. 20-17. First, transforming to random variables $X \equiv x - \bar{x}$ and $Y \equiv y - \bar{y}$, which corresponds to a translation of the coordinate axes so that their new origin is at the centroid of the distribution, the mean values $\bar{X}$ and $\bar{Y}$ are clearly equal to zero. Next, transforming to random variables

$$u \equiv X \cos \theta + Y \sin \theta$$
$$v \equiv -X \sin \theta + Y \cos \theta \tag{20-84}$$

which corresponds to a counter-clockwise rotation of the axes about the new origin, the covariance of u and v can also be brought to zero by properly specifying the rotation angle θ. Note that $\bar{u} = \bar{v} = 0$.

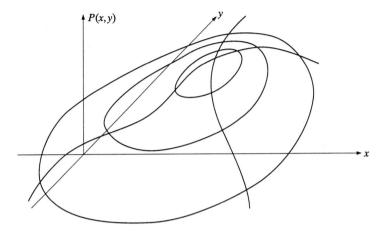

FIGURE 20-16
Arbitrary joint probability density function for random variables x and y.

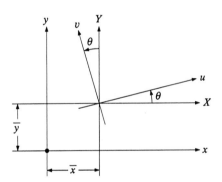

FIGURE 20-17
Coordinate transformations.

Expressing Eqs. (20-63), (20-64), and (20-66) in terms of u and v, rather than x and y, one finds upon substitution of Eqs. (20-84) that

$$\sigma_u^2 = \sigma_X^2 \cos^2 \theta + 2\mu_{XY} \cos \theta \sin \theta + \sigma_Y^2 \sin^2 \theta$$

$$\sigma_v^2 = \sigma_X^2 \sin^2 \theta - 2\mu_{XY} \cos \theta \sin \theta + \sigma_Y^2 \sin^2 \theta \tag{20-85}$$

$$\mu_{uv} = \mu_{XY}(\cos^2 \theta - \sin^2 \theta) - (\sigma_X^2 - \sigma_Y^2) \cos \theta \sin \theta$$

These equations are identical in form to those obtained in the transformation of two-dimensional plane stress as seen by substituting normal stresses for corresponding variances and shear stress for the covariance. Therefore the transformation of random variables X and Y to the new set u and v follows the same procedure used in the transformation of plane stress.

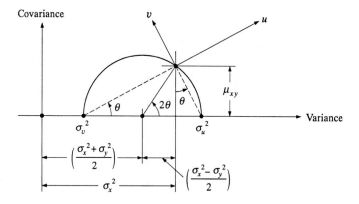

FIGURE 20-18
Mohr's circle for locating principal axes of joint probability.

Setting covariance μ_{uv} equal to zero in the third of Eqs. (20-85) and solving for θ gives the desired rotation angle

$$\theta = \frac{1}{2} \tan^{-1} \left(\frac{2\mu_{XY}}{\sigma_X^2 - \sigma_Y^2} \right)$$ (20-86)

which upon substitution into the first and second of Eqs. (20-85) yields

$$\sigma_{u,v}^2 = \left(\frac{\sigma_X^2 - \sigma_Y^2}{2} \right) \pm \sqrt{\left(\frac{\sigma_X^2 - \sigma_Y^2}{2} \right)^2 + \mu_{XY}^2}$$ (20-87)

The corresponding Mohr's circle for locating the principal axes u and v is shown in Fig. 20-18.

Subsequently in Section 20-10, it is shown that a linear transformation of normally distributed random variables yields a new set of random variables which are also normally distributed; therefore, the above transformation applied to the normal distribution on x and y as represented by Eq. (20-70) yields

$$p(u, v) = \frac{1}{2\pi \, \sigma_u \, \sigma_v} \exp \left[-\frac{1}{2} \left(\frac{u^2}{\sigma_u^2} + \frac{v^2}{\sigma_v^2} \right) \right]$$ (20-88)

in which σ_u and σ_v are given by Eqs. (20-87) through substitution of $\sigma_X^2 = \sigma_x^2$, $\sigma_Y^2 = \sigma_y^2$, and $\mu_{XY} = \mu_{xy}$.

Example   E20-9.   Consider the random variables x and y defined in Example E20-8 having the joint probability density function

$$p(x, y) = \frac{1}{6\sqrt{3}\,\pi} \exp \left[-\frac{2}{3} \left(\frac{x^2}{4} - \frac{xy}{6} + \frac{y^2}{9} \right) \right]$$ (a)

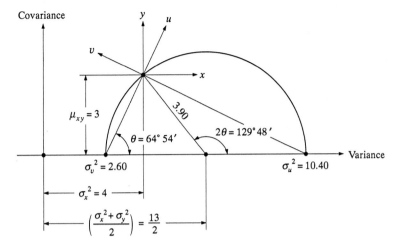

FIGURE E20-3
Mohr's circle for joint probability.

Defining new random variables u and v through the transformations

$$u = x \cos\theta + y \sin\theta$$
$$v = -x \sin\theta + y \cos\theta$$

(b)

find the angle θ which will uncouple u and v statistically and for this particular angle find the variances of u and v.

As shown in Example E20-8 the variances of x and y are 4 and 9, respectively, and their covariance is 3. These numerical values can be used to construct Mohr's circle for locating the principal axes of joint probability as shown in Fig. E20-3. From this circle it is readily seen that

$$\theta = \frac{1}{2}\tan^{-1}\left(\frac{2\mu_{xy}}{\sigma_x^2 + \sigma_y^2}\right) = 64° \, 54'$$

(c)

and that

$$\sigma_v^2 = 2.60 \qquad \sigma_u^2 = 10.40$$

(d)

20-8 RAYLEIGH PROBABILITY DENSITY FUNCTION

Consider random variables u and v as represented by the normal joint probability density function of Eq. (20-88) for the special case $\sigma_u = \sigma_v = \sigma$; thus

$$p(u, v) = \frac{1}{2\pi\,\sigma^2}\exp\left[-\frac{1}{2\sigma^2}(u^2 + v^2)\right]$$

(20-89)

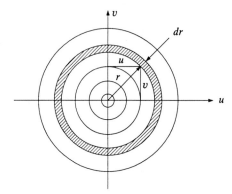

FIGURE 20-19
Equal probability contours as defined by
Eq. (20-89).

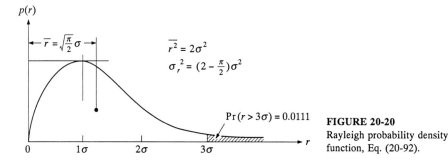

FIGURE 20-20
Rayleigh probability density
function, Eq. (20-92).

The contours of equal joint probability for this distribution are circles as shown in Fig. 20-19. If one is interested only in the absolute value of the vector sum of u and v, i.e.,

$$r \equiv (u^2 + v^2)^{1/2} \tag{20-90}$$

its probability density function $p(r)$ can be obtained easily since

$$p(r)\, dr = 2\pi r\, p(u, v)\, dr \tag{20-91}$$

Upon substitution of Eqs. (20-89) and (20-90), this relation gives

$$p(r) = \frac{r}{\sigma^2} \exp\left(-\frac{r^2}{2\sigma^2}\right) \qquad r \geq 0 \tag{20-92}$$

which is known as the Rayleigh distribution. This equation is plotted in Fig. 20-20.

Example E20-10. Consider random variable r having the Rayleigh distribution of Eq. (20-92). Find its most probable value, mean value, mean square value, variance, and probability distribution function. What is the probability that r will exceed 1σ, 3σ, and 5σ?

The most probable value of r is that value which maximizes Eq. (20-92); thus, differentiating with respect to r, setting the resulting equation to zero, and solving for r give $r = \sigma$. The mean value is found using Eq. (20-9); thus

$$\bar{r} = \int_0^\infty \frac{r^2}{\sigma^2} \exp\left(-\frac{r^2}{2\sigma^2}\right) dr = \sqrt{\frac{\pi}{2}}\, \sigma \tag{a}$$

In accordance with Eq. (20-10), the mean square value is given by

$$\overline{r^2} = \int_0^\infty \frac{r^3}{\sigma^2} \exp\left(-\frac{r^2}{2\sigma^2}\right) dr = 2\,\sigma^2 \tag{b}$$

The variance is found using

$$\sigma_r^2 = \overline{r^2} - \bar{r}^2 = \left(2 - \frac{\pi}{2}\right)\sigma^2 = 0.4292\,\sigma^2 \tag{c}$$

From Eq. (20-5), the probability distribution function is found to be

$$P(r) = 1 - \exp\left(-\frac{r^2}{2\sigma^2}\right) \tag{d}$$

giving the corresponding probability of exceedance function

$$Q(r) \equiv 1 - P(r) = \exp\left(-\frac{r^2}{2\sigma^2}\right) \tag{e}$$

from which $Q(1\sigma) = 0.6065$, $Q(3\sigma) = 0.0111$, and $Q(5\sigma) = 0.0000037$.

20-9 m RANDOM VARIABLES

Assume a single set of random variables $x_1, x_2, \cdots, x_m$ is obtained by spinning separately m disks of the type shown in either Fig. 20-1a or 20-1b. After obtaining n such sets in the limit as $n \to \infty$, a multivariate probability density function $p(x_1, x_2, \cdots, x_m)$ can be obtained as defined by

$$p(X_1, X_2, \cdots, X_m)\, dx_1\, dx_2 \cdots dx_m$$
$$\equiv \Pr(X_1 < x_1 < X_1 + dx_1, X_2 < x_2 < X_2 + dx_2,$$
$$\cdots, X_m < x_m < X_m + dx_m) \tag{20-93}$$

This probability density function will be of discrete form with disks of the type shown in Fig. 20-1a and of continuous form with disks of the type shown in Fig. 20-1b.

For a general experiment involving m random variables, the probability density function defined by Eq. (20-93) may be discrete, continuous, or a combination of these forms. However, because of the central-limit theorem referred to in Section 20-3, the normal distribution given by

$$p(x_1, x_2, \cdots, x_m) = \frac{1}{(2\pi)^{m/2} |\boldsymbol{\mu}|^{1/2}} \exp\left\{ -\frac{1}{2}\left[\mathbf{x} - \bar{\mathbf{x}}\right]^T \boldsymbol{\mu}^{-1}\left[\mathbf{x} - \bar{\mathbf{x}}\right]\right\} \qquad (20\text{-}94)$$

will often give reasonably good results in engineering applications. In this expression $\mathbf{x}$ and $\bar{\mathbf{x}}$ denote vectors

$$\mathbf{x} \equiv \left\{ \begin{matrix} x_1 \\ x_2 \\ \vdots \\ x_m \end{matrix} \right\} \qquad\qquad \bar{\mathbf{x}} \equiv \left\{ \begin{matrix} \bar{x}_1 \\ \bar{x}_2 \\ \vdots \\ \bar{x}_m \end{matrix} \right\} \qquad (20\text{-}95)$$

and $\boldsymbol{\mu}$ is the $m \times m$ matrix

$$\boldsymbol{\mu} \equiv \begin{bmatrix} \mu_{11} & \mu_{12} & \cdots & \mu_{1m} \\ \mu_{21} & \mu_{22} & \cdots & \mu_{2m} \\ \cdots\cdots\cdots\cdots\cdots\cdots \\ \mu_{m1} & \mu_{m2} & \cdots & \mu_{mm} \end{bmatrix} \qquad (20\text{-}96)$$

containing individual coefficients defined by

$$\mu_{ij} \equiv \overline{(x_i - \bar{x}_i)(x_j - \bar{x}_j)} \qquad i, j = 1, 2, \cdots, m \qquad (20\text{-}97)$$

These coefficients are covariances for $i \neq j$ and variances for $i = j$. Usually, however, $\boldsymbol{\mu}$ is simply referred to as the covariance matrix. The correlation coefficients are given by

$$\rho_{ij} \equiv \frac{\mu_{ij}}{\sqrt{\mu_{ii}\,\mu_{jj}}} = \frac{\mu_{ij}}{\sigma_{ii}\,\sigma_{jj}} \qquad (20\text{-}98)$$

The reader can easily verify that Eq. (20-94) reduces to the form of Eq. (20-70) when $m = 2$. If the random variables are statistically independent, the above covariance matrix will be a diagonal matrix and all m variables will appear in an uncoupled form in Eq. (20-94).

The multivariate probability density function for a new set of random variables $y_1, y_2, \cdots, y_m$ as defined by

$$y_1 = y_1(x_1, x_2, \cdots, x_m)$$

$$y_2 = y_2(x_1, x_2, \cdots, x_m)$$

$$\cdots\cdots\cdots\cdots\cdots\cdots\cdots \qquad (20\text{-}99)$$

$$y_m = y_m(x_1, x_2, \cdots, x_m)$$

can be obtained using the Jacobian transformation

$$p(y_1, y_2, \cdots, y_m) = \begin{vmatrix} \dfrac{\partial x_1}{\partial y_1} & \dfrac{\partial x_2}{\partial y_1} & \cdots & \dfrac{\partial x_m}{\partial y_1} \\ \dfrac{\partial x_1}{\partial y_2} & \dfrac{\partial x_2}{\partial y_2} & \cdots & \dfrac{\partial x_m}{\partial y_2} \\ \cdots\cdots\cdots\cdots\cdots\cdots\cdots \\ \dfrac{\partial x_1}{\partial y_m} & \dfrac{\partial x_2}{\partial y_m} & \cdots & \dfrac{\partial x_m}{\partial y_m} \end{vmatrix} p(x_1, x_2, \cdots, x_m) \qquad (20\text{-}100)$$

provided that Eqs. (20-99) and their inverse relations

$$x_1 = x_1(y_1, y_2, \cdots, y_m)$$

$$x_2 = x_2(y_1, y_2, \cdots, y_m)$$

$$\cdots\cdots\cdots\cdots\cdots\cdots \qquad (20\text{-}101)$$

$$x_m = x_m(y_1, y_2, \cdots, y_m)$$

are all single-valued functions. This procedure is a straightforward extension of the two-dimensional case treated earlier.

The statistical average of random variable $r = r(x_1, x_2, \cdots, x_m)$ can be obtained using the relation

$$\bar{r} = \int_{-\infty}^{\infty} \int_{-\infty}^{\infty} \cdots \int_{-\infty}^{\infty} r(x_1, x_2, \cdots, x_m)\, p(x_1, x_2, \cdots, x_m)\, dx_1\, dx_2 \, \cdots \, dx_m \quad (20\text{-}102)$$

which is a generalization of the simpler two-dimensional form given by Eq. (20-61).

20-10 LINEAR TRANSFORMATIONS OF NORMALLY DISTRIBUTED RANDOM VARIABLES

If Eqs. (20-99) are of the linear form

$$y_1 \equiv a_{11} x_1 + a_{12} x_2 + \cdots + a_{1m} x_m$$

$$y_2 \equiv a_{21} x_1 + a_{22} x_2 + \cdots + a_{2m} x_m$$

$$\cdots\cdots\cdots\cdots\cdots\cdots\cdots\cdots\cdots\cdots \qquad (20\text{-}103)$$

$$y_m \equiv a_{m1} x_1 + a_{m2} x_2 + \cdots + a_{mm} x_m$$

variables $y_1, y_2, \cdots, y_m$ will always have a gaussian distribution when variables $x_1, x_2, \cdots, x_m$ are normally distributed. To prove this important characteristic of linear transformations, substitute the matrix form of Eqs. (20-103), namely

$$\mathbf{y} = \mathbf{a}\,\mathbf{x} \qquad (20\text{-}104)$$

into the right hand side of Eq. (20-94) and apply the Jacobian transformation given by Eq. (20-100) to obtain

$$p(y_1, y_2, \cdots, y_m) = \frac{|\mathbf{a}^{-1}|}{(2\pi)^{m/2} |\boldsymbol{\mu}|^{1/2}} \exp\left\{ -\frac{1}{2}[\mathbf{y} - \bar{\mathbf{y}}]^T [\mathbf{a}^T]^{-1} \boldsymbol{\mu}^{-1} \mathbf{a}^{-1} [\mathbf{y} - \bar{\mathbf{y}}] \right\}$$

(20-105)

or

$$p(y_1, y_2, \cdots, y_m) = \frac{1}{(2\pi)^{m/2} |\mathbf{a}\,\boldsymbol{\mu}\,\mathbf{a}^T|^{1/2}} \exp\left\{ -\frac{1}{2}[\mathbf{y} - \bar{\mathbf{y}}]^T [\mathbf{a}\,\boldsymbol{\mu}\,\mathbf{a}^T]^{-1} [\mathbf{y} - \bar{\mathbf{y}}] \right\}$$

(20-106)

Evaluating the individual covariance terms

$$\nu_{ij} \equiv \overline{(y_i - \bar{y}_i)(y_j - \bar{y}_j)} \qquad i, j = 1, 2, \cdots, m \tag{20-107}$$

directly from Eqs. (20-103) shows the covariance matrix for random variables y_1, y_2, $\cdots$, y_m to be

$$\nu = \mathbf{a}\,\boldsymbol{\mu}\,\mathbf{a}^T \tag{20-108}$$

When Eq. (20-108) is substituted into Eq. (20-106), the desired probability density function becomes

$$p(y_1, y_2, \cdots, y_m) = \frac{1}{(2\pi)^{m/2} |\nu|^{1/2}} \exp\left\{ -\frac{1}{2}[\mathbf{y} - \bar{\mathbf{y}}]^T \nu^{-1} [\mathbf{y} - \bar{\mathbf{y}}] \right\} \tag{20-109}$$

which is obviously gaussian when compared with Eq. (20-94).

PROBLEMS

20-1. A random variable x has the probability density function

$$p(x) = \begin{cases} 1 - |x| & 0 \le |x| \le 1 \\ 0 & |x| \ge 1 \end{cases}$$

If a new random variable y is defined by the relation $y = ax^2$, find and plot the probability density function $p(y)$.

20-2. The probability density function for random variable x has the exponential form

$$p(x) = a \, \exp(-b|x|)$$

where a and b are constants. Determine the required relation between constants a and b and, for $a = 1$, find the probability distribution function $P(X)$.

20-3. Consider the one-dimensional random walk when the probability density function for a single step length is

$$p(L) = 0.6\delta\,(L - \Delta L) + 0.4\delta\,(L + \Delta L)$$

Find the probability density function for random variable x_4 as defined by

$$x_4 = \sum_{j=1}^{4} L_j$$

which represents distance from the origin after four steps.

20-4. Consider the one-dimensional random walk when the probability density function for a single step length is

$$p(L) = 0.1\delta(L + \Delta L) + 0.3\delta(L) + 0.5\delta(L - \Delta L) + 0.1\delta(L - 2\Delta L)$$

Approximately what is the probability of being a tlocation $6\Delta L$ after 10 steps?

20-5. Let x and y represent two statistically independent random variables and define a third random variable z as the product of x and y; that is, $z = xy$. Derive an expression for the probability density function $p(z)$ in terms of the probability density functions $p(x)$ and $p(y)$.

20-6. Two statistically independent random variables x and y have identical probability density functions:

$$p(x) = \begin{cases} \dfrac{1}{2} & -1 < x < 1 \\ 0 & x < -1; \; x > 1 \end{cases} \qquad\qquad p(y) = \begin{cases} \dfrac{1}{2} & -1 < y < 1 \\ 0 & y < -1; \; y > 1 \end{cases}$$

What is the probability density function for random variable z in the range $0 < z < 1$ when z is defined by the relation $z = yx^{-2}$?

20-7. The joint probability density function for two random variables x and y is

$$p(x, y) = \begin{cases} \dfrac{y}{\pi\sqrt{1 - x^2}} \exp\left(-\dfrac{y^2}{2}\right) & y \geq 0; \; |x| < 1 \\ 0 & \text{otherwise} \end{cases}$$

What are the expressions for the marginal probability density function $p(y)$ and the conditional probability density function $p(x \mid y)$, and what is the mean value of x? Are random variables x and y statistically independent?

20-8. Prove the validity of Eqs. (20-69).

20-9. The probability density function for the random variables x and y is

$$p(x, y) = \begin{cases} a \, \exp(-x - y) & x > 0; \; y > 0 \\ 0 & x < 0; \; y < 0 \end{cases}$$

Find the numerical value of a so that this function is properly normalized. What is the probability that x will be in the range $0 < x < 1$ when $y = 1$? Are random variables x and y statistically indenpendent? What is the probability that x and y will fall outside the square $OABC$ of unit area as shown in Fig. P20-1? Find the probability distribution function $P(X, Y)$.

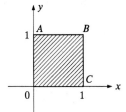

FIGURE P20-1
Region OABC in the xy plane of Prob. 20-9.

20-10. Random variables x and y are statistically independent and can be sampled in accordance with the marginal probability density functions

$$p(x) = \begin{cases} 2(1-x) & 0 < x < 1 \\ 0 & x < 0;\ x > 1 \end{cases} \qquad p(y) = \begin{cases} 2(1-y) & 0 < y < 1 \\ 0 & y < 0;\ y > 1 \end{cases}$$

Sketch the joint probability density function $p(x, y)$ and find mean values $\bar{x}$ and $\bar{y}$, mean square values $\overline{x^2}$ and $\overline{y^2}$, covariance μ_{xy}, and the mean value $\overline{x+y}$.

20-11. The joint probability density function for two random variables x and y equals a constant C over the region shown in Fig. P20-2 and equals zero outside that region.

(a) Find the numerical value of C so that $p(x, y)$ is properly normalized.

(b) Plot the marginal probability density functions $p(x)$ and $p(y)$.

(c) Plot the conditional probability density functions $p(x\,|\,y = 0.5)$ and $p(y\,|\,x = 1.5)$.

(d) Are random variables x and y statistically independent?

(e) Find mean values $\bar{x}$ and $\bar{y}$, variances σ_x^2 and σ_y^2, and the covariance μ_{xy}.

(f) Consider sampling values of x and y, say x_1, x_2, x_3, $\cdots$ and y_1, y_2, y_3, $\cdots$, respectively. If two new random variables r and s are defined as

$$r_n = x_1 + x_2 + x_3 + \cdots + x_n$$

$$s_n = y_1 + y_2 + y_3 + \cdots + y_n$$

find an appropriate expression for the joint probability density function $p(r_n, s_n)$ when $n = 20$.

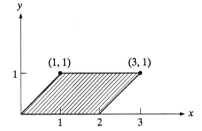

FIGURE P20-2
Region of nonzero joint probability in the xy plane of Prob. 20-11.

20-12. Consider again random variables x and y as defined in Prob. 20-11. Defining two new random variables u and v through the transformation

$$u = (y - A) \, \sin \theta + (x - B) \, \cos \theta$$

$$v = (y - A) \, \cos \theta - (x - B) \, \sin \theta$$

find the values of A and B which will give zero-mean values for u and v and find the angle θ which will uncouple u and v statistically. For this particular angle find the variances of u and v.

CHAPTER
21

RANDOM
PROCESSES

21-1 DEFINITION

A random process is a family, or ensemble, of n random variables related to a similar phenomenon which may be functions of one or more independent variables. For example, suppose n accelerometers are mounted on the frames of n automobiles for the purpose of measuring vertical accelerations as these automobiles travel over a rough country road. The recorded accelerometer signals $x_i(t)$ $(i = 1, 2, \cdots, n)$, which are functions of one independent variable, namely, time t, might look something like the waveforms shown in Fig. 21-1. Each waveform in such a process differs from all other waveforms; that is, $x_r(t) \neq x_s(t)$ for $r \neq s$. To characterize this process $x(t)$ in a probabilistic sense, it is necessary to establish the multivariate probability density function $p(x_1, x_2, \cdots, x_m)$ as defined by the relation

$$p(X_1, X_2, \cdots, X_m)dx_1\, dx_2 \cdots dx_m$$

$$\equiv \Pr(X_1 < x_1 < X_1 + dx_1, X_2 < x_2 < X_2 + dx_2, \cdots, X_m < x_m < X_m + dx_m)$$

$$(21\text{-}1)$$

for $m = 1, 2, \cdots$, where x_i is the random variable consisting of sample values $x_{i1}, x_{i2}, \cdots, x_{im}$ across the ensemble at time t_i. Usually in engineering fields, it is sufficient to establish only the first two of these functions, that is, $p(x_1)$ and $p(x_1, x_2)$ but with t_1 and t_2 treated as variables.

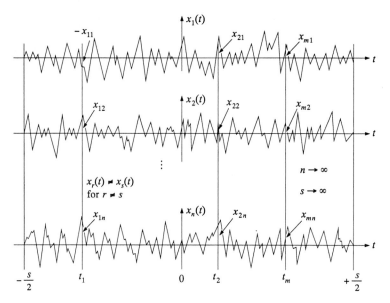

FIGURE 21-1
Random process (one independent variable).

The number of members n in the ensemble required to characterize a random process depends upon the type of process and the accuracy desired. Should it be necessary to establish the probability density functions statistically by sampling values of the random variables across the ensemble, exact results are obtained only in the limit as n approaches infinity. In practice, however, sufficient accuracy can be obtained using a finite number of members.

For some random processes, the desired probability density functions can be determined from an analysis of a single member of each process, in which case their exact characterizations are obtained only in the limit as the duration s approaches infinity. In practice these processes are always limited in duration; therefore, the characterizations obtained can only be approximate; however, engineering accuracy can usually be obtained with relatively short-duration sample waveforms.

In the above example, time t happens to be the independent variable, but it should be recognized that in general the independent variable can be any quantity.

As a second example of a random process, consider the wind drag force per unit height $p(x, t)$ acting on a tall industrial smokestack during a strong windstorm. This forcing function will contain a large steady-state or static component but will in addition contain a significant random component due to air turbulence. Clearly such turbulence produces drag forces which are not only random with respect to time t but are random with respect to the vertical space coordinate x as well. This process

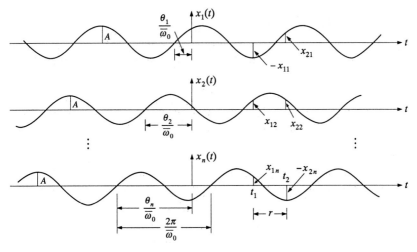

FIGURE 21-2
Random process of harmonic waveform.

therefore involves two independent variables.

The pressure fluctuations over the surface of an aircraft during flight are an example of a random process involving three independent variables, namely, time and two surface coordinates.

Obviously, the larger the number of independent variables involved in a random process, the more difficult it is to characterize the process.

21-2   STATIONARY AND ERGODIC PROCESSES

A specific random process will now be described in detail to help the reader develop a better understanding of random processes involving one independent variable. Consider the random process $x(t)$ shown in Fig. 21-2, which is defined by the relation

$$x_r(t) = A\sin(\overline{\omega}_0 t + \theta_r) \qquad r = 1, 2, \cdots, \infty \qquad (21\text{-}2)$$

where $x_r(t)$ = rth member of the ensemble

A = fixed amplitude for each harmonic waveform

$\overline{\omega}_0$ = fixed circular frequency

θ_r = rth sampled value of a random phase angle θ having a uniform probability density function in the range $0 < \theta < 2\pi$ of intensity $1/2\pi$

This process shows that waveforms need not be irregular, that is, contain many frequency components, to be classified as random. Harmonic, periodic, or aperiodic waveforms may or may not be random, depending upon whether they are fully pre-scribed or not. If known in a probabilistic sense only, they are defined as random. From this definition it is clear that once a random signal has been sampled, that par-ticular waveform immediately becomes fully known and can no longer by itself be considered random; however, it still is considered part of the random process from which it was sampled. By statistically studying a sufficient number of sampled wave-forms, the probability density functions for the process can be estimated, in which case any unsampled waveform becomes known in a probabilistic sense.

To establish the probability density function for random variable $x_1 \equiv x(t_1)$, a transformation relation similar to that given by Eq. (20-4) is used, namely,

$$p(x_1) = 2p(\theta)\left|\frac{d\theta}{dx_1}\right| \tag{21-3}$$

This equation differs slightly from Eq. (20-4) since the latter is valid only when $x_1 = x_1(\theta)$ and $\theta = \theta(x_1)$, its inverse relation, are single-valued functions. In this example, however, as random variable θ is allowed to change over its full range $0 < \theta < 2\pi$, random variable x_1 changes not once but twice over the range $-A < x_1 < +A$, which explains why the factor of 2 appears in Eq. (21-3). When Eq. (21-2) is substituted into Eq. (21-3) and the known information

$$p(\theta) = \begin{cases} \dfrac{1}{2\pi} & 0 < \theta < 2\pi \\ 0 & \theta < 0 \;;\; 0 > 2\pi \end{cases} \tag{21-4}$$

is used, the probability density function $p(x_1)$ becomes

$$p(x_1) = \begin{cases} \dfrac{1}{\pi\sqrt{A^2 - x_1^2}} & -A < x_1 < A \\ 0 & x_1 < -A \;;\; x_1 > A \end{cases} \tag{21-5}$$

Equations (21-4) and (21-5) are plotted in Fig. 21-3.

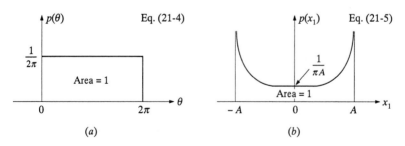

FIGURE 21-3
Probability density functions for θ and x_1, where $x_1 = A\sin(\overline{\omega}_0 t_1 + \theta)$.

The joint probability density function $p(x_1, x_2)$, where $x_1 \equiv x(t_1)$ and $x_2 \equiv x(t_2)$, can be obtained for the above process in the following manner. First, by using the appropriate trigonometric identity, x_2 can be expressed in the form

$$x_2 \equiv x(t_2) = x_1 \cos \overline{\omega}_0 \tau \pm \sqrt{A^2 - x_1^2} \sin \overline{\omega}_0 \tau \qquad -A \leq x_1 \leq A \qquad (21\text{-}6)$$

Clearly this relation shows that for any sampled value of x_1, random variable x_2 has only two possible values with equal chances of occurring. In other words, for a given time interval $\tau = t_2 - t_1$, the conditional probability density function $p(x_2|x_1)$ consists of two Dirac delta functions, namely,

$$p(x_2|x_1) = \frac{1}{2} \left[\delta \left(x_2 - x_1 \cos \overline{\omega}_0 \tau + \sqrt{A^2 - x_1^2} \sin \overline{\omega}_0 \tau \right) \right.$$

$$\left. + \delta \left(x_2 - x_1 \cos \overline{\omega}_0 \tau - \sqrt{A^2 - x_1^2} \sin \overline{\omega}_0 \tau \right) \right] \qquad (21\text{-}7)$$

Substituting Eqs. (21-5) and (21-7) into the following form of Eq. (20-53)

$$p(x_1, x_2) = p(x_1) \, p(x_2|x_1) \qquad (21\text{-}8)$$

leads to

$$p(x_1, x_2) = \frac{1}{2\pi \sqrt{A^2 - x_1^2}}$$

$$\times \left[\delta(x_2 - x_1 \cos \overline{\omega}_0 \tau + \sqrt{A^2 - x_1^2} \sin \overline{\omega}_0 \tau) \right.$$

$$\left. + \delta(x_2 - x_1 \cos \overline{\omega}_0 \tau - \sqrt{A^2 - x_1^2} \sin \overline{\omega}_0 \tau) \right] \qquad (21\text{-}9)$$

which is valid in the range $-A < x_2 < +A$ and $-A < x_1 < +A$. Outside this range $p(x_1, x_2)$ equals zero.

Example E21-1. Consider the single harmonic random process defined by Eq. (21-2), namely,

$$x_r(t) = A \sin(\overline{\omega}_0 t + \theta_r) \qquad r = 1, 2, \cdots, \infty \qquad (a)$$

where A is a fixed amplitude, $\overline{\omega}_0$ is a fixed circular frequency, and θ_r is the rth sampled value of a random phase angle θ having a uniform probability density function over the range $0 < \theta < 2\pi$. Defining the random variables x_1 and x_2 as

$$x_1 \equiv x(t) \qquad x_2 \equiv x(t + \tau) \qquad (b)$$

characterize the form of the scatter diagram for variables x_1 and x_2 and plot the diagram for $\overline{\omega}_0\tau = 0,\ \pi/4,\ \pi/2,\ 3\pi/4,$ and π.

The form of the scatter diagram can easily be obtained from Eq. (21-9) by noting that sample pairs of random variables x_1 and x_2 must satisfy the condition

$$x_2 - x_1 \cos\overline{\omega}_0\tau = \pm\sqrt{A^2 - x_1^2}\,\sin\overline{\omega}_0\tau \tag{c}$$

Squaring both sides of Eq. (c) gives

$$x_2^2 - 2\cos\overline{\omega}_0\tau\, x_1 x_2 + x_1^2 = A^2 \sin^2\overline{\omega}_0\tau \tag{d}$$

This equation represents an ellipse with its major and minor axes at 45° from the x_1 and x_2 axes. To determine the dimensions of the ellipse along the major and minor axes, transform Eq. (d) to a new set of orthogonal axes u and v located on the principal axes of the ellipse; that is, use the linear transformation

$$u = \frac{1}{\sqrt{2}}(x_1 + x_2) \qquad v = \frac{1}{\sqrt{2}}(x_2 - x_1) \tag{e}$$

to obtain

$$\frac{u^2}{a^2} + \frac{v^2}{b^2} = 1 \tag{f}$$

where

$$a^2 = \frac{\sin^2\overline{\omega}_0\tau}{1 - \cos\overline{\omega}_0\tau}A^2 \qquad b^2 = \frac{\sin^2\overline{\omega}_0\tau}{1 + \cos\overline{\omega}_0\tau}A^2 \tag{g}$$

Thus it is shown that the scatter diagram is in the form of an ellipse with its principal axes at 45° from the x_1 and x_2 axes and with the ellipse dimensions along its principal axes being

$$2a = \frac{2\sin\overline{\omega}_0\tau}{\sqrt{1 - \cos\overline{\omega}_0\tau}}A \qquad 2b = \frac{2\sin\overline{\omega}_0\tau}{\sqrt{1 + \cos\overline{\omega}_0\tau}}A \tag{h}$$

as shown in Fig. E21-1. Substituting the values $0,\ \pi/4,\ \pi/2,\ 3\pi/4,$ and π, separately, into Eqs. (h) for $\overline{\omega}_0\tau$ gives the corresponding values $\sqrt{2}\,A,\ 1.31\,A,$ $1.00\,A,\ 0.54\,A,$ and 0 for a and $0,\ 0.54\,A,\ 1.00\,A,\ 1.31\,A,$ and $\sqrt{2}\,A$ for b. Plots of the scatter diagrams for each of these five cases are shown in Fig. E21-2. Note from the figure that the ellipse degenerates into a straight line for $\overline{\omega}_0\tau = 0$ and π. From the above it is clear that a straight line with positive slope of 1 will occur for $\overline{\omega}_0\tau = 0,\ 2\pi,\ 4\pi,\ 6\pi,\cdots$, a straight line with negative slope of 1 will occur for $\overline{\omega}_0\tau = \pi,\ 3\pi,\ 5\pi,\cdots$, a circle will occur for $\overline{\omega}_0\tau = \pi/2,\ 3\pi/2,$ $5\pi/2,\cdots$, and an ellipse will occur for all other values of $\overline{\omega}_0\tau$.

Usually of main interest are the mean values, mean square values, variances, the covariance, and the correlation coefficient for random variables x_1 and x_2. Using

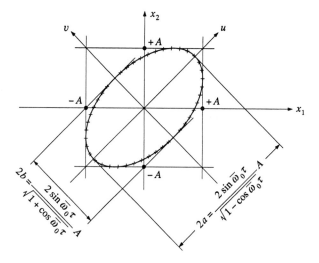

FIGURE E21-1
Scatter diagram for random variables x_1 and x_2 derived from single
harmonic process, Eq. (21-2).

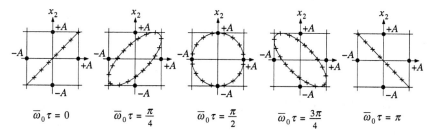

FIGURE E21-2
Scatter diagrams for five cases of the more general diagram in Fig. E21-1.

Eqs. (20-62) to (20-68) and (21-9) gives the following ensemble averages for the
process:

Mean values: $\qquad\qquad E(x_1) = E(x_2) = 0$

Mean square values: $\qquad E(x_1^2) = E(x_2^2) = \dfrac{A^2}{2}$

Variances: $\qquad\qquad\quad \sigma_{x_1}^2 = \sigma_{x_2}^2 = \dfrac{A^2}{2}$

Covariance: $\qquad\qquad\quad \mu_{x_1 x_2} = \dfrac{A^2}{2} \cos \overline{\omega}_0 \tau$

Correlation coefficient: $\qquad \rho_{x_1 x_2} = \cos \overline{\omega}_0 \tau \qquad\qquad$ (21-10)

The letter E has been introduced as a substitute for the bar previously placed above the random variable. It indicates that the variable has been averaged across the ensemble.

It is significant to note that all ensemble averages for this example process are independent of time t. Processes having this characteristic are defined as *stationary processes*.

It is also significant that for this process, any average obtained with respect to time t along any member r of the ensemble is exactly equal to the corresponding average across the ensemble at an arbitrary time t. Mathematically, this statement can be expressed in the form

$$\langle f(x_r) \rangle_{r=1,2,\cdots} \equiv \lim_{s \to \infty} \frac{1}{s} \int_{-s/2}^{s/2} f(x_r) \, dt = E\left[f(x_i)\right]_{i=1,2,\cdots} \tag{21-11}$$

where $f(x_r)$ is any function of the variable $x_r(t)$, $x_i = x(t_i)$, and where the angle brackets indicate time average. Processes having this characteristic are defined as *ergodic processes*.

It is suggested that the reader check the results given by Eq. (21-10) using Eq. (21-11) to show that the example process being considered, Eq. (21-2), is indeed ergodic; that is, show

$$\langle x_r \rangle = \lim_{s \to \infty} \frac{1}{s} \int_{-s/2}^{s/2} x_r(t) \, dt = 0$$

$$\langle x_r^2 \rangle = \lim_{s \to \infty} \frac{1}{s} \int_{-s/2}^{s/2} x_r(t)^2 \, dt = \frac{A^2}{2}$$

$$\sigma_{x_r}^2 = \frac{A^2}{2} \qquad r = 1, 2, \cdots \tag{21-12}$$

$$\mu(\tau) = \frac{A^2}{2} \cos \overline{\omega}_0 \tau$$

$$\rho(\tau) = \cos \overline{\omega}_0 \tau$$

According to the above definitions, an ergodic process must always be stationary; however, a stationary process may or may not be ergodic.

21-3 AUTOCORRELATION FUNCTION FOR STATIONARY PROCESSES

Consider again the general random process $x(t)$ shown in Fig. 21-1, which involves one independent variable. Assume for this discussion that this process is stationary (but not necessarily ergodic) and that it has a zero ensemble mean value, that is, $E(x) = 0$.

The covariance function $E[x(t)x(t+\tau)]$ in this case, like all ensemble averages, will be independent of time t and therefore will be a function of τ only. This function of τ will be referred to subsequently as the *autocorrelation function* and will be expressed in the form

$$R_x(\tau) = E[x(t)\,x(t+\tau)] \tag{21-13}$$

Certain important properties of the autocorrelation function should be noted, namely,

$$R_x(0) = \sigma_x^2 \qquad R_x(\tau) = R_x(-\tau) \qquad |R_x(\tau)| \leq R_x(0) \tag{21-14}$$

The first of Eqs. (21-14) is obvious since $R_x(0) = E[x(t)x(t)]$ is the variance when $E[x] = 0$. The second equation is a direct result of the assumed stationarity of the process, and the third equation can readily be proved using the fact that the following mean square average must always be greater than or equal to zero:

$$E\{[x(t) \pm x(t+\tau)]^2\} = R_x(0) \pm 2R_x(\tau) + R_x(0) \geq 0 \tag{21-15}$$

or

$$|R_x(\tau)| \leq R_x(0) \tag{21-16}$$

For most stationary processes, the autocorrelation function decays rapidly with increasing values of τ, thus showing a similar rapid loss of correlation of the two random variables as they are separated with respect to time. One notable exception, however, is the random process consisting of discrete harmonic waveforms, as shown in Fig. 21-2. This process has the autocorrelation function

$$R_x(\tau) = E(x_1\,x_2) = \frac{A^2}{2}\cos\overline{\omega}_0\tau \tag{21-17}$$

Clearly, regardless of the process, the two random variables $x(t)$ and $x(t+\tau)$ approach each other numerically as the time separation τ approaches zero. Therefore, these variables correlate completely in the limit as reflected by the correlation coefficient

$$\rho_x(0) = \frac{R_x(0)}{\sigma_x^2} = 1 \tag{21-18}$$

It is very significant to note that if the general process $x(t)$ being considered is stationary, has a zero mean value $E[x(t)] = 0$, and has the gaussian distribution given by Eq. (20-94), the autocorrelation function $R_x(\tau)$ completely characterizes the process. This fact is evident since all variance and covariance functions given by Eq. (20-97) are directly related to the autocorrelation function as follows:

$$\mu_{ik} = \begin{cases} R_x(0) & i = k \\ R_x(\tau) & i \neq k \end{cases} \qquad \tau = t_k - t_i \qquad (21\text{-}19)$$

For an ergodic process, the ensemble average given by Eq. (21-13) can be obtained by averaging along any single member (x_r) of the ensemble, in which case the autocorrelation function is more easily obtained using the relation

$$R_x(\tau) = \lim_{s \to \infty} \frac{1}{s} \int_{-s/2}^{s/2} x_r(t) x_r(t + \tau)\, dt \qquad r = 1, 2, \cdots \qquad (21\text{-}20)$$

It should now be obvious to the reader why a gaussian ergodic process is so easily characterized in a probabilistic sense.

Example E21-2.   A sample function $x_r(t)$ of random process $x(t)$ is established by assigning statistically independent sampled values of a random variable x to successive ordinates spaced at equal intervals along the time abscissa and by assuming a linear variation of the ordinates over each interval as shown in Fig. E21-3. A complete ensemble of such sample functions $(r = 1, 2, \cdots)$ can be obtained in a similar manner.

   If the probability density function for x is prescribed arbitrarily, except that its mean value $\bar{x}$ is held equal to zero, and if the ordinate x_{1r} occurs at time $t = \alpha_r$, where α_r is a sampled value of a random variable α uniformly distributed over the range $0 < \alpha < \Delta\varepsilon$, determine the mean value, mean square value, and variance of $x(t)$ and the covariance of $x(t)$ and $x(t+\tau)$. What kind of random process is $x(t)$?

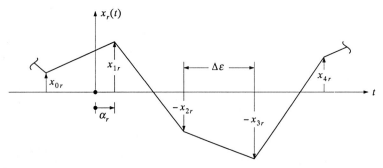

FIGURE E21-3
Sample function $x_r(t)$ from random process $x(t)$.

First, consider the above process but with all values of $\alpha_r (r = 1, 2, \cdots)$ set equal to zero, thus forcing all ordinates $x_{ir}(i, r = 1, 2, \cdots)$ to occur at time $t = (i - 1)\Delta\varepsilon$. The linear variation of ordinates shown in Fig. E21-3 leads to

$$x_r(t) = \left(1 - \frac{t}{\Delta\varepsilon}\right)x_{1r} + \frac{t}{\Delta\varepsilon}x_{2r} \qquad 0 < t < \Delta\varepsilon$$

$$x_r(t + \tau) = \begin{cases} \left(1 - \dfrac{t + \tau + \Delta\varepsilon}{\Delta\varepsilon}\right)x_{0r} + \dfrac{t + \tau + \Delta\varepsilon}{\Delta\varepsilon}x_{1r} & -\Delta\varepsilon < t + \tau < 0 \\[2mm] \left(1 - \dfrac{t + \tau}{\Delta\varepsilon}\right)x_{1r} + \dfrac{t + \tau}{\Delta\varepsilon}x_{2r} & 0 < t + \tau < \Delta\varepsilon \\[2mm] \left(1 - \dfrac{t + \tau - \Delta\varepsilon}{\Delta\varepsilon}\right)x_{2r} + \dfrac{t + \tau - \Delta\varepsilon}{\Delta\varepsilon}x_{3r} & \Delta\varepsilon < t + \tau < 2\Delta\varepsilon \end{cases}$$

$$\text{(a)}$$

Taking the ensemble average of the first of Eqs. (a) gives

$$E[x(t)] = \left(1 - \frac{t}{\Delta\varepsilon}\right)E(x_1) + \frac{t}{\Delta\varepsilon}E(x_2)$$

However, when it is noted that

$$E(x_i) = \overline{x} = \int_{-\infty}^{\infty} x\, p(x)\, dx \qquad i = 1, 2, \cdots \tag{b}$$

the result is

$$E[x(t)] = \overline{x} = 0 \tag{c}$$

Squaring the first of Eqs. (a) and taking the ensemble average gives

$$E[x(t)^2] = \left(1 - \frac{t}{\Delta\varepsilon}\right)^2 E(x_1^2) + 2\left(1 - \frac{t}{\Delta\varepsilon}\right)\frac{t}{\Delta\varepsilon}E(x_1 x_2) + \left(\frac{t}{\Delta\varepsilon}\right)^2 E(x_2^2)$$

Making use of the relations

$$E[x_i^2] = \overline{x^2} = \int_{-\infty}^{\infty} x^2\, p(x)\, dx \qquad i, j = 1, 2, \cdots$$

$$\tag{d}$$

$$E[x_i x_j] = 0 \qquad i \neq j$$

results in

$$E[x(t)^2] = \overline{x^2}\left(1 - \frac{2t}{\Delta\varepsilon} + \frac{2t^2}{\Delta\varepsilon^2}\right) \tag{e}$$

Therefore,

$$\sigma^2_{x(t)} = \overline{x^2}\left(1 - \frac{2t}{\Delta\varepsilon} + \frac{2t^2}{\Delta\varepsilon^2}\right) \tag{f}$$

From Eqs. (a) and (d)

$$
E[x(t)x(t+\tau)] =
\begin{cases}
\left[\left(-\dfrac{1}{\Delta\varepsilon^2}\right)t^2 + \left(-\dfrac{\tau}{\Delta\varepsilon^2}\right)t + \left(\dfrac{\tau}{\Delta\varepsilon}+1\right)\right]\overline{x^2} \\[2mm]
\qquad 0 \le t \le \Delta\varepsilon \qquad -\Delta\varepsilon \le t+\tau \le 0 \\[4mm]
\left[\dfrac{2}{\Delta\varepsilon^2}t^2 + \left(\dfrac{2\tau}{\Delta\varepsilon^2}-\dfrac{2}{\Delta\varepsilon}\right)t + \left(1-\dfrac{\tau}{\Delta\varepsilon}\right)\right]\overline{x^2} \\[2mm]
\qquad 0 \le t \le \Delta\varepsilon \qquad 0 \le t+\tau \le \Delta\varepsilon \\[4mm]
\left[\left(-\dfrac{1}{\Delta\varepsilon^2}\right)t^2 + \left(\dfrac{2}{\Delta\varepsilon}-\dfrac{\tau}{\Delta\varepsilon^2}\right)t\right]\overline{x^2} \\[2mm]
\qquad 0 \le t \le \Delta\varepsilon \qquad \Delta\varepsilon \le t+\tau \le 2\Delta\varepsilon
\end{cases}
\tag{g}
$$

Note that the covariance of $x(t)$ and $x(t+\tau)$ as given by Eq. (g) is time dependent; therefore, the random process treated above is nonstationary. Further, note that this covariance equals zero for values of τ outside the ranges indicated for Eqs. (g). The ranges indicated for the first, second, and third of Eqs. (g) are shown by the shaded regions 1, 2, and 3, respectively, in Fig. E21-4. If the origin of time $t = 0$ had been selected coincident with $x_{ir}(r = 1, 2, \cdots)$ rather than x_{1r}, as above, Eqs. (a) would obviously be of exactly the same form except that x_{0r}, x_{1r}, x_{2r}, and x_{3r} would be replaced by $x_{i-1,r}$, x_{ir}, $x_{i+1,r}$, and $x_{i+2,r}$, respectively. Thus, the covariance function $E[x(t)x(t+\tau)]$ must be periodic in time with period $\Delta\varepsilon$. This periodic behavior is also indicated in Fig. E21-4 by a repetition of the shaded regions in each interval along the time t axis.

If the probability density function $p(x)$ used in sampling values of x were gaussian in form, then the entire process $x(t)$ would be gaussian, in which case

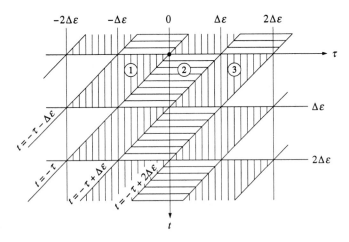

FIGURE E21-4
Regions of nonzero covariance for random variables $x(t)$ and $x(t + \tau)$.

Eqs. (a) would completely characterize the process in a probabilistic sense even though it is nonstationary.

The restriction placed on $\alpha_r (r = 1, 2, \cdots)$ above is now removed, and it is sampled from a uniform distribution over the range $0 < \alpha < \Delta\varepsilon$ as originally stated. Since any arbitrary time t will now occur uniformly over the intervals $\Delta\varepsilon$ looking across the ensemble, the process must be stationary and the covariance function $E[x(t)x(t + \tau)]$ is obtained by simply averaging that function as given by Eqs. (g) over time. Since the resulting function is independent of time and depends only upon the time difference τ, it becomes the autocorrelation function $R_x(\tau)$ for the process. Carrying out this averaging procedure gives

$$
R_x(\tau) = \begin{cases}
\dfrac{\overline{x^2}}{\Delta\varepsilon} \displaystyle\int_{-\tau-\Delta\varepsilon}^{\Delta\varepsilon} \left[\left(-\dfrac{1}{\Delta\varepsilon^2} \right) t^2 + \left(-\dfrac{\tau}{\Delta\varepsilon^2} \right) t + \left(\dfrac{\tau}{\Delta\varepsilon} + 1 \right) \right] dt \\
\qquad\qquad -2\Delta\varepsilon < \tau < -\Delta\varepsilon \\[2ex]
\dfrac{\overline{x^2}}{\Delta\varepsilon} \left\{ \displaystyle\int_0^{-\tau} \left[\left(-\dfrac{1}{\Delta\varepsilon^2} \right) t^2 + \left(-\dfrac{\tau}{\Delta\varepsilon^2} \right) t + \left(\dfrac{\tau}{\Delta\varepsilon} + 1 \right) \right] dt \right. \\
\qquad \left. + \displaystyle\int_{-\tau}^{\Delta\varepsilon} \left[\dfrac{2}{\Delta\varepsilon^2} t^2 + \left(\dfrac{2\tau}{\Delta\varepsilon^2} - \dfrac{2}{\Delta\varepsilon} \right) t + \left(1 - \dfrac{\tau}{\Delta\varepsilon} \right) \right] dt \right\} \\
\qquad\qquad -\Delta\varepsilon < \tau < 0 \\[2ex]
\dfrac{\overline{x^2}}{\Delta\varepsilon} \left\{ \displaystyle\int_0^{-\tau+\Delta\varepsilon} \left[\dfrac{2}{\Delta\varepsilon^2} t^2 + \left(\dfrac{2\tau}{\Delta\varepsilon^2} - \dfrac{2}{\Delta\varepsilon} \right) t + \left(1 - \dfrac{\tau}{\Delta\varepsilon} \right) \right] dt \right. \\
\qquad \left. + \displaystyle\int_{-\tau+\Delta\varepsilon}^{\Delta\varepsilon} \left[\left(-\dfrac{1}{\Delta\varepsilon^2} \right) t^2 + \left(\dfrac{2}{\Delta\varepsilon} - \dfrac{\tau}{\Delta\varepsilon^2} \right) t \right] dt \right\} \\
\qquad\qquad 0 < \tau < \Delta\varepsilon \\[2ex]
\dfrac{\overline{x^2}}{\Delta\varepsilon} \displaystyle\int_0^{-\tau+2\Delta\varepsilon} \left[\left(-\dfrac{1}{\Delta\varepsilon^2} \right) t^2 + \left(\dfrac{2}{\Delta\varepsilon} - \dfrac{\tau}{\Delta\varepsilon^2} \right) t \right] dt \\
\qquad\qquad \Delta\varepsilon < \tau < 2\Delta\varepsilon
\end{cases}
\tag{h}
$$

When the above integrals are completed and terms are collected, the result is

$$
R_x(\tau) = \begin{cases}
\left(\dfrac{4}{3} + \dfrac{2\tau}{\Delta\varepsilon} + \dfrac{\tau^2}{\Delta\varepsilon^2} + \dfrac{\tau^3}{6\Delta\varepsilon^3} \right) \overline{x^2} & -2\Delta\varepsilon \leq \tau \leq -\Delta\varepsilon \\[2ex]
\left(\dfrac{2}{3} - \dfrac{\tau^2}{\Delta\varepsilon^2} - \dfrac{\tau^3}{2\Delta\varepsilon^3} \right) \overline{x^2} & -\Delta\varepsilon \leq \tau \leq 0 \\[2ex]
\left(\dfrac{2}{3} - \dfrac{\tau^2}{\Delta\varepsilon^2} + \dfrac{\tau^3}{2\Delta\varepsilon^3} \right) \overline{x^2} & 0 \leq \tau \leq \Delta\varepsilon \\[2ex]
\left(\dfrac{4}{3} - \dfrac{2\tau}{\Delta\varepsilon} + \dfrac{\tau^2}{\Delta\varepsilon^2} - \dfrac{\tau^3}{6\Delta\varepsilon^3} \right) \overline{x^2} & \Delta\varepsilon \leq \tau \leq 2\Delta\varepsilon
\end{cases}
\tag{i}
$$

Because of the second of Eqs. (d), $R_x(\tau) = 0$ for $\tau \leq -2\Delta\varepsilon$ and $\tau \geq 2\Delta\varepsilon$.

If the random variable x has a normal distribution, the entire process is gaussian, in which case it is completely characterized by Eqs. (i).

21-4   POWER SPECTRAL DENSITY FUNCTION FOR STATIONARY PROCESSES

As demonstrated in Chapter 6, any sample waveform $x_r(t)$ taken from a real stationary random process having a zero mean value, that is, $E[x(t)] = 0$, can be separated into its frequency components using a standard Fourier analysis. If this waveform is represented only over the finite interval $-s/2 < t < +s/2$, the Fourier series representation can be used, namely,

$$x_r(t) = \sum_{n=-\infty}^{\infty} C_{nr} \exp(in\,\overline{\omega}_0 t) \tag{21-21}$$

where

$$C_{nr} = \frac{1}{s} \int_{-s/2}^{s/2} x_r(t) \exp(-in\,\overline{\omega}_0 t)dt$$

and where $\overline{\omega}_0 \equiv 2\pi/s$. If $x_r(t)$ is periodic, Eqs. (21-21) give an exact representation of the entire waveform provided the integration interval s is taken as one full period. Such periodic waveforms consist of discrete harmonics having circular frequencies $\overline{\omega}_0, 2\overline{\omega}_0, 3\overline{\omega}_0, \cdots$, with corresponding finite amplitudes $A_{1r} = 2|C_{1r}|$, $A_{2r} = 2|C_{2r}|$, $A_{3r} = 2|C_{3r}|, \cdots$, provided, of course, corresponding negative and positive frequency components are combined.

Usually the quantity of most interest when analyzing stationary random processes is the mean square value of $x_r(t)$ over the interval $-s/2 < t < +s/2$, which can be obtained by substituting the first of Eqs. (21-21) into the relation

$$\langle x_r(t)^2 \rangle = \frac{1}{s} \int_{-s/2}^{s/2} x_r(t)^2 \, dt \tag{21-22}$$

to obtain

$$\langle x_r(t)^2 \rangle = \sum_{n=-\infty}^{\infty} |C_{nr}|^2 = \sum_{n=1}^{\infty} \frac{A_{nr}^2}{2} \tag{21-23}$$

When $\Delta\overline{\omega}$ represents the frequency spacing of the discrete harmonics, that is,

$$\Delta\overline{\omega} = \overline{\omega}_0 = \frac{2\pi}{s} \tag{21-24}$$

and the second of Eqs. (21-21) is used, Eq. (21-23) becomes

$$\langle x_r(t)^2 \rangle = \sum_{n=-\infty}^{\infty} \frac{\left| \int_{-s/2}^{s/2} x_r(t) \exp(-in\,\overline{\omega}_0 t)dt \right|^2}{2\pi s} \Delta\overline{\omega} \tag{21-25}$$

If s is now allowed to approach infinity, $\Delta\bar{\omega} \to d\bar{\omega}$, $n\bar{\omega}_0 \to \bar{\omega}$, and the summation becomes an integral; thus, Eq. (21-25) is converted into the form

$$\langle x_r(t)^2 \rangle = \int_{-\infty}^{\infty} S_{x_r}(\bar{\omega}) \, d\bar{\omega} \tag{21-26}$$

where the function

$$S_{x_r}(\bar{\omega}) \equiv \lim_{s \to \infty} \frac{\left| \int_{-s/2}^{s/2} x_r(t) \exp(-i\bar{\omega}t) dt \right|^2}{2\pi s} \tag{21-27}$$

is defined as the *power spectral density function* for waveform $x_r(t)$ provided a limit actually exists. According to this definition, the power spectral density function is an even function when $x_r(t)$ is a real function, is positive and finite for all values of $\bar{\omega}$, and yields the mean square value of $x_r(t)$ when integrated over the entire range $-\infty < \bar{\omega} < +\infty$.

The power spectral density function for the entire stationary process $x(t)$ is obtained by simply averaging the power spectral density functions for individual members across the ensemble as follows:

$$S_x(\bar{\omega}) = \lim_{n \to \infty} \frac{1}{n} \sum_{r=1}^{n} S_{x_r}(\bar{\omega}) \tag{21-28}$$

The ensemble average of the mean square value of $x(t)$ can now be obtained by integrating $S_x(\bar{\omega})$ over the entire range $-\infty < \bar{\omega} < +\infty$.

If the random process is ergodic, each member of the ensemble will yield the same power spectral density function, in which case it is unnecessary to average across the ensemble. It is sufficient simply to generate the power spectral density function using one member. For most ergodic processes encountered in engineering, the power spectral density function given by Eq. (21-27) approaches its limit rapidly with increasing values of s, so that sufficient accuracy can usually be obtained with a relatively short sample of the waveform.

21-5   RELATIONSHIP BETWEEN POWER SPECTRAL DENSITY AND AUTOCORRELATION FUNCTIONS

Let a function $F_{x_r}(\bar{\omega})$ be defined as the Fourier transform of the time average $\langle x_r(t)x_r(t+\tau) \rangle$; that is, let

$$F_{x_r}(\bar{\omega}) \equiv \int_{-\infty}^{\infty} \left[\lim_{s \to \infty} \frac{1}{s} \int_{-s/2}^{s/2} x_r(t)x_r(t+\tau) dt \right] \exp(-i\bar{\omega}\tau) d\tau \tag{21-29}$$

Assuming that the function $F_{x_r}(\overline{\omega})$ does indeed exist, Fourier transform theory requires that the quantity in square brackets in Eq. (21-29), which is a function of τ only, decay with increasing values of $|\tau|$ so that the integral

$$I \equiv \int_{-\infty}^{\infty} \left| \lim_{s \to \infty} \frac{1}{s} \int_{-s/2}^{s/2} x_r(t) x_r(t+\tau) dt \right| d\tau \tag{21-30}$$

exists. When Eq. (21-29) is expressed in its equivalent form

$$\frac{1}{2\pi} F_{x_r}(\overline{\omega}) = \lim_{s \to \infty} \frac{1}{2\pi s} \int_{-s/2}^{s/2} \int_{-s/2}^{s/2} x_r(t)\, x_r(t+\tau)\, \exp(-i\overline{\omega}\tau)\, d\tau\, dt \tag{21-31}$$

and a change of variable as defined by

$$\theta \equiv t + \tau \tag{21-32}$$

is substituted, Eq. (21-31) becomes

$$\frac{1}{2\pi} F_{x_r}(\overline{\omega}) = \lim_{s \to \infty} \frac{1}{2\pi s} \int_{-s/2}^{s/2} x_r(t) \exp(i\overline{\omega}t) dt \int_{t-s/2}^{t+s/2} x_r(\theta) \exp(-i\overline{\omega}\theta) d\theta \tag{21-33}$$

The expanding domain of integration given by Eq. (21-33) is shown in Fig. 21-4a. Since the function $F_{x_r}(\overline{\omega})$ can exist only when the total integrand of this equation decays rapidly with increasing values of $|\tau|$, it is valid to change the limits of the second integral as shown by the relation

$$\frac{1}{2\pi} F_{x_r}(\overline{\omega}) = \lim_{s \to \infty} \frac{1}{2\pi s} \int_{-s/2}^{s/2} x_r(t) \exp(i\overline{\omega}t) dt \int_{-s/2}^{s/2} x_r(\theta) \exp(-i\overline{\omega}\theta) d\theta \tag{21-34}$$

which simply changes the expanding domain of integration to that shown in Fig. 21-4b. At this point θ can be changed to t since it is serving only as a dummy time

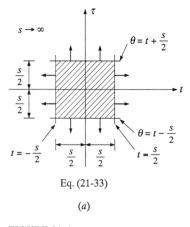

Eq. (21-33)

(a)

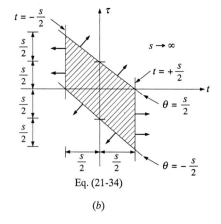

Eq. (21-34)

(b)

FIGURE 21-4
Expanding domains of integration.

variable. Equation (21-34) then can be expressed in the form

$$\frac{1}{2\pi}F_{x_r}(\overline{\omega}) = \lim_{s\to\infty}\frac{\left|\int_{-s/2}^{s/2}x_r(t)\exp(-i\overline{\omega}t)dt\right|^2}{2\pi s} \tag{21-35}$$

When Eq. (21-35) is compared with Eq. (21-27), it is clear that

$$\frac{1}{2\pi}F_{x_r}(\overline{\omega}) = S_{x_r}(\overline{\omega}) \tag{21-36}$$

If the stationary process being considered is ergodic, $F_{x_r}(\overline{\omega})$ is simply the Fourier transform of the autocorrelation function $R_x(\tau)$, and $S_{x_r}(\overline{\omega})$ equals the power spectral density for the process $S_x(\overline{\omega})$. Thus, it has been shown that for an ergodic process, the autocorrelation and power spectral density functions for the process are related through the Fourier integrals given by

$$S_x(\overline{\omega}) = \frac{1}{2\pi}\int_{-\infty}^{\infty}R_x(\tau)\exp(-i\overline{\omega}\tau)\,d\tau$$
$$\tag{21-37}$$
$$R_x(\tau) = \int_{-\infty}^{\infty}S_x(\overline{\omega})\exp(i\overline{\omega}\tau)\,d\overline{\omega}$$

If the stationary process being considered is nonergodic, an additional step must be taken by averaging Eq. (21-36) across the ensemble as expressed by the relation

$$\frac{1}{2\pi}\lim_{n\to\infty}\frac{1}{n}\sum_{r=1}^{n}F_{x_r}(\overline{\omega}) = \lim_{n\to\infty}\frac{1}{n}\sum_{r=1}^{n}S_{x_r}(\overline{\omega}) \tag{21-38}$$

When Eq. (21-31) is used, it is observed that the left hand side of Eq. (21-38) is equal to $1/2\pi$ times the Fourier transform of $R_x(\tau)$. Since the right side of this same equation is $S_x(\overline{\omega})$, Eqs. (21-37) must also be valid for a nonergodic stationary process.

It was previously demonstrated that if a stationary process having a zero mean value is gaussian, it is completely characterized by the autocorrelation function. Now that it has been shown that the power spectral density function can be obtained by a Fourier transformation of the autocorrelation function, that function must also completely characterize such a process.

Example E21-3. Derive the power spectral density function for random process $x(t)$ as given in stationary form by Example E21-2.

Substituting Eqs. (i) of Example E21-2 into the first of Eqs. (21-37), namely,

$$S_x(\overline{\omega}) = \frac{1}{2\pi}\int_{-\infty}^{\infty}R_x(\tau)\exp(-i\overline{\omega}\tau)d\tau \tag{a}$$

gives

$$S_x(\bar{\omega}) = \frac{\overline{x^2}}{2\pi}\left[\int_{-2\Delta\varepsilon}^{-\Delta\varepsilon}\left(\frac{4}{3} + \frac{2\tau}{\Delta\varepsilon} + \frac{\tau^2}{\Delta\varepsilon^2} + \frac{\tau^2}{6\Delta\varepsilon^3}\right)\exp(-i\bar{\omega}\tau)d\tau\right.$$

$$+ \int_{-\Delta\varepsilon}^{0}\left(\frac{2}{3} - \frac{\tau^2}{\Delta\varepsilon^2} - \frac{\tau^3}{2\Delta\varepsilon^3}\right)\exp(-i\bar{\omega}\tau)d\tau$$

$$+ \int_{0}^{\Delta\varepsilon}\left(\frac{2}{3} - \frac{\tau^2}{\Delta\varepsilon^2} + \frac{\tau^3}{2\Delta\varepsilon^3}\right)\exp(-i\bar{\omega}\tau)d\tau$$

$$\left.+ \int_{\Delta\varepsilon}^{2\Delta\varepsilon}\left(\frac{4}{3} - \frac{2\tau}{\Delta\varepsilon} + \frac{\tau^2}{\Delta\varepsilon^2} - \frac{\tau^3}{6\Delta\varepsilon^3}\right)\exp(-i\bar{\omega}\tau)d\tau\right] \quad \text{(b)}$$

After integrating and collecting all terms, the result is

$$S_x(\bar{\omega}) = \frac{\overline{x^2}}{2\pi}\left\{\frac{1}{\bar{\omega}^4\,\Delta\varepsilon^3}[6 - 4\exp(-i\bar{\omega}\tau) - 4\exp(i\bar{\omega}\tau)\right.$$

$$\left.+ \exp(-2i\bar{\omega}\tau) + \exp(2i\bar{\omega}\tau)]\right\} \quad \text{(c)}$$

which can be converted to the trigonometric form

$$S_x(\bar{\omega}) = \frac{\overline{x^2}}{2\pi}\left[\frac{6 - 8\cos\bar{\omega}\Delta\varepsilon + 2\cos 2\bar{\omega}\Delta\varepsilon}{\bar{\omega}^4\Delta\varepsilon^3}\right] \quad -\infty < \bar{\omega} < \infty \quad \text{(d)}$$

21-6   POWER SPECTRAL DENSITY AND AUTOCORRELATION FUNCTIONS FOR DERIVATIVES OF PROCESSES

When the power spectral density and autocorrelation functions for the random variable $x(t)$ are known, these same functions can easily be obtained for time derivatives of this variable such as $\dot{x}(t)$ and $\ddot{x}(t)$. To illustrate the method, consider the autocorrelation function for $x(t)$ in its most basic form, that is,

$$R_x(\tau) \equiv E[x(t)\,x(t+\tau)] \qquad (21\text{-}39)$$

Differentiating with respect to τ gives

$$R_x'(\tau) = \frac{dR_x(\tau)}{d\tau} = E[x(t)\,\dot{x}(t+\tau)] \qquad (21\text{-}40)$$

Since the process $x(t)$ is stationary, Eq. (21-40) can also be expressed in the form

$$R_x'(\tau) = E[x(t-\tau)\,\dot{x}(t)] \qquad (21\text{-}41)$$

Differentiating once more with respect to τ gives

$$R_x''(\tau) = -E[\dot{x}(t-\tau)\,\dot{x}(t)] = -E[\dot{x}(t)\,\dot{x}(t+\tau)]$$ (21-42)

Since the ensemble average in Eq. (21-42) is by definition the autocorrelation function for $\dot{x}(t)$, it becomes apparent that

$$R_{\dot{x}}(\tau) = -R_x''(\tau)$$ (21-43)

Differentiating in the same manner two more times shows that

$$R_{\ddot{x}}(\tau) = -R_{\dot{x}}''(\tau) = R_x^{iv}(\tau)$$ (21-44)

The above autocorrelation functions can be expressed in the form of the second of Eqs. (21-37), namely

$$R_x(\tau) = \int_{-\infty}^{\infty} S_x(\overline{\omega}) \exp(i\overline{\omega}\tau)\, d\overline{\omega}$$

$$R_{\dot{x}}(\tau) = \int_{-\infty}^{\infty} S_{\dot{x}}(\overline{\omega}) \exp(i\overline{\omega}\tau)\, d\overline{\omega}$$ (21-45)

$$R_{\ddot{x}}(\tau) = \int_{-\infty}^{\infty} S_{\ddot{x}}(\overline{\omega}) \exp(i\overline{\omega}\tau)\, d\overline{\omega}$$

Substituting the first of Eq. (21-45) into Eqs. (21-43) and (21-44) gives

$$R_{\dot{x}}(\tau) = \int_{-\infty}^{\infty} \overline{\omega}^2 S_x(\overline{\omega}) \exp(i\overline{\omega}\tau)\, d\overline{\omega}$$

$$R_{\ddot{x}}(\tau) = \int_{-\infty}^{\infty} \overline{\omega}^4 S_x(\overline{\omega}) \exp(i\overline{\omega}\tau)\, d\overline{\omega}$$ (21-46)

Comparing Eqs. (21-46) with the second and third of Eqs. (21-45) shows that

$$S_{\dot{x}}(\overline{\omega}) = \overline{\omega}^2 S_x(\overline{\omega}) \qquad S_{\ddot{x}}(\overline{\omega}) = \overline{\omega}^4 S_x(\overline{\omega})$$ (21-47)

Example   E21-4.   If random process $x(t)$ has the autocorrelation function

$$R_x(\tau) = (1 - \tau^2)\, e^{-\tau^2}$$ (a)

find the corresponding autocorrelation functions for random processes $\dot{x}(t)$ and $\ddot{x}(t)$.

Taking derivatives of Eq. (a) gives

$$R'_x(\tau) = (2\tau^3 - 4\tau)\ e^{-\tau^2}$$

$$R''_x(\tau) = (-4\tau^4 + 14\tau^2 - 4)\ e^{-\tau^2}$$

$$R'''_x(\tau) = (8\tau^5 - 44\tau^3 + 36\tau)\ e^{-\tau^2}$$

$$R^{iv}_x(\tau) = (-16\tau^6 + 128\tau^4 - 204\tau^2 + 36)\ e^{-\tau^2}$$

(b)

Thus from Eqs. (21-43) and (21-44)

$$R_{\dot{x}}(\tau) = (4\tau^4 - 14\tau^2 + 4)\ e^{-\tau^2} \tag{c}$$

$$R_{\ddot{x}}(\tau) = (-16\tau^6 + 128\tau^4 - 204\tau^2 + 36)\ e^{-\tau^2} \tag{d}$$

21-7   SUPERPOSITION OF STATIONARY PROCESSES

Consider a stationary process $q(t)$ which is defined as the sum of three separate stationary processes $x(t)$, $y(t)$, and $z(t)$ all of which have zero mean values. To find the autocorrelation function for this process, namely,

$$R_q(\tau) \equiv E[q(t)\,q(t+\tau)] \tag{21-48}$$

substitute the relation

$$q(t) = x(t) + y(t) + z(t) \tag{21-49}$$

into Eq. (21-48) to obtain

$$
\begin{aligned}
R_q(\tau) =\ & E[x(t)x(t+\tau)] + E[y(t)y(t+\tau)] + E[z(t)z(t+\tau)] \\
& + E[x(t)y(t+\tau)] + E[y(t)z(t+\tau)] + E[x(t)z(t+\tau)] \\
& + E[y(t)x(t+\tau)] + E[z(t)y(t+\tau)] + E[z(t)x(t+\tau)]
\end{aligned} \tag{21-50}
$$

The first three ensemble averages on the right hand side of this equation are the autocorrelation functions for processes $x(t)$, $y(t)$, and $z(t)$, respectively, and the last six ensemble averages are *cross-correlation functions* (or covariance functions), which will be designated as

$$
\begin{aligned}
R_{xy}(\tau) &= E[x(t)y(t+\tau)] & R_{yx}(\tau) &= E[y(t)x(t+\tau)] \\
R_{yz}(\tau) &= E[y(t)z(t+\tau)] & R_{zy}(\tau) &= E[z(t)y(t+\tau)] \\
R_{xz}(\tau) &= E[x(t)z(t+\tau)] & R_{zx}(\tau) &= E[z(t)x(t+\tau)]
\end{aligned} \tag{21-51}
$$

Thus, the autocorrelation function for process $q(t)$ can be expressed in terms of the autocorrelation and cross-correlation functions for $x(t)$, $y(t)$, and $z(t)$ as follows:

$$R_q(\tau) = R_x(\tau) + R_y(\tau) + R_z(\tau) + R_{xy}(\tau) + R_{yz}(\tau)$$

$$+ R_{xz}(\tau) + R_{yx}(\tau) + R_{zy}(\tau) + R_{zx}(\tau) \tag{21-52}$$

If random processes $x(t)$, $y(t)$, and $z(t)$ are uncorrelated with each other, their cross-correlation functions will equal zero, in which case

$$R_q(\tau) = R_x(\tau) + R_y(\tau) + R_z(\tau) \tag{21-53}$$

It should be noted that for real stationary processes

$$R_{xy}(\tau) = R_{yx}(-\tau) \qquad R_{yz}(\tau) = R_{zy}(-\tau) \qquad R_{xz}(\tau) = R_{zx}(-\tau) \tag{21-54}$$

The power spectral density function for process $q(t)$ is obtained using the first of Eqs. (21-37), that is,

$$S_q(\varpi) = \frac{1}{2\pi} \int_{-\infty}^{\infty} R_q(\tau) \exp(-i\varpi\tau) d\tau \tag{21-55}$$

Substituting Eq. (21-52) into Eq. (21-55) gives

$$S_q(\varpi) = S_x(\varpi) + S_y(\varpi) + S_z(\varpi) + S_{xy}(\varpi) + S_{yz}(\varpi)$$

$$+ S_{xz}(\varpi) + S_{yx}(\varpi) + S_{zy}(\varpi) + S_{zx}(\varpi) \tag{21-56}$$

where $S_{xy}(\varpi)$, $S_{yz}(\varpi)$, $\cdots$ are *cross-spectral density functions* which are related to their respective cross-correlation functions through the Fourier transform relation

$$S_{xy}(\varpi) = \frac{1}{2\pi} \int_{-\infty}^{\infty} R_{xy}(\tau) \exp(-i\varpi\tau) d\tau \tag{21-57}$$

Note that $S_{yx}(\varpi)$ is the complex conjugate of $S_{xy}(\varpi)$. The inverse of Eq. (21-57) is, of course,

$$R_{xy}(\tau) = \int_{-\infty}^{\infty} S_{xy}(\varpi) \exp(i\varpi\tau) d\varpi \tag{21-58}$$

When the procedure of Section 21-4 is followed, the time average of the product $x_r(t)y_r(t)$ becomes

$$\langle x_r(t)y_r(t) \rangle = \int_{-\infty}^{\infty} S_{x_r y_r}(\varpi) d\varpi \tag{21-59}$$

where

$$S_{x_r y_r}(\varpi) \equiv \lim_{s \to \infty} \frac{\left[\int_{-s/2}^{s/2} x_r(t) \exp(-i\varpi t) dt \right] \left[\int_{-s/2}^{s/2} y_r(t) \exp(+i\varpi t) dt \right]}{2\pi s} \tag{21-60}$$

Note that $S_{x_r y_r}(-\overline{\omega})$ is the complex conjugate of $S_{x_r y_r}(\overline{\omega})$. Therefore, only the real part of $S_{x_r y_r}(\overline{\omega})$ contributes to the integral in Eq. (21-59). If processes $x(t)$ and $y(t)$ are ergodic, $S_{x_r y_r}(\overline{\omega})$ as given by Eq. (21-60) represents the cross-spectral density for these processes. However, if processes $x(t)$ and $y(t)$ are nonergodic, the cross-spectral density function for these processes must be obtained by averaging across the ensemble, that is,

$$S_{xy}(\overline{\omega}) = \lim_{n \to \infty} \frac{1}{n} \sum_{r=1}^{n} S_{x_r y_r}(\overline{\omega}) \tag{21-61}$$

21-8 STATIONARY GAUSSIAN PROCESSES: ONE INDEPENDENT VARIABLE

In engineering it is common practice to assume a gaussian, or normal, distribution for random processes. To help in establishing a rational basis for this assumption, consider a real stationary zero-mean random process $x(t)$ of the form

$$x_{jr}(t) = \sum_{n=-j}^{j} C_{nr} \exp(in\,\overline{\omega}_0 t) \qquad r = 1, 2, \cdots \tag{21-62}$$

where $x_{jr}(t)$ is the rth member of the ensemble which contains j discrete harmonics having frequencies $\overline{\omega}_0, 2\overline{\omega}_0, \cdots, j\overline{\omega}_0$, and where C_{nr} represents random complex constants. For the process to have a zero-mean value, it is necessary, of course, that coefficients C_{0r} equal zero; and since it is assumed that the process contains real functions only, it is necessary that complex coefficients C_{nr} and C_{mr} be conjugate pairs when $n = -m$.

To define the randomness of coefficients C_{nr}, assume first that $|C_{nr}| = C$ (a constant) for all permissible values of n and r but that their corresponding phase angles α_{nr} are sampled values of a random variable α which has a uniform probability density function of intensity $1/2\pi$ in the range $0 < \alpha < 2\pi$. Under these conditions Eq. (21-62) can be written in the form

$$x_{jr}(t) = \sum_{n=-j}^{j} |C_{nr}| \exp\left[i(n\,\overline{\omega}_0 t + \alpha_{nr})\right] \tag{21-63}$$

or

$$x_{jr} = 2C \sum_{n=1}^{j} \sin(n\,\overline{\omega}_0 t + \theta_{nr}) \qquad r = 1, 2, \cdots \tag{21-64}$$

where $\theta_{nr} = +(\pi/2) + \alpha_{nr}$. Since this process contains discrete harmonics at frequency intervals of $\overline{\omega}_0$, each ensemble member will be periodic with a period $s = 2\pi/\overline{\omega}_0$. When a new random variable $L(t)$ is defined so that

$$L_{nr}(t) = 2C \sin(n\,\overline{\omega}_0 t + \theta_{nr}) \tag{21-65}$$

Eq. (21-64) can be written in the form of the one-dimensional random walk:

$$x_{jr}(t) = \sum_{n=1}^{j} L_{nr}(t) \tag{21-66}$$

From Eqs. (21-5) and (21-10), it is clear that

$$p[L(t)] = \begin{cases} \dfrac{1}{\pi\sqrt{4C^2 - L^2}} & -2C < L < 2C \\ 0 & L < -2C \; ; \; L > 2C \end{cases} \tag{21-67}$$

and that

$$\overline{L(t)} = 0 \qquad \sigma^2_{L(t)} = 2C^2 \tag{21-68}$$

When the one-dimensional random-walk relations given by Eq. (20-34) are used, it follows that

$$\overline{x_j(t)} = 0 \qquad \sigma^2_{x(t)} = 2jC^2 \tag{21-69}$$

At this point, apply the same limiting procedure previously used in the one-dimensional random-walk development, that is, let $\overline{\omega}_0 \rightarrow 0$, $j \rightarrow \infty$, and $C^2 \rightarrow 0$, but in such a manner that

$$n\overline{\omega}_0 \rightarrow \overline{\omega} \; \text{(a variable)} \quad j\,\overline{\omega}_0 \rightarrow \overline{\omega}_1 \; \text{(a constant)} \quad C^2/\overline{\omega}_0 \rightarrow S_0 \; \text{(a constant)} \tag{21-70}$$

Since $\overline{\omega}_0 = 2\pi/s$, period $s \rightarrow \infty$ by this limiting procedure.

When the above relations and the second of Eqs. (21-21) are used, one finds that

$$|C_{nr}| = C = \frac{1}{s} \left| \int_{-s/2}^{s/2} x_{jr}(t) \exp(-in\,\overline{\omega}_0 t)dt \right| \tag{21-71}$$

It is now evident that in the limit

$$S_0 = \lim_{s \rightarrow \infty} \frac{\left| \int_{-s/2}^{s/2} x_r(t) \exp(-i\overline{\omega}t)dt \right|^2}{2\pi s} \qquad r = 1, 2, \cdots \tag{21-72}$$

where

$$x_r(t) = \lim_{\substack{j \rightarrow \infty \\ \overline{\omega}_0 \rightarrow 0}} x_{jr}(t) \tag{21-73}$$

A comparison of Eq. (21-72) with Eq. (21-27) and recognition of the limiting conditions given by the first and second of Eqs. (21-70) lead to the conclusion that ensemble member $x_r(t)$ has a uniform power spectral density function $S_{x_r}(\overline{\omega})$ of intensity S_0 over the frequency range $-\overline{\omega}_1 < \overline{\omega} < \overline{\omega}_1$ and of intensity zero outside this range and that since this power spectral density function is invariant with r, the process is ergodic; thus, the power spectral density for the entire process $x(t)$ is that function

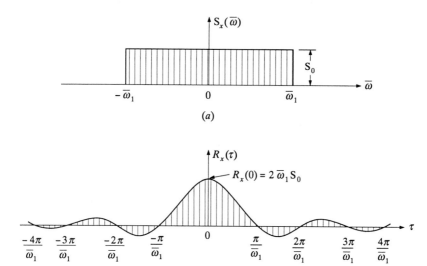

FIGURE 21-5
Power spectral density and autocorrelation functions for random process $x(t)$.

shown in Fig. 21-5a. Further, the earlier one-dimensional random-walk development leads to the conclusion that this random process is gaussian and that its variance [see the second of Eqs. (21-69)] is given by

$$\sigma^2_{x(t)} = 2\bar{\omega}_1 S_0 \tag{21-74}$$

Substituting the power spectral density function shown in Fig. 21-5a into the second of Eqs. (21-37), the autocorrelation function for random process $x(t)$ is found to be

$$R_x(\tau) = \frac{2S_0}{\tau} \sin\bar{\omega}_1\tau \qquad -\infty < \tau < \infty \tag{21-75}$$

This relation is plotted in Fig. 21-5b.

Note that when the power spectral density function for this process becomes uniform over the entire frequency range, that is, when $\bar{\omega}_1 \rightarrow \infty$, the variance $\sigma^2_x \rightarrow \infty$ and the autocorrelation function $R_x(\tau) \rightarrow 2\pi S_0 \delta(\tau)$, where $\delta(\tau)$ is a Dirac delta function located at the origin. This process, which is commonly referred to as a *white* process or simply *white noise*, can be considered as totally random since $x(t)$ is completely independent of $x(t+\tau)$ for all values of $\tau \neq 0$.

Consider again random process $x_r(t)$ but this time assume that coefficients $|C_{nr}|$ equal zero for all values of n in the range $-k < n < +k$, where $k < j$, and that they equal a constant C for all values of n in the ranges $-j \leq n \leq -k$ and

$+k \leq n \leq +j$. The same procedures as before are followed, but this time $\bar{w}_0 \to 0$, $k \to \infty$, $j \to \infty$, and $C^2 \to 0$ in such a manner that $n\bar{w}_0 \to \bar{w}$, $k\bar{w}_0 \to \bar{w}_1$, $j\bar{w}_0 \to \bar{w}_2$, and $C^2/\bar{w}_0 \to S_0$; again the process becomes gaussian in the limit and its power spectral density and autocorrelation functions are of the form

$$S_x(\bar{w}) = \begin{cases} S_0 & -\bar{w}_2 < \bar{w} < -\bar{w}_1 \ ; \ \bar{w}_1 < \bar{w} < \bar{w}_2 \\ 0 & \bar{w} < -\bar{w}_2 \ ; \ -\bar{w}_1 < \bar{w} < \bar{w}_1 \ ; \ \bar{w} > \bar{w}_2 \end{cases} \tag{21-76}$$

$$R_x(\tau) = \frac{2S_0}{\tau}(\sin\bar{w}_2\tau - \sin\bar{w}_1\tau) \qquad -\infty < \tau < \infty$$

To generalize one step further, consider a random process $z(t)$ defined as the sum of the statistically independent gaussian ergodic processes $x(t)$ and $y(t)$, both of which are developed separately from Eq. (21-62) using the same limiting procedure as before. From the proof given in Section 20-10, process $z(t)$ will also have a gaussian distribution.

Finally, once more use the process given by Eq. (21-63) as expressed in the equivalent form

$$x_{jr}(t) = \sum_{n=1}^{j} 2|C_{nr}| \ \sin(n\bar{w}_0 t + \theta_{nr}) \qquad r = 1, 2, \cdots \tag{21-77}$$

For this process assume that phase angles θ_{nr} are sampled values of random variable θ which has the uniform probability density function shown in Fig. 21-6a and that coefficients $|C_{nr}|$ are sampled values of a second random variable C which has an arbitrary, but prescribed, probability density as shown in Fig. 21-6b. When $L_{nr}(t)$ is defined by the relation

$$L_{nr}(t) \equiv 2|C_{nr}| \ \sin(n\bar{w}_0 t + \theta_{nr}) \tag{21-78}$$

$x_r(t)$ can again be expressed in the form

$$x_{jr}(t) \equiv \sum_{n=1}^{j} L_{nr}(t) \tag{21-79}$$

When the forms of probability density functions $p(\theta)$ and $p(C)$ are known, the probability density function for $L(t)$, as defined by Eq. (21-78), can be established if

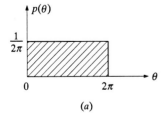
(a)

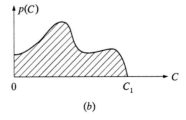
(b)

FIGURE 21-6
Probability density functions for random variables ϕ and C.

desired. For this process, however, this step is unnecessary since the mean value $\overline{L}(t)$ and the variance $\sigma^2_{L(t)}$ are the only quantities required in the random-walk development and they can be obtained without establishing the function $p[L(t)]$. From the form of Eq. (21-78), it can be reasoned that

$$\overline{L(t)} = 0 \qquad \sigma^2_{L(t)} = 2\overline{C^2} = 2\int_{-\infty}^{\infty} C^2 p(C)\, dC \tag{21-80}$$

This process is stationary since the variance for $L(t)$ is independent of time t.

When the one-dimensional random-walk relations given by Eq. (20-34) are used again, it follows that

$$\overline{x}_j(t) = 0 \qquad \sigma^2_{x_j(t)} = 2j\overline{C^2} \tag{21-81}$$

When Eq. (21-81) is compared with Eq. (21-69), it is clear that the same limiting procedures used previously can once again be used provided C^2 is replaced by $\overline{C^2}$. In this case

$$S_0 \equiv \frac{\overline{C^2}}{\overline{\omega}_0} = \lim_{n \to \infty} \frac{1}{n} \sum_{r=1}^{n} S_{x_r}(\overline{\omega}) \tag{21-82}$$

where

$$S_{x_r}(\overline{\omega}) = \lim_{s \to \infty} \frac{\left| \int_{-s/2}^{s/2} x_r(t) \exp(-i\overline{\omega}t)\, dt \right|^2}{2\pi s} \tag{21-83}$$

Although coefficients $|C_{nr}|$ for this process are random in accordance with Fig. 21-bb, the power spetral density function as defined by Eq. (21-83) will be independent of r when the limiting procedure is applied, i.e., when $j \to \infty$ and $\overline{\omega}_0 \to 0$. If S_0 as given by Eq. (21-82) is to be finite, it is necessary that C_1 in Fig. 21-bb approach zero in such a way that $\overline{C^2}/\overline{\omega}_0 = S_0$ is finite. If the process given by Eq. (21-77) is to have a nonuniform power spectral density function over the frequency range $j\,\overline{\omega}_0$, the coefficients $|C_{nr}|$ would have to be dependent upon n; and, if the process is to be nonergodic, they would have to be dependent upon r.

The earlier development which restricted the nonzero-frequency components to the range $\overline{\omega}_1 < \overline{\omega} < \overline{\omega}_2$ and the development which presented the principle of superposition obviously both apply equally well to the present process involving two random variables. Therefore, it may be concluded that any stationary process $x(t)$ (whether ergodic or not) will be gaussian when its power spectral density function $S_x(\overline{\omega})$ truly exists and when all the phase angles between frequency components which are randomly distributed in a uniform manner over 360° are statistically independent of each other.

When the phase angles between frequency components are not uniformly distributed over the full 360°, gaussian processes will still result in the limit; however, stationarity will no longer be maintained. For example, if the random phase angle θ for the process defined by Eq. (21-2) has a uniform probability density function of intensity $1/\theta_1$ over the range $0 < \theta < \theta_1$, where $\theta_1 < 2\pi$, the ensemble mean square value $E[x(t)^2]$ (or variance in this case) will be time dependent. To prove this statement, substitute Eqs. (21-2) and (20-4) into Eq. (20-10) to obtain

$$E[x(t)^2] = \int_{A\sin\overline{\omega}_0 t}^{A\sin(\overline{\omega}_0 t + \theta_1)} \frac{x^2}{\theta_1\sqrt{A^2 - x^2}}\, dx \qquad (21\text{-}84)$$

After the integration is completed, this equation becomes

$$E[x(t)^2] = \frac{A^2}{2}\left\{1 - \frac{1}{2\theta_1}[\sin 2\theta_1 \cos 2\overline{\omega}_0 t - (1 - \cos 2\theta_1)\sin 2\overline{\omega}_0 t]\right\} \qquad (21\text{-}85)$$

which clearly shows the time dependency. Note that as $\theta_1 \to 2\pi$, the time dependency is gradually removed; that is, $E[x(t)^2] \to A^2/2$, and as $\theta_1 \to 0$, the random character of the process is gradually lost, so that $E[x(t)^2] \to A^2\sin^2\overline{\omega}_0 t$.

It is important to recognize that gaussian processes result only when the random variables involved are statistically independent.

Example E21-5. Assume random variables r_1 and r_2 as defined in Example E20-7 are used as successive discrete ordinates for all members of random process $x(t)$ given in Example E21-2. What is the joint probability density function for random variables $x(t)$ and $x(t + \tau)$?

First it should be recognized that random variables $x(t)$ and $x(t + \tau)$ are linearly related to random variables r_1 and r_2 in accordance with the first and second of Eqs. (a) in Example E21-2. Since random variables r_1 and r_2 have the normal distribution given by Eq. (f) of Example E20-7, random variables $x(t)$ and $x(t + \tau)$ must also have a normal distribution in accordance with the principle of linear transformation treated in Section 20-9. Thus the probability density functions must be of the form

$$p(x_1) = \frac{1}{\sqrt{2\pi}\,\sigma_{x_1}} \exp\left[-\frac{(x_1 - \overline{x}_1)^2}{2\sigma_{x_1}^2}\right]$$

$$p(x_1, x_2) = \frac{1}{2\pi\sigma_{x_1}\sigma_{x_2}\sqrt{1 - \rho_{x_1 x_2}^2}} \times \exp\left\{-\frac{1}{2(1 - \rho_{x_1 x_2}^2)}\right.$$

$$\left. \times \left[\frac{(x_1 - \overline{x}_1)^2}{\sigma_{x_1}^2} - \frac{2\rho_{x_1 x_2}(x_1 - \overline{x}_1)(x_2 - \overline{x}_2)}{\sigma_{x_1}\sigma_{x_2}} + \frac{(x_2 - \overline{x}_2)^2}{\sigma_{x_2}^2}\right]\right\}$$

(a)

where $x_1 \equiv x(t)$ and $x_2 \equiv x(t + \tau)$. With the results in Examples E20-7 and E21-2, it is shown that

$$\overline{x}_1 = \overline{x}_2 = 0$$

$$\sigma_{x_1}^2 = \sigma_{x_2}^2 = R_x(0) = \frac{2}{3}\,\overline{x_2^2} = \frac{2}{3} \qquad\qquad\qquad\text{(b)}$$

$$\rho_{x_1\,x_2}(\tau) = \frac{R_x(\tau)}{R_x(0)}$$

$$= \begin{cases} 1 - \dfrac{3\tau^2}{2\Delta\varepsilon^2} + \dfrac{3|\tau|^3}{4\Delta\varepsilon^3} & -\Delta\varepsilon \le \tau \le \Delta\varepsilon \\[2mm] 2 - \dfrac{3|\tau|}{\Delta\varepsilon} + \dfrac{3\tau^2}{2\Delta\varepsilon^2} - \dfrac{|\tau|^3}{4\Delta\varepsilon^3} & \begin{array}{l} -2\Delta\varepsilon \le \tau \le -\Delta\varepsilon \\ \Delta\varepsilon \le \tau \le 2\Delta\varepsilon \end{array} \\[2mm] 0 & \tau \le -2\Delta\varepsilon\,;\ \tau \ge 2\Delta\varepsilon \end{cases} \qquad\text{(c)}$$

Substituting Eqs. (b) into Eqs. (a) gives the desired probability density function.

21-9   STATIONARY WHITE NOISE

In the previous discussion on stationary gaussian processes, white noise was defined as a process having a uniform power spectral density function of intensity S_0 over the entire frequency range $-\infty < \overline{\omega} < \infty$, which corresponds to a Dirac delta function of intensity $2\pi S_0$ at the origin for the autocorrelation function. By this definition it is clear that such processes contain frequency components of equal intensity (based on squared amplitude as a measure of intensity) over the entire frequency range thus the random variables at time t and $t + \tau$ are uncorrelated for all $\tau \ne 0$.

In subsequent developments, it will be found desirable to express white-noise processes in an equivalent but quite different manner. To develop this new type of representation, consider the random process

$$x_r(t) = \lim_{N \to \infty} \sum_{k=-N}^{N-1} a_{kr}\eta(t - k\,\Delta t - \varepsilon_r) \qquad r = 1, 2, \cdots \qquad (21\text{-}86)$$

where coefficients a_{kr} are statistically independent random variables having a zero mean value and sampled in accordance with the arbitrary but prescribed probability density function $p(a)$ shown in Fig. 21-7a, Δt is a constant time interval, variables ε_r are statistically independent random phase parameters having the uniform probability density function shown in Fig. 21-7b, and $\eta(t)$ is the function defined in Fig. 21-7c. The rth member of this ensemble is shown in Fig. 21-7d. The uniformly random phase shift ε over a full interval Δt is a necessary condition for the process to be stationary.

The power spectral density function for member $x_r(t)$ can be derived by using Eq. (21-27) in its equivalent form

$$S_{x_r}(\overline{\omega}) = \lim_{N \to \infty} \frac{|Q_{x_r}(i\overline{\omega})|^2}{4\pi N\,\Delta t} \qquad\qquad (21\text{-}87)$$

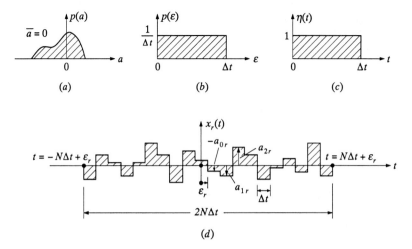

FIGURE 21-7
White-noise process, Eq. (21-86).

where

$$Q_{x_r}(i\varpi) = \int_{-N\,\Delta t+\varepsilon_r}^{N\,\Delta t+\varepsilon_r} \left[\sum_{k=-N}^{N-1} a_{kr}\eta(t - k\,\Delta t - \varepsilon_r) \right] \exp(-i\varpi t)dt \qquad (21\text{-}88)$$

When the change of variable $\theta \equiv t - k\,\Delta t - \varepsilon_r$ is substituted into this equation and the order of summation and integration is changed, it becomes

$$Q_{x_r}(i\varpi) = \sum_{k=-N}^{N-1} a_{kr} \exp[-i\varpi(k\,\Delta t + \varepsilon_r)] \int_{-(N+k)\Delta t}^{(N-k)\Delta t} \eta(\theta)\exp(-i\varpi\theta)d\theta$$

$$(21\text{-}89)$$

or

$$Q_{x_r}(i\varpi) = \frac{i}{\varpi}[\exp(-i\varpi\,\Delta t) - 1] \sum_{k=-N}^{N-1} a_{kr}\exp[-i\varpi(k\,\Delta t + \varepsilon_r)] \qquad (21\text{-}90)$$

Substituting this equation into Eq. (21-87) gives

$$S_{x_r}(\varpi) = \lim_{N\to\infty} \frac{1}{4\pi N\,\Delta t\,\varpi^2}[\exp(-i\varpi\,\Delta t) - 1][\exp(i\varpi\,\Delta t) - 1]$$

$$\times \sum_{k=-N}^{N-1} \sum_{j=-N}^{N-1} a_{kr}\,a_{jr}\exp[-i\varpi(k - j)\Delta t] \qquad (21\text{-}91)$$

Since the process as defined is stationary but nonergodic, the power spectral density function for the process must be obtained by averaging Eq. (21-91) across the ensemble.

Since random variables a_{kr} and a_{jr} $(r = 1, 2, \cdots, \infty)$ are statistically indepen-dent, their covariances, that is, $E(a_{kr}\, a_{jr})$ for $j \neq k$, must all equal zero. Therefore, the double summation in Eq. (21-91) reduces to a single summation, which obviously equals $2N\sigma_a^2$ when averaged with respect to r across the ensemble. Thus, the power spectral density function for the process becomes

$$S_x(\overline{\omega}) = \frac{\sigma_a^2}{2\pi\,\Delta t\,\overline{\omega}^2}[\exp(-i\overline{\omega}\,\Delta t) - 1][\exp(i\overline{\omega}\,\Delta t) - 1] \qquad (21\text{-}92)$$

or

$$S_x(\overline{\omega}) = \frac{\sigma_a^2\,\Delta t}{2\pi}\frac{\sin^2[(\overline{\omega}\,\Delta t)/2]}{[(\overline{\omega}\,\Delta t)/2]^2} \qquad (21\text{-}93)$$

When $\sigma_a^2 \to \infty$ and $\Delta t \to 0$ in such a way that $\sigma_a^2\,\Delta t = C$ (a constant), this equation becomes

$$S_x(\overline{\omega}) = \frac{C}{2\pi} = S_0 \qquad (21\text{-}94)$$

showing that the process becomes white noise in the limit.

As a special case of the above process, let the probability density function $p(a)$ consist of two Dirac delta functions of intensity 1/2 located at $a = \pm A$. This process becomes white noise having a uniform power spectral density function of intensity $S_0 = C/2\pi$ when $A^2 \to \infty$, $\Delta t \to 0$, and $A^2\,\Delta t \to C$.

Example E21-6. For the stationary random process defined in Exam-ple E20-5, (1) show that this process approaches white noise in the limit as $\Delta\varepsilon \to 0$ and (2) find the normalization factor C which would force this limiting process to have a constant power spectral density equal to S_0.

From the form of the autocorrelation function given in Example E21-2

$$\lim_{\Delta\varepsilon \to 0} R_x(\tau) \begin{cases} = 0 & \tau \neq 0 \\ \neq 0 & \tau = 0 \end{cases} \qquad (a)$$

which suggests the form of a Dirac delta function. Integrating $R_x(\tau)$ over the infinite τ domain gives

$$\int_{-\infty}^{\infty} R_x(\tau)d\tau = 2\overline{x^2}\left\{ \int_0^{\Delta\varepsilon}\left(\frac{2}{3} - \frac{\tau^2}{\Delta\varepsilon^2} + \frac{\tau^3}{2\Delta\varepsilon^3}\right)d\tau \right.$$

$$\left. + \int_{\Delta\varepsilon}^{2\Delta\varepsilon}\left(\frac{4}{3} - \frac{2\tau}{\Delta\varepsilon} + \frac{\tau^2}{\Delta\varepsilon^2} - \frac{\tau^3}{6\Delta\varepsilon^3}\right)d\tau\right\}$$

or

$$\int_{-\infty}^{\infty} R_x(\tau)d\tau = \overline{x^2}\,\Delta\varepsilon = \Delta\varepsilon \tag{b}$$

Multiplying all discrete ordinates of process $x(t)$ by the constant $(2\pi S_0/\Delta\varepsilon)^{1/2}$ giving a new process $a(t)$, following the above procedures, would demonstrate that

$$\lim_{\Delta\varepsilon \to 0} R_a(\tau) \begin{cases} = 0 & \tau \neq 0 \\ \neq 0 & \tau = 0 \end{cases} \tag{c}$$

$$\int_{-\infty}^{\infty} R_a(\tau)d\tau = 2\pi S_0$$

thus showing that

$$R_a(\tau) \to 2\pi S_0\,\delta(\tau) \tag{d}$$

which means that process $a(t)$ approaches white noise of intensity S_0. Therefore the normalization factor C is given by

$$C = \left(\frac{2\pi S_0}{\Delta\varepsilon}\right)^{1/2} \tag{e}$$

The solution to this example can be obtained more easily by noting that the power spectral density function for the process $x(t)$, as given by Eq. (b) in Example E21-3, becomes in the limit

$$\lim_{\Delta\varepsilon \to 0} S_x(\overline{\omega}) = \frac{\overline{x^2}\,\Delta\varepsilon}{2\pi} = \frac{\Delta\varepsilon}{2\pi} \tag{f}$$

Likewise the limiting power spectral density function for process $a(t)$ would be

$$\lim_{\Delta\varepsilon \to 0} S_a(\overline{\omega}) = S_0 \tag{g}$$

again showing that the normalization factor is given by Eq. (e).

21-10 PROBABILITY DISTRIBUTION FOR MAXIMA[1]

Consider a zero-mean stationary gaussian process $x(t)$ having an arbitrary power spectral density function $S_x(p)$. A sample function taken from this process (Fig. 21-8) shows positive and negative maxima and positive and negative minima. From

---

[1] D. E. Cartwright and M. S. Longuet-Higgins, "The Statistical Distributions of the Maxima of a Random Function," *Proc. R. Soc.*, Ser. A, Vol. 237, pp. 212-232, 1956; A. G. Davenport, "Note on the Distribution of the Largest Value of a Random Function with Application to Gust Loading," *Proc. Inst. Civ. Eng.*, Vol. 28, pp. 187-196, 1964.

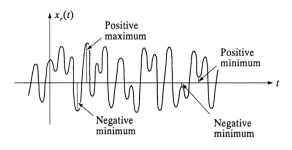

Positive
maximum

Positive
minimum

Negative
minimum

Negative
minimum

FIGURE 21-8
Sample function of process $x(t)$.

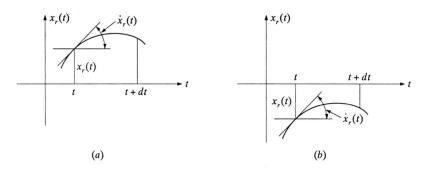

(a)

(b)

FIGURE 21-9
Maxima occurring in the time interval $(t, t + dt)$: (a) positive maxima; (b) negative maxima.

Fig. 21-9, it is clear that for a maximum (+ or −) to occur in the time interval $(t, t + dt)$, it is necessary that $\dot{x}_r(t)$ be positive and $\ddot{x}_r(t)$ be negative and that

$$0 < \dot{x}_r(t) < |\ddot{x}_r(t)| \, dt \qquad (21\text{-}95)$$

With the definition of three new random variables $\zeta_1 \equiv x(t)$, $\zeta_2 \equiv \dot{x}(t)$, and $\zeta_3 \equiv \ddot{x}(t)$, the probability density function $p(\zeta_1, \zeta_2, \zeta_3)$ can be written in its normal form

$$p(\zeta_1, \zeta_2, \zeta_3) = \frac{1}{(2\pi)^{3/2} |\mu|^{1/2}} \exp\left\{ -\frac{1}{2} [\zeta - \bar{\zeta}]^T \mu^{-1} [\zeta - \bar{\zeta}] \right\} \qquad (21\text{-}96)$$

where ζ is the vector $[\zeta_1, \zeta_2, \zeta_3]^T$, $\bar{\zeta}$ is the vector $[\bar{\zeta}_1, \bar{\zeta}_2, \bar{\zeta}_3]^T = 0$, and μ is the covariance matrix

$$\mu = \begin{bmatrix} \mu_{11} & \mu_{12} & \mu_{13} \\ \mu_{21} & \mu_{22} & \mu_{23} \\ \mu_{31} & \mu_{32} & \mu_{33} \end{bmatrix} \qquad (21\text{-}97)$$

where

$$\mu_{ik} = E(\zeta_i \, \zeta_k) \qquad (21\text{-}98)$$

When

$$m_n = \int_{-\infty}^{\infty} \overline{\omega}^n \, S_x(\overline{\omega}) \, d\overline{\omega} \tag{21-99}$$

it is easily shown, using the techniques of derivation in Section 21-6, that

$$\mu = \begin{bmatrix} m_0 & 0 & -m_2 \\ 0 & m_2 & 0 \\ -m_2 & 0 & m_4 \end{bmatrix} \tag{21-100}$$

Thus, Eq. (21-96) becomes

$$p(\zeta_1, \zeta_2, \zeta_3) = \frac{1}{(2\pi)^{3/2} \, (m_2 \, \triangle)^{1/2}}$$

$$\times \exp\left[-1/2\left(\frac{\zeta_2^2}{m_2} + \frac{m_4 \, \zeta_1^2 + 2m_2 \, \zeta_1 \, \zeta_3 + m_0 \, \zeta_3^2}{\triangle}\right)\right] \tag{21-101}$$

where

$$\triangle \equiv m_0 \, m_4 - m_2^2 \tag{21-102}$$

From Fig. 21-10 it becomes apparent that the probability of a maximum (+ or −) occurring in the range $(\zeta_1, \zeta_1 + d\zeta_1)$ during the time interval $(t, t + dt)$ is expressed by the relation

$$F(\zeta_1) d\zeta_1 \, dt = \left[\int_{-\infty}^{0} p(\zeta_1, 0, \zeta_3) \, |\zeta_3| \, d\zeta_3\right] d\zeta_1 \, dt \tag{21-103}$$

Thus, it follows that the mean frequency of occurrence of maxima (+ or −) over the complete range $-\infty < \zeta_1 < \infty$ is given by

$$N_1 \equiv \int_{-\infty}^{\infty} \left[\int_{-\infty}^{0} p(\zeta_1, 0, \zeta_3) \, |\zeta_3| \, d\zeta_3\right] d\zeta_1 \tag{21-104}$$

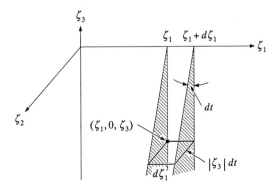

FIGURE 21-10
Shaded region satisfies the conditions for a maximum (+ or −) occurring in the range $(\zeta_1, \zeta_1 + d\zeta_1)$ and in the time interval $(t, t + dt)$.

Substituting Eq. (21-101) for $\zeta_2 = 0$ into this relation and carrying out the double integration leads to Rice's equation

$$N_1 = \frac{1}{2\pi}\left(\frac{m_4}{m_2}\right)^{1/2} \tag{21-105}$$

Since the probability density function for maxima is by definition the ratio $F(\zeta_1)/N_1$, it can be obtained using Eqs. (21-103) and (21-105). In doing so, it is convenient to express maxima in the nondimensional form

$$\eta \equiv \frac{\zeta_1}{m_0^{1/2}} \tag{21-106}$$

allowing its probability density function to be expressed in the form

$$p(\eta) = \frac{1}{(2\pi)^{1/2}}\left[\varepsilon e^{-\eta^2/2\varepsilon^2} + (1 - \varepsilon^2)^{1/2}\,\eta e^{-\eta^2/2}\int_{-\infty}^{[\eta(1-\varepsilon^2)^{1/2}]/\varepsilon} e^{-x^2/2}dx\right] \tag{21-107}$$

where

$$\varepsilon^2 \equiv \frac{m_0\,m_4 - m_2^2}{m_0\,m_4} = \frac{\Delta}{m_0\,m_4} \tag{21-108}$$

From Eq. (21-99) it can easily be shown that Δ is always positive; therefore, ε, as defined by Eq. (21-108), must always be in the range

$$0 < \varepsilon < 1 \tag{21-109}$$

Equation (21-107) is plotted in Fig. 21-11 for different values of ε throughout this range. Note that for a narrow-band process approaching the single harmonic process given by Eq. (21-2), $\varepsilon \to 0$, in which case Eq. (21-107) reduces to the form of a Rayleigh distribution, Eq. (20-92). When the process is white noise or band-limited white noise, as given by Eq. (21-75), $\varepsilon = 2/3$. The limiting case $\varepsilon = 1$ can be approached by superposition of a single harmonic process $y(t)$ at frequency $\bar{\omega}_2$ and a band-limited process $z(t)$ within the frequency range $-\bar{\omega}_1 < \bar{\omega} < \bar{\omega}_1$, provided that $\bar{\omega}_2/\bar{\omega}_1 \to \infty$ and $\sigma_y^2/\sigma_z^2 \to 0$. This is equivalent to placing a very high-frequency, low-amplitude "dither" signal on top of a low-frequency band-limited signal. The resulting distribution of maxima as given by Eq. (21-107) approaches the form of a gaussian distribution.

If the value of ε is to be estimated using a single sample waveform from process $x(t)$, this can easily be accomplished by first counting the total number of maxima ($+$ and $-$) N and the number of negative maxima N^- occurring in a sample function of reasonable duration. Dividing N^- by N gives the proportion r of negative maxima present in the total, which must be equal to the area under the probability density

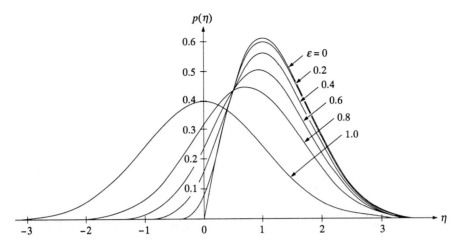

$p(\eta)$

FIGURE 21-11
Probability density function for maxima for different values of ε.

function $p(\eta)$ to the left of the origin in Fig. 21-11. It can be shown that ε is approximately related to this area by

$$\varepsilon^2 = 4r(1 - r) \tag{21-110}$$

Thus after $r = N^-/N$ has been determined, ε can immediately be estimated by this relation.

Example E21-7. Compute the numerical value of ε for stationary process $x(t)$ which has a uniform power spectral density function of intensity S_0 over the ranges $-\bar{\omega}_2 < \bar{\omega} < -\bar{\omega}_1$ and $\bar{\omega}_1 < \bar{\omega} < \bar{\omega}_2$ as given by Eqs. (21-76).

Substituting the first of Eqs. (21-76) into Eq. (21-99) and completing the integral for $n = 0$, 2, and 4 gives, respectively,

$$m_0 = 2S_0(\bar{\omega}_2 - \bar{\omega}_1)$$

$$m_2 = \frac{2S_0}{3}(\bar{\omega}_2^3 - \bar{\omega}_1^3) \tag{a}$$

$$m_4 = \frac{2S_0}{5}(\bar{\omega}_2^5 - \bar{\omega}_1^5)$$

Substituting these relations into Eq. (21-108) yields

$$\varepsilon^2 = 1 - \frac{5}{9}\frac{(1 - \gamma^3)^2}{(1 - \gamma)(1 - \gamma^5)} \tag{b}$$

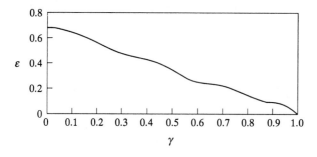

FIGURE E21-5
Parameter ε versus frequency
ratio $\bar{\omega}_1/\bar{\omega}_2$.

where γ is the dimensionless frequency parameter

$$\gamma = \frac{\bar{\omega}_1}{\bar{\omega}_2} \qquad (c)$$

Equation (b) is plotted in Fig. E21-5, showing that ε varies in an approximately linear fashion from a value of $2/3$ at $\gamma = 0$ to a value of zero at $\gamma = 1$, thus (from Fig. 21-11) showing how the probability density function for maxima approaches the Rayleigh distribution as the frequency bandwidth narrows.

21-11 PROBABILITY DISTRIBUTION FOR EXTREME VALUES[2]

Consider N independently observed maxima having the probability density function $p(\eta)$ given by Eq. (21-107). The probability (Pr) that all N maxima will be less than η is given by

$$\text{Pr (all } N \text{ maxima} < \eta) = P(\eta)^N \qquad (21\text{-}111)$$

where $P(\eta)$ is the probability distribution function for maxima as defined by

$$P(\eta) \equiv \int_{-\infty}^{\eta} p(\eta) \, d\eta \qquad (21\text{-}112)$$

Obviously, the probability distribution function for the largest maximum (extreme value) must also be given by Eq. (21-111), that is,

$$P_e(\eta) = P(\eta)^N \qquad (21\text{-}113)$$

[2] D. E. Cartwright and M. S. Longuet-Higgins, ''The Statistical Distribution of the Maxima of a Random Function,'' loc. cit.; A. G. Davenport, ''Note on the Distribution of the Largest Value of a Random Function with Application to Gust Loading,'' loc. cit.

Taking the derivative of Eq. (21-113) gives the probability density function for the extreme value in the form

$$P_e(\eta) = N\, P(\eta)^{N-1} p(\eta) \tag{21-114}$$

For large values of N, it is quite apparent that relatively large values of η_e (extreme value) are of interest; therefore the accuracy with which the extreme-value distribution $P_e(\eta)$ can be defined depends very much on the accuracy of the function $P(\eta)$ as it approaches unity asymptotically with increasing values of η.

Using Eqs. (21-107), (21-112), and (21-113), Davenport has shown, relying in part on earlier work by Cartwright and Lonquet-Higgins, that the probability distribution function for extreme values η_e is

$$P(\eta_e) = \exp\left[-\nu\, T \exp\left(-\frac{\eta_e^2}{2}\right)\right] \tag{21-115}$$

in which ν, the mean frequency of occurrence of zero crossings with positive slope, is given by

$$\nu \equiv \frac{1}{2\pi}\left(\frac{m_2}{m_0}\right)^{1/2} \tag{21-116}$$

The corresponding probability density function $p(\eta_e)$ can easily be obtained by differentiating Eq. (21-115) with respect to η_e.

Using the extreme-value probability distribution function given by Eq. (21-115), it has been shown by Davenport that the mean extreme value is given by the approximate relation

$$\bar{\eta}_e \doteq (2\ln\nu\, T)^{1/2} + \frac{\gamma}{(2\ln\nu\, T)^{1/2}} \tag{21-117}$$

in which γ is Euler's constant, equal to 0.5772, and that the standard deviation of the extreme values is given by

$$\sigma_{\eta_e} = \frac{\pi}{\sqrt{6}}\frac{1}{(2\ln\nu\, T)^{1/2}} \tag{21-118}$$

Figure 21-12 shows a plot of the probability density function for process $x(t)$, a plot of the probability density function for maxima $\eta(\varepsilon = 2/3)$, and plots of the extreme-value probability density function for four different values of $\nu\, T(10^2,\ 10^3,\ 10^4,\ 10^5)$. It should be noted that the probability density functions for extreme values are sharply peaked and that the degree of peaking increases with increasing values of $\nu\, T$. Because of this characteristic, engineering designs can often be based on the mean extreme value $\bar{\eta}_e$ as expressed by Eq. (21-117), which is plotted in Fig. 21-13. It

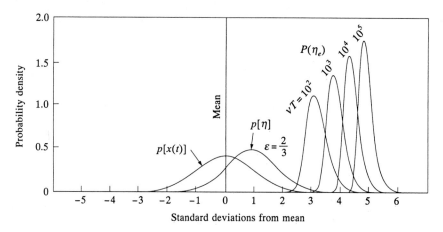

FIGURE 21-12
Probability density functions for $x(t)$, η, and η_e.

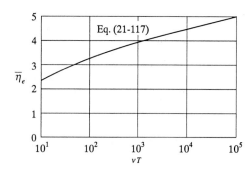

FIGURE 21-13
Normalized mean extreme-value vs. νT.

is clear from this figure that arbitrarily assuming $\overline{\eta}_e$ equal to 3, as is often done in practice, can be considerably on the unconservative side for large values of νT.

Since the general form of the probability distribution function for largest maxima $P_e(\eta)$ closely depends on the accuracy of the probability distribution function for maxima $P(\eta)$ as it nears unity with increasing values of η, other forms of $P_e(\eta)$ have been derived by making various assumptions regarding the manner in which $P(\eta)$ approaches unity. One such assumption is that $P(\eta)$ approaches unity in the manner

$$P(\eta) = 1 - e^{-\eta} \tag{21-119}$$

With this asymptotic form, the extreme-value distribution (Gumbel Type I)[3] can be

[3] E. J. Gumbel and P. G. Carlson, "Extreme Values in Aeronautics," *Jour. of Aero. Sci.*, pp. 389-398, June, 1954;   E. J. Gumbel, "Probability Tables for the Analysis of Extreme-Value Data," *Natl. Bur. Stds. Appl. Math. Ser.* 22, July, 1953.

expressed as

$$P(\eta_e) = \exp\{-\exp[-\alpha(\eta_e - u)]\} \qquad (21\text{-}120)$$

where α and u are constants. Since the second derivative of Eq. (21-120) vanishes for $\eta_e = u$, constant u must equal the most probable value of η_e. Equation (21-120) gives the mean and standard deviation for the extreme values in the forms

$$\bar{\eta}_e = u + \frac{\gamma}{\alpha} \qquad (21\text{-}121)$$

$$\sigma_{\eta_e} = \frac{\pi}{\sqrt{6}\,\alpha} \qquad (21\text{-}122)$$

where γ is Euler's constant (0.5772). From Eq. (21-122) it is clear that constant α is a measure of the dispersion of the extreme values.

If a very large number of experimental extreme values are known, the mean and standard deviation can be calculated fairly accurately, whereupon Eqs. (21-121) and (21-122) can be used to solve for α and u. However, if the number of extreme values is relatively small, a correction should be made using the procedure reported by Gumbel.

The lognormal probability density function given previously by Eq. (20-40) is often used as the probability density function for extreme values. Its use requires finding not only the mean and standard deviation of the extreme values but their median value (50 percentile) as well.

Example E21-8. The extreme values of 50 sample members of random process $x(t)$ have been measured giving the following numerical values, the absolute values of which have been arranged in order of rank:

0.82	1.14	-1.54	1.97	2.67
-0.90	1.16	1.60	1.99	-2.74
0.98	-1.20	-1.64	-2.02	-2.98
-1.03	1.29	-1.67	-2.09	3.33
-1.06	-1.39	1.70	2.11	3.50
1.08	-1.44	1.75	2.13	-3.63
1.10	1.46	-1.77	-2.23	3.85
-1.11	1.48	1.84	2.37	-4.07
-1.12	-1.50	-1.90	-2.51	-4.18
-1.13	1.51	-1.93	2.60	4.33

Assuming the positive and negative extreme values have the same Gumbel Type I probability distribution, generate its proper relation in the form of Eq. (21-120).

An approximation of the distribution function can be obtained using Eqs. (21-121) and (21-122). When the signs of the measured extreme values are ignored, the result is

$$\bar{x}_e = \frac{1}{50} \sum_{i=1}^{50} x_{e_i} = 1.97 \qquad \sigma_{x_e}^2 = \frac{1}{50} \sum_{i=1}^{50} (x_{e_i} - \bar{x}_e)^2 = 0.839 \qquad \text{(a)}$$

Using Eqs. (21-121) and (21-122) gives

$$\alpha = \frac{\pi}{\sqrt{6}\, \sigma_{x_e}} = 1.40 \qquad u = \bar{x}_e - \frac{0.577}{\alpha} = 1.56 \qquad \text{(b)}$$

Substituting Eqs. (b) into Eq. (21-120) results in

$$P(x_e) \doteq \exp\left\{ - \exp\left[- 1.40(x_e - 1.56)\right]\right\} \qquad \text{(c)}$$

Using the correction as given by Gumbel for the case of 50 sample values gives the more accurate expression

$$P(x_e) = \exp\left\{ - \exp\left[- 1.27(x_e - 1.54)\right]\right\} \qquad \text{(d)}$$

21-12   NONSTATIONARY GAUSSIAN PROCESSES

A stationary process has previously been defined as one for which all ensemble averages are independent of time; therefore, a nonstationary process is one for which these same ensemble averages are time dependent. Thus the ensemble average $E[x(t)x(t+\tau)]$, which completely characterizes a nonstationary gaussian process $x(t)$, will be dependent upon time t as well as the time interval τ.

In engineering, a nonstationary process $x(t)$ can often be represented fairly well using the quasi-stationary form

$$x(t) = \zeta(t)\, z(t) \qquad (21\text{-}123)$$

where $\zeta(t)$ is a fully prescribed function of time and $z(t)$ is a stationary process. If $z(t)$ is a gaussian process, $x(t)$ will also be gaussian, in which case the covariance function

$$E[x(t)x(t + \tau)] = \zeta(t)\, \zeta(t + \tau)\, R_z(\tau) \qquad (21\text{-}124)$$

completely characterizes the process.

The above characterization of nonstationary gaussian processes involving one independent variable can be extended directly to processes involving more than one independent variable.

21-13 STATIONARY GAUSSIAN PROCESS: TWO OR MORE INDEPENDENT VARIABLES

All the stationary gaussian processes characterized previously involved one independent variable which was considered to be time t. The basic concepts developed for these processes will now be extended to stationary gaussian processes involving two or more independent variables. To illustrate this extension, suppose the variable of interest is random not only with respect to time but with respect to certain space coordinates as well. For example, consider the wind drag force per unit height acting on a tall industrial smokestack during a strong windstorm, as described in Section 21-1. This loading involves two independent variables, x and t.

To characterize the random component of drag $p(x, t)$ in a probabilistic sense, it is necessary to establish probability density functions involving random variables $p(x, t)$ and $p(\alpha, t + \tau)$, where α and τ are dummy space and time variables, respectively. If the process is gaussian, these probability density functions will be completely known provided the covariance function as given by the ensemble average $E[p(x, t)p(\alpha, t + \tau)]$ can be defined. If the process is stationary, this ensemble average will be independent of time but will depend upon the time difference τ, in which case the covariance function defined by the relation

$$R_p(x, \alpha, \tau) \equiv E[p(x, t)p(\alpha, t + \tau)] \tag{21-125}$$

completely characterizes the process.

Assuming the above process is ergodic, that is, the mean wind velocity remains constant for all members of the ensemble, the cross-spectral density function for the rth member, that is,

$$S_{p_r}(x, \alpha, \overline{\omega}) \equiv \lim_{s \to \infty} \frac{\left[\int_{-s/2}^{s/2} p_r(x, t) \exp(-i\overline{\omega}t) dt \right] \left[\int_{-s/2}^{s/2} p_r(\alpha, t) \exp(+i\overline{\omega}t) dt \right]}{2\pi s} \tag{21-126}$$

will also characterize the process. This cross-spectral density function is related to the covariance function through the Fourier transform relations

$$S_p(x, \alpha, \overline{\omega}) = \frac{1}{2\pi} \int_{-\infty}^{\infty} R_p(x, \alpha, \tau) \exp(-i\overline{\omega}\tau) \, d\tau$$

$$R_p(x, \alpha, \tau) = \int_{-\infty}^{\infty} S_p(x, \alpha, \overline{\omega}) \exp(i\overline{\omega}\tau) \, d\overline{\omega} \tag{21-127}$$

Extending the above characterizations to stationary gaussian processes involving more than two independent variables is straightforward. For example, to characterize a field potential $\Phi(x, y, z, t)$ which is random with respect to time and each space coordinate, one must establish either the covariance function

$$R_\Phi(x, y, z, \alpha, \beta, \gamma, \tau) \equiv E[\Phi(x, y, z, t)\Phi(\alpha, \beta, \gamma, t + \tau)] \tag{21-128}$$

or the corresponding cross-spectral density function $S_\Phi(x, y, z, \alpha, \beta, \gamma, \bar{\omega})$. Terms α, β, and γ are dummy variables for x, y, and z, respectively.

If the field potential $\Phi(x, y, z, t)$ is homogeneous, the covariance and cross-spectral density functions depend only on the differences in coordinates, that is, on

$$ X \equiv x - \alpha \qquad Y \equiv y - \beta \qquad Z \equiv z - \gamma \tag{21-129} $$

The process is then characterized either by the function $R_\Phi(X, Y, Z, \tau)$ or by the function $S_\Phi(X, Y, Z, \bar{\omega})$.

If the potential function $\Phi(x, y, z, t)$ happens to be isotropic as well as homogeneous, the covariance and cross-spectral density functions will depend only upon the distance between points, that is, the distance

$$ \rho \equiv [(x - \alpha)^2 + (y - \beta)^2 + (z - \gamma)^2]^{1/2} \tag{21-130} $$

in which case the process will be characterized by either $R_\Phi(\rho, \tau)$ or $S_\Phi(\rho, \bar{\omega})$.

PROBLEMS

21-1. Show that the Fourier transform of an even function and of an odd function are real and imaginary, respectively.

21-2. Find the Fourier transform of each function $x(t)$ shown in Fig. P21-1.

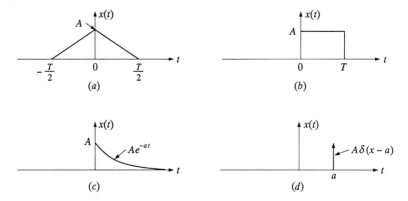

FIGURE P21-1
Functions $x(t)$ referred to in Prob. 21-2.

21-3. Consider the function $x(t) = A \cos at$ in the range $-T/2 < t < T/2$ and $x(t) = 0$ outside this range. Find and sketch the Fourier transform $X(\bar{\omega})$ when (a) $T = \pi/a$, (b) $T = 3\pi/a$, (c) $T = 5\pi/a$, and (d) $T \to \infty$.

21-4. Evaluate the integral

$$I = \int_1^\infty \left[\int_1^\infty \frac{x^2 - y^2}{(x^2 + y^2)^2} dy \right] dx$$

by integrating first with respect to y and then with respect to x. Then reverse the order of integration and reevaluate the integral. Finally evaluate the limit L of integral I by integrating over the finite domain and then taking the limit as follows:

$$L = \lim_{T \to \infty} \left\{ \int_1^T \left[\int_1^T \frac{x^2 - y^2}{(x^2 + y^2)^2} dy \right] dx \right\}$$

Noting that the integrand in integral I is antisymmetric about the line $x = y$, which form of integration would you recommend for engineering applications?

21-5. Evaluate the integral

$$I = \int_{-\infty}^\infty \frac{\sin^2 x}{x^2} dx$$

21-6. Consider the stationary random process $x(t)$ defined by

$$x_r(t) = \sum_{n=1}^{10} A_{nr} \, \cos(n\bar{\omega}_0 t + \theta_{nr}) \qquad r = 1, 2, \cdots$$

where $x_r(t)$ = rth member of ensemble
A_{nr} = sample values of random variable A
$\bar{\omega}_0$ = fixed circular frequency
θ_{nr} = sample values of random phase angle θ having a uniform probability density function in range $0 < \theta < 2\pi$ of intensity $1/2\pi$

If random variable A is gaussian having a known mean value $\bar{A}$ and a known variance σ_A^2, find the ensemble mean value of $x(t)$ and the ensemble variance of $x(t)$. Is process $x(t)$ a gaussian process?

21-7. Derive the autocorrelation function for the stationary random process $x(t)$ defined in Prob. 21-6.

21-8. Derive the power spectral density function for the stationary random process $x(t)$ defined in Prob. 21-6, assuming that Dirac delta functions are permitted in the answer.

21-9. A stationary random process $x(t)$ has the autocorrelation function

$$R_x(\tau) = A \, \exp(-a|\tau|)$$

where A and a are real constants. Find the power spectral density function for this process.

21-10. Consider a random process $x(t)$ which takes the value $+A$ or $-A$, with equal probability, throughout each interval $n\Delta\varepsilon < t < (n+1)\Delta\varepsilon$ of each member of the process, where n is an integer running from $-\infty$ to $+\infty$. Find and plot the ensemble covariance function $E[x(t)x(t+\tau)]$. Is this process stationary or nonstationary?

21-11. If the origin of time, that is, $t = 0$, for each member of process $x(t)$ defined in Prob. 21-10 is selected randomly over an interval $\Delta\varepsilon$ with uniform probability of occurrence, what is the covariance function $E[x(t)x(t+\tau)]$? Is this process stationary or nonstationary?

21-12. Assuming that you find the process in Prob. 21-11 stationary, what are the autocorrelation and power spectral density functions for this process? Use Eqs. (21-35) and (21-38) in finding the power spectral density function.

21-13. Show that the autocorrelation and power spectral density functions obtained in Prob. 21-12 are Fourier transform pairs in accordance with Eqs. (21-37).

21-14. Each member of a stationary random process $x(t)$ consists of a periodic infinite train of triangular pulses, as shown in Fig. P21-2. All members of the process are identical except for phase, which is a random variable uniformly distributed over the interval $(0, T)$. Assuming that the period T is not less than $2a$, where a is the duration of a single pulse, find the autocorrelation function for this process.

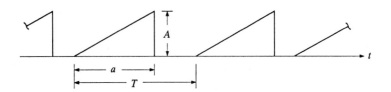

FIGURE P21-2
One sample member of process $x(t)$ referred to in Prob. 21-14.

21-15. Each member of a random process $x(t)$ consists of the superposition of rectangular pulses of duration $\Delta\varepsilon$ and of constant intensity A which are located in a random fashion with respect to time as shown in Fig. P21-3a. Each value of ε_n is independently sampled in accordance with the uniform probability density function $p(\varepsilon)$ given in Fig. P21-3b. What is the ensemble value $\overline{x(t)}$ for this process? Is this process stationary or nonstationary?

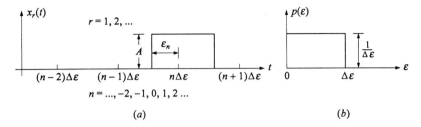

FIGURE P21-3
One sample pulse in member $x_r(t)$ of process $x(t)$ and probability density function for random variable ε referred to in Prob. 21-15.

21-16. Assume the autocorrelation and power spectral density functions $R_{xx}(\tau)$ and $S_{xx}(\overline{w})$ for a stationary random process $x(t)$ are known. Derive the expressions for $S_{x\dot{x}}(\overline{w})$, $S_{\dot{x}x}(\overline{w})$, $S_{x\ddot{x}}(\overline{w})$, $S_{\ddot{x}x}(\overline{w})$, $S_{\dot{x}\ddot{x}}(\overline{w})$, $S_{\ddot{x}\dot{x}}(\overline{w})$ and $R_{x\dot{x}}(\tau)$, $R_{\dot{x}x}(\tau)$, $R_{x\ddot{x}}(\tau)$, $R_{\ddot{x}x}(\tau)$, $R_{\dot{x}\ddot{x}}(\tau)$, $R_{\ddot{x}\dot{x}}(\tau)$ in terms of $S_{xx}(\overline{w})$ and $R_{xx}(\tau)$, respectively.

21-17. Considering two stationary random processes $x(t)$ and $y(t)$, show that $S_{yx}(\overline{w})$ is the complex conjugate of $S_{xy}(\overline{w})$.

21-18. Two stationary random processes $x(t)$ and $y(t)$ have the joint probability density function

$$p[x(t)y(t+\tau)] = \frac{1}{2ab\pi\sqrt{1-c^2}} \exp\left[-\frac{1}{2(1-c^2)}\left(\frac{x^2}{a^2} - \frac{2cxy}{ab} + \frac{y^2}{b^2}\right)\right]$$

Define a, b and c in terms of the appropriate autocorrelation and/or cross-correlation functions for processes $x(t)$ and $y(t)$. What is the corresponding joint probability density function $p[\dot{x}(t)\dot{y}(t+\tau)]$? Define the coefficients in this function in terms of the appropriate autocorrelation and/or cross-correlation functions for processes $x(t)$ and $y(t)$.

CHAPTER

22

STOCHASTIC RESPONSE OF LINEAR SDOF SYSTEMS

22-1 TRANSFER FUNCTIONS

This chapter develops the appropriate input-output relationships for stable linear SDOF systems having constant coefficients and characterizes the stationary output processes of such systems in terms of their corresponding stationary input processes and their transfer functions.

Suppose that a stationary gaussian process $p(t)$ is the input to a linear SDOF system and that $v(t)$ is the desired output process, as shown in Fig. 22-1, where TF_1, $TF_2, \cdots, TF_n$ represent the transfer functions of systems 1, 2, $\cdots$, n, respectively. Since uncontrollable random variables are always present during construction of real systems (even though of identical design), these transfer functions will also have random characteristics. Usually, however, in vibration analysis, the randomness of these characteristics is small in comparison with the randomness of the input $p(t)$ and therefore can be neglected, in which case $TF_1 = TF_2 = \cdots = TF_n = TF$. Thus, in the subsequent treatment of linear systems, the coefficients appearing in all mathematical representations will be considered as fixed constants; that is, the transfer function (or functions) TF_r will in each case be treated as independent of r. When the transfer function TF and either the autocorrelation function $R_p(\tau)$ or

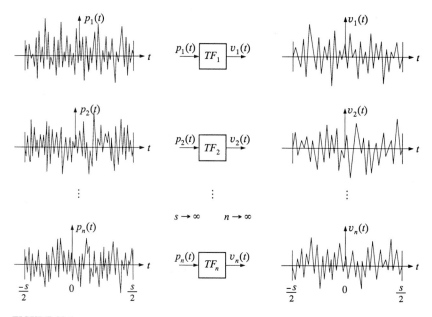

FIGURE 22-1
Input and output processes of a stable linear SDOF system.

the power spectral density function $S_p(\overline{\omega})$ are known, the output process $v(t)$ can be completely characterized. The transfer functions used here will be the unit-impulse response function $h(t)$ and the complex frequency response function $H(i\overline{\omega})$.

22-2   RELATIONSHIP BETWEEN INPUT AND OUTPUT AUTOCORRELATION FUNCTIONS

The output or response function $v_r(t)$ shown in Fig. 22-1 is related to its corresponding input function $p_r(t)$ through the convolution integral relation

$$v_r(t) = \int_{-\infty}^{t} p_r(\tau)\, h(t - \tau)\, d\tau \qquad r = 1, 2, \cdots, \infty \tag{22-1}$$

If the input process is assumed to have a zero mean value, that is,

$$E[p(t)] = 0 \tag{22-2}$$

the mean value for the output process can be obtained by averaging Eq. (22-1) across the ensemble, which gives

$$E[v(t)] = E\left[\int_{-\infty}^{t} p(\tau)\, h(t - \tau)\, d\tau\right] = \int_{-\infty}^{t} E[p(\tau)]\, h(t - \tau)\, d\tau = 0 \tag{22-3}$$

Thus it is shown that if the input ensemble has a zero mean value, the output ensemble will also have a zero mean value.

Consider now the ensemble average $E[v(t)\,v(t+\tau)]$, which can be evaluated by using Eq. (22-1) as shown in the relation

$$E[v(t)\,v(t+\tau)]$$

$$= E\left[\int_{-\infty}^{t} p(\theta_1)\,h(t-\theta_1)\,d\theta_1 \int_{-\infty}^{t+\tau} p(\theta_2)\,h(t+\tau-\theta_2)\,d\theta_2\right] \quad (22\text{-}4)$$

where θ_1 and θ_2 are dummy time variables. When a change of variables is introduced in accordance with the definitions

$$u_1 \equiv t - \theta_1 \qquad \theta_1 = t - u_1$$
$$u_2 \equiv t + \tau - \theta_2 \qquad \theta_2 = t + \tau - u_2 \quad (22\text{-}5)$$

Eq. (22-4) becomes

$$E[v(t)\,v(t+\tau)]$$

$$= E\left[\int_{t+\infty}^{0} p(t-u_1)\,h(u_1)\,du_1 \int_{t+\tau+\infty}^{0} p(t+\tau-u_2)\,h(u_2)\,du_2\right] \quad (22\text{-}6)$$

When the limits of both integrals are inverted and use is made of the fact that $h(u_1)$ and $h(u_2)$ damp out for stable systems, Eq. (22-6) can be written in the form

$$E[v(t)\,v(t+\tau)] = E\left[\int_{0}^{\infty}\int_{0}^{\infty} p(t-u_1)\,p(t+\tau-u_2)\,h(u_1)\,h(u_2)\,du_1\,du_2\right] \quad (22\text{-}7)$$

Since only the functions $p(t-u_1)$ and $p(t+\tau-u_2)$ change across the ensemble, Eq. (22-7) becomes

$$E[v(t)\,v(t+\tau)] = \int_{0}^{\infty}\int_{0}^{\infty} E[p(t-u_1)\,p(t+\tau-u_2)]\,h(u_1)\,h(u_2)\,du_1\,du_2 \quad (22\text{-}8)$$

The ensemble average on the right hand side of Eq. (22-8) is the autocorrelation function for stationary process $p(t)$ and is independent of time; therefore, the ensemble average on the left hand side must also be independent of time. This shows that the output process is stationary and that its autocorrelation function is given by the relation

$$R_v(\tau) = \int_{0}^{\infty}\int_{0}^{\infty} R_p(\tau - u_2 + u_1)\,h(u_1)\,h(u_2)\,du_1\,du_2 \quad (22\text{-}9)$$

If the input process $p(t)$ is gaussian, the output process $v(t)$ for a linear stable system will also be gaussian; therefore, in such cases the autocorrelation function given by Eq. (22-9) completely characterizes the process. To prove the first part of this statement, consider Eq. (21-1) in the limiting form

$$v_r(t) = \lim_{\Delta\tau \to 0} \sum_{i=-\infty}^{t/\Delta\tau} p_r(\tau_i)\, h(t-\tau_i)\, \Delta\tau \qquad r = 1, 2, \cdots, \infty \qquad (22\text{-}10)$$

Since all terms $h(t-\tau_i)$ are known constants for fixed values of t, Eqs. (22-10) are identical in form with the linear transformations given by Eqs. (20-103), which have already been shown to retain the gaussian distribution.

To illustrate the application of Eq. (22-9), assume that the excitation $p(t)$ of a viscously damped SDOF system is white noise; that is, its power spectral density function equals a constant S_0 which corresponds to the autocorrelation function

$$R_p(\tau) = 2\pi\, S_0\, \delta(\tau) \qquad (22\text{-}11)$$

Further, assume that the system is undercritically damped, in which case the unit-impulse function $h(t)$ is given by Eq. (6-51). Substituting this relation along with Eq. (22-11) into Eq. (22-9) gives

$$R_v(\tau) = \frac{2\pi\, S_0}{\omega_D^2\, m^2} \int_0^\infty \int_0^\infty \delta(\tau - u_2 + u_1)$$

$$\times \exp\big[-\omega\,\xi(u_1 + u_2)\big] \, \sin \omega_D\, u_1 \, \sin \omega_D\, u_2 \, du_1 \, du_2 \qquad (22\text{-}12)$$

Completing the double integration and introducing the relation $k = m\omega^2$ lead to

$$R_v(\tau) = \frac{\pi\omega S_0}{2k^2\,\xi} \left(\cos \omega_D\, |\tau| + \frac{\xi}{\sqrt{1-\xi^2}} \sin \omega_D\, |\tau| \right) \exp(-\omega\,\xi\, |\tau|)$$

$$-\infty < \tau < \infty \qquad (22\text{-}13)$$

If the system is overcritically damped, the unit-impulse function $h(t)$ is given by the second of Eqs. (f) in Example E12-5 and the above integration procedure leads to the relation

$$R_v(\tau) = \frac{\pi\omega S_0}{2k^2\,\xi} \left[\phi \exp(\omega\sqrt{\xi^2-1}\,|\tau|) - \theta \exp(-\omega\sqrt{\xi^2-1}\,|\tau|) \right] \exp(-\omega\,\xi|\tau|)$$

$$-\infty < \tau < \infty \qquad (22\text{-}14)$$

where

$$\phi \equiv \frac{1}{2\big[\xi\sqrt{\xi^2-1} - (\xi^2-1)\big]} \qquad \theta \equiv \frac{1}{2\big[\xi\sqrt{\xi^2-1} + (\xi^2-1)\big]} \qquad (22\text{-}15)$$

If the white-noise input is gaussian, the output response $v(t)$ will also be gaussian, in which case the autocorrelation functions given by Eqs. (22-13) and (22-14) completely characterize the processes they represent. Plots of these functions are shown in Fig. 22-2.

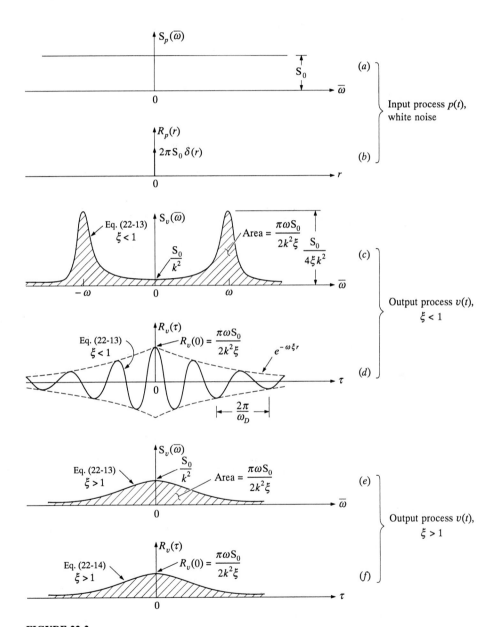

FIGURE 22-2
Output relations for SDOF system subjected to white-noise excitation.

Example   E22-1.   Consider the SDOF system

$$m\ddot{v} + c\dot{v} + kv = p(t) \tag{a}$$

excited by a zero mean ergodic random process $p(t)$ having constant power spectral density equal to S_0 over the range $-\infty < \overline{\omega} < \infty$. Determine the average rate of energy dissipation in the system with time.

Since the damping force $c\dot{v}$ is the only nonconservative force in the system, the instantaneous rate of energy dissipation is given by $c\dot{v}^2$. Therefore, the average rate of energy dissipation P_{avg} can be expressed as

$$P_{\text{avg}} = c\langle \dot{v}^2 \rangle = c\,R_{\dot{v}}(0) \tag{b}$$

Using Eq. (21-43) gives

$$R_{\dot{v}}(\tau) = -R_v''(\tau) \tag{c}$$

Substituting Eq. (22-13) into Eq. (c) gives

$$R_{\dot{v}}(\tau) = -R_v''(\tau) = \frac{\pi\omega S_0}{2k^2\,\xi}\left(\omega^2\,\cos\omega_D\,|\tau| - \frac{\omega^2\,\xi}{\sqrt{1-\xi^2}}\sin\omega_D\,|\tau|\right)$$

$$\times \exp(-\omega\,\xi\,|\tau|) \tag{d}$$

from which

$$R_{\dot{v}}(0) = -R_v''(0) = \frac{\pi\omega^3\,S_0}{2k^2\,\xi} \tag{e}$$

Substituting Eq. (e) into Eq. (b) and making use of the relations $k^2 = m^2\,\omega^4$ and $c = 2m\omega\xi$ lead to

$$P_{\text{avg}} = \frac{\pi S_0}{m} \tag{f}$$

Note that the average rate of energy dissipation is independent of the damping ratio ξ.

22-3   RELATIONSHIP BETWEEN INPUT AND OUTPUT POWER SPECTRAL DENSITY FUNCTIONS

The power spectral density function for the output process $v(t)$ is related to its autocorrelation function through the Fourier transform relation

$$S_v(\overline{\omega}) = \frac{1}{2\pi}\int_{-\infty}^{\infty} R_v(\tau)\,\exp(-i\,\overline{\omega}\tau)\,d\tau \tag{22-16}$$

Substituting Eq. (22-9) into Eq. (22-16) gives

$$S_v(\overline{\omega}) = \frac{1}{2\pi} \int_{-\infty}^{\infty} \left[\int_0^\infty \int_0^\infty R_p(\tau - u_2 + u_1)\, h(u_1)\, h(u_2)\, du_1\, du_2 \right] \exp(-i\overline{\omega}\tau)\, d\tau$$

(22-17)

Interchanging the order of integration and introducing expanding limits of integration lead to

$$S_v(\overline{\omega}) = \frac{1}{2\pi} \lim_{s \to \infty} \left[\int_0^s h(u_1)\, du_1 \int_0^s h(u_2)\, du_2 \int_{-s}^s R_p(\tau - u_2 + u_1) \right.$$

$$\left. \times \exp(-i\overline{\omega}\tau)\, d\tau \right] \qquad (22\text{-}18)$$

When a change of variable $\theta \equiv \tau - u_2 + u_1$ is substituted, Eq. (22-18) changes to the form

$$S_v(\overline{\omega}) = \frac{1}{2\pi} \lim_{s \to \infty} \left[\int_0^s h(u_1)\, \exp(i\overline{\omega}\, u_1)\, du_1 \int_0^s h(u_2)\, \exp(-i\overline{\omega}\, u_2)\, du_2 \right.$$

$$\left. \times \int_{-s+u_1-u_2}^{s+u_1-u_2} R_p(\theta)\, \exp(-i\overline{\omega}\theta)\, d\theta \right] \qquad (22\text{-}19)$$

Since the unit-impulse-response functions $h(u_1)$ and $h(u_2)$ equal zero for $u_1 < 0$ and $u_2 < 0$, respectively, the lower limits of the first two integrals can be changed from zero to $-s$. Also since these functions must damp out with increasing values of u_1 and u_2 for the system to be stable, these terms can be dropped from the limits of the third integral. When use is made of the first of Eqs. (21-37) and (6-53), Eq. (22-19) reduces to the form

$$S_v(\overline{\omega}) = H(-i\overline{\omega})\, H(i\overline{\omega})\, S_p(\overline{\omega}) = |H(i\overline{\omega})|^2\, S_p(\overline{\omega}) \qquad (22\text{-}20)$$

in which $H(i\overline{\omega})$ is the frequency-domain transfer function between loading and response.

When the viscously-damped SDOF system is subjected to a zero-mean white-noise excitation $p(t)$, that is, $S_p(\overline{\omega}) = S_0$, substitution of Eq. (6-52) into (22-20) gives

$$S_v(\overline{\omega}) = \frac{S_0}{k^2[1 + (4\xi^2 - 2)(\overline{\omega}/\omega)^2 + (\overline{\omega}/\omega)^4]} \qquad (22\text{-}21)$$

Power spectral density functions for both the under- and overcritically damped cases are shown in Figs. 22-2c and e, respectively.

Example E22-2.   Derive Eq. (22-13) directly from Eq. (22-21) making use of the Fourier transform relation

$$R_v(\tau) = \int_{-\infty}^{\infty} S_v(\bar{\omega}) \exp(i\bar{\omega}\tau) \, d\bar{\omega} \tag{a}$$

Substituting Eq. (22-21) into Eq. (a) gives

$$R_v(\tau) = \frac{\omega S_0}{k^2} \int_{-\infty}^{\infty} \frac{\exp(i\omega\beta\tau)}{(\beta - r_1)(\beta - r_2)(\beta + r_1)(\beta + r_2)} \, d\beta \tag{b}$$

after introducing

$$\beta = \frac{\bar{\omega}}{\omega} \qquad r_1 = i\xi + \sqrt{1 - \xi^2} \qquad r_2 = i\xi - \sqrt{1 - \xi^2} \tag{c}$$

The integrand in Eq. (b) is an analytic function everywhere in the complex β plane except at points $\beta = r_1$, $\beta = r_2$, $\beta = -r_1$, and $\beta = -r_2$, where poles of order 1 exist. Points $\beta = r_1$ and $\beta = r_2$ are in the upper half plane, while points $\beta = -r_1$ and $\beta = -r_2$ are in the lower half plane. For positive values of τ, contour integration is carried out in the upper half plane; for negative values of τ, integration is carried out in the lower half plane. When Cauchy's residue theorem is used, the integral in Eq. (b) is easily carried out by procedures similar to those in Example E6-3, resulting in the relation

$$R_v(\tau) = \begin{cases} \dfrac{\pi\omega S_0}{2k^2\,\xi} \left(\cos \omega_D\tau + \dfrac{\xi}{\sqrt{1 - \xi^2}} \sin \omega_D\tau \right) \exp(-\omega\xi\tau) & \tau > 0 \\[4mm] \dfrac{\pi\omega S_0}{2k^2\,\xi} \left(\cos \omega_D\tau - \dfrac{\xi}{\sqrt{1 - \xi^2}} \sin \omega_D\tau \right) \exp(\omega\xi\tau) & \tau < 0 \end{cases} \tag{d}$$

Thus the validity of Eq. (22-13) is verified.

22-4   RESPONSE CHARACTERISTICS FOR NARROWBAND SYSTEMS

Most structural systems have reasonably low damping ($\xi < 0.1$) and therefore are classified as narrowband systems. This classification results because the area under the response power spectral density function is highly concentrated near the natural frequency of the system, as shown in Fig. 22-2c. Such a concentration indicates that the predominant frequency components in a sample response function $v_r(t)$ will be contained in a relatively narrow band centered on the undamped natural frequency ω. Because of the beat phenomenon associated with two harmonics whose frequencies are close together, the response envelope for a narrowband system can be expected to show similar characteristics; however, since the predominant frequencies are spread

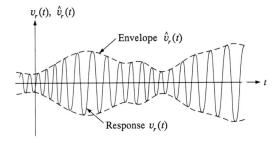

FIGURE 22-3
r th sample function of a narrowband process $v(t)$.

over a narrow band, the beat behavior will be random in character, as shown in Fig. 22-3. Thus, it is correctly reasoned that the response will locally appear as a slightly distorted sine function with a frequency near the natural frequency of the system and with amplitudes that vary slowly in a random fashion. This same type of response can also be predicted from the autocorrelation function, Eq. (22-13), as plotted in Fig. 22-2d, since this function approaches the autocorrelation function for the single harmonic process presented in Fig. 21-2 as damping approaches zero.

When it is noted that the sharply peaked output power spectral density function $S_v(\overline{\omega})$ shown in Fig. 22-2c is obtained by multiplying the similarly peaked transfer function $|\mathrm{H}(i\,\overline{\omega})|^2$ by the constant power spectral density function S_0 of the white-noise input, it becomes clear that the response $v(t)$ is caused primarily by those frequency components in the input process $p(t)$ which are near the natural frequency of the system. Therefore in those cases when the input power spectral density function $S_p(\overline{\omega})$ is not a constant but is a slowly varying function of $\overline{\omega}$ in the vicinity of the natural frequency ω, a white-noise input process can be assumed with little loss in predicting response provided the constant power spectral density S_0 is set equal to the intensity of $S_p(\overline{\omega})$ at $\overline{\omega} = \omega$; that is, let $S_p(\overline{\omega}) = S_p(\omega)$; thus, the output power spectral density function can be approximated by the relation

$$S_v(\overline{\omega}) = \frac{S_p(\omega)}{k^2[1 + (4\xi^2 - 2)(\overline{\omega}/\omega)^2 + (\overline{\omega}/\omega)^4]} \qquad \xi \ll 1 \qquad (22\text{-}22)$$

Note that as the damping ratio ξ approaches zero, the area under this function becomes more and more concentrated at the natural frequency ω and approaches infinity in the limit. This means that the stationary mean square response of an undamped SDOF system is infinite when subjected to white-noise excitation of finite intensity. Such systems are classified as unstable systems.

To clarify further the response characteristics of narrowband linear systems subjected to a stationary gaussian excitation having zero mean values, consider the response conditional probability density function

$$p(v_2|v_1) = \frac{p(v_1, v_2)}{p(v_1)} \qquad (22\text{-}23)$$

where $v_2 \equiv v(t+\tau)$ and $v_1 \equiv v(t)$. When use is made of the standard relations

$$p(v_1, v_2) = \frac{1}{2\pi\sigma_v^2\sqrt{1-\rho_v^2}}$$

$$\times \exp\left[-\frac{1}{2\sigma_v^2(1-\rho_v^2)}(v_1^2 - 2\rho_v\, v_1\, v_2 + v_2^2)\right] \qquad (22\text{-}24)$$

$$p(v_1)\frac{1}{\sqrt{2\pi}\,\sigma_v}\exp\left(-\frac{v_1^2}{2\sigma_v^2}\right) \qquad (22\text{-}25)$$

where

$$\sigma_v = R_v(0)^{1/2} \qquad \rho_v(\tau) = \frac{R_v(\tau)}{R_v(0)} \qquad (22\text{-}26)$$

Eq. (22-23) becomes

$$p[v_2|v_1] = \frac{1}{\sqrt{2\pi}\,\sigma_v\sqrt{1-\rho_v^2}}\exp\left[-\frac{(v_2-\rho_v\,v_1)^2}{2(1-\rho_v^2)\,\sigma_v^2}\right] \qquad (22\text{-}27)$$

This equation shows that when $v(t)$ is fixed, the expected value of $v(t+\tau)$ is $\rho_v(\tau)\,v(t)$ and its variance is $[1-\rho_v(\tau)^2]\,\sigma_v^2$, thus lending support to the previously described response characteristics of narrowband systems.

Finally consider the joint probability density function

$$p[v(t),\,\dot{v}(t)] = \frac{1}{2\pi\sigma_v\,\sigma_{\dot v}}\exp\left[-\frac{1}{2}\left(\frac{v^2}{\sigma_v^2} + \frac{\dot{v}^2}{\sigma_{\dot v}^2}\right)\right] \qquad (22\text{-}28)$$

in which

$$\sigma_{\dot v} = R_{\dot v}(0)^{1/2} = -R_v''(0)^{1/2} \qquad (22\text{-}29)$$

Since $R_v'(0) = 0$, the covariance of random variables v and $\dot{v}$ will also equal zero, which explains the uncoupled form of Eq. (22-28). This equation can now be used to find the probability that response $v(t)$ will cross a fixed level $\hat{v}$ with positive velocity within the time limits t and $t+dt$. To satisfy this condition, $v(t)$ must conform to the relation

$$\Pr\left[v(t) < \hat{v} < v(t+dt)\right] = \Pr\left[0 < [\hat{v}-v(t)] < \dot{v}(t)\,dt\right] \qquad (22\text{-}30)$$

as illustrated graphically in Fig. 22-4a for one member of the ensemble. From the $v\,\dot{v}$ plane shown in Fig. 22-4b, it is clear that those ensemble members which are favorable to this condition must have values of $v(t)$ and $\dot{v}(t)$ which fall within the shaded region. Therefore, if $Q(\hat{v})\,dt$ represents the probability condition given by Eq. (22-30), this term can be evaluated by simply integrating the joint probability density function given by Eq. (22-28) over the shaded region, that is,

$$Q(\hat{v})\,dt = \int_0^\infty \int_{\hat{v}-\dot{v}\,dt}^{\hat{v}} \frac{1}{2\pi\sigma_v\,\sigma_{\dot v}}\exp\left[-\frac{1}{2}\left(\frac{v^2}{\sigma_v^2} + \frac{\dot{v}^2}{\sigma_{\dot v}^2}\right)\right]d\dot{v}\,dv \qquad (22\text{-}31)$$

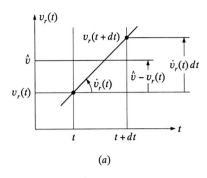
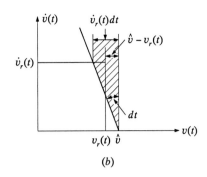

FIGURE 22-4
Velocity-displacement relations for positive slope crossings at level $\hat{v}$ within time limits t and $t + dt$.

Substituting $dv = \dot{v}\, dt$ into this equation, completing the integration, and dividing the result by dt give Rice's relation[1]

$$Q(\hat{v}) = \frac{1}{2\pi} \frac{\sigma_{\dot{v}}}{\sigma_v} \exp\left(-\frac{1}{2} \frac{\hat{v}^2}{\sigma_v^2}\right) \tag{22-32}$$

which is the number of crossings of response $v(t)$ at level $\hat{v}$ with positive velocity per unit of time. Setting $\hat{v}$ equal to zero, the number of zero crossings with positive slope per unit time becomes

$$Q(0) = \frac{1}{2\pi} \frac{\sigma_{\dot{v}}}{\sigma_v} = \frac{1}{2\pi} \sqrt{-\frac{R_v''(0)}{R_v(0)}} = \frac{1}{2\pi} \sqrt{\frac{m_2}{m_0}} \tag{22-33}$$

in which m_2 and m_0 are defined by Eq. (21-99); this same relation was previously given, without derivation, by Eq. (21-116). Making use of Eqs. (22-13) and (22-33) one obtains the desired result

$$Q(0) = \frac{\omega}{2\pi} \tag{22-34}$$

thus indicating that low-damped SDOF systems have the same average number of zero crossings when excited by white noise as when vibrating in a free undamped state. This result is further evidence of the type of response which characterizes narrowband systems.

To approximate the probability density function for the distribution of maxima in the response function $v(t)$ for narrowband systems, assume that on the average one maximum exists for each zero crossing, that is, one for each time interval $T =$

---

[1] S. O. Rice, "Mathematical Analysis of Random Noise," in N. Wax (ed.), *Selected Papers on Noise and Stochastic Processes*, Dover, New York, 1954.

$2\pi/\omega = 2\pi\,(\sigma_v/\sigma_{\dot{v}})$. The possibility of having negative maxima exists, of course, but it is highly improbable for this class of SDOF systems. As a direct result of this assumption, the probability density function for the single maximum occurring in each member of the response ensemble $v(t)$ during a specified time period $2\pi/\omega$ can be approximated by the differential relation

$$p(\hat{v}) = -\frac{dQ(\hat{v})}{d\hat{v}}\left(2\pi\,\frac{\sigma_v}{\sigma_{\dot{v}}}\right) \tag{22-35}$$

Substituting Eq. (22-32) into this equation gives the Rayleigh distribution

$$p(\hat{v}) = \frac{\hat{v}}{\sigma_v^2}\,\exp\left(-\frac{1}{2}\,\frac{\hat{v}^2}{\sigma_v^2}\right) \tag{22-36}$$

which can be considered the approximate probability density function for the upper response envelope shown in Fig. 22-3. This distribution is a special case of that given previously by Eq. (21-107) for $\varepsilon = 0$; see Fig. 21-11.

22-5   NONSTATIONARY MEAN SQUARE RESPONSE RESULTING FROM ZERO INITIAL CONDITIONS

The response characteristics previously defined for output processes are based on steady-state conditions, that is, those conditions which result when input processes are assumed to start at time $t = -\infty$. In actual practice, however, input processes must be assumed to start at time $t = 0$. While such input processes may be assumed as stationary for $t > 0$, the resulting output processes will be nonstationary due to the usual zero initial conditions which are present at $t = 0$. To illustrate the type of nonstationarity which results, consider the input process $p(t)$ to the viscously-damped SDOF system as stationary white noise of intensity S_0 starting at $t = 0$. This input process, as previously demonstrated, can be represented [see Eq. (21-86) and Fig. 21-7] by the relation

$$p_r(t) = \lim_{N\to\infty}\sum_{k=0}^{N-1} a_{kr}\,\eta(t - k\,\Delta t - \varepsilon_r) \qquad r = 1, 2, \cdots \tag{22-37}$$

which is plotted in Fig. 22-5 for one value of r, provided $A^2 \to \infty$ and $\Delta t \to 0$ in such a way that $A^2\,\Delta t \to C$ (a constant). As shown in Eq. (21-94), constant C equals $2\pi\,S_0$. Assuming an undercritically damped system ($\xi < 1$), response can be obtained by superposition through the time domain to give

$$v_r(j\,\Delta t) = \lim_{\Delta t\to 0}\sum_{k=0}^{j} \frac{a_{kr}\,\Delta t}{\omega_D\,m}\,\exp\left[-\omega\xi(j - k)\Delta t\right]\,\sin\omega_D(j - k)\Delta t \tag{22-38}$$

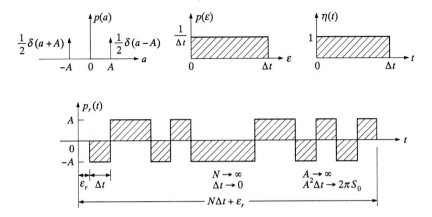

FIGURE 22-5
Stationary white-noise input process $p(t)$ for $t > 0$, Eq. (22-37).

and

$$v_r(j \, \Delta t)^2 = \lim_{\Delta t \to 0} \sum_{k=0}^{j} \sum_{g=0}^{j} \frac{a_{kr} \, a_{gr} \, \Delta t^2}{w_D^2 \, m^2} \exp[-\omega\xi(2j - k - g)\Delta t]$$

$$\times \sin w_D(j - k)\Delta t \, \sin w_D(j - g)\Delta t \qquad (22\text{-}39)$$

The ensemble average of the double summation term in the equation can be reduced immediately to the ensemble average of a single summation term since all covariances of random variables a_k and a_g $(k \ne g)$ equal zero; therefore, the ensemble average is

$$E[v(j \, \Delta t)^2] = \left[\lim_{\Delta t \to 0} \frac{\Delta t^2}{w_D^2 \, m^2} \sum_{k=0}^{j} \exp[-2\omega\xi(j - k) \, \Delta t] \right.$$

$$\left. \times \sin^2 w_D(j - k)\Delta t \right] \left[\lim_{n \to \infty} \frac{1}{n} \sum_{r=1}^{n} a_{kr}^2 \right] \qquad (22\text{-}40)$$

However, since the second square-bracket term in this equation equals A^2 for all values of k, one obtains in the limit as $j \, \Delta t \to t$ and $k \, \Delta t \to \tau$ the relation

$$E[v(t)^2] = \frac{A^2 \, \Delta t}{w_D^2 \, m^2} \exp(-2\omega\xi\tau) \int_0^t \exp(2\omega\xi\tau) \sin^2 w_D(t - \tau) \, d\tau \qquad (22\text{-}41)$$

After substituting $2\pi \, S_0$ for $A^2 \Delta t$ and k^2 for $\omega^4 \, m^2$ and completing the integration, this equation yields

$$E[v(t)^2] = \frac{\pi\omega S_0}{2\xi k^2} \left\{ 1 - \frac{\exp(-2\xi\omega t)}{w_D^2} \left[w_D^2 + \frac{(2\xi\omega)^2}{2} \sin^2 2w_D \, t + \xi\omega w_D \, \sin 2w_D \, t \right] \right\}$$

$$\xi < 1 \qquad (22\text{-}42)$$

which is plotted in Fig. 22-6 for various values of damping ratio ξ. Note the manner and relatively rapid rate at which the ensemble mean square value $E[v(t)^2]$ or variance $\sigma_v(t)^2$ approaches its steady-state value

$$R_v(0) = \frac{\pi S_0 \, \omega}{2\xi k^2} \tag{22-43}$$

If the power spectral density function for the stationary input process starting at time $t = 0$ is nonuniform but varies reasonably slowly in the vicinity of the undamped frequency ω, the variance of the output process for low-damped systems ($\xi < 0.1$) can be approximated reasonably well using Eq. (22-42) provided the power spectral density intensity at $\overline{\omega} = \omega$ is substituted for S_0, that is,

$$E[v(t)^2] \doteq \frac{\pi \omega S_p(\omega)}{2\xi k^2} \left\{ 1 - \frac{\exp(-2\omega\xi t)}{\omega_D^2} \right.$$

$$\left. \left[\omega_D^2 + \frac{(2\omega\xi)^2}{2} \sin^2 2\omega_D t + \xi\omega\,\omega_D \sin 2\omega_D t \right] \right\} \quad \xi < 1 \tag{22-44}$$

The harmonic terms in this equation are relatively small and can be dropped with little loss of accuracy giving the approximate relation

$$E[v(t)^2] \doteq \frac{\pi \omega S_p(\omega)}{2\xi k^2} \left[1 - \exp(-2\omega\xi t) \right] \quad \xi < 1 \tag{22-45}$$

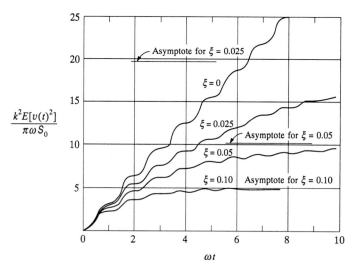

FIGURE 22-6
Nonstationary mean square response resulting from zero initial conditions, Eq. (22-42).

As the damping ratio approaches zero, this equation becomes in the limit

$$E[v(t)^2] \doteq \frac{\pi\omega^2 S_p(\omega)}{k^2} t \qquad \xi = 0 \qquad (22\text{-}46)$$

Assuming that the stationary input process is gaussian, the nonstationary output process will also be gaussian, in which case the probability density function $p[v(t)]$ will be given by

$$p[v(t)] = \frac{1}{\sqrt{2\pi E[v(t)^2]}} \exp\left\{ -\frac{v(t)^2}{2E[v(t)^2]} \right\} \qquad (22\text{-}47)$$

Example E22-3. Consider the SDOF system

$$m\ddot{v} + c\dot{v} + kv = p(t) \qquad (a)$$

excited by a zero-mean ergodic random process $p(t)$ having constant power spectral density equal to S_0 over the range $-\infty < \bar{\omega} < \infty$ and assume zero initial conditions are imposed on the system; that is, $v_r(0) = \dot{v}_r(0) = 0$ $(r = 1, 2, \cdots)$. Calculate the ratio of the variance of $v(t)$ to the steady-state variance for $t/T = t\omega/2\pi = 2, 5, 10,$ and 15 and $\xi = 0.02$ and 0.05.

Using Eq. (22-45) gives

$$\frac{\sigma^2_{v(t)}}{\sigma^2_{v(\infty)}} = 1 - \exp(-2\omega\xi t) = 1 - \exp\left(-4\pi\xi\frac{t}{T} \right) \qquad (b)$$

Substituting the above numerical values into this equation gives the results shown in Table 22-1 which indicate the very rapid rate at which the response process $v(t)$ approaches its steady-state condition.

Table 22-1 Ratio $\sigma^2_{v(t)}/\sigma^2_{v(\infty)}$

ξ	t/T			
	2	5	10	15
0.02	0.395	0.714	0.919	0.987
0.05	0.715	0.957	0.998	0.99995

22-6   FATIGUE PREDICTIONS FOR NARROWBAND SYSTEMS

The fatigue life of a narrowband SDOF system can easily be determined provided the material follows a prescribed S-N relationship (S = harmonic stress amplitude, N = number of cycles to failure) and provided that Miner's linear-accumulative-damage criterion applies.[2]

Proceeding on this basis, assume that the S-N relationship has a known form

$$N = N(S) \tag{22-48}$$

The accumulative damage (AD) can then be expressed in the discrete form

$$AD = \frac{n_1}{N(S_1)} + \frac{n_2}{N(S_2)} + \frac{n_3}{N(S_3)} + \cdots \tag{22-49}$$

where $n_1, n_2, n_3, \cdots$ are the numbers of harmonic stress cycles applied to the material at amplitudes $S_1, S_2, S_3, \cdots$, respectively. Failure occurs when the accumulative damage reaches unity, that is, $AD = 1$.

If the system is responding as a narrowband system, the accumulative damage can be expressed in the continuous form

$$AD = \int_0^\infty \frac{n(S)}{N(S)} \, dS \tag{22-50}$$

where $n(S) \, dS$ represents the number of harmonic stress cycles with amplitudes between S and $S + dS$. If a stationary response process of duration T is assumed, the total number of stress cycles will be equal to $\omega T / 2\pi$, in which case

$$n(S) \, dS = \frac{\omega T}{2\pi} p(S) \, dS \tag{22-51}$$

where $p(S)$ is the probability density function for stress amplitude S. Substituting Eq. (22-51) into Eq. (22-50) gives the accumulative damage in the form

$$AD = \frac{\omega T}{2\pi} \int_0^\infty \frac{p(S)}{N(S)} \, dS \tag{22-52}$$

If $p(S)$ is of the Rayleigh form as represented by a narrowband process, that is,

$$p(S) = \frac{S}{\sigma_s^2} \exp\left(-\frac{S^2}{2\sigma_s^2}\right) \tag{22-53}$$

---

[2] M. A. Miner, "Cumulative Damage in Fatigue," *J. Appl. Mech.*, Ser. A, Vol. 12, No. 1, pp. 159-164, 1945.

where σ_s^2 is the variance of the critical stress $s(t)$, and if $N(S)$ takes on the familiar form

$$N(S) = \left(\frac{S_1}{S}\right)^b N_1 \tag{22-54}$$

where S_1 and N_1 represent a convenient point on the S-N curve and b is an even integer (usually $b > 10$), Eq. (22-52) becomes, after substituting Eqs. (22-53) and (22-54),

$$AD = \frac{\omega T}{2\pi N_1} \left(\frac{\sigma_s}{S_1}\right)^b 2^{b/2} \left(\frac{b}{2}\right)! \tag{22-55}$$

Setting the accumulative damage equal to unity and solving for T gives the expected time to failure

$$T_{\text{failure}} = \frac{2\pi N_1}{\omega} \left(\frac{S_1}{\sigma_s}\right)^b \frac{2^{-b/2}}{(b/2)!} \tag{22-56}$$

For a linear system, the critical stress is related to displacement $v(t)$ by the relation

$$s(t) = C\, v(t) \tag{22-57}$$

where C is a known constant. It follows therefore that

$$\sigma_s^2 = C^2\, \sigma_v^2 \tag{22-58}$$

Thus, if $p(t)$ is a white-noise process of intensity S_0, as shown in Fig. 22-2a,

$$\sigma_s^2 = \frac{\pi \omega C^2 S_0}{2k^2 \xi} \tag{22-59}$$

If the above SDOF system is excited by a nonstationary zero mean process $p(t)$ for which the ensemble average $E[p(t)\, p(t+\tau)]$ is varying slowly with time, the response process $v(t)$ will be essentially a quasi-stationary process. In this case, one must treat σ_s as time dependent. In engineering applications, this time dependency is usually random; therefore, one must establish the probability distribution function $P[\sigma_s(t)]$ from a statistical analysis of available data controlling the excitation over a long period of time assuming process $\sigma_s(t)$ to be ergodic. Having established $P[\sigma_s(t)]$, the accumulative damage is then given by

$$(AD)_L = \frac{\omega T_L}{2\pi N_1} \left(\frac{1}{S_1}\right)^b 2^{b/2} \left(\frac{b}{2}\right)! \int_0^\infty \sigma_s^b \frac{dP[\sigma_s]}{d\sigma_s}\, d\sigma_s \tag{22-60}$$

in which $(AD)_L$ denotes accumulative damage over the life of the structure T_L.

Example E22-4.   As represented by Eq. (2-17), a viscously-damped SDOF system excited by random support excitation $\ddot{v}_g(t)$ results in the equation of motion

$$m\ddot{v} + c\dot{v} + kv = -m\ddot{v}_g(t) = p_e(t) \tag{a}$$

Stationary random process $\ddot{v}_g(t)$ has a uniform spectral density equal to $0.2\ ft^2/sec^3$ over the frequency range $2 < \bar{\omega} < 100$ and $-100 < \bar{\omega} < -2$. The system has a natural frequency of $10\ Hz$ and damping equal to 2 percent of critical, and the critical stress from a fatigue standpoint is given by

$$s(t) = 2 \times 10^5\ v(t) \tag{b}$$

where units are pounds and inches. If the material at the critical location satisfies the fatigue relation

$$N(S) = \left(\frac{60,000}{s}\right)^{12} \times 10^5 \tag{c}$$

find the expected time to failure caused by excitation $\ddot{v}_g(t)$.

From Eq. (22-13) it is seen that the variance of $v(t)$ is

$$\sigma^2_{v(t)} = R_v(0) = \frac{\pi \omega S_0}{2k^2\ \xi} \tag{d}$$

where S_0 is the power spectral density for a white-noise excitation $p_e(t)$. Since the natural frequency of the system ω falls within the bandwidth of support excitation, white-noise excitation can be assumed here of intensity

$$S_{p_e}(\bar{\omega}) = S_0 = m^2\ S_{\ddot{v}_g(t)}(\bar{\omega}) \tag{e}$$

Substituting Eq. (e) into Eq. (d) and making use of the notation $k^2 = \omega^4 m^2$ give

$$\sigma^2_{v(t)} = \frac{\pi S_{\ddot{v}_g(t)}(\omega)}{2\omega^3\ \xi} \tag{f}$$

Making use of Eq. (22-58) gives

$$\sigma^2_{s(t)} = \frac{\pi C^2\ S_{\ddot{v}_g(t)}(\omega)}{2\omega^3\ \xi} \tag{g}$$

where $C = 2 \times 10^5\ lb/in^3$, $S_{\ddot{v}(t)}(\omega) = 0.2\ ft^2/sec^3$, $\xi = 0.02$, and $\omega = 2\pi f = 62.8\ rad/sec$. Thus one obtains $\sigma_{s(t)} = 1.91 \times 10^4\ psi$. When it is noted from Eq. (c) that $b = 12$, $N_1 = 10^5$, and $S_1 = 60,000\ psi$, Eq. (22-56) yields the expected time to failure

$$T_{\text{failure}} = 1.99 \times 10^5\ sec = 55.5\ hr \tag{h}$$

PROBLEMS

22-1. Consider the SDOF system represented by

$$m\,\ddot{v} + c\,\dot{v} + k\,v = p(t)$$

when excited by a gaussian zero-mean stationary process $p(t)$ having a constant power spectral density $S_0 = 2 \times 10^4\ lb^2 \times sec$ over two wide-frequency bands centered on $\pm\omega$, where ω is the natural circular frequency $\sqrt{k/m}$. The system is characterized by a mass m equal to $100\ lb \cdot sec^2/ft$, a natural frequency ω equal to $62.8\ rad/sec$, and a damping ratio ξ equal to 2 percent of critical.
 (a) Find the numerical value for the mean square displacement $E[v(t)^2]$.
 (b) Find the numerical value for the mean square velocity $E[\dot{v}(t)^2]$.
 (c) What is the joint probability density function for $v(t)$ and $\dot{v}(t)$? Find the numerical values for all constants in this function.
 (d) What is the probability density function for the maxima of response process $v(t)$? Find the numerical values for all constants in this function. (Assume a Rayleigh distribution in this case since the parameter ε appearing in Fig. 21-11 is nearly equal to zero.)
 (e) What is the numerical value of the mean extreme value of the process $v(t)$ as given by Eq. (21-117) when the duration T of the process is $30\ sec$? [For this low-damped system, ν as given by Eq. (21-116) is approximately equal to $\omega/2\pi$.]
 (f) What is the numerical value of the standard deviation of the extreme values for process $v(t)$?

22-2. Approximately how much will the numerical values found in Prob. 22-1 change if the power spectral density function for the process $p(t)$ is changed from $S_p(\overline{\omega}) = S_0$ to

$$S_p(\overline{\omega}) = S_0\ \exp(-0.0111\ |\overline{\omega}|) \qquad -\infty < \overline{\omega} < \infty$$

22-3. The one-mass system shown in Fig. P22-1 is excited by support displacement $x(t)$. The spring and viscous dashpot are linear, having constant k and c, respectively. Let $\omega^2 = k/m$ and $\xi = c/2m\omega$.
 (a) Obtain the unit-impulse-response function $h(t)$ for spring force $f_S(t) = k[y(t) - x(t)]$ when $x(t) = \delta(t)$.
 (b) Obtain the complex-frequency-response function $H(i\overline{\omega})$ for force $f_S(t)$ which is the ratio of the complex amplitude of $f_S(t)$ to the complex amplitude of $x(t)$ when the system is performing simple harmonic motion at frequency $\overline{\omega}$.

(c) Verify that $h(t)$ and $H(i\bar\omega)$ are Fourier transform pairs in accordance with Eqs. (12-73).

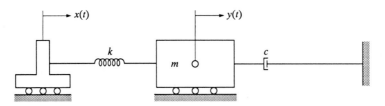

FIGURE P22-1
One-mass system of Prob. 22-3.

22-4. If the input $x(t)$ to the system defined in Prob. 22-3 is a stationary random process having a constant power spectral density S_0 over the entire frequency range $-\infty < \bar\omega < \infty$, derive the power spectral density and autocorrelation functions for response process $f_s(t)$ using Eqs. (22-20) and (22-9), respectively.

22-5. Show that the power spectral density and autocorrelation functions obtained in Prob. 22-4 are indeed Fourier transform pairs in accordance with Eqs. (21-37).

22-6. If $x(t)$ is the stationary random input to a linear system and $y(t)$ is the corresponding stationary random output, express the cross-correlation function $R_{xy}(\tau)$ in terms of $R_x(\tau)$ and $h(t)$.

22-7. If the input $x(t)$ to a linear system and the corresponding output $y(t)$ are given by

$$x(t) = \begin{cases} e^{-t} & t > 0 \\ 0 & t < 0 \end{cases} \qquad y(t) = \begin{cases} \dfrac{1}{a-b}\left[e^{-(b/a)t} - e^{-t}\right] & t \ge 0 \\ 0 & t \le 0 \end{cases}$$

what is the complex-frequency-response function $H(i\bar\omega)$ for the system?

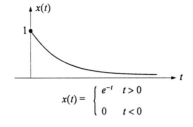
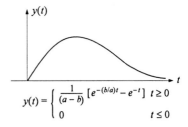

$$x(t) = \begin{cases} e^{-t} & t > 0 \\ 0 & t < 0 \end{cases}$$

$$y(t) = \begin{cases} \dfrac{1}{(a-b)}\left[e^{-(b/a)t} - e^{-t}\right] & t \ge 0 \\ 0 & t \le 0 \end{cases}$$

FIGURE P22-2
Input and output functions of Prob. 22-7.

22-8. If a rectangular input $x(t)$ to a linear system produces a single sine-wave output $y(t)$, as shown in Fig. P22-3,

$$x(t) = \begin{cases} 1 & 0 < t < T \\ 0 & t < 0;\ t > T \end{cases} \qquad y(t) = \begin{cases} \sin \dfrac{2\pi t}{T} & 0 \le t \le T \\ 0 & t < 0;\ t > T \end{cases}$$

what would be the power spectral density function for the output process when the input is a stationary white-noise process of intensity S_0?

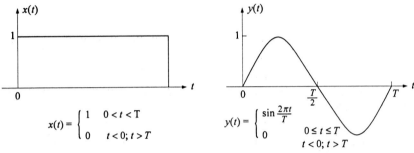

FIGURE P22-3
Input and output functions of Prob. 22-8.

22-9. Consider two stationary processes $x(t)$ and $y(t)$ related through the differential equation

$$\ddot{x}_r(t) + A\,\dot{x}_r(t) + B\,y_r(t) + C\,\dot{y}_r(t) = 0 \qquad r = 1, 2, \cdots$$

Express the power spectral density function for random process $x(t)$ in terms of the power spectral density function for process $y(t)$ and real constants A, B, and C.

CHAPTER

23

STOCHASTIC RESPONSE OF LINEAR MDOF SYSTEMS

23-1 TIME-DOMAIN RESPONSE FOR LINEAR SYSTEMS USING NORMAL MODES

As shown in Chapter 13, the dynamic response of linear viscously-damped MDOF systems (discrete or continuous) can be determined by solving the normal equations of motion

$$\ddot{Y}_n(t) + 2\omega_n\xi_n\dot{Y}_n(t) + \omega_n^2 Y_n(t) = \frac{P_n(t)}{M_n} \quad n = 1, 2, \cdots \tag{23-1}$$

where n is the mode number. Any response quantity $z(t)$ linearly related to the normal coordinates can be found using the relation

$$z(t) = \sum_n B_n Y_n(t) \tag{23-2}$$

where coefficients B_n ($n = 1, 2, \cdots$) are obtained by standard methods of analysis. Usually, the rapid convergence rate of the series means that only a limited number of lower modes need be considered.

If random excitations on the system are assumed, each generalized forcing function $P_n(t)$ should be considered as a separate stochastic process. If the excitations are stationary, the response processes will also be stationary, in which case one's interest is in obtaining the autocorrelation function for response $z(t)$, that is,

$$R_z(\tau) = E[z(t)\, z(t+\tau)] \tag{23-3}$$

Substituting Eq. (23-2) into Eq. (23-3) gives

$$R_z(\tau) = E\left[\sum_m \sum_n B_m\, B_n\, Y_m(t)\, Y_n(t+\tau)\right] \tag{23-4}$$

Solving for response through the time domain, one obtains

$$Y_n(t) = \int_{-\infty}^{t} P_n(\tau)\, h_n(t-\tau)\, d\tau \tag{23-5}$$

in which, for undercritically damped systems,

$$h_n(t) = \frac{1}{\omega_{D_n} M_n} \exp(-\xi_n \omega_n t) \sin \omega_{D_n} t \qquad \omega_{D_n} = \omega_n (1 - \xi_n^2)^{1/2} \tag{23-6}$$

Substituting Eq. (23-5) into (23-4) gives

$$R_z(\tau) = E\left[\sum_m \sum_n \int_{-\infty}^{t} \int_{-\infty}^{t+\tau} B_m\, B_n\, P_m(\theta_1)\, P_n(\theta_1) \right.$$

$$\left. \times\, h_m(t-\theta_1)\, h_n(t+\tau-\theta_2)\, d\theta_1\, d\theta_2 \right] \tag{23-7}$$

where θ_1, θ_2, and τ are dummy time variables. With the change of variables

$$u_1 \equiv t - \theta_1 \qquad\qquad u_2 \equiv t + \tau - \theta_2$$

$$du_1 = -d\theta_1 \qquad\qquad du_2 = -d\theta_2 \tag{23-8}$$

and recognition that $h_n(t)$ damps out for stable systems, Eq. (23-7) can be written in the form

$$R_z(\tau) = \sum_m \sum_n R_{z_m z_n}(\tau) \tag{23-9}$$

where

$$R_{z_m z_n}(\tau) = \int_0^\infty \int_0^\infty B_m\, B_n\, R_{P_m P_n}(\tau - u_2 + u_1)\, h_m(u_1)\, h_n(u_2)\, du_1\, du_2 \tag{23-10}$$

$R_{P_m P_n}(\tau)$ is the covariance function for random variables $P_m(t)$ and $P_n(t+\tau)$, and $R_{z_m z_n}(\tau)$ is the covariance function for modal responses $z_m(t)$ and $z_n(t+\tau)$.

From this derivation it is clear that if the covariance function $R_{P_m P_n}(\tau)$ is known for all combinations of m and n, the integrations in Eq. (23-10) and the summations in Eq. (23-9) can be completed to obtain the desired autocorrelation function for response $z(t)$.

For systems that are lightly damped and have well-separated modal frequencies, as is usually the case in structural engineering, response process $z_m(t)$ produced by mode m is almost statistically independent of response $z_n(t)$ produced by mode n; that is, the cross-terms in Eq. (23-9) are nearly equal to zero. Therefore, the autocorrelation function for total response can usually be approximated by the relation

$$R_z(\tau) \doteq \sum_m R_{z_m z_m}(\tau) \qquad (23\text{-}11)$$

where $R_{z_m z_m}(\tau)$ is the autocorrelation function for process $z_m(t)$. When τ is made equal to zero, Eq. (23-11) can be written in terms of standard deviations, that is,

$$\sigma_z = (\sigma_{z1}^2 + \sigma_{z2}^2 + \sigma_{z3}^2 + \cdots)^{1/2} \qquad (23\text{-}12)$$

Since the mean extreme values of response for processes $z(t)$ and $z_m(t)$ ($m = 1, 2, \cdots$) are proportional to their respective standard deviations σ_z and σ_{z_m}, Eq. (23-12) lends support to the common square-root-of-the-sum-of-squares (SRSS) method of weighting the maximum normal mode responses when estimating maximum total response.

When the system has very closely spaced pairs of frequencies, the corresponding cross-terms in Eq. 23-9 must be retained. Doing so leads to the complete-quadratic-combination (CQC) method[1] of weighting the maximum normal mode contributions to response; see Eq. (26-116).

23-2 FREQUENCY-DOMAIN RESPONSE FOR LINEAR SYSTEMS USING NORMAL MODES

The power-spectral-density function for response $z(t)$ is obtained by taking the Fourier transform of the autocorrelation function, that is,

$$S_z(\overline{\omega}) = \frac{1}{2\pi} \int_{-\infty}^{\infty} R_z(\tau) \, \exp(-i\overline{\omega}\tau) \, d\tau \qquad (23\text{-}13)$$

Substituting Eq. (23-10) into Eq. (23-9) and then Eq. (23-9) into Eq. (23-13) gives

$$S_z(\overline{\omega}) = \frac{1}{2\pi} \int_{-\infty}^{\infty} \left\{ \sum_m \sum_n \int_0^{\infty} \int_0^{\infty} B_m \, B_n \, R_{P_m P_n}(\tau - u_2 + u_1) \right.$$

$$\left. \times \, h_m(u_1) \, h_n(u_2) \, du_1 \, du_2 \right\} \exp(-i\overline{\omega}\tau) \, d\tau \qquad (23\text{-}14)$$

---

[1] A. Der Kiureghian, "Structural Response to Stationary Excitation," *Jour. of Engineering Mechanics Division*, ASCE, December 1980.

or

$$S_z(\overline{\omega}) = \frac{1}{2\pi} \sum_m \sum_n B_m B_n$$

$$\times \left[\lim_{T \to \infty} \int_0^T h_m(u_1) \, du_1 \int_0^T h_n(u_2) \, du_2 \right.$$

$$\left. \times \int_{-T}^T R_{P_m P_n}(\tau - u_2 + u_1) \, \exp(-i\overline{\omega}\tau) \, d\tau \right] \qquad (23\text{-}15)$$

Since $h(u_1)$ and $h(u_2)$ equal zero for u_1 and u_2 less than zero, the lower limits of the first two integrals in Eq. (23-15) can be changed from zero to $-T$. After substituting the change of variable

$$\gamma \equiv \tau - u_2 + u_1 \qquad (23\text{-}16)$$

Equation (23-15) becomes

$$S_z(\overline{\omega}) = \frac{1}{2\pi} \sum_m \sum_n B_m B_n$$

$$\times \left[\lim_{T \to \infty} \int_{-T}^T h_m(u_1) \, \exp(i\overline{\omega}u_1) \, du_1 \right.$$

$$\left. \times \int_{-T}^T h_n(u_2) \, \exp(-i\overline{\omega}u_2) \, du_2 \times \int_{-T-u_2+u_1}^{T-u_2+u_1} R_{P_m P_n}(\gamma) \, \exp(-i\overline{\omega}\gamma) \, d\gamma \right] \tag{23-17}$$

Since $R_{P_r P_s}(\tau)$ damps out with increasing values of $|\tau|$, the limits of the last integral in Eq. (23-17) can be changed to $\int_{-T}^T$. With use of Eq. (6-53) and the first of Eqs. (21-37), Eq. (23-17) becomes

$$S_z(\overline{\omega}) = \sum_m \sum_n S_{z_m z_n}(\overline{\omega}) \qquad (23\text{-}18)$$

in which

$$S_{z_m z_n}(\overline{\omega}) \equiv B_m \, B_n \, H_m(-i\overline{\omega}) \, H_n(i\overline{\omega}) \, S_{P_m P_n}(\overline{\omega}) \qquad (23\text{-}19)$$

is the cross-spectral density function for modal responses $z_m(t)$ and $z_n(t)$, $S_{P_m P_n}(\overline{\omega})$ is the cross-spectral density function for processes $P_m(t)$ and $P_n(t)$, and

$$H_m(-i\overline{\omega}) = \frac{1}{K_m \left[1 - 2i\xi_m(\overline{\omega}/\omega_m) - (\overline{\omega}/\omega_m)^2\right]}$$

$$H_n(i\overline{\omega}) = \frac{1}{K_n \left[1 + 2i\xi_n(\overline{\omega}/\omega_n) - (\overline{\omega}/\omega_n)^2\right]}$$

$$(23\text{-}20)$$

For lightly damped systems with well-separated modal frequencies, the cross-terms in Eq. (23-18) contribute very little to the mean-square response, $\int_{-\infty}^{\infty} S_z(\overline{\omega}) \, d\overline{\omega}$, in which case $S_z(\overline{\omega})$ can be simplified to the approximate form

$$S_z(\overline{\omega}) \doteq \sum_m S_{z_m z_m}(\overline{\omega}) \tag{23-21}$$

where

$$S_{z_m z_m}(\overline{\omega}) = B_m^2 \, |H_m(i\overline{\omega})|^2 \, S_{P_m P_m}(\overline{\omega}) \tag{23-22}$$

$$|H_m(i\overline{\omega})|^2 = \frac{1}{K_m^2 \left[1 + (4\xi_m^2 - 2)(\overline{\omega}/\omega_m)^2 + (\overline{\omega}/\omega_m)^4 \right]} \tag{23-23}$$

and $S_{P_m P_m}(\overline{\omega})$ is the power spectral density function for process $P_m(t)$. As previously pointed out, when two important normal modes have nearly the same frequencies, the corresponding cross-term in Eq. (23-9) must be retained. Likewise the corresponding cross-term in Eq. (23-18) must be retained.

If all input processes are gaussian, response $z(t)$ will also be gaussian, in which case $S_z(\overline{\omega})$ completely characterizes the process.

23-3 NORMAL MODE FORCING FUNCTION DUE TO DISCRETE LOADINGS

It was shown in Chapter 12 that if a linear structure is subjected to discrete applied loadings $p_i(t)$ $(i = 1, 2, \cdots)$, the generalized forcing function for the nth mode becomes

$$P_n(t) = \sum_i \phi_{in} p_i(t) = \boldsymbol{\phi}_n^T \, \mathbf{p}(t) \tag{23-24}$$

where constants ϕ_{in} are the components of nth-modal displacements at points i in the directions of corresponding forces $p_i(t)$. If each discrete force $p_i(t)$ is a stationary gaussian process as defined by $S_{p_i}(\overline{\omega})$ or $R_{p_i}(\tau)$, the cross-spectral density and covariance functions for $P_m(t)$ and $P_n(t)$ become

$$S_{P_m P_n}(\overline{\omega}) = \sum_i \sum_k \phi_{im} \, \phi_{kn} \, S_{p_i p_k}(\overline{\omega}) = \boldsymbol{\phi}_m^T \, \mathbf{S}_p \, \boldsymbol{\phi}_n \tag{23-25}$$

$$R_{P_m P_n}(\tau) = \sum_i \sum_k \phi_{im} \, \phi_{kn} \, R_{p_i p_k}(\tau) = \boldsymbol{\phi}_m^T \, \mathbf{R}_p \, \boldsymbol{\phi}_n \tag{23-26}$$

The power spectral density and autocorrelation functions for response $z(t)$ are now obtained by substituting Eqs. (23-25) and (23-26) into Eqs. (23-19) and (23-10), respectively.

Example E23-1. A uniform inverted L-shaped member of mass $\overline{m}$ per unit length and flexural stiffness EI in its plane is discretized as shown in

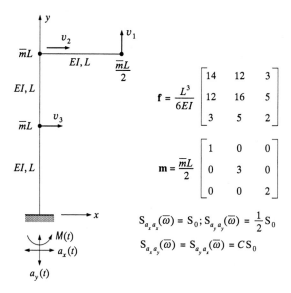

$$S_{a_x a_x}(\bar{\omega}) = S_0; \quad S_{a_y a_y}(\bar{\omega}) = \frac{1}{2}S_0$$

$$S_{a_x a_y}(\bar{\omega}) = S_{a_y a_x}(\bar{\omega}) = CS_0$$

FIGURE E23-1
Discrete model of uniform
inverted L-shaped member.

Fig. E23-1. This model is subjected to simultaneous stationary random base accelerations $a_x(t)$ and $a_y(t)$ having power spectral and cross-spectral densities given by

$$S_{a_x a_x}(\bar{\omega}) = S_0 \qquad S_{a_y a_y}(\bar{\omega}) = \frac{1}{2}S_0 \qquad S_{a_x a_y}(\bar{\omega}) = S_{a_y a_x}(\bar{\omega}) = C\,S_0 \quad \text{(a)}$$

where C is a real constant. Assuming modal damping of the uncoupled form, where the damping ratio in each normal mode equals ξ, determine the variance of base moment $M(t)$ expressed in terms of $\overline{m}$, L, EI, ξ, S_0, and C. What is the range of possible numerical values for C?

Using the flexibility and mass matrices shown in Fig. E23-1 gives the following mode shapes and frequencies:

$$\phi_1^T = [-0.807 \quad 1.000 \quad 0.307] \qquad \omega_1^2 = 0.197\,\frac{EI}{\overline{m}L^4}$$

$$\phi_2^T = [-1.000 \quad -0.213 \quad -0.280] \qquad \omega_2^2 = 2.566\,\frac{EI}{\overline{m}L^4} \qquad \text{(b)}$$

$$\phi_3^T = [-0.309 \quad -0.304 \quad 1.000] \qquad \omega_3^2 = 21.858\,\frac{EI}{\overline{m}L^4}$$

From Eq. (12-12a), the normal-coordinate generalized masses are

$$M_1 = 3.84\,\frac{\overline{m}L}{2} \qquad M_2 = 1.29\,\frac{\overline{m}L}{2} \qquad M_3 = 2.41\,\frac{\overline{m}L}{2} \qquad \text{(c)}$$

and from Eq. (12-12c), the corresponding generalized forces are

$$P_n(t) = \boldsymbol{\phi}_n^T \mathbf{p}_{\text{eff}}(t) = -\boldsymbol{\phi}_n^T \mathbf{m} \begin{Bmatrix} a_y \\ a_x \\ a_x \end{Bmatrix} \qquad n = 1, 2, 3$$

From this equation it follows that

$$P_m(t)\, P_n(t) = \boldsymbol{\phi}_m^T \mathbf{m} \begin{Bmatrix} a_y \\ a_x \\ a_x \end{Bmatrix} \langle a_y \quad a_x \quad a_x \rangle \mathbf{m}\, \boldsymbol{\phi}_n$$

from which the power and cross-spectral density functions become

$$S_{P_m P_n}(\overline{\omega}) = \boldsymbol{\phi}_m^T \mathbf{m} \begin{bmatrix} S_{a_y a_y}(\overline{\omega}) & S_{a_y a_x}(\overline{\omega}) & S_{a_y a_x}(\overline{\omega}) \\ S_{a_x a_y}(\overline{\omega}) & S_{a_x a_x}(\overline{\omega}) & S_{a_x a_x}(\overline{\omega}) \\ S_{a_x a_y}(\overline{\omega}) & S_{a_x a_x}(\overline{\omega}) & S_{a_x a_x}(\overline{\omega}) \end{bmatrix} \mathbf{m}\boldsymbol{\phi}_n$$

or

$$S_{P_m P_n}(\overline{\omega}) = \boldsymbol{\phi}_m^T \mathbf{m} \begin{bmatrix} \frac{1}{2}S_0 & CS_0 & CS_0 \\ CS_0 & S_0 & S_0 \\ CS_0 & S_0 & S_0 \end{bmatrix} \mathbf{m}\boldsymbol{\phi}_n$$

$$m, n = 1, 2, 3 \qquad \text{(d)}$$

To find the moment $M_n(t)$ contributed by the nth normal mode, apply the force vector

$$\begin{Bmatrix} f_{1n} \\ f_{2n} \\ f_{3n} \end{Bmatrix} = \omega_n^2 \mathbf{m}\boldsymbol{\phi}_n Y_n(t)$$

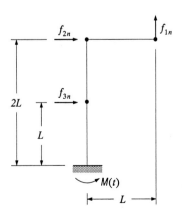

FIGURE E23-2

to the member as shown in Fig. E23-2. Summing the moments about the base gives

$$M_n(t) = \omega_n^2 \langle -L \quad 2L \quad L \rangle \, \mathbf{m} \boldsymbol{\phi}_n \, Y_n(t) \qquad n = 1, 2, 3$$

Thus for this case, Eq. (23-2) becomes

$$M(t) = \sum_{n=1}^{3} B_n \, Y_n(t) \tag{e}$$

where

$$B_n = \omega_n^2 \langle -L \quad 2L \quad L \rangle \, \mathbf{m} \, \boldsymbol{\phi}_n \qquad n = 1, 2, 3 \tag{f}$$

When Eqs. (23-18) and (23-19) are used, the spectral density for base moment becomes

$$S_M(\overline{\omega}) = \sum_{m=1}^{3} \sum_{n=1}^{3} B_m B_n H_m(-i\overline{\omega}) \, H_n(i\overline{\omega}) \, S_{P_m P_n}(\overline{\omega}) \tag{g}$$

where $H_m(-i\overline{\omega})$ and $H_n(i\overline{\omega})$ are given by Eqs. (23-20). Integrating this expression with respect to $\overline{\omega}$ from $-\infty$ to $+\infty$ gives the variance for base moment as

$$\sigma_{M(t)}^2 = \sum_{m=1}^{3} \sum_{n=1}^{3} \left[B_m \, B_n \, S_{P_m P_n} \int_{-\infty}^{\infty} H_m(-i\overline{\omega}) \, H_n(i\overline{\omega}) \, d\overline{\omega} \right] \tag{h}$$

where $S_{P_m P_n}$ has the frequency-invariant form given by Eq. (d). By using Eqs. (23-20), the integral in Eq. (h) can be completed to give

$$\sigma_{M(t)}^2 = \sum_{m=1}^{3} \sum_{n=1}^{3} \frac{4\pi\xi \, B_m \, B_n \, S_{P_m P_n}}{M_n M_m (\omega_n + \omega_m) \left[(\omega_n - \omega_m)^2 + 4\xi^2 \omega_n \omega_m \right]} \tag{i}$$

Assuming a low-damped system, that is, $\xi < 0.1$, the cross-terms in this equation will be relatively small and can be neglected, which leads to the approximate expression

$$\sigma_{M(t)}^2 \doteq \frac{\pi}{2\xi} \sum_{n=1}^{3} \frac{B_n^2 \, S_{P_n P_n}}{M_n^2 \omega_n^3} \tag{j}$$

Upon substitution from Eqs. (b) to (d) and (f), Eq. (j) becomes

$$\sigma_{M(t)}^2 \doteq \frac{\overline{m}^2 L^2}{\xi} \sqrt{\frac{EI}{\overline{m}}} (9.3 + 3.2 \, C) \, S_0 \tag{k}$$

If excitations $a_x(t)$ and $a_y(t)$ are statistically independent, $S_{a_x a_y} = C S_0 = 0$. Therefore, $C = 0$ in this case. However, if $a_x(t)$ and $a_y(t)$ are fully correlated statistically, then $a_y(t) = \alpha a_x(t)$, where α is a real constant. From the definitions of power and cross-spectral densities given by Eqs. (21-27) and (21-60), respectively, it follows that

$$S_{a_y a_y} = \alpha^2 S_{a_x a_x} \qquad S_{a_x a_y} = \alpha S_{a_x a_x}$$

From these relations and Eqs. (a), $C = \alpha = \pm 1/\sqrt{2}$. Thus the range of possible numerical values for C is

$$-\frac{1}{\sqrt{2}} \le C \le +\frac{1}{\sqrt{2}} \tag{l}$$

Substituting the maximum and minimum values of C into Eq. (k) gives

$$\sigma^2_{M(t)} = \begin{cases} 7.0 \dfrac{\overline{m}^2 L^2}{\xi} \sqrt{\dfrac{EI}{\overline{m}}} S_0 & C = -\dfrac{1}{\sqrt{2}} \\[3mm] 11.6 \dfrac{\overline{m}^2 L^2}{\xi} \sqrt{\dfrac{EI}{\overline{m}}} S_0 & C = +\dfrac{1}{\sqrt{2}} \end{cases} \tag{m}$$

23-4 NORMAL MODE FORCING FUNCTION DUE TO DISTRIBUTED LOADINGS

If the distributed loading $p(x,t)$ applied on a linear structure is random with respect to both x and t, the generalized forcing function for the nth mode is of the form

$$P_n(t) = \int_{-\infty}^{\infty} \phi_n(x)\, p(x,t)\, dx \tag{23-27}$$

where $\phi_n(x)$ is simply the continuous form of ϕ_{in} defined in Section 23-3. If $p(x,t)$ is a stationary gaussian process defined by $S_p(x, \alpha, \overline{\omega})$ or $R_p(x, \alpha, \overline{\tau})$, the cross-spectral density and covariance functions for $P_m(t)$ and $P_n(t)$ become

$$S_{P_m P_n}(\overline{\omega}) = \int_{-\infty}^{\infty} \int_{-\infty}^{\infty} \phi_m(x)\, \phi_n(\alpha)\, S_p(x, \alpha, \overline{\omega})\, dx\, d\alpha \tag{23-28}$$

$$R_{P_m P_n}(\overline{\omega}) = \int_{-\infty}^{\infty} \int_{-\infty}^{\infty} \phi_m(x)\, \phi_n(\alpha)\, R_p(x, \alpha, \overline{\omega})\, dx\, d\alpha \tag{23-29}$$

where α is a dummy space variable. The power spectral density and autocorrelation functions for response $z(t)$ are now obtained by substituting Eqs. (23-28) and (23-29) into Eqs. (23-19) and (23-10), respectively.

It should now be quite apparent to the reader how one determines the stochastic response of linear structures subjected to stationary gaussian excitations which are random with respect to all three space coordinates as well as time.

23-5   FREQUENCY-DOMAIN RESPONSE FOR LINEAR SYSTEMS HAVING FREQUENCY-DEPENDENT PARAMETERS AND/OR COUPLED NORMAL MODES

In many engineering applications, the governing linear discrete parameter equations of motion of the form

$$\mathbf{m}\,\ddot{\mathbf{v}}(t) + \mathbf{c}\,\dot{\mathbf{v}}(t) + \mathbf{k}\,\mathbf{v}(t) = \mathbf{p}(t) \tag{23-30}$$

cannot be solved by the normal mode procedures described above because the normal modes are coupled due to the nonorthogonal nature of the damping matrix $\mathbf{c}$. In other applications, the damping and stiffness matrices have frequency-dependent coefficients, thus preventing standard time-domain solutions. For either of these cases, an alternate frequency-domain solution can be used which solves Eq. (23-30) directly without the use of normal modes. To formulate this approach, consider Eq. (23-30) in its frequency-domain form as given by

$$\left[(\mathbf{k} - \overline{\omega}^2\,\mathbf{m}) + i(\overline{\omega}\mathbf{c})\right]\mathbf{V}(i\overline{\omega}) = \mathbf{P}(i\overline{\omega}) \tag{23-31}$$

where $\mathbf{V}(i\overline{\omega})$ and $\mathbf{P}(i\overline{\omega})$ are the Fourier transforms of vectors $\mathbf{v}(t)$ and $\mathbf{p}(t)$, respectively. For the moment, assume vectors $\mathbf{v}(t)$ and $\mathbf{p}(t)$ are of finite durations so that their Fourier transforms do indeed exist.

Assuming an NDOF system, solve for the system's $N \times N$ complex frequency response matrix containing complex frequency response functions $H_{jk}(i\overline{\omega})$ ($j, k = 1, 2, \cdots, N$) where each function is the complex response in displacement coordinate j due to a unit harmonic loading in coordinate k. This step is accomplished by solving for response vectors using Eqs. (23-31), i.e., using

$$\mathbf{H}_k(i\overline{\omega}) \equiv \langle H_{1k}(i\overline{\omega})\ H_{2k}(i\overline{\omega})\ \cdots\ H_{Nk}(i\overline{\omega})\rangle^T = \mathbf{I}(i\overline{\omega})^{-1}\langle 0\ 0\ \cdots\ 0\ 1\ 0\ \cdots\ 0\rangle^T \tag{23-32}$$

where matrix $\mathbf{I}(i\overline{\omega})$ is the system's known impedance matrix given by the square-bracket quantity on the left hand side of Eq. (23-31) and where the kth component in the load vector on the right hand side of this same equation equals unity with all other components set equal to zero. Vectors $\mathbf{H}_k(i\overline{\omega})$ ($k = 1, 2, \cdots, N$) make up the columns in the desired $N \times N$ complex frequency response matrix $\mathbf{H}(i\overline{\omega})$.

This matrix is generated for widely spaced discrete values of $\overline{\omega}$ using Eq. (23-32) and then fifth-order interpolation, as described in Section 12-9, is used to generate it for the other closely-spaced discrete values of frequency used in the fast Fourier transforms of the components in vector $\mathbf{p}(t)$. Having the complete transfer matrix $\mathbf{H}(i\overline{\omega})$ for closely-spaced discrete values of $\overline{\omega}$, response vector $\mathbf{V}(i\overline{\omega})$ is obtained by superposition using

$$\mathbf{V}(i\overline{\omega}) = \mathbf{H}(i\overline{\omega})\,\mathbf{P}(i\overline{\omega}) \tag{23-33}$$

Assume now the vector $\mathbf{p}(t)$ represents a stationary random process characterized by its spectral density matrix

$$\mathbf{S}_p(i\overline{\omega}) = \begin{bmatrix} S_{p_1 p_1}(\overline{\omega}) & S_{p_1 p_2}(i\overline{\omega}) & \cdots & S_{p_1 p_N}(i\overline{\omega}) \\ S_{p_2 p_1}(i\overline{\omega}) & S_{p_2 p_2}(\overline{\omega}) & \cdots & S_{p_2 p_N}(i\overline{\omega}) \\ \cdots\cdots\cdots\cdots\cdots\cdots\cdots\cdots\cdots\cdots\cdots \\ S_{p_N p_1}(i\overline{\omega}) & S_{p_N p_2}(i\overline{\omega}) & \cdots & S_{p_N p_N}(\overline{\omega}) \end{bmatrix} \qquad (23\text{-}34)$$

where

$$S_{p_j p_k}(i\overline{\omega}) \equiv \lim_{s\to\infty} \frac{\left[\int_{-s/2}^{s/2} p_j(t)\,\exp(-i\overline{\omega}t)\,dt\right]\left[\int_{-s/2}^{s/2} p_k(t)\,\exp(i\overline{\omega}t)\,dt\right]}{2\pi\,s} \qquad (23\text{-}35)$$

Consistent with this stationary condition, (1) postmultiply each side of Eq. (23-33) by the transpose of its own complex conjugate, (2) divide both sides of the resulting equation by $2\pi\,s$ where s represents duration of the process, and (3) take the limit as $s \to \infty$; thus, one obtains

$$\lim_{s\to\infty} \frac{\mathbf{V}(i\overline{\omega})\,\mathbf{V}^T(-i\overline{\omega})}{2\pi\,s} = \lim_{s\to\infty} \frac{\mathbf{H}(i\overline{\omega})\,\mathbf{P}(i\overline{\omega})\,\mathbf{P}(-i\overline{\omega})^T\,\mathbf{H}(-i\overline{\omega})^T}{2\pi\,s} \qquad (23\text{-}36)$$

Using the definition of cross-spectral density given by Eq. (23-35), Equation (23-36) becomes

$$\mathbf{S}_v(i\overline{\omega}) = \mathbf{H}(i\overline{\omega})\,\mathbf{S}_p(i\overline{\omega})\,\mathbf{H}(-i\overline{\omega})^T \qquad (23\text{-}37)$$

where $\mathbf{S}_v(i\overline{\omega})$ is the spectral density matrix for response given by

$$\mathbf{S}_v(i\overline{\omega}) = \begin{bmatrix} S_{v_1 v_1}(\overline{\omega}) & S_{v_1 v_2}(-i\overline{\omega}) & \cdots & S_{v_1 v_N}(i\overline{\omega}) \\ S_{v_2 v_1}(i\overline{\omega}) & S_{v_2 v_2}(\overline{\omega}) & \cdots & S_{v_2 v_N}(i\overline{\omega}) \\ \cdots\cdots\cdots\cdots\cdots\cdots\cdots\cdots\cdots\cdots\cdots \\ S_{v_N v_1}(i\overline{\omega}) & S_{v_N v_2}(i\overline{\omega}) & \cdots & S_{v_N v_N}(\overline{\omega}) \end{bmatrix} \qquad (23\text{-}38)$$

If one is interested in an r-component response vector $\mathbf{z}(t)$ as given in the time and frequency domains, respectively, by

$$\mathbf{z}(t) = \mathbf{A}\,\mathbf{v}(t) \qquad \mathbf{Z}(i\overline{\omega}) = \mathbf{A}\,\mathbf{V}(i\overline{\omega}) \qquad (23\text{-}39)$$

where $\mathbf{A}$ is a known $r \times N$ coefficient matrix, the $r \times r$ spectral density matrix for vector $\mathbf{z}(t)$ is then given by

$$\mathbf{S}_z(i\overline{\omega}) = \mathbf{A}\,\mathbf{H}(i\overline{\omega})\,\mathbf{S}_p(i\overline{\omega})\,\mathbf{H}(-i\overline{\omega})^T\,\mathbf{A}^T \qquad (23\text{-}40)$$

This matrix fully characterizes random vector $\mathbf{z}(t)$ if the input process $\mathbf{p}(t)$ is gaussian.

PROBLEMS

23-1. Consider the linear system with two degrees of freedom shown in Fig. P23-1, where $p_1(t)$ and $p_2(t)$ are two different zero-mean stationary random processes. The system has discrete masses and springs as indicated and may be assumed to be undercritically damped, with linear viscous damping of the uncoupled form yielding modal damping ratios $\xi_1 = \xi_2 = \xi$. If the power and cross-spectral density functions for processes $p_1(t)$ and $p_2(t)$ are

$$S_{p_1 p_1}(\overline{\omega}) = S_0 \quad S_{p_2 p_2}(\overline{\omega}) = A\,S_0 \quad S_{p_1 p_2}(\overline{\omega}) = (B + iC)\,S_0$$

over the entire frequency range $-\infty < \overline{\omega} < \infty$, where A, B, and C are real constants (A must be positive, but B and C may be positive or negative), express the power spectral density function for spring force $f_s(t)$ in terms of constants k, m, ξ, S_0, A, B, and C. Write an expression, involving constants A, B, and C, which gives the ranges of possible values for constants B and C. *Note:* The range of possible values for B cannot be expressed independently of the range of possible values for A.

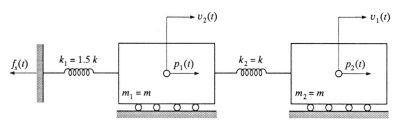

FIGURE P23-1
Two-mass system of Prob. 23-1.

23-2. A uniform simple beam of length L, stiffness EI, and mass $\overline{m}$ per unit length is subjected to zero-mean stationary random vertical-support motions at each end. Let $v(x, t)$ represent the total vertical displacement of the member; the vertical displacements are given by $v(0, t)$ and $v(L, t)$. Assume viscous damping of the uncoupled form with all normal modes having the same damping ratio ξ ($0 < \xi < 1$). If the power and cross-sepctral density functions for vertical-support accelerations are given by

$$S_{a_1 a_1}(\overline{\omega}) = S_0 \quad S_{a_2 a_2}(\overline{\omega}) = 0.5\,S_0 \quad S_{a_1 a_2}(\overline{\omega}) = (0.4 + 0.2i)\,S_0$$

over the entire frequency range $-\infty < \overline{\omega} < \infty$, where subscripts a_1 and a_2 are used to represent $\ddot{v}(0, t)$ and $\ddot{v}(L, t)$, respectively, find the power spectral density functions for (a) displacement $v(x,t)$, (b) moment $M(x,t)$, and (c) shear $V(x,t)$. Give your answers in series form expressed in terms of L, EI, $\overline{m}$, ξ, S_0, and $\overline{\omega}$. Discuss the relative rates of convergence of these series.

23-3. The tapered vertical cantilever member shown in Fig. P23-2 is subjected to a distributed zero-mean gaussian stationary random loading $p(x,t)$ having the power spectral density function

$$S_p(x, \alpha, \overline{\omega}) = S(\overline{\omega}) \, \exp\left[- (A/L)|x - \alpha| \right]$$

where $S(\overline{\omega})$ is a known function of $\overline{\omega}$ with units of $lb^2 \cdot sec/ft^2$ and A is a known positive real constant. The functions $m(x)$ and $EI(x)$ are known, and viscous damping can be assumed of the uncoupled form yielding the same damping ratio ξ in each normal mode. *Outline* how you would make an analysis of the stochastic response of this structure. Explain in sufficient detail to show that you could actually obtain correct numerical results if requested to do so.

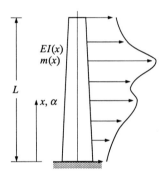

FIGURE P23-2
Cantilever member of Prob. 23-3.

EARTHQUAKE ENGINEERING

CHAPTER
24

SEISMOLOGICAL BACKGROUND

24-1 INTRODUCTORY NOTE

As was mentioned in Chapter 1, the dynamic loadings that act on structural systems may result from a wide range of input mechanisms. One important class of loading involves vehicular systems such as ships, airplanes, automobiles, etc., for which the dynamic loading is a result of the motion of the vehicle. The other basic class concerns fixed civil engineering structures such as bridges, buildings, dams, etc., to which the dynamic loading is applied externally. Of the many sources of external load that must be considered in the design of fixed structures, the most important by far in terms of its potential for disastrous consequences is the earthquake.

The degree of importance of earthquake loading in any given region is related, of course, to its probable intensity and likelihood of occurrence — that is, to the seismicity of the region. However, the importance of the earthquake problem, in general, was amplified greatly some years ago by the advent of the nuclear power industry because stringent seismic criteria were adopted that had to be considered in the design of nuclear power stations to be built in any part of the United States, and similar criteria also are applicable in most well-developed areas of the world. For this reason alone it would be desirable to use the field of earthquake engineering as the framework on which to demonstrate the application of the theories and techniques presented in Parts One through Four of this text.

In addition, however, it is evident that the design of economic and attractive structures which can successfully withstand the forces induced by a severe ground motion is a challenge demanding the best in structural engineering, art, and science. Furthermore, to paraphrase the comment made by Newmark and Rosenblueth:[1] "Earthquakes systematically bring out the mistakes made in design and construction — even the most minute mistakes; it is this aspect of earthquake engineering that makes it challenging and fascinating, and gives it an educational value far beyond its immediate objectives."

The essential background for study or practice in the field of earthquake engineering is, of course, knowledge about the earthquake itself. The detailed study of earthquakes and earthquake mechanisms lies in the province of seismology, but in his or her studies the earthquake engineer must take a different point of view than the seismologist. Seismologists have focused their attention primarily on the global or long-range effects of earthquakes and therefore are concerned with very small-amplitude ground motions which induce no significant structural responses. Engineers, on the other hand, are concerned mainly with the local effects of large earthquakes, where the ground motions are intense enough to cause structural damage. These so-called *strong-motion earthquakes* are too violent to be recorded by the very sensitive seismographs typically used by seismologists and have necessitated the development of special types of strong-motion seismographs. Nevertheless, even though the objectives of an earthquake engineer differ from those of a seismologist, there are many topics in seismology which are of immediate engineering interest. A brief summary of the more important topics is presented in this chapter.

24-2 SEISMICITY

The seismicity of a region determines the extent to which earthquake loadings may control the design of any structure planned for that location, and the principal indicator of the degree of seismicity is the historical record of earthquakes that have occurred in the region. Because major earthquakes often have had disastrous consequences, they have been noted in chronicles dating back to the beginnings of civilization. In China, records have been kept that are thought to include every major destructive seismic event for a time span of nearly 3000 years; thus it is evident that considerable knowledge about the seismicity of China is available. Also there are reports of major earthquake damages that occurred in the Middle East as much as two thousand years ago, though the record for that region is not as complete as it is in China.

In general, however, information about global seismicity is much less extensive

---

[1] N. M. Newmark and E. Rosenblueth, *Fundamentals of Earthquake Engineering*, Prentice-Hall Inc., Englewood Cliffs, N. J., 1971.

than might be inferred from these examples because long historic records have not been maintained in most regions of the earth. Even though data about earthquake occurrences increased rapidly with the spread of civilization, the record remained very incomplete and inconsistent until the 1960's when a Worldwide Standardized Network of seismological stations was installed by the United States. This network consisted of about 120 stations equipped with standard seismograph units and was distributed among 60 countries. Therefore, when this new earthquake recording capability was added to the existing networks installed by many other countries, it became possible to study seismicity on a truly global scale.

Earthquake occurrence information compiled from these seismograph networks typically is presented in maps such as that shown in Fig. 24-1, which indicates the location and magnitude of all earthquakes recorded during 1977 through 1986. The most obvious conclusion to be drawn from this map is that earthquake occurrences are not distributed uniformly over the surface of the earth; instead they tend to be concentrated along well-defined lines which are known to be associated with the boundaries of segments or "plates" of the earth's crust. The mechanism of earthquake generation along plate boundaries is discussed in Section 24-5 under the heading of "plate tectonics," following brief descriptions of earthquake faults and waves, and of the structure of the earth, that are presented below in Sections 24-3 and 24-4, respectively.

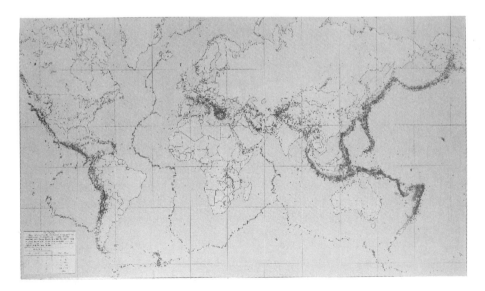

FIGURE 24-1
Global distribution of seismicity: 1977–1986.
[*Susan K. Goter, U.S. Geological Survey.*]

24-3 EARTHQUAKE FAULTS AND WAVES

From the study of geology, it has become apparent that the rock near the surface of the earth is not as rigid and motionless as it appears to be. There is ample evidence in many geological formations that the rock was subjected to extensive deformations at a time when it was buried at some depth. Apparently, when it is subjected to great pressure from the overburden, the rock can be bent like elasto-plastic metals or squeezed into new shapes like soft clay. In addition, the geological structures show that numerous ruptures have occurred within the rock masses, presumably when they were strained beyond the deformational capacity of the type of material involved. When such ruptures occurred, relative sliding motions were developed between the opposite sides of the rupture surface creating what is called a geological fault. The orientation of the fault surface is characterized by its "strike," the orientation from north of its line of intersection with the horizontal ground surface, and by its "dip," the angle from horizontal of a line drawn on the fault surface perpendicular to this intersection line.

The type of displacement that occurs on a fault depends on the state of stress in the rock that led to the rupture. Horizontal shear stress leads to lateral motion in the direction of the fault strike, thus this is called a "strike-slip" or lateral fault; it may be classified as either left-lateral or right-lateral according to whether the rock mass on the side of the fault opposite the observer moved toward the left or right relative to the rock on the near side. The relative sliding motion also may be in the direction of the fault dip, in which case it is called a "dip-slip" fault. This is designated as a "normal" fault if the rock mass on the upper side of the fault has a relative downward movement, and it is a "reverse" fault if the upper rock mass moves in the upward direction relative to the lower side. In general, normal faults are associated with a state of tensile stress in the surficial rock layer, while a reverse fault may be induced by compressive stress in the surface rock layer if the dip is small; in this case the result is called a thrust fault. All of this fault terminology is depicted in Fig. 24-2.

The important fact about any fault rupture is that the fracture occurs when the deformations and stresses in the rock reach the breaking strength of the material. Accordingly it is associated with a sudden release of strain energy which then is transmitted through the earth in the form of vibratory elastic waves radiating outward in all directions from the rupture point. These displacement waves passing any specified location on the earth constitute what is called an earthquake. The point on the fault surface where the rupture first began is called the earthquake focus, and the point on the ground surface directly above the focus is called the epicenter.

Two types of waves may be identified in the earthquake motions that are propagated deep within the earth: "P" waves, in which the material particles move along the path of the wave propagation inducing an alternation between tension and compression deformations, and "S" waves, in which the material particles move in a

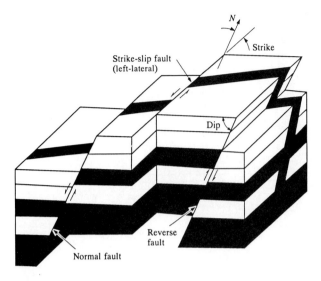

FIGURE 24-2
Definition of fault orientation, and of the basic types of fault displacement.
[*Adapted from Earthquake by Bruce A. Bolt, W. H. Freeman and Company 1988.*]

direction perpendicular to the wave propagation path, thus inducing shear deformations. The "P" or Primary wave designation refers to the fact that these normal stress waves travel most rapidly through the rock and therefore are the first to arrive at any given point. The "S" or Secondary wave designation refers correspondingly to the fact that these shear stress waves travel more slowly and therefore arrive after the "P" waves.

When the vibratory wave energy is propagating near the surface of the earth rather than deep in the interior, two other types of waves known as Rayleigh and Love waves can be identified. The Rayleigh surface waves are tension-compression waves similar to the "P" waves except that their amplitude diminishes with distance below the surface of the ground. Similarly the Love waves are the counterpart of the "S" body waves; they are shear waves that diminish rapidly with distance below the surface. Figure 24-3 illustrates the nature of these four types of elastic earthquake waves.

24-4 STRUCTURE OF THE EARTH

Much of present knowledge about the structure of the earth has been deduced from information about the relative length of time required for an earthquake wave to propagate from the point of rupture (the focus) to observation points (seismographs) which may be located at many points on the earth's surface. The relative arrival times of the P and S waves can be interpreted in terms of the distance of the observatory from

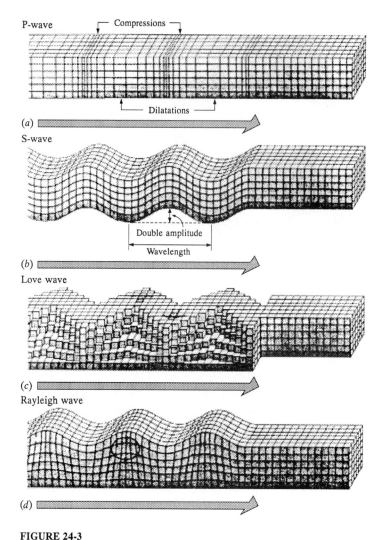

P-wave

Compressions

Dilatations

(a)

S-wave

Double amplitude

Wavelength

(b)

Love wave

(c)

Rayleigh wave

(d)

FIGURE 24-3
Diagram illustrating the forms of ground motion near the ground surface in four types of earthquake waves. [*From Bruce A. Bolt, Nuclear Explosions and Earthquakes: The Parted Veil (San Francisco: W. H. Freeman and Company. Copyright © 1976).*]

the focus if the properties of the materials through which the waves travel are known; also this relative delay time provides evidence regarding reflection and refraction of the earthquake waves from the boundaries between concentric layers of rock having different moduli of elasticity and densities. The paths of some P-type earthquake waves are shown schematically in Fig. 24-4. Based on this kind of information, it has been deduced that the earth consists of several discrete concentric layers as illustrated

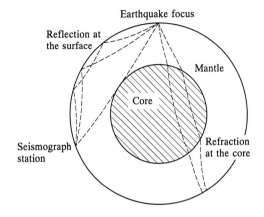

FIGURE 24-4
Paths of some P-type earthquake waves from the focus.

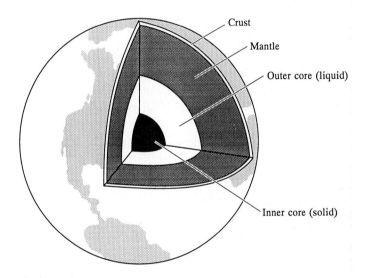

FIGURE 24-5
Zonation of the earth's interior. The crust, which includes continents at the surface of the earth, rests on the mantle. The mantle, in turn, rests on the core. The outer core is liquid, but the inner core is solid. [*After W. J. Kauffman, Planets and Moons, W. H. Freeman and Company, New York, 1979.*]

in Fig. 24-5 and described in the following paragraphs.

   The central sphere (called the inner core) is a very dense solid thought to consist mainly of iron; surrounding that is a layer of similar density, but thought to be a liquid because shear waves are not transmitted through it. Next is a solid thick envelope of lesser density around the core that is called the mantle, and finally is the rather thin layer at the earth's surface called the crust. Seismic waves propagating from the mantle into the crust exhibit a very abrupt decrease of velocity known as the Mohorovicic discontinuity (or Moho) which identifies the base of the crust at that location. From such seismic information, it has been noted that the crust is relatively thin under the oceans but that it is correspondingly thick beneath high mountain ranges. This relationship is believed to be a demonstration of the principle of isostasy indicating that the crust is floating on the mantle. In this circumstance, a deeper layer of crust is needed to provide buoyancy for a high mountain range as compared with regions where the topography reaches only to sea level or below.

Based on this concept, the mantle is considered to consist of two distinct layers. The upper mantle together with the crust form a rigid layer called the lithosphere. Below that, a zone with low velocity has been identified; this layer, called the asthenosphere, is thought to be partially molten rock consisting of solid particles incorporated within a liquid component. Although the asthenosphere represents only a small fraction of the total thickness of the mantle, it is because of its highly plastic character that the lithosphere can act as if it is floating on a liquid and thus can be subjected to large crustal deformations. The lithosphere does not move as a single unit, however; instead it is divided into a pattern of plates of various sizes, and it is the relative movements along the plate boundaries that cause the earthquake occurrence patterns seen in Fig. 24-1. A simplified map of the Earth's crustal plates is shown in Fig. 24-6. The detailed description of the motions of these plates is a subject called plate tectonics; development of understanding of this subject is one of the great advances of geology and seismology during the present century as described briefly in Section 24-5, below.

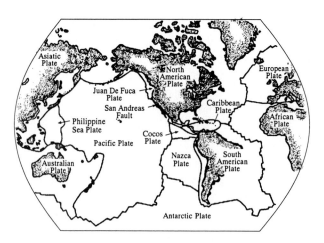

FIGURE 24-6
Simplified map of the Earth's crustal plates.

24-5 PLATE TECTONICS

The basic concept embodied in plate tectonics was recognized several centuries ago by persons working with world maps who noted that the outline of the west coast of Africa is a good match for the east coast of South America and suggested that the two continents might at one time have been joined together. Although this fact was a matter of great curiosity, it was not until much later that a Frenchman, Antonio Snider-Pellegrini made the first attempt to develop the concept of substantial continental movement; in a book published in 1858 he hypothesized that the Atlantic Ocean had been formed by the tearing apart and separation of the two continents. The author supported this postulation of "continental drift" (as it came to be known) by presenting extensive evidence that the geological structure on the South American coast was very similar to that on the African coast and also by identifying similarities between the ancient plant and animal life in the two regions as indicated by their fossil specimens.

Most geologists completely rejected the continental drift idea at that time, but it was proposed again in 1908 by an American geologist, F. B. Taylor, who included in his argument the idea that the Mid-Atlantic Ridge (which had been identified and mapped in the 1870's) is the line along which the single supercontinent ruptured to form the present two continents. A much more comprehensive argument for the concept of continental drift was made in 1915 by a German meteorologist, Alfred Wegener, in his book *On the Origin of Continents and Oceans*, in which he claimed that essentially all the world's large land masses at one time had been joined in a single supercontinent. He identified the sequence in which these great land areas had split apart and moved to their present locations, and following the lead of Snider-Pellegrini he also based his argument for the previous interconnection of the system on the similarities of the geological structure and of the fossil flora and fauna of the related components. In the following years numerous other scientists began to accept the idea of continental drift, but most geophysicists rejected it because they could not imagine how major continental units could be driven through the oceanic crust for any significant distance.

Evidence for continental drift continued to be compiled in the decades following publication of Wegener's work, but no satisfactory explanation of the mechanism that caused it was proposed until the publication in 1962 of the paper "History of Ocean Basins" by the American geologist H. H. Hess. This paper, which opened the door to our present understanding of plate tectonics, was based on extensive studies of the sea floor including investigations made by Hess during World War II while he was serving as captain of a U. S. Navy vessel. In the course of his normal duties he was able to survey the bottom topography using the ship's depth sounding equipment and this led to his discovery of flat topped features called seamounts that rose from the sea floor. From their size and shape he concluded that these at one time had been islands, and that their flat tops had been caused by wave erosion near sea level. The basic question over which he puzzled was how they had subsided to their present great depths, and he eventually formed the hypothesis that this was the result of extensive lateral motion having a downward component rather than simple vertical motion. Combining this concept with knowledge of the position of the mid-ocean ridges and also with the idea of convective circulation of the earth's mantle that had been proposed by others previously, Hess conceived a general hypothesis for motions of the earth's crust. This continual movement is thought to be driven by heating of the mantle due to radioactivity, which causes molten rock to well upwards along the ocean ridges. As this material reaches the surface and cools, it forms the crust at the surface of the lithosphere; the entire lithosphere floating on the plastic asthenosphere is caused to spread outwards in both directions by the continued upwelling of additional molten rock. This new crust then sinks beneath the surface of the sea as it cools and spreads outward, and the motion continues until eventually the lithosphere reaches a deep sea trench where it plunges downward into the asthenosphere in a process called subduction.

This concept of sea floor spreading is supported by many types of physical evidence, including the presence of seamounts at great depths as was mentioned earlier, but the most striking proof of the theory is given by patterns of magnetic orientation as shown by the zebra stripe patterns that have been observed in maps of sea floor magnetism, as shown in Fig. 24-7. As the crust is first formed, it is magnetized in accordance with the polarity of the earth's magnetic field at that time, and it maintains that polarity as it spreads outward from the ridge. However, when

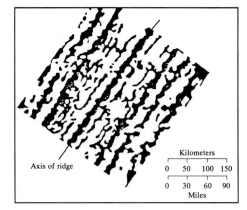

FIGURE 24-7
Magnetic-anomaly pattern of the North Atlantic sea floor. Symmetrical striping is revealed by measurement of the strength of the magnetic field at many locations from a ship. The position of the area represented in the lower diagram is shown in the map above.
[*From A. Cox et al., "Reversals of the Earth's Magnetic Field." Copyright © 1967 by Scientific American, Inc. All rights reserved.*]

the earth's field changes polarity, as has happened at intervals of one-half to one million years during the present Cenozoic era (65 million years), the crust that has just developed on both sides of the ridge shows the newly reversed polarity in maps such as Fig. 24-7. Using such data, the rate of sea floor spreading may be evaluated by relating the spacing of these polarity stripes to the chronology of the polarity changes.

Other convincing evidence of the movement of the earth's crust is provided by the alignment of volcanic island groups such as the Hawaiian Islands in the Pacific Ocean. These islands were produced by plumes of molten rock welling up from the base of the mantle at the times when each island location was positioned in sequence over an isolated point of volcanic activity in the mantle base, known as a "hot spot." These hot spots are relatively stationary in the earth's sphere, so when the lithosphere moves in accordance with its convective circulation, the plumes rise into it at successively differing locations. In the area of the Hawaiian Island chain, the crust is moving in a northwesterly direction, so the island which was formed first (Kawai) is located at the northwest end of the chain. The other islands are positioned according to their relative ages, with the youngest (Hawaii) located in its chronological sequence at the southeast end of the chain.

From this description, it is evident that the Hess concept of sea floor spreading provides answers for the principal questions which originally had led to rejection of the idea of continental drift. Because the continents are embedded in the lithosphere and are transported with it like on a conveyor belt, they do not have to be driven through the crust as they move. However, it is apparent from the seismicity map of Fig. 24-1 that the continental motions are associated with a variety of different circulation patterns; hence relative motions are induced at the plate boundaries and it is these relative motions that are the cause of most earthquakes that have occurred. For example, the interaction between the Pacific Plate and the continental plates all around its periphery has produced the great majority of the earthquakes indicated in Fig. 24-1. The well-known San Andreas fault is the juncture between the Pacific Plate and the North American Plate. This fault and the subsidiary faults branching from it, shown in Fig. 24-8, has been the source of most major California earthquakes during historic times, including the great San Francisco earthquake of 1906. It is one of the most active as well as the most studied fault systems in the world; its location is apparent in topographic features (Fig. 24-9) over nearly its entire extent in California. Research on this fault zone and the earthquakes associated with it has contributed greatly to present knowledge of earthquake mechanisms and earthquake-motion characteristics. An interesting fact is that relative movements along this fault, corresponding with the counterclockwise rotation of the Pacific basin mentioned above, have been observed both in fault breaks occurring during earthquakes and in continual creep deformations measured by geodetic surveys. These measurements show the geological structure

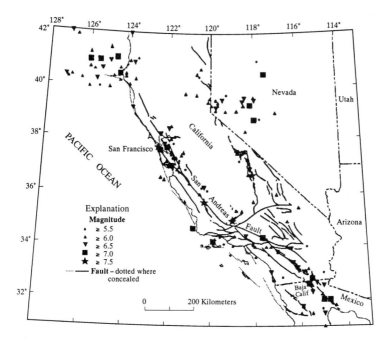

FIGURE 24-8
Principal earthquakes and faults in California.

FIGURE 24-9
San Andreas fault on east edge of Carrizo Plain, Central California.
[*R. E. Wallace, U.S. Geological Survey.*]

west of the fault in California to be moving northward relative to the east side at a rate of several centimeters per year.

24-6   ELASTIC-REBOUND THEORY OF EARTHQUAKES

It was from study of the rupture which occurred along the San Andreas fault during the San Francisco earthquake of 1906 that H. F. Reid first put into clear focus the elastic-rebound theory of earthquake generation. Many seismologists had already surmised that earthquakes result somehow from fractures, or faulting, of the earth's crust. However, Reid's investigation of the large-amplitude shearing displacements which resulted from this earthquake for dozens of miles along the fault (a typical displacement is shown in Fig. 24-10) led him to conclude that the specific source of the earthquake vibration energy is the release of accumulated strain in the earth's crust, the release itself resulting from the sudden shear-type rupture.

The essential concept of this elastic-rebound mechanism, which still provides the most satisfactory explanation for the types of earthquakes causing intense, potentially damaging surface motions, is portrayed in Fig. 24-11. The active fault zone is shown in the center of the sketches, and, as with the San Andreas system, the geological structure to the left is assumed to be moving northward at a constant rate. If a series of fences were built perpendicularly across the fault (Fig. 24-11a), this continual northward drift would gradually distort the fence lines as shown in Fig. 24-11b. Also shown in the sketch is a road which is assumed to have been built after the fence-line deformations developed. Eventually, the continuing deformation of the crustal

FIGURE 24-10
Fence offset by San Andreas fault slip, San Francisco earthquake, April 18, 1906.

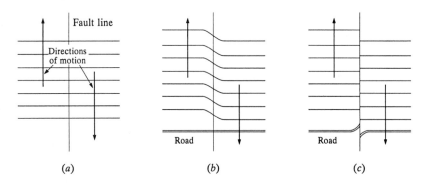

FIGURE 24-11
Elastic-rebound theory of earthquake generation: (*a*) before straining; (*b*) strained (before earthquake); (*c*) after earthquake.

structure will lead to stresses and strains which exceed the material strength. A rupture will then initiate at some critical point in the fault zone and will propagate rapidly throughout the length of the highly stressed material. The resulting release of strain and the corresponding displacements lead to the conditions depicted in Fig. 24-11*c*, with large offsets visible across the fault on both the road and the fence lines. With the release of strain, the fence lines would become straight, but the road (which was built over strained basement rock) would be locally curved.

Fault displacements such as those pictured in Figs. 24-9 and 24-10 are local manifestations of a major earthquake. Displacements like these undoubtedly will cause significant damage to any structure that happens to be founded directly over the fault break. However, the strong ground shaking initiated by the rupture and radiating outward in all directions from the focus is the more hazardous aspect of the earthquake because of its great area of influence. In contrast to the motions recorded at great distances which are too small even to be felt, the ground motions resulting at close range from an important earthquake are so strong that they cannot be recorded on a teleseismic instrument, as was mentioned earlier, and special ''strong-motion'' seismographs have been developed to obtain detailed information about such intense local shaking. Within this potentially damaging range the wave generation mechanism may be described conveniently by reference to a penny-shaped crack located on the fault surface, as shown in Fig. 24-12, following a general line of reasoning due to G. W. Housner. Suppose that the state of stress in the area of this crack zone has reached the rupture point. When the rupture occurs, the release of strain adjacent to the crack surface will be accompanied by a sudden relative displacement of the two sides. This displacement initiates a displacement wave which propagates radially from the source; a record of this simple displacement pulse as it passes a recording station located at a moderate distance from the focus would be like that shown in

Fig. 24-13. The velocity and acceleration records corresponding to this displacement pulse are also shown. Comparison of these idealized ground-motion records with the

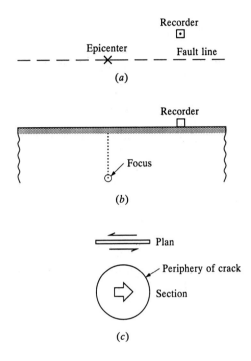

FIGURE 24-12
Idealized point-source earthquake rupture (after G. W. Housner): (*a*) plan; (*b*) section along fault line; (*c*) penny-shaped crack at focus.

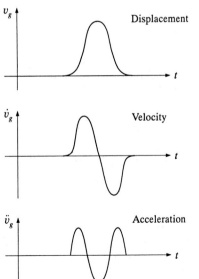

FIGURE 24-13
Idealized ground motion from point source.

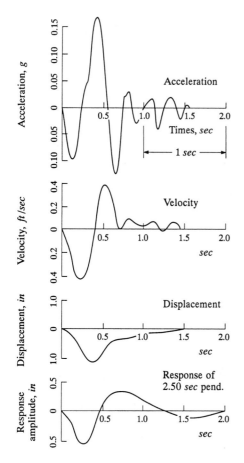

FIGURE 24-14
Accelerogram from Port Hueneme earthquake,
March 18, 1957 (*NS* component).

actual seismograph acceleration record obtained from the Port Hueneme earthquake of 1957 (shown in Fig. 24-14, together with the velocity and displacement diagrams obtained by integration of the accelerogram) demonstrates that the Port Hueneme earthquake was essentially a single-displacement pulse; thus it may be inferred that the mechanism which generated this earthquake was similar to the rupture of the simple penny-shaped crack.

Records of typical earthquakes, such as the El Centro earthquake of 1940 (one component of which is shown in Fig. 24-15), are much more complicated than the Port Hueneme record, and it is probable that the generating mechanism is correspondingly more complex. A hypothesis which provides a satisfactory explanation of the typical record is that the earthquake involves a sequence of ruptures along the fault surface. Each successive rupture is the source of a simple earthquake wave of the Port Hueneme type, but because they occur at different locations and times, the motions

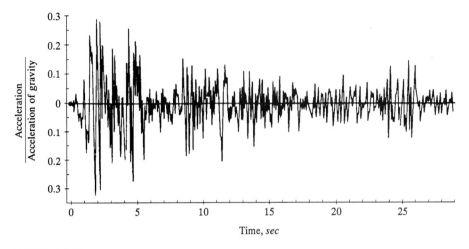

FIGURE 24-15
Accelerogram from El Centro earthquake, May 18, 1940 (*NS* component).

observed at a nearby station will be a random combination of simple records which could look much like Fig. 24-15.

In this context, it is important to recall that the focus of the earthquake is the point within the earth at which the *first* rupture of the fault surface takes place and the *epicenter* is the point at the surface directly above the focus. If the earthquake results from a sequence of ruptures along the fault line, it is evident that the focus may not coincide with the center of energy release. In a major earthquake, which may be associated with a fault break hundreds of miles in length, the distance of a building from the epicenter may be of little importance; the significant factor is the distance to the nearest point along the rupture surface.

24-7 MEASURES OF EARTHQUAKE SIZE

To an earthquake engineer, the most important aspect of an earthquake's ground motions is the effect they will have on structures, that is, the stresses and deformations or the amount of damage they would produce. This damage potential is, of course, at least partly dependent on the "size" of the earthquake, and a number of measures of size are used for different purposes. The most important measure of size from a seismological point of view is the amount of strain energy released at the source, and this is indicated quantitatively as the *magnitude*. By definition, Richter magnitude is the (base 10) logarithm of the maximum amplitude, measured in micrometers (10^{-6} m) of the earthquake record obtained by a Wood-Anderson seismograph, corrected to a distance of 100 km. This magnitude rating has been related empirically to the amount of earthquake energy released E by the formula

$$\log E = 11.8 + 1.5\,M \qquad (24\text{-}1)$$

in which M is the magnitude. By this formula, the energy increases by a factor of 32 for each unit increase of magnitude. More important to engineers, however, is the empirical observation that earthquakes of magnitude less than 5 are not expected to cause structural damage, whereas for magnitudes greater than 5, potentially damaging ground motions will be produced.

The magnitude of an earthquake by itself is not sufficient to indicate whether structural damage can be expected. This is a measure of the size of the earthquake at its source, but the distance of the structure from the source has an equally important effect on the amplitude of its response. The severity of the ground motions observed at any point is called the *earthquake intensity*; it diminishes generally with distance from the source, although anomalies due to local geological conditions are not uncommon. The oldest measures of intensity are based on observations of the effects of the ground motions on natural and man-made objects. In the United States, the standard measure of intensity for many years has been the Modified Mercalli (MM) scale. This is a 12-point scale ranging from I (not felt by anyone) to XII (total destruction). Results of earthquake-intensity observations are typically compiled in the form of isoseismal maps like that shown in Fig. 24-16. Although such subjective intensity ratings are very valuable in the absence of any instrumented records of an earthquake, their deficiencies in providing criteria for the design of earthquake-resistant structures are obvious.

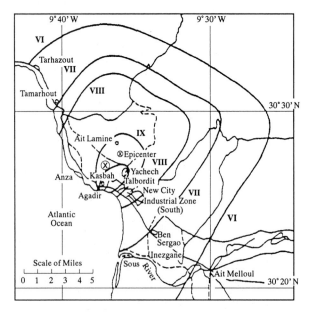

FIGURE 24-16
Isoseismal map of Agadir earthquake, 1960 (Modified Mercalli intensity scale).

Basic information on the characteristics of earthquake motions which could be used for earthquake engineering purposes did not become available until the first strong-motion-recording accelerographs were developed and a network of such instruments was installed by the United States Coast and Geodetic Survey. The accelerogram of Fig. 24-15 (together with the perpendicular horizontal component and the vertical component) was recorded by one of these early instruments. The rate at which such instrumental information was gathered was very slow for many years because the number and distribution of the instruments were very limited. Gradually more extensive networks have been installed in Japan, Mexico, the most active seismic regions of the United States, and various other parts of the world, and much new and significant information is now being obtained. Unfortunately, however, the distribution of instruments is still quite limited, and destructive earthquakes in most parts of the world provide no strong-motion records. Consequently, basic data concerning the influence of such factors as magnitude, distance, and local soil conditions on the characteristics of earthquake motions are still very scarce.

The three components of ground motion recorded by a strong-motion accelerograph provide a complete description of the earthquake which would act upon any structure at that site. However, the most important features of the record obtained in each component (such as Fig. 24-15), from the standpoint of its effectiveness in producing structural response, are the amplitude, the frequency content, and the duration. The amplitude generally is characterized by the peak value of acceleration or sometimes by the number of acceleration peaks exceeding a specified level. (It is worth noting that the ground velocity may be a more significant measure of intensity than the acceleration, but it generally is not available without supplementary calculations.) The frequency content can be represented roughly by the number of zero crossings per second in the accelerogram and the duration by the length of time between the first and the last peaks exceeding a given threshold level. It is evident, however, that all these quantitative measures taken together provide only a very limited description of the ground motion and certainly do not quantify its damage-producing potential adequately. The quantitative description of earthquake motions is the subject of Chapter 25.

FREE-FIELD SURFACE GROUND MOTIONS

25-1 FOURIER AND RESPONSE SPECTRA

In designing structures to perform satisfactorily under earthquake conditions, the engineer needs a much more precise characterization of ground shaking than is provided by the Modified Mercalli intensity. For this purpose, the response of a simple oscillator, such as the SDOF frame shown in Fig. 25-1, has proved to be invaluable. The relative displacement response of this frame to a specified single component of ground acceleration $\ddot{v}_g(t)$ may be expressed in the time domain by

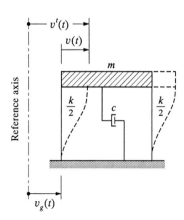

FIGURE 25-1
Basic SDOF dynamic system.

means of the Duhamel integral, Eq. (6-7), if it is noted that the effective loading is given by Eq. (2-17), i.e., $p_{\text{eff}}(t) = -m\,\ddot{v}_g(t)$; thus,

$$v(t) = \frac{1}{m\,\omega_D} \int_0^t -m\,\ddot{v}_g(\tau)\,\sin\omega_D(t-\tau)\exp[-\xi\omega(t-\tau)]\,d\tau \tag{25-1}$$

When the difference between the damped and the undamped frequency is neglected, as is permissible for small damping ratios usually representative of real structures (say $\xi < 0.10$), and when it is noted that the negative sign has no real significance with regard to earthquake excitation, this equation can be reduced to

$$v(t) = \frac{1}{\omega} \int_0^t \ddot{v}_g(\tau)\,\sin\omega(t-\tau)\exp[-\xi\,\omega(t-\tau)]\,d\tau \tag{25-2}$$

Taking the first time derivative of Eq. (25-2), one obtains the corresponding relative velocity time-history

$$\dot{v}(t) = \int_0^t \ddot{v}_g(\tau)\,\cos\omega(t-\tau)\exp[-\xi\omega(t-\tau)]\,d\tau$$

$$- \xi \int_0^t \ddot{v}_g(\tau)\,\sin\omega(t-\tau)\exp[-\xi\omega(t-\tau)]\,d\tau \tag{25-3}$$

Further, substituting Eqs. (25-2) and (25-3) into the forced-vibration equation of motion, written in the form

$$\ddot{v}^t(t) = -2\,\omega\,\xi\,\dot{v}(t) - \omega^2\,v(t) \tag{25-4}$$

one obtains the total acceleration relation

$$\ddot{v}^t(t) = \omega\,(2\,\xi^2 - 1) \int_0^t \ddot{v}_g(\tau)\,\sin\omega(t-\tau)\exp[-\xi\omega(t-\tau)]\,d\tau$$

$$- 2\,\omega\,\xi \int_0^t \ddot{v}_g(\tau)\,\cos\omega(t-\tau)\exp[-\xi\omega(t-\tau)]\,d\tau \tag{25-5}$$

The absolute maximum values of the response given by Eqs. (25-2), (25-3), and (25-5) are called the spectral relative displacement, spectral relative velocity, and spectral absolute acceleration, respectively; these will be denoted herein as $S_d(\xi, \omega)$, $S_v(\xi, \omega)$, and $S_a(\xi, \omega)$, respectively.

As will be shown subsequently, it is usually necessary to calculate only the so-called pseudo-velocity spectral response $S_{pv}(\xi, \omega)$ defined by

$$S_{pv}(\xi, \omega) \equiv \left[\int_0^t \ddot{v}_g(\tau)\,\sin\omega(t-\tau)\exp\left[-\xi\omega(t-\tau)\right]\,d\tau \right]_{\max} \tag{25-6}$$

where the subscript max refers to the maximum absolute value of response over the entire time-history. Now from Eq. (25-2), it is seen that

$$S_d(\xi, \omega) = \frac{1}{\omega} S_{pv}(\xi, \omega) \tag{25-7}$$

and from Eqs. (25-3) and (25-6) that (for $\xi = 0$)

$$S_v(0, \omega) = \left[\int_0^t \ddot{v}_g(\tau) \cos \omega(t - \tau) \, d\tau \right]_{\max} \tag{25-8}$$

$$S_{pv}(0, \omega) = \left[\int_0^t \ddot{v}_g(\tau) \sin \omega(t - \tau) \, d\tau \right]_{\max} \tag{25-9}$$

which are identical except for the trigonometric terms. It has been demonstrated by Hudson[1,2] that $S_v(0, \omega)$ and $S_{pv}(0, \omega)$ differ very little numerically, except in the case of very long period oscillators, i.e., very small values of ω. For damped systems, the difference between S_v and S_{pv} is considerably larger and can differ by as much as 20 percent for $\xi = 0.20$. Note also from Eq. (25-5) for $\xi = 0$ that

$$S_a(0, \omega) = \left[\omega \int_0^t \ddot{v}_g(t) \sin \omega(t - \tau) \, d\tau \right]_{\max} \tag{25-10}$$

thus, from Eq. (25-6),

$$S_a(0, \omega) = \omega \, S_{pv}(0, \omega) \tag{25-11}$$

It can be shown that Eq. (25-11) is very nearly satisfied for damping values over the range $0 < \xi < 0.20$; therefore, we are able to use the approximate relation

$$S_a(\xi, \omega) \doteq \omega \, S_{pv}(\xi, \omega) \tag{25-12}$$

with little error being introduced. The entire quantity on the right hand side of Eq. (25-12) is called the pseudo-acceleration spectral response and is denoted herein as $S_{pa}(\xi, \omega)$. This quantity is particularly significant since it is a measure of the maximum spring force developed in the oscillator, i.e.,

$$f_{s,\max} = k \, S_d(\xi, \omega) = \omega^2 \, m \, S_d(\xi, \omega) = m \, S_{pa}(\xi, \omega) \tag{25-13}$$

[1] D. E. Hudson, "Response Spectrum Techniques in Engineering Seismology," *Proc. 1st World Conference on Earthquake Engineering*, Earthquake Engineering Research Institute, Berkeley, CA, 1956.

[2] D. E. Hudson, "Some Problems in the Application of Spectrum Techniques to Strong Motion Earthquake Analysis," *Bull. Seismological Society of America*, Vol. 52, No. 2, April, 1962.

Clearly, from the above treatment, only the pseudo-velocity response spectrum as defined by Eq.(25-6) need be generated for any prescribed single component of earthquake ground motion. The other desired response spectra can be easily obtained therefrom using the relations

$$S_d(\xi, \omega) = \frac{1}{\omega} \, S_{pv}(\xi, \omega) \tag{25-14}$$

$$S_{pa}(\xi, \omega) = \omega \, S_{pv}(\xi, \omega) \tag{25-15}$$

As indicated above these response quantities depend not only on the ground motion time-history but also on the natural frequency and damping ratio of the oscillator. Thus for any given earthquake accelerogram, by assuming discrete values of damping ratio and natural frequency, it is possible to calculate the corresponding discrete values of $S_{pv}(\xi, \omega)$ using Eq. (25-6) and to calculate corresponding values of $S_d(\xi, \omega)$ and $S_{pa}(\xi, \omega)$ using Eqs. (25-14) and (25-15), respectively.

Graphs of the values for $S_{pv}(\xi, \omega)$, $S_d(\xi, \omega)$, and $S_{pa}(\xi, \omega)$ plotted as functions of frequency (or functions of period $T = \frac{2\pi}{\omega}$) for discrete values of damping ratio are called pseudo-velocity response spectra, displacement response spectra, and pseudo-acceleration response spectra, respectively. If plotted in linear form, each type of spectra must be plotted separately similar to the set of $S_{pv}(\xi, T)$ shown in Fig. 25-2 for the El Centro, California, earthquake of May 18, 1940 (NS component). However, due to the simple relationships existing among the three types of spectra as given by Eqs. (25-14) and (25-15), it is possible to present them all in a single plot. This may be accomplished by taking the log (base 10) of Eqs. (25-14) and (25-15) to obtain

$$\log S_d(\xi, \omega) = \log S_{pv}(\xi, \omega) - \log \omega \tag{25-16}$$

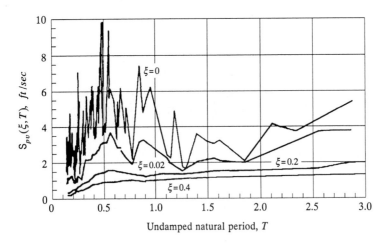

FIGURE 25-2
Pseudo-velocity response spectra, El Centro, California earthquake, May 18, 1940 (NS component).

$$\log S_{pa}(\xi, \omega) = \log S_{pv}(\xi, \omega) + \log \omega \qquad (25\text{-}17)$$

From these relations, it is seen that when a plot is made with $\log S_{pv}(\xi, \omega)$ as the ordinate and $\log \omega$ as the abscissa, Eq. (25-16) is a straight line with slope of $+45°$ for a constant value of $\log S_d(\xi, \omega)$ and Eq. (25-17) is a straight line with slope of $-45°$ for a constant value of $\log S_{pa}(\xi, \omega)$. Thus, a four-way log plot as shown in Fig. 25-3 allows all three types of spectra to be illustrated on a single graph. When interpreting such plots, it is important to note the following limiting values:

$$\lim_{\omega \to 0} S_d(\xi, \omega) = \left[v_g(t)\right]_{\max} \qquad (25\text{-}18)$$

$$\lim_{\omega \to \infty} S_{pa}(\xi, \omega) = \left[\ddot{v}_g(t)\right]_{\max} \qquad (25\text{-}19)$$

These limiting conditions mean that all response spectrum curves on the four-way log plot, as illustrated in Fig. 25-3, approach asymptotically the maximum ground displacement with increasing values of oscillator period (or decreasing values of frequency) and the maximum ground acceleration with decreasing values of oscillator period (or increasing values of frequency) for typical values of damping ratio, say $\xi < 0.20$.

It is evident that the above response spectra provide a much more meaningful characterization of earthquake ground motion, as related to structural response, than

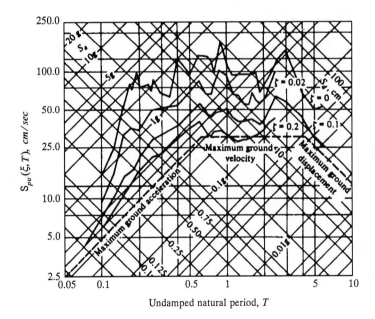

FIGURE 25-3
Pseudo-velocity response spectra for El Centro, California earthquake, May 18, 1940
(*NS* component).

does any single quantity such as Modified Mercalli intensity or peak ground acceleration (PGA). In fact, these response spectra show directly the extent to which real SDOF structures (with specific values of damping ratio and natural period) respond to the input ground motion. The only limitation in their application is that the response must be linear elastic because linear response is inherent in the Duhamel integral of Eq. (25-6). Therefore, such response spectra cannot accurately represent the extent of damage to be expected from a given earthquake excitation, inasmuch as damage involves inelastic (nonlinear) deformations. Nevertheless, the maximum amount of elastic deformation produced by an earthquake is a very meaningful indication of ground motion intensity. Moreover, such response spectra indicate the maximum deformations for all structures having periods within the range for which they were evaluated; hence, the integral of a single response spectrum over an appropriate period range can be used as an effective measure of ground motion intensity. Housner[3] originally introduced such a measure of ground motion intensity when he suggested defining the integral of the pseudo-velocity response spectrum over the period range $0.1 < T < 2.5$ *sec* as the spectrum intensity:

$$SI(\xi) \equiv \int_{0.1}^{2.5} S_{pv}(\xi, T) \, dT \tag{25-20}$$

As indicated, this integral can be evaluated for any desired damping ratio; however, Housner recommended using $\xi = 0.20$.

Structural engineers normally use the previously defined response spectra to characterize earthquake ground motion in terms of its maximum influence on simple oscillators; however, often it is helpful in understanding the characteristic features of such response to generate the Fourier spectrum for the ground motion as given by

$$\ddot{V}_g(i\overline{\omega}) \equiv \int_{-\infty}^{\infty} \ddot{v}_g(t) \, \exp[-i\overline{\omega}t] \, dt \tag{25-21}$$

This allows one to express the ground acceleration through the superposition of a full spectrum of harmonics as indicated by the inverse relation

$$\ddot{v}_g(t) = \frac{1}{2\pi} \int_{-\infty}^{\infty} \ddot{V}_g(i\overline{\omega}) \, \exp[i\overline{\omega}t] \, d\overline{\omega} \tag{25-22}$$

Assuming that the ground motion is nonzero only in the range $0 < t < t_1$, Eq. (25-21) can be separated into its real and imaginary parts as follows:

$$\ddot{V}_g(i\overline{\omega}) = \int_{0}^{t_1} \ddot{v}_g(t) \, \cos\overline{\omega}t \, dt - i \int_{0}^{t_1} \ddot{v}_g(t) \, \sin\overline{\omega}t \, dt \tag{25-23}$$

[3] G. W. Housner, "Spectrum Intensities of Strong Motion Earthquakes," *Proc. of the Symposium on Earthquake and Blast Effects on Structures*, Earthquake Engineering Research Institute, Los Angeles, 1952.

Usually, one is primarily interested in the Fourier amplitude spectrum $|\ddot{V}_g(i\overline{\omega})|$ defined by

$$|\ddot{V}_g(i\overline{\omega})| \equiv \left[C(\overline{\omega})^2 + S(\overline{\omega})^2 \right]^{1/2} \tag{25-24}$$

where $C(\overline{\omega})$ and $S(\overline{\omega})$ are the cosine and sine transforms of $\ddot{v}_g(t)$ defined by the first and second integrals, respectively, on the right hand side of Eq. (25-23). The quantity $|\ddot{V}_g(i\overline{\omega})|$ in Eq. (25-24), when divided by 2π, is the ground acceleration amplitude intensity at frequency $\overline{\omega}$ per unit of $\overline{\omega}$. Of lesser importance than the Fourier amplitude spectrum, but sometimes of interest, is the Fourier phase spectrum defined by

$$\theta(\overline{\omega}) \equiv -\tan^{-1} \left[\frac{S(\overline{\omega})}{C(\overline{\omega})} \right] \tag{25-25}$$

Note that the Fourier amplitude spectrum [Eq. (25-24)] alone does not uniquely define a ground motion time-history since the phase angles between pairs of harmonics have been lost through its definition. However, the complex Fourier spectrum, Eq. (25-21), does uniquely define the ground motion time-history as indicated by Eq. (25-22). Seismologists often find both the Fourier amplitude and phase spectra to be useful in interpreting the various phenomena associated with the transmission of energy from the earthquake source to distant locations.

25-2 FACTORS INFLUENCING RESPONSE SPECTRA

If one generates sets of response spectrum curves, as illustrated in Fig. 25-3, for ground motions recorded at different locations during past earthquakes, large variations will be observed in both the response spectral values and the shapes of the spectrum curves from one set to another. These variations depend upon many factors, such as energy release mechanism in the vicinity of the focus or "hypocenter" and along fault interfaces, epicentral distance and focal depth, geology and variations in geology along energy transmission paths, Richter magnitude, and local soil conditions at the recording station.[4,5,6,7] Thus, the response spectral values S (S_{pv}, S_{pa}, and S_d) for earthquake ground motions should be thought of in the form

$$S = S(SM, ED, FD, GC, M, SC, \xi, T) \tag{25-26}$$

---

[4] H. B. Seed and I. M. Idriss, "Ground Motions and Soil Liquefaction During Earthquakes," Monograph published by the Earthquake Engineering Research Institute, 1982.

[5] N. M. Newmark and W. J. Hall, "Earthquake Spectra and Design," Monograph published by the Earthquake Engineering Research Institute 1982.

[6] G. W. Housner, "Design Spectrum," *Earthquake Engineering*, Chapter 5, Ed. R. L. Wiegel, Prentice-Hall, Englewood Cliffs, N. J., 1970.

[7] G. W. Housner, "Properties of Strong Ground Motion Earthquakes," *Bull. Seismological Society of America*, Vol. 45, No. 3, July, 1955.

where the independent variables denote source mechanism, epicentral distance, focal depth, geological conditions, Richter magnitude, soil conditions, damping ratio, and period, respectively. The effects of SM and GC on both spectral values and shapes of the response spectrum curves are not well understood; therefore, such effects cannot be quantified when defining response spectra for design purposes. The effects of ED, FD, and M are usually taken into consideration when specifying the intensity levels of the design response spectra; however, they are often ignored when specifying the shapes of these spectra because of lack of knowledge as to their influences. On the other hand, the effects of SC on both the intensities and shapes of response spectra are now being considered widely when defining design response spectra. Because of the above considerations, modern design response spectrum curves, when normalized to a fixed intensity level, are usually specified in terms of only two parameters as indicated by

$$S = S(SC, \xi) \tag{25-27}$$

Therefore, the subsequent discussion in this section will concentrate on the influence of local soil conditions on the shapes of response spectrum curves.

For many years, the shear beam model shown in Fig. 25-4 has been used to deterministically study the influence of local soil conditions on the characteristics of horizontal free-field surface ground motions. This is the shear deformation equivalent of the normal stress wave propagation model described in Section 19-5; it assumes that strain energy is transmitted in the form of pure shear waves propagating in the vertical direction. In the earlier years, it was common practice to assume a fixed lower boundary condition for this model; however, in recent years the more realistic viscous or impedance boundary condition has been introduced, as is discussed later in Chapter 27.

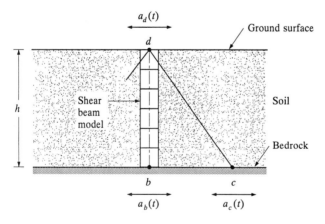

FIGURE 25-4
The shear beam model used for soil response analyses.

Referring to Fig. 25-4, if the soil layer is of reasonably uniform thickness extending over a large horizontal distance and if the bedrock motion is horizontal with no significant variations over a distance bc, where $bc \gg h$, then the shear beam assumption described above is quite valid. However, when there are significant phase differences in the bedrock motions from point b to point c and when distance bc is of the same order of magnitude as h, or when Rayleigh waves and horizontally propagating shear waves, for example, dominate the free-field surface ground motions, then the shear beam model is not realistic. Because of these and other complicating factors, one cannot rigorously justify, in general, the shear beam model for studying the influence of local soil conditions on the characteristics of free-field surface motions.

An alternative approach to characterizing the influence of local soil conditions on free-field surface motions is the direct statistical approach whereby the characteristics of numerous recorded motions are correlated with soil conditions at the recording stations. To carry out such studies effectively, one would like to have many recorded accelerograms for each soil type (hard to soft) with all other influencing factors, such as source mechanism, magnitude, epicentral distance, etc., held constant. Unfortunately, sufficient strong motion records are not available to meet this requirement; therefore, rigor must be relaxed in order to obtain usable results.

A very valuable study was carried out by H. B. Seed et al. using the above mentioned statistical approach,[8] and some of their most revealing findings are shown in Figs. 25-5 and 25-6. As indicated in these figures, Seed and his colleagues analyzed 104 strong motion accelerograms, 15 recorded on soft to medium clay and sand, 30 recorded on deep cohesionless soils, 31 recorded on stiff soil conditions, and 28 recorded on rock. Figure 25-5 shows the average pseudo-acceleration response spectra for each of the four soil conditions, normalized with respect to peak ground acceleration. Figure 25-6 shows the corresponding 84 percentile (median plus one standard deviation) spectra. All spectra in these two figures are for 5 percent critical damping, i.e., $\xi = 0.05$.

While the above described spectrum curves definitely show a correlation of spectral values with type of soil condition, one should be careful in judging this correlation. First, even if one ignores the variations caused by such influencing factors as fault mechanism, epicentral distance, and magnitude on spectral shapes, which are indeed present in the results of Figs. 25-5 and 25-6, the statistical correlations between spectral shapes and soil types show large dispersions of the spectral values.[9] To

---

[8] H. B. Seed, C. Ugas, and J. Lysmer, "Site-Dependent Spectra for Earthquake Resistant Design," *Bull. of the Seismological Society of America*, Vol. 66, No. 1, February, 1976.

[9] J. Penzien, "Statistical Nature of Earthquake Ground Motions and Structural Response," Proc. U.S.-Southeast Asia Symposium on Engineering for Natural Hazards Protection, Manila, Philippines, September, 1977.

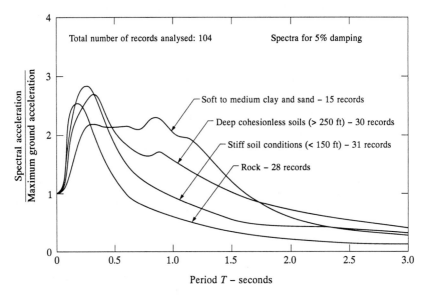

FIGURE 25-5
Average pseudo-acceleration spectra for different site conditions (by Seed et al.).

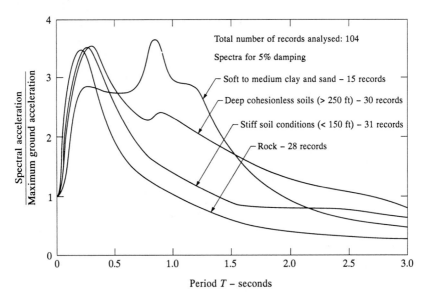

FIGURE 25-6
84 percentile pseudo-acceleration spectra for different site conditions (by Seed et al.).

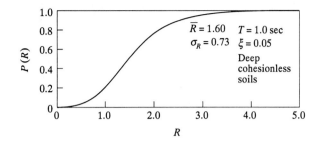

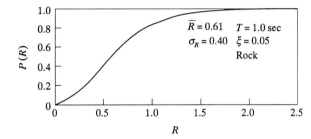

FIGURE 25-7
Gumbel Type I probability distribution functions for response ratio R for two soil types.

elaborate on this point, let us examine the probability distribution function (cumulative probability) for the acceleration spectral value in two particular cases: (1) the case for $T = 1.0$ sec, $\xi = 0.05$, and a deep cohesionless soil type and (2) the case for $T = 1.0$ sec, $\xi = 0.05$, and rock as the soil type. Fig. 25-7 shows the Gumbel Type I extreme-value distribution $P(R)$ for both cases where R is the normalized pseudo-acceleration spectral value. (Note the factor of 2 difference in scale for R between the two plots.) It is significant to recognize the large overlapping dispersions of the spectral values for these two cases. The second reason for being cautious when interpreting the results of Figs. 25-5 and 25-6 is that the effects of other factors, such as magnitude and epicentral distance, have been averaged out. Housner[10] has pointed out that both magnitude and epicentral distance can significantly influence the shapes of the response spectrum curves. Source mechanism differences can also influence these shapes. For example, the response spectrum shapes for earthquake ground motions recorded in California are significantly different from the response spectrum shapes for earthquake ground motions recorded in the SMART-1 strong motion array located in Lotung, Taiwan.[11] Undoubtedly, this difference is, at least in part, caused by the differences in source mechanisms. The shallow lateral fault slippage along the San Andreas fault is obviously quite different from the slippage

[10] G. W. Housner, "Design Spectrum," *Earthquake Engineering*, Chapter 5, Ed. R. L. Weigel, Prentice-Hall, Englewood Cliffs, N. J., 1970.

[11] C. H. Loh, J. Penzien, and Y. B. Tsai, "Engineering Analysis of SMART-1 Array Accelerograms," *Jour. of Earthquake Engineering and Structural Dynamics*, Vol. 10, 1982.

caused by deep thrusting of the Philippine crustal plate under Taiwan.

Clearly, there is need for additional strong ground motion data and for additional studies to further clarify the effects of the various factors mentioned above on the intensity and shapes of response spectrum curves. In developing response spectra for design application, one should place emphasis on strong ground motions recorded in the region of the site where the spectra are to be applied. If such data are lacking, they must be supplemented by data collected for near-similar conditions in other regions.

25-3 DESIGN RESPONSE SPECTRA

Dual Strategy of Seismic Design

The design of structures to perform satisfactorily under expected seismic conditions requires that realistic earthquake loadings be specified and that the structural components be proportioned to resist these and other combined loadings within the limits of certain design requirements. In regions of high seismicity, earthquake loading is often critical among the types of loading that must be considered because a great earthquake will usually cause greater stresses and deformations in the various critical components of a structure than will all other loadings combined; yet, the probability of such an earthquake occurring within the life of the structure is very low. In order to deal effectively with this combination of extreme loading and low probability of occurrence, a strategy based on the following dual criteria has usually been adopted:

(1) A moderate earthquake which reasonably may be expected to occur once at the site of a structure during its lifetime is taken as the basis of design. The structure should be proportioned to resist the intensity of ground motion produced by this earthquake without significant damage to the basic system.

(2) The most severe earthquake which could possibly ever be expected to occur at the site is applied as a test of structural safety. Because this earthquake is very unlikely to occur within the life of the structure, the designer is economically justified in permitting it to cause significant structural damage; however, collapse and serious personal injury or loss of life must be avoided.

Currently, the trend is to strengthen the second of these criteria for critical and expensive structures by calling for limited repairable damage, thus, focusing not only on life safety but on protection of financial investment as well.

In order to establish the characteristics of the design earthquake and the maximum probable earthquake for any given site, it is necessary to first study all seismological and geophysical data (instrumental and historical) available for the region of the site. From these data, supplemented by similar data for other regions as needed, earthquake loadings for both the design earthquake and the maximum probable earthquake can be developed. These loadings are usually specified in the form of prescribed response spectra as defined in Section 25-1. Such spectra will be referred to herein as design response spectra.

Usually, it is assumed that the shapes of the design spectra are the same for both the design and maximum probable earthquakes but that they differ in intensity as measured by peak ground acceleration. Thus, it has been common practice to first normalize the intensity of these design spectra to the 1 g peak acceleration level so that Eq. (25-19) becomes

$$\lim_{\omega \to \infty} S_{pa}(\xi, \omega) = 1 \ g \tag{25-28}$$

and then later to scale them down to the appropriate peak acceleration levels representing the design and maximum probable earthquakes. Once the shapes of these common normalized spectra have been developed, taking into consideration local soil conditions, appropriate scaling factors are applied representing the intensity levels of the peak free-field surface ground accelerations (PGA) produced by the design and maximum probable earthquakes.

Peak Ground Accelerations

For a site located in a region of moderate to high seismicity, the following deterministic procedure has often been used to establish the peak free-field surface ground acceleration (PGA) values for the maximum probable and design earthquakes:

(1) The locations of known active faults in the near region of the site are established. In some cases, extensive investigations are required to assess whether or not a known fault is indeed active or capable.

(2) All instrumental and historical seismicity data available for past earthquakes occurring on these active faults are carefully studied, along with the associated geological and geophysical conditions, to establish the maximum Richter magnitude possible for future earthquakes occurring along each fault. If the past history of observation of earthquakes covers a very long period of time, say many hundreds of years, the maximum magnitude possible for a given fault will most likely not greatly exceed the magnitude of the strongest earthquake experienced over the same period of time. Usually, however, the subjective judgments of specialists must be relied upon in setting the maximum probable magnitude for each fault. This is especially the case when a very long period is specified for the mean return period as in the case of a nuclear power plant where the mean return period for the maximum probable earthquake is commonly specified in the range 10^3 to 10^4 years.

(3) Empirical attenuation relations, expressing the median (50 percentile) or mean PGA as functions of source-to-site distance R and local Richter magnitude M, are established for the region of the site. Since the true attenuation relation is unknown, several possible forms of attenuation judged appropriate to the local

region are selected. For example, the Campbell relation of the form[12]

$$a = b_1 \left[R + b_4 \exp(b_5 \, M) \right]^{-b_3} \exp(b_2 \, M) \qquad (25\text{-}29)$$

may be selected along with other forms. The constants appearing in such relations are determined through nonlinear regression analyses using, to the extent possible, strong motion data recorded in the local region of the site. If such data are lacking, supplementary data for other similar regions must be used. Having established a possible set of attenuation relations, the subjective judgments of specialists are then used in assigning relative weights to the individual forms selected.

(4) Using the shortest distance from the site to each active fault as source-to-site distance R, the corresponding maximum possible magnitude established in Step 2 above as M, and the attenuation relations with their respective relative weights as selected in Step 3, a single numerical value is obtained for the median or mean PGA expected at the site. Using the largest of such values obtained for all active faults under consideration, the design PGA is established for the maximum possible earthquake. This PGA is then used to scale down the normalized (1 g) response spectra to the appropriate level representing the maximum probable earthquake.

(5) For each active fault, the Richter magnitude is estimated corresponding to the mean frequency of occurrence of earthquakes having magnitudes greater than M being equal to one event per life span of the structure. Using this magnitude for each active fault, instead of the corresponding maximum probable magnitude, Step 4 above is repeated to obtain the PGA to be used as the scale factor for the design earthquake in applying the normalized (1 g) response spectra.

While the predominantly deterministic approach described above for setting the PGA values for the maximum probable and design earthquakes has often been used in the past, the recent emergence into engineering practice of Probability Risk Assessment (PRA) methodologies makes it possible to use a consistent probabilistic approach. The final result of this approach is a seismic hazard curve which is simply a plot expressing annual mean frequency of exceedance as a function of PGA for the site under consideration. The analytical procedure used to generate this function consists of the following steps:[13,14]

---

[12] K. W. Campbell, "Near-Source Attenuation of Peak Horizontal Acceleration," *Bull. Seismological Society of America*, Vol. 76, No. 6, December, 1981.

[13] C. A. Cornell, "Engineering Seismic Risk Analysis," *Bull. of the Seismological Society of America*, Vol. 58, No. 5, October, 1968.

[14] A. Der Kiureghian and A. H-S. Ang, "A Fault-Rupture Model for Seismic Risk Analysis," *Bull. of the Seismological Society of America*, Vol. 67, No. 4, August, 1977.

(1) The near-region around the site is divided into surface seismotectonic zones, each having similar tectonic characteristics over its area. Volume zoning below the surface may also be required to properly represent zones of high seismic activity, e.g., subduction zones. Uniform seismicity is usually assumed over each zone and geologically appropriate source (point and/or fault-rupture) models are specified. Nonuniform seismicity can however be assigned in the volume zones if sufficiently supported by field evidence. Several possible zoning schemes are normally considered.

(2) A magnitude recurrence relation of the Gutenberg-Richter type $\log N = a - b M$ is established for each zone. Constant b in this relation is obtained by a least-square fitting of an appropriate seismicity data subset obtained instrumentally, which is judged to be complete; constant a is then obtained, using both historical and instrumental data over a long period of time, by specifying a relatively large value of M for which the corresponding value of N is believed to be known accurately.

(3) An upper bound magnitude is established for each zone based primarily on available recorded and historical seismicity data but supplemented by the subjective judgments of specialists.

(4) Attenuation relations are selected for the region of the site by the procedure previously described and lognormal distributions of PGA are assumed. A rupture-length/magnitude relation is sometimes established, using all pertinent data available for the near-region of the site, representing mean and upper and lower bound values. This relation should always be compared with similar relations based on world-wide data so as to judge its appropriateness.[15]

(5) An upper bound acceleration, which represents some reasonable physical upper bound value for the site, should be specified using the subjective judgments of experts. This acceleration is used to truncate the seismic hazard curves.

(6) Then a set of seismic hazard curves is generated by a consistent probabilistic approach using the different zoning schemes and corresponding distributions of events, the rupture-length/magnitude relation, the frequency of occurrence and attenuation relations, and the upper-bound values set for magnitudes and acceleration. After giving relative weights to the resulting seismic hazard curves following the subjective judgments of specialists, a final single seismic hazard curve is established along with its 10 and 90 percent uncertainty bounds.

(7) Once the mean annual frequencies of exceedance have been specified for the PGA values of the design and maximum probable earthquakes, their values can

[15] M. G. Bonilla, "A Review of Recently Active Faults in Taiwan," U.S. Department of the Interior, Geological Survey, Open-File Report 75-41, 1975.

be taken directly from the final seismic hazard curve for use in scaling the normalized (1 g) response spectrum curves. When normalized response spectra are used for nuclear power plant design, they are scaled down to the appropriate design and maximum probable intensity levels (PGA) known as the OBE (Operating Basis Earthquake) and SSE (Safe Shutdown Earthquake) g-levels, respectively. In the past, it has been common practice to establish the design SSE g-level and then reduce it by one-half to set the corresponding OBE g-level. Recent trends, however, indicate that the OBE g-level may be reduced somewhat below this level in the future so that the SSE, rather than the OBE, governs the design of piping and equipment. When using the consistent probabilistic approach, both the SSE and OBE g-levels can be taken directly from the seismic hazard curve after specifying their annual frequencies of exceedance. A value in the range 10^{-3} to 10^{-4} is generally considered acceptable for the SSE while a value of 10^{-2} is reasonable for the OBE. It has also been common practice in the past to set the vertical design OBE and SSE g-levels at $2/3$ their corresponding values for horizontal motions. However, in some cases, they have been set at equal levels consistent with available ground motion data representative of the site. Thrust faulting very near the site is usually the reason for increasing the relative vertical component acceleration in this way.

Response Spectrum Shapes

To establish the shapes of the normalized (PGA=1 g) response spectrum curves, the Newmark-Hall approach (or some variation thereof) has often been used.[16] Comparing the solid response spectrum curves in Fig. 25-3 with the dashed curve showing maximum values of ground displacement, velocity, and acceleration, it appears that the smoothed response spectrum curves are essentially scalar amplifications of the maximum values of ground displacement, velocity, and acceleration in their respective frequency (or period) ranges as indicated. This observation led Newmark and Hall to suggest using piecewise linear plots for each response spectrum curve as shown in Fig. 25-8 where the three linear portions of the curve below $f = f_2$ are simply amplifications of the maximum values of ground displacement, velocity, and acceleration in their respective frequency ranges. Following this suggestion, site-dependent design response spectrum curves can be established using the following steps:

(1) Establish the expected PGA values for the design and maximum probable earthquakes following one of the two procedures previously outlined.

(2) Calculate the corresponding expected peak values of ground velocity and dis-

---

[16] N. M. Newmark and W. J. Hall, "Earthquake Spectra and Design," loc. cit.

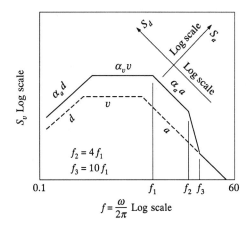

FIGURE 25-8
Design response spectrum.

placement using the relations[17,18]

$$v = c_1 \frac{a}{g} \qquad\qquad d = c_2 \frac{v^2}{a} \qquad (25\text{-}30)$$

where a, v, and d represent the expected maximum values of ground acceleration (PGA), velocity, and displacement, respectively; g is the acceleration of gravity; and constants c_1 and c_2 are selected appropriately for the known site conditions based on the results of statistical analyses of numerous strong ground motion accelerograms, giving preference to those accelerograms recorded in the local region of the site. In his 1982 paper, Hall suggested using for horizontal ground motions $c_1 = 48$ *in/sec* and 36 *in/sec* for competent soil and rock sites, respectively. Further, he suggested using $c_2 = 6$ for both soil conditions in this case. Mohraz in 1976 reported mean, 50 percentile, and 84.1 percentile values for c_1 and c_2 (assuming lognormal distributions) for different site conditions based primarily on California earthquake data.

(3) Having established numerical values for a, v, and d, multiply them by their corresponding amplification factors α_a, α_v, and α_d, respectively. Then plot the results on four-way log plots as indicated in Fig. 25-8. Since the amplification factors for both horizontal and vertical motions are random quantities, they must be obtained through statistical analyses of strong ground motion data representative of the local site conditions. Mohraz has published numerous results of

---

[17] W. J. Hall, "Observations on Some Current Issues Pertaining to Nuclear Power Plant Seismic Design," *Nuclear Engineering and Design*, North-Holland Publishing Co., Vol. 69, 1982.

[18] B. Mohraz, "A Study of Earthquake Response Spectra for Different Geological Conditions," *Bull. of the Seismological Society of America*, Vol. 66, No. 3, June, 1976.

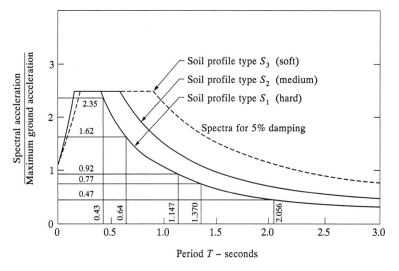

Spectral acceleration / Maximum ground acceleration vs Period T – seconds

FIGURE 25-9
ATC-3 normalized response spectra recommended for use in building code.

this type, generated under the assumption of a lognormal distribution. It has been common practice to use amplification factors at the 84.1 percentile level in combination with mean or 50 percentile (median) values of maximum ground acceleration, velocity, and displacement. Assuming a lognormal distribution, the 84.1 percentile level represents the median-plus-one-standard-deviation level.

(4) Knowing that any response spectrum curve must approach asymptotically the peak ground acceleration as the SDOF system frequency $f = \omega/2\pi$ increases toward large values, as previously indicated by Eq. (25-19), the design response spectrum curve as shown in Fig. 25-8 is forced to approach the maximum ground acceleration in a linear fashion when going from frequency f_2 to f_3. Statistical analyses show that these frequencies are often reasonably well predicted using the approximate relations

$$f_2 \doteq 4 \, f_1 \qquad\qquad f_3 \doteq 10 \, f_1 \qquad\qquad (25\text{-}31)$$

where f_1 is that frequency on the response spectrum curve corresponding to the intersection point of the amplified ground acceleration and amplified ground velocity.

This general approach has been used to establish normalized design response spectra for building codes, including those recommended by ATC-3 as shown in Fig. 25-9.[19]

---

[19] Applied Technology Council (ATC), "Tentative Provisions for the Development of Seismic Regulations for Buildings," ATC Publication ATC3-06, NBS Special Publication 510, and NSF Publication 78-8, 1978.

It should be emphasized that when developing such spectra, every effort should be made to use values of a, c_1, c_2, α_a, α_v, and α_d established from strong ground motion data representative of the region they represent. Often to finalize these numerical values, subjective judgments must be made based on rational assessments of all factors involved.

When setting design response spectra for regions having very deep soft soil conditions, such as those present in the Mexico City and Taipei, Taiwan, basins, the very narrow-band type of ground motions which occur should be included in the data under consideration. An example demonstrating the influence of such narrow-band motion on the shapes of response spectra is shown in Fig. 25-10. This figure shows the normalized response spectrum curves for 5 percent damping using eight accelerograms recorded in Taipei during the earthquake of November 14, 1986. The recorded accelerograms had predominate frequency contents centered around the period $T = 1.65$ *sec* which is reflected in the unique peaking of the spectra in the near neighborhood of this period. Using these eight spectrum curves to set a smooth design spectrum, a curve similar to that shown in Fig. 25-11 would result. Note that the segment of the spectrum usually controlled by velocity has disappeared.

In the above discussion, the general focus has been on procedures for setting the shapes of the response spectra for horizontal free-field surface motions representing the design and maximum probable earthquake events. These same procedures can, of course, be used for the vertical free-field surface motions if sufficient vertical strong motion data are available. In doing so, it should be recognized that the shapes of

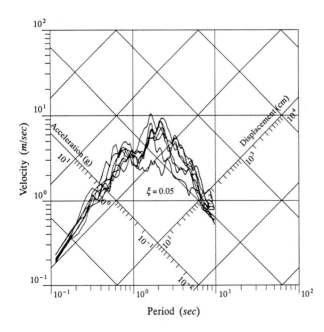

FIGURE 25-10
Normalized response spectrum curves for $\xi = 0.05$ using eight accelerograms recorded in Taipei, Taiwan during the earthquake of November 14, 1986.

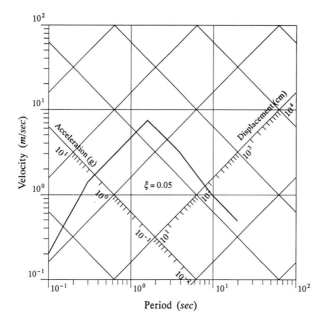

FIGURE 25-11
Normalized design response
spectrum curve for $\xi = 0.05$
representing Taipei, Taiwan,
soft soil conditions.

the design spectra for vertical motions differ from those for horizontal motions. For example, note the shape differences between the two sets of normalized (PGA=1 g) USNRC (United States Nuclear Regulatory Commission) design spectra shown in Fig. 25-12 which have been used extensively in the USA and many other countries of the world in designing nuclear power plants.[20,21] Both sets of curves in this figure represent a variation on the above described Newmark-Hall procedure for developing design spectra in that the straight lines in the velocity controlled region of frequencies are not horizontal. Negative slopes have been assigned to these lines in this case to provide a better fit to the family of normalized real spectra used in the statistical analysis. The shapes of all curves in Fig. 25-12 represent spectral values at the mean-plus-one-standard-deviation levels.

When using piecewise linear design spectra, as represented on 4-way log plots, one should avoid taking spectral values directly from such plots as the nonlinear interpolation required along the log scales cannot be done visually with acceptable accuracy. It is recommended therefore that the desired spectral values for specified periods be calculated directly from the equations representing the straight line segments. For each straight line segment having one of the three spectra as a constant,

[20] N. M. Newmark, J. A. Blume, and K. K. Kapur, "Seismic Design Spectra for Nuclear Power Plants," *Jour. Power Division, ASCE*, Vol. 99, No. PO2, November, 1973.

[21] U. S. Atomic Energy Commission, Regulatory Guide 1.60, "Design Response Spectra for Seismic Design of Nuclear Power Plants," Revision 1, December, 1973.

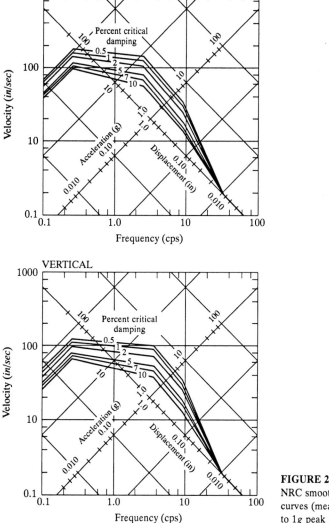

FIGURE 25-12
NRC smooth design response spectrum curves (mean $+1\sigma$ levels) normalized to $1g$ peak ground acceleration.

its conversions to representations of the other two spectra are easily carried out using Eqs. (25-14) and (25-15).

Uniform-Hazard Site-Specific Response Spectra

It should be recognized that response spectra of the type shown in Fig. 25-12 have two significant deficiencies: (1) they are independent of local site conditions and (2) they do not represent the same probability of exceedance over the full frequency range of interest. Because of these deficiencies, every effort should be made to establish site-specific spectra which do represent a uniform probability of exceedance over the entire frequency range. To establish these spectra, a statistical analysis is required using sets of response spectra generated for a family of accelerograms characterizing the expected ground motions at the site. Since the number of available accelerograms recorded in the near-region of the site is usually quite limited, the family should be supplemented using recorded accelerograms obtained in other regions having similar geological, seismological, and geophysical (GSG) conditions. All recorded accelerograms selected should be scaled individually in an appropriate manner so that they are representative of the single magnitude and single source-to-site distance of the earthquake under consideration and the specific site being represented.[22,23,24,25] To adjust for magnitude and site condition differences, the scaling is accomplished by Fourier transforming each accelerogram and then multiplying the real and imaginary parts of each transformed result by the appropriate frequency dependent scaling factors. At this point, a similar multiplication by a third scaling factor, which is frequency independent, is made to adjust for differences in the source-to-site distance. This scaling factor follows the attenuation relation expressing the mean (or median) PGA as a function of magnitude and source-to-site distance. Having applied all three scaling factors, the results are inverse Fourier transformed back to the time domain giving the adjusted accelerograms. If the number of recorded accelerograms in the family are insufficient for statistical analysis purposes, synthetic accelerograms generated from a stochastic model can be used provided the model is carefully selected to represent the design earthquake and local site conditions. Having the final full set of accelerograms representing the expected motions at the specific site, response spectra for each accelerogram are generated. Statistical analyses are then carried out on all sets of spectra to obtain the probability distributions on PGA and on spectral values at discrete frequencies over the frequency range of interest. Spectra selected at a specified percentile level from these results represent uniform probability of exceedance over the entire frequency range. The development of site-specific design spectra, by the above procedure, should be carried out separately for the vertical and horizontal ground motions.

22 H. B. Seed, C. Ugas, and J. Lysmer, "Site-Dependent Spectra for Earthquake Resistant Design," loc. cit.

23 R. K. McGuire, "A Simple Model for Estimating Fourier Amplitude Spectra of Horizontal Ground Acceleration," *Bull. of the Seismological Society of America*, Vol. 68, No. 3, June, 1978.

24 M. D. Trifunac, "Preliminary Empirical Model for Scaling Fourier Amplitude Spectra of Strong Ground Acceleration in Terms of Earthquake Magnitude, Source-to-Site Distance, and Recording Site Conditions," *Bull. of the Seismological Society of America*, Vol. 66, No. 4, August, 1976.

25 D. M. Boore, "Stochastic Simulation of High-Frequency Ground Motions Based on Seismological Models of the Radiated Spectra," *Bull. of the Seismological Society of America*, Vol. 73, No. 6, December, 1983.

An alternative approach to generating site-specific response spectra representing a uniform probability of exceedance over the entire frequency range is similar to that described above with one exception. Instead of using the attenuation relation to adjust for differences in the source-to-site distances, all accelerograms obtained after adjustments for magnitude and site condition differences are normalized to the $1\,g$ PGA level. Sets of response spectra are then generated for these normalized accelerograms and statistical analyses are carried out on all sets to obtain probability distributions on the spectral values at discrete frequencies, or simply to obtain mean and 84 percentile levels. However, because the commonly used 84 percentile curves converge to the peak ground acceleration in the high-frequency range, such curves do not represent uniform probabilities of exceedance over the entire frequency range when scaled down using the mean (or median) PGA for the design or maximum probable earthquake. For this reason, it is more rational to use the mean normalized (PGA $= 1\,g$) response spectra and then scale them down using the 84 percentile PGA. This latter approach leads to reasonably uniform probabilities of exceedance over the entire frequency range.

Recently the trend is to establish uniform-hazard site-specific response spectra by first establishing sets of seismic hazard curves, each of which expresses annual mean frequency of exceedance as a function of response spectral value for a specified discrete value of frequency (or period) and a specified discrete value of damping. The procedure for doing this is the same as that described previously for generating seismic hazard curves for PGA; however, the computational effort is much greater due to the large number of discrete values of frequency and damping involved. Having the sets of hazard curves, response spectra for a specified probability of exceedance over the entire frequency range are easily established.

Two Horizontal Components of Motion

If a structure is to be designed using two simultaneous orthogonal components of horizontal free-field ground acceleration, the normalized design spectra previously described for single component horizontal motion can be used for both components; however, consistent with strong ground motion data, the design and maximum probable intensity levels of one component should be reduced about 15 percent below the corresponding intensity levels of the other component. The component of larger intensity should be directed along the critical axis of the structure.

25-4 DESIGN ACCELEROGRAMS

As explained in the previous section, the design and maximum probable earthquake ground motions are usually specified in terms of design response spectra. Assuming linear structural systems, these spectra can be used to obtain corresponding maximum response levels through standard modal analyses which are discussed in

Chapters 26 and 27. In many cases, however, time-history dynamic analyses must be carried out in predicting maximum response levels. Various reasons exist for having to do this. For example, under maximum probable earthquake conditions, most structures will experience damage which means that such structures respond in a nonlinear manner. Thus, the linear modal analysis approach does not apply, and time-history nonlinear analyses may be required. In other cases, where linear response analyses are acceptable, the complexity and/or nature of structural modeling may be such as to require time-history dynamic analyses. Extreme complexities in structural geometries, causing difficulty in combining modal contributions to response, is one such case. Modeling, which contains critical frequency dependent parameters, is another. Regardless of the reason for requiring any particular time-history dynamic analysis, the earthquake inputs must be specified in terms of free-field ground motion accelerograms. Since the design and maximum probable earthquake free-field ground motions are usually specified in terms of smooth design response spectra, accelerograms used for time-history dynamic analyses should be compatible with these spectra.

Response Spectrum Compatible Accelerograms

To generate a synthetic ground motion accelerogram compatible with a response spectrum, the following steps can be used:

(1) By computer, generate a set of random numbers, denoted by x_1, x_2, x_3, $\cdots$, x_n, consistent with a uniform probability density function of intensity 1 over the range zero to $+1$. Computer programs are readily available for carrying out this step.

(2) Convert consecutive pairs of these random numbers to corresponding consecutive pairs of new random numbers using the relations

$$y_i = (-2 \ln x_i)^{1/2} \cos 2\pi x_{i+1}$$

$$(25\text{-}32)$$

$$y_{i+1} = (-2 \ln x_i)^{1/2} \sin 2\pi x_{i+1}$$

As shown in Example E20-7, numbers y_1, y_2, y_3, $\cdots$, y_n so obtained have a gaussian distribution with zero mean value and a variance of unity.

(3) A sample time function $y(t)$ can now be established by assigning the discrete values y_1, y_2, y_3, $\cdots$, y_n, obtained in Step 2 to n successive ordinates spaced at equal time intervals Δt along a time abscissa and by assuming a linear variation of ordinates over each interval. The initial ordinate y_0 at $t = 0$ is set equal to zero.

As shown in Example E21-2, the autocorrelation function for waveform $y(t)$, as defined by

$$R_y(\tau) \equiv \lim_{n \to \infty} \frac{1}{n \, \Delta t} \int_0^{n \, \Delta t} y(t) \, y(t + \tau) \, dt \qquad (25\text{-}33)$$

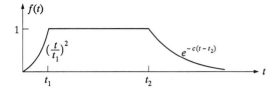

FIGURE 25-13
Intensity function $f(t)$ for nonstationary process $a(t)$.

is given by the relations

$$
R_y(\tau) = \begin{cases}
\frac{2}{3} - \left(\frac{\tau}{\Delta t}\right)^2 + \frac{1}{2}\left(\frac{|\tau|}{\Delta t}\right)^3 & -\Delta t \le \tau \le \Delta t \\[2mm]
\frac{4}{3} - 2\frac{|\tau|}{\Delta t} + \left(\frac{\tau}{\Delta t}\right)^2 - \frac{1}{6}\left(\frac{|\tau|}{\Delta t}\right)^3 & \begin{array}{l} -2\Delta t \le \tau \le -\Delta t; \\ \Delta t \le \tau \le 2\Delta t \end{array} \\[2mm]
0 & \begin{array}{l} \tau \le -2\Delta t; \\ \tau \ge 2\Delta t \end{array}
\end{cases} \qquad (25\text{-}34)
$$

and the corresponding power spectral density function, as defined by

$$
S_y(\varpi) \equiv \lim_{n\to\infty} \frac{|\int_0^{n\Delta t} y(t)\exp(-i\varpi t)\,dt|^2}{2\pi\, n\,\Delta t} \qquad (25\text{-}35)
$$

is given by

$$
S_y(\varpi) = \frac{\Delta t}{\pi}\left[\frac{6 - 8\cos\varpi\,\Delta t + 2\cos 2\varpi\,\Delta t}{(\varpi\,\Delta t)^4}\right] \qquad \varpi \ge 0 \qquad (25\text{-}36)
$$

As reported by Ruiz and Penzien,[26] this latter function is flat to within 5 percent error for $\varpi\,\Delta t < 0.57$ and to within 10 percent error for $\varpi\,\Delta t < 0.76$. It drops to 50 percent of its initial value at $\varpi\,\Delta t = 2$. For generating synthetic earthquake accelerograms, it is usually sufficient to let $\Delta t = 0.01\ sec$.

(4) Multiply the stationary-type waveform $y(t)$ obtained in Step 3 by a suitable deterministic time function $f(t)$ to convert it to a nonstationary form $z(t)$ appropriate to the magnitude and source-to-site distance of the design or maximum probable earthquake being considered. One function commonly used for this purpose is shown in Fig. 25-13.[27] Constants t_1, t_2, and c in this figure should be assigned numerical values only after considering such factors as earthquake magnitude and epicentral distance.[28] Another form which has been used for this purpose is

$$
f(t) = a_1\, t\, \exp(-a_2\, t) \qquad (25\text{-}37)
$$

[26] P. Ruiz and J. Penzien, "Probabilistic Study of Behavior of Structures during Earthquakes," University of California, Berkeley, Earthquake Engineering Research Center, Rept. 69-3, 1969.

[27] P. C. Jennings, G. W. Housner, and N. C. Tsai, "Simulated Earthquake Motions," Rept., Earthquake Engineering Research Laboratory, California Institute of Technology, April, 1968.

[28] G. W. Housner, "Design Spectrum," loc. cit.

where again the constants involved are assigned values after considering such factors as earthquake magnitude and epicentral distance.[29] For the general class of accelerograms recorded during the San Fernando, California, earthquake, statistical studies show that constants a_1 and a_2 can be assigned the values 0.45 and 1/6, respectively, giving $f(0) = 0$, $f(6) = 1$, $f(12) = 0.74$, $f(20) = 0.32$. Other forms which have been used for $f(t)$ can be found in the literature.[30]

(5) Fast Fourier Transform (FFT) the wave form $z(t)$ obtained in Step 4 to get $Z(i\overline{\omega})$ which should then be multiplied by filter functions $H_1(i\overline{\omega})$ and $H_2(i\overline{\omega})$ to get

$$B(i\overline{\omega}) \equiv Z(i\overline{\omega}) \, H_1(i\overline{\omega}) \, H_2(i\overline{\omega}) \qquad (25\text{-}38)$$

where

$$H_1(i\overline{\omega}) = \frac{\left[1 + 2\,i\,\xi_1\left(\frac{\overline{\omega}}{\omega_1}\right)\right]}{\left[\left(1 - \frac{\overline{\omega}^2}{\omega_1^2}\right) + 2\,i\,\xi_1\left(\frac{\overline{\omega}}{\omega_1}\right)\right]}$$

and $\qquad\qquad\qquad\qquad\qquad\qquad\qquad\qquad\qquad\qquad\qquad (25\text{-}39)$

$$H_2(i\overline{\omega}) = \frac{\left(\frac{\overline{\omega}}{\omega_2}\right)^2}{\left[\left(1 - \frac{\overline{\omega}^2}{\omega_2^2}\right) + 2\,i\,\xi_2\left(\frac{\overline{\omega}}{\omega_2}\right)\right]}$$

The first of Eqs. (25-39) is the well-known Kanai/Tajimi filter function which amplifies the frequency content in $Z(i\overline{\omega})$ in the neighborhood of $\overline{\omega} = \omega_1$ and increasingly attenuates the frequency content above $\overline{\omega} = \omega_1$ as $\overline{\omega} \to \infty$.[31] The second of Eqs. (25-39) greatly attenuates the very low frequencies in $Z(i\overline{\omega})$ which is necessary to correct possible drifting in time of the first and second integral functions of $z(t)$. Parameters ω_1 and ξ_1 appearing in $H_1(i\overline{\omega})$ may be thought of as some characteristic ground frequency and characteristic damping ratio, respectively. Kanai has suggested 15.6 r/sec for ω_1 and 0.6 for ξ_1 as being representative of firm soil conditions. As the soil conditions become softer, ω_1 and ξ_1 should both be adjusted appropriately to reflect changes in

---

[29] J. L. Bogdanoff, J. E. Goldberg, and M. C. Bernard, "Response of a Simple Structure to a Random Earthquake-Type Disturbance," *Bull. of the Seismological Society of America*, Vol. 51, No. 2, April, 1961.

[30] T. Kubo and J. Penzien, "Time and Frequency Domain Analyses of Three-Dimensional Ground Motions, San Fernando Earthquake," University of California, Berkeley, Earthquake Engineering Research Center, Report No. 76-6, March, 1976.

[31] K. Kanai, "Semi-empirical Formula for the Seismic Characteristics of the Ground," University of Tokyo, Bull., Earthquake Research Institute, Vol. 35, pp. 309-325, 1957; H. Tajimi, "A Statistical Method of Determining the Maximum Response of a Building Structure during an Earthquake," Proc. 2nd World Conference on Earthquake Engineering, Tokyo and Kyoto, Vol. II, pp. 781-798, July, 1960.

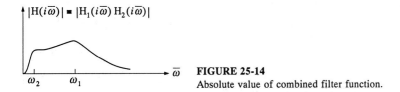

FIGURE 25-14
Absolute value of combined filter function.

the frequency content of the ground motions. Parameters ω_2 and ξ_2 appearing in $H_2(i\overline{\omega})$ must be set appropriately to produce the desired filtering of the very low frequencies. The absolute value of the product of $H_1(i\overline{\omega})$ and $H_2(i\overline{\omega})$ will have the general appearance shown in Fig. 25-14.

(6) Inverse FFT the complex function $B(i\overline{\omega})$ obtained in Step 5 to get the corresponding waveform $b(t)$ which should then be normalized by a scale factor α so that its PGA equals 1 g. Let this normalized waveform represent an accelerogram $a(t)$.

(7) Generate the pseudo-velocity response spectrum for the normalized accelerogram $a(t)$ obtained in Step 6 using a damping ratio ξ consistent with the representative structural damping ratio, say a damping ratio ξ_s. Let us denote this spectrum as $S_{pv}^a(\xi_s, T)$. The superscript a is used here to indicate that the generated spectrum is for the normalized accelerogram $a(t)$.

(8) Compare $S_{pv}^a(\xi_s, T)$ with the specified design response spectrum $S_{pv}(\xi_s, T)$ which has been normalized to the 1 g peak acceleration level. Since these two spectra will not match, except at very low periods approaching zero, an adjustment must be made to accelerogram $a(t)$ to make it spectrum compatible. After dividing the entire frequency range of interest into narrow frequency bands, each containing a number of frequency intervals as used in the FFT of $a(t)$, this adjustment is easily made by multiplying the discrete values of the real and imaginary parts of $A(i\overline{\omega})$ over each frequency band by the corresponding average of the ratios of design spectral value, $S_{pv}(\xi_s, T)$, to response spectral value of $a(t)$, $S_{pv}^a(\xi_s, T)$. Now inverse FFT the resulting adjusted complex function, to get the corresponding adjusted accelerogram. The response spectrum of this adjusted accelerogram will more closely match the specified design spectrum. An even better fitting can be achieved, if desired, by repeating this step using the response spectrum for the adjusted accelerogram instead of $S_{pv}^a(\xi_s, T)$ as indicated above. Note that the maximum or peak value of the adjusted accelerogram will not equal 1 g. Nevertheless, this adjusted accelerogram should not be normalized to that level as it already has been modified to better represent the entire smooth design spectrum.

One's success in using the above procedure is very much dependent upon the number

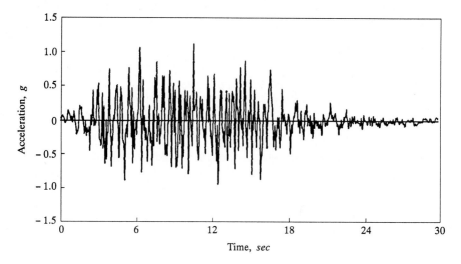

FIGURE 25-15
Synthetic accelerogram adjusted to be compatible with smooth design spectrum.

of FFT frequency intervals specified for the narrow frequency band in Step 8. Even with an optimum choice of this number, full convergence to spectrum compatibility is not achievable; however, the degree of spectrum compatibility is usually satisfactory. To illustrate, consider the spectrum compatible accelerogram generated by this procedure as shown in Fig. 25-15. Its actual response spectrum is shown in Fig. 25-16 where it can be compared with the specified smooth design spectrum. The degree of spectrum compatibility achieved in this case is quite good; however, should one desire even better spectrum compatibility, an improved procedure allowing greater convergence is required.[32]

Accelerograms generated by the above procedure are compatible with the design response spectrum for one value of damping only. Should one wish to have each accelerogram compatible with the response spectra for two values of damping, say $\xi = 0.02$ and 0.05, a more complex system of adjustments must be made to each accelerogram. Procedures and associated computer programs for making these adjustments have been developed.[32,33]

Spectrum compatible accelerograms can also be obtained by modifying real accelerograms. This is easily done by Fourier transforming each real accelerogram,

---

[32] K. Lilhanand and W. S. Tseng, "Development and Application of Realistic Earthquake Time Histories Compatible with Multiple-Damping Design Spectra," Proceedings of the Ninth World Conference on Earthquake Engineering, Tokyo/Kyoto Japan (Vol. II), August 29, 1988.

[33] M. Watabe, "Characteristics and Synthetic Generation of Earthquake Ground Motions," Proc., Canadian Earthquake Engineering Conference, July, 1987.

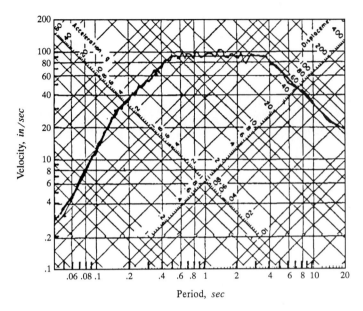

FIGURE 25-16
Smooth design response spectrum and response spectrum for adjusted synthetic accelerogram; $\xi = 0.05$.

using the FFT method, to obtain a complex function $B(i\bar{w})$. Using this complex function, carry out Steps 6, 7, and 8 exactly as indicated above to obtain the desired spectrum compatible accelerogram. Figure 25-17 shows a normalized accelerogram representing the recorded $N21°E$ component of motion recorded during the Taft, California, earthquake of 1952. Its response spectrum is shown in Fig. 25-19. After adjusting this accelerogram by the above procedure, the accelerogram shown in Fig. 25-18 is obtained. Its response spectrum is shown in Fig. 25-19 where it can be compared with the specified design spectrum. Very close agreement has been achieved.

Principal Axes of Motion

Let us now consider the generation of multiple components of spectrum compatible accelerograms representing motion at a fixed location. It is important that these components be realistically correlated with each other. To illustrate this requirement, let us consider three orthogonal components of ground acceleration $a_x(t)$, $a_y(t)$, and $a_z(t)$. A 3×3 covariance matrix μ can be generated for these components using the relation

$$\mu_{ij} \equiv \frac{1}{t_d} \int_0^{t_d} a_i(t)\, a_j(t)\, dt \qquad i, j = x, y, z \tag{25-40}$$

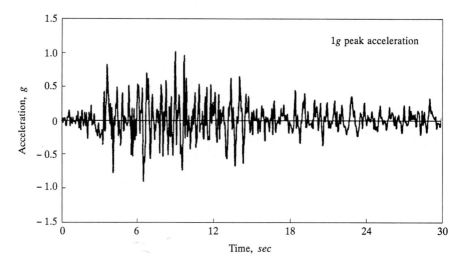

FIGURE 25-17
Normalized accelerogram – Taft California N21°E, 1952.

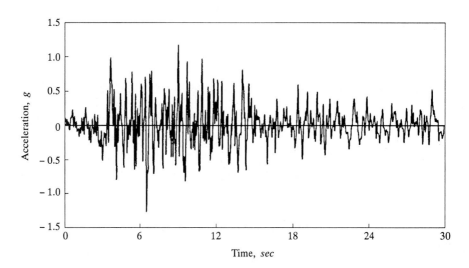

FIGURE 25-18
Taft accelerogram adjusted to be compatible with smooth design spectrum.

where t_d is a specified duration of motion which may be selected to represent any portion or all of the total duration of motion. The diagonal terms of this matrix represent the mean square intensities of motion and the off-diagonal terms represent cross correlations of the various components. Components $a_x(t)$, $a_y(t)$, and $a_z(t)$ can

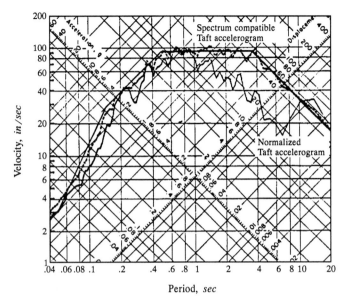

FIGURE 25-19
Smooth design response spectrum and response spectra for normalized Taft
accelerogram and adjusted Taft accelerogram; $\xi = 0.05$.

easily be transformed to a new orthogonal set of axes x', y', z' giving components of
motion $a_{x'}(t)$, $a_{y'}(t)$, and $a_{z'}(t)$ for which a covariance matrix μ' can be generated
using the corresponding relation to Eq. (25-40). It is easily shown that covariance
matrices μ and μ' are related through the orthogonal transformation given by[34]

$$\mu' = \mathbf{A}^T \, \mu \, \mathbf{A} \qquad (25\text{-}41)$$

which is identical to the transformation of the three-dimensional stress matrix from
axes x, y, z to axes x', y', z'; thus by analogy this demonstrates the existence of
principal axes for which the covariance matrix is diagonal in form. The procedure
for finding the principal axes of motion is identical to the procedure for finding the
principal axes of stress, i.e., both cases require a solution to the same eigenproblem.

It has been found that the directions of the major and minor principal axes of
recorded earthquake ground motions correlate to limited degrees with the direction
to the epicenter and the vertical direction, respectively.[33] Figure 25-20 shows the

---

[34] J. Penzien and M. Watabe, "Simulation of 3-Dimensional Earthquake Ground Motions," *Bull. of the
International Institute of Seismology and Earthquake Engineering*, Vol. (1974), and *J. of Earthquake
Engineering and Structural Dynamics*, Vol. 3, No. 4, April-June, 1975.

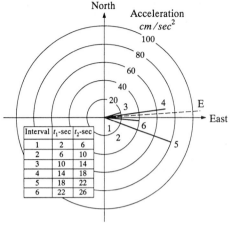

Interval	t_1-sec	t_2-sec
1	2	6
2	6	10
3	10	14
4	14	18
5	18	22
6	22	26

Tokachi-Oki, Japan
(Hachinoe Station)
May 16, 1968.

FIGURE 25-20
Directions of major principal axis of ground
motion – Tokachi-Oki, Japan, earthquake
(Hachinoe Station) May 16, 1968.

directions of the major principal axis for different time intervals $t_1 < t < t_2$ during the May 16, 1968, Tokachi-Oki, Japan, earthquake. The lengths of the solid arrows in this figure represent mean square intensities of the major principal motions over the corresponding time intervals. In this case, correlation of the major principal direction with the epicenter direction (dashed line) is quite good. While a correlation of this type usually exists for most recorded earthquake motions, it is often quite weak. Nevertheless, the above transformation and resulting correlation, even though often weak, suggest that components of ground motion generated from a stochastic model should be uncorrelated with each other, with the major axis directed towards the expected epicenter and the minor axis directed vertically.

To generate multiple components of synthetic uncorrelated acceleration representing motion at a fixed location by the previously described procedure outlined in eight steps, it is necessary that the corresponding sets of random numbers x_1, x_2, x_3, $\cdots$, x_n as defined in Step 1 be statistically independent. Since this requirement is easily satisfied, no difficulty arises in generating uncorrelated components of acceleration. For engineering purposes, one may wish to transform the uncorrelated components to a new set of directions, e.g., directions corresponding to principal axes of a structure, using the orthogonal coordinate transformation. The new set of motions obtained will possess nonzero cross correlations, if the mean square intensities of motion along the principal axes are assigned different values. However, the numerical values of these cross correlations will always be quite small compared with the mean square intensities, regardless of the changes introduced in coordinate directions, because the mean square intensities of the horizontal components of motion do not differ greatly. Consider, for example, two horizontal principal components of motion $a_{x'}(t)$ and $a_{y'}(t)$ where the mean square intensity of $a_{x'}(t)$ has been normalized to unity and the corresponding mean square intensity of $a_{y'}(t)$ is 0.72. This roughly corresponds to the intensity of $a_{y'}(t)$ being 85 percent of the intensity of $a_{x'}(t)$, i.e., $0.72 = (0.85)(0.85)$, as previously suggested for use in generating synthetic accelerograms. Transforming motions $a_{x'}(t)$ and $a_{y'}(t)$ to $a_x(t)$ and $a_y(t)$, where the x, y axes are at $45°$ to the x', y' axes, one obtains the maximum possible cross correlation between $a_x(t)$ and $a_y(t)$. Assuming mean square intensities of $a_{x'}(t)$ and $a_{y'}(t)$ to be 1.00 and 0.72, respectively, the corresponding maximum possible cross correlation of $a_x(t)$ and $a_y(t)$ is $(1.00 - 0.72)/2 = 0.14$. It is clear therefore that one should never use highly correlated components to represent earthquake ground motion at a particular point, as such components are not realistic.

Spatially Correlated Motions

If one wishes to generate multiple components of earthquake motion to represent free-field surface motions at a number of points, as might be desired when performing a time-history dynamic analysis of a very large structure such as a long multiple-span bridge, spatial variations of the ground motions should be taken into consideration. In such a case, the same set of smooth design spectra may apply to all points; however, in generating synthetic components of motion, pairs of components in the same direction representing the motions at two different points must be properly cross correlated. If the two points under consideration are very far apart, then this pair of motions will be almost totally uncorrelated in which case they can be generated statistically independent of each other by the method previously described. On the other hand, as the distance between the two points decreases, the cross correlation will increase. In the limit as the distance approaches zero, the cross correlation will approach the mean square intensity of motion in the direction under consideration. Cross correlations of pairs of components in orthogonal directions representing motions at two different points will always be very low in comparison with the corresponding mean square intensities. As indicated above, this cross correlation is always relatively low even when the distance between the two points tends to zero.

Because of the recent installation of dense strong motion arrays in various countries of the world, much data are being collected which are now allowing researchers to characterize the spatial variations of ground motions. Most of the instruments in these arrays, e.g., the SMART-1 array in Lotung, Taiwan, and the El Centro differential array in California, are located on the ground surface; thus, the progress made in characterizing spatial variations is restricted primarily to the free-field surface motions. Since these variations in surface motions are important to the seismic analysis of extended structures, spatially correlated accelerograms may be required. To develop such accelerograms, assume the free-field motions at different locations on the ground surface to be random in nature as represented by the previously described stochastic model

$$\ddot{\mathbf{v}}_g(t) = f(t)\,\mathbf{a}(t) \tag{25-42}$$

where $\ddot{\mathbf{v}}_g(t)$ is an n-component vector containing nonstationary free-field ground accelerations, $f(t)$ is an appropriate deterministic time intensity function, and $\mathbf{a}(t)$ is an n-component vector containing stationary random accelerations. These stationary random time-histories can be characterized by their $n \times n$ spectral density matrix given by

$$\mathbf{S}_a(i\overline{\omega}) = \begin{bmatrix} S_{11}(\overline{\omega}) & S_{12}(i\overline{\omega}) & \cdots & S_{1n}(i\overline{\omega}) \\ S_{21}(i\overline{\omega}) & S_{22}(\overline{\omega}) & \cdots & S_{2n}(i\overline{\omega}) \\ \vdots & \vdots & \ddots & \vdots \\ S_{n1}(i\overline{\omega}) & S_{n2}(i\overline{\omega}) & \cdots & S_{nn}(\overline{\omega}) \end{bmatrix} \tag{25-43}$$

If the random process is ergodic, as normally assumed, each function in this matrix is given by

$$S_{ij}(i\overline{\omega}) \equiv \lim_{s \to \infty} \frac{\int_{-s/2}^{s/2} a_i^r(t)\exp[-i\overline{\omega}t]\,dt \ \int_{-s/2}^{s/2} a_j^r(t)\exp[i\overline{\omega}t]\,dt}{2\,\pi\,s} \tag{25-44}$$

$$i,j = 1,2,\cdots,n \qquad\qquad r = 1,2,\cdots$$

where superscript r denotes the rth member of the random process. Since this process is ergodic, $S_{ij}(i\overline{\omega})$ is independent of r; thus, the superscript r will be dropped in the subsequent development.

It has been shown in Chapter 21 that

$$< a_i(t) \quad a_j(t) > = \int_{-\infty}^{\infty} S_{ij}(i\overline{\omega})\,d\overline{\omega}$$

$$S_{ij}(i\overline{\omega}) = S_{ij}^*(-i\overline{\omega}) \tag{25-45}$$

$$S_{ij}(i\overline{\omega}) = S_{ji}^*(i\overline{\omega})$$

where $\langle -- \rangle$ denotes time average and superscript * denotes complex conjugate. The spectral density matrix of Eq. (25-43) can be reasonably well represented in the form

$$\mathbf{S}_a(i\overline{\omega}) = \begin{bmatrix} 1 & \gamma_{12}(i\overline{\omega}) & \gamma_{13}(i\overline{\omega}) & \cdots & \gamma_{1n}(i\overline{\omega}) \\ \gamma_{21}(i\overline{\omega}) & 1 & \gamma_{23}(i\overline{\omega}) & \cdots & \gamma_{2n}(i\overline{\omega}) \\ \vdots & \vdots & \vdots & \ddots & \vdots \\ \gamma_{n1}(i\overline{\omega}) & \gamma_{n2}(i\overline{\omega}) & \gamma_{n3}(i\overline{\omega}) & \cdots & 1 \end{bmatrix} S_0(\overline{\omega}) \tag{25-46}$$

where

$$S_0(\overline{\omega}) = \frac{\left[1 + 4\,\xi_1^2(\overline{\omega}/\omega_1)^2\right](\overline{\omega}/\omega_2)^4}{\left\{\left[1 - (\overline{\omega}/\omega_1)^2\right]^2 + 4\,\xi_1^2(\overline{\omega}/\omega_1)^2\right\}\left\{\left[1 - (\overline{\omega}/\omega_2)^2\right]^2 + 4\,\xi_2^2(\overline{\omega}/\omega_2)^2\right\}}\,S_0 \tag{25-47}$$

is a frequency dependent power-spectral density function expressed in terms of a constant power spectral density function S_0, representing white noise, multiplied by the transfer function $H_1(i\overline{\omega})$, $H_1(-i\overline{\omega})$, $H_2(i\overline{\omega})$, $H_2(-i\overline{\omega})$ in which $H_1(i\overline{\omega})$ and $H_2(-i\overline{\omega})$ are given by Eqs. (25-39). As previously explained, parameters ξ_1 and ω_1 must be chosen appropriately to represent the local site condition and parameters ω_2 and ξ_2 must be set appropriately to produce the desired filtering of the very low frequencies. Further, the dimensionless coherency γ-functions must be defined consistent with the degree of correlation between acceleration time-histories $a_i(t)$ and $a_j(t)$ for $i \neq j$. If $a_i(t)$ and $a_j(t)$ represent components of acceleration in the same direction at stations i and j, the simplest coherency function one could hypothesize would be

$$\gamma(i\overline{\omega}) = \exp\left[-i\overline{\omega}\,\frac{d_{ij}}{V_a}\right] \tag{25-48}$$

which corresponds to ground motions produced by a single wave train moving with apparent wave velocity V_a without attenuation. The quantity d_{ij} in this relation is the projected distance along the direction of wave propagation between stations i and j. Such motions are fully correlated with the introduction of a time lag $\tau_{ij} = d_{ij}/V_a$.

The coherency function of Eq. (25-48) is unrealistic, however, because the free-field ground motions are produced by multiple waves having numerous reflections and refractions along their paths of propagation. Wave scattering and other unknown effects are also experienced; thus, an improved coherency relation would be

$$\gamma_{ij}(i\overline{\omega}) = \exp\left[-\alpha\,|d_{ij}|\right]\,\exp\left[-i\overline{\omega}\,\frac{d_{ij}}{V_a(\overline{\omega})}\right] \tag{25-49}$$

where the first exponential term on the right hand side represents loss of correlation with distance $|d_{ij}|$ due to unknown random effects; α is a parameter reflecting the rate of loss of correlation; and $V_a(\overline{\omega})$ is the apparent wave velocity which is frequency dependent. Numerous coherency functions, which are superior to that given by Eq. (25-49), have been developed recently using the SMART-1 array data,[35,36,37,38]

[35] C. H. Loh and J. Penzien, "Identification of Wave-Types, Directions, and Velocities Using SMART-1 Strong Motion Array Data," Proc., 8th World Conference on Earthquake Engineering, San Francisco, Ca., July 21-28, 1984.

[36] R. Harada, "Probabilistic Modeling of Spatial Variation of Strong Earthquake Ground Displacements," Proc., 8th World Conference on Earthquake Engineering, San Francisco, Ca., July 21-28, 1984.

[37] N. A. Abrahamson and B. A. Bolt, "The Spatial Variation of the Phasing of Seismic Strong Ground Motion," *Bull. of the Seismological Society of America*, Vol. 75, No. 5, October, 1985.

[38] R. S. Harichandran and E. H. Vanmarke, "Stochastic Variation of Earthquake Ground Motion in Space and Time," *ASCE, J. of Engineering Mechanics*, February, 1986.

e.g., the relation[39]

$$\gamma_{ij}(d_{ij}^L, d_{ij}^T, \overline{\omega}) = \exp\left[-\beta_1 \, d_{ij}^L - \beta_2 \, d_{ij}^T\right] \times$$

$$\exp\left[-\left(\alpha_1 \, d_{ij}^{L\ 1/2} + \alpha_2 \, d_{ij}^{T\ 1/2}\right) \overline{\omega}^2\right] \times \exp\left[i \, 2 \, \pi \, \overline{\omega} \, \frac{d_{ij}^L}{V_a}\right] \qquad (25\text{-}50)$$

where $\overline{\omega}$ is circular frequency; β_1 and β_2 are constant parameters; α_1 and α_2 are frequency dependent parameters; and d_{ij}^L and d_{ij}^T are projected distances between stations i and j in the dominant direction of wave propagation (towards the epicenter) and transverse to it, respectively. Coherency models, such as Eq. (25-50), are site-specific in their rigorous applicability, i.e., they cannot reliably be considered applicable to all sites. Nevertheless, they are very useful in estimating the spatial variations in free-field surface motions.

Should one wish to have available free-field surface acceleration time-histories $a_1(t)$, $a_2(t)$, $\cdots$, $a_n(t)$ representing motions in the same direction at discrete points over an extended distance, then their loss of coherency with distance between discrete points should be considered. If an appropriate coherency model for the site, such as given by Eq. (25-50), is established, then accelerograms which are coherency compatible can be generated using the time-domain relation[40]

$$a_i(t) = \sum_{j=1}^{i} \sum_{k=1}^{N} A_{ij}(\overline{\omega}_k) \, \cos\left[\overline{\omega}_k t + \beta_{ij}(\overline{\omega}_k) + \phi_{jk}(\overline{\omega}_k)\right] \qquad i = 1, 2, \cdots, n$$

$$(25\text{-}51)$$

for values of i and j representing components of acceleration in the same direction. In this relation

$$\phi_{jk}(\overline{\omega}_k) = \begin{cases} 0 & \text{for } j \neq i \\ \text{uniformly random} & \text{for } j = i \end{cases} \qquad (25\text{-}52)$$

and

$$\beta_{ij}(\overline{\omega}_k) = 0 \qquad \text{for} \qquad j = i$$

For each discrete harmonic in Eq. (25-51) at frequency $\overline{\omega}_k$, i.e., for harmonic

$$a_{ik}(t) = \sum_{j=1}^{i} A_{ij}(\overline{\omega}_k) \, \cos\left[\overline{\omega}_k \, t + \beta_{ij}(\overline{\omega}_k) + \phi_{jk}(\overline{\omega}_k)\right] \qquad i = 1, 2, \cdots \qquad (25\text{-}53)$$

---

[39] H. Hao, C. S. Oliveira, and J. Penzien, "Multiple-Station Ground Motion Processing and Simulation Based on SMART-1 Array Data," *Nuclear Engineering and Design 111*, North-Holland, Amsterdam, 1989.

[40] E. Samaras, M. Shinozuka, and A. Tsurui, "Time Series Generation Using the Auto-Regressive Moving-Average Model," Technical Report, Department of Civil Engineering, Columbia University, New York, May, 1983.

one must solve for the $2i$ unknowns $A_{i1}, A_{i2}, \cdots, A_{ii}$ and $\beta_{i1}, \beta_{i2}, \cdots, \beta_{ii}$ for each value of i using the specified coherency function $\gamma_{ij}(i\bar{\omega})$ and the complex relations

$$S_0(\bar{\omega}_k) \left[\gamma_{ij}(i\bar{\omega}_k) + \gamma_{ij}(-i\bar{\omega}_k)\right] \Delta\bar{\omega} = \langle a_{ik}(t) \quad a_{jk}(t) \rangle$$

$$i, j = 1, 2, \cdots, n \qquad j \leq i \qquad (25\text{-}54)$$

where $\Delta\bar{\omega}$ is the constant frequency interval between the discrete harmonics in $a_i(t)$. The complete solution is achieved by successively solving these relations in the sequential order of i starting with station 1 $(i = 1)$ and proceeding to the last station n $(i = n)$. By expressing Eqs. (25-54) in a matrix equation form, one can easily see that the right side of the resulting equation represents a lower-triangular matrix containing the A_{ij} and β_{ij} coefficients and the solution procedure described above is equivalent to performing a Choleski factorization of the complex coefficient matrix γ_{ij}, $i, j = 1, \cdots, n$, which is a hermitian matrix, i.e., the conjugate and transpose of the matrix equals the matrix itself. Having obtained $a_1(t), a_2(t), \cdots$ in that sequential order by the above procedure, each stationary acceleration time-history $a_i(t)$ $(i = 1, 2, \cdots)$ is made nonstationary using Eq. (25-42), and finally, each nonstationary component so obtained is made response spectrum compatible by the method described above whereby the amplitudes of the harmonics in the motions are iteratively adjusted while the phase angles, which totally control their coherencies, are held to their original fixed values. By this procedure, one can obtain the time-histories of free-field surface accelerations in vector $\ddot{\mathbf{v}}_g(t)$ which are both properly correlated spatially with each other and are response spectrum compatible.

CHAPTER
26

DETERMINISTIC EARTHQUAKE RESPONSE: SYSTEMS ON RIGID FOUNDATIONS

26-1 TYPES OF EARTHQUAKE EXCITATION

A special feature of earthquake excitation of structures, compared with most other forms of dynamic excitation, is that it is applied in the form of support motions rather than by external loads; thus, the effective seismic loadings must be established in terms of these motions. Defining the support motions is the most difficult and uncertain phase of the problem of predicting structural response to earthquakes. When these input motions have been established, however, the calculation of the corresponding stresses and deflections in any given structure is a standard problem of structural dynamics which can be carried out by the techniques described earlier in Parts One through Four. This chapter will discuss deterministic methods of earthquake response analysis, which provide valuable insights into the seismic behavior. In some cases, however, it may be desirable to carry out a stochastic seismic analysis which describes response in probabilistic terms. The methodologies used in conducting that type of analysis are described in Chapter 28.

As discussed in Chapter 25, the earthquake excitation considered to act on a structure is the free-field ground motion at support points, expressed in terms of three components of translational acceleration and usually characterized through design response spectra and corresponding spectrum compatible accelerograms. It should be recognized, however, that rotational components are also present. Unfortunately, little is known regarding the magnitude and character of these rotational components; consequently, their effects are usually not taken into account when carrying out seismic analyses. This neglect is not serious, however, as, in most cases, response to the translational components greatly exceeds response to the rotational components. Nevertheless, in the interest of completeness, dynamic response to support rotational input also will be treated herein.

The response of any linear system to multiple components of input can be computed by superposing the responses calculated separately for each component; thus, the analytical problem can be reduced to separate evaluations of structural response to each single component of input. For instructional purposes, it is desirable to proceed in this way as it gives better understanding of the overall problem; however, in engineering practice, it may well be desirable to carry out a single solution using all components of input simultaneously when it is more efficient computationally to do so.

Inherent in the usual treatment of earthquake excitations is the assumption that the same free-field ground motion acts simultaneously at all support points of the structure with its foundation. If rotational motions are neglected, this assumption is equivalent to considering the foundation soil or rock to be rigid. This hypothesis clearly is not consistent with the concept of earthquake waves propagating through the earth's crust from the source of energy release; however, if the base dimensions of the structure are small relative to the predominant wavelengths in the basement rock motions, the assumption is acceptable. For example, if the apparent velocity of wave propagation is $6,000 \ ft/sec$, a wave of $3 \ Hz$ frequency will have a length of $2,000 \ ft$; thus, a building with a maximum base dimension of $100 \ ft$ would be subjected to essentially the same motions over its entire length. On the other hand, a suspension bridge or a dam having a length of, say, $1,500 \ ft$ would obviously be subjected to drastically differing motions along its length. Since such differences contribute significantly to the dynamic response mechanism, it is important to develop analysis procedures capable of dealing with multiple support excitations, i.e., different displacement histories at the different points of support.

When specifying input motions at the base of a structure, it should be recognized that the actual structure base motions during an earthquake may be significantly different from the corresponding free-field motions that would have occurred without the structure being present. This "soil-structure interaction" effect will be of slight importance if the foundation is relatively stiff and the structure is relatively flexible;

in this case, the structure transmits little energy into the foundation and the free-field motions are adequate measures of the actual foundation displacements. On the other hand, if a heavy stiff structure (such as a nuclear power plant containment building) is supported on a deep, relatively soft layer of soil, considerable energy will be transferred from the structure to the soil and the base motions will differ drastically from those experienced by the soil under free-field conditions. This soil-structure interaction mechanism is independent of, and in addition to, the effect of local soil conditions on the free-field motions as discussed in Chapter 25. In general, both effects can be important and should be accounted for in conducting earthquake-response analyses.

It is the purpose of this chapter to discuss the methodologies of deterministic response analysis of various types of structural systems to earthquake excitations, considering cases in which the soil is assumed to be rigid so it does not interact with the dynamic response of the structure. Chapter 27 which follows is concerned with excitation through a flexible soil medium, i.e., it deals with soil-structure interaction effects.

26-2   RESPONSE TO RIGID-SOIL EXCITATIONS

Lumped SDOF Elastic Systems, Translational Excitation

The simplest form of earthquake response problem involves a SDOF lumped-mass system subjected to identical single-component translations of all support points. An example of such a system is shown in Fig. 26-1 which is identical to the system used in Chapter 25 to define earthquake response spectra. It also is identical to the structure shown in Chapter 2 (Fig. 2-3), where it was used in the formulation of the

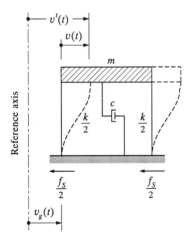

FIGURE 26-1
Lumped SDOF system subjected to rigid-base translation.

equation of motion for a system subjected to base translation, Eq. (2-14), which is repeated here for convenience:

$$m \, \ddot{v}^t(t) + c \, \dot{v}(t) + k \, v(t) = 0 \tag{26-1}$$

The superscript t in the first term of this equation denotes total displacement, and as explained in Chapter 2, it is this term that makes possible expressing the equation of motion in terms of an effective loading as follows [see Eq. (2-17)]:

$$m \, \ddot{v}(t) + c \, \dot{v}(t) + k \, v(t) = p_{\text{eff}}(t) \tag{26-2}$$

where

$$p_{\text{eff}}(t) = -m \, \ddot{v}_g(t) \tag{26-3}$$

As was noted in Chapter 2, in this expression $\ddot{v}_g(t)$ represents the free-field input acceleration applied at the base of the structure; the negative sign has little significance in earthquake response analysis and generally is ignored.

As was explained before, it also is possible to express the effective earthquake force in terms of the free-field velocity and displacement, $\dot{v}_g(t)$ and $v_g(t)$, if the equation of motion is formulated in terms of total rather than relative motion [see Eq. (2-18)]. However, this form of the equation of motion seldom is used because it is so much simpler to evaluate the base acceleration expression for effective loading, Eq. (26-3).

For purposes of this discussion, it is convenient to express the earthquake displacement response given by the solution of Eq. (26-2) in terms of the Duhamel integral expression for low damped systems, as follows:

$$v(t) = \frac{1}{\omega} \int_0^t \ddot{v}_g(t) \, \exp[-\xi\omega(t - \tau)] \, \sin\omega(t - \tau) \, d\tau \equiv \frac{1}{\omega} V(t) \tag{26-4}$$

as given previously by Eq. (25-2); however, it should be noted that the response could equally well be calculated by step-by-step integration or by a frequency-domain analysis rather than by the Duhamel integral. As was defined in Chapter 25, the maximum absolute value over the entire earthquake history of the earthquake response integral $V(t)$ in Eq. (26-4) is the pseudo-velocity spectral response $S_{pv}(\xi, \omega)$. The related displacement and pseudo-acceleration spectral responses, as previously given by Eqs. (25-14) and (25-15), are

$$S_d(\xi, \omega) = \frac{1}{\omega} \, S_{pv}(\xi, \omega) \tag{26-5}$$

$$S_{pa}(\xi, \omega) = \omega \, S_{pv}(\xi, \omega) \tag{26-6}$$

Since all aspects of these spectral response expressions for the lumped SDOF system were discussed in Chapter 25, it will suffice here to simply give an example of their use.

Example E26-1. Assuming that the structure of Fig. 26-1 has the following properties:

$$m = 2 \; kips \cdot sec^2/in \qquad k = 60 \; kips/in \qquad c = 1.10 \; kips \cdot sec/in$$

compute the maximum relative displacement and maximum base shear produced in this structure by an earthquake of 0.3 g peak acceleration having the acceleration response spectrum shown in Fig. 25-9 for hard soil conditions (Type S_1). The first step in this analysis is to determine the vibration period and damping ratio of the structure:

$$\omega = \sqrt{k/m} = \sqrt{60/2} = 5.48 \; rad/sec$$

$$T = 2\pi/\omega = 2\pi/5.48 = 1.147 \; sec \tag{a}$$

$$\xi = c/2m\omega = 1.10/(2)(2)(5.48) = 0.05$$

From Fig. 25-9, the spectral acceleration for this period and damping ratio is $S_{pa} = (0.3)(32.2)(0.92) = 8.89 \; ft/sec^2$. Hence the maximum relative displacement produced by this earthquake is

$$v_{max} = S_{pa}/\omega^2 = \pm 8.89/(5.48)^2 = \pm 0.296 \; ft \tag{b}$$

The maximum base shear force may be computed in either of two ways:

$$V_{max} = k \, v_{max} = (60)(0.296)(12) = 213 \; kips$$
$$V_{max} = m \, S_{pa} = (2)(12)(8.89) = 213 \; kips \tag{c}$$

This base shear corresponds to 27.5 percent of the weight represented by mass m.

Generalized-Coordinate SDOF Elastic Systems, Translational Excitation

Any structure of arbitrary form can be treated as a SDOF system if it is assumed that its displacements are restricted to a single shape, as explained in Chapter 8. This generalized-coordinate approach can be used effectively in earthquake engineering; the only special problem to be considered in this case is the evaluation of the generalized effective force resulting from the support excitation.

The formulation of the generalized-coordinate equation of motion will be explained with reference to the tower structure of Fig. 26-2. The equilibrium of this

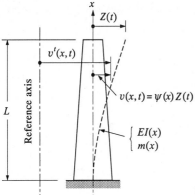

FIGURE 26-2
Generalized SDOF system with rigid-base translation.

system involves inertial, damping, and elastic forces, which are distributed along the axis and may be expressed as

$$f_I(x,t) + f_D(x,t) + f_S(x,t) = 0 \tag{26-7}$$

The basic assumption of the SDOF approximation is that the displacements are given by the product of a single shape function $\psi(x)$ and a generalized-coordinate amplitude $Z(t)$, that is,

$$v(x,t) = \psi(x)\,Z(t) \tag{26-8}$$

When a virtual displacement of the form $\delta v = \psi(x)\,\delta Z$ is applied, the principle of virtual work leads to the SDOF equilibrium relationship

$$f_I^*\,\delta Z + f_D^*\,\delta Z + f_S^*\,\delta Z = 0 \tag{26-9}$$

in which

$$f_I^* = \int_0^L f_I(x,t)\,\psi(x)\,dx$$

$$f_D^* = \int_0^L f_D(x,t)\,\psi(x)\,dx \tag{26-10}$$

$$f_S^* = \int_0^L f_S(x,t)\,\psi(x)\,dx$$

Because the distributed damping and elastic forces are assumed to depend only on the relative motions, the corresponding generalized forces here are the same as for the situation discussed in Chapter 8, where the dynamic load was applied externally, that is,

$$f_D^* = c^*\,\dot{Z} \qquad f_S = k^*\,Z$$

where c^* and k^* are given by expressions of the type shown in Eqs. (8-14). However, the local inertial forces depend on the total acceleration, that is,

$$f_I(x,t) = m(x)\,\ddot{v}^t(x,t)$$

Thus since

$$v^t(x,t) = v(x,t) + v_g(t) = \psi(x)\,Z(t) + v_g(t)$$

the generalized inertial force is found to be

$$f_I^* = \ddot{Z}(t)\int_0^L m(x)\,[\psi(x)]^2\,dx + \ddot{v}_g\int_0^L m(x)\,\psi(x)\,dx$$

Substituting all these generalized-force expressions into Eq. (26-9) then leads to the final equation of motion

$$m^*\,\ddot{Z}(t) + c^*\,\dot{Z}(t) + k^*\,Z(t) = -\mathcal{L}\,\ddot{v}_g(t) \tag{26-11}$$

in which

$$m^* = \int_0^L m(x)\,[\psi(x)]^2\,dx \tag{26-12}$$

$$\mathcal{L} = \int_0^L m(x)\,\psi(x)\,dx \tag{26-13}$$

Equation (26-12) is the same generalized-mass expression shown in Eqs. (8-14), while the quantity $\mathcal{L}$ given by Eq. (26-13) is the earthquake-excitation factor representing the extent to which the earthquake motion tends to excite response in the assumed shape $\psi(x)$.

Ignoring the sign of the effective earthquake force in Eq. (26-11) and dividing by the generalized mass leads to

$$\ddot{Z}(t) + 2\,\xi\,\omega\,\dot{Z}(t) + \omega^2\,Z(t) = \frac{\mathcal{L}}{m^*}\,\ddot{v}_g(t) \tag{26-14}$$

By analogy with the foregoing analysis of the lumped SDOF system, the solution of Eq. (26-14) may now be written

$$Z(t) = \frac{\mathcal{L}}{m^*\,\omega}\,V(t) \tag{26-15}$$

and hence the local displacements [from Eq. (26-8)] are

$$v(x,t) = \frac{\psi(x)\,\mathcal{L}}{m^*\,\omega}\,V(t) \tag{26-16}$$

It will be noted by comparison of Eqs. (26-4) and (26-15) that the factor $\mathcal{L}/m^*$ characterizes the difference between the lumped and the generalized SDOF response;

this factor depends on the mass distribution of the structure as well as its assumed shape function and generally is significantly different from unity.

In principle, the elastic forces produced by the earthquake motions can be evaluated from the structural displacements of Eq. (26-16) acting on the structural stiffness properties. However, when expressed in this way, the forces in this generalized-coordinate analysis depend on derivatives of the displacements or, in other words, on derivatives of the assumed shape functions $\psi(x)$. Thus the local forces obtained from such an analysis usually are less accurate than the displacements because the derivatives of the assumed shapes are poorer approximations than the shapes themselves. A more dependable formulation of the elastic forces can be obtained by expressing them in terms of the inertial forces of free vibration, following the general approach described previously for the lumped-mass case. The equilibrium condition in undamped free vibration is obtained by omitting the damping term from Eq. (26-7); thus since the inertial force in free harmonic motion is

$$f_I(x,t) = m(x)\,\ddot{v}(x,t) = -\omega^2\,m(x)\,v(x,t)$$

the resulting equation may be written

$$-\omega^2\,m(x)\,v(x,t) + f_S(x,t) = 0 \qquad (26\text{-}17)$$

Now if the the displacements are assumed to be of the form given by Eq. (26-8), the force balance implied by Eq. (26-17) will not be satisfied at all points along the span, in general; that is, the assumed shape will not satisfy equilibrium locally. However, if a virtual displacement of this form is introduced, the virtual-work principle can be used to obtain an approximate global-equilibrium relationship

$$\delta Z \int_0^L \left[-\omega^2\,m(x)\,v(x,t) + f_S(x,t)\right]\,\psi(x)\,dx = 0$$

Thus, even though it is valid only in an integrated or weighted-average sense, Eq. (26-17) provides the best available estimate of the elastic forces developed during the earthquake response, that is,

$$f_S(x,t) = \omega^2\,m(x)\,v(x,t) = m(x)\,\psi(x)\,\frac{\mathcal{L}}{m^*}\,\omega\,V(t) \qquad (26\text{-}18)$$

From these distributed elastic forces, which are depicted in Fig. 26-3, any desired force resultant can be obtained by standard methods of statics. For example, the base shear V_0 is given by

$$V_0(t) = \int_0^L f_S(x,t)\,dx = \frac{\mathcal{L}}{m^*}\,\omega\,V(t) \int_0^L m(x)\,\psi(x)\,dx$$

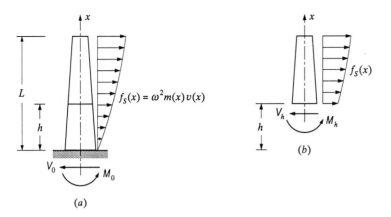

FIGURE 26-3
Elastic-force response of generalized SDOF system: (a) base forces; (b) section forces.

that is,

$$V_0(t) = \frac{\mathcal{L}^2}{m^*} \, \omega \, V(t) \qquad (26\text{-}19)$$

Similarly the base moment is given by $M_0(t) = \int_0^L f_S(x,t) \, x \, dx$, that is,

$$M_0(t) = \frac{\mathcal{L}}{m^*} \, \omega \, V(t) \int_0^L m(x) \, \psi(x) \, x \, dx \qquad (26\text{-}20)$$

Expressions for moment and shear at any arbitrary section h can be written similarly:

$$V_h(t) = \frac{\mathcal{L}}{m^*} \, \omega \, V(t) \int_h^L m(x) \, \psi(x) \, dx$$

$$(26\text{-}21)$$

$$M_h(t) = \frac{\mathcal{L}}{m^*} \, \omega \, V(t) \int_h^L m(x) \, \psi(x) \, (x - h) \, dx$$

Of course, the evaluation of time-varying response expressions such as Eqs. (26-16) and (26-18) requires the numerical integration of the earthquake response integral $V(t)$ as defined by Eq. (26-4). However, maximum response values can be determined readily from the corresponding earthquake response spectra, as explained in the discussion of the lumped SDOF systems, by merely selecting spectral values appropriate to the period of vibration and damping of the structure. For example, the maximum local displacements and local elastic forces are given, respectively, by

$$v_{\max}(x) = \psi(x) \, \frac{\mathcal{L}}{m^*} \, S_d(\xi, T)$$

$$(26\text{-}22)$$

$$f_{S,\max}(x) = m(x) \, \psi(x) \, \frac{\mathcal{L}}{m^*} \, S_{pa}(\xi, T)$$

Since the signs in these relations have no significance, they can always be taken as positive.

Example E26-2.   The earthquake response analysis of a generalized SDOF structure having 5 percent of critical damping will be demonstrated by subjecting the uniform cantilever column of Fig. E26-1 to a base motion $v_g(t)$ corresponding to an earthquake of 0.3 g peak acceleration having the acceleration response spectrum shown in Fig. 25-9 for a hard soil condition (Type S_1). It will be assumed that the displaced shape of the column is given by $\psi(x) = 1 - \cos(\pi\,x\,/\,2\,L)$; hence the generalized properties of this structure are as given in Example E8-3. For the numerical values shown in Fig. E26-1, these properties are

$$m^* = 0.228\,\overline{m}\,L = 0.456\ kips \cdot sec^2/ft$$

$$k^* = \frac{\pi^4}{32}\frac{EI}{L^3} = 4.26\ kips/ft \tag{a}$$

$$\mathcal{L} = 0.364\,\overline{m}\,L = 0.728\ kips \cdot sec^2/ft$$

From these values, the circular frequency of the column is

$$\omega = \sqrt{\frac{k^*}{m^*}} = 3.056\ rad/sec \tag{b}$$

Hence the period is

$$T = \frac{2\,\pi}{\omega} = 2.056\ sec \tag{c}$$

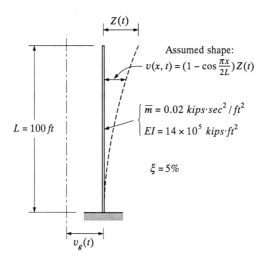

FIGURE E26-1
SDOF idealization of uniform cantilever column.

With this period and a damping ratio $\xi = 5$ percent, the spectral acceleration shown by Fig. 25-9 is $S_{pa} = (0.47)\,(32.2)\,(0.30) = 4.54 \; ft/sec^2$. Hence the maximum generalized-coordinate displacement [by analogy with Eq. (26-15)] is

$$Z_{\max} = \frac{\mathcal{L}}{m^* \omega^2} \, S_{pa} = 0.776 \; ft \tag{d}$$

and so the maximum displacements of the column are

$$v_{\max} = 0.776 \left(1 - \cos \frac{\pi x}{2\,L} \right) ft \tag{e}$$

Similarly the maximum base shear [by analogy with Eq. (26-19)] is

$$V_{0,\max} = \frac{\mathcal{L}^2}{m^*} \, S_{pa} = 5.27 \; kips \tag{f}$$

which is approximately 8.2 percent of the column's total weight. The maximum distributed earthquake forces acting on the column are given by

$$f_{S,\max} = \frac{\overline{m}\,\psi(x)}{\mathcal{L}} \, V_{0,\max} = 0.145 \left(1 - \cos \frac{\pi\,x}{2\,L} \right) kips/ft \tag{g}$$

Lumped MDOF Elastic Systems, Translational Excitation

The formulation of the earthquake response analysis of a lumped MDOF system can be carried out in matrix notation in a manner entirely analogous to the foregoing development of the lumped SDOF equations. Thus the equations of motion of the multistory shear building shown in Fig. 26-4 can be written by analogy with Eq. (26-1)

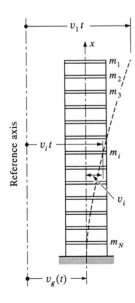

FIGURE 26-4
Discretized MDOF system with rigid-base translation.

as

$$\mathbf{m}\,\ddot{\mathbf{v}}^t(t) + \mathbf{c}\,\dot{\mathbf{v}}(t) + \mathbf{k}\,\mathbf{v}(t) = \mathbf{0} \tag{26-23}$$

and again the effective earthquake force can be derived by expressing the total displacements as the sum of the relative motions plus the displacements resulting directly from the support motions. For the system of Fig. 26-4 this relationship may be written

$$\mathbf{v}^t(t) = \mathbf{v}(t) + \{\mathbf{1}\}\,v_g(t) \tag{26-24}$$

in which $\{\mathbf{1}\}$ represents a column of ones. This vector expresses the fact that a unit static translation of the base of this structure produces directly a unit displacement of all degrees of freedom. Of course, this simple relationship is a consequence of the type of support displacement which has been applied as well as of the structural configuration; for other forms of structures or support motions this static-displacement vector would be different. Thus, the structure shown in Fig. 26-4 should be considered as a special case, even though a large number of practical analyses are assumed to be of this type.

Substituting Eq. (26-24) into (26-23) leads to the relative-response equations of motion

$$\mathbf{m}\,\ddot{\mathbf{v}}(t) + \mathbf{c}\,\dot{\mathbf{v}}(t) + \mathbf{k}\,\mathbf{v}(t) = \mathbf{p}_{\text{eff}}(t) \tag{26-25}$$

in which

$$\mathbf{p}_{\text{eff}}(t) = -\mathbf{m}\,\{\mathbf{1}\}\,\ddot{v}_g(t) \tag{26-26}$$

Equations (26-25) could be solved directly through the frequency domain or by numerical integration of the coupled equations in the time domain; however, in analyzing the earthquake response of linear structures, it generally is much more efficient to transform to a system of normal (modal) coordinates because the support motions tend to excite strongly only the lowest modes of vibration. Thus good approximations of the earthquake response of systems having dozens or even hundreds of degrees of freedom can often be obtained by carrying out the analysis for only a few normal coordinates.

The transformation to normal coordinates has been described in adequate detail in Chapter 12. If it is assumed that the damping matrix is of a form which satisfies the same orthogonality conditions as the mass and stiffness matrices, the result is a set of N uncoupled modal equations of the form

$$M_n\,\ddot{Y}_n + C_n\,\dot{Y}_n + K_n\,Y_n = P_n(t) \tag{26-27}$$

in which M_n, C_n, and K_n are the generalized properties associated with mode n [see Eqs. (12-12) and (12-15a)], Y_n is the amplitude of this modal response, and the generalized force resulting from the earthquake excitation [neglecting the negative sign in Eq. (26-26)] is given by

$$P_n(t) = \boldsymbol{\phi}_n^T\,\mathbf{p}_{\text{eff}}(t) = \mathcal{L}_n\,\ddot{v}_g(t) \tag{26-28}$$

in which, for the structure of Fig. 26-4, the modal earthquake-excitation factor is given by

$$\mathcal{L}_n \equiv \boldsymbol{\phi}_n^T \, \mathbf{m}\{1\} \qquad (26\text{-}29)$$

It will be recognized that this is the matrix equivalent of Eq. (26-13), which was derived for the generalized SDOF system; of course the modal excitation factor is different for each mode because it contains the mode shape $\boldsymbol{\phi}_n$.

By analogy with the derivation of the generalized SDOF response, it may be seen that the response of each mode of the MDOF system is given by

$$Y_n(t) = \frac{\mathcal{L}_n}{M_n \, \omega_n} V_n(t) \qquad (26\text{-}30)$$

where the modal earthquake response integral is of the form defined by Eq. (26-4) and is dependent on the damping ratio ξ_n and frequency ω_n of the nth mode of vibration. The relative-displacement vector produced in this mode then is given by

$$\mathbf{v}_n(t) = \boldsymbol{\phi}_n \frac{\mathcal{L}_n}{M_n \, \omega_n} V_n(t) \qquad (26\text{-}31)$$

Finally, the relative-displacement vector due to all modal responses is obtained by superposition, that is,

$$\mathbf{v}(t) = \boldsymbol{\Phi} \, \mathbf{Y}(t) = \boldsymbol{\Phi} \left\{ \frac{\mathcal{L}_n}{M_n \, \omega_n} V_n(t) \right\} \qquad (26\text{-}32)$$

in which $\boldsymbol{\Phi}$ is made up of all mode shapes for which the modal response is excited significantly by the earthquake, and the term in braces represents a vector of response terms defined for each mode considered in the analysis.

The elastic forces associated with the relative displacements can be obtained directly by premultiplying $\mathbf{v}(t)$ by the stiffness matrix $\mathbf{k}$ as given by

$$\mathbf{f}_S(t) = \mathbf{k} \, \mathbf{v}(t) = \mathbf{k} \, \boldsymbol{\Phi} \, \mathbf{Y}(t) \qquad (26\text{-}33)$$

However, as mentioned in the discussion of the SDOF systems, it frequently is more effective to express these forces in terms of the equivalent inertial forces developed in undamped free vibrations. The equivalence of the elastic and inertial forces is expressed by the eigenproblem relationship, which may be written

$$\mathbf{k} \, \boldsymbol{\Phi} = \mathbf{m} \, \boldsymbol{\Phi} \, \boldsymbol{\Omega}^2 \qquad (26\text{-}34)$$

in which $\boldsymbol{\Omega}^2$ is a diagonal matrix of the squared modal frequencies ω_n^2. Substituting Eq. (26-34) into Eq. (26-33) results in the alternate expression for the elastic forces

$$\mathbf{f}_S(t) = \mathbf{m} \, \boldsymbol{\Phi} \, \boldsymbol{\Omega}^2 \, \mathbf{Y}(t) = \mathbf{m} \, \boldsymbol{\Phi} \left\{ \frac{\mathcal{L}_n}{M_n} \omega_n \, V_n(t) \right\} \qquad (26\text{-}35)$$

It will be noted that the elastic-force vector associated with each mode in this equation, that is,

$$\mathbf{f}_{Sn}(t) = \mathbf{m}\, \boldsymbol{\phi}_n \frac{\mathcal{L}_n}{M_n} \omega_n \, V_n(t) \tag{26-36}$$

is given by the matrix equivalent of the generalized SDOF expression of Eq. (26-18). It must be emphasized that Eq. (26-35) is a completely general expression for the elastic forces developed in a damped structure subjected to arbitrarily varying ground motions; the fact that it was derived from an expression for undamped free vibrations does not limit its applicability.

When the distribution of these effective elastic forces at any time t during the earthquake has been determined, as illustrated, for example, in Fig. 26-5, the value of any desired force resultant at that same time can be computed by standard statics procedures. For example, the base shear force $V_0(t)$ of the system in Fig. 26-5 is given by the sum of all the story forces, that is,

$$V_0(t) = \sum_{i=1}^{N} f_{Si}(t) = \langle \mathbf{1} \rangle \, \mathbf{f}_S(t)$$

where $\langle \mathbf{1} \rangle$ represents a row vector of ones. Substituting Eq. (26-35) into this expression leads to

$$V_0(t) = \sum_{n=1}^{N} \frac{\mathcal{L}_n^2}{M_n} \omega_n \, V_n(t) \tag{26-37}$$

in which it can be noted from Eq. (26-29) that

$$\langle \mathbf{1} \rangle \, \mathbf{m} \, \boldsymbol{\Phi} = \langle \mathcal{L}_1 \quad \mathcal{L}_2 \quad \cdots \quad \mathcal{L}_N \rangle$$

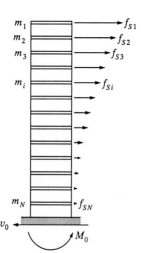

FIGURE 26-5
Elastic forces in lumped MDOF system.

Similarly, the resultant overturning moment at the base of the building is

$$M_0(t) = \sum_{i=1}^{N} x_i\, f_{Si}(t) = \langle \mathbf{x} \rangle\, \mathbf{f}_S(t)$$

in which x_i is the height of mass i above the base and $\langle \mathbf{x} \rangle$ is a row vector of these heights. Substituting Eq. (26-35) into this yields the expression for the base moment

$$M_0(t) = \langle \mathbf{x} \rangle\, \mathbf{m}\, \boldsymbol{\Phi}\, \boldsymbol{\Omega}^2\, \mathbf{Y}(t) = \langle \mathbf{x} \rangle\, \mathbf{m}\, \boldsymbol{\Phi}\, \left\{ \frac{\mathcal{L}_n}{M_n} \omega_n\, V_n(t) \right\} \qquad (26\text{-}38)$$

The quantity $\mathcal{L}^2 / M_n$ in Eq. (26-37) has the dimensions of mass and is sometimes called the *effective modal mass* of the structure because it can be interpreted as the part of the total mass responding to the earthquake in each mode. This interpretation of the expression is valid only for structures of the type shown in Fig. 26-4, having masses lumped along a vertical axis; for such structures, the total mass M_T is given by

$$M_T = \langle \mathbf{1} \rangle\, \mathbf{m}\, \{\mathbf{1}\} \qquad (26\text{-}39)$$

Now it can be proved that the sum of all effective modal masses is equal to the total mass by expressing the vector of ones $\{\mathbf{1}\}$ in modal coordinates as

$$\{\mathbf{1}\} = \boldsymbol{\Phi}\, \mathbf{Y}$$

where each modal amplitude Y_n can be evaluated by multiplying both sides by $\boldsymbol{\phi}_n^T\, \mathbf{m}$ and applying the mass orthogonality relationship, that is,

$$\boldsymbol{\phi}_n^T\, \mathbf{m}\, \{\mathbf{1}\} = \boldsymbol{\phi}_n^T\, \mathbf{m}\, \boldsymbol{\Phi}\, \mathbf{Y} = M_n\, Y_n$$

Since the left hand triple matrix product is $\mathcal{L}_n$, each modal amplitude can be expressed as $Y_n = \mathcal{L}_n / M_n$ and the ones vector is given by

$$\{\mathbf{1}\} = \boldsymbol{\Phi}\, \left\{ \frac{\mathcal{L}_n}{M_n} \right\} \qquad (26\text{-}40)$$

Substituting this into Eq. (26-39) gives

$$M_T = \langle \mathbf{1} \rangle\, \mathbf{m}\, \boldsymbol{\Phi} \left\{ \frac{\mathcal{L}_n}{M_n} \right\}$$

$$= \langle \mathcal{L}_1\ \ \mathcal{L}_2\ \ \cdots\ \ \mathcal{L}_N \rangle \left\{ \frac{\mathcal{L}_n}{M_n} \right\} = \sum_{i=1}^{N} \frac{\mathcal{L}_n^2}{M_n} \qquad \text{Q.E.D.} \qquad (26\text{-}41)$$

Hence each modal contribution $V_{0n}(t)$ to the base shear of Eq. (26-37) may be looked upon as the reaction of the effective modal mass to the effective modal acceleration of the ground $\omega_n\, V_n(t)$.

Example E26-3. To demonstrate the earthquake response analysis of a MDOF structure, the three-story building shown in Fig. E26-2 will be considered. This is the structure of Example E12-1 with its stiffness reduced by a factor of 10 to provide frequencies typical of a taller building in which the higher modes would contribute more to the response. The vibration properties, generalized masses, and modal excitation factors are also shown in the figure; in addition it is assumed that the damping is 5 percent critical in each mode.

From the given frequency and damping values, the first-mode response integral $V_1(t)$ was calculated for the entire history of a certain earthquake motion $v_g(t)$; the maximum value of this integral occurred at $t_1 = 3.08$ sec. Then the second- and third-mode response integrals were evaluated at the same time; the values of the three modal response integrals at this time were

$$\mathbf{V}(t_1) = - \begin{bmatrix} 1.74 \\ 1.22 \\ 0.77 \end{bmatrix} \; ft/sec \tag{a}$$

When these values and the other modal properties are introduced into Eq. (26-30), the normal coordinates at this time are

$$\mathbf{Y}(t_1) = \left\{ \frac{\mathcal{L}_n}{M_n \, \omega_n} V_n(t_1) \right\} = \begin{bmatrix} 0.541 \\ 0.0635 \\ 0.00475 \end{bmatrix} \; ft \tag{b}$$

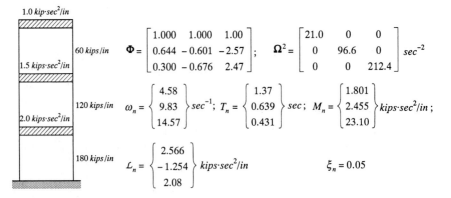

1.0 kip·sec²/in

60 kips/in

1.5 kips·sec²/in

120 kips/in

2.0 kips·sec²/in

180 kips/in

$$\mathbf{\Phi} = \begin{bmatrix} 1.000 & 1.000 & 1.00 \\ 0.644 & -0.601 & -2.57 \\ 0.300 & -0.676 & 2.47 \end{bmatrix} ; \qquad \mathbf{\Omega}^2 = \begin{bmatrix} 21.0 & 0 & 0 \\ 0 & 96.6 & 0 \\ 0 & 0 & 212.4 \end{bmatrix} sec^{-2}$$

$$\omega_n = \left\{ \begin{matrix} 4.58 \\ 9.83 \\ 14.57 \end{matrix} \right\} sec^{-1} ; \; T_n = \left\{ \begin{matrix} 1.37 \\ 0.639 \\ 0.431 \end{matrix} \right\} sec ; \; M_n = \left\{ \begin{matrix} 1.801 \\ 2.455 \\ 23.10 \end{matrix} \right\} kips \cdot sec^2/in ;$$

$$\mathcal{L}_n = \left\{ \begin{matrix} 2.566 \\ -1.254 \\ 2.08 \end{matrix} \right\} kips \cdot sec^2/in \qquad \qquad \xi_n = 0.05$$

FIGURE E26-2
Building frame and its vibration properties.

and the resulting displacements are

$$\mathbf{v}(t_1) = \boldsymbol{\Phi}\mathbf{Y}(t_1) = \begin{bmatrix} 0.541 + 0.064 + 0.005 \\ 0.348 - 0.038 - 0.012 \\ 0.162 - 0.043 + 0.012 \end{bmatrix} \{\mathbf{1}\} = \begin{bmatrix} 0.610 \\ 0.298 \\ 0.131 \end{bmatrix} \; ft \quad \text{(c)}$$

where the individual modal displacements have been shown as a matter of interest. Similarly, the elastic-force vector at this time is

$$\mathbf{f}_S(t_1) = \mathbf{m}\,\boldsymbol{\Phi}\left\{ \frac{\mathcal{L}_n}{M_n}\omega_n\, V_n(t_1) \right\}$$

$$= \begin{bmatrix} 11.35 + 6.13 + 1.01 \\ 10.95 - 5.53 - 3.90 \\ 6.80 - 8.29 + 5.00 \end{bmatrix} \{\mathbf{1}\} = \begin{bmatrix} 18.49 \\ 1.52 \\ 3.51 \end{bmatrix} \; kips \quad \text{(d)}$$

and the base shear force is given by the sum of the story forces:

$$V_0(t_1) = 23.52 \; kips \quad \text{(e)}$$

To evaluate the earthquake response of a lumped MDOF system at any time t using Eq. (26-32) or (26-35) involves the evaluation of the earthquake response integral at that time for each significant response mode. Hence, the evaluation of the maximum response requires that each modal response be computed in this way for each time during the earthquake history, in order that the maximum value can be identified. This obviously constitutes a major computational task and makes an approximate analysis based on the ground motion response spectra an attractive alternative.

For each individual mode of the structure, the maximum response can be obtained directly from the response spectrum as described for the SDOF systems. For example, from Eq. (26-31) the maximum displacement vector in mode n is given by

$$\mathbf{v}_{n,\text{max}} = \boldsymbol{\phi}_n\, \frac{\mathcal{L}_n}{M_n}\, S_d(\xi_n, T_n) \quad (26\text{-}42)$$

where $S_d(\xi_n, T_n)$ is the spectral displacement corresponding to the damping and period of the nth mode of vibration. Similarly, from Eq. (26-36) the maximum elastic-force vector in mode n is given by

$$\mathbf{f}_{Sn,\text{max}} = \mathbf{m}\,\boldsymbol{\phi}_n\, \frac{\mathcal{L}_n}{M_n}\, S_{pa}(\xi_n, T_n) \quad (26\text{-}43)$$

where $S_{pa}(\xi_n, T_n)$ is the spectral acceleration for the nth mode.

Maximum total response cannot be obtained, in general, by merely adding the modal maxima because these maxima usually do not occur at the same time. In most cases, when one mode achieves its maximum response, the other modal responses are less than their individual maxima. Therefore, although the superposition of the modal spectral values obviously provides an upper limit to the total response, it generally over estimates this maximum by a significant amount. A number of different formulas have been proposed to obtain a more reasonable estimate of the maximum response from the spectral values. The simplest and most popular of these is the square root of the sum of the squares (SRSS) of the maximum modal responses. Thus if the maximum modal displacements are given by Eq. (26-42), the SRSS approximation of the maximum total displacements is given by

$$\mathbf{v}_{\max} \doteq \sqrt{(\mathbf{v}_1)_{\max}^2 + (\mathbf{v}_2)_{\max}^2 + \cdots} \tag{26-44}$$

where the terms under the radical sign represent vectors of the maximum modal displacements squared. Similarly the maximum story forces could be approximated from the modal maxima of Eq. (26-43) as follows:

$$\mathbf{f}_{S,\max} \doteq \sqrt{(\mathbf{f}_{S1})_{\max}^2 + (\mathbf{f}_{S2})_{\max}^2 + \cdots} \tag{26-45}$$

Theoretical justification of using the SRSS method is presented in the following Section 26-3, where it is shown that this method is fundamentally sound when the modal frequencies are well separated. However, when the frequencies of major contributing modes are very close together, the SRSS method can give poor results, in which case the more general complete quadratic combination (CQC) method should be used. This method also is developed in Section 26-3.

Example E26-4. The response spectrum analysis of a 5 percent critically damped MDOF structure will be illustrated by evaluating the response of the building of Example E26-3 to an earthquake of 0.3 g peak acceleration having the acceleration response spectrum shown in Fig. 25-9 for the hard soil condition (Type S_1). When the periods given in Example E26-3 are used, the modal spectral accelerations are

$$\mathbf{S}_{pa} = (0.3)\,(32.2) \begin{Bmatrix} 0.77 \\ 1.62 \\ 2.35 \end{Bmatrix} = \begin{Bmatrix} 7.40 \\ 15.6 \\ 22.7 \end{Bmatrix} ft/sec^2 = \begin{Bmatrix} 88.8 \\ 187.2 \\ 272.4 \end{Bmatrix} in/sec^2 \tag{a}$$

Hence the modal maximum displacements, given by

$$\mathbf{v}_{n,\max} = \boldsymbol{\phi}_n \frac{L_n}{M_n} \frac{S_{pa,n}}{\omega_n^2} \tag{b}$$

are

$$\mathbf{v}_{1,\max} = \begin{bmatrix} 0.503 \\ 0.324 \\ 0.151 \end{bmatrix} ft \quad \mathbf{v}_{2,\max} = \begin{bmatrix} 0.083 \\ -0.050 \\ -0.056 \end{bmatrix} ft \quad \mathbf{v}_{3,\max} = \begin{bmatrix} 0.010 \\ -0.025 \\ 0.024 \end{bmatrix} ft \tag{c}$$

Superposing the modal maxima by the SRSS procedure gives the approximate total maximum displacements

$$\mathbf{v}_{\max} \doteq \begin{bmatrix} 0.510 \\ 0.329 \\ 0.163 \end{bmatrix} ft \tag{d}$$

Similarly, the modal maximum forces given by

$$\mathbf{f}_{Sn,\max} = \mathbf{m}\, \boldsymbol{\phi}_n \frac{L_n}{M_n} S_{pa,n} \tag{e}$$

are

$$\mathbf{f}_{S1,\max} = \begin{bmatrix} 127 \\ 122 \\ 76 \end{bmatrix} kips \quad \mathbf{f}_{S2,\max} = \begin{bmatrix} 96 \\ -86 \\ -129 \end{bmatrix} kips \quad \mathbf{f}_{S3,\max} = \begin{bmatrix} 25 \\ -94 \\ 121 \end{bmatrix} kips \tag{f}$$

which, when summed cumulatively from top to bottom, give the modal maximum story shears

$$V_{1,m} = \begin{bmatrix} 127 \\ 249 \\ 325 \end{bmatrix} kips \quad V_{2,m} = \begin{bmatrix} 96 \\ 10 \\ -119 \end{bmatrix} kips \quad V_{3,m} = \begin{bmatrix} 25 \\ -69 \\ 52 \end{bmatrix} kips \tag{g}$$

Superposing these forces and story shears by the SRSS method gives

$$\mathbf{f}_{S,\max} = \begin{bmatrix} 16 \\ 17 \\ 192 \end{bmatrix} kips \qquad V_{\max} = \begin{bmatrix} 161 \\ 258 \\ 350 \end{bmatrix} kips \tag{h}$$

This example demonstrates clearly that the maximum story shears cannot be obtained by simply summing the maximum story forces; also it is evident that taking the SRSS must always be the last step in evaluating the maximum value of any response quantity.

It is interesting to note in this example that the effective modal masses for this structure are

$$\left\{ \frac{\mathcal{L}_n^2}{M_n} \right\} = \begin{bmatrix} 3.66 \\ 0.64 \\ 0.18 \end{bmatrix} \tag{i}$$

and their sum is 4.48 which except for a slight error due to round-offs is the same as the sum of the story masses which is equal to 4.50. This equality applies to all building-type structures, as was noted earlier.

In both Eqs. (26-44) and (26-45) only the significant modal contributions need be included, and because each term is squared, the lesser terms have little effect so that very few modes need be considered in most cases. Again let it be emphasized that this approximate superposition procedure must be applied directly to the response quantity in question. As was shown in Example E26-4, to estimate the maximum story shears, it was necessary to compute the modal story shears and superpose them using

$$V_{max} \doteq \sqrt{(V_1)_{max}^2 + (V_2)_{max}^2 + \cdots} \tag{26-46}$$

The base shear could not be found by summing the maximum forces $\mathbf{f}_{S,max}$ over the height of the building because the signs of the local force quantities were lost in the squaring process.

It was pointed out at the beginning of this discussion of lumped MDOF systems that the type of system shown in Fig. 26-4, having a vertical axis and subjected to horizontal excitation, represents a special class of earthquake problem for which the relationship between the total and relative motions takes the simple form of Eq. (26-24). In a more general case, where the relative displacements are not all measured parallel to the ground motion, an example of which is shown in Fig. 26-6, the total displacement may be expressed as the sum of the relative displacement and the quasi-static displacements $\mathbf{v}^s$ that would result from a static-support displacement, that is,

$$\mathbf{v}^t(t) = \mathbf{v}(t) + \mathbf{v}^s(t) \tag{26-47}$$

The quasi-static displacements can be expressed conveniently by an influence co-efficient vector $\mathbf{r}$ which represents the displacements resulting from a unit support displacement; thus $\mathbf{v}^s = \mathbf{r}\, v_g$ and

$$\mathbf{v}^t(t) = \mathbf{v}(t) + \mathbf{r}\, v_g(t) \tag{26-48}$$

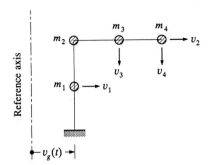

FIGURE 26-6
General lumped MDOF system with rigid-base translation.

From comparison of Eqs. (26-24) and (26-48) it is evident that $\mathbf{r}$ is a vector of ones for the structure of Fig. 26-4; however, for the system of Fig. 26-6 it would be given by $\mathbf{r}^T = \langle 1 \quad 1 \quad 0 \quad 0 \rangle$.

This generalization affects only the effective-force vector generated by the earthquake motion; that is, in place of Eq. (26-26), which was derived for the special static-displacement influence vector, the general expression is

$$\mathbf{P}_{\text{eff}}(t) = -\mathbf{m}\,\mathbf{r}\,\ddot{v}_g(t) \tag{26-49}$$

Similarly, the general form of the modal earthquake-excitation factor, replacing Eq. (26-29), would be

$$\mathcal{L}_n = \boldsymbol{\phi}_n^T \mathbf{m}\,\mathbf{r} \tag{26-50}$$

With this general definition of $\mathcal{L}_n$, the response equations [Eqs. (26-30) to (26-36)] are now fully applicable to general forms of lumped-mass structures. Of course it must be noted that the elastic forces act in the directions of the corresponding displacements $\mathbf{v}$; hence new expressions for the force resultants (such as base shear or moment) would have to be derived, appropriate to the given structural configuration.

Example E26-5. The earthquake response of a structure for which a unit static support movement does not cause a unit displacement of each degree of freedom will be demonstrated by analysis of the structure shown in Fig. E26-3. The mass and stiffness matrices defined for the two specified degrees of freedom are shown, as are the eigenvectors and eigenvalues describing its free vibration.

From these data, the modal earthquake response parameters are

$$\begin{bmatrix} M_1 \\ M_2 \end{bmatrix} = \begin{bmatrix} 2.557 \\ 3.834 \end{bmatrix} m \qquad \begin{bmatrix} \mathcal{L}_1 \\ \mathcal{L}_2 \end{bmatrix} = \begin{bmatrix} 1.293 \\ 3.000 \end{bmatrix} m$$

$$\begin{bmatrix} \omega_1 \\ \omega_2 \end{bmatrix} = \begin{bmatrix} 5.49 \\ 16.86 \end{bmatrix} rad/sec \qquad \begin{bmatrix} T_1 \\ T_2 \end{bmatrix} = \begin{bmatrix} 1.144 \\ 0.373 \end{bmatrix} sec \qquad \text{(a)}$$

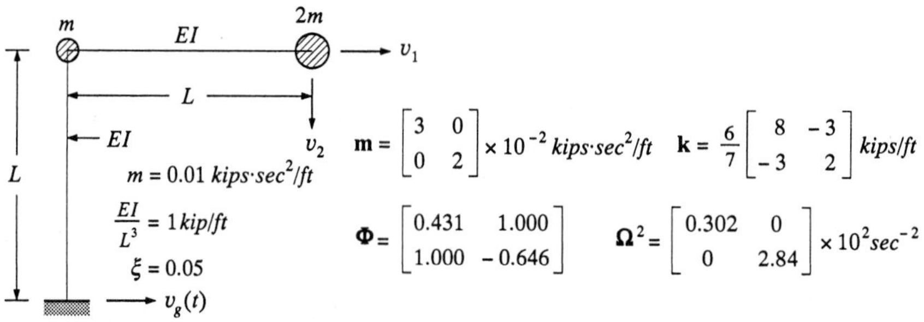

FIGURE E26-3
Two-DOF frame and its vibration properties.

$$\mathbf{m} = \begin{bmatrix} 3 & 0 \\ 0 & 2 \end{bmatrix} \times 10^{-2} \; kips \cdot sec^2/ft \qquad \mathbf{k} = \tfrac{6}{7} \begin{bmatrix} 8 & -3 \\ -3 & 2 \end{bmatrix} \; kips/ft$$

$$\mathbf{\Phi} = \begin{bmatrix} 0.431 & 1.000 \\ 1.000 & -0.646 \end{bmatrix} \qquad \mathbf{\Omega}^2 = \begin{bmatrix} 0.302 & 0 \\ 0 & 2.84 \end{bmatrix} \times 10^2 \; sec^{-2}$$

$$\text{(b)}$$

When it is assumed that this structure is subjected to an earthquake motion corresponding to the hard site (Type S_1) spectrum of Fig. 25-9 having a peak acceleration of 0.3 g (note the 5 percent critical modal damping ratios), the following modal spectral accelerations are obtained:

$$\begin{bmatrix} S_{pa,1} \\ S_{pa,2} \end{bmatrix} = \begin{bmatrix} 8.89 \\ 24.15 \end{bmatrix} \; ft/sec^2 \qquad \text{(c)}$$

Accordingly, from Eq. (26-43) the maximum modal response forces are

$$\mathbf{f}_{S1,max} = \{\mathbf{m}\ \phi_1\} \frac{\mathcal{L}_1}{M_1} S_{pa,1}$$

$$= \begin{bmatrix} 1.293 \\ 2.000 \end{bmatrix} \frac{1.293}{2.557} (8.89) \times 10^{-2} = \begin{bmatrix} 5.81 \\ 8.89 \end{bmatrix} \times 10^{-2} \; kips \qquad \text{(d)}$$

$$\mathbf{f}_{S2,max} = \begin{bmatrix} 3.000 \\ -1.292 \end{bmatrix} \frac{3.000}{3.834} (24.15) \times 10^{-2} = \begin{bmatrix} 56.69 \\ -24.39 \end{bmatrix} \times 10^{-2} \; kips$$

Applying the SRSS method to these modal results gives the approximate maximum response forces

$$\mathbf{f}_{S,max} = \begin{bmatrix} 57.0 \\ 26.0 \end{bmatrix} \times 10^{-2} \; kips \qquad \text{(e)}$$

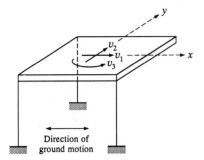

FIGURE 26-7
Rigid slab subjected to base translation.

Similar comments apply to the system of Fig. 26-7, which consists of a rigid rectangular slab supported by identical columns at three corners. When the degrees of freedom of this system are defined as the x and y translations of the center of mass, together with the rotation about that center, that is $\mathbf{v}^T = \langle v_1 \quad v_2 \quad v_3 \rangle$, and when it is assumed that the earthquake motions act in the direction of the x axis, the influence vector for this case is given by $\mathbf{r}^T = \langle 1 \quad 0 \quad 0 \rangle$. The modal earthquake-excitation factors for this structure are then obtained by substituting this vector into Eq. (26-50), and the response is given finally by Eqs. (26-30) to (26-36).

Example E26-6. Because the earthquake response analysis of a rigid slab structure of this type involves several features of special interest, the example structure of Fig. E26-4 will be discussed in some detail. It is assumed that the three columns supporting the slab are rigidly attached to the foundation and to the slab, so that the resistance at the top of each column to lateral displacement in any direction is $12EI/L^3 = 5 \; kips/ft$. The torsional stiffness of each individual column is negligible. Damping is assumed to be 5 percent of critical in all modes.

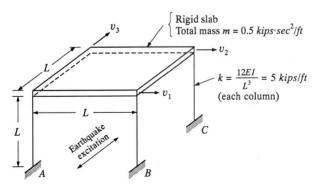

Rigid slab
Total mass $m = 0.5 \; kips \cdot sec^2/ft$

$k = \dfrac{12EI}{L^3} = 5 \; kips/ft$
(each column)

FIGURE E26-4
Slab supported by three columns.

For the purpose of this example, the three degrees of freedom of the slab are represented by the displacement components of the corners as shown. The total mass of the slab is $m = 0.5 \; kip \cdot sec^2/ft$ and is distributed uniformly over the area. The structure is subjected to a 0.3 g peak acceleration earthquake having the hard site response spectrum of Fig. 25-9 and acting in the direction parallel with coordinate v_3. It is desired to determine the maximum displacements of the slab due to this earthquake.

The mass and stiffness matrices of this system can be evaluated by direct application of the definitions of the influence coefficients. Considering first the stiffness matrix, a unit displacement $v_1 = 1$ is applied while the other coordinates are constrained, as shown in Fig. E26-5a. The forces exerted by the columns in resisting this displacement are shown in this sketch, and the equilibrating forces corresponding to the degrees of freedom are shown in Fig. E26-5b. By applying separately unit displacements of the other two coordinates, the remaining stiffness coefficients can be determined similarly.

The mass matrix is evaluated by applying a unit acceleration separately to each degree of freedom and determining the resulting inertial forces in the slab. For example, Fig. E26-6a shows the unit acceleration $\ddot{v}_2 = 1$ and the slab inertial forces resisting this acceleration, while Fig. E26-6b shows the mass influence coefficients which equilibrate these inertial forces. The other mass coefficients can be found by applying separately unit accelerations of the other two coordinates. The complete stiffness and mass matrices for the system are

$$\mathbf{k} = \frac{12EI}{L^3} \begin{bmatrix} 4 & -2 & 2 \\ -2 & 4 & -2 \\ 2 & -2 & 3 \end{bmatrix} \qquad \mathbf{m} = \frac{m}{6} \begin{bmatrix} 4 & -1 & 3 \\ -1 & 4 & -3 \\ 3 & -3 & 6 \end{bmatrix} \qquad (a)$$

When the eigenproblem $[\mathbf{k} - \omega^2\mathbf{m}]\mathbf{v} = \mathbf{0}$ is solved, the mode shapes and frequencies of the system are found to be

$$\mathbf{\Phi} = \begin{bmatrix} 0.366 & 1.000 & -1.366 \\ 1.000 & 1.000 & 1.000 \\ 1.000 & -1.000 & 1.000 \end{bmatrix} \qquad \omega^2 = \begin{bmatrix} 25.36 \\ 30.00 \\ 94.64 \end{bmatrix} \; (rad/sec)^2 \qquad (b)$$

Study of these mode shapes reveals that the first and third represent rotations about points on the symmetry diagonal while the second is simple translation along this diagonal. Obviously these motions could have been identified more easily by a more appropriate coordinate system; translation of the center of mass in the direction of the two diagonals plus rotation about the center of mass would have been a better choice.

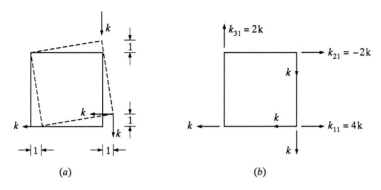

FIGURE E26-5
Evaluation of stiffness coefficients for $v_1 = 1$: (a) displacement $v_1 = 1$ and resisting
column forces; (b) column forces and equilibrating stiffness coefficients.

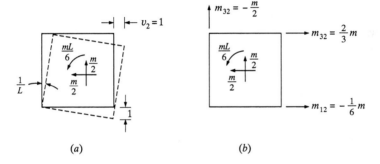

FIGURE E26-6
Evaluation of mass coefficients for $\ddot{v}_2 = 1$. (a) Acceleration $\ddot{v}_2 = 1$ and resisting
inertial forces; (b) slab inertial forces and equilibrating mass coefficients.

The frequencies, periods of vibration, and spectral accelerations given by
Fig. 25-9 (assuming 5 percent damping) for the three modes of this structure
are

$$
\boldsymbol{\omega} = \begin{bmatrix} 5.036 \\ 5.477 \\ 9.464 \end{bmatrix} rad/sec \quad \mathbf{T} = \begin{bmatrix} 1.25 \\ 1.15 \\ 0.65 \end{bmatrix} sec \quad S_{pa} = \begin{bmatrix} 8.5 \\ 8.9 \\ 15.6 \end{bmatrix} ft/sec^2 \quad (c)
$$

Also the generalized masses M_n and modal earthquake-excitation factors
$\mathcal{L}_n = \boldsymbol{\phi}_n^T \mathbf{m} \, \mathbf{r}$ where $\mathbf{r}^T = \langle 0 \quad 0 \quad 1 \rangle$ are

$$
\boldsymbol{M} = \begin{bmatrix} 0.5 \\ 1.0 \\ 0.5 \end{bmatrix} kips \cdot sec^2/ft \quad \boldsymbol{\mathcal{L}} = \begin{bmatrix} 0.3415 \\ -0.5000 \\ -0.0915 \end{bmatrix} kips \cdot sec^2/ft \quad (d)
$$

Hence the maximum modal displacements are found from

$$\mathbf{v}_{n,\text{max}} = \phi_n \frac{L_n}{M_n} \frac{S_{pa}}{\omega_n^2} \tag{e}$$

to be

$$\mathbf{v}_{1,\text{max}} = \begin{bmatrix} 0.084 \\ 0.229 \\ 0.229 \end{bmatrix} ft \quad \mathbf{v}_{2,\text{max}} = \begin{bmatrix} 0.148 \\ 0.148 \\ -0.148 \end{bmatrix} ft \quad \mathbf{v}_{3,\text{max}} = \begin{bmatrix} -0.044 \\ 0.032 \\ 0.032 \end{bmatrix} ft \tag{f}$$

An approximation of the maximum displacement in each coordinate could be determined from these results by the SRSS method.

The reader is reminded that although the response spectral values, as used in the above examples, are always positive by definition, the actual maximum responses they represent may be either positive or negative.

Comparison with ATC-3 Recommended Code Provisions

It is of interest to compare the foregoing formulation of expressions for seismically-induced forces in a multistory building, such as in Fig. 26-4 with the seismic design forces specified in a typical building code. For example, the ATC-3 recommended provisions define the effective intensity of the design earthquake in terms of the maximum shear force it produces at the base of the building as given by[1]

$$V_0 = \frac{1.2\, A_v\, S}{R\, T^{2/3}}\, W \tag{26-51}$$

in which $A_v\, g$ is the velocity-related peak ground acceleration having an annual probability of exceedance equal to 0.002 (usually taken the same as the effective peak ground acceleration $A_a\, g$), S is a site characteristic coefficient, R is a response modification factor depending upon the type of structural framing used, T is the fundamental period of the building, and W is its total weight. The site characteristic coefficient S is assigned a value of 1.0, 1.2, or 1.5 depending on whether the site condition is hard, medium, or soft, respectively, while the response modification factor R is assigned a value in the range $2 < R < 8$ depending upon the inelastic energy-absorption capacity of the framing system.

---

[1] Applied Technology Council (ATC), "Tentative Provisions for the Development of Seismic Regulations for Buildings," loc. cit.

The ATC-3 provisions specify that the equivalent static seismic loadings producing the above base shear be calculated in accordance with

$$f_{Si} = \frac{w_i\, x_i^k}{\sum w_i\, x_i^k}\, V_0 \tag{26-52}$$

in which f_{Si} is the lateral force at level i, w_i is the weight at level i, x_i^k is the height of level i above the building base with exponent k being related to the building's fundamental period (T) in accordance with $k = 1$ for $T \le 0.5$ sec and $k = 2$ for $T \ge 2.5$ sec. For buildings having a period between 0.5 and 2.5 sec, k may be taken as 2 or may be determined by linear interpolation between 1 and 2.

A corresponding analytical expression can be obtained for the first-mode dynamic solution by substituting the first-mode form of Eq. (26-37) into Eq. (26-36), making use of Eq. (26-29), and selecting only the ith component in the resulting vector. Taking this action, one obtains

$$f_{Si}(t) = \frac{m_i\, \phi_{i1}}{\sum m_i\, \phi_{i1}}\, V_{01}(t) \tag{26-53}$$

Comparison of Eqs. (26-52) and (26-53) shows that the code expression represents a loading distribution equivalent in shape to the dynamic loading of a lumped-mass system which is constrained to deflect with the shape $\phi_{i1} = x_i^k / L$. This shape has been incorporated into the code because observations of the vibrations of a large number of tall buildings demonstrate that the first-mode shape generally is quite close to a straight line $(k = 1)$ when the fundamental period is 0.5 sec or less; and is quite close to a parabola $(k = 2)$ when the fundamental period is 2.5 sec or more. An intermediate shape usually exists for a period in the range $0.5 < T < 2.5$ sec.

Let us now make a comparison between the design seismic forces obtained through Eqs. (26-51) and (26-52) above and the corresponding maximum dynamic forces obtained through Eqs. (26-43) and (26-45) using a response spectrum for a specified design earthquake. This comparison can best be accomplished through an example solution as follows:

Example E26-7. Assume the three-story building of Fig. E26-2, having a ductile moment resisting steel frame $(R = 8)$, is subjected to the same horizontal rigid-base motion specified in Example E26-4. In that example, it was determined that the maximum dynamic story shears are given by

$$\mathbf{V} \text{ (dynamic)} = \begin{bmatrix} 161 \\ 258 \\ 350 \end{bmatrix} \ kips \tag{a}$$

Using Eq. (26-51) for the case where $A_v = A_a = 0.3$, $S = S_1 = 1.0$, $R = 8$, $W = 1737$ $kips$ ($M = 4.5$ $kips \cdot sec^2/in$), and $T = T_1 = 1.37$ sec, the design base shear for this structure is 63.4 $kips$. Distributing this base shear in accordance with Eq. (26-52) using $k = 0.75 + 0.5\,T = 1.43$ and summing cumulatively from top to bottom gives the design story shears

$$
\mathbf{V}\,(\text{design}) = \begin{bmatrix} 28.1 \\ 51.7 \\ 63.4 \end{bmatrix} \; kips \tag{b}
$$

Following standard design procedures and code requirements, full interstory yielding will take place when the interstory shear forces reach values about twice their design values, i.e., at the levels

$$
\mathbf{V}\,(\text{yield}) = \begin{bmatrix} 56.2 \\ 103.4 \\ 126.8 \end{bmatrix} \; kips \tag{c}
$$

Since the predicted elastic dynamic shears in the first, second, and third stories exceed their respective yield levels by factors of about 2.8, 2.5, and 2.7, respectively, it is clear that such a structure would deform inelastically under the specified earthquake loading and that the large predicted elastic dynamic shears could not actually develop.

Based on the results of Example E26-7, it is evident that buildings designed in accordance with any modern building code will experience rather large inelastic deformations under maximum probable earthquake conditions. Every effort should be made, however, to insure that these deformations do not exceed dangerous levels because collapse of a building is unacceptable within the standard design strategy. It is therefore important to develop a basic understanding of the inelastic response of such systems to earthquake excitations, and for this reason a brief discussion of the calculated behavior of an elastic-plastic MDOF system is presented in the last subsection of this Section 26-2.

Distributed-Parameter Elastic Systems, Translational Excitation

The formulation of the earthquake response equations for systems having continuously distributed properties can be carried out by procedures which are completely analogous to those described previously. The decoupled normal-coordinate equations of motion take the same form as for the lumped-mass system and may be expressed as

$$\ddot{Y}_n(t) + 2\,\xi_n\,\omega_n\,\dot{Y}_n(t) + \omega_n^2\,Y_n(t) = \frac{P_n(t)}{M_n} = \frac{\mathcal{L}_n}{M_n}\,\ddot{v}_g(t) \qquad (26\text{-}54)$$

However, the generalized mass associated with the distributed mass $m(x)$ is given by

$$M_n = \int_0^L \phi_n^2(x)\,m(x)\,dx \qquad (26\text{-}55)$$

and the modal earthquake-excitation factor takes the integral form equivalent to the previous triple matrix product

$$\mathcal{L}_n = \int_0^L \phi_n(x)\,m(x)\,r(x)\,dx \qquad (26\text{-}56)$$

In this equation, $r(x)$ is the static-displacement influence function representing the displacements resulting from a unit displacement of the ground $v_g = 1$; thus

$$v^s(x) = r(x)\,v_g \qquad (26\text{-}57)$$

By using these expressions for M_n and $\mathcal{L}_n$ in Eq. (26-30) the amplitude of each modal response can be evaluated. Then the total displacement response can be obtained by superposition, using the continuous equivalent of Eq. (26-32) as follows:

$$v(x,t) = \sum_{n=1}^M \phi_n(x)\,Y_n(t) = \sum_{n=1}^M \phi_n(x)\,\frac{\mathcal{L}_n}{M_n\,\omega_n}\,V_n(t) \qquad (26\text{-}58)$$

In practice only the significant modal responses (up to mode M) are included in the superposition even though in principle an infinite number of modes might be considered. The elastic-force distribution is given similarly by an expression analogous to Eq. (26-35):

$$f_S(x,t) = \sum_{n=1}^M m(x)\,\phi_n(x)\,\omega_n^2\,Y_n(t) = \sum_{n=1}^M m(x)\,\phi_n(x)\,\frac{\mathcal{L}_n}{M_n}\,\omega_n\,V_n(t) \qquad (26\text{-}59)$$

Equations (26-58) and (26-59) express the time-history of earthquake response for arbitrary distributed-parameter systems. The procedure for approximating the maximum earthquake response of this type of structure by response-spectrum superposition is entirely equivalent to that described earlier for the lumped-mass systems and need not be discussed further.

Although the analysis procedure for structures with distributed properties outlined above is completely general in principle, its use in practice is limited by the fact that the vibration mode shapes and frequencies can be obtained only for very simple systems. For this reason, the more complicated distributed-parameter systems usually are discretized by the finite-element method so that their analysis can be carried out in matrix form. The matrix equations for the earthquake response analysis of structures that have been idealized as finite-element systems are identical in form to the lumped-mass equations described above, except that where the consistent-mass formulation is used, mass coupling exists between the degrees of freedom so the mass matrix no longer is diagonal. If the column of coefficients in the mass matrix which introduces coupling between the response degrees of freedom and the support displacement is denoted as vector $\mathbf{m}_g$, the equations of motion become

$$\mathbf{m} \, \ddot{\mathbf{v}}^t(t) + \mathbf{m}_g \, \ddot{v}_g(t) + \mathbf{c} \, \dot{\mathbf{v}}(t) + \mathbf{k} \, \mathbf{v}(t) = \mathbf{0} \qquad (26\text{-}60)$$

With the total acceleration expressed in terms of the relative and the quasi-static components by Eq. (26-48), this equation can be written in the form of Eq. (26-25) but with the effective force now given by

$$\mathbf{p}'_{\text{eff}}(t) = -(\mathbf{m} \, \mathbf{r} + \mathbf{m}_g) \, \ddot{v}_g(t) \qquad (26\text{-}61)$$

The corresponding modal earthquake-excitation factor then becomes

$$\mathcal{L}_n = \boldsymbol{\phi}_n^T \, \mathbf{m} \, \mathbf{r} + \boldsymbol{\phi}_n^T \, \mathbf{m}_g \qquad (26\text{-}62)$$

Once this factor has been evaluated, the rest of the analysis is carried out exactly as for a lumped-mass system. In most cases there are few nonzero terms in the mass-coupling vector $\mathbf{m}_g$, and when present they generally are relatively small, hence the second term in Eq. (26-62) usually contributes little to the earthquake-excitation factor; however, it should be included in the formulation for completeness.

Lumped MDOF Elastic Systems, Rotational Excitation

In the preceding discussions and example solutions of this Section 26-2, the earthquake excitation has consisted of a single translational component of rigid-soil motion. Now will be considered lumped MDOF systems subjected to a simple component of rotation applied by the rigid-soil support. In this case, the total displacements are expressed as the sum of the relative displacements and the quasi-static displacements that would result from a static support rotation, i.e.,

$$\mathbf{v}^t(t) = \mathbf{v}(t) + \mathbf{r} \, \theta_g(t) \qquad (26\text{-}63)$$

where θ_g is the applied base rotation and $\mathbf{r}$ is a vector containing the displacements resulting from a unit base rotation. Obviously this equation is equivalent to Eq. (26-48) for the case of base translation.

The governing equation for this system can now be written in the standard form

$$\mathbf{m}\,\ddot{\mathbf{v}}(t) + \mathbf{c}\,\dot{\mathbf{v}}(t) + \mathbf{k}\,\mathbf{v}(t) = -\mathbf{m}\,\mathbf{r}\,\ddot{\theta}_g(t) \tag{26-64}$$

and solved in the usual way. The development of the vector $\mathbf{r}$ is illustrated with reference to the lumped MDOF system shown in Fig. 26-6, which is shown again in Fig. 26-8, subjected to a unit static base rotation. The displacements denoted by v_1, v_2, v_3, and v_4 are the components in the vector $\mathbf{r}$, given by

$$\mathbf{r} = \langle h_1 \quad h_2 \quad x_3 \quad x_4 \rangle^T \tag{26-65}$$

When this vector is used in expressing the effective earthquake forces,

$$\mathbf{p}_{\text{eff}}(t) = -\mathbf{m}\,\mathbf{r}\,\ddot{\theta}_g(t) \tag{26-66}$$

and in the modal earthquake-excitation factor given by Eq. (26-50), the response resulting from the base rotational acceleration $\ddot{\theta}_g(t)$ is calculated in exactly the same way as described previously for systems subjected to rigid-soil translations. It must be noted in this structure that the mass corresponding to displacement v_2 is $m_2 + m_3 + m_4$ because the girders interconnecting these mass lumps are assumed to be inextensible.

In the above illustration it was assumed that the masses were point masses having no rotational inertia. If, however, they have significant rotational inertia, then effective earthquake moments are induced in proportion to the rotational inertias of the masses, which act simultaneously with the effective translational forces resulting from translational inertia. For example, if it is assumed that the tower shown in Fig. 26-9 has rotational inertias J_1 and J_2 in addition to the translational inertias m_1 and m_2, then the rotational degrees of freedom (designated here as v_3 and v_4) must be included in the quasi-static influence coefficient vector representing the motions resulting from rigid-soil rotation as follows: $r = \langle h_1 \quad h_2 \quad 1 \quad 1 \rangle$.

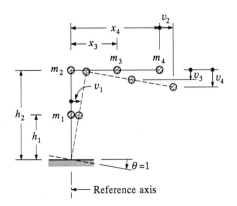

FIGURE 26-8
Lumped MDOF system with rigid-base rotation.

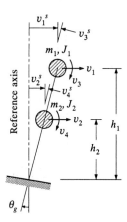

FIGURE 26-9
Tower with lumped masses having rotational inertias subjected to rigid-base rotation.

Lumped MDOF Elastic Systems, Multiple Excitation

In the case when a linear elastic structure is supported at more than one point and is subjected to different — possibly even multicomponent — input components the formulation of the response to each input component is somewhat different from that described above for a system having only one point of support. The difference is that when the multiple supports move independently of each other, they induce quasi-static stresses that must be considered in addition to the dynamic response effects resulting from inertial forces.

To formulate the equations of motion for this general case of earthquake excitation, the multistory frame shown in Fig. 26-10 is considered. This represents a completely general finite-element model for which all the superstructure nodal response components are listed in the vector $\mathbf{v}^t$, where the superscript t denotes that these are total nodal displacements. Similarly all components of support displacement are listed in the vector $\mathbf{v}_g$; these independent input components express the seismic excitation to which the structure is subjected. For this rigid-base input, these are the total displacements of the support points, and no superscript is needed to indicate this fact.

The equilibrium equation expressing the motion of the response degrees of freedom now is written in partitioned matrix form as follows:

$$\begin{bmatrix} \mathbf{m} & \mathbf{m}_g \end{bmatrix} \left\{ \begin{matrix} \ddot{\mathbf{v}}^t(t) \\ \ddot{\mathbf{v}}_g(t) \end{matrix} \right\} + \begin{bmatrix} \mathbf{c} & \mathbf{c}_g \end{bmatrix} \left\{ \begin{matrix} \dot{\mathbf{v}}^t(t) \\ \dot{\mathbf{v}}_g(t) \end{matrix} \right\} + \begin{bmatrix} \mathbf{k} & \mathbf{k}_g \end{bmatrix} \left\{ \begin{matrix} \mathbf{v}^t(t) \\ \mathbf{v}_g(t) \end{matrix} \right\} = 0 \qquad (26\text{-}67)$$

in which the motion vectors have been partitioned to separate the response quantitities from the input, and the property matrices have been partitioned to correspond. The coupling matrices that express forces in the response degrees of freedom due to motions of the supports are denoted here with the subscript g. It will be noted that

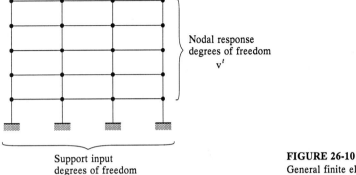

Nodal response
degrees of freedom
v^t

Support input
degrees of freedom
v_g

FIGURE 26-10
General finite element earthquake
response model.

Eq. (26-67) expresses the equilibrium of forces in the response degrees of freedom only and that there are no external loads corresponding to these displacements.

Now an expression for the effective seismic loading is obtained by separating the support motion effects from the response quantities and transferring these input terms to the right hand side; thus

$$\mathbf{m}\,\ddot{\mathbf{v}}^t(t) + \mathbf{c}\,\dot{\mathbf{v}}^t(t) + \mathbf{k}\,\mathbf{v}^t(t)$$

$$= -\mathbf{m}_g\,\ddot{\mathbf{v}}_g(t) - \mathbf{c}_g\,\dot{\mathbf{v}}_g(t) - \mathbf{k}_g\,\mathbf{v}_g(t) = \mathbf{p}_{\text{eff}}(t) \qquad (26\text{-}68)$$

However, the solution for the response to this input can be simplified if the total response motions are expressed as the combination of a quasi-static displacement vector $\mathbf{v}^s(t)$ plus a dynamic response vector $\mathbf{v}(t)$ as has been described for the previous earthquake response formulations; thus

$$\mathbf{v}(t) = \mathbf{v}^s(t) + \mathbf{v}(t) \qquad (26\text{-}69)$$

To evaluate the quasi-static displacements, the static equivalent of Eq. (26-60) is obtained by setting all time-derivative terms to zero and noting that the total displacements then are merely the quasi-static motions (i.e., $\mathbf{v}^t \equiv \mathbf{v}^s$ for this static case). The result of this process is $\mathbf{k}\mathbf{v}^s(t) = -\mathbf{k}_g\,\mathbf{v}_g(t)$ which may be solved for the quasi-static displacements as follows:

$$\mathbf{v}^s(t) = -\mathbf{k}^{-1}\,\mathbf{k}_g\,\mathbf{v}_g(t) \equiv \mathbf{r}\,\mathbf{v}_g(t) \qquad (26\text{-}70)$$

Here it is evident that the influence coefficient matrix $\mathbf{r}$ which expresses that response in all degrees of freedom due to unit support motions is given by

$$\mathbf{r} = -\mathbf{k}^{-1}\,\mathbf{k}_g \qquad (26\text{-}71)$$

Finally, introducing Eq. (26-70) into Eq. (26-69), substituting the result into Eq. (26-68), and transferring all terms associated with the input support motions to the right side leads to

$$\mathbf{m}\,\ddot{\mathbf{v}}(t) + \mathbf{c}\,\dot{\mathbf{v}}(t) + \mathbf{k}\,\mathbf{v}(t) = -\left[\mathbf{mr} + \mathbf{m}_g\right]\ddot{\mathbf{v}}_g(t) - \left[\mathbf{cr} + \mathbf{c}_g\right]\dot{\mathbf{v}}_g(t) \qquad (26\text{-}72)$$

It will be noted that there is no stiffness term in the effective forces on the right side; it drops out because of the definition of the quasi-static displacement matrix given by Eq. (26-71). Also, it may be recognized that this relationship will eliminate any effective input associated with a stiffness-proportional component of the viscous damping. In fact, it can be demonstrated by numerical experiment that the entire velocity-dependent part of this effective input is negligible in comparison to the contribution due to inertia if the viscous damping ratio has any reasonable value. Consequently, Eq. (26-72) may be written in the following approximate form:

$$\mathbf{m}\,\ddot{\mathbf{v}}(t) + \mathbf{c}\,\dot{\mathbf{v}}(t) + \mathbf{k}\,\mathbf{v}^t(t) = -\left[\mathbf{mr} + \mathbf{m}_g\right]\ddot{\mathbf{v}}_g(t) \qquad (26\text{-}73)$$

Comparing this with Eq. (26-61), it is apparent that the effective force vector derived previously for rigid-base translation input is merely a special case of the general effective force expression shown in the right side of Eq. (26-73).

Since this effective force vector is fully known, the dynamic earthquake response can be found by mode superposition in the usual way, first solving the undamped free-vibration equations of motion

$$\mathbf{m}\,\ddot{\mathbf{v}}(t) + \mathbf{k}\,\mathbf{v}(t) = \mathbf{0} \qquad (26\text{-}74)$$

for the normal mode shapes and frequencies, and then using the orthogonality properties of these coordinates to obtain uncoupled modal coordinate equations of motion of the form

$$\ddot{Y}_n(t) + 2\xi_n\omega_n\dot{Y}_n(t) + \omega_n^2\,Y_n(t) = \frac{P_n(t)}{M_n} \equiv \frac{\mathcal{L}_n}{M_n}\,\ddot{v}_g(t) \qquad (26\text{-}75)$$

in which

$$\mathcal{L}_n = \boldsymbol{\phi}_n^T\left[\mathbf{m}\,\mathbf{r} + \mathbf{m}_g\right]$$
$$M_n = \boldsymbol{\phi}_n^T\,\mathbf{m}\,\boldsymbol{\phi}_n \qquad (26\text{-}76)$$

Solution of these modal equations of motion may be obtained either through the time domain or the frequency domain. The dynamic response then is given by standard mode superposition

$$\mathbf{r}(t) = \sum_{n=1}^{M}\boldsymbol{\phi}_n\,Y_n(t) = \mathbf{K}\,\boldsymbol{\phi}\,Y(t) \qquad (26\text{-}77)$$

where only as many modes as are required for engineering accuracy are included in the mode shape matrix ϕ and the modal response vector $Y(t)$.

As was discussed previously, the elastic forces in the superstructure degrees of freedom may be obtained from these dynamic displacements either by premultiplying them by the stiffness matrix $\mathbf{k}$ [as in Eq. (26-33)] or from the equivalent mass matrix formulation, Eq. (26-36). In the stiffness matrix formulation, it is important to note that the support displacements have no effect on the structure nodal forces even though it appears from Eq. (26-67) that the coupling stiffness matrix, $\mathbf{k}_g$, would have this effect. This may be demonstrated by considering the general elastic force expression in Eq. (26-67)

$$\mathbf{f}_s(t) = \mathbf{k}\,\mathbf{v}^t(t) + \mathbf{k}_g\,\mathbf{v}_g(t) \tag{26-78}$$

and using Eqs. (26-69) and (26-70) to express the total response motions, $\mathbf{v}^t(t)$; thus

$$\mathbf{f}_s(t) = \mathbf{k}\,\mathbf{v}(t) + \mathbf{k}\,\mathbf{v}^s(t) + \mathbf{k}_g\,\mathbf{v}_g(t)$$

$$= \mathbf{k}\,\mathbf{v}(t) + \left[\mathbf{k}_g - \mathbf{k}\,\mathbf{k}^{-1}\,\mathbf{k}_g\right]v_g(t) = \mathbf{k}\,\mathbf{v}(t) \tag{26-79}$$

The bracketed term in Eq. (26-79) vanishes because it was set to zero in solving for the quasi-static displacments. On the other hand, elastic forces are developed at the supports by coupling with displacements of the superstructure nodes as well as by the support motions. A general expression for the elastic support forces may be written as follows:

$$\mathbf{f}_g(t) = \mathbf{k}_g^T\,\mathbf{v}(t) + \mathbf{k}_{gg}\,\mathbf{v}_g(t) \tag{26-80}$$

in which $\mathbf{k}_g^T$ expresses the support forces due to superstructure displacements and $\mathbf{k}_{gg}$ represents the support forces due to support motions.

When all of the system elastic nodal forces have been evaluated using Eq. (26-79) [or its mass equivalent Eq. (26-36)] and Eq. (26-80), internal forces of interest may be evaluated by standard methods of statics. However, in typical finite-element programs, the internal element stresses are evaluated directly from the element nodal displacements, and the elastic nodal forces generally are not employed in the analysis.

Lumped SDOF Elastic-Plastic Systems, Translational Excitation

To develop an insight into the seismic behavior of nonlinear yielding systems, consider again the lumped SDOF system of Fig. 26-1 subjected to rigid-base earthquake excitation; however, in this case, let us assume that the columns respond in an elastic-plastic manner such that their base shear forces combine to follow the force-displacement relation shown in Fig. 26-11.

The governing nonlinear equation of motion for this system is

$$\ddot{v} + 2\,\omega\,\xi\,\dot{v} + \frac{f_s(v)}{m} = -\ddot{v}_g(t) \tag{26-81}$$

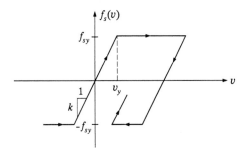

FIGURE 26-11
Elastic-plastic force-displacement
relation.

where $\omega = \sqrt{k/m}$ and $\xi = c/2m\omega$ are its natural frequency and damping ratio, respectively, in the elastic range. This equation can be solved for any prescribed set of parameters and earthquake excitation using the step-by-step integration procedures described earlier in Chapter 7. Let the maximum value of relative displacement so obtained be expressed in terms of a displacement ductility factor μ defined by

$$\mu \equiv \frac{|v(t)|_{\text{max}}}{v_y} \tag{26-82}$$

where v_y is the displacement at which yielding is initiated, as shown in Fig. 26-11. Clearly, by this definition, the entire time-history of response is elastic if $\mu \leq 1$, in which case Eq. (26-81) is identical to Eq. (26-2); however, if $\mu > 1$, the system will have responded into the inelastic range within certain intervals during the time-history. In this latter case, the maximum force developed in the system will equal the yield force f_{sy}, consistent with the relation shown in Fig. 26-11.

Although a step-by-step analysis may be performed easily for a SDOF elastic-plastic system such as this, it requires much more computational effort than is needed for a response spectrum analysis of a linear system. Consequently, it is of practical interest to have comparisons of the maximum values of response $|v(t)|_{\text{max}}$ and $|f_s(t)|_{\text{max}}$ (or $|\ddot{v}_t(t)|_{\text{max}} = |f_s(t)|_{\text{max}}/m$) obtained for an elastic system and for an elastic-plastic system having the same values of parameters ω, ξ, and k and subjected to the same earthquake excitation $\ddot{v}_g(t)$. The objective of the comparisons is to obtain an approximation of the response of the nonlinear system by appropriate interpretation of the response of the linear system. For this purpose, let us denote these maximum values by v_{el}, $f_{s,\text{el}}$, v_{elpl}, and $f_{s,\text{elpl}}$ where subscripts "el" and "elpl" refer to the elastic and elastic-plastic values, respectively. The elastic values are obtained through Eq. (26-2) while the elastic-plastic values are obtained by a step-by-step analysis through Eq. (26-81). As reported in the literature, these comparisons are best made in terms of the above defined ductility factor μ.[2]

---

[2] N. M. Newmark and W. J. Hall, "Earthquake Spectra and Design," loc. cit.

Consider first flexible systems having very low values of natural frequency $f = \omega/2\pi$. Certainly as $f \to 0$, both v_{el} and v_{elpl} will approach the maximum ground displacement $|v_g(t)|_{max}$ because the structural resistance vanishes. In the approximate frequency range $0 < f < 0.3\ Hz$, this condition is nearly true; thus, as seen in Fig. 26-12a,

$$\left.\begin{array}{c} v_{el} \doteq v_{elpl} \doteq |v_g(t)|_{max} \\[2mm] f_{s,y} \doteq f_{s,el}/\mu \doteq k\,|v_g(t)|_{max}/\mu \end{array}\right\} \qquad 0 < f < 0.3\ Hz \qquad (26\text{-}83)$$

In the approximate frequency range $0.3 < f < 2\ Hz$, v_{el} and v_{elpl} will no longer equal $|v_g|_{max}$, but solutions to Eqs. (26-2) and (26-81) for numerous earthquake excitations show that the relative displacement is fairly well preserved, i.e.,

$$\left.\begin{array}{c} v_{el} \doteq v_{elpl} \\[2mm] f_{s,y} \doteq f_{s,el}/\mu \end{array}\right\} \qquad 0.3 < f < 2\ Hz \qquad (26\text{-}84)$$

as indicated in Fig. 26-12b. In the approximate frequency range $2 < f < 8$, the results of many solutions indicate that the deformation energy is fairly well preserved, i.e., the areas under the elastic and the elastic-plastic curves in Fig. 26-12c are nearly equal. Expressing these two areas first in terms of $f_{s,y}$, $f_{s,el}$, k, and μ and second in terms of v_{elpl}, v_{el}, k and μ and then equating the areas in each case, one obtains

$$\left.\begin{array}{c} v_{elpl} = \dfrac{\mu}{\sqrt{2\mu-1}}\,v_{el} \\[4mm] f_{s,y} = \dfrac{1}{\sqrt{2\mu-1}}\,f_{s,el} \end{array}\right\} \qquad 2 < f < 8\ Hz \qquad (26\text{-}85)$$

As $f \to \infty$ ($k \to \infty$), finite yielding of the rigid system will occur with the slightest reduction of $f_{s,y}$ from $f_{s,el}$. Since $v_y \to 0$ in this case, finite yielding corresponds to $\mu = \infty$; thus, for finite values of μ, the force must be preserved. Results have shown that this is nearly true for frequencies $f > 33\ Hz$; thus

$$f_{s,y} \doteq f_{s,el} \doteq m\,|\ddot{v}_g(t)|_{max} \qquad f > 33\ Hz \qquad (26\text{-}86)$$

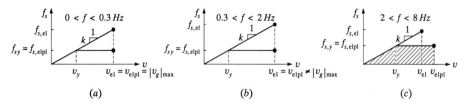

FIGURE 26-12
Elastic and elastic-plastic force-displacement relations.

Substituting $k\,v_y = k\,v_{\text{elpl}}\,/\,\mu$ for $f_{s,y}$ and $k\,v_{\text{el}}$ for $f_{s,\text{el}}$, Eq. (26-86) gives

$$v_{\text{elpl}} \doteq \mu\,v_{\text{el}} \qquad f > 33\ Hz \tag{26-87}$$

In the approximate frequency range $8 < f < 33$, the response is transitional between the state represented by Eqs. (26-85) and that represented by Eqs. (26-86) and (26-87) in which case some form of interpolation between the two cases is required. Fortunately, the fundamental frequencies of most structures, which one might wish to model in the simple elastic-plastic form, are below $8\ Hz$; thus, Eqs. (26-83) through (26-85) will suffice in estimating maximum force (or total acceleration) and relative displacement responses of most SDOF yielding systems subjected to strong earthquake excitations.

26-3   COMBINING MAXIMUM MODAL RESPONSES

As stated in Section 26-2, the square root of the sum of squares (SRSS) method of combining maximum modal responses is fundamentally sound when the modal frequencies are well separated; however, when the frequencies of major contributing modes are very close together, this method will give poor results. Examples of when this situation arises are (1) a tall building with its fundamental lateral vibration mode being very close to its fundamental torsional mode and (2) a complex 3-D nuclear power plant piping system which can have many very closely spaced normal mode frequencies. In cases such as these, the more general Complete Quadratic Combination (CQC) method should be used. Derivations of these combination rules as well as the so-called 30 percent rule for combining the responses to two components of excitation are presented in this section.

Mean Square Response of a Single Mode

The response of a MDOF system in its nth normal mode to a single component of earthquake input $\ddot{v}_g(t)$ is represented by

$$\ddot{Y}_n + 2\,\omega_n\,\xi_n\,\dot{Y}_n + \omega_n^2\,Y_n = -\frac{\mathcal{L}_n}{M_n}\,\ddot{v}_g(t) \tag{26-88}$$

which is the same as Eq. (26-54) except that the sign of the earthquake acceleration has been retained. Taking the direct Fourier transform of Eq. (26-88) gives

$$\mathrm{Y}_n(i\overline{\omega}) = -\frac{\mathcal{L}_n}{K_n}\,\mathrm{H}_n(i\overline{\omega})\,\ddot{\mathrm{V}}_g(i\overline{\omega}) \tag{26-89}$$

where

$$\mathrm{H}_n(i\overline{\omega}) \equiv \frac{1}{\left[\left(1 - \frac{\overline{\omega}^2}{\omega_n^2}\right) + 2i\xi_n\left(\frac{\overline{\omega}}{\omega_n}\right)\right]} \tag{26-90}$$

The inverse Fourier transform of Eq. (26-89) is

$$Y_n(t) = -\frac{\mathcal{L}_n}{2\pi K_n} \int_{-\infty}^{\infty} H_n(i\bar{\omega}) \, \ddot{V}_g(i\bar{\omega}) \, \exp(i\bar{\omega}t) \, d\bar{\omega} \qquad (26\text{-}91)$$

Let us now calculate the mean square intensity of $Y_n(t)$, over the effective duration of the earthquake t_d, as defined by

$$\langle Y_n(t)^2 \rangle \equiv \frac{1}{t_d} \int_0^{t_d} Y_n(t)^2 \, dt \qquad (26\text{-}92)$$

Using Eq. (26-91) in its discrete form, consistent with the FFT of $\ddot{v}_g(t)$, Eq. (26-92) becomes

$$\langle Y_n(t)^2 \rangle = \frac{1}{t_d} \int_0^{t_d} \left[\frac{\mathcal{L}_n^2 \Delta\bar{\omega}^2}{4\pi^2 K_n^2} \sum_{j=-\infty}^{\infty} \sum_{k=-\infty}^{\infty} H_n(i\bar{\omega}_j) \, H_n(i\bar{\omega}_k) \right.$$

$$\left. \ddot{V}_g(i\bar{\omega}_j) \, \ddot{V}_g(i\bar{\omega}_k) \, \exp(i\bar{\omega}_j t) \, \exp(i\bar{\omega}_k t) \right] dt \qquad (26\text{-}93)$$

where $\Delta\bar{\omega} = 2\pi / T_d$ with T_d being the total duration used in the FFT. Note that T_d normally includes a period of time when $\ddot{v}_g(t) = 0$; therefore, $T_d > t_d$. Since $t_d \gg (2\pi / \omega_n)$ for the usual structural frequencies of interest, the time integrals in Eq. (26-93) will be given with sufficient accuracy by the approximate expression

$$\frac{1}{t_d} \int_0^{t_d} \exp(i\omega_j t) \, \exp(i\omega_k t) \, dt \doteq \begin{cases} 0 & |j| \neq k; \ \ k = j \\ 1 & k = -j \end{cases}$$

$$j, k = 1, 2, 3, \cdots \qquad (26\text{-}94)$$

thus, Eq. (26-93) becomes, after converting back to the continuous form,

$$\langle Y_n(t)^2 \rangle = \frac{1}{t_d} \frac{\mathcal{L}_n^2}{4\pi^2 K_n^2} \int_{-\infty}^{\infty} \left| H_n(i\bar{\omega}) \right|^2 \left| \ddot{V}_g(i\bar{\omega}) \right|^2 d\bar{\omega} \qquad (26\text{-}95)$$

Note that for low damped systems, say $\xi < 0.10$, the term $\left| H_n(i\bar{\omega}) \right|^2$ is very highly peaked in the close neighborhood of $\bar{\omega} = \omega_n$ while the term $\left| \ddot{V}_g(i\bar{\omega}) \right|^2$ is not highly peaked; therefore, Eq. (26-95) can be expressed approximately as

$$\langle Y_n(t)^2 \rangle \doteq \frac{C_{nn} \mathcal{L}_n^2}{K_n^2} \int_{-\infty}^{\infty} \left| H_n(i\bar{\omega}) \right|^2 d\bar{\omega} \qquad (26\text{-}96)$$

where

$$C_{nn} \equiv \frac{\left| \ddot{V}_g(i\omega_n) \right|^2}{4\pi^2 t_d} \qquad (26\text{-}97)$$

Making use of Eq. (26-90), it can be shown using contour integration that the integral in Eq. (26-96), denoted as $I_{nn}(\xi_n, \omega_n)$, is given by

$$I_{nn}(\xi_n, \omega_n) = \frac{\pi \omega_n}{2\xi_n} \tag{26-98}$$

thus, the desired mean square response becomes

$$\langle Y_n(t)^2 \rangle = \frac{\pi \omega_n C_{nn} \mathcal{L}_n^2}{2\xi_n K_n^2} \tag{26-99}$$

Covariance of Response Produced by Two Modes

Letting the covariance of response produced by modes m and n be defined by the time average

$$\langle Y_n(t) \, Y_m(t) \rangle \equiv \frac{1}{t_d} \int_0^{t_d} Y_n(t) \, Y_m(t) \, dt \tag{26-100}$$

upon substitution of Eq. (26-91) and following the same steps described above, the corresponding relation to Eq. (26-95) is found to be

$$\langle Y_n(t) \, Y_m(t) \rangle = \frac{1}{t_d} \frac{\mathcal{L}_n \mathcal{L}_m}{4\pi^2 K_n K_m} \int_{-\infty}^{\infty} H_n(i\bar{\omega}) \, H_m(-i\bar{\omega}) \, \left| \ddot{V}_g(i\bar{\omega}) \right|^2 d\bar{\omega} \tag{26-101}$$

Note that for low damped systems, say ξ_n and $\xi_m < 0.10$, the terms $\left| H_n(i\bar{\omega}) \right|$ and $\left| H_m(i\bar{\omega}) \right|$ are very highly peaked in the close neighborhoods of $\bar{\omega} = \omega_n$ and $\bar{\omega} = \omega_m$, respectively. When frequencies ω_n and ω_m are well separated, the narrow peaks of $\left| H_n(i\bar{\omega}) \right|$ and $\left| H_m(i\bar{\omega}) \right|$ do not overlap. In this case, the numerical value of the integral in Eq. (26-101) is relatively small; thus, the covariance given by this equation is very small compared to the mean square intensities of $Y_n(t)$ and $Y_m(t)$. However, when the frequencies ω_n and ω_m are very close together, the narrow peaks of $\left| H_n(i\bar{\omega}) \right|$ and $\left| H_m(i\bar{\omega}) \right|$ overlap sufficiently so that the covariance given by Eq. (26-101) becomes of similar order of magnitude to the mean square intensities. Since the frequencies ω_n and ω_m must become very close to each other for this to happen, the value of $\left| \ddot{V}_g(i\bar{\omega}) \right|^2$ will not vary greatly in the neighborhood of these closely spaced frequencies. Thus, Eq. (26-101) can be written in the approximate form

$$\langle Y_n(t) \, Y_m(t) \rangle \doteq \frac{C_{nm} \mathcal{L}_n \mathcal{L}_m}{K_n K_m} \int_{-\infty}^{\infty} H_n(i\bar{\omega}) \, H_m(-i\bar{\omega}) \, d\bar{\omega} \tag{26-102}$$

where

$$C_{nm} \equiv \text{Re} \, \frac{[V_g(i\omega_n) \, V_g(-i\omega_m)]}{4\pi^2 t_d} \tag{26-103}$$

Making use of Eq. (26-90), it can be shown using contour integration that the integral in Eq. (26-102), denoted as $I_{nm}(\xi_n, \xi_m, \omega_n, \omega_m)$, is[3]

$$I_{nm}(\xi_n, \xi_m, \omega_n, \omega_m)$$

$$= \frac{\pi}{2} \sqrt{\frac{\omega_n \omega_m}{\xi_n \xi_m}} \left[\frac{8\sqrt{\xi_n \xi_m} \, (\xi_n + r\xi_m) \, r^{3/2}}{(1 - r^2)^2 + 4\xi_n \xi_m \, r(1 + r^2) + 4(\xi_n^2 + \xi_m^2) \, r^2} \right] \qquad (26\text{-}104)$$

where

$$r \equiv \omega_n / \omega_m \qquad\qquad \omega_m > \omega_n$$

Thus, the covariance given by Eq. (26-102) becomes

$$\langle Y_n(t) \quad Y_m(t) \rangle \doteq \frac{\pi C_{nm} \mathcal{L}_n \mathcal{L}_m}{2K_n K_m} \sqrt{\frac{\omega_n \omega_m}{\xi_n \xi_m}} \, \rho_{nm} \qquad (26\text{-}105)$$

where

$$\rho_{nm} \equiv \frac{8\sqrt{\xi_n \xi_m} \, (\xi_n + r\xi_m) \, r^{3/2}}{(1 - r^2)^2 + 4\xi_n \xi_m \, r \, (1 + r^2) + 4(\xi_n^2 + \xi_m^2) \, r^2} \qquad (26\text{-}106)$$

When $\xi_n = \xi_m = \xi$, Eq. (26-106) simplifies to the form

$$\rho_{nm} = \rho_{mn} = \frac{8\xi^2 \, (1 + r) \, r^{3/2}}{(1 - r^2)^2 + 4\xi^2 \, r(1 + r)^2} \qquad (26\text{-}107)$$

Note that

$$0 \le \rho_{nm} \le 1 \qquad (26\text{-}108)$$

and

$$\rho_{nn} = \rho_{mm} = 1 \qquad (26\text{-}109)$$

SRSS and CQC Combination of Modal Responses

Consider a response $z(t)$ which has contributions from all N normal modes as indicated by

$$z(t) = \sum_{n=1}^{N} A_n \, Y_n(t) \qquad (26\text{-}110)$$

where coefficients A_n are known for the structural system under consideration. The corresponding mean square response is then given by

$$\sigma_z^2 \equiv \langle z(t)^2 \rangle = \sum_{n=1}^{N} \sum_{m=1}^{N} A_n A_m \, \langle Y_n(t) \quad Y_m(t) \rangle \qquad (26\text{-}111)$$

---

[3] A. Der Kiureghian, "Structural Response to Stationary Excitation," loc. cit.

Making use of Eqs. (26-99) and (26-105) and recognizing that all quantities in these equations are positive, except for $\mathcal{L}_n$ and $\mathcal{L}_m$ which may be either positive or negative in accordance with Eq. (26-29), Eq. (26-111) can be written in the form

$$\sigma_z^2 = \sum_{n=1}^{N} \sum_{m=1}^{N} A_n\, A_m \frac{C_{nm}}{\sqrt{C_{nn}C_{mm}}} \frac{\mathcal{L}_n\mathcal{L}_m}{|\mathcal{L}_n|\,|\mathcal{L}_m|} \rho_{nm} \langle Y_n^2(t)\rangle^{1/2} \langle Y_m^2(t)\rangle^{1/2}$$

(26-112)

Since $\rho_{nn} = \rho_{mm} = 1$ and $\rho_{nm} = \rho_{mn} \ll 1$ when frequencies ω_n and ω_m are well separated, only those cross terms in Eq. (26-112) with ω_n and ω_m close together contribute significantly to σ_z^2. The corresponding values of $C_{nm}/\sqrt{C_{nn}C_{mm}}$ for these terms are nearly equal to unity, and since the values of $C_{nn}/\sqrt{C_{nn}C_{nn}}$ are identically equal to unity, Eq. (26-112) can be written in the form

$$\sigma_z^2 = \sum_{n=1}^{N} \sum_{m=1}^{N} \alpha_{nm}\, A_n\, A_m\, \rho_{nm} \langle Y_n^2(t)\rangle^{1/2} \langle Y_m^2(t)\rangle^{1/2}$$

(26-113)

where

$$\alpha_{nm} \equiv \frac{\mathcal{L}_n\mathcal{L}_m}{|\mathcal{L}_n|\,|\mathcal{L}_m|}$$

(26-114)

Note that α_{nm} is either $+1$ or -1 for $m \neq n$ depending upon the signs of $\mathcal{L}_n$ and $\mathcal{L}_m$. It is, of course, always $+1$ for $m = n$.

It has been shown in Chapter 21 that the maximum values of modal response over duration t_d are proportional to their respective root mean square values, i.e.,

$$|Y_n(t)|_{\max} = B_n\langle Y_n^2(t)\rangle^{1/2} \qquad |Y_m(t)|_{\max} = B_m\langle Y_m^2(t)\rangle^{1/2}$$

(26-115)

The numerical values of B_n and B_m depend upon the ratios $t_d\omega_n/2\pi$ and $t_d\omega_m/2\pi$, respectively, as shown in Fig. 21-13; however, they do not differ greatly in magnitude unless the ratios just mentioned differ greatly, i.e., by an order of magnitude or more. Often in engineering practice, coefficients B_n and B_m are assigned the numerical value 3.

If it is assumed that all separate maximum contributions of modal response, and even the maximum of their combined responses, can be obtained from their corresponding root mean square values using the same proportionality factor B, then it follows from Eq. (26-113) that

$$|z(t)|_{\max} = \left[\sum_{n=1}^{N} \sum_{m=1}^{N} \alpha_{nm}\, A_n\, A_m\, \rho_{nm}\, |Y_n(t)|_{\max}\, |Y_m(t)|_{\max}\right]^{1/2}$$

(26-116)

This method of evaluating maximum total response from the individual maxima of modal responses is known as the complete quadratic combination (CQC) method.[4]

---

[4] A. Der Kiureghian, "Structural Response to Stationary Excitation," loc. cit.

When major contributing modes have frequencies close together, the corresponding cross terms in Eq. (26-116), which may be plus or minus, can be very significant; thus, they should be retained. If, however, the frequencies of the contributing modes are well separated, the cross terms in Eq. (26-116) are negligible, in which case this equation reduces to

$$|z(t)|_{max} = \left[\sum_{n=1}^{N} A_n^2 \, |Y_n(t)|_{max}^2\right]^{1/2} \tag{26-117}$$

which is the SRSS method given previously by Eqs.(26-44) and (26-45) and used in example solutions E26-4 and E26-5.

Expressing maximum values of modal response through their corresponding response spectral values for the specified earthquake motion, Eqs. (26-116) and (26-117) become

$$|z(t)|_{max} = \left[\sum_{n=1}^{N}\sum_{m=1}^{N} A_n \, A_m \, \frac{\mathcal{L}_n \mathcal{L}_m}{M_n M_m} \, \rho_{nm} \, S_d(\xi_n, \omega_n) \, S_d(\xi_m, \omega_m)\right]^{1/2} \tag{26-118}$$

and

$$|z(t)|_{max} = \left[\sum_{n=1}^{n} A_n^2 \, \frac{\mathcal{L}_n^2}{M_n^2} \, S_d(\xi_n, \omega_n)^2\right]^{1/2} \tag{26-119}$$

respectively.

Example E26-8. The three-dimensional structure shown in Fig. E26-7 is subjected to a single component of base excitation in the x-direction which corresponds to a 0.3 g peak acceleration earthquake having the acceleration response spectrum shown in Fig. 25-9 for the hard-soil condition (Type S_1). Damping is assumed to be 5 percent of critical in each normal mode.

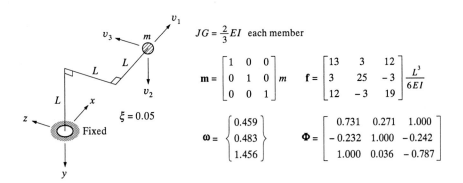

FIGURE E26-7
3-DOF system subjected to rigid-base translation.

Assume all mass is lumped at the single point indicated and that the flexibility of the system consists of flexure and torsion only. If each of the three segments of length L has a solid circular cross section, the torsional stiffness JG will equal 2/3 of the flexural stiffness EI assuming $G = E/3$. Under these conditions, the mass and stiffness matrices for the coordinates shown will be as indicated in the figure. The corresponding eigenvectors and frequencies for the system are also indicated. Note the closeness of the first and second mode frequencies.

Using the hard site response spectrum in Fig. 25-9 and periods corresponding to the natural frequencies given in Fig. E26-7, the acceleration spectral values are found to be

$$S_{pa} = \begin{bmatrix} 7.44 \\ 8.04 \\ 22.9 \end{bmatrix} ft/sec^2 \tag{a}$$

and the modal parameters M_n and $\mathcal{L}_n$ are found to be

$$M = \begin{bmatrix} 1.588 \\ 1.075 \\ 1.678 \end{bmatrix} m \qquad \mathcal{L} = \begin{bmatrix} 0.731 \\ 0.271 \\ 1.000 \end{bmatrix} m \tag{b}$$

From these results, the maximum modal displacements are given by

$$\mathbf{v}_{1,max} = \begin{bmatrix} 0.119 \\ -0.038 \\ 0.162 \end{bmatrix} ft \quad \mathbf{v}_{2,max} = \begin{bmatrix} 0.039 \\ 0.143 \\ 0.005 \end{bmatrix} ft \quad \mathbf{v}_{3,max} = \begin{bmatrix} 0.064 \\ -0.016 \\ -0.055 \end{bmatrix} ft \tag{c}$$

Combining the modal maxima by the SRSS procedure, i.e., using Eq. (26-44), one obtains

$$\mathbf{v}_{max} = \begin{bmatrix} 0.140 \\ 0.149 \\ 0.171 \end{bmatrix} ft \qquad \text{SRSS} \tag{d}$$

However, if one combines the modal maxima using the CQC method, the equivalent to Eq. (26-44) is, in this case [see Eq. (26-116)],

$$\mathbf{v}_{max} \doteq \sqrt{\left[(\mathbf{v}_1)^2_{max} + 2\rho_{12}(\mathbf{v}_1)_{max}(\mathbf{v}_2)_{max} + (\mathbf{v}_2)^2_{max} \right.}$$

$$\left. +2\rho_{23}(\mathbf{v}_2)_{max}(\mathbf{v}_3)_{max} + (\mathbf{v}_3)^2_{max} + 2\rho_{13}(\mathbf{v}_1)_{max}(\mathbf{v}_3)_{max} \right] \tag{e}$$

assuming all damping ratios are the same so that Eq. (26-107) can be used to find the cross-correlation coefficients for all modes. In this case, one finds that $\rho_{12} = \rho_{21} = 0.792$, $\rho_{23} = \rho_{32} = 0.006$, $\rho_{13} = \rho_{31} = 0.006$; thus, using the above equation, one obtains

$$\mathbf{v}_{\max} = \begin{bmatrix} 0.165 \\ 0.117 \\ 0.175 \end{bmatrix} \; ft \qquad\qquad \text{CQC} \qquad\qquad (f)$$

Comparing Eqs. (d) and (f), it is seen that the SRSS method under-estimates the first term while it overestimates the second term; the third term remains nearly the same. It must be remembered that the cross-terms in the CQC method carry a sign, i.e., individual terms are either plus or minus.

In concluding this section on combining modal responses, the reader is reminded that when using Eqs. (26-118) and (26-119), the response quantity of interest must be expressed in the form of Eq. (26-110). While this is the obvious form to be used when $z(t)$ represents an internal force component or deformation, it may not be obvious for all response quantities of interest. For example, suppose $z(t)$ represents the total absolute acceleration of mass m_i in a lumped-mass system as represented by Eq. (26-24), i.e., $z(t) \equiv \ddot{v}_i^t(t)$. As seen by this equation, this response has a contribution not only from the relative motion $\ddot{v}_i(t)$ but from the ground acceleration $\ddot{v}_g(t)$ as well. Nevertheless, it can be expressed in the form of Eq. (26-110) since

$$\ddot{v}_i^t(t) = -\frac{[f_{S_i}(t) + f_{D_i}(t)]}{m_i} \qquad\qquad (26\text{-}120)$$

At the instant $\ddot{v}_i^t(t)$ reaches its maximum absolute value, the damping force $F_{D_i}(t)$ will be very small compared with the spring force $F_{S_i}(t)$; thus, Eq. (26-120) can be expressed in the approximate form

$$\ddot{v}_i^t(t) \doteq -\frac{f_{S_i}(t)}{m_i} = \sum_{n=1}^{N} \omega_n^2 \phi_{in} Y_n(t) \qquad\qquad (26\text{-}121)$$

This equation expresses the response quantity of interest in the form of Eq. (26-110) with $A_n = \omega_n^2 \phi_{in}$.

Combining Two-Component-Excitation Responses

In the previous section, the SRSS and CQC methods have been developed for combining maximum modal responses produced by a single component of horizontal earthquake excitation. Let us now develop procedures for combining the two maximum responses produced by two horizontal components of excitation. If $z(t)$ is the response quantity of interest, it will have two contributions as given by

$$z(t) = z_x(t) + z_y(t) \qquad (26\text{-}122)$$

where $z_x(t)$ and $z_y(t)$ are the contributions produced by horizontal earthquake excitations in the x- and y-directions, respectively. Usually, the x and y axes are taken along principal axes of the structure. Since, as described in Chapter 25, the input earthquake excitations in the x- and y-directions will have very low cross-correlation, the cross-correlation of $z_x(t)$ and $z_y(t)$ will likewise be very low so that it can be neglected. Thus, it is statistically sound to use the SRSS method for weighting the maximum values of $z_x(t)$ and $z_y(t)$, i.e.,

$$\left| z(t) \right|_{\max} = \left[\left| z_x(t) \right|^2_{\max} + \left| z_y(t) \right|^2_{\max} \right]^{1/2} \qquad (26\text{-}123)$$

Suppose $\left| z_x(t) \right|_{\max}$ and $\left| z_y(t) \right|_{\max}$ are obtained using the same design response spectrum and using either Eq. (26-118) or (26-119), as judged appropriate, giving responses that are proportional to each other as expressed by

$$\left| z_y(t) \right|_{\max} = B \left| z_x(t) \right|_{\max} \qquad (26\text{-}124)$$

If the directions of the x and y axes are chosen so that

$$\left| z_x(t) \right|_{\max} \geq \left| z_y(t) \right|_{\max} \qquad (26\text{-}125)$$

then constant B will have a value somewhere in the range $0 \leq B \leq 1$ depending upon the transfer functions between the x and y components of excitation and the particular response $z(t)$ under consideration. Since, however, the design response spectrum represents the stronger component of horizontal excitation and since, as discussed in Chapter 25, the intensity of the weaker component can be taken as 85 percent of the intensity of the stronger component, it is more realistic to use

$$\left| z_y(t) \right|_{\max} = 0.85 \, B \left| z_x(t) \right|_{\max} \qquad (26\text{-}126)$$

Substituting Eq. (26-126) into Eq. (26-123) gives

$$\left| z(t) \right|_{\max} = (1 + 0.723 \, B^2)^{1/2} \left| z_x(t) \right|_{\max} \qquad (26\text{-}127)$$

Let us now compare this statistically rational equation with the "30 percent rule" commonly used in building design as given by

$$\left| z(t) \right|_{\max} = \left| z_x(t) \right|_{\max} + 0.3 \left| z_y(t) \right|_{\max} \qquad (26\text{-}128)$$

in which $\left|z_x(t)\right|_{\max}$ and $\left|z_y(t)\right|_{\max}$ are evaluated through either Eq. (26-118) or Eq. (26-119), as judged appropriate, using the same response spectrum without reduction in the direction of weaker intensity, i.e., using Eq. (26-124); in this case, Eq. (26-128) becomes

$$\left|z(t)\right|_{\max} = (1 + 0.3\,B)\left|z_x(t)\right|_{\max} \tag{26-129}$$

Plots of the ratio $\left|z(t)\right|_{\max}/\left|z_x(t)\right|_{\max}$, as given by Eqs. (26-127) and (26-129), as functions of B are shown in Fig. 26-13. Comparing the two curves in this figure, it is seen that the 30 percent rule, Eq. (26-129), gives identical results for $B = 0$, almost identical results for $B = 1$, and somewhat higher values for $0 < B < 1$; the largest difference is approximately 5 percent.

In the above development $z(t)$ as expressed in Eq. (26-122) represents a specific response quantity, e.g., stress at a particular critical point in a structure or a particular force component in an individual member. Usually, estimating separately the maximum values of specific responses by one of the above procedures is sufficient; however, in some cases this procedure is deficient due to the multiplicity of possible critical responses. One such case is the response of a vertical cantilever structure, such as a smoke stack, or an intake tower having a circular cross section.

Suppose a structure of this type is subjected to earthquake excitations in the x- and y-directions at its base as described above. In attempting to use either of the above procedures, one might possibly select one of the two outer-fiber locations on the x-axis of the base cross section as representing the most critical-stress location. Letting $z(t)$ be the bending stress at this location, one finds that $|z(t)|_{\max}$ given by either Eq. (26-127) or Eq. (26-129) equals $|z_x(t)|_{\max}$ since $B = 0$. This suggests that the circular cantilever can be designed on the basis of a single earthquake input representing the major principal component of ground motion. It needs to be recognized, however, that at the critical time t_c when the bending moment about the y-axis produces $|z_x(t)|_{\max} = |z_x(t_c)|$, the resultant bending moment on the cross section

$$M(t_c) = \sqrt{M_x(t_c)^2 + M_y(t_c)^2} \tag{26-130}$$

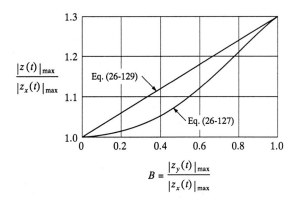

FIGURE 26-13
Statistical approach versus 30% rule in combining two components of horizontal response.

will be greater than $|M_y(t_c)|$ thus producing a bending stress greater than $|z_x(t)|_{\max}$. Further, it should be recognized that the maximum value of the resultant moment $M(t)$ occurs at a time different from t_c and that its absolute value is greater than $M(t_c)$; thus, the absolute value of maximum stress on the cross section is even greater than the maximum stress produced by $M(t_c)$. Therefore, rather than focusing on stress at a fixed point, one should focus on the maximum stress due to $M(t)$ even though its location is changing with time. Since the maximum stress is proportional to the resultant moment, the subsequent analytical treatment will be concerned with predicting the maximum absolute value of the resultant moment $M(t)$ which can then be compared with the maximum absolute value of moment $M_y(t)$.

Let $x \equiv M_x(t)$ and $y \equiv M_y(t)$ represent bending moments produced by input ground motions in the y- and x-directions, respectively. The resultant moment $r \equiv M(t)$ is given by $r = (x^2 + y^2)^{1/2}$. Making the usual assumptions that the ground motions in the x- and y-directions are uncorrelated and have normal distributions, the marginal and joint probability density functions are given by

$$p(x) = \frac{1}{\sqrt{2\pi}\sigma_x} \exp\left[-\frac{1}{2}\frac{x^2}{\sigma_x^2}\right]$$

$$p(y) = \frac{1}{\sqrt{2\pi}\sigma_y} \exp\left[-\frac{1}{2}\frac{y^2}{\sigma_y^2}\right] \qquad (26\text{-}131)$$

$$p(x,y) = \frac{1}{2\pi\sigma_x\sigma_y} \exp\left[-\frac{1}{2}\left(\frac{x^2}{\sigma_x^2} + \frac{y^2}{\sigma_y^2}\right)\right]$$

Letting θ be the angle between the resultant moment vector r and the x-axis, random variables x and y can be transformed to random variables r and θ using

$$x = r\,\cos\theta$$
$$y = r\,\sin\theta \qquad (26\text{-}132)$$

Using the jacobian transformation given by Eq. (20-47), one obtains

$$p(r,\theta) = \frac{r}{2\pi\sigma_x\sigma_y} \exp\left[-\frac{r^2}{2}\left(\frac{\cos^2\theta}{\sigma_x^2} + \frac{\sin^2\theta}{\sigma_y^2}\right)\right] \qquad r \geq 0 \quad 0 \leq \theta \leq 2\pi \quad (26\text{-}133)$$

From this joint probability density function, the probability that r will exceed some prescribed value s is given by

$$\Pr[r > s] = 1 - \int_{r=0}^{s}\int_{\theta=0}^{2\pi} p(r,\theta)\,d\theta\,dr \qquad (26\text{-}134)$$

Substituting $p(r,\theta)$ from Eq. (26-133) into this equation and carrying out the double integration gives the desired result $\Pr[r > s]$.

Consider first the case for which $\sigma_x = \sigma_y = \sigma$. The joint probability density function $p(r, \theta)$ given by Eq. (26-133) reduces to a form independent of θ as given by

$$p(r, \theta) = \frac{r}{2\pi\sigma^2} \exp\left[-\frac{r^2}{2\sigma^2}\right] \qquad r \geq 0 \quad 0 \leq \theta \leq 2\pi \qquad (26\text{-}135)$$

and the probability exceedance function given by Eq. (26-134) becomes

$$\Pr[r > s] = \exp\left[-\frac{s^2}{2\sigma^2}\right] \qquad (26\text{-}136)$$

Letting $s = 3\sigma$, as is often done in estimating the mean extreme value of response, one finds from Eq. (26-136) that $\Pr[r > 3\sigma] = 0.01111$. Using the normal distribution given by the second of Eqs. (26-131) with σ substituted for σ_y, one finds that $\Pr[|y| > 2.535\,\sigma] = 0.0111$. This shows that the resultant moment $M(t)$ at the 3σ level has the same probability of exceedance as does $|M_y(t)|$ at the $2.535\,\sigma$ level. Therefore, when the intensities of the ground motions in the x- and y-directions are the same, i.e., when $\sigma_x = \sigma_y$, the expected maximum stress on the cross section due to both components of input acting simultaneously will be approximately 18 percent higher than the expected maximum stress on the same cross section due to only one component of input.

As pointed out earlier, the intensity of motion in the y-direction is usually about 85 percent of the intensity of the motion in the x-direction, in which case $\sigma_x = 0.85\,\sigma_y$. Substituting this relation into Eq. (26-133) and evaluating Eq. (26-134) for $r = 3\sigma_y$, one obtains $\Pr[r > 3\sigma_y] = 0.00734$. Using the normal distribution given by the second of Eqs. (26-131), one finds that $\Pr[|y| > 2.685\sigma_y] = 0.00734$. This shows that the resultant moment $M(t)$ at the $3\sigma_y$ level has the same probability of exceedance as does $|M_y(t)|$ at the $2.685\sigma_y$ level. Therefore, when the intensity of ground motion in the y-direction is 85 percent of the intensity in the x-direction, i.e., $\sigma_x = 0.85\sigma_y$, the expected maximum stress on the cross section due to both components of input acting simultaneously will be approximately 12 percent higher than the expected maximum stress on the same cross section due to the stronger x-component of ground motion acting alone.

Since the above case with $\sigma_x = 0.85\sigma_y$ is more realistic than the previous case with $\sigma_x = \sigma_y = \sigma$, the design maximum stress level for two simultaneous components of input can be estimated by multiplying the expected maximum stress level for one component of input by the factor 1.12.

PROBLEMS

26-1. Assume that the structure of Fig. 26-1 has the following properties:

$$m = 3.2 \; kips \cdot sec^2/in \qquad k = 48 \; kips/in \qquad \xi = 0.05$$

Determine the maximum displacement and base shear force caused by an earthquake of 0.3 g peak acceleration having the type S_2 response spectrum of Fig. 25-9.

26-2. Repeat Prob. 26-1 assuming that the stiffness of the structure is increased to $k = 300 \; kips/in$. Comment on the effectiveness of increasing stiffness as a means of increasing earthquake resistance.

26-3. Assume that the uniform cantilever column of Fig. E26-2 has the properties $\overline{m} = 0.016 \; kips \cdot sec^2/ft^2$ and $EI = 10^6 \; kips \cdot ft^2$, and that its deflected shape is $\psi(x) = 1 - \cos(\pi x/2L)$. If this structure is subjected to the ground motion of 0.3 g peak acceleration having the type S_1 response spectrum of Fig. 25-9.

(a) Determine the maximum tip displacement, base moment, and base shear.

(b) Determine the maximum displacement, moment, and shear at mid-height.

26-4. Repeat Prob. 26-3 assuming the same response spectrum shape but considering the following nonuniform mass and stiffness properties:

$$m(x) = 0.01 \, (2 - x/L) \; kips \cdot sec^2/ft^2$$

$$EI(x) = 5 \times 10^5 \, (1 - x/L)^2 \; kips \cdot ft^2$$

Use Simpson's rule with $\Delta x = L/2$ to evaluate the generalized property integrals.

26-5. A building similar to that shown in Fig. E26-3 has the following mass matrix and vibration properties:

$$\mathbf{m} = 2 \begin{bmatrix} 1 & 0 & 0 \\ 0 & 1 & 0 \\ 0 & 0 & 1 \end{bmatrix} kips \cdot sec^2/ft \qquad \Phi = \begin{bmatrix} 1.000 & 1.000 & 1.00 \\ 0.548 & -1.522 & -6.26 \\ 0.198 & -0.872 & 12.10 \end{bmatrix}$$

$$\omega = \begin{Bmatrix} 3.88 \\ 9.15 \\ 15.31 \end{Bmatrix} rad/sec$$

Determine the displacement and overturning moment at each floor level and the shear force within each story at a time t_1 during an earthquake when the response integrals for the three modes are

$$\mathbf{V}(t_1) = \left\{ \begin{array}{c} 1.38 \\ -0.50 \\ 0.75 \end{array} \right\} \; ft/sec$$

The height of each story is 12 ft.

26-6. For the structure and earthquake of Prob. 26-5, the acceleration response spectrum values for the three modes are

$$\mathbf{S}_a = \left\{ \begin{array}{c} 9.66 \\ 5.15 \\ 12.88 \end{array} \right\} \; ft/sec^2$$

(*a*) For each mode of vibration, calculate the maximum values of displacement and overturning moment at each floor level and the maximum shear force within each story.

(*b*) By the SRSS method, determine approximate total maximums for each of the response quantities of part *a*.

26-7. For preliminary design purposes, the tall building in Fig. P26-1 will be assumed to behave as a uniform shear beam, the vibration properties of which are completely analogous to those of the uniform bar in axial deformation discussed in Section 18-5. To express this correspondence, it may be noted that the axial rigidity EA and mass per unit length $\overline{m}$ of Section 18-5 are replaced respectively by $(12 \sum EI)/h^2$ and m_j/h to represent the shear building (where $\sum EI$ denotes the sum of the flexural rigidities of all columns

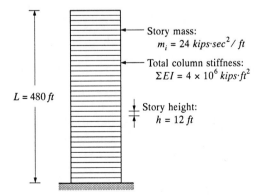

Story mass:
$m_i = 24 \; kips \cdot sec^2/ft$

Total column stiffness:
$\Sigma EI = 4 \times 10^6 \; kips \cdot ft^2$

$L = 480 \, ft$

Story height:
$h = 12 \, ft$

FIGURE P26-1
Uniform shear building.

within one story). Thus the building mode shapes and frequencies are given by

$$\phi_n(x) = \sin \frac{2n-1}{2}\left(\frac{\pi x}{L}\right)$$

$$\omega_n = \frac{2n-1}{2}\pi\left(\frac{12\sum EI}{m_i h L^2}\right)^{1/2}$$

where the values of the properties are shown in the figure above.

(a) Determine the effective modal mass $\mathcal{L}_n^2/M_n$ for each of the first five modes. What fraction of the total mass is associated with each mode?

(b) Compute the approximate maximum top displacement, base shear, and base overturning moment by the root-sum-square method, assuming that the velocity response spectrum value for each mode is 1.6 ft/sec.

26-8. A structure is idealized as the two-degree-of-freedom system shown in Fig. P26-2; also shown are its vibration mode shapes and frequencies. Assuming $\xi = 5$ percent in each mode, using the response spectrum of type S_2 in Fig. 25-9, compute the approximate (SRSS) maximum moment at the column base assuming the direction of the earthquake motions is

(a) Horizontal.

(b) Vertical.

(c) Along the inclined axis ZZ.

Assume the ground motions are caused by an earthquake of 0.3 g peak acceleration.

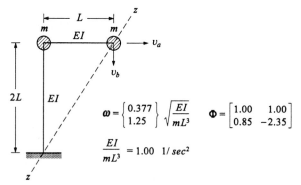

FIGURE P26-2
2DOF plane frame.

26-9. A 6-in concrete slab is supported by four $W8 \times 40$ columns which are located and oriented as shown in Fig. P26-3. Also shown are the structure's mass matrix and vibration properties, based on the assumption that the slab is rigid, that the columns are weightless, and that the clear height of the columns is 12 ft. The mass matrix and mode shapes are expressed in terms of the slab centroid coordinates that are shown.

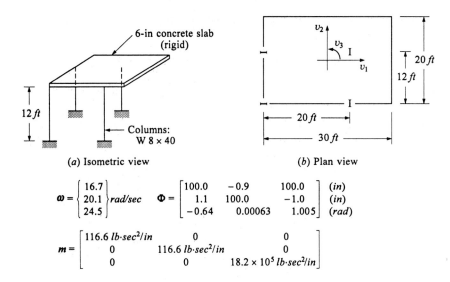

$$\boldsymbol{\omega} = \left\{ \begin{array}{c} 16.7 \\ 20.1 \\ 24.5 \end{array} \right\} rad/sec \qquad \boldsymbol{\Phi} = \left[\begin{array}{ccc} 100.0 & -0.9 & 100.0 \\ 1.1 & 100.0 & -1.0 \\ -0.64 & 0.00063 & 1.005 \end{array} \right] \begin{array}{c} (in) \\ (in) \\ (rad) \end{array}$$

$$\boldsymbol{m} = \left[\begin{array}{ccc} 116.6\ lb\!\cdot\!sec^2/in & 0 & 0 \\ 0 & 116.6\ lb\!\cdot\!sec^2/in & 0 \\ 0 & 0 & 18.2 \times 10^5\ lb\!\cdot\!sec^2/in \end{array} \right]$$

FIGURE P26-3
Rigid deck frame.

Assuming that an earthquake of 0.3 g peak acceleration, having the type S_2 response spectrum of Fig. 25-9, acts in the direction of coordinate v_1, determine the maximum dynamic displacement at the top of each column in the *first* mode of vibration.

26-10. A uniform bridge deck is simply supported with an 80-ft span, as shown below. Also shown are the mass and stiffness properties as well as an idealized earthquake-velocity-response spectrum. Assuming that this same earthquake acts simultaneously on *both* end supports in the vertical direction,

   (a) Compute the maximum moment at midspan for each of the first three modes of vibration.

   (b) Compute the approximate (SRSS) maximum midspan moment due to these three modes.

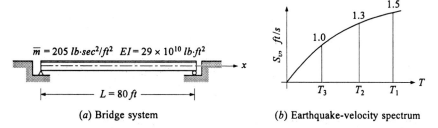

$\overline{m} = 205\ lb\!\cdot\!sec^2/ft^2 \quad EI = 29 \times 10^{10}\ lb\!\cdot\!ft^2$

(a) Bridge system

(b) Earthquake-velocity spectrum

FIGURE P26-4
Bridge subjected to vertical earthquake motions.

26-11. Repeat Prob. 26-10, assuming that only the right hand support is subjected to this vertical motion. Note that $r(x) = x/L$ in this case.

26-12. The service platform for a space rocket is idealized as a lumped mass tower, as shown below. Also shown are the shapes and frequencies of its first two modes of vibration. Determine the maximum moment developed at the base of this tower due to a harmonic horizontal ground acceleration $\ddot{v}_g = A \sin \bar{\omega} t$ where $A = 5 \ ft/sec^2$ and $\bar{\omega} = 8 \ rad/sec$. Consider only the steady-state response of the first two modes, and neglect damping.

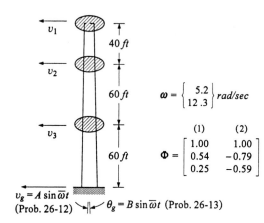

FIGURE P 26-5
Lumped mass tower subjected to earthquake.

26-13. Repeat Prob. 26-12 assuming that the harmonic ground motion applied at the base is a rotation θ_g rather than horizontal translation. In this case $\ddot{\theta}_g = B \sin \bar{\omega} t$ where $B = 0.06 \ rad/sec^2$ and $\bar{\omega} = 8 \ rad/sec$.

26-14. A rigid bar of length L and total uniformly distributed mass m has an additional lumped mass $m/2$ at each end. This bar is rigidly attached to the top of a weightless column of length L and has a lateral spring support at midheight, as shown in Fig. P26-6. The mass matrix for the rigid bar and the stiffness matrix for the entire system including the support degrees of freedom are shown in the figure, together with the vibration properties. This system is subjected to a ground motion for which the spectral velocity at the first mode period is 2.7 ft/sec. Determine the *first* mode maximum response of coordinate v_2 if the earthquake motion is applied:

(a) At both support points simultaneously.

(b) Only at the column base (coordinate v_{gb}), while the spring support (v_{ga}) is fixed against motion.

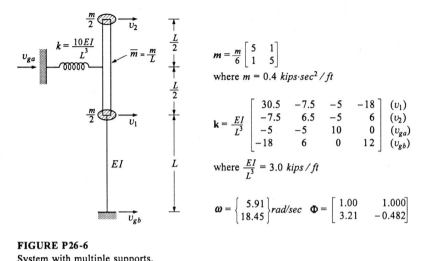

$$m = \frac{m}{6}\begin{bmatrix} 5 & 1 \\ 1 & 5 \end{bmatrix}$$

where $m = 0.4 \ kips\cdot sec^2/ft$

$$\mathbf{k} = \frac{EI}{L^3}\begin{bmatrix} 30.5 & -7.5 & -5 & -18 \\ -7.5 & 6.5 & -5 & 6 \\ -5 & -5 & 10 & 0 \\ -18 & 6 & 0 & 12 \end{bmatrix}\begin{matrix} (v_1) \\ (v_2) \\ (v_{ga}) \\ (v_{gb}) \end{matrix}$$

where $\dfrac{EI}{L^3} = 3.0 \ kips/ft$

$$\omega = \begin{Bmatrix} 5.91 \\ 18.45 \end{Bmatrix} rad/sec \quad \Phi = \begin{bmatrix} 1.00 & 1.000 \\ 3.21 & -0.482 \end{bmatrix}$$

FIGURE P26-6
System with multiple supports.

CHAPTER
27

DETERMINISTIC EARTHQUAKE RESPONSE: INCLUDING SOIL-STRUCTURE INTERACTION

27-1 SOIL-STRUCTURE INTERACTION BY DIRECT ANALYSIS

In the preceding discussions of structural response to earthquakes, it has been assumed that the foundation medium is very stiff and that the seismic motions applied at the structure support points are the same as the free-field earthquake motions at those locations; in other words, the effects of soil-structure interaction (SSI) have been neglected. In actuality, however, the structure always interacts with the soil to some extent during earthquakes, imposing soil deformations that cause the motions of the structure-soil interface to differ from those that would have been observed in the free field.

The nature and amount of this interaction depends not only on the soil stiffness, but also on the stiffness and mass properties of the structure. The interaction effect associated with the stiffness of the structure is termed kinematic interaction and the corresponding mass-related effect is called inertial interaction. In this presentation it will be possible to give only a brief explanation of the concepts of SSI, starting with

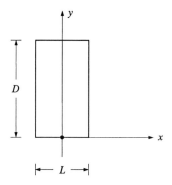

FIGURE 27-1
Rigid rectangular basemat of a large structure.

a derivation of the kinematic interaction effect for a rigid, massless foundation slab — the so-called "Tau effect." This is followed by a brief description of the direct analysis concept for a combined soil-structure system in which the soil underlying the structure is represented as a "bounded" finite-element model. Finally, the substructure approach to SSI analysis is described, in which the structure to be analyzed (and possibly a portion of the adjacent soil) is considered as one substructure while the second substructure is a differential equation representation of the remaining soil domain. For a more comprehensive treatment of this important subject, the reader is referred to the excellent books by Dr. John P. Wolf.[1]

Kinematic Interaction for Translational Excitation; the Tau Effect

Kinematic interaction due to translational excitation is discussed here with reference to the rigid, rectangular basemat shown in Fig. 27-1. When the free-field earthquake motions vary significantly within the area where this mat is located, it is apparent that they will be constrained to some extent by the rigid mat. If the dimensions D and L of the mat are small compared with the apparent wave lengths in the free-field motions over the frequency range of interest, the slab will exert little constraint on the soil and the slab motions will be essentially the same as the free-field motions at that location. But if the mat dimensions are of the same order as the wave lengths, then the resulting slab motions will be only some average of the free-field motions in that area.

Suppose, for example, that these free-field motions act in the x-direction only and are independent of x, as denoted by the ground acceleration function $\ddot{v}_{gx}(y,t)$. If these motions are caused by a single wave train moving in the y-direction with

---

[1] *Dynamic Soil-Structure Interaction* (1985) and *Soil-Structure Interaction Analysis in the Time Domain* (1988), both published by Prentice-Hall, Inc., Englewood Cliffs, N. J.

apparent velocity V_a, they can be expanded into an orthogonal series

$$\ddot{v}_{gx}(y,t) = \sum_i a_{ix}(t)\,\gamma_i(y) \tag{27-1}$$

where the wave shape functions $\gamma_i(y)$ satisfy the orthogonality condition

$$\int_0^D \gamma_i(y)\,\gamma_k(y)\,dy = 0 \qquad (i \neq k) \tag{27-2}$$

A reasonable set of assumed dimensionless displacement functions γ_i $(i = 1, 2, \cdots)$ is

$$\left.\begin{aligned} \gamma_1(y) &= 1 \\ \gamma_2(y) &= 1 - \frac{2y}{D} \\ \gamma_3(y) &= \cdots \end{aligned}\right\} \qquad 0 < y < D \tag{27-3}$$

which express the vibration mode shapes of a uniform free-free beam. The first two functions represent the rigid-body modes while the others correspond to the flexural modes. The acceleration coefficients a_{ix} may be evaluated by multiplying both sides of Eq. (27-1) by $\gamma_k(y)$, integrating with respect to y from zero to D, and using the orthogonality relation Eq. (27-2). The first two coefficients obtained in this way are

$$a_{1x}(t) = \frac{1}{D}\int_0^D \ddot{v}_{gx}(y,t)\,dy$$

$$a_{2x}(t) = \frac{3}{D}\int_0^D \left(1 - \frac{2y}{D}\right)\ddot{v}_{gx}(y,t)\,dy \tag{27-4}$$

The first term in Eq. (27-1) represents uniform rigid-body translation in the x-direction over the entire base area, the second term represents rigid-body rotation about the vertical z-axis, and the remaining terms represent motions which would be filtered out by a rigid foundation. Since only the first term is effective as a translational input to the structure, the resulting acceleration of the rigid base can be taken as

$$a_x(t) = \frac{1}{D}\int_0^D \ddot{v}_{gx}(y,t)\,dy \tag{27-5}$$

Now if the free-field ground acceleration at a selected value of y, as produced by the wave train, say at $y = 0$, is designated as $\ddot{v}_{gx}(t)$, it can be expressed as the combination of a sequence of harmonic terms, i.e., by the Fourier integral

$$\ddot{v}_{gx}(t) = \frac{1}{2\pi}\int_{-\infty}^{\infty} A(i\bar{\omega})\,\exp(i\bar{\omega}t)\,d\bar{\omega} \tag{27-6}$$

where

$$A(i\overline{\omega}) = \int_{-\infty}^{\infty} \ddot{v}_{gx}(t) \exp(-i\overline{\omega}t) \, dt \qquad (27\text{-}7)$$

The motions at any other value of y, $\ddot{v}_{gx}(y,t)$, can then be expressed as

$$\ddot{v}_{gx}(y,t) = \frac{1}{2\pi} \int_{-\infty}^{\infty} A(i\overline{\omega}) \exp\left[i\overline{\omega}(t - \frac{y}{V_a})\right] d\overline{\omega} \qquad (27\text{-}8)$$

Substituting Eq. (27-8) into Eq. (27-5) and integrating over y give

$$a_x(t) = \frac{1}{2\pi} \int_{-\infty}^{\infty} A(i\overline{\omega}) \left[\frac{\exp(-i\frac{\overline{\omega}D}{V_a}) - 1}{(-i\frac{\overline{\omega}D}{V_a})}\right] \exp(i\overline{\omega}t) \, d\overline{\omega} \qquad (27\text{-}9)$$

This equation shows that the modified rigid-base translational input acceleration $a_x(t)$ is obtained by Fourier transforming the specified single-point acceleration $\ddot{v}_{gx}(t)$, Eq. (27-7), multiplying the resulting function $A(i\overline{\omega})$ by the complex square bracket term in Eq. (27-9), and then inverse Fourier transforming in accordance with the same equation.

If a τ-factor is now defined as the ratio of amplitudes of the harmonics in the rigid-base translational motion, Eq. (27-9), to the corresponding free-field amplitudes in Eq. (27-6), one obtains

$$\tau = \frac{1}{\alpha} \sqrt{2\left(1 - \cos\alpha\right)} \qquad (27\text{-}10)$$

where

$$\alpha \equiv \frac{\overline{\omega}\,D}{V_a} = \frac{2\pi\,D}{\lambda(\overline{\omega})} \qquad (27\text{-}11)$$

and the wave lengths, denoted by $\lambda(\overline{\omega})$, equal $2\pi\,V_a\,/\,\overline{\omega}$. A plot of this τ-factor over the range $0 < \alpha < 5\pi/2$ is shown in Fig. 27-2. Note that it diminishes from unity for an infinite wave length ($\lambda = \infty$) at $\alpha = 0$, to zero at $\alpha = 2\pi$, where the wave length is equal to the base dimension ($\lambda = D$).

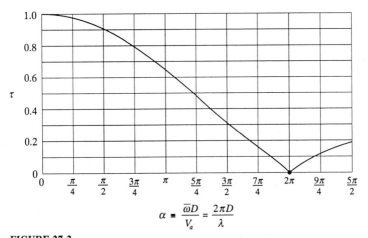

FIGURE 27-2
τ-factor as a function of frequency and apparent wave velocity.

Consider, for example, an offshore gravity tower having a base dimension $D = 400\ ft$ resting on relatively soft soil conditions having a shear wave velocity equal to $1,000\ ft/sec$. For a wave train moving horizontally with $V_a = 1,000\ ft/sec$, the τ-factor for a frequency of 1 Hz would be equal to 0.757 showing a considerable reduction in the excitation at 1 Hz. Since a tower such as this may have a fundamental frequency similar to this value, the τ-factor could significantly lower the excitation at its fundamental frequency, thus reducing response accordingly.

Direct Inclusion of a Bounded Soil Layer

It is evident from the foregoing description that the "Tau Effect" accounts for only a very limited aspect of soil-structure interaction: just the kinematic effect of earthquake motions that vary within the contact zone of the structure. To deal fully with the SSI mechanism, the soil must be represented explicitly in the analytical model, and in principle it appears that this could be done by merely combining a layer of soil with the model of the structure; in fact, some of the earliest attempts to deal with SSI treated the problem in this way.

Unfortunately, this direct approach has the major deficiency that the bounded soil model does not allow vibration energy in the structure and soil to propagate away, and thus it ignores an effective damping mechanism. For this reason a bounded soil layer model should be used only in cases where the soil supporting the structure is underlain by a very stiff rock layer as depicted by Fig. 27-3. If the soil is modelled as a finite-element assemblage, as suggested there, a direct analysis may be performed for

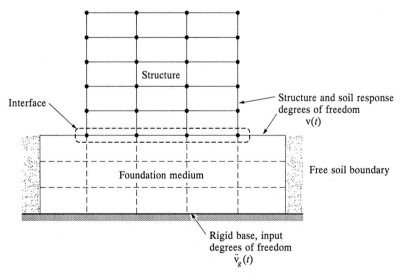

FIGURE 27-3
Finite-element model of combined structure and supporting soil.

the combined soil-structure system by means of Eq. (26-67), if the nodal displacements of both soil and structure are considered as response displacements and the input is represented by prescribed motions of any selected number of nodes at the rock base.

One deficiency of this formulation is that the earthquake excitation is applied at the base of the soil layer, whereas the seismic input usually is expressed in terms of accelerograms recorded at the free-field soil surface. Consequently it is useful to reformulate the response equation so that the effective input is expressed in terms of the free-field motions of the soil layer. Such a reformulation is described in the first edition of this text[2] and need not be restated here. It must be recognized, however, that the free-field input considered in that formulation is associated with a bounded soil layer; therefore the analysis procedure developed there should be used only if the soil is resting on rigid-base rock.

27-2 SUBSTRUCTURE ANALYSIS OF SSI RESPONSE

As was explained above, in a substructure analysis of SSI the foundation mechanism and the structure are represented as two independent mathematical models — or substructures. The connection between them is provided by interaction forces of equal amplitude but acting in opposite directions on the two substructures. The total motions developed at the interface are the sum of the free-field motions at the interface of the soil without the added structure plus the additional motions resulting from the interaction. The dynamic equilibrium relationships for the interface degrees of freedom are written in terms of these motions and then are solved to determine the resulting displacements. Because the stiffness and damping properties of the soil substructure are frequency dependent, it is most convenient to carry out the earthquake response analysis in the frequency domain and then to obtain the response history by transforming back to the time domain. In this presentation, this method of analysis is first described in detail for the very simple case of a SDOF structure supported by a rigid foundation slab resting on an elastic half-space. The concepts are then extended to the analysis of a MDOF structure resting on a flexible base system. In this case, the excitation may include any number of prescribed independent free-field components of motion.

Lumped SDOF System on Rigid Mat Foundation

Consider now the lumped SDOF elastic system shown in Fig. 27-4a which is supported on a rigid basemat of mass m_0 and mass moment of inertia J_0 (about the x-axis) which is, in turn, supported on an elastic half-space. For this case, assume the horizontal dimensions of the basemat are sufficiently small so that the τ-effect described in Section 27-1 is negligible. The uniform free-field ground acceleration

---

[2] *Dynamics of Structures*, McGraw-Hill Book Company, New York (1975), pp. 584–588.

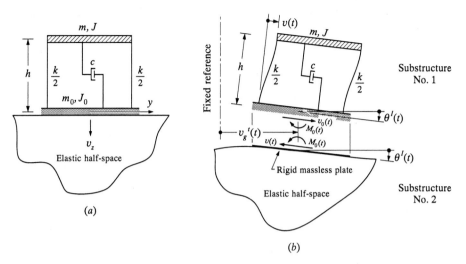

FIGURE 27-4
Lumped SDOF elastic system on rigid mat foundation.

$\ddot{v}_g(t)$ in the y-direction at the half-space surface will then cause foundation forces to develop at the interface between the basemat and the half-space, forcing the basemat to translate and rotate as indicated in Fig. 27-4b. Since substructuring, with the total structure as Substructure No. 1 and the soil foundation as Substructure No. 2, will be used in this formulation, a rigid massless plate is assumed to be present on the surface of the half-space to insure its displacement compatibility with the lower surface of the rigid basemat. Note that the total base displacement $v_g^t(t)$ shown in Fig. 27-4b equals the ground free-field surface displacement $v_g(t)$ plus the added displacement $v_g^I(t)$ caused by soil-structure interaction, i.e.,

$$v_g^t(t) = v_g(t) + v_g^I(t) \qquad (27\text{-}12)$$

The basemat rotation $\theta^I(t)$ is, of course, totally caused by soil-structure interaction since no free-field ground rotation is being considered in this case.

The substructure method will now be used to develop the governing equilibrium equations of motion expressed in terms of the parameters of the overall system and the three unknown displacements $v(t)$, $v_g^I(t)$, and $\theta^I(t)$ where $v(t)$ is the motion of mass m relative to the base, as shown in the figure. Because of soil-structure interaction which allows displacements $v_g^I(t)$ and $\theta^I(t)$ to take place, the overall system has 3DOF; otherwise it would be a SDOF system as treated previously.

To obtain the equations of motion for Substructure No. 1, first the mass m is isolated to get its horizontal force equilibrium equation:

$$m\,\ddot{v} + 2\,m\,\omega\,\xi\,\dot{v} + k\,v + m\,h\,\ddot{\theta}^I + m\,\ddot{v}_g^I + m\,\ddot{v}_g = 0 \qquad (27\text{-}13)$$

where

$$\xi = c \, / \, 2 m \, \omega \qquad \omega = \sqrt{k/m} \qquad (27\text{-}13a)$$

Next, the entire structure (Substructure No. 1) is isolated from the elastic half-space (Substructure No. 2) to get the substructure horizontal force equilibrium equation, as follows:

$$m \, \ddot{v} + m \, h \, \ddot{\theta}^I + (m + m_0) \, \ddot{v}_g^I + (m + m_0) \, \ddot{v}_g = V_0(t) \qquad (27\text{-}14)$$

where $V_0(t)$ denotes the base interaction shear force. Finally, moments about the centroidal x-axis of the basemat are summed for Substructure No. 1 to obtain

$$m \, h \, \ddot{v} + (m \, h^2 + J + J_0) \, \ddot{\theta}^I + m \, h \, \ddot{v}_g^I + m \, h \, \ddot{v}_g = M_0(t) \qquad (27\text{-}15)$$

in which $M_0(t)$ is the base interaction moment. Fourier transforming Eqs.(27-13), (27-14), and (27-15) gives the corresponding equations of motion for Substructure No. 1 in the frequency domain as

$$(-\overline{\omega}^2 m + 2 i \overline{\omega} \omega \xi m + k) \, \mathrm{V}(i\overline{\omega}) - \overline{\omega}^2 m h \, \Theta^I(i\overline{\omega})$$

$$- \overline{\omega}^2 m \, \mathrm{V}_g^I(i\overline{\omega}) + m \, \ddot{\mathrm{V}}_g(i\overline{\omega}) = 0$$

$$-\overline{\omega}^2 m \, \mathrm{V}(i\overline{\omega}) - \overline{\omega}^2 m h \, \Theta^I(i\overline{\omega}) - \overline{\omega}^2 (m + m_0) \, \mathrm{V}_g^I(i\overline{\omega})$$

$$+ (m + m_0) \, \ddot{\mathrm{V}}_g(i\overline{\omega}) = V_0(i\overline{\omega}) \qquad (27\text{-}16)$$

$$-\overline{\omega}^2 m h \, \mathrm{V}(i\overline{\omega}) - \overline{\omega}^2 (m h^2 + J + J_0) \, \Theta^I(i\overline{\omega})$$

$$- \overline{\omega}^2 m h \, \mathrm{V}_g^I(i\overline{\omega}) + m h \, \ddot{\mathrm{V}}_g(i\overline{\omega}) = M_0(i\overline{\omega})$$

The equations of motion for Substructure No. 2 involve only the two soil-structure interaction DOF, $v_g^I(t)$ and $\theta^I(t)$. These equations for a circular rigid basemat are given by the solutions of Veletsos and Wei[3] and/or Luco and Westman[4] which provide the complex frequency dependent compliance functions (dynamic flexibility functions) for a rigid massless circular plate resting on the surface of an isotropic homogeneous elastic half-space. By subjecting the rigid plate separately to each of the four unit amplitude harmonic forces shown in Fig. 27-5, and treating the half-space as a continuum, these authors solved the governing 3-D equations of motion to obtain the corresponding compliance functions which are coupled only in the translational

---

[3] A. S. Veletsos and Y. T. Wei, "Lateral and Rocking Vibrations of Footings," *Jour. of the Soil Mechanism and Foundations Division*, ASCE, Vol. 97, 1971.

[4] J. E. Luco and R. A. Westman, "Dynamic Response of Circular Footings," *Jour. of the Engineering Mechanism Division*, ASCE, Vol. 97, (EM5), 1971.

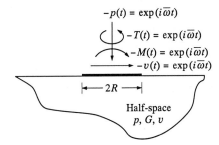

FIGURE 27-5
Rigid massless circular plate on half-space.

and rocking modes. Inverting this set of compliance functions gives the corresponding complex frequency dependent impedance functions (dynamic stiffness functions) in the form

$$G(ia_0) = G^R(a_0) + i\, G^I(a_0) \tag{27-17}$$

where superscripts R and I refer to the real and imaginary parts, respectively, and a_0 is a dimensionless frequency defined by

$$a_0 \equiv R\overline{\omega} / V_s \tag{27-18}$$

in which V_s is the shear wave velocity for the material of the uniform half-space and R is the radius of the circular plate. Plots of $G^R(a_0)$ and $G^I(a_0)$ in nondimensional form are shown subsequently in Fig. 27-6 for each excitation case, namely (a) vertical translation, (b) torsion, (c) lateral translation, (d) rocking, and (e) coupled lateral translation and rocking. Each nondimensional graph of $G^R(a_0)$ and $G^I(a_0)$ in this figure is expressed in terms of plate radius R, and the shear modulus and Poisson's ratio of the half-space material are denoted by G and ν, respectively.

Using these impedance functions, the interaction forces acting on Substructure No. 2 in Fig. 27-4b are given in the frequency domain by

$$\left\{ \begin{array}{c} -V_0(i\overline{\omega}) \\ -M_0(i\overline{\omega}) \end{array} \right\} = \left[\begin{array}{cc} G_{v_g^I v_g^I}(i\overline{\omega}) & G_{v_g^I \theta^I}(i\overline{\omega}) \\ G_{\theta^I v_g^I}(i\overline{\omega}) & G_{\theta^I \theta^I}(i\overline{\omega}) \end{array} \right] \left\{ \begin{array}{c} \mathrm{V}_g^I(i\overline{\omega}) \\ \Theta^I(i\overline{\omega}) \end{array} \right\} \tag{27-18}$$

where impedances $G_{v_g^I v_g^I}(i\overline{\omega})$, $G_{\theta^I \theta^I}(i\overline{\omega})$, and $G_{v_g^I \theta^I}(i\overline{\omega}) = G_{\theta^I v_g^I}(i\overline{\omega})$ are given by the dimensionless plots shown in Figs. 27-6c, d, and e, respectively.

After writing Eqs. (27-16) in matrix form and substituting Eq. (27-19) for the interaction forces, one obtains

$$\left[\begin{array}{ccc} G_{11} & G_{12} & G_{13} \\ G_{21} & G_{22} & G_{23} \\ G_{31} & G_{32} & G_{33} \end{array} \right] \left\{ \begin{array}{c} \mathrm{V}(i\overline{\omega}) \\ \mathrm{V}_g^I(i\overline{\omega}) \\ \Theta^I(i\overline{\omega}) \end{array} \right\} = \left\{ \begin{array}{c} -m \\ -(m+m_0) \\ -mh \end{array} \right\} \ddot{\mathrm{V}}_g(i\overline{\omega}) \tag{27-20}$$

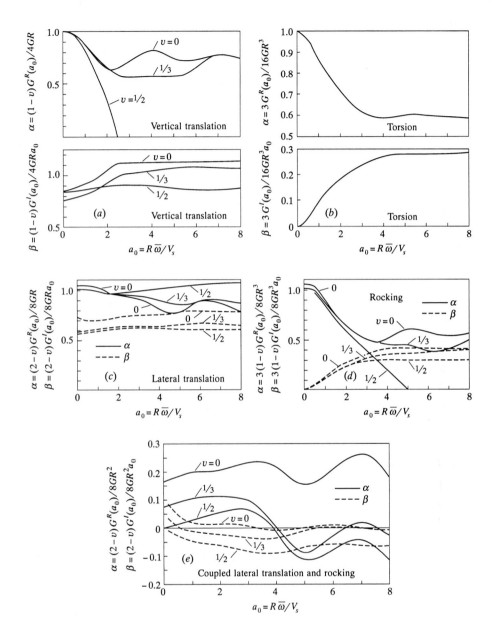

FIGURE 27-6
Rigid massless circular plate impedances.

where

$$G_{11} = -\overline{\omega}^2 m + 2 i \overline{\omega} \omega \xi m + k$$

$$G_{12} = G_{21} = -\overline{\omega}^2 m$$

$$G_{13} = G_{31} = -\overline{\omega}^2 m h$$

$$G_{22} = -\overline{\omega}^2 (m + m_0) + G_{v_g^I v_g^I}(i\overline{\omega}) \qquad (27\text{-}21)$$

$$G_{23} = G_{32} = -\overline{\omega}^2 m h + G_{v_g^I \theta^I}(i\overline{\omega})$$

$$G_{33} = -\overline{\omega}^2 (m h^2 + J + J_0) + G_{\theta^I \theta^I}(i\overline{\omega})$$

Having the prescribed free-field ground acceleration $\ddot{v}_g(t)$, it can be Fourier transformed to give $\ddot{V}_g(i\overline{\omega})$; then Eqs. (27-20) can be solved for discrete values of $\overline{\omega}$ giving responses $V(i\overline{\omega})$, $V_g^I(i\overline{\omega})$, and $\Theta^I(i\overline{\omega})$ in the frequency domain. Inverse Fourier transforming these responses gives $v(t)$, $v_g^I(t)$, and $\theta(t)$.

Direct use of the impedances in Fig. 27-6 implies that the basemat for the above system is rigid and circular. If it is indeed rigid but square of dimension $2b$ rather than circular of diameter $2R$, these circular plate impedances still can be used to get an approximate solution by simply substituting a circular basemat of radius $R = 1.13b$ which has the same area. When the basemat is rigid and rectangular of area $2b_1 \times 2b_2$, this equivalent area approach is not valid if the two dimensions b_1 and b_2 differ significantly. Regardless of the shape of the basemat, the above approach cannot be used when the basemat is flexible rather than rigid. The engineer should check to be sure it is sufficiently rigid to justify the above type of solution.

General MDOF System with Multiple Support Excitation

The substructure method previously applied to the system shown in Fig. 27-4 will now be used to formulate the governing equations for a general system subjected to multiple earthquake excitations. As examples, suppose one wishes to consider the structures shown in Fig. 27-7, namely a tall smoke stack, an earth dam, and a nuclear power plant containment building with associated heavy equipment. Application of

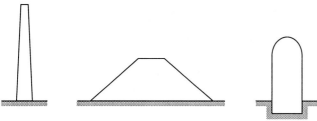

FIGURE 27-7
Example structures for soil-structure interaction analysis.

this method requires that the two substructures be defined specifically. Substructure No. 1 will consist of the structure itself and it may or may not include a portion of the foundation soil near its base. This substructure will be modelled appropriately using the lumped-parameter or finite-element method. It will be subjected to the forces which develop at its interface with Substructure No. 2, which in turn is subjected to these same forces acting in the opposite directions. Because this substructure is modeled as a continuum, the interaction forces are defined in terms of a matrix of impedance functions. While this general case is more complex in its formulation than the simple case of Fig. 27-4, the basic principles involved are the same.

Suppose the substructuring of the systems of Fig. 27-7 is adopted as shown in Fig. 27-8. Since the smoke stack is essentially surface supported, the stack itself could be taken as Substructure No. 1 and the half-space of soil below as Substructure No. 2. In this case, a 3-D analysis could be carried out using the impedance functions for a rigid massless circular plate resting on an elastic half-space. For the embankment dam, one would most likely choose to carry out a 2-D analysis using a dam cross section of unit thickness as Substructure No. 1 and the corresponding 2-D half-space below as Substructure No. 2. Depending upon possible variations in soil conditions below the dam, one may decide to also include a portion of the foundation soil as part of Substructure No. 1 as shown in Fig. 27-8c. Finite-element modeling would undoubtedly be used for Substructure No. 1 and the impedances representing Substructure No. 2 would probably be of a rather simple form as described later. In the case of the nuclear power plant containment building shown in Fig. 27-8d, a 3-D analysis can be carried out using the building, its associated heavy equipment, and a portion of soil within a hemispherical boundary as Substructure No. 1, while a half-space with a hemispherical cavity is used for Substructure No. 2. The con-

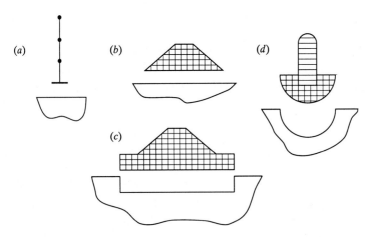

FIGURE 27-8
Substructures Nos. 1 and 2 for the systems shown in Fig. 26-7.

tainment structure and soil in the near field could be modelled using axisymmetric finite elements while the heavy equipment would be modelled appropriately using lumped-parameter systems. The far field, Substructure No. 2, would be modelled through a set of impedance functions yet to be defined, which would be compatible with the generalized DOF chosen for the near-field nodal rings at the hemispherical boundary.

To formulate the governing equations for the general soil-structure system, let n_b represent the number of DOF at the soil-structure interface, n_d the number of DOF at the interface of Substructure Nos. 1 and 2, n_a the number of DOF in the structure (including heavy equipment, if present) but excluding the n_b DOF, and n_c the number of DOF in the soil region but excluding the n_b and n_d DOF. The dynamic equilibrium equations of motion for Substructue No. 1 can be written in the form

$$
\begin{bmatrix}
\mathbf{m}_{aa} & \mathbf{m}_{ab} & 0 & \vdots & 0 \\
\mathbf{m}_{ba} & \mathbf{m}_{bb} & \mathbf{m}_{bc} & \vdots & 0 \\
0 & \mathbf{m}_{cb} & \mathbf{m}_{cc} & \vdots & \mathbf{m}_{cd} \\
\cdots & \cdots & \cdots & \cdots & \cdots \\
0 & 0 & \mathbf{m}_{dc} & \vdots & \mathbf{m}_{dd}
\end{bmatrix}
\begin{Bmatrix}
\ddot{\mathbf{v}}_a^t \\ \ddot{\mathbf{v}}_b^t \\ \ddot{\mathbf{v}}_c^t \\ \cdots \\ \ddot{\mathbf{v}}_d^t
\end{Bmatrix}
+
\begin{bmatrix}
\mathbf{c}_{aa} & \mathbf{c}_{ab} & 0 & \vdots & 0 \\
\mathbf{c}_{ba} & \mathbf{c}_{bb} & \mathbf{c}_{bc} & \vdots & 0 \\
0 & \mathbf{c}_{cb} & \mathbf{c}_{cc} & \vdots & \mathbf{c}_{cd} \\
\cdots & \cdots & \cdots & \cdots & \cdots \\
0 & 0 & \mathbf{c}_{dc} & \vdots & \mathbf{c}_{dd}
\end{bmatrix}
\begin{Bmatrix}
\dot{\mathbf{v}}_a^t \\ \dot{\mathbf{v}}_b^t \\ \dot{\mathbf{v}}_c^t \\ \cdots \\ \dot{\mathbf{v}}_d^t
\end{Bmatrix}
$$

$$
+
\begin{bmatrix}
\mathbf{k}_{aa} & \mathbf{k}_{ab} & 0 & \vdots & 0 \\
\mathbf{k}_{ba} & \mathbf{k}_{bb} & \mathbf{k}_{bc} & \vdots & 0 \\
0 & \mathbf{k}_{cb} & \mathbf{k}_{cc} & \vdots & \mathbf{k}_{cd} \\
\cdots & \cdots & \cdots & \cdots & \cdots \\
0 & 0 & \mathbf{k}_{dc} & \vdots & \mathbf{k}_{dd}
\end{bmatrix}
\begin{Bmatrix}
\mathbf{v}_a^t \\ \mathbf{v}_b^t \\ \mathbf{v}_c^t \\ \cdots \\ \mathbf{v}_d^t
\end{Bmatrix}
=
\begin{Bmatrix}
0 \\ 0 \\ 0 \\ \cdots \\ \mathbf{p}_d^t
\end{Bmatrix}
\tag{27-22}
$$

where the displacement vector represents total displacements from a fixed reference and where $\mathbf{p}_d^t$ represents total nodal forces developed between Substructure Nos. 1 and 2. Partitioning this equation as indicated, it can be written in the simplified form

$$
\begin{bmatrix}
\mathbf{m} & \mathbf{m}_g \\
\mathbf{m}_g^T & \mathbf{m}_{dd}
\end{bmatrix}
\begin{Bmatrix}
\ddot{\mathbf{v}}^t \\ \ddot{\mathbf{v}}_d^t
\end{Bmatrix}
+
\begin{bmatrix}
\mathbf{c} & \mathbf{c}_g \\
\mathbf{c}_g^T & \mathbf{c}_{dd}
\end{bmatrix}
\begin{Bmatrix}
\dot{\mathbf{v}}^t \\ \dot{\mathbf{v}}_d^t
\end{Bmatrix}
+
\begin{bmatrix}
\mathbf{k} & \mathbf{k}_g \\
\mathbf{k}_g^T & \mathbf{k}_{dd}
\end{bmatrix}
\begin{Bmatrix}
\mathbf{v}^t \\ \mathbf{v}_d^t
\end{Bmatrix}
=
\begin{Bmatrix}
0 \\ \mathbf{p}_d^t
\end{Bmatrix}
\tag{27-23}
$$

To assist in the solution of this equation, the total displacement vector is separated into two quasi-static vectors and a dynamic vector as follows:

$$
\begin{Bmatrix}
\mathbf{v}^t \\ \mathbf{v}_d^t
\end{Bmatrix}
=
\begin{Bmatrix}
\tilde{\mathbf{v}}^s \\ \tilde{\mathbf{v}}_d^s
\end{Bmatrix}
+
\begin{Bmatrix}
\hat{\mathbf{v}}^s \\ \hat{\mathbf{v}}_d^s
\end{Bmatrix}
+
\begin{Bmatrix}
\mathbf{v}^d \\ \mathbf{v}_d^d
\end{Bmatrix}
\tag{27-24}
$$

The first of these components $\langle \widetilde{\mathbf{v}}^s{}^T \quad \widetilde{\mathbf{v}}_d^s{}^T \rangle^T$, which can be written out as $\langle \widetilde{\mathbf{v}}_a^s{}^T \quad \widetilde{\mathbf{v}}_b^s{}^T \quad \widetilde{\mathbf{v}}_c^s{}^T \quad \widetilde{\mathbf{v}}_d^s{}^T \rangle^T$, is defined such that the vectors $\widetilde{\mathbf{v}}_b^s$, $\widetilde{\mathbf{v}}_c^s$, and $\widetilde{\mathbf{v}}_d^s$ have the values of the seismic free-field ground displacements $\mathbf{v}_{b_g}$, $\mathbf{v}_{c_g}$, and $\mathbf{v}_{d_g}$, respectively. Vector $\widetilde{\mathbf{v}}_a^s$ is the quasi-static displacements that result from imposing the free-field ground motions at the interface of the added structure while its other nodes are allowed to move freely, i.e., letting $\widetilde{\mathbf{p}}_a^s = 0$. These displacements may be calculated from the first of Eqs. (27-22) after excluding dynamic effects by setting the velocity and acceleration terms to zero, with the result

$$\widetilde{\mathbf{v}}_a^s = -\mathbf{k}_{aa}^{-1}\mathbf{k}_{ab}\, \widetilde{\mathbf{v}}_b^s \equiv -\mathbf{k}_{aa}^{-1}\mathbf{k}_b\, \mathbf{v}_{b_g} \tag{27-25}$$

However, restraint forces $\widetilde{\mathbf{p}}_b^s$ must be applied to the added structure to maintain its required interface displacements $\mathbf{v}_{b_g}$ while these quasi-static motions are developed. The values of these restraints may be determined from the static version of the second of Eqs. (27-22); thus considering only the stiffness coefficients of the added structure

$$\mathbf{k}_{ba}\, \widetilde{\mathbf{v}}_a^s + \mathbf{k}_{bb}^{(2)}\, \widetilde{\mathbf{v}}_b^s = \widetilde{\mathbf{p}}_b^s \tag{27-26}$$

where the superscript (2) indicates that only those quantities contributed by the structure are included. Substituting the displacement expression of Eq. (27-25), this may be written

$$\widetilde{\mathbf{p}}_b^s = \left[\mathbf{k}_{bb}^{(2)} - \mathbf{k}_{ba}\mathbf{k}_{aa}^{-1}\mathbf{k}_{ab} \right] \mathbf{v}_{bg} \tag{27-27}$$

Since these quasi-static forces do not actually exist, they are removed from the analysis by applying to the complete Substructure No. 1 a corresponding vector acting in the reverse direction. The results of this action are determined in a separate quasi-static analysis for which the governing equation may be written

$$\begin{bmatrix} \mathbf{k} & \mathbf{k}_g \\ \mathbf{k}_g^T & \mathbf{k}_{dd} \end{bmatrix} \begin{Bmatrix} \hat{\mathbf{v}}^s \\ \hat{\mathbf{v}}_d^s \end{Bmatrix} = \begin{Bmatrix} -\widetilde{\mathbf{p}}_s \\ \hat{\mathbf{p}}_d^s \end{Bmatrix} \tag{27-28}$$

The interaction forces $\hat{\mathbf{p}}_d^s$ shown in this equation act both on this substructure as well as in the opposite direction on Substructure No. 2. The values of these (negative) forces acting on this continuum substructure depend on the frequency of excitation and are expressed most conveniently in the frequency domain as follows:

$$-\hat{\mathbf{P}}_d^s(i\overline{\omega}) = \mathbf{G}_{dd}(i\overline{\omega})\, \hat{\mathbf{V}}_d^s(i\overline{\omega}) \tag{27-29}$$

in which the impedance matrix $\mathbf{G}_{dd}(i\overline{\omega})$ expresses the dynamic resistance of Substructure No. 2 and $\hat{\mathbf{V}}_d^s(i\overline{\omega})$ is the Fourier transform of the displacements $\hat{\mathbf{v}}_d^s$.

The total forces associated with these interface displacements are obtained by combining the contribution of Substructure No. 2 [Eq. (27-29)] with that from the first substructure given by the Fourier transformed version of Eq. (27-28) using the concept of direct stiffness assembly, with the following result:

$$
\begin{bmatrix}
\mathbf{k} & \mathbf{k}_g \\
\mathbf{k}_g^T & [\mathbf{k}_{dd} + G_{dd}(i\overline{\omega})]
\end{bmatrix}
\begin{Bmatrix}
\hat{\mathbf{V}}^s(i\overline{\omega}) \\
\hat{\mathbf{V}}_d^s(i\overline{\omega})
\end{Bmatrix}
=
\begin{Bmatrix}
-\tilde{\mathbf{P}}_s(i\overline{\omega}) \\
0
\end{Bmatrix}
\tag{27-30}
$$

in which it will be noted that the interface forces have cancelled each other. In this equation, $\tilde{\mathbf{P}}^s(i\overline{\omega}) = < 0 \quad \mathbf{P}_b^s(i\overline{\omega}) \quad 0 >^T$ in which $\tilde{\mathbf{P}}_b^s(i\overline{\omega})$ may be obtained from the Fourier transformed version of Eq. (27-27):

$$
\tilde{\mathbf{P}}_b^s(i\overline{\omega}) = -\frac{1}{\overline{\omega}^2}\left[\mathbf{k}_{bb}^{(2)} - \mathbf{k}_{ba}\mathbf{k}_{aa}^{-1}\mathbf{k}_{ab}\right]\ddot{\mathbf{V}}_{b_g}(i\overline{\omega})
\tag{27-31}
$$

Substituting this into the solution of Eq. (27-30), given by

$$
\begin{Bmatrix}
\hat{\mathbf{V}}^s(i\overline{\omega}) \\
\hat{\mathbf{V}}_d^s(i\overline{\omega})
\end{Bmatrix}
=
\begin{bmatrix}
\mathbf{k} & \mathbf{k}_g \\
\mathbf{k}_g^T & [\mathbf{k}_{dd} + G_{dd}(i\omega)]
\end{bmatrix}^{-1}
\begin{Bmatrix}
-\tilde{\mathbf{P}}_s(i\overline{\omega}) \\
0
\end{Bmatrix}
\tag{27-32}
$$

leads finally to a frequency-domain expression for this quasi-static displacement vector stated in terms of the free-field accelerations at the interface degrees of freedom.

Now to obtain an expression for the dynamic response to the seismic input, $\mathbf{v}^d$, Eq. (27-24) is substituted into the equation of dynamic equilibrium, Eq. (27-23), and the effective load terms resulting from the quasi-static motions are transferred to the right hand side, with the following result:

$$
\begin{bmatrix}
\mathbf{m} & \mathbf{m}_g \\
\mathbf{m}_g^T & \mathbf{m}_{dd}
\end{bmatrix}
\begin{Bmatrix}
\ddot{\mathbf{v}}^d \\
\ddot{\mathbf{v}}_d^d
\end{Bmatrix}
+
\begin{bmatrix}
\mathbf{c} & \mathbf{c}_g \\
\mathbf{c}_g^T & \mathbf{c}_{dd}
\end{bmatrix}
\begin{Bmatrix}
\dot{\mathbf{v}}^d \\
\dot{\mathbf{v}}_d^d
\end{Bmatrix}
+
\begin{bmatrix}
\mathbf{k} & \mathbf{k}_g \\
\mathbf{k}_g^T & \mathbf{k}_{dd}
\end{bmatrix}
\begin{Bmatrix}
\mathbf{v}^d \\
\mathbf{v}_d^d
\end{Bmatrix}
$$

$$
= -
\begin{bmatrix}
\mathbf{m} & \mathbf{m}_g \\
\mathbf{m}_g^T & \mathbf{m}_{dd}
\end{bmatrix}
\begin{Bmatrix}
\ddot{\tilde{\mathbf{v}}}^s + \ddot{\hat{\mathbf{v}}}^s \\
\ddot{\tilde{\mathbf{v}}}_d^s + \ddot{\hat{\mathbf{v}}}_d^s
\end{Bmatrix}
-
\begin{bmatrix}
\mathbf{c} & \mathbf{c}_g \\
\mathbf{c}_g^T & \mathbf{c}_{dd}
\end{bmatrix}
\begin{Bmatrix}
\dot{\tilde{\mathbf{v}}}^s + \dot{\hat{\mathbf{v}}}^s \\
\dot{\tilde{\mathbf{v}}}_d^s + \dot{\hat{\mathbf{v}}}_d^s
\end{Bmatrix}
$$

$$
-
\begin{bmatrix}
\mathbf{k} & \mathbf{k}_g \\
\mathbf{k}_g^T & \mathbf{k}_{dd}
\end{bmatrix}
\begin{Bmatrix}
\tilde{\mathbf{v}}^s + \hat{\mathbf{v}}^s \\
\tilde{\mathbf{v}}_d^s + \hat{\mathbf{v}}_d^s
\end{Bmatrix}
+
\begin{Bmatrix}
0 \\
\mathbf{p}_d^t
\end{Bmatrix}
\tag{27-33}
$$

This equation may be simplified by incorporating into it the governing equation for free-field response of the soil and also the two equations for the quasi-static response motions. The governing equation for the free-field soil response is obtained by removing the structure's contribution to the system property matrices in Eq. (27-23) (noting that $\tilde{\mathbf{v}}^s$ represents motions of the soil only) and denoting the interface forces for this soil model by $\mathbf{p}_{dg}$, with the following result:

$$
\begin{bmatrix}
\mathbf{m}^{(1)} & \mathbf{m}_g \\
\mathbf{m}_g^T & \mathbf{m}_{dd}
\end{bmatrix}
\begin{Bmatrix}
\ddot{\tilde{\mathbf{v}}}^s \\
\ddot{\mathbf{v}}_{dg}
\end{Bmatrix}
+
\begin{bmatrix}
\mathbf{c}^{(1)} & \mathbf{c}_g \\
\mathbf{c}_g^T & \mathbf{c}_{dd}
\end{bmatrix}
\begin{Bmatrix}
\dot{\tilde{\mathbf{v}}}^s \\
\dot{\mathbf{v}}_{dg}
\end{Bmatrix}
+
\begin{bmatrix}
\mathbf{k}^{(1)} & \mathbf{k}_g \\
\mathbf{k}_g^T & \mathbf{k}_{dd}
\end{bmatrix}
\begin{Bmatrix}
\tilde{\mathbf{v}}^s \\
\mathbf{v}_{dg}
\end{Bmatrix}
=
\begin{Bmatrix}
0 \\
\mathbf{p}_{dg}
\end{Bmatrix}
\tag{27-34}
$$

in which the superscript (1) indicates that the matrix contains only those coefficients contributed by the soil. The static force-displacement relationship for the added structure subjected to the free-field ground motions that was used previously in deriving Eqs. (27-26) and (27-27) may be written here as

$$
\begin{bmatrix} \mathbf{k}^{(2)} & 0 \\ 0 & 0 \end{bmatrix} \begin{Bmatrix} \tilde{\mathbf{v}}^s \\ \mathbf{v}_{dg} \end{Bmatrix} = \begin{Bmatrix} \tilde{\mathbf{p}}_s \\ 0 \end{Bmatrix}
\tag{27-35}
$$

where the superscript (2) indicates that the matrix only contains coefficients from the structure. The corresponding relationships for the combined system subjected to these same restraint forces but acting in the opposite direction was given previously by Eq. (27-28).

If the force vector at the right side of each of Eqs. (27-34), (27-35), and (27-28) is transferred to the left side, the result in each case is a zero vector which can be added to the right side of Eq. (27-33) without affecting the equilibrium. Taking this step, it is found that Eq. (27-33) may be simplified to the following form:

$$
\begin{bmatrix} \mathbf{m} & \mathbf{m}_g \\ \mathbf{m}_g^T & \mathbf{m}_{dd} \end{bmatrix} \begin{Bmatrix} \ddot{\mathbf{v}}^d \\ \ddot{\mathbf{v}}_d^d \end{Bmatrix} + \begin{bmatrix} \mathbf{c} & \mathbf{c}_g \\ \mathbf{c}_g^T & \mathbf{c}_{dd} \end{bmatrix} \begin{Bmatrix} \dot{\mathbf{v}}^d \\ \dot{\mathbf{v}}_d^d \end{Bmatrix} + \begin{bmatrix} \mathbf{k} & \mathbf{k}_g \\ \mathbf{k}_g^T & \mathbf{k}_{dd} \end{bmatrix} \begin{Bmatrix} \mathbf{v}^d \\ \mathbf{v}_d^d \end{Bmatrix}
$$

$$
= - \begin{bmatrix} \mathbf{m}^{(2)} & 0 \\ 0 & 0 \end{bmatrix} \begin{Bmatrix} \ddot{\tilde{\mathbf{v}}}^s \\ \ddot{\tilde{\mathbf{v}}}_d^s \end{Bmatrix} - \begin{bmatrix} \mathbf{m} & \mathbf{m}_g \\ \mathbf{m}_g^T & \mathbf{m}_{dd} \end{bmatrix} \begin{Bmatrix} \ddot{\hat{\mathbf{v}}}^s \\ \ddot{\hat{\mathbf{v}}}_d^s \end{Bmatrix} - \begin{bmatrix} \mathbf{c}^{(2)} & 0 \\ 0 & 0 \end{bmatrix} \begin{Bmatrix} \dot{\tilde{\mathbf{v}}}^s \\ \dot{\tilde{\mathbf{v}}}_d^s \end{Bmatrix}
$$

$$
- \begin{bmatrix} \mathbf{c} & \mathbf{c}_g \\ \mathbf{c}_g^T & \mathbf{c}_{dd} \end{bmatrix} \begin{Bmatrix} \dot{\hat{\mathbf{v}}}^s \\ \dot{\hat{\mathbf{v}}}_d^s \end{Bmatrix} + \begin{Bmatrix} 0 \\ \mathbf{p}_d^t - \mathbf{p}_{dg} - \hat{\mathbf{p}}_d^s \end{Bmatrix}
\tag{27-36}
$$

The damping terms on the right hand side of this equation make little contribution to the effective load of a relatively low damped system, say $\xi < 0.1$, and can be neglected. Thus defining the combined interface force terms in the equation as follows:

$$
\mathbf{p}_d^d = \mathbf{p}_d^t - \mathbf{p}_{dg} - \hat{\mathbf{p}}_d^s
\tag{27-37}
$$

the Fourier transformed version of Eq. (27-35) becomes

$$
\begin{bmatrix} -\overline{\omega}^2 \begin{bmatrix} \mathbf{m} & \mathbf{m}_g \\ \mathbf{m}_g^T & \mathbf{m}_{dd} \end{bmatrix} + i\overline{\omega} \begin{bmatrix} \mathbf{c} & \mathbf{c}_g \\ \mathbf{c}_g^T & \mathbf{c}_{dd} \end{bmatrix} + \begin{bmatrix} \mathbf{k} & \mathbf{k}_g \\ \mathbf{k}_g^T & \mathbf{k}_{dd} \end{bmatrix} \end{bmatrix} \begin{Bmatrix} \mathbf{V}^d(i\overline{\omega}) \\ \mathbf{V}_d^d(i\overline{\omega}) \end{Bmatrix}
$$

$$
= \overline{\omega}^2 \begin{bmatrix} \mathbf{m}^{(2)} & 0 \\ 0 & 0 \end{bmatrix} \begin{Bmatrix} \tilde{\mathbf{V}}^s(i\overline{\omega}) \\ \tilde{\mathbf{V}}_d^s(i\overline{\omega}) \end{Bmatrix} + \overline{\omega}^2 \begin{bmatrix} \mathbf{m} & \mathbf{m}_g \\ \mathbf{m}_g^T & \mathbf{m}_{dd} \end{bmatrix} \begin{Bmatrix} \hat{\mathbf{V}}^s(i\overline{\omega}) \\ \hat{\mathbf{V}}_d^s(i\overline{\omega}) \end{Bmatrix} + \begin{Bmatrix} 0 \\ \mathbf{P}_d^d(i\overline{\omega}) \end{Bmatrix}
\tag{27-38}
$$

Before Eq. (27-38) can be used in practical dynamic response analysis, it is necessary to simplify the effective load expression on the right hand side. First the vector $\mathbf{P}_d^d(i\bar{\omega})$ in the third term is expressed using the impedance matrix of Substructure No. 2, similar to Eq. (27-29) for the quasi-static motions; thus

$$-\mathbf{P}_d^d(i\bar{\omega}) = \mathbf{G}_{dd}(i\bar{\omega})\,\mathbf{V}_d^d(i\bar{\omega}) \qquad (27\text{-}39)$$

and this expression is combined with the stiffness matrix terms on the left side of Eq. (27-38). Next, it is apparent that displacements $\widetilde{\mathbf{V}}^s(i\bar{\omega})$ and $\widetilde{\mathbf{V}}_d^s(i\bar{\omega})$ contribute nothing to the first term of the effective loading because of the zero submatrices in $\mathbf{m}^{(2)}$. In addition, because $\widetilde{\mathbf{V}}_a^s(i\bar{\omega})$ can be expressed in terms of $\mathbf{V}_{bg}(i\bar{\omega})$ by Eq. (27-25), only the free-field motions in the n_b DOF remain to contribute to this term. Likewise, Eqs. (27-31) and (27-32) show that only these same free-field motions contribute to the second term. Making use of all these relationships, Eq. (27-38) can be put in the following form:

$$\left[-\bar{\omega}^2\begin{bmatrix}\mathbf{m} & \mathbf{m}_g \\ \mathbf{m}_g^T & \mathbf{m}_{dd}\end{bmatrix} + i\bar{\omega}\begin{bmatrix}\mathbf{c} & \mathbf{c}_g \\ \mathbf{c}_g^T & \mathbf{c}_{dd}\end{bmatrix} + \begin{bmatrix}\mathbf{k} & \mathbf{k}_g \\ \mathbf{k}_g^T & [\mathbf{k}_{dd} + \mathbf{G}_{dd}(i\bar{\omega})]\end{bmatrix}\right]\begin{Bmatrix}\mathbf{V}^d(i\bar{\omega}) \\ \mathbf{V}_d^d(i\bar{\omega})\end{Bmatrix}$$
$$= \boldsymbol{\mathcal{K}}(i\bar{\omega})\,\ddot{\mathbf{V}}_{bg}(i\bar{\omega}) \qquad (27\text{-}40)$$

where $\boldsymbol{\mathcal{K}}(i\bar{\omega})$ is an $n_b \times n_b$ matrix given by

$$\boldsymbol{\mathcal{K}}(i\bar{\omega}) = \begin{bmatrix}\mathbf{m}^{(2)} & \mathbf{0} \\ \mathbf{0} & \mathbf{0}\end{bmatrix}\begin{Bmatrix}\widetilde{\boldsymbol{\phi}}^s \\ \mathbf{0}\end{Bmatrix} + \begin{bmatrix}\mathbf{m} & \mathbf{m}_g \\ \mathbf{m}_g^T & \mathbf{m}_{dd}\end{bmatrix}\begin{bmatrix}\mathbf{k} & \mathbf{k}_g \\ \mathbf{k}_g^T & [\mathbf{k}_{dd} + \mathbf{G}_{dd}(i\bar{\omega})]\end{bmatrix}^{-1}\begin{Bmatrix}\hat{\boldsymbol{\phi}}^s \\ \mathbf{0}\end{Bmatrix}$$
$$(27\text{-}41)$$

in which

$$\widetilde{\boldsymbol{\phi}}^s \equiv \begin{Bmatrix}-\mathbf{r} \\ -\mathbf{I} \\ \mathbf{0}\end{Bmatrix} \qquad \hat{\boldsymbol{\phi}}^s \equiv \begin{Bmatrix}\mathbf{0} \\ [\mathbf{k}_{bb}^{(2)} + \mathbf{k}_{ab}^T\mathbf{r}] \\ \mathbf{0}\end{Bmatrix} \qquad (27\text{-}42)$$

Matrix $\mathbf{r} = -\mathbf{k}_{aa}^{-1}\,\mathbf{k}_{ab}$ contains the static displacement influence coefficients previously defined by Eq. (26-71) and matrix $\mathbf{I}$ is an $n_b \times n_b$ identity matrix. Matrix $\boldsymbol{\mathcal{K}}(i\bar{\omega})$ is complex and frequency dependent because it involves the impedance matrix $\mathbf{G}_{dd}(i\bar{\omega})$.

If the free-field soil accelerations at the location of the soil-structure interface are known, then Eq. (27-40) can be solved for discrete values of $\bar{\omega}$ by the standard procedures of frequency-domain analysis giving the dynamic displacement vectors $\mathbf{V}^d(i\bar{\omega})$ and $\mathbf{V}_d^d(i\bar{\omega})$. Combining these vectors with vectors $\widetilde{\mathbf{V}}^s(i\bar{\omega})$, $\widetilde{\mathbf{V}}_d^s(i\bar{\omega})$, $\hat{\mathbf{V}}^s(i\bar{\omega})$, and $\hat{\mathbf{V}}_d^s(i\bar{\omega})$, consistent with Eq. (27-24), one obtains the total displacement vectors $\mathbf{V}^t(i\bar{\omega})$ and $\mathbf{V}_d^t(i\bar{\omega})$ which can be inverse Fourier transformed to give the corresponding time-domain vectors $\mathbf{v}^t(t)$ and $\mathbf{v}_d^t(t)$. The internal stresses and deformations in the complete near-field soil-structure system (Substructure No. 1) can then be obtained by standard procedures of static analysis.

At this point, it should be emphasized that the complete solution as described above requires that the free-field ground motions $\ddot{\mathbf{V}}_{bg}(i\overline{\omega})$, $\ddot{\mathbf{V}}_{cg}(i\overline{\omega})$, and $\ddot{\mathbf{V}}_{dg}(i\overline{\omega})$ be known. For a structure having embedment, such as the nuclear power plant containment building shown in Fig. 27-8d, these motions include wave scattering effects due to the presence of the surface cavity left upon removal of the structure.

To account for such wave scattering effects, as is necessary to obtain a rigorous solution, the full foundation half-space is substructured as shown in Fig. 27-9; in this case, the soil in the cavity region is used as a substitute for the structure in the previous formulation. Using standard finite-element modeling for the soil in this region results in $n_A + n_b$ DOF where the n_b DOF should be identical to those previously defined for the soil-structure system. The n_A DOF within the soil region shown in Fig. 27-9 are, of course, different from the n_a DOF previously defined for the structure. All of the equations previously developed for the soil-structure system, Eqs. (27-22) through (27-42), can be used in exactly the same forms to represent the full half-space soil system by simply changing all subscripts "a" appearing in these equations to "A" to adjust for substituting the soil in the cavity region for the structure. In this case, the second and third vectors on the right hand side of the modified Eq. (27-24), as expressed in the frequency domain, and the modified vector $\widetilde{\mathbf{V}}_A^s(i\overline{\omega})$ in this same equation can be expressed in terms of vector $\ddot{\mathbf{V}}_{bg}(i\overline{\omega}) = -\overline{\omega}^2 \mathbf{V}_{bg}(i\overline{\omega})$ using the modified Eqs. (27-32), (27-33), (27-34), (27-40), (27-41), and (27-42). The vector on the left hand side of the modified Eq. (27-24) represents total free-field displacements in the full foundation half-space with no cavity or structure present. By hypothesizing the types of waves traveling in the foundation half-space, this vector can be stated a priori. Then using the modified Eq. (27-24) in its frequency-domain form, the desired vectors $\mathbf{V}_{bg}(i\overline{\omega})$, $\mathbf{V}_{cg}(i\overline{\omega})$, and $\mathbf{V}_{dg}(i\overline{\omega})$ can be determined. These vectors are

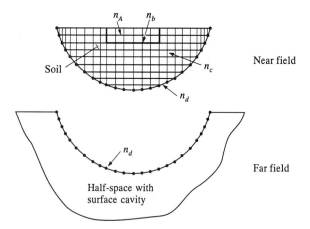

Near field

Far field

Soil

n_A n_b

n_c

n_d

n_d

Half-space with
surface cavity

FIGURE 27-9
Modeling of foundation full
half-space.

identical to the free-field motions previously expressed for the soil-structure system because they include the wave scattering effects due to presence of the surface cavity; therefore, they can be used in carrying out the previously described complete solution for the soil-structure system.

If the structure under consideration has relatively small dimensions at its inter-face with the soil region so that the wave lengths of the free-field seismic motions in the critical frequency range of interest are much larger in comparison, wave scattering effects will be quite small, allowing them to be neglected, and the spatial variations of motions within these small distances in the full half-space foundation will also be small. In this case, one can assume the components of free-field acceleration in vector $\ddot{\mathbf{v}}_{bg}(t)$ to be reasonably similar; thus, the approximate relation

$$\ddot{\mathbf{v}}_{bg}(t) \doteq \mathbf{r}_{bg} \ \ddot{\mathbf{v}}_g(t) \tag{27-43}$$

can be used where $\mathbf{r}_{bg}$ is an $n_b \times 3$ rigid-body static displacement influence coefficient matrix and $\ddot{\mathbf{v}}_g(t)$ is a three-component vector containing the average over the cavity surface of the free-field accelerations $\ddot{v}_{gx}(t)$, $\ddot{v}_{gy}(t)$, and $\ddot{v}_{gz}(t)$ in the x-, y-, and z-directions, respectively. Using the corresponding frequency-domain expression,

$$\ddot{\mathbf{V}}_{bg}(i\overline{\omega}) = \mathbf{r}_{bg} \ \ddot{\mathbf{V}}_g(i\overline{\omega}) \tag{27-44}$$

the previously described solution for the soil-structure system is greatly simplified because vectors $\tilde{\mathbf{V}}_a(i\overline{\omega})$ and $\tilde{\mathbf{V}}_b(i\overline{\omega})$ correspond to rigid-body-type displacements of the structure that cause no deformations and vectors $\hat{\mathbf{V}}^s(i\overline{\omega})$ and $\hat{\mathbf{V}}_d^s(i\overline{\omega})$ become zero vectors. Therefore, we are only interested in the dynamic solution given by Eq. (27-40) which now has a simplified form on its right hand side, i.e.,

$$\boldsymbol{K}(i\overline{\omega}) \ \ddot{\mathbf{V}}_{bg}(i\overline{\omega}) = \boldsymbol{K} \ \ddot{\mathbf{V}}_g(i\overline{\omega}) \tag{27-45}$$

where $\boldsymbol{K}$ is an $n \times 3$ matrix given by

$$\boldsymbol{K} = \begin{bmatrix} \mathbf{m}^{(2)} & \mathbf{0} \\ \mathbf{0} & \mathbf{0} \end{bmatrix} \begin{Bmatrix} \tilde{\boldsymbol{\phi}}^s \\ \mathbf{0} \end{Bmatrix} \tag{27-46}$$

in which

$$\tilde{\boldsymbol{\phi}}^s \equiv \begin{Bmatrix} -\mathbf{r}_{ag} \\ -\mathbf{r}_{bg} \\ \mathbf{0} \end{Bmatrix} \tag{27-47}$$

and where $\mathbf{r}_{ag}$ and $\mathbf{r}_{bg}$ are $n_a \times 3$ and $n_b \times 3$ matrices, respectively, containing the rigid-body displacement influence coefficients corresponding to unit displacements in coordinates v_{gx}, v_{gy}, and v_{gz}. Note that $\boldsymbol{K}$ in this case is frequency independent.

Upon substitution of Eq. (27-45), Eq. (27-40) can be solved by the standard procedure of frequency-domain analysis, and the stresses and deformations in the entire near-field soil-structure system due to $\mathbf{v}^d(t)$ can be determined using basic finite-element methods of static analysis. It should be noted, however, that these stresses and deformations are only those produced by soil-structure interaction. To obtain total stresses and deformations, these results must be superimposed on the free-field stresses produced by the traveling seismic waves. Also it should be noted that all accelerations in the structure given by $\ddot{\mathbf{v}}^d(t)$ represent only accelerations relative to the input accelerations in $\ddot{\mathbf{v}}_g(t)$; thus, the rigid-body accelerations produced by $\ddot{\mathbf{v}}_g(t)$ must be added to those produced by $\ddot{\mathbf{v}}^d(t)$ to obtain absolute values.

In all of the above linear formulations of this section, material damping of the viscous form has been used for Substructure No. 1. However, since the frequency-domain form of solution is being used, it is more efficient and effective to use material damping of the hysteretic form, that is, to use the complex stiffness matrix of the form

$$\mathbf{k}_c = \mathbf{k} + i\,\mathbf{c} \tag{27-48}$$

where

$$\mathbf{c} = 2\,\xi\,\mathbf{k} \tag{27-49}$$

in place of the corresponding real stiffness matrix $\mathbf{k}$, with the damping ratio ξ selected appropriately for the material of each finite element of the system. Using this formulation, all previously indicated viscous damping terms can be removed from the equations.

Even though all of the foregoing formulations for treating soil-structure interaction are linear, they can still be used when nonlinear soil behavior is experienced close to the base of the structure by using the equivalent linearization procedure commonly implemented into soil-structure interaction computer programs.[5,6,7,8]

---

[5] H. B. Seed, R. T. Wong, I. M. Idriss, and K. Tokimatsu, "Moduli and Damping Factors for Dynamic Analysis for Cohesionless Soils," University of California, Berkeley, Earthquake Engineering Research Center, Report No. EERC 84-14, 1984.

[6] B. O. Hardin and V. P. Drnevich, "Shear Modulus and Damping in Soils: Design Equations and Curves," *Jour. of the Soil Mechanics and Foundation Division, ASCE,* Vol. 98, No. SM7, July, 1972.

[7] J. Lysmer, T. Udaka, C. F. Tsai, and H. B. Seed, "Flush — A Computer Program for Approximate 3-D Analysis of Soil-Structure Interaction Problem," University of California, Berkeley, Earthquake Engineering Research Center, Report No. EERC 75–30, 1975.

[8] I. Katayama, C. H. Chen, and J. Penzien, "Near-Field Soil-Structure Interaction Analysis Using Nonlinear Hybrid Modeling," Proc. SMIRT Conference, Anaheim, Ca., 1989.

Generation of Boundary Impedances

In the preceding treatment of the seismic response of soil-structure systems by the substructure method of analysis, an impedance matrix relating the boundary forces of Substructure No. 2 to its corresponding boundary displacements has been used, e.g., matrix $\mathbf{G}_{dd}(i\bar{\omega})$ in Eq. (27-29). In the past, analysts have often implicitly assumed the impedances in this matrix to have infinite numerical values, i.e., rigid boundary conditions have been assumed. In these cases, the interface between Substructures 1 and 2 is usually chosen to be some distance away from the base of the structure itself, as described above with reference to the direct SSI analysis of a system with a bounded soil layer. However, this rigid-boundary assumption may lead to erroneous results, and in most cases an appropriate impedance matrix that represents the actual foundation conditions is preferable to a bounded soil layer model. In the following, some commonly used procedures for generating the impedance matrix are described.

One-Dimensional Plane Waves — In many cases, the interaction between Substructures 1 and 2 produce traveling waves in the far field which can be modelled reasonably well using one-dimensional plane waves. For example, consider the 2-D model of an earth dam shown in Fig. 27-8b. If the foundation below the dam is reasonably uniform in the horizontal directions, the dam-foundation interaction forces produced by horizontal earthquake ground motions will be primarily shear forces causing shear waves propagating downward into the half-space and the interaction forces caused by vertical ground motions will be primarily normal forces producing downward-propagating compression waves. It is reasonable, in this case, to generate the boundary impedance matrix assuming such waves to be one-dimensional in form.

To illustrate the type of motion induced by horizontal earthquake excitations, let us first consider a semi-infinite shear-beam column of unit cross-sectional area located in a uniform foundation half-space as indicated in Fig. 27-10a. The equation of horizontal motion for this uniform column is equivalent to that discussed in Chapter

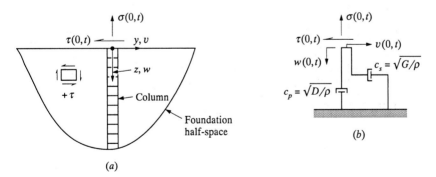

(a)

(b)

FIGURE 27-10
Substructure No. 2 as a uniform half-space having viscous boundary elements as its equivalent.

19 [Eq. (19-25)], but here the motions are denoted $v(z,t)$ so the equation becomes

$$\ddot{v}(z,t) - V_s^2\, v''(z,t) = 0 \tag{27-50}$$

in which

$$V_s = \sqrt{\frac{G}{\rho}} \tag{27-51}$$

is the shear wave propagation velocity and where G is the shear modulus. The general solution of this equation is equivalent to Eq. (19-27), and thus it may be written as the sum of two traveling waves.

The dynamic behavior of the shear-beam column depicted in Fig. 27-10a can also be expressed in terms of its shear stress distribution since

$$\tau(z,t) = G\,\gamma(z,t) = G\,\frac{\partial v(z,t)}{\partial z} \tag{27-52}$$

Thus by analogy with Eq. (19-29), when the stress wave functions $G\,\partial f_1/\partial z$ and $G\,\partial f_2/\partial z$ are designated as g_1 and g_2, this equation may be written as

$$\tau(z,t) = g_1\,(z - V_s t) + g_2\,(z + V_s t) \tag{27-53}$$

showing both a forward- and a backward-propagating wave.

To establish the forms of the displacement functions f_1, f_2 and their corresponding stress functions g_1, g_2, one must make use of prescribed boundary conditions. For example, suppose the top of the shear-beam column is subjected to a harmonic displacement

$$v(0,t) = \bar{v}\,\exp(i\bar{\omega}t) \tag{27-54}$$

Then, the steady-state $v(z,t)$ takes the similar traveling waveform

$$v(z,t) = \bar{v}\,\exp\!\left[i\bar{\omega}\left(t - \frac{z}{V_s}\right)\right] = \bar{v}\,\exp(i\bar{\omega}t)\,\exp\!\left(-i\bar{\omega}\frac{z}{V_s}\right) \tag{27-55}$$

in order to satisfy this top boundary condition. The corresponding downward-traveling shear wave is given by

$$\tau(z,t) = G\,\frac{\partial v(z,t)}{\partial z} = \left(\frac{-G i\bar{\omega}}{V_s}\right)\bar{v}\,\exp(i\bar{\omega}t)\,\exp\!\left(-i\bar{\omega}\frac{z}{V_s}\right) \tag{27-56}$$

which shows that

$$\tau(0,t) = \left(\frac{-G i\bar{\omega}}{V_s}\right)\bar{v}\,\exp(i\bar{\omega}t) \tag{27-57}$$

Since the upper-boundary impedance function for horizontal motion has been defined through the relation

$$\tau(0,t) = -G_y(i\bar{\omega})\,v(0,t) \tag{27-58}$$

it is seen by substitution of Eqs. (27-54) and (27-57) that

$$G_y(i\overline{\omega}) = i\,G\overline{\omega}\,/\,V_s \qquad (27\text{-}59)$$

Also, because the boundary shear stress of Eq. (27-57) is 90 degrees out-of-phase with the corresponding boundary displacement of Eq. (27-54), the same boundary impedance would be provided by the dash pot equivalent shown in Fig. 27-10b provided its coefficient is given by $c_s = G\,/\,V_s = \sqrt{G\rho}$.

Using the above plane shear wave model for Substructure No. 2, the impedance matrix $G_{dd}(i\overline{\omega})$ representing the horizontal boundary DOF would be diagonal in form with each coefficient simply being the product of $i\,G\overline{\omega}\,/\,V_s$ and the tributary area associated with the corresponding boundary node.

Let us now consider the case of Substructure No. 2 being a uniform layer of depth H_1 resting on a uniform half-space as shown in Fig. 27-11. In certain cases, one can again generate an uncoupled boundary impedance matrix based on vertically propagating plane waves traveling through both the layer and the half-space in this case. Suppose the top of a shear-beam column of unit cross-sectional area as shown in Fig. 27-11 is subjected to the horizontal harmonic displacement

$$v_1(0,t) = \overline{v}\,\exp(i\overline{\omega}t) \qquad (27\text{-}60)$$

Under steady-state conditions, the horizontal displacements in the layer will be given by

$$v_1(z_1,t) = \left[A_1\,\exp\!\left(-i\overline{\omega}\,\frac{z_1}{V_{s1}}\right) + B_1\,\exp\!\left(i\overline{\omega}\,\frac{z_1}{V_{s1}}\right) \right] \exp(i\overline{\omega}t) \qquad (27\text{-}61)$$

which represents both an upward-traveling wave and a downward-traveling wave. Both types of waves must be present in order to satisfy the boundary conditions at

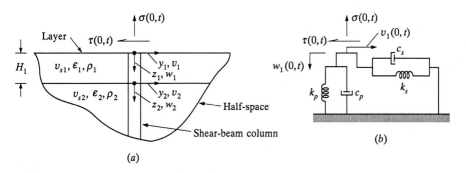

FIGURE 27-11
Substructure No. 2 as a uniform layer on a uniform half-space having viscous and spring boundary elements as its equivalent.

both top and bottom of the layer. The shear stress distribution in the layer is given by the corresponding relation

$$\tau_1(z_1, t) = G_1 \frac{i\bar{\omega}}{V_{s1}} \left[-A_1 \exp\left(-i\bar{\omega}\frac{z_1}{V_{s1}}\right) + B_1 \exp\left(i\bar{\omega}\frac{z_1}{V_{s1}}\right) \right] \exp(i\bar{\omega}t) \quad (27\text{-}62)$$

The steady-state displacements in the half-space below will be given by

$$v_2(z_2, t) = \left[A_2 \exp\left(-i\bar{\omega}\frac{z_2}{V_{s2}}\right) \right] \exp(i\bar{\omega}t) \quad (27\text{-}63)$$

which represents a downward-traveling wave only, and the corresponding shear stress distribution is given by

$$\tau_2(z_2, t) = G_2 \frac{i\bar{\omega}}{V_{s2}} \left[-A_2 \exp\left(-i\bar{\omega}\frac{z_2}{V_{s2}}\right) \right] \exp(i\bar{\omega}t) \quad (27\text{-}64)$$

Making use of Eqs. (27-61) and (27-62) for $x_1 = 0$ to satisfy the upper boundary conditions of the layer, i.e., the displacement condition given by Eq. (27-60) and the corresponding stress condition given by Eq. (27-58), one obtains

$$A_1 = \frac{1}{2} \left[1 + \frac{V_{s1} G_y(i\bar{\omega})}{G_1 i\bar{\omega}} \right] \bar{v} \quad (27\text{-}65)$$

and

$$B_1 = \frac{1}{2} \left[1 - \frac{V_{s1} G_y(i\bar{\omega})}{G_1 i\bar{\omega}} \right] \bar{v} \quad (27\text{-}66)$$

Using Eqs. (27-61) and (27-63) for $z_1 = H_1$ and $z_2 = 0$, respectively, to satisfy the displacement compatibility condition at the layer/half-space interface, one obtains

$$A_2 = \left[A_1 \exp\left(-i\bar{\omega}\frac{H_1}{V_{s1}}\right) + B_1 \exp\left(i\bar{\omega}\frac{H_1}{V_{s1}}\right) \right] \bar{v} \quad (27\text{-}67)$$

Finally, using Eqs. (27-62) and (27-64) for $z_1 = H_1$ and $z_2 = 0$, respectively, to satisfy the stress compatibility condition at the layer/half-space interface and using Eqs. (27-65), (27-66), and (27-67), the relation

$$G_y(i\bar{\omega}) = \frac{i G_1 \bar{\omega}}{V_{s1}} \left[\frac{(1 + \alpha_s) \exp(i\beta_s) + (1 - \alpha_s) \exp(-i\beta_s)}{(1 + \alpha_s) \exp(i\beta_s) - (1 - \alpha_s) \exp(-i\beta_s)} \right] \quad (27\text{-}68)$$

is obtained, where

$$\alpha_s \equiv \sqrt{\frac{G_1 \rho_1}{G_2 \rho_2}} \qquad\qquad \beta_s = \frac{\bar{\omega} H_1}{V_{s1}} \quad (27\text{-}69)$$

Introducing the Euler equations to convert Eq. (27-68) from exponential form to trigonometric form leads finally to

$$G_y(i\overline{\omega}) = \frac{i\,G_1\overline{\omega}}{V_{s1}} \left[\frac{\alpha_s + i\,[\alpha_s^2 - 1]\,\cos\beta_s\,\sin\beta_s}{\alpha_s^2\,\cos^2\beta_s + \sin^2\beta_s} \right] \qquad (27\text{-}70)$$

Note that when $\alpha_s = 1$, i.e., when the layer and the half-space have the same properties, Eq. (27-70) reduces to the imaginary form of Eq. (27-59) as it should. Also, when $\alpha_s = 0$, representing a flexible layer resting on a rigid half-space, Eq. (27-70) reduces to the real form

$$G_y(i\overline{\omega}) = \frac{G_1\overline{\omega}}{V_{s1}}\,\cot\beta_s = G_y(\overline{\omega}) \qquad (27\text{-}71)$$

indicating that no energy can be transmitted downward through the lower boundary of the layer into the rigid half-space. This impedance has zero values when the nondimensional parameter β_s equals $\pi/2$, $3\pi/2$, $5\pi/2$, etc., and infinite values when it equals π, 2π, 3π, etc. Such values of impedance at these frequencies are caused by matching of input frequencies $\overline{\omega}$ with normal shear mode frequencies in the layer, causing pure resonance to develop. The upper and lower boundary conditions of the layer which produce these normal modes are free and fixed, respectively, for those cases yielding zero impedances and fixed and fixed, respectively, for those cases yielding infinite impedances.

When $\alpha_s = \infty$, representing a flexible layer resting on an inviscid fluid, Eq. (27-70) reduces to the real form

$$G_y(i\overline{\omega}) = \frac{G_1\overline{\omega}}{V_{s1}}\,\tan\beta_s = G_y(\overline{\omega}) \qquad (27\text{-}72)$$

again indicating that no energy can be transmitted downward through the lower boundary of the layer into the half-space. This impedance has zero values when parameter β_s equals π, 2π, 3π, etc., and infinite values when it equals $\pi/2$, $3\pi/2$, $5\pi/2$, etc. Again such values are caused by matching of input frequencies $\overline{\omega}$ with normal shear mode frequencies in the layer causing pure resonance to develop. The upper and lower boundary conditions on the layer which produce these normal modes are free and free, respectively, for those cases yielding zero impedances and fixed and free, respectively, for those cases yielding infinite impedances.

It is interesting to note that $G_y(i\overline{\omega})$ has a real part only when $\alpha_s = 0$ and ∞, an imaginary part only when $\alpha_s = 1$, and is complex having both real and imaginary parts when α_s takes on other values in the range $0 < \alpha_s < \infty$.

Making use of Eqs. (27-54) and (27-58) when $G_y(i\overline{\omega})$ as given by Eq. (27-70) has both real and imaginary parts, i.e., $G_y(i\overline{\omega}) = G_y^R(\overline{\omega}) + iG_y^I(\overline{\omega})$, it is easily shown that the horizontal boundary impedance spring/dashpot equivalent of Fig. 27-11a is the system shown in Fig. 27-11b having frequency dependent parameters

$$c_s = G_y^I(\bar{\omega}) / \bar{\omega}$$

$$k_s = G_y^R(\bar{\omega})$$

(27-73)

The imaginary part of the boundary impedance represents partial radiation of energy downward into the half-space while the real part represents partial reflection of energy back into the layer.

Let us now consider the case of vertically propagating compression waves in the columns of Figs. 27-10a and 27-11a produced by vertical seismic motions. The compression wave equation equivalent to Eq. (27-50) is

$$\ddot{w}(z,t) - V_p^2 \, w''(z,t) = 0$$

(27-74)

in which V_p represents the vertical compression wave velocity given by

$$V_p = \sqrt{D / \rho}$$

(27-75)

The compression modulus D for the plane strain condition assumed here (i.e., where no strain is permitted in the x- and y-directions) is given by

$$D = E \left[\frac{1 - \nu}{(1 + \nu)(1 - 2\nu)} \right]$$

(27-76)

in which E and ν denote Young's modulus and Poisson's ratio, respectively. For values of ν typical of foundation media, this compression modulus is considerably increased over the modulus E used in Chapter 19 for analysis of plane stress compression waves in rods.

Following the same procedure used in deriving Eq. (27-70), one finds the vertical boundary impedance $G_z(i\bar{\omega})$ for Substructure No. 2 to be of the equivalent form

$$G_z(i\bar{\omega}) = \frac{i D_1 \bar{\omega}}{V_{p1}} \left[\frac{\alpha_p + i \left[\alpha_p^2 - 1\right] \cos \beta_p \sin \beta_p}{\alpha_p^2 \cos^2 \beta_p + \sin^2 \beta_p} \right]$$

(27-77)

where

$$\alpha_p \equiv \sqrt{\frac{D_1 \rho_1}{D_2 \rho_2}} \qquad\qquad \beta_p = \frac{\bar{\omega} H_1}{V_{p1}}$$

(27-78)

The corresponding spring/dashpot system shown in Fig. 27-11b has the frequency dependent parameters

$$c_p = G_z^I(\bar{\omega}) / \bar{\omega}$$

$$k_p = G_z^R(\bar{\omega})$$

(27-79)

which are equivalent to the parameters shown in Eqs. (27-73) for horizontal motion. The unit impedances given by Eq. (27-77) must be multiplied by tributary areas to obtain the corresponding uncoupled impedances in matrix $\mathbf{G}_{dd}(i\overline{w})$.

Assuming one-dimensional vertically propagating plane waves as in the single layer case of Fig. 27-11a, the uniform surface impedances $G_y(i\overline{w})$ and $G_z(i\overline{w})$ can easily be generated for a multiple horizontally layered system as well using the same harmonic traveling wave solutions for each layer and for the half-space and satisfying the same surface boundary conditions and the same compatibility conditions at each interface. Having these uniform surface impedances per unit area, they can be discretized as before by multiplying by nodal tributary areas to obtain the uncoupled impedances in matrix $\mathbf{G}_{dd}(i\overline{w})$.

Example E27-1.   A long uniform point bearing pile is being driven by dropping a rigid hammer on its head through a massless SDOF cushion as shown in Fig. E27-1. Using the substructure method of analysis, determine the downward-traveling wave produced in the pile by a single hammer impact at initial velocity V_h.

Using the displacements and system properties defined in Fig. E27-1a and the pile impedance function of Eq. (27-77) for $\alpha_p = 1$ with the plane stress compression modulus $D = E$, the complete system can be modelled as shown in Fig. E27-1b. The axial force acting on the top of the pile, which equals the combined dashpot and spring forces in Fig. E27-1b, can be expressed using the three conditions

$$N(0,t) = -\frac{AE}{V_p}\,\dot{w}(0,t) \qquad (a)$$

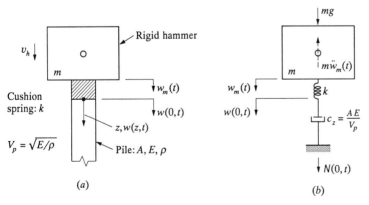

FIGURE E27-1
Hammer-cushion-pile system.

$$N(0,t) = k\left[w(0,t) - w_m(t)\right] \tag{b}$$

$$N(0,t) = m\,\ddot{w}_m(t) - m\,g \tag{c}$$

Substituting (a) into (b) and differentiating twice give

$$-\frac{AE}{V_p}\,\ddot{w}(0,t) = k\left[\ddot{w}(0,t) - \ddot{w}_m(t)\right] \tag{d}$$

Substituting (a) into (c), solving for $\ddot{w}_m(t)$, and in turn substituting this acceleration into (d) give

$$\dddot{V}_0(t) + \frac{k\,V_p}{AE}\,\ddot{V}_0(t) + \frac{k}{m}\,\dot{V}_0(t) = \frac{k\,g\,V_p}{AE} \tag{e}$$

where

$$V_0(t) \equiv \dot{w}(0,t) \tag{f}$$

The term on the right hand side of (e) results from the weight mg in (c); that is, it represents that part of the contact force $N(0,t)$ contributed by the hammer weight which is normally a small fraction of the inertial-force contribution and therefore may be neglected.

The initial conditions on $w_m(t)$ are

$$w_m(t) = 0 \qquad\qquad \dot{w}_m(0) = V_h \tag{g}$$

where V_h is the hammer velocity at the instant of initial contact with the cushion. Since $w(0,0)$ equals zero, combining (a) and (b) and using the first of (g) give

$$V_0(0) = \dot{w}(0,0) = 0 \tag{h}$$

Substituting (a) into (b), differentiating the resulting equation once and setting $t = 0$, and then making use of the second of (g) together with (h) give

$$\dot{V}_0(0) = \ddot{w}(0,0) = \frac{k\,V_p\,V_h}{AE} \tag{i}$$

Using the initial conditions as given by (h) and (i), the solution to the homogeneous form of (e) is

$$V_0(t) = \frac{2\,\xi\,\omega\,V_h}{\omega_D}\,\exp(-\xi\omega t)\cdot\sin\omega_D t \tag{j}$$

where

$$\omega = \sqrt{k/m} \qquad\qquad \omega_D = \omega\sqrt{1-\xi^2} \qquad\qquad \xi = \frac{k\,V_p}{2\,\omega\,AE} \tag{k}$$

Using (a) and (f), the axial force in the pile at its top end is

$$N(0,t) = -\frac{k V_h}{\omega_D} \exp(-\xi\omega t) \cdot \sin \omega_D t \tag{l}$$

To satisfy this top-end condition, the downward axial-force traveling wave in the pile is

$$N(z,t) = -\frac{k V_h}{\omega_D} \exp\left[-\xi\omega\left(t - \frac{z}{V_p}\right)\right] \cdot \sin \omega_D \left(t - \frac{z}{V_p}\right) \tag{m}$$

Upon rebound from its first impact with the pile head, separation occurs between the hammer and the cushion; thus, the above impact force and corresponding traveling wave expressions are valid only in the range

$$0 < \omega_D \left(t - \frac{z}{V_p}\right) < \pi \tag{n}$$

when describing the single hammer impact.

Example E27-2. To provide a numerical demonstration of the analysis of the axial force developed by a pile-driving hammer, a concrete pile having a Young's modulus $E = 3 \times 10^6$ psi, an area $A = 400$ in^2, and a unit weight $\gamma = 150$ pcf is considered. The velocity of wave propagation in the pile is given by

$$V_p = \sqrt{E/\rho} = \sqrt{Eg/\gamma} = (1.15)(10^5) \ in/sec$$

Assuming the hammer weight $W = 2,000$ lb, the cushion spring constant $k = (2,054)(10^3)$ lb/in, and the hammer velocity upon initial impact $V_h = 184$ in/sec, one finds

$$\omega = \sqrt{\frac{k g}{W}} = 628 \ r/sec \qquad\qquad \xi = \frac{V_p k}{2\omega AE} = 0.156$$

$$\omega_D = \omega\sqrt{1 - \xi^2} = 620 \ r/sec \qquad\qquad \frac{k V_h}{\omega_D} = (606)(10^3) \ lb$$

Eq. (m) then becomes

$$P(z,t) = -(606)(10^3)$$
$$\times \exp\left[-97.97\left(t - \frac{z}{13.8 \times 10^5}\right)\right] \cdot \sin 620 \left(t - \frac{z}{13.8 \times 10^5}\right) \ lb$$

where t is measured in seconds and z is measured in feet. A plot of this axial force against coordinate z is shown in Fig. E27-2 for $t = \pi / 620 = 0.00507 \ sec$,

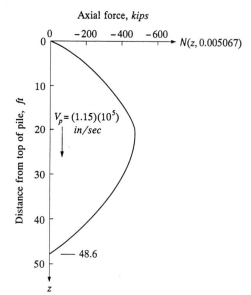

FIGURE E27-2
Axial-force distribution in concrete pile
0.00507 *sec* after initial hammer impact
with cushion.

which corresponds to the instant of hammer separation from the cushion. At this time, the traveling wave has advanced a distance of $V_p t = 48.6$ *ft* down the pile.

<u>Two-Dimensional Waves</u> — Soil-structure interaction effects of long narrow structures subjected to transverse seismic excitation often can be modelled in two-dimensional form. If they are surface supported on rigid continuous footings, one may find it appropriate to generate the impedances for Substructure No. 2 by simply using the inverted compliances shown in Fig. 27-12 for the massless infinitely-long rigid strip resting on the surface of an elastic half-space.[9]

If the long narrow structure under consideration has embedment, one can select a half-cylindrical interface between Substructures 1 and 2 and then use 2-D radial and tangential far-field impedances per unit area over the surface of the half-cylindrical cavity as indicated in Fig. 27-13. If this cavity is contained within a layer of depth H having a shear wave velocity V_{s1} which, in turn, is resting on an elastic homogeneous isotropic half-space having a shear wave velocity V_{s2}, it has been shown[10] that the

---

[9] T. J. Tzong, S. Gupta, and J. Penzien, "Two-Dimensional Hybrid Modeling of Soil-Structure Interaction," Report No. UC-EERC 81/11, Earthquake Engineering Research Center, University of California, Berkeley, August, 1981.

[10] T. J. Tzong and J. Penzien, "Hybrid-Modeling of a Single Layer Half-Space System in Soil-Structure Interaction," *Earthquake Engineering and Structural Dynamics*, Vol. 14, 1986.

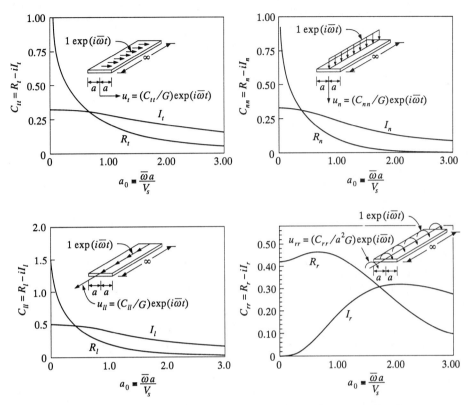

FIGURE 27-12
Compliances of infinite rigid massless strip of width $2a$; G = shear modulus, V_s = shear-wave velocity.

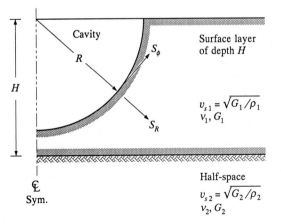

FIGURE 27-13
Continuous far-field impedance functions S_p and S_R along half-cylindrical cavity surface.

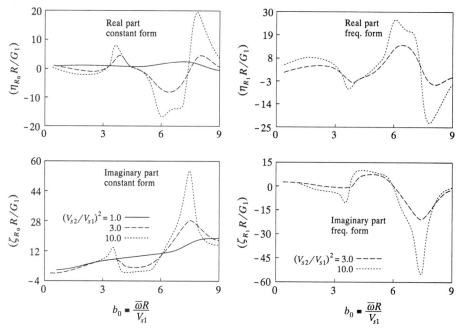

FIGURE 27-14
Parameters defining impedance S_R along half-cylindrical cavity surface.

radial and tangential impedances, per unit area, can be expressed in the approximate forms

$$S_R \doteq \eta_{R_0} + i\,\zeta_{R_0} + (\eta_{R_1} + i\,\zeta_{R_1})\,\cos\phi$$

$$S_\phi \doteq \eta_{\phi_0} + i\,\zeta_{\phi_0} + (\eta_{\phi_1} + i\,\zeta_{\phi_1})\,\cos\phi \tag{27-80}$$

where parameters η_{R_0}, ζ_{R_0}, η_{R_1}, and ζ_{R_1} have numerical values as indicated in Fig. 27-14 and parameters η_{ϕ_0}, ζ_{ϕ_0}, η_{ϕ_1}, and ζ_{ϕ_1} have numerical values as indicated in Fig. 27-15. Note that all parameters in these figures are plotted in their nondimensional forms as functions of the nondimensional frequency $b_0 \equiv \overline{\omega}R\,/\,V_{s1}$. The first plots in Figs. 27-14 and 27-15 are for three different numerical values of the ratio $V_{s1}^2\,/\,V_{s2}^2$, namely 1.0, 3.0, and 10.0; while the second plots in these figures are for only two different numerical values, namely 3.0 and 10.0. The numerical values for all four coefficients in the trigonometric terms of Eqs. (27-80) equal zero for $V_{s1}^2\,/\,V_{s2}^2 = 1.0$ because they are not needed for the nonlayered system. In all plots of Figs. 27-14 and 27-15, the results have been generated using Poisson ratio values $\nu_1 = \nu_2 = 1/3$ and $H/R = 4/3$.

The continuous impedances, per unit area, as given by Eqs. (27-80) can now be discretized using standard procedures to obtain the half-cylindrical boundary impedances for Substructure No. 2 consistent with the DOF chosen for the near-

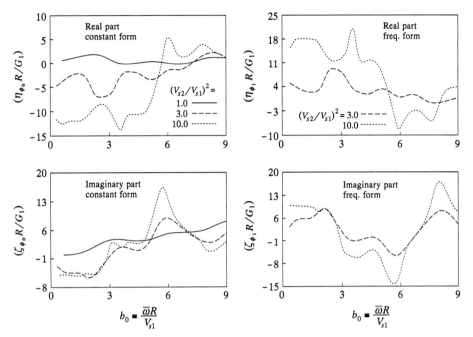

FIGURE 27-15
Parameters defining impedance S_ϕ along half-cylindrical cavity surface.

field/far-field interface; thus, the desired impedance matrix $\mathbf{G}_{dd}(i\bar{\omega})$ to be used in the substructure method of analysis is obtained.

Three-Dimensional Waves — Soil-structure interaction effects of structure surfaces supported on circular or square rigid mats or footings can often be modelled in three-dimensional form using the known impedances for a rigid massless circular plate resting on the surface of an isotropic homogeneous elastic half-space as indicated earlier in treating the system shown in Fig. 27-4; see Fig. 27-5, and Eqs. (27-17) and (27-18). These impedances are shown in their nondimensional forms in Figs. 27-6a through 27-6e.[11,12,13]

As pointed out earlier, one should be careful in using the impedances shown in these figures for mat foundations of large dimensions as the rigid-plate assumption may not be valid leading to erroneous solutions. One should also be careful in using

[11] A. S. Veletsos and Y. T. Wei, "Lateral and Rocking Vibrations of Footings," loc. cit.

[12] J. E. Luco and R. A. Westman, "Dynamic Response of Circular Footings," loc. cit.

[13] A. S. Veletsos and V. V. D. Nair, "Torsional Vibration of Foundations," Structural Research at Rice, Report No. 19, Department of Civil Engineering, Rice University, June, 1973.

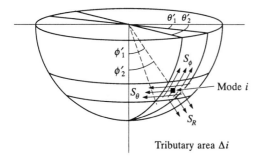

Tributary area Δi

FIGURE 27-16
Far-field impedances over the hemispherical
cavity surface in spherical co-ordinates.

these impedances when the foundation is layered as they are strictly valid only for a uniform half-space. A limited number of similar impedances have been generated for a layered half-space which can be used when found applicable.[14,15]

If the structure under consideration has a flexible mat foundation, or it has embedment as shown for the nuclear power plant containment building in Fig. 27-8d, one can obtain an appropriate three-dimensional model by introducing a hemispherical interface between Substructures 1 and 2 and then using the continuous radial, tangential, and circumferential far-field impedances per unit area over the surface of the hemispherical cavity as indicated in Fig. 27-16. It has been shown that these impedances can be approximated reasonably well for engineering purposes using the simple relations

$$S_R \doteq \eta_R + i\,\zeta_R$$

$$S_\phi \doteq \eta_\phi + i\,\zeta_\phi \qquad (27\text{-}81)$$

$$S_\theta \doteq \eta_\theta + i\,\zeta_\theta$$

which are complex and frequency dependent but are constant over the hemispherical surface of radius R. The three η and three ζ parameters in these equations have been generated for a uniform half-space of shear modulus G and Poisson's ratio $\nu = 1/3$, with results as shown in Fig. 27-14.[16] In each case, the parameter is nondimensionalized using the ratio R/G and it is plotted as a function of the dimensionless frequency $b_0 \equiv \overline{\omega} R/V_s$. If Substructure No. 2 is layered horizontally even within the depth R, the impedances of Fig. 27-17 can still be used but the impedances at

[14] E. Kausel, "Forced Vibrations of Circular Footings on Layered Media," MIT Research Report R74-11, Mass. Inst. of Tech., Cambridge, Mass., 1974.

[15] J. E. Luco, "Impedance Functions for a Rigid Foundation on a Layered Medium," *Nuclear Engineering and Design*, Vol. 31, No. 2, 1979.

[16] S. Gupta, T. W. Lin, J. Penzien, and C. S. Yeh, "Three-Dimensional Hybrid Modeling of Soil-Structure Interaction," *Int. Jour. of Earthquake Eng. and Struct. Dyn.*, Vol. 10, No. 1, Jan. – Feb., 1982.

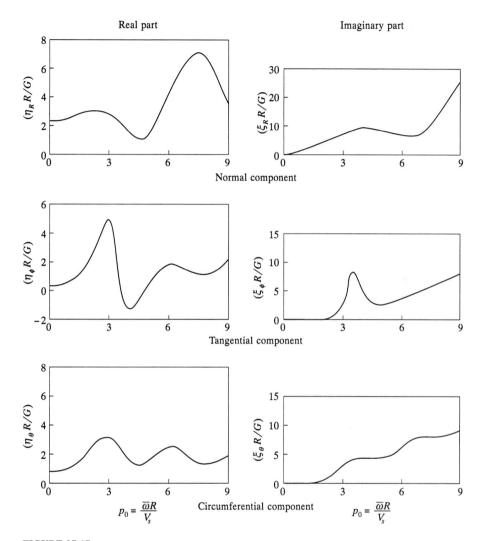

FIGURE 27-17
Far-field impedance functions over the hemispherical cavity surface.

any level corresponding to $R \cos \phi$ must be consistent with the shear wave velocity representative of the material at that depth.

These continuous impedances, per unit area, can now be discretized using standard tributary area procedures to obtain the hemispherical boundary impedances for Substructure No. 2 consistent with the DOF chosen for the near-field/far-field interface, thus yielding the desired impedance matrix $\mathbf{G}_{dd}(i\overline{\omega})$ to be used in the substructure method of analysis.

27-3   RESPONSE OF UNDERGROUND STRUCTURES

The earthquake response of above ground structures, including those having embedment, is produced primarily by effective seismic forces which are independent of the response they produce; therefore, such structures are basically under force control. However, the response of underground structures, such as subway stations, tunnels, and pipelines, is produced primarily by the ground deformations under free-field conditions; and consequently, such structures are basically under deformation control. The soil-structure interaction of an underground structure can be treated in a quasi-static fashion since the mass of the soil displaced by the presence of the structure is very large compared with the mass of the structure itself. Only when heavy equipment items are present do significant inertial forces develop.

Free-Field Ground Motions Due To Propagating Plane Waves

As was mentioned above, it is common practice to calculate free-field ground motions assuming they are produced by vertically propagating plane waves, i.e., with shear waves producing the horizontal components and compression waves producing the vertical component. Considering a linear horizontally-layered soil system resting on a uniform elastic half-space, the same form of harmonic traveling wave solutions used previously in generating boundary impedances for the one-dimensional plane wave case, i.e., equations of the form of Eqs. (27-61) through (27-64), can be used to describe the free-field ground motions produced by vertically propagating plane waves.

Assuming upward-traveling incident waves in the uniform half-space, which can be expressed in terms of a series of harmonics, one can calculate the corresponding upward- and downward-traveling harmonics in each layer, and the corresponding downward-traveling harmonics in the half-space resulting from its interaction with the layered system. Decomposing the specified components of free-field control motion (usually acceleration) into their harmonics by the FFT procedure, satisfying the surface zero stress condition and all interface stress and displacement compatibility conditions, one can calculate the amplitude and phase angle of each upward- and each downward-traveling harmonic in each layer and in the half-space; thus time-histories of ground motion (acceleration, velocity, and displacement) are obtained at all levels which are compatible with the soil properties and the specified free-field control motions. Usually these control motions are specified at the surface; however, the analytical procedure allows them to be specified at any level. The soil properties (shear-modulus and damping-ratio values) should be adjusted iteratively so that they are compatible with the resulting shear strain levels reached under free-field conditions. Soil material damping can easily be included in the basic analytical formulations [Eqs. (27-61) through (27-64)] by simply substituting the complex shear modulus $G_c = G(1 + 2i\xi)$ for the real modulus G.

Free-field ground motions produced by other than vertically propagating plane waves also can be formulated for use in seismic analyses of underground structures, e.g., motions produced by Rayleigh waves or nonvertically propagating shear waves.

Racking Deformations of Cross Sections

As a result of vertically propagating plane waves, the cross sections of underground structures are subjected to racking (shear-type) deformations. To illustrate, suppose the cross section of tunnel lining shown in Fig. 27-18 is subjected to a free-field soil environment produced by vertically propagating shear waves producing horizontal displacements $w(y,t)$. The instant at which the cross section experiences maximum racking deformations can be taken as that time when the difference in the free-field soil displacements at depths corresponding to the top and bottom elevations of the tunnel cross section reaches a maximum, i.e., when $\left| w(y_t, t) - w(y_b, t) \right|$ becomes a maximum. To find this maximum difference, it is necessary to calculate the entire time-history of the difference through the free-field site analysis and then observe its maximum (or critical) value.

The racking analysis of the cross section can then be carried out in two steps by a procedure that is entirely equivalent to the analysis of quasi-static motions in the SSI analysis of a MDOF structure with multiple support excitation. First, a vector set of forces $\mathbf{F}$ applied in all degrees of freedom of the outer-boundary cross-sectional nodes is calculated which will deform the outer boundary so as to be totally compatible with the critical free-field soil-displacement and stress patterns as indicated in Fig. 27-18a. This set of forces is composed of a vector $\mathbf{F}_1$ required to deform the cross section (when removed from the soil) into a shape compatible with the free-field soil displacements and a vector $\mathbf{F}_2$ representing, in discrete form, the free-field shear stress distribution acting on the outside boundary of the rectangular soil element to be displaced by the cross section; thus, $\mathbf{F} = \mathbf{F}_1 - \mathbf{F}_2$. Since the applied forces in vector $\mathbf{F}$ do not exist, they must be removed by applying them in opposite directions to the soil-structure system shown in Fig. 27-18b. The complete solution is then obtained by

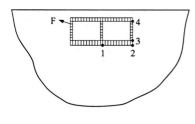

(a) Free-field deformations

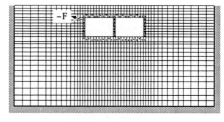

(b) SSI deformations

FIGURE 27-18
Modeling for cross-section racking analyses.

superposition of the two solutions, the first solution representing the compatible free-field deformations only, Fig. 27-18a, and the second solution representing the quasi-static soil-structure interaction deformations. The second solution requires plane-strain finite-element modeling of the soil within prescribed boundaries as indicated in Fig. 27-18b. The side and lower boundaries of this region should be sufficiently far away from the cross section so that the soil stresses due to pure racking soil-structure interaction (i.e., due to the pure racking components of **F** which are in self-equilibrium) have decayed sufficiently. Note that $\mathbf{F}_1$ contains only pure racking forces which are in self-equilibrium but $\mathbf{F}_2$ contains both pure racking and other forces which are not in self-equilibrium.

Results of racking analyses of rectangular cross sections using the above procedure have been reported in the literature.[17] They clearly show the importance of including soil-structure interaction effects.

Overall Axial and Flexural Deformations

Consider a pure harmonic shear wave moving under free-field conditions at velocity V_{ff} in the X-direction as shown in Fig. 27-19. The lateral displacement produced by this wave in the transverse Y-direction can be expressed in the form

$$V_n(X,t) = -\frac{a(i\overline{\omega}_n)}{\overline{\omega}_n^2} \exp\left[i\overline{\omega}_n\left(t - \frac{X}{V_{ff}}\right)\right]$$

which corresponds to the following velocity and acceleration expressions

$$\dot{V}_n(X,t) = -\frac{i\,a(i\overline{\omega}_n)}{\overline{\omega}_n} \exp\left[i\overline{\omega}_n\left(t - \frac{X}{V_{ff}}\right)\right]$$

$$\ddot{V}_n(X,t) = a(i\overline{\omega}_n) \exp\left[i\overline{\omega}_n\left(t - \frac{X}{V_{ff}}\right)\right]$$

(27-82)

In addition, the displacement components in the x- and y-directions may be written as

$$u_n(x,t) = \frac{a(i\overline{\omega}_n)}{\overline{\omega}_n^2} \sin\theta \, \exp\left[i\overline{\omega}_n\left(t - \frac{x\cos\theta}{V_{ff}}\right)\right]$$

$$v_n(x,t) = -\frac{a(i\overline{\omega}_n)}{\overline{\omega}_n^2} \cos\theta \, \exp\left[i\overline{\omega}_n\left(t - \frac{x\cos\theta}{V_{ff}}\right)\right]$$

(27-83)

---

[17] J. Penzien, C. H. Chen, W. Y. Jean, and Y. J. Lee, "Seismic Analysis of Rectangular Tunnels in Soft Ground," Proceedings of the Tenth World Conference on Earthquake Engineering, Madrid, Spain, July, 1992.

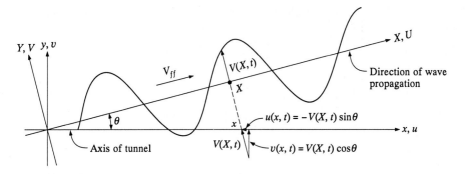

FIGURE 27-19
Shear wave moving in the X-direction at velocity V_{ff}.

Treating the soil-structure interaction of the tunnel elastic axis in a quasi-static fashion, the governing equations for axial and bending deformations are given by

$$\frac{\partial^2 u_{tn}(x,t)}{\partial x^2} - \frac{k_x}{AE} u_{tn}(x,t) = -\frac{k_x}{AE} u_n(x,t)$$

$$\frac{\partial^4 v_{tn}(x,t)}{\partial x^4} + \frac{k_y}{EI} v_{tn}(x,t) = +\frac{k_y}{EI} v_n(x,t) \tag{27-84}$$

where u_{tn} and v_{tn} represent total structural displacements, k_x and k_y are subgrade moduli in the x- and y-directions, and AE and EI are the axial and flexural stiffnesses of the tunnel. It can easily be shown that the flexural deformations are small relative to the axial deformations when the wave acts in the critical direction in each case, i.e., $\theta_{cr} = 0°$ for flexure and $\theta_{cr} = 45°$ for axial deformation.[16]

Therefore, only the axial deformation caused by $\theta = \theta_{cr} = 45°$ will be considered here, and in this case, the solution of the first of Eqs. (27-84) upon substitution of the first of Eqs. (27-83) is

$$u_{tn}(x,t) = \frac{a(i\bar{\omega}_n)}{\sqrt{2}\,\bar{\omega}_n^2\,(1+\Theta_{un})} \exp\left[i\bar{\omega}_n\left(t - \frac{x}{\sqrt{2}\,V_{ff}}\right)\right] \tag{27-85}$$

where

$$\Theta_{un} = \frac{AE\,\bar{\omega}_n^2}{2\,k_x\,V_{ff}^2} \tag{27-86}$$

The tunnel axial strain is given by

$$\epsilon_a(x,t) = \frac{\partial u_{tn}(x,t)}{\partial x} = \frac{-i\,a(i\bar{\omega}_n)}{2\,\bar{\omega}_n\,V_{ff}\,(1+\Theta_{un})} \exp\left[i\bar{\omega}_n\left(t - \frac{x}{\sqrt{2}\,V_{ff}}\right)\right] \tag{27-87}$$

---

[16] S. Gupta, T. W. Lin, J. Penzien, and C. S. Yeh, "Three-Dimensional Hybrid Modeling of Soil-Structure Interaction," loc. cit.

The axial strain produced by all discrete harmonics in $\ddot{V}(X,t)$ is then given by superposition, yielding

$$\epsilon_a(x,t) = \sum_n \frac{-i\,a(i\bar{\omega}_n)}{2\,\bar{\omega}_n\,V_{ff}\,(1 + \Theta_{un})}\ \exp\left[i\bar{\omega}_n\left(t - \frac{x}{\sqrt{2}\,V_{ff}}\right)\right] \qquad (27\text{-}88)$$

in which $a(i\bar{\omega}_n)$ are the Fourier coefficients of the prescribed accelerogram $\ddot{V}(t)$ obtained by the FFT procedure, V_{ff} is the free-field shear wave velocity, and Θ_{un} is given by Eq. (27-86).

For numerical analysis purposes, it is convenient to introduce the constant

$$\phi \equiv \frac{2\,k_x\,V_{ff}^2}{AE}\quad sec^{-2} \qquad (27\text{-}89)$$

into Eq. (27-88) and solve the equation for $V_{ff}\,\epsilon_a(x,t)$ giving

$$V_{ff}\,\epsilon_a(x,t) = \sum_n \frac{-i\,a(i\bar{\omega}_n)}{2\,\bar{\omega}_n\,(1 + \bar{\omega}_n^2/\phi)}\ \exp\left[i\bar{\omega}_n\left(t - \frac{x}{\sqrt{2}\,V_{ff}}\right)\right] \qquad (27\text{-}90)$$

This relation gives directly the corresponding functions

$$V_{ff}\,\epsilon_a(x,0) = \sum_n \frac{-i\,a(i\bar{\omega}_n)}{2\,\bar{\omega}_n\,(1 + \bar{\omega}_n^2/\phi)}\ \exp\left[-\frac{i\bar{\omega}_n\,x}{\sqrt{2}\,V_{ff}}\right]$$

$$V_{ff}\,\epsilon_a(0,t) = \sum_n \frac{-i\,a(i\bar{\omega}_n)}{2\,\bar{\omega}_n\,(1 + \bar{\omega}_n^2/\phi)}\ \exp\left[i\bar{\omega}_n\,t\right] \qquad (27\text{-}91)$$

for any prescribed accelerogram $\ddot{V}(t)$ from which the coefficients $a(i\bar{\omega}_n)$ can be obtained. The functions shown in Eqs. (27-91) are identical, except that the corresponding independent variables are $\frac{x}{\sqrt{2}V_{ff}}$ and t; thus, they can be represented on the same plot. The maximum absolute value of either function gives the quantity of primary interest.

The subgrade modulus k_x appearing in Eq. (27-89) cannot be evaluated rigorously for a given soil/cross section system; however, for practical solutions it can be approximated using $k_x \doteq 3G = 3\rho V_{ssi}^2$ in which G is the effective shear modulus in the dominant region controlling soil-structure interaction and V_{ssi} is the corresponding shear wave velocity. Note that this velocity may be considerably less than the velocity V_{ff} controlling the train of traveling free field waves approaching the tunnel alignment. Obviously, considerable judgement must be used in assigning the numerical values of both V_{ff} and V_{ssi}, after taking into consideration known factors such as geometry of soil layers relative to the tunnel alignment, results of soil tests, and levels of soil shear strain produced by the free-field motions and by the soil-structure interaction. In considering the latter factor, one should note that the shear strains produced by the free-field ground motions and by the axial soil-structure interaction, at the outer-boundary location of the cross section are given, respectively, by the approximate relations

$$\gamma_{ff}(x,t) \doteq \frac{i}{V_{ff}} \sum_n \frac{a(i\overline{\omega}_n)}{\overline{\omega}_n} \exp\left[i\overline{\omega}_n\left(t - \frac{x}{\sqrt{2}\,V_{ff}}\right)\right]$$

$$\gamma_{ssi}(x,t) \doteq \frac{-AE}{2\sqrt{2}\,p\rho\,V_{ssi}^2\,V_{ff}^2} \sum_n \frac{a(i\overline{\omega}_n)}{(1+\Theta_{un})} \exp\left[i\overline{\omega}_n\left(t - \frac{x}{\sqrt{2}\,V_{ff}}\right)\right]$$

(27-92)

in which p is the outside perimeter dimension of the cross section.

The above procedure for evaluating strains in a tunnel is extremely conservative due to the two assumptions made regarding the free-field soil motions: (1) the ground motion is produced by a single shear wave train moving at velocity V_{ff} and (2) the wave train impinges upon the tunnel at the most critical angle θ which is approximately 45°. Therefore, judgement should be used in reducing these strains accordingly.

Influence of Transverse Joints on Axial Deformations

The above evaluation of axial strains in the cross section assumes no transverse joints are present. Placement of open joints in the cross section will, however, reduce these strains to zero at the joint locations and will also reduce them at intermediate locations, with the smallest reductions occurring at locations midway between adjacent pairs of joints. Using the first of Eqs. (27-84), it can be shown that the ratio of the axial strain at this midway point, ϵ_{mp}, to the maximum axial strain (without joints), $\left|\epsilon_a\right|_{\max}$, is given by

$$\left(\epsilon_{mp}/\left|\epsilon_a\right|_{\max}\right) = 1 - \frac{2}{\exp(\frac{\beta L}{2}) + \exp(-\frac{\beta L}{2})}$$

(27-93)

where L is the distance between two adjacent joints and

$$\beta \equiv \sqrt{k_x/AE}$$

(27-94)

in which k_x is the associated subgrade modulus.

While the placement of multiple joints will reduce axial strains, they also allow joint separations to occur. Using the first of Eqs. (27-84), and assuming the maximum strain $\left|\epsilon_a\right|_{\max}$ to be reasonably uniform over the distance L when no joints are present as indicated in Fig. 27-20, it can be shown that the joint separation $\triangle_j$ is given by

$$\triangle_j = \frac{2}{\beta}\left[\frac{\exp(\frac{\beta L}{2}) - \exp(-\frac{\beta L}{2})}{\exp(\frac{\beta L}{2}) + \exp(-\frac{\beta L}{2})}\right] \left|\epsilon_a\right|_{\max}$$

(27-95)

Joint separation estimates should be made to insure that the joint details can accommodate the gap displacements during an earthquake without producing undesirable effects, e.g., leakage of water-tight seals.

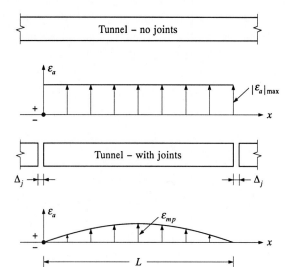

FIGURE 27-20
Tunnel axial strains with and without joints.

CHAPTER 28

STOCHASTIC STRUCTURAL RESPONSE

28-1 MODELING OF STRONG GROUND MOTIONS

Since seismic waves are initiated by irregular slippage along faults followed by numerous random reflections, refractions, and attenuations within the complex ground formations through which they pass, stochastic modeling of strong ground motions seems appropriate. The generation of synthetic accelerograms to reflect such stochastic behavior has already been discussed in Section 25-4. The procedure presented there consists of generating stationary waveforms having nearly constant power spectral density functions over the frequency range of interest, converting them to nonstationary waveforms by multiplying each by an appropriate deterministic time intensity function, and then filtering the resulting waveforms appropriately in the frequency domain; see Eqs. (25-32) through (25-39). It is the purpose of this chapter to discuss the stochastic response of structural systems to these excitations.

28-2 STOCHASTIC RESPONSE OF LINEAR SYSTEMS

SDOF Systems

If a stationary white-noise process of intensity S_0 is assumed for ground acceleration $\ddot{v}_g(t)$, the response of a linear SDOF system to this support acceleration is governed by the equation

$$\ddot{v} + 2\,\xi\,\omega\,\dot{v} + \omega^2\,v = -\ddot{v}_g(t) \tag{28-1}$$

where v is the displacement of the mass relative to the moving support. The principles

711

set forth in Chapter 22 give for an undercritically damped system

$$R_v(\tau) = \frac{\pi S_0}{2\,\omega^3\,\xi} \left(\cos\omega_D|\tau| + \frac{\xi}{\sqrt{1-\xi^2}}\,\sin\omega_D|\tau|\right)\exp(-\omega\,\xi\,|\tau|) \qquad (28\text{-}2)$$

$$R_{\dot v}(\tau) = \frac{\pi S_0}{2\,\omega\,\xi} \left(\cos\omega_D|\tau| - \frac{\xi}{\sqrt{1-\xi^2}}\,\sin\omega_D|\tau|\right)\exp(-\omega\,\xi\,|\tau|) \qquad (28\text{-}3)$$

$$S_v(\overline\omega) = \frac{S_0\,\omega^{-4}}{[1-(\overline\omega/\omega)^2]^2 + 4\,\xi^2\,(\overline\omega/\omega)^2} \qquad (28\text{-}4)$$

$$S_{\dot v}(\overline\omega) = \frac{S_0\,(\overline\omega/\omega)^2\,\omega^{-2}}{[1-(\overline\omega/\omega)^2]^2 + 4\,\xi^2\,(\overline\omega/\omega)^2} \qquad (28\text{-}5)$$

$$\sigma_v^2 = \frac{\pi S_0}{2\,\omega^3\,\xi} \qquad (28\text{-}6)$$

$$\sigma_{\dot v}^2 = \frac{\pi S_0}{2\,\omega\,\xi} \qquad (28\text{-}7)$$

If a stationary filtered process having power spectral density $S_{\ddot v_g}(\overline\omega)$ is assumed for ground acceleration $\ddot v_g(t)$, Eqs. (28-4) and (28-5) are still valid provided $S_{\ddot v_g}(\overline\omega)$ is substituted for S_0. Means and standard deviations of extreme values can be estimated using the procedures given in Section 21-11.

MDOF Systems

The linear response of discrete MDOF systems subjected to the same stationary acceleration $\ddot v_g(t)$ at all support points can be determined using normal-mode superposition as described in Chapter 23. The generalized forcing function $P_n(t)$ shown in Eq. (23-1) and defined by Eq. (23-24) becomes

$$P_n(t) = -\ddot v_g(t) \sum_i m_i\,\phi_{in} = \boldsymbol\phi_n^T\,\mathbf{m}\,\{1\}\,\ddot v_g(t) \qquad n = 1,2,\cdots \qquad (28\text{-}8)$$

and Eqs. (23-25) and (23-26) can be expressed in the forms

$$S_{P_m P_n}(\overline\omega) = S_{\ddot v_g}(\overline\omega) \sum_i \sum_k m_i\,m_k\,\phi_{im}\,\phi_{kn} = S_{\ddot v_g}(\overline\omega)\,\boldsymbol\phi_m^T\,\mathbf{m}\,\{1\}\,\{1\}^T\,\mathbf{m}\,\boldsymbol\phi_n$$
$$(28\text{-}9)$$

$$R_{P_m P_n}(\tau) = R_{\ddot v_g}(\tau) \sum_i \sum_k m_i\,m_k\,\phi_{im}\,\phi_{kn} = R_{\ddot v_g}(\tau)\,\boldsymbol\phi_m^T\,\mathbf{m}\,\{1\}\,\{1\}^T\,\mathbf{m}\,\boldsymbol\phi_n$$
$$(28\text{-}10)$$

For a distributed-mass system with $m = m(x)$, Eqs. (28-8) through (28-10) become

$$P_n(t) = -\ddot v_g(t) \int m(x)\,\phi_n(x)\,dx \qquad (28\text{-}11)$$

$$S_{P_m P_n}(\overline\omega) = S_{\ddot v_g}(\overline\omega) \int \int m(x)\,m(\alpha)\,\phi_m(x)\,\phi_n(\alpha)\,dx\,d\alpha \qquad (28\text{-}12)$$

$$R_{P_m P_n}(\tau) = R_{\ddot v_g}(\tau) \int \int m(x)\,m(\alpha)\,\phi_m(x)\,\phi_n(\alpha)\,dx\,d\alpha \qquad (28\text{-}13)$$

where α is a dummy space coordinate. The power spectral density and autocorrelation functions for response $z(t)$ defined by Eq. (23-2) are now obtained by substituting Eqs. (28-9) and (28-10) or (28-12) and (28-13) into Eqs. (23-19) and (23-10), respectively.

28-3 EXTREME-VALUE RESPONSE OF NONLINEAR SYSTEMS

The stochastic response of nonlinear SDOF or MDOF systems cannot be obtained by the methods previously presented, which employ the principle of superposition. For these complex systems, which are often history dependent due to hysteresis effects, one is usually forced to generate an ensemble of ground motion accelerograms by the techniques of Section 25-4 to determine deterministically the time-history response of the nonlinear system to each input accelerogram and then to examine the output response process using Monte Carlo methods. Usually, one is interested primarily in the mean and standard deviation values of extreme response.

This general method of stochastic analysis can also be used for linear systems; however, in this case, the direct method previously described is usually preferable.

SDOF Systems

Linear Models — Consider the SDOF linear system represented by Eq. (28-1) subjected to earthquake ground motion $\ddot{v}_g(t)$. Figure 28-1 shows Housner's pseudo-velocity design-spectrum curves for this system for different damping ratios, that is, $\xi = 0, 0.02, 0.05, 0.10$.[1] Since these curves were obtained by normalizing eight components of recorded ground accelerations (two components each of El Centro 1940, El Centro 1934, Olympia 1949, and Taft 1952) to a common intensity level and by averaging the eight pseudo-velocity response spectra derived therefrom, one can consider the ordinates in Fig. 28-1 as representing mean extreme values of relative pseudo-velocity. The multiplication factors given in this figure increase the ordinates to intensity levels corresponding to the earthquakes indicated.

Using an analog computer, Bycroft studied the possibility of using a white-noise process to represent earthquake ground motions at a given intensity level.[2] In these studies, he noted the extreme values of response for a SDOF system using 20 separate bursts of stationary white-noise input of 25 *sec* duration each. It was necessary in these studies to limit the input bandwidth having constant power spectral density to the range 0 to 35 Hz. To compare his mean extreme values with Housner's earlier published

[1] G. W. Housner, "Behavior of Structures During Earthquakes," *Proc. ASCE*, Vol. 85, No. EM-4, October, 1959.

[2] G. N. Bycroft, "White Noise Representation of Earthquakes," Proc. Paper 2434, *J. Eng. Mech. Div., ASCE*, Vol. 86, EM2, April, 1960.

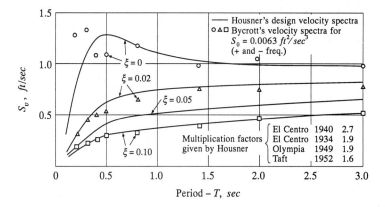

FIGURE 28-1
Mean extreme values of pseudo-relative velocity for linear SDOF systems
(stationary white-noise excitation).

velocity spectra, Bycroft normalized his results to that power spectral density of
input S_0 which would give full agreement with Housner's results for $T_n = 3$ sec and
$\xi = 0.20$. This normalization criterion resulted in a value of S_0 equal to 0.75 ft^2/Hz
over the frequency range $0 < f < \infty$. A further normalization of these same results
so that they can be compared with Housner's design velocity spectra requires that
$S_0 = 0.0063$ $ft^2/rad \cdot sec^3$ over the entire frequency range. Bycroft's mean extreme
values normalized to this intensity level are shown in Fig. 28-1. These results would
appear to indicate that even white noise is a reasonable representation of earthquake
ground accelerations.

Elastic-Plastic and Bilinear Stiffness-Degrading Models — Many investigators have
used stationary filtered white noise to simulate earthquake ground accelerations. In
one of these investigations, Liu and Penzien[3] used a single filter having the transfer
function given by the first of Eqs. (25-39) with $\omega_g = 15.6$ rad/sec and $\xi_g = 0.6$. Fifty
sample functions of band-limited white noise were generated by the digital-computer
methods of Section 25-4 with $S_0 = 0.00614$ ft^2/sec^3 and $\Delta t = 0.025$ sec. These
sample functions, each having 30 sec duration, were then filtered by digital-computer
techniques to provide an ensemble of 50 artificial accelerograms.

Complete time-histories of response for the linear SDOF system when subjected
separately to each of the 50 input accelerations were established by deterministic
methods. The extreme values of relative displacement were noted in each case and

---

[3] J. Penzien and S. C. Liu, "Nondeterministic Analysis of Nonlinear Structures Subjected to Earthquake
Excitations," Proc. 4th World Conf. Earthquake Eng., Santiago, Chile, Vol. I, Sec. A-1, January, 1969.

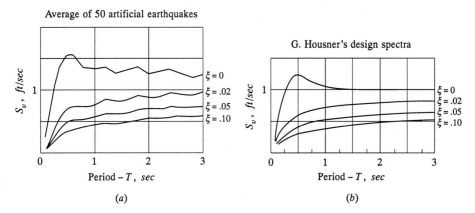

FIGURE 28-2
Mean extreme values of pseudo-relative velocity for linear SDOF systems (filtered stationary white-noise excitation).

were averaged to obtain mean values. These mean values of displacement were then converted to mean extreme values of pseudo-velocity by multiplying by ω. These values are plotted in Fig. 28-2a, where they may be compared with Housner's design-spectrum curves in Fig. 28-2b. The close agreement of these two sets of curves lends support to using filtered stationary white noise in the simulation of strong-earthquake ground motions.

Using numerical integration procedures, complete time-histories of relative-displacement response $v(t)$ were established also for SDOF elastic-plastic and stiffness-degrading models having the static-force-deflection relations shown in Figs. 28-3b and c, respectively. Each model was separately subjected to support accelerations $\ddot{v}_g(t)$ corresponding to the filtered process described above but after normalizing by a factor of $(2.90)^2$ so that the process intensity S_0 would represent the intensity of the NS component of the 1940 El Centro earthquake $[S_0 = (2.90)^2 (0.00614) = 0.0516 \ ft^2/sec^3]$.

The basic parameters of these nonlinear models, which are comparable to those used for the linear models, are shown in Fig. 28-3. In all cases T and ξ represent the period of vibration and viscous-damping ratio, respectively, in the initial elastic range. The strength ratio B and ductility demand μ_d are defined for these models in accordance with the relations $B \equiv V_y/W$ and $\mu_d \equiv |v(t)|_{max}/v_y$. It is significant that in addition to loss of stiffness following any yielding, the stiffness-degrading model permits hysteresis loops to be formed even at very low amplitudes of oscillation. Therefore, this model dissipates more energy in the lower-amplitude ranges of response than does the equivalent elastic-plastic model.

The response of the elastic-plastic and stiffness-degrading models are considered

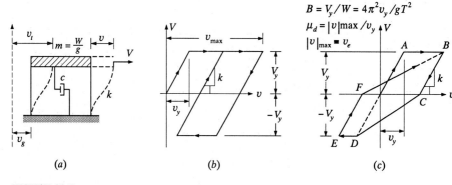

FIGURE 28-3
Nonlinear SDOF models.

TABLE 28-1

Case No.	Structural type *	Period T sec	Damping ratio, ξ	Strength ratio, B	Yield displ. v_y in
1	E	0.3	0.02	—	—
2	EP	0.3	0.02	0.10	0.088
3	SD	0.3	0.02	0.10	0.088
4	E	0.3	0.10	—	—
5	EP	0.3	0.10	0.10	0.088
6	SD	0.3	0.10	0.10	0.088
7	E	2.7	0.02	—	—
8	EP	2.7	0.02	0.048	3.42
9	SD	2.7	0.02	0.048	3.42
10	E	2.7	0.10	—	—
11	EP	2.7	0.10	0.048	3.42
12	SD	2.7	0.10	0.048	3.42

* E – Elastic   EP – Elastic-plastic   SD – Stiffness degrading

for two different periods, $T = 0.3$ and 2.7 sec, and for two different damping ratios, $\xi = 0.02$ and 0.10; thus, the response of eight different nonlinear models as presented in Table 28-1 are discussed. Strength ratios B are based on the assumption that the yield resistance V_y equals twice the design load as specified in the 1973 edition of the Uniform Building Code for moment-resisting frames, that is, $B = 2KC = (2 \times 0.667)\,(0.05)\,T^{-1/3}$.

Probability distribution functions $P(|v|_{\max})$ based on the 50 extreme values for each of the eight nonlinear models are shown in Fig. 28-4. For comparison, probability distribution functions are also presented for the four corresponding linear elastic models, that is, models having the same corresponding initial stiffnesses and

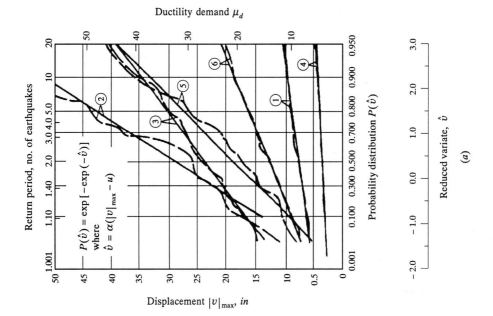

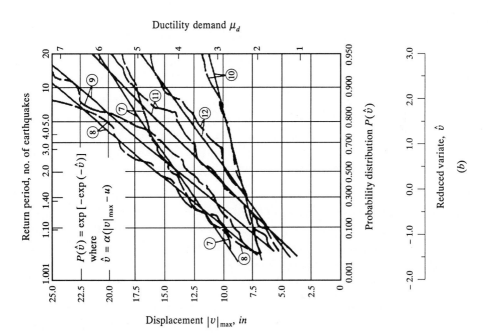

FIGURE 28-4
Probability distributions for extreme values of relative displacement.

viscous-damping ratios. These twelve models, as identified by the arabic numerals 1 through 12 in Fig. 28-4, have the properties listed in Table 28-1.

Two probability distribution functions are shown in Fig. 28-4 for each of the 12 structural models, namely, a wavy-line function, which is a plot of the actual extreme-value distribution as determined numerically for process $v(t)$, and a straight-line function, which is the theoretical Gumbel Type I distribution corresponding to Eq. (21-120), namely

$$P(|v|\text{max}) = \exp\left[-\exp(-\hat{v})\right] \qquad (28\text{-}14)$$

in which $\hat{v}$ is the reduced extreme value defined by

$$\hat{v} \equiv \alpha\,(|v|\text{max} - u) \qquad (28\text{-}15)$$

The numerical values of α and u for each of the twelve cases were determined by the "more accurate" procedure mentioned at the end of Example E21-8.[4]

The probability distribution scale on Gumbel extreme-value charts as shown in Fig. 28-4 varies in such a manner that Eq. (28-14) plots as a straight line with its ordinate at the origin ($\hat{v} = 0$) representing the most probable extreme value and with its slope being proportional to the standard deviation of the extreme values. Note that the extreme values in this figure for the nonlinear models can be measured also in terms of the ductility demand μ_d and that the probability distribution can be measured in terms of the return period, that is, the expected number of earthquakes required to produce a single extreme value having the magnitude shown by the ordinate scale.

The significant features to be noted in Fig. 28-4 are the following:

(1) The most probable extreme values of response for short-period structures as represented in Fig. 28-4a are much greater for the elastic-plastic and stiffness-degrading models than for their corresponding linear models, are appreciably greater for the elastic-plastic models than for their corresponding stiffness-degrading models, and are considerably greater for those models having 2 percent of critical damping than for their corresponding models having 10 percent of critical damping.

(2) The most probable extreme values of response for long-period structures as represented at $\hat{v} = 0$ in Fig. 28-4b are considerably greater for those models having 2 percent of critical damping than for their corresponding models having 10 percent of critical damping; however, these values differ very little from one model to another.

(3) The standard deviations of extreme-value response for the short-period structures are considerably larger for the elastic-plastic and stiffness-degrading models

---

[4] E. J. Gumbel and P. G. Carlson, *Extreme-Values in Aeronautics*, op. cit.; E. J. Gumbel, *Probability Tables for the Analysis of Extreme-Value Data*, op. cit.

than for their corresponding linear models and are appreciably larger for the elastic-plastic models than for their corresponding stiffness-degrading models.

(4) The standard deviations of extreme-value response for long-period structures correlate in a manner quite similar to short-period structures except that the differences are not so great.

(5) Increasing the viscous-damping ratio decreases the standard deviations of extreme-value response for each model type.

(6) The theoretical extreme-value functions as represented by Eq. (28-14) and plotted as straight lines in Fig. 28-4 show very good correlations with the actual distributions.

The probability distribution functions for extreme values shown in Fig. 28-4 result from an input process $\ddot{v}_g(t)$ having a duration of 30 sec. The corresponding extreme values will, of course, be less for processes of shorter duration. To illustrate this time-duration effect, a ratio of the ensemble average of extreme values for an input process of duration T_0 to the ensemble average of extreme values for an input process of 30 sec is plotted in Fig. 28-5 as a function of the duration ratio $T_0/30$.

It is quite evident, from curve No. 2 in Fig. 28-5a, that the mean peak response of a typically damped, linear short-period structure ($T = 0.3$ sec) increases very slowly with duration beyond approximately 6 sec. A long-period structure is, of course, more sensitive to duration, as shown by curve No. 2 in Fig. 28-5b. This latter curve indicates that the magnitude of mean peak response for a 15 sec duration process is approximately 95 percent of the magnitude observed for a 30 sec duration process. As shown in Figs. 28-5a and b, elastic-plastic and stiffness-degrading structures are much more sensitive to duration than elastic structures are; thus, it is apparent that

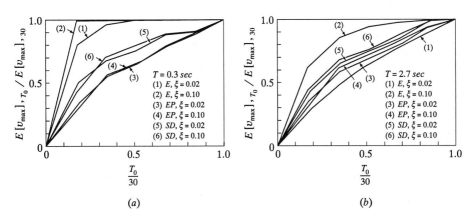

FIGURE 28-5
Duration effect of stationary process on mean peak response of linear and nonlinear structures.

realistic durations should be used for stationary inputs when investigating the response of nonlinear structures.

As demonstrated above, stationary processes of relatively short duration can be used quite effectively to establish the probabilistic peak response of both linear and nonlinear systems to strong-motion earthquakes of a given intensity level. However, as the true dynamic characteristics of real structures become better known, damage will likely be measured using various accumulative-damage criteria, in which case, it becomes desirable to use appropriate nonstationary processes for the excitation.

Trilinear Stiffness-Degrading Models — To further illustrate the variability of structural response to be expected from strong ground motion excitation, let us now examine some selected results reported by Murakami and Penzien[5] for an improved stiffness-degrading single-degree-of-freedom model[6] representing reinforced concrete structures where the nonlinear deformations and failure characteristics are controlled primarily by flexure. The force-displacement hysteretic relation for this model is shown in Fig. 28-6.

The nonstationary stochastic model used in this investigation for the single component of ground acceleration input was of the form

$$\ddot{v}_g(t) = f(t)\, a(t) \tag{28-16}$$

where $f(t)$ is a deterministic time intensity function and $a(t)$ is a stationary filtered process. In generating the filtered process, both the high- and low-frequency filters of Eq. (25-39) were used. In the high-frequency filter ω_1 and ξ_1 were assigned the numerical values 15.6 rad/sec and 0.6, respectively; while at the same time, parameters ω_2 and ξ_2 in the low-frequency filter were assigned the numerical values 0.897 rad/sec and $1/\sqrt{2}$, respectively. The intensity of the stationary process S_0 was assigned a fixed value in generating the sample ground motions; however, a normalization was carried out so that the final response results were independent of this value. The time intensity function used was of the form

$$f(t) = \begin{cases} t^2/16 & 0 \le t \le 4 \ sec \\ 1 & 4 \le t \le 35 \ sec \\ \exp[-0.0357(t-35)] & t \ge 35 \ sec \end{cases} \tag{28-17}$$

[5] M. Murakami and J. Penzien, "Nonlinear Response Spectra for Probabilistic Seismic Design and Damage Assessment of Reinforced Concrete Structures," Univ. of Calif., Berkeley, Earthquake Engineering Research Center, Report No. 75–38, 1975.

[6] H. Umemura et. al., *Earthquake Resistant Design of Reinforced Concrete Buildings Accounting for the Dynamic Effects of Earthquakes*, Giko-do, Tokyo, Japan, 1973 (in Japanese).

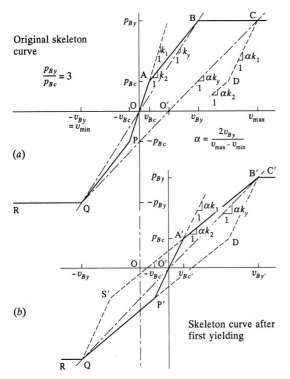

Original skeleton
curve

$$\frac{p_{By}}{p_{Bc}} = 3$$

$$\alpha = \frac{2v_{By}}{v_{max} - v_{min}}$$

(a)

(b)

Skeleton curve after
first yielding

FIGURE 28-6
Trilinear stiffness-degrading hysteretic
model.

which represents roughly the upper bound motions to be expected in the vicinity of the causative fault of an earthquake having a Richter magnitude 8 or greater.[7]

The hysteretic form of the force-displacement relation is characterized by p_{Bc}, p_{By}, v_{Bc}, and v_{By} which represent the load at which the concrete cracks due to flexure, the load at which the main reinforcing steel starts to yield due to flexure, the relative displacement produced by p_{Bc}, and the relative displacement produced by p_{By}, respectively. Linear elastic behavior (without hysteretic loops) always takes place for oscillatory displacements where the corresponding loads are in the range $-p_{Bc} < p < p_{Bc}$; however, hysteretic behavior occurs with each cycle of deformation which has a load level above p_{Bc} or below $-p_{Bc}$. During the period of time between the initiation of loading and that instant at which the relative displacement first increases above v_{By} or decreases below $-v_{By}$, the trilinear model behaves exactly like the standard bilinear hysteretic model having stiffnesses k_1 and k_2 (QPOAB; Fig. 28-6). However, as soon as the relative displacement increases above v_{By} or decreases below $-v_{By}$, a new bilinear hysteretic relation controls the response. For example, suppose the relative displacement for the first time increases above v_{By} to

---

[7] P. C. Jennings, G. W. Housner, and N. C. Tsai, "Simulated Earthquake Motions," loc. cit.

level $v_{\max}$ as represented by C in Fig. 28-6a. Upon decreasing the displacement from this level, the corresponding load decreases along path CD which has a slope equal to αk_1, where

$$\alpha \equiv \frac{2\, v_{By}}{v_{\max} + v_{By}} \tag{28-18}$$

As soon as the load drops by the amount $2p_{Bc}$ reaching point D, any further drop in load will follow the continuing path shown having a slope αk_2. It should be noted that point D is located at load level p_{Bc} in Fig. 28-6a but only because the particular trilinear model represented in that figure is for $p_{By}/p_{Bc} = 3.0$. If this ratio had been assigned a different numerical value, the load level at point D would be different from p_{Bc}.

The new bilinear hysteretic model controlling the continuing motion is shown in Fig. 28-6b. Note that the origin of the skeleton curve is shifted from point O, the origin of the original bilinear hysteretic model, to point O'. This point is the intersection point of line QC and the abscissa axis in Fig. 28-6a; therefore OO' is equal to $BC/2$. The stiffnesses of the new bilinear model are αk_1, and αk_2.

If during the period of response controlled by the second bilinear model of Fig. 28-6b, the relative displacement should increase beyond $v_{\max}$ ($v_{\max} = v_{By'}$) as represented by point B' to a new level as represented by C', the continuing response would be controlled by a third bilinear hysteretic model whose characteristics could be obtained in exactly the same manner as the characteristics of the second model. Also if yielding of the trilinear model had taken place at load level $-p_{By}$ rather than load level p_{By}, the new bilinear model controlling the continuing motion would be obtained by a similar procedure.

The trilinear stiffness-degrading model is completely characterized by any four of the seven parameters k_1, k_2, k_y, p_{Bc}, p_{By}, v_{Bc}, and v_{By} shown in Fig. 28-6. It is convenient to use k_1, k_y, p_{Bc}, and p_{By} for this purpose but substituting period parameters T_1 and T_2, defined by $T_1 \equiv 2\pi\sqrt{m/k_1}$ and $T_2 \equiv 2\pi\sqrt{m/k_y}$, for parameters k_1 and k_y, respectively. For reinforced concrete members, k_1 and p_{By} usually fall in the ranges $2k_y < k_1 < 4k_y$ and $2p_{Bc} < p_{By} < 3p_{Bc}$, respectively; therefore, for a given design, one can specify the numerical values for ratios k_1/k_y and p_{By}/p_{Bc}, which reduces the number of independent parameters to two. For the example case presented herein, $k_1/k_y = 2$ and $p_{By}/p_{Bc} = 2$ in which case $T_2 = \sqrt{2}T_1$. Period T_1 and strength p_{By} can now be used as the two independent model parameters. It is convenient to normalize strength parameter p_{By} by dividing by the force $m\bar{\bar{v}}_{g0}$ where $\bar{\bar{v}}_{g0}$ is the mean peak acceleration of the ground motion excitation; thus the final results are independent of the numerical value of S_0.

Viscous damping in the structural model is controlled by prescribing the numerical value of the damping ratio ξ_1 defined by $\xi_1 \equiv c/2\sqrt{mk_1}$. A value of 2% of critical, i.e., $\xi_1 = 0.02$, has been selected for the example presented herein.

To illustrate the probabilistic nature of maximum structural response to a class of seismic excitations of fixed intensity, consider the trilinear stiffness-degrading model defined above subjected to a family of ground motion excitations defined by Eq. (28-16). Subjecting this model to 20 separate ground motions for each set of fixed parameters T_1 and p_{By}, the mean and coefficient of variation (ratio of standard deviation to mean value) of the corresponding 20 maximum values of response are obtained. Measuring maximum response in terms of ductility demand (μ_d) defined by

$$\mu_d \equiv v(t)_{\max} / v_{By} \tag{28-19}$$

the mean value of μ_d ($\overline{\mu}_d$) and its coefficient of variation (standard deviation/mean value) have distributions as shown in Fig. 28-7. For simplicity, the subscript "B" has been dropped from the terms p_{Bc}, p_{By}, v_{Bc}, and v_{By} in this figure.

As shown in Fig. 28-7, the mean ductility demands generally increase with decreasing period (T_1) and the spread of ductility demands over the full-strength range $0.50 < (p_y/m\overline{v}_{g0}) < 1.75$ increases with decreasing period. Also note that the ductility demands for a fixed period increase with decreasing structural strength. The trends of the coefficients of variation with period are similar to the trends for mean ductility demand, particularly regarding strength level and strength variation. It is most significant to note that the coefficients of variation are low when the response is essentially elastic ($\mu_d < 1$) but they become very large with increasing inelastic deformations.

When using the results in Fig. 28-7, one can calculate mean values of maximum response using the relation

$$\overline{v}(t)_{\max} = T_2^2 \, \beta_f \, \overline{\mu} \left[\frac{g}{4\pi^2} \right] \left[\frac{\overline{\overline{v}}_{g0}}{g} \right] \tag{28-20}$$

where

$$\beta_f \equiv p_y / m\overline{\overline{v}}_{g0} \tag{28-21}$$

MDOF Systems

Using a nonstationary process to represent strong ground motions, Ruiz[8] studied the probabilistic response of multistory shear buildings. Selected results of his investigation are presented here to provide an example of the stochastic response of MDOF systems.

Ruiz generated a ground-acceleration process $\ddot{v}_g(t)$ to simulate the expected ground motions on firm soil at a distance of about 45 mi from the epicenter of

---

[8] P. Ruiz and J. Penzien, "Probabilistic Study of Behavior of Structures during Earthquakes," University of California, Berkeley, Earthquake Engineering Research Center, Rept. 69-3, 1969.

Type A earthquake

$$\xi_1 = 0.02 \qquad T_2 = \sqrt{2}\, T_1 \qquad p_y = 2p_c$$

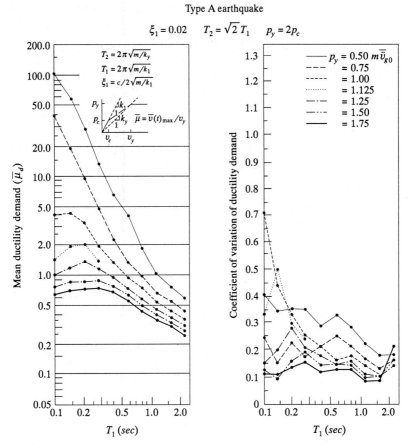

FIGURE 28-7

Mean ductility demands and corresponding coefficients of variation versus period T_1 for the trilinear stiffness degrading model having different strength levels p_y.

a magnitude 8.3 earthquake. Twenty sample functions of band-limited stationary white noise were generated by digital-computer methods. These sample functions were then multiplied by the deterministic intensity function $f(t)$ shown in Fig. 25-13 with $t_1 = 0$, $t_2 = 11.5$ sec, and $c = 0.155$ sec^{-1}. The resulting nonstationary waveforms were then filtered using the first of Eqs. (25-39) as the filter function with $\omega_g = 15.7$ rad/sec and $\xi_g = 0.6$. The process was normalized to an intensity level corresponding to an expected peak acceleration of 0.3 g.

The complete time-history of the elastic-plastic response of an eight-story shear building was calculated deterministically for each of the 20 input ground accelerations. The eight lumped masses of this building were of equal magnitude and equally spaced, and the relative story elastic spring constants were adjusted so that the fundamental mode shape of the building was linear. The lateral drift of each story was related to its shear force through a bilinear hysteretic-force-deflection relation independent of axial forces acting in the columns. Yielding in all stories was assumed to start simultaneously as the static lateral loading, distributed in accordance with the 1973 Uniform Building Code, increased monotonically to a level twice as great as the design loading. The yielding stiffness in each story was set at 10 percent of its initial elastic stiffness. Viscous damping was introduced into the coupled nonlinear equations of motion to provide specified damping in the elastic range of response.

Probability distribution functions based on the 20 extreme values of ductility demand in each of the eight stories are presented in the form of Gumbel plots (Type I) in Fig. 28-8 for two different shear-type buildings having fundamental periods of 0.5 and 2.0 *sec*; thus, they represent a stiff building and a flexible building. Both buildings are assigned damping ratios of 5 percent in all elastic modes. These probability distribution functions are similar to those shown in Fig. 28-4 for the SDOF systems described in Section 28-4; therefore, no additional description of the meaning of these plots is necessary.

From the results of Fig. 28-8a, the following observations are made with regard to stiff shear buildings:

(1) The most probable ductility demands given at 0.368 on the abscissa scale decrease toward the top of the structure.

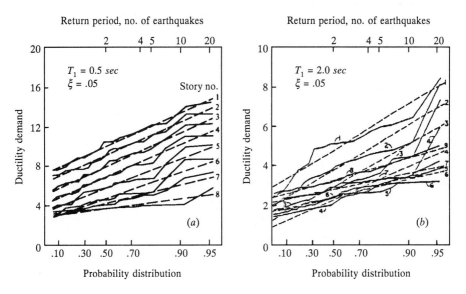

FIGURE 28-8
Probability distributions for story ductility demands.

(2) The standard deviations of the ductility demands are almost the same for all stories except the top story, where a large reduction is observed.

(3) The estimated probability distribution functions show very good agreement with the corresponding theoretical distributions represented by straight lines.

Likewise from the results in Fig. 28-8b, the following observations are made with regard to flexible shear buildings:

(1) The most probable ductility demands $P(0.368)$ decrease toward the middle stories and then increase toward the top story.

(2) The standard deviations of the ductility demands decrease toward the upper stories but with a slight increase in the top story.

(3) The agreement between the estimated probability distributions and the corresponding theoretical straight-line distributions is only fair in this case.

Comparing the results of Figs. 28-8a and b, it is clear that the most probable ductility demands and their standard deviations are significantly higher for the stiffer building.

28-4 DESIGN CONSIDERATIONS

The generally accepted philosophy of seismic design has been that only minor damage is acceptable under moderate earthquake conditions and that total damage or complete failure should be avoided under maximum probable earthquake conditions. It is implied in this statement that economical considerations permit a certain level of risk with regard to damage of structures in high seismic regions. To minimize total costs (initial costs, repair costs after earthquakes, etc.), damage is often permitted to limited degrees under moderate to severe earthquake conditions. It should be understood that permitting some damage to occur in a well-designed structure has the beneficial effect of limiting damage to that same structure. This is due to the fact that the energy absorption associated with damage is effective in limiting the maximum levels of oscillatory motion in the structure. Therefore, a good seismic resistant structure should be designed to possess high-energy absorption capacity so that it will experience controlled damage under moderate to severe earthquake conditions. In terms of the hysteretic structural models presented previously, this concept means that the ductility demands should be limited to certain values, well below their corresponding ductility capacities, consistent with the basic design philosophy.

Assume for the moment that one prescribes two numerical values of ductility demand for a given structural model, the smaller value chosen to be consistent with light damage under moderate earthquake conditions and the larger value chosen to be consistent with much greater damage (but not complete failure) under the most severe condition which could occur. Two questions come to mind (1) "What is the probability of these ductility demands being exceeded during a single earthquake?" and (2) "What ductility demands should be specified, consistent with the design philosophy?" To fully answer these questions, one needs the appropriate probability density or distribution functions.

In the previous section, it was shown that the probability distribution function for maximum (or extreme) response for a single earthquake follows closely the Gumbel Type I distribution. Using here the trilinear SDOF model and measuring the extreme values of response in terms of ductility demand (μ_d), the probability distribution function for 20 sample values can be expressed in the Gumbel Type I form

$$P(q) = \exp\left\{-\exp\left[-\frac{1.063}{c}(q-1+0.493\,c)\right]\right\} \qquad (28\text{-}22)$$

in which q is the ductility demand normalized by its mean value, i.e.,

$$q \equiv \frac{\mu_d}{\overline{\mu}_d} \qquad (28\text{-}23)$$

and c is the coefficient of variation of μ_d, i.e.,

$$c \equiv \frac{\sigma_{\mu_d}}{\overline{\mu}_d} \qquad (28\text{-}24)$$

This probability distribution function is plotted in Fig. 28-9 over a range of discrete values of c, $0 < c < 1.5$. Since the probability distribution function is defined such that

$$P(x) \equiv \text{Probability } [q < x] \qquad (28\text{-}25)$$

the probability exceedance function is given by

$$Q(x) \equiv \text{Probability } [q > x] = 1 - P(x) \qquad (28\text{-}26)$$

The first question raised, namely, "What is the probability of these ductility factors being exceeded during a single earthquake?" can be easily answered using Eq. (28-26), Fig. 28-9, and data of the type presented in Fig. 28-7. The second question raised, i.e., "What ductility factors are required consistent with the design philosophy?" is more difficult to answer. Before attempting to answer this question, one must realize that the basic design criteria cannot be met in absolute terms, i.e., with 100% confidence. This complication is due to the scatter of extreme values of response for each family of earthquake excitations. The best one can do is to reduce the probability of exceedance associated with each of the two ductility factors to an acceptable level. Deciding on an acceptable level is, of course, complex as it involves economic, social, and political considerations; however, suppose for example, it was decided that a 15

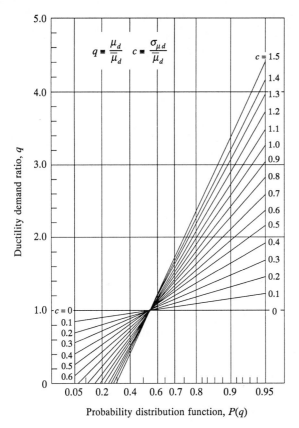

FIGURE 28-9
Probability distribution functions for ductility demand ratios on Gumbel Type I plot.

percent probability of exceedance was acceptable, i.e., $Q(q) = 0.15$ corresponding to $P(q) = 0.85$. Using Fig. 28-9 and data similar to that shown in Fig. 28-7, one can easily establish that ductility factor μ_d^{85} associated with $P(q) = 0.85$. Figure 28-10 presents the results for the trilinear case previously described, i.e., the case represented in Fig. 28-7.

To establish the required strength level of a structural system, one must first prescribe basic criteria consistent with the basic design philosophy. For example, consider the case where the design and maximum probable PGA levels have been set at 0.30 g and 0.45 g, respectively, and the two ductility demands, consistent with light and heavy (but controlled) damage for the trilinear stiffness-degrading model, have been set at 2 and 4, respectively. Having set these criteria, one can use the established data on acceptable probability of exceedance, for example $\mu_d^{85} = \mu_d^{85}(T_1, p_y, \text{etc.})$ as shown in Fig. 28-10, to obtain the corresponding required strength ratios μ_d^{85} ($\beta \equiv p_y/m\bar{\bar{v}}_{g0}$) for any discrete value of T_1 under consideration. Considerable

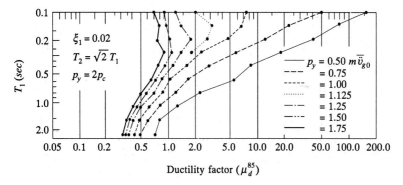

FIGURE 28-10
Response ductility demands for the 85% level on probability distribution functions.

information of this type has been generated and reported in the literature.[9,10,11]

In this discussion on design considerations, probabilistic concepts have been presented for assessing inelastic structural performance under seismic conditions. While the numerical results given apply to the trilinear SDOF model only, they do provide considerable insight into the inelastic performance of MDOF structures as well. A thorough treatment of the probabilistic response of such systems is beyond the scope of this book.

28-5 ALLOWABLE DUCTILITY DEMAND VERSUS DUCTILITY CAPACITY

When setting the allowable ductility demand of the maximum credible earthquake for a particular SDOF or MDOF structure, its overall ductility capacity should be taken into consideration. Ductility capacity is defined in general terms as the maximum ductility ratio (v_{max}/v_y) the structure can withstand under a specified number of complete cycles of deformation to that level without experiencing significant loss of structural integrity (strength and stiffness). This specified number should correspond to one's estimate of the number of complete cycles of inelastic deformation the structure will undergo during a single maximum credible earthquake event. To

---

[9] M. Murakami and J. Penzien, loc. cit.

[10] J. Penzien, "Predicting the Performance of Structures on Regions of High Seismicity," Proc. 2nd Canadian Conf. on Earthquake Engineering, McMasters Univ., Hamilton, Untario, Canada, June, 1975.

[11] H. A. Sawyer, "Comprehensive Design of Reinforced Concrete Frames by Plasticity Factors," 9th Plenary Session of the Comite European du Beton Symposium "Hyperstatique," Ankara, Turkey, September, 1964.

provide for a proper margin of safety of performance, ductility capacity should exceed the allowable ductility demand by an appropriate factor. A factor of 2 is commonly used for this purpose.

INDEX

AUTHOR INDEX